HARNESSING
AutoCAD® 2006

HARNESSING
AutoCAD® 2006

THOMAS A. STELLMAN
G.V. KRISHNAN

Autodesk

Australia • Canada • Mexico • Singapore • Spain • United Kingdom • United States

Harnessing AutoCAD® 2006
Thomas A. Stellman / G.V. Krishnan

Vice President, Technology and Trades SBU:
Alar Elken

Editorial Director:
Sandy Clark

Senior Acquisitions Editor:
James DeVoe

Senior Development Editor:
John Fisher

Marketing Director:
Dave Garza

Channel Manager:
Dennis Williams

Marketing Coordinator:
Stacey Wiktorek

Production Director:
Mary Ellen Black

Production Manager:
Andrew Crouth

Production Editor:
Jennifer Hanley

Art & Design Specialist:
Mary Beth Vought

Technology Project Manager:
Kevin Smith

Technology Project Specialist:
Linda Verde

Editorial Assistant
Tom Best

COPYRIGHT © 2006 Thomson Delmar Learning. Thomson, the Star Logo, and Delmar Learning are trademarks used herein under license. Autodesk, AutoCAD, and the AutoCAD logo are registered trademarks of Autodesk. Delmar Learning uses "Autodesk Press" with permission from Autodesk for certain purposes.

Printed in the United States
1 2 3 4 5 XX 07 06 05 04

For more information contact Thomson Delmar Learning
Executive Woods, 5 Maxwell Drive, PO Box 8007, Clifton Park, NY 12065-8007

Or find us on the World Wide Web at www.delmarlearning.com

ALL RIGHTS RESERVED. No part of this work covered by the copyright hereon may be reproduced in any form or by any means—graphic, electronic, or mechanical, including photocopying, recording, taping, Web distribution, or information storage and retrieval systems—without the written permission of the publisher.

For permission to use material from the text or product, contact us by
Tel.   (800) 730-2214
Fax   (800) 730-2215
www.thomsonrights.com

Library of Congress Cataloging-in-Publication Data:
Stellman, Thomas A.
  Harnessing AutoCAD 2006 / Thomas A. Stellman, G.V. Krishnan.
    p. cm.
  ISBN 1-4180-2035-4
  1. Computer graphics. 2. AutoCAD. I. Krishnan, G.V. II. Title.

T385.S75186 2005
620'.0042'0285536--dc22

2005020011

NOTICE TO THE READER

Publisher does not warrant or guarantee any of the products described herein or perform any independent analysis in connection with any of the product information contained herein. Publisher does not assume, and expressly disclaims, any obligation to obtain and include information other than that provided to it by the manufacturer.

The reader is expressly warned to consider and adopt all safety precautions that might be indicated by the activities herein and to avoid all potential hazards. By following the instructions contained herein, the reader willingly assumes all risks in connection with such instructions.

The publisher makes no representation or warranties of any kind, including but not limited to, the warranties of fitness for particular purpose or merchantability, nor are any such representations implied with respect to the material set forth herein, and the publisher takes no responsibility with respect to such material. The publisher shall not be liable for any special, consequential, or exemplary damages resulting, in whole or part, from the readers' use of, or reliance upon, this material.

CONTENTS

Introduction xxi
HARNESSING THE POWER OF AUTOCAD 2006 xxi
HIGHLIGHTS AND FEATURES OF THIS NEW EDITION xxii
HOW TO USE THIS BOOK xxiii
    Overview xxiii
    Fundamentals xxiii
    Intermediate xxiii
    Advanced xxiii
    Appendices xxiii
STYLE CONVENTIONS xxiv
HOW TO INVOKE COMMANDS xxiv
COORDINATE INPUT – WHAT'S THE POINT? xxv
DRAWING EXERCISES AND BONUS CHAPTER ON VISUAL LISP xxv
EXERCISE ICONS xxvi
HARNESSING AUTOCAD 2006 EXERCISE MANUAL xxvi
ONLINE COMPANION™ xxvi
E-RESOURCE xxvi
ABOUT THE AUTHORS xxvii
ACKNOWLEDGMENTS xxviii

Chapter 1 Getting Started 1
STARTING AUTOCAD 1
    The Startup Window 1
AUTOCAD SCREEN 4
    Graphics Window 5
    Status Bar and Status Bar Tray 5
    Title Bar 7
    Toolbars 7
    Tool Palettes Window 12
    Menu Bar 14
    Model Tab and Layout Tab 15
    Focusing on the Design — Dynamic Input 15
    Command Window 16
AUTOCAD COMMANDS AND INPUT METHODS 18
    Input Methods 19
GETTING HELP 25
    Info Palette 25
    Traditional Help 26

New Features Workshop	28
Developer Help	28
Online Resources	28

BEGINNING A NEW DRAWING 28
OPENING AN EXISTING DRAWING 30
SETTING UP THE DRAWING ENVIRONMENT 33
    Setting Drawing Units 33
    Setting Drawing Limits 37
BEGINNING A NEW DRAWING BY USING THE CREATE NEW DRAWING
    DIALOG BOX (STARTUP SYSTEM VARIABLE SET TO 1) 39
    Starting a New Drawing by Using Wizards 40
    Starting a New Drawing with a Template 45
    Starting a New Drawing from Scratch 47
WORKING WITH MULTIPLE DRAWINGS 48
SAVING A DRAWING 49
EXITING AUTOCAD 51
REVIEW QUESTIONS 52

Chapter 2 Fundamentals I 55

CONSTRUCTING GEOMETRIC FIGURES 55
    Drawing Lines 56
    Drawing Rectangles 60
    Drawing Wide Lines 62
    Coordinate System 63
    Three Types of Coordinate Systems 67
    Methods to Specify Points 70
    Overrides for Absolute/Relative Coordinate Entry 73
    Coordinates Display 73
    Drawing Circles 74
    Drawing Arcs 78
OBJECT SELECTION 89
    Selection by Window 89
    Selection by Crossing 90
    Previous Selection 91
    Last Selection 91
MODIFYING OBJECTS 92
    Erasing Objects 92
    Getting It Back 94
REVIEW QUESTIONS 95

Chapter 3 Fundamentals II 99

DRAFTING SETTINGS 99
    SNAP Command 99
    GRID Command 105
    ORTHO Command 110
    Object Snap 111

Drawing Objects Using Tracking	118
Drawing Objects Using Direct Distance	121
POLAR TRACKING and OBJECT SNAP TRACKING	123
Dynamic Input, On-Screen Prompts, and Geometric Values Display	131

DISPLAY CONTROL 136
- ZOOM Command 136
- PAN Command 143
- Aerial View 144
- Controlling Display with IntelliMouse 145
- REDRAW Command 145
- REGEN Command 145

SETTING MULTIPLE VIEWPORTS 146
- Creating Tiled Viewports 148

CREATING AND MODIFYING LAYERS 149
- Creating and Managing Layers with the Layer Properties Manager Dialog Box 152
- Making an Object's Layer Current 164
- Undoing Layer Settings 164

SETTING THE LINETYPE SCALE FACTOR 165
WILD CARDS AND NAMED OBJECTS 166
U, UNDO, AND REDO COMMANDS 167
- U Command 167
- UNDO Command 168
- REDO Command 170

REVIEW QUESTIONS 171

Chapter 4 Fundamentals III 177
DRAWING CONSTRUCTION LINES 177
- XLINE Command 177
- RAY Command 179

DRAWING POLYGONS 179
- Polygon Options 180

DRAWING ELLIPSES 182
- Drawing an Ellipse by Specifying Axis End Points 182
- Isometric Circles (or Isocircles) 184

DRAWING POLYLINES 186
DRAWING TEXT 191
- Creating a Single Line of Text 191
- Creating Multiline Text 195

EDITING TEXT 205
- Finding and Replacing Text 206
- Justifying Text 208
- Changing Text from One Space to Another 208
- Spell-Checking 208
- Line Spacing 209
- Controlling the Display of Text 210

CREATING AND MODIFYING TEXT STYLES 210

- CREATING AND MODIFYING TABLES 213
  - Inserting Tables 213
  - Editing Text in a Cell 215
  - Modifying Tables 216
  - Creating and Modifying Table Styles 216
- CREATING OBJECTS FROM EXISTING OBJECTS 219
  - Copying Objects 219
  - Creating a Pattern of Copies 221
  - Creating Parallel Lines, Parallel Curves, and Concentric Circles 226
  - Creating a Mirror Copy of Objects 229
  - Creating a Fillet Between Two Objects 231
  - Creating a Chamfer Between Two Objects 234
- MODIFYING OBJECTS 236
  - Moving Objects 236
  - Trimming Objects 238
  - Erasing Parts of Objects 240
  - Extending Objects to Meet Another Object 241
- REVIEW QUESTIONS 244

Chapter 5 Fundamentals IV 249
- CONSTRUCTING GEOMETRIC FIGURES 249
  - Drawing Solid-Filled Circles 249
  - Drawing Solid-Filled Polygons 250
  - Drawing Point Objects 252
  - Drawing Sketch Line Segments 254
- OBJECT SELECTION 255
  - WPolygon (WP) Selection 257
  - CPolygon (CP) Selection 258
  - Fence (F) Selection 258
  - All Selection 258
  - Multiple Selections 258
  - Box Selection 259
  - Auto Selection 259
  - Undo Selection 259
  - Add Selection 259
  - Remove Selection 259
  - Single Selection 260
- OBJECT SELECTION MODES 260
- MODIFYING OBJECTS 262
  - Lengthening Objects 262
  - Stretching Objects 264
  - Rotating Objects 265
  - Scaling Objects 267
  - Modifying Polylines 269
  - Joining Similar Objects 275
  - Matching Properties 276
- REVIEW QUESTIONS 278

Chapter 6 Fundamentals V 283
MULTILINES 283
- Drawing Multiple Parallel Lines 283
- Editing Multiple Parallel Lines 286
- Creating and Modifying Multiline Styles 292
SPLINE CURVES 297
- Editing Spline Curves 299
WIPEOUT 301
REVISION CLOUD 302
EDITING WITH GRIPS 304
- Using Grips 306
SELECTING OBJECTS BY QUICK SELECT 310
SELECTION SET BY FILTER TOOL 311
CHANGING PROPERTIES OF SELECTED OBJECTS 313
GROUPING OBJECTS 315
INFORMATION ABOUT OBJECTS 319
- LIST Command 319
- DBLIST Command 320
- AREA Command 320
- ID Command 322
- DIST Command 323
SYSTEM VARIABLES 323
REVIEW QUESTIONS 326

Chapter 7 Dimensioning 333
DIMENSION TERMINOLOGY 336
- Dimension Line 337
- Arrowhead 337
- Extension Line 337
- Dimension Text 337
- Leader 337
- Center Mark 337
ASSOCIATIVE, NON-ASSOCIATIVE & EXPLODED DIMENSIONS .. 337
- Associative 337
- Non-Associative 339
- Exploded 339
DIMENSIONING COMMANDS 339
- Linear Dimensioning 339
- Linear Dimensioning by Selecting an Object 343
- Aligned Dimensioning 345
- Ordinate Dimensioning 346
- Radius Dimensioning 348
- Diameter Dimensioning 349
- Arc Length Dimensioning 350
- Angular Dimensioning 351

  Baseline Dimensioning..354
  Continue Dimensioning..356
  Quick Dimensioning...358
  Qleader and Leader...358
  Tolerances..362
  Drawing Cross Marks for Arcs or Circles...368
  Oblique Dimensioning...368
EDITING DIMENSIONS AND DIMENSION TEXT...369
  DIMEDIT Command..369
  DIMTEDIT Command..370
  Editing Dimensions with Grips..371
  DIMDISASSOCIATE Command..371
  DIMREASSOCIATE Command...371
  DIMREGEN Command..372
  Editing Dimensions Using the Shortcut Menu...372
DIMENSION STYLES..373
  Dimension Style Manager Dialog Box...374
OVERRIDING THE DIMENSION FEATURE...391
UPDATING DIMENSIONS..392
REVIEW QUESTIONS..393

Chapter 8 Plotting and Layouts..397
PLANNING THE PLOTTED SHEET...397
  Example of Computing Plot Scale, Plot Size, and Limits..........................399
  Setting for LTSCALE..402
  Setting for DIMSCALE...402
  Scaling Annotations and Symbols...402
WORKING IN MODEL SPACE AND PAPER SPACE..404
PLOTTING FROM MODEL SPACE...404
  Plot Settings...405
PLOTTING FROM PAPER SPACE (WYSIWYG)..414
  Planning to Plot...415
  Setting Up Layouts..417
  Working with Floating Viewports...421
  Scaling Views Relative to Paper Space...431
  Centering Model Space Objects in a Viewport..432
  Hiding Viewport Borders..433
  Controlling the Visibility of Layers within Viewports................................434
  Plotting from Layout..436
RECONFIGURING THE LAYOUT WITH PAGE SETUP.....................................437
  Changing the Current Page Setup..439
  Modifying the Page Setup...439
CREATING A LAYOUT BY LAYOUT WIZARD..441
MAKING THINGS LOOK RIGHT FOR PLOTTING...444
  Setting Paper Space Linetype Scaling..444
  Dimensioning in Model Space and Paper Space.......................................444

CREATING A PLOT STYLE TABLE 445
    Creating a New Plot Style Table 446
    Modifying a Plot Style Table 448
    Changing Plot Style Property for an Object or Layer 452
    Configuring Plotters 454
REVIEW QUESTIONS 459

Chapter 9 Hatching, Gradients, and Boundaries 463

WHAT IS HATCHING? 463
WHAT IS GRADIENT FILL? 464
DEFINING THE HATCH OR GRADIENT BOUNDARY 465
    Selecting Objects Versus Picking a Point 466
HATCH AND GRADIENT FILL WITH THE HATCH COMMAND 467
    Hatch—Related Settings 468
    Gradient—Related Settings 471
HATCH PATTERNS USING TOOL PALETTES 479
EDITING HATCHES AND GRADIENTS 481
CONTROLLING THE VISIBILITY OF HATCH PATTERNS 484
REVIEW QUESTIONS 485

Chapter 10 Block References and Attributes 487

CREATING BLOCKS 488
    Creating a Block Definition 489
INSERTING BLOCK REFERENCES 493
NESTED BLOCKS 495
EXPLODE COMMAND 496
    Possible Changes Caused by the EXPLODE Command 496
    Exploding Block References with Nested Elements 497
BASE COMMAND 497
ATTRIBUTES 498
    A Definition Within a Definition 498
    Visibility and Plotting 499
    Attribute Components 499
    Attribute Commands 500
    Redefining a Block and Its Associated Attributes 515
    Block Attribute Manager 515
DIVIDING OBJECTS 519
MEASURING OBJECTS 520
DYNAMIC BLOCKS 521
    Three Steps to Create a Simple Dynamic Block 522
    Using the Dynamics of a Dynamic Block 525
    The Block Editor—Making a Block Dynamic 527
    Parameters 531
    Action, Parameter, and Grip Properties 542

  Using Actions 543
  Adding Visibility States to Dynamic Blocks 576
  Advanced Dynamic Block Utilities and Features 579
REVIEW QUESTIONS 583

Chapter 11   External References and Images 587
EXTERNAL REFERENCES 587
EXTERNAL REFERENCES AND DEPENDENT SYMBOLS 589
ATTACHING AND MANIPULATING XREFS WITH THE XREF MANAGER 591
  Attaching External Reference Drawings 593
  Detaching External Reference Drawings 596
  Reloading External Reference Drawings 596
  Unloading External Reference Drawings 596
  Binding External Reference Drawings 597
  Opening the External Reference 598
  Changing the Path 598
  Saving the Path 598
ADDING DEPENDENT SYMBOLS TO THE CURRENT DRAWING 598
CONTROLLING THE DISPLAY OF EXTERNAL REFERENCES 599
EDITING REFERENCE FILES/XREF EDIT CONTROL 601
  Adding/Removing Objects from the Working Set 604
MANAGING EXTERNAL REFERENCES 605
IMAGES 607
  Controlling the Display of the Image Objects 610
  Adjusting the Image Settings 611
  Adjusting the Display Quality of Images 612
  Controlling the Transparency of an Image 612
  Controlling the Frame of an Image 613
REVIEW QUESTIONS 614

Chapter 12   AutoCAD DesignCenter 617
DesignCenter WINDOW 618
  Content 618
  Content Type 618
  Container 618
OPENING THE DesignCenter WINDOW 619
POSITIONING THE DesignCenter WINDOW 620
WORKING WITH THE DesignCenter 622
  Folders 622
  Toolbar 622
VIEWING CONTENT 630
  Using the Tree View 631
  Using the Content Area 633
  Viewing Images 635
  Open Drawings 636

History 637
DC Online 638
ADDING CONTENT TO DRAWINGS 643
Layers, Linetypes, Text Styles, and Dimension Styles 643
Blocks 644
Raster Images 645
External References 645
REVIEW QUESTIONS 646

Chapter 13 Utility Commands 649
TOOL PALETTES 649
Tool Palettes Window Shortcut Menus 652
Insert Blocks/Hatch Patterns from a Tool Palette 659
Block Tool Properties 659
Pattern Tool Properties 660
Creating and Populating Tool Palettes 661
PARTIAL LOAD 662
DRAWING PROPERTIES 663
QUICKCALC 665
MANAGING NAMED OBJECTS 668
DELETING UNUSED NAMED OBJECTS 669
COMMAND MODIFIER—MULTIPLE 671
UTILITY DISPLAY COMMANDS 671
Saving Views 671
Controlling the Regeneration 673
Controlling the Drawing of Objects 674
Controlling the Display of Marker Blips 674
CHANGING THE DISPLAY ORDER OF OBJECTS 674
OBJECT PROPERTIES 675
Setting an Object's Color 675
Setting an Object's Linetype 677
LINEWEIGHT Command 681
X, Y, AND Z FILTERS—AN ENHANCEMENT TO OBJECT SNAP 683
Filters with @ 683
Filters with Object Snap 685
SHELL COMMAND 686
SETTING UP A DRAWING 687
LAYER TRANSLATOR 691
TIME COMMAND 693
AUDIT COMMAND 694
OBJECT LINKING AND EMBEDDING (OLE) 695
SECURITY, PASSWORDS, AND ENCRYPTION 703
Passwords 704
Encryption 705
Digital Signature 705

- CUSTOM SETTINGS WITH THE OPTIONS DIALOG BOX 706
  - Files 707
  - Display 708
  - Open And Save 709
  - Plot and Publish 712
  - System 715
  - User Preferences 717
  - Drafting 722
  - Selection 722
  - Profiles 723
- SAVING OBJECTS IN OTHER FILE FORMATS (EXPORTING) 725
- IMPORTING VARIOUS FILE FORMATS 726
- STANDARDS 727
- SLIDES AND SCRIPTS 730
  - Making a Slide 730
  - Viewing a Slide 731
  - Scripts 731
  - Workspaces 736
- REVIEW QUESTIONS 738

Chapter 14 Internet Utilities and Drawing Sets 743
- LAUNCHING THE DEFAULT WEB BROWSER 744
- COMMUNICATION CENTER 745
- OPENING AND SAVING DRAWINGS FROM THE INTERNET 748
- WORKING WITH HYPERLINKS 749
- DESIGN WEB FORMAT 753
  - Using ePlot to create DWF files 753
  - Viewing DWF Files 758
- PUBLISHING AUTOCAD DRAWINGS TO THE WEB 759
- PUBLISH 764
- eTRANSMIT UTILITY 768
- DRAWING SETS 770
  - Creating a New Sheet Set 770
  - Sheet Set Manager 776
  - Viewing and Modifying a Sheet Set 777
  - Placing a View on a Sheet 780
  - Creating a Sheet List Table 783
  - Creating a Transmittal Package 785
  - Creating an Archive of the Sheet Set 786
  - Plotting the Sheet Set and Publishing to DWF 786
- REVIEW QUESTIONS 788

Chapter 15 AutoCAD 3D 791
- WHAT IS 3D? 791
- COORDINATE SYSTEMS 793
  - Right-hand Rule 793

　　　　Setting the Display of the UCS Icon 794
　　　　Defining a New UCS 797
　　　　Selecting a Predefined Orthographic UCS 804
　　　　Viewing a Drawing from Plan View 807
　VIEWING IN 3D 808
　　　　Viewing a Model by Means of the VPOINT Command 808
　　　　Viewing a Model by Means of the DVIEW Command 811
　　　　Using 3DORBIT 818
　　　　Working with Multiple Viewports in 3D 821
　CREATING 3D OBJECTS 822
　　　　2D Drawing Commands in 3D Space 823
　　　　Setting Elevation and Thickness 823
　　　　Creating a Region Object 824
　　　　Drawing 3D Polylines 825
　　　　Creating 3D Faces 825
　CREATING MESHES 828
　　　　Creating a Free-form Polygon Mesh 829
　　　　Creating a 3D Polyface Mesh 830
　　　　Creating a Ruled Surface Between Two Objects 832
　　　　Creating a Tabulated Surface 834
　　　　Creating a Revolved Surface 835
　　　　Creating an Edge Surface with Four Adjoining Sides 836
　　　　Editing Polymesh Surfaces 837
　EDITING IN 3D 837
　　　　Aligning Objects 837
　　　　Rotating Objects About a 3D Object 838
　　　　Mirroring About a Plane 839
　　　　Creating a 3D Array 840
　　　　Extending and Trimming in 3D 841
　CREATING SOLID SHAPES 842
　　　　Creating a Solid Box 843
　　　　Creating a Solid Cone 844
　　　　Creating a Solid Cylinder 845
　　　　Creating a Solid Sphere 846
　　　　Creating a Solid Torus 847
　　　　Creating a Solid Wedge 848
　　　　Creating Solids from Existing 2D Objects 849
　　　　Creating Solids by Means of Revolution 851
　CREATING COMPOSITE SOLIDS 852
　　　　Union Operation 852
　　　　Subtraction Operation 853
　　　　Intersection Operation 855
　EDITING 3D SOLIDS 856
　　　　Chamfering Solids 856
　　　　Filleting Solids 858
　　　　Sectioning Solids 859
　　　　Slicing Solids 860
　　　　Solid Interference 861

Editing Faces of *3D* Solids	862
Editing Edges of *3D* Solids	869
Imprinting Solids	871
Separating Solids	871
Shelling Solids	872
Cleaning Solids	873
Checking Solids	873
MASS PROPERTIES OF A SOLID	874
Hiding Objects	875
PLACING MULTIVIEWS IN PAPER SPACE	875
GENERATING VIEWS IN VIEWPORTS	878
GENERATING PROFILES	878
REVIEW QUESTIONS	880

Chapter 16 Rendering ... 887

SHADING A MODEL	887
RENDERING A MODEL	889
SETTING UP LIGHTS	894
Creating a New Light	895
Modifying a Light	898
Deleting a Light	898
Selecting a Light	898
SETTING UP A SCENE	898
Creating a New Scene	899
Modifying an Existing Scene	900
Deleting a Scene	900
MATERIALS	900
SETTING PREFERENCES FOR RENDERING	903
SAVING AN IMAGE	904
VIEWING AN IMAGE	904
STATISTICS	905
REVIEW QUESTIONS	907

Chapter 17 Customizing AutoCAD ... 911

CUSTOMIZING THE USER INTERFACE	912
Workspaces	913
Toolbars	917
Custom Commands (Macros)	920
Menus	922
Shortcut Menus	924
Keyboard Shortcuts	925
Mouse Buttons	927
Legacy	928
Partial CUI Files	928
EXTERNAL COMMANDS AND ALIASES	929

CUSTOMIZING MENUS WITH MACROS 930
    Menu Macro Syntax 930
CUSTOMIZING HATCH PATTERNS 932
    Custom Hatch Patterns and Trigonometry 935
    Repeating Closed Polygons 938
CUSTOMIZING SHAPES AND TEXT FONTS 942
    Pen Movement Distances and Directions 943
    Special Codes 945
    Text Fonts 949
CUSTOM LINETYPES 951
EXPRESS TOOLS 952
CUSTOMIZING AND PROGRAMMING LANGUAGE 952
REVIEW QUESTIONS 953

Chapter 18 The Tablet and Digitizing 957

TABLET OPERATION 957
    Tablet Configuration 958
CUSTOM MENUS 959
TABLET COMMAND 960
    Transformation Options 964
    The Calibration Table 966
TABLET MODE AND SKETCHING 967
    Editing Sketches 967
    Sketching in Polylines 967
    Linetypes in Sketching 967
REVIEW QUESTIONS 968

Chapter 19 Visual LISP 971

BACKGROUND 971
    VLISP 971
    Interfaces 971
    Integrated Development Environment 972
A TEST DRIVE OF VISUAL LISP 972
    Simple Example 972
    Launch Visual LISP 972
    Toolbars 973
    Developing and Application 974
    Text Editor 974
    Save Your Program 975
    Load Your Program 976
    Run Your Program 976
VLISP FUNDAMENTALS 978
    Project File 978
    Project Window 979
    Compiling VLISP Application (or, So Long Kelvinator) 982

MAKING AN APPLICATION 982
    Simple Wizard...983
    Project File Selections 984
    Review Selections 984
    Expert Wizard..985
PROJECT DEFINITION 986
    Project File...986
    Project Properties 986
VIEW TOOLBAR 988
    Activate AutoCAD..988
    Select Window 988
    Visual LISP Console 988
    Inspect...988
    Trace..988
    Apropos...989
    Watch Window 989
AUTOLISP BASICS 989
    Expressions and Variables 989
    Terminology and Fundamental Concepts 991
    Example 994
    Exercises from the Keyboard 995
LISTS...997
    Types of Arguments 997
    Creating a List..998
EXERCISES 1000
PAUSING FOR USER INPUT 1000
    AutoLISP Functions—The COMMAND Function 1000
    Pause Symbol...1001
    Symbology Used in this Chapter to Describe Functions 1002
ELEMENTARY FUNCTIONS...1002
    Rules of Promotion of an Integer to a Real 1002
    More Elementary Functions...1002
    Trigonometry Functions 1004
LIST HANDLING FUNCTIONS 1004
    CAR, CDR, and Combinations..1006
    CAR and CADR Mainly for Graphics..1010
    Review..1010
TYPE-CHANGING FUNCTIONS..1012
INPUT FUNCTIONS..1013
CONDITIONAL AND LOGIC FUNCTIONS 1017
    Test Expressions 1018
EXERCISES 1020
CUSTOM COMMANDS AND FUNCTIONS..1022
    Arguments and Local Symbols 1022
    Defined Functions and Commands 1023
    Writing a Defined Function 1023
    Writing and Storing a Defined Function...1024

WHAT IS THE DATABASE? 1028
    Entity Names, Entity Data, and Selection Sets 1028
    Entity Name Functions 1028
    Walking Through the Data Base 1028
    Entity Data Functions 1029
    The Association List 1031
    Extracting Data from a List 1033
    Other Entity Data Functions 1034
    Selection Sets 1034
    Advanced Association List and (Scanning) Functions 1036
DEBUG YOUR PROGRAM 1038
    Load Project 1038
    Set Break Point 1039
    Debug Toolbar 1039
    Step Into 1039
    Step Over 1040
    Step Out 1040
    Continue 1040
    Quit 1040
    Reset 1040
    Add Watch 1040
AUTOCAD DEVELOPMENT SYSTEM (ADS) 1040
REVIEW QUESTIONS 1041

Appendix A Hardware Requirements 1043
RECOMMENDED CONFIGURATION 1043

Appendix B Listing of AutoCAD Commands
(Menu and Command Prompt) 1045
DIMENSIONING COMMANDS 1072
DIMENSIONING VARIABLES 1073
OBJECT SELECTION 1077

Appendix C AutoCAD Toolbars 1079

Appendix D System Variables 1083

Appendix E Hatch and Fill Patterns 1109
HATCH PATTERNS 1110
POSTSCRIPT FILL PATERNS 1115

Appendix F Fonts 1117
USING SHX AND PFB FILES 1117

Appendix G Linetypes and Lineweights 1121

Appendix H Command Aliases 1123

Appendix I Express Tools 1129

Index 1133

INTRODUCTION

HARNESSING THE POWER OF AUTOCAD 2006

The key phrase in AutoCAD 2006 is "Focus on the Design." The software engineers at Autodesk recognize the need for the designer/drafter to maintain focus on the task at hand, that is, communicating the design by the creation and manipulation of objects on the drawing screen. In a manner similar to the surgeon who holds out his hand during an operation and says: "scalpel", AutoCAD 2006 puts the necessary design tools in the designer/drafter's hands with minimal distractions, thereby helping the user to continually focus on the center of work, the cursor.

AutoCAD 2006, with its new Dynamic Block feature, has made the powerful BLOCK command even more impressive. Previous versions of AutoCAD allowed you to insert a Block (a pre-drawn combination of objects) into your drawing and change its size or stretch it in the *X* or *Y* direction. The new Dynamic Block capabilities allow changes in size, shape, orientation and visibility of individual objects after the block is inserted. Now, instead of having to maintain a library of hundreds of blocks, a single block can be used to represent hundreds of variations of similar geometries. *Harnessing AutoCAD 2006* contains a comprehensive section describing the new Dynamic Block feature, along with many examples which demonstrate the application of new functions.

It is now easier to customize the User Interface in AutoCAD 2006. A new dialog box has been introduced to allow a single location to customize Menus, Toolbars, Shortcuts and the Workspace.

Harnessing AutoCAD 2006 brings you comprehensive descriptions, explanations, and examples of the basics of AutoCAD 2006 and its new innovations. It has been written to be used both in the classroom as a textbook, and in industry by the professional CADD designer/drafter as a reference and learning tool. Early chapters point out with explanations and examples how the new "Focus on the Design" innovations have been implemented in AutoCAD 2006. In a carryover from AutoCAD 2005, a chapter is dedicated to the application of the Drawing Set feature. Whether you're new to AutoCAD or a seasoned user upgrading your skills, *Harnessing AutoCAD 2006* will show you how to rein in the power of AutoCAD to improve your professional skills and increase your productivity.

HIGHLIGHTS AND FEATURES OF THIS NEW EDITION

The improvements that AutoCAD 2006 brings include the following:

- Viewing and entering data at your point of focus
- Control of dynamic input
- Dynamic Blocks
- Smooth view transitions when panning and zooming
- Dynamic feedback to show which objects have been selected
- Customizable scale list
- Mathematical calculations with the Quick Calculator
- Editing text in place
- Creating numbered and bulleted lists
- Enhancements to dimensions include varying dimension Linetypes, assigned fixed-length extension lines, arc length dimensioning, dimensioning large radii curves, flipping arrows and specifying initial lengths.
- Block improvements to Attribute data extraction
- Ability to perform calculations on table data
- Enhanced control of creating Hatching and Boundaries
- Finding the area of Hatching
- General enhancements to the COPY, MOVE, STRETCH, ROTATE, SCALE, OFFSET, CHAMFER, FILLET, TRIM, EXTEND and RECTANGLE commands
- Improvements to the MLINE creating and editing commands
- Accessing Object Snaps on 3D geometry
- Customizing the user interface with a new dialog box, temporarily overriding settings and enhancing Tool Palettes
- Utilizing workspaces
- Locking Toolbars and Windows
- Finding AutoCAD files from Windows Explorer
- Improved recovery of damaged drawing
- External Reference Bubble Notification
- Preview merged objects
- Buzzsaw Integration
- Enhanced DWF

HOW TO USE THIS BOOK

OVERVIEW
The first chapter of this text provides an overview of the AutoCAD program, its interface, the commands, special features and warnings, and AutoCAD 2006 enhancements. Specific commands are described in detail throughout the book, along with lessons on how to use them.

FUNDAMENTALS
Harnessing AutoCAD 2006 contains five chapters devoted to teaching the fundamentals of AutoCAD 2006. Fundamentals I introduces some of the basic commands and concepts, and Fundamentals II through V continue to build logically on that foundation, until the student has a reasonable competency in the most basic functions of AutoCAD.

INTERMEDIATE
After mastering the fundamentals, you move on to the intermediate topics, which include dimensioning, plotting and printing, hatching and boundaries, blocks and attributes, external references, and drawing environments. Other chapters teach students to make the most of AutoCAD using utility commands, scripts and slides, 3D commands, rendering, and the digitizing tablet.

ADVANCED
For the advanced AutoCAD user, this book offers a chapter on customizing AutoCAD 2006 (including toolbar customization) and Visual LISP. These two chapters teach you to make AutoCAD 2006 more individualized and powerful as you tailor it to your special needs.

APPENDICES
There are nine appendices in the back of this book. Appendix A is an introduction to hardware and software requirements of AutoCAD. Appendix B is a quick reference of AutoCAD commands with a brief description of their basic functions, and Appendix C provides a visual reference of AutoCAD toolbars. Appendix D lists system variables, including default setting, type, whether or not it is read-only, and an explanation of the system variables. To see hatch and fill patterns, fonts, and linetypes provided with AutoCAD, refer to Appendices E, F, and G respectively. Appendix H lists AutoCAD command aliases. Appendix I addresses Express Tools.

STYLE CONVENTIONS

In order to make this text easier for you to use, we have adopted certain typographic conventions that are used throughout the book:

Convention	Example
Command names are in small caps	The MOVE command
Shortcut menu names and Option names italics	Choose *Close* from the shortcut menu are in
Menu names appear with the first letter capitalized	Draw pulldown menu
Toolbar menu names appear with the first letter capitalized	Standard toolbar
Toolbar buttons and icons apper in boldface	ORTHO
Command sequences are indented. User inputs are indicated by boldface. Instructions are indicated by italics and are enclosed in parentheses	**move** Enter variable name or [?]: **snapmode** Enter group name: *(enter group name)*

HOW TO INVOKE COMMANDS

Like most Windows based programs, AutoCAD offers more that one method of invoking a command or accomplishing a particular task. As you progress through the different concepts and skill levels of using AutoCAD and as you use the program in your job, you will want to determine which method best suits your applications.

In early DOS-based versions, interfacing with AutoCAD usually involved typing command names at the "Command line", in the "Command Window" or using the cursor to navigate through the nested levels of the "Side Screen Menu". The Side Screen Menu became obsolete with the migration from DOS to Windows and its Menus and Toolbars for selecting commands. Now, with the new "Cursor-focused" interface, the Command line has taken a back seat as the prime method of interface and command entry.

Almost all commands offer options to the default sequence of prompts and responses. For example, the CIRCLE command's default method of drawing a circle is to specify the center and then the radius. You can override the default by using the center-diameter option or the tangent-tangent option. These options have previously been displayed in the Command Window and also available in the Shortcut menu accessed by right-clicking after invoking the command. This means you can turn off the display of the Command Window, but still access the list of options by right-clicking during a command.

For purposes of expediency, explanations and examples in *Harnessing AutoCAD 2006* assume that the user is entering commands at the On-Screen cursor, rather than entering them at the Command line.

COORDINATE INPUT – WHAT'S THE POINT?

AutoCAD, like all bona fide Computer-Aided-Drafting programs, uses points in a coordinate system when drawing and creating objects. When you select a point on the screen with your pointing device (like you do in a paint program) that point contains highly accurate coordinate information. You can also type in the coordinates of the point you wish to specify, when prompted. In AutoCAD 2006, the new default method of doing this is to type in the coordinates and see the values reported in text boxes adjacent to the new on-screen prompt, referred to as Dynamic Input.

Caution! When you enter coordinates under certain conditions, AutoCAD 2006 automatically prefixes them with a symbol that forces them to be relative to the last point entered. If you want the point coordinates you enter to be absolute (relative to the coordinate system origin), you must either prefix them with the proper symbol or reset the appropriate system variable. See the explanations in Chapter 2 in the section on METHODS TO SPECIFY POINTS and in Chapter 3 in the section on DYNAMIC INPUT, ON-SCREEN PROMPTS, AND GEOMETRIC VALUES DISPLAY.

DRAWING EXERCISES AND BONUS CHAPTER ON VISUAL LISP

Harnessing AutoCAD 2006, like its predecessors, still offers comprehensive learning exercises associated with the lessons in each chapter. These exercises are representative of the types of discipline-related drawing problems found in the design industry today. They can be found on the CD in the back of this book and in the companion exercise book: *Harnessing AutoCAD 2006 Exercise Manual*. The CD contains PDF files of each chapter in the *Harnessing AutoCAD 2006 Exercise Manual*. These chapters correspond to the text in this book and contain a project exercise, as well as exercises developed specifically for the following disciplines: mechanical, architectural, civil, electrical, and piping. Exercise icons and tabs identify the discipline sections (refer to the following table of exercise icons). In addition, a bonus chapter entitled "Visual LISP" is included on the CD in *.pdf* format.

EXERCISE ICONS

Step-by-step Project Exercises are identified by the special icon shown in the following table. Exercises that give you practice with types of drawings that are often found in a particular discipline are identified by the icons shown in the following table.

Type of Exercise	Icon	Type of Exercise	Icon
Project Exercises		Civil	
Electrical		Mechanical	
Piping		Architectural	

HARNESSING AUTOCAD 2006 EXERCISE MANUAL

This printed exercise manual contains project exercises and discipline-specific exercises for Chapters 2 through 11 and Chapter 15 of the core text.

ISBN 1-4180-2036-2.

ONLINE COMPANION™

The Online Companion™ is your link to AutoCAD on the Internet. Updates are posted monthly, including a command of the month, tutorials, and FAQs. To access the Online Companions, go to the following URL:

http://www.autodeskpress.com/resources/olcs/index.aspx

E-RESOURCE

E-resource is an educational resource that creates a truly electronic Classroom. It is a CD-ROM containing tools and instructional resources that enrich your classroom and make your preparation time shorter. The elements of e-resource link directly to the text and tie together to provide a unified instructional system. Spend your time teaching, not preparing to teach.

Features contained in e-resource include:

Syllabus: Lesson plans created by chapter that list goals and discussion topics. You have the option of using these lesson plans with your own course information.

Chapter Hints: Objectives and teaching hints that provide direction on how to present the material and coordinate the subject matter with student projects.

Answers to Review Questions: These solutions enable you to grade and evaluate end of chapter tests.

PowerPoint® Presentation: These slides provide the basis for a lecture outline that helps you to present concepts and material. Key points and concepts can be graphically highlighted for student retention.

Exam View Computerized Test Bank: Over 800 questions of varying levels of difficulty are provided in true/false and multiple-choice formats. Exams can be generated to assess student comprehension, or questions can be made available to the student for self evaluation.

Animations: These .AVI files graphically depict the execution of key concepts and commands in drafting, design, and AutoCAD and let you bring multimedia presentations into the classroom.

Spend your time teaching, not preparing to teach!

ISBN 1-4018-5081-2.

ABOUT THE AUTHORS

Thomas A. Stellman received a B.A. degree in Architecture from Rice University and has over 20 years of experience in the architecture, engineering, and construction industries. He has taught at the college level for over ten years and has been teaching courses in AutoCAD since the introduction of version 2.0 in 1984. He conducts seminars covering both introductory and advanced AutoLISP. In addition, he develops and markets third-party software for AutoCAD.

G.V. Krishnan is director of the Applied Business and Technology Center, University of Houston-Downtown, a Premier Autodesk Training Center. He has used AutoCAD since the introduction of version 1.4 and writes about AutoCAD from the standpoint of a user, instructor, and general CADD consultant to area industries. Since 1985 he has taught courses ranging from basic to advanced levels of AutoCAD, including customization, 3D AutoCAD, solid modeling, and AutoLISP programming.

ACKNOWLEDGMENTS

We would like to thank and acknowledge the many professionals who reviewed the manuscript to help us publish this *Harnessing AutoCAD 2006* text. The authors would like to acknowledge and thank the following staff members of Delmar Thomson Learning:

Publisher: Alar Elken

Senior Acquisitions Editor: James DeVoe

Production Manager: Andrew Crouth

Senior Developmental Editor: John Fisher

Production Editor: Jennifer Hanley

Editorial Assistant: Tom Best

The authors also would like to acknowledge and thank the following:

Composition: John Shanley and Phoenix Creative Graphics

Copyeditor: Margaret Berson

Reviewer: Alex Lepeska, Renton Technical College/Pierce College, Renton, WA

CHAPTER 1

Getting Started

INTRODUCTION

This chapter covers starting AutoCAD, entering commands, and finding your way around the AutoCAD screen.

After completing this chapter, you will be able to do the following:

- Start AutoCAD
- Identify the various parts on the screen
- Use various methods of command and data input
- Obtain help about the AutoCAD commands and features
- Start a new drawing
- Open an existing drawing
- Set up the drawing environment

STARTING AUTOCAD

Designing and drafting is what AutoCAD (and this book) is all about. So how do you get into AutoCAD? Choose the Start button (Windows 2000/ME, Windows NT 4.0, and XP operating systems), navigate to the Autodesk/AutoCAD 2006 program group, and then select the AutoCAD 2006 program. Or on some systems you can double-click on the AutoCAD 2006 startup icon on the Windows desktop to start AutoCAD.

THE STARTUP WINDOW

By default, when AutoCAD is started, it displays a blank drawing window surrounded by menus and toolbars, as shown in Figure 1–1.

Figure 1–1 *AutoCAD 2006 OTB (Out of the Box) Startup window*

The window shown in Figure 1–1 is one of the possible windows that might appear when AutoCAD is opened. This one appears when you start the program for the first time. The layout of the graphics area conforms to a particular set of drawing parameters. Other arrangements of the startup window and layout area are possible, depending on how AutoCAD has been configured. This arrangement along with other settings is referred to as the drawing environment. Individual configurations and arrangements surrounding the drawing area can be named and saved as unique workspaces and then recalled when needed. If you double-click on an AutoCAD drawing file icon in the Windows Explorer window, the layout and settings affecting the drawing area will conform to that drawing's environment within the current workspace arrangement. As covered later in this chapter, it is possible to configure AutoCAD to open with a Startup dialog box to assist you in setting up parameters different from the default opening layout.

Within the AutoCAD 2006 program window, you can create drawings for viewing, printing (referred to as plotting in the trade), solving geometry and engineering problems, accumulating data, creating three-dimensional views of objects, and various other design, graphics, and engineering applications. Whatever your objective is, you will very likely have to make changes in the layout and drawing parameters, or you can configure the startup configuration to suit your needs.

Note: The difference between a workspace environment and a particular drawing environment is that the workspace controls the tools set up on your particular station that you use to make a drawing (menus, toolbars, palettes, etc.), while the drawing environment controls features that are configured for the drawing file itself (units, limits, layers, styles, etc.) and travel with the drawing from station to station.

The Drawing Layout

In this chapter, the significance of the width and height of the graphics area on the Startup window and how they correlate to a final plotted sheet will be dealt with. The three key elements of drafting are location, direction, and distance. When you draw objects on a paper sheet, you choose a starting point on the drawing sheet for the location of a point on the object to be drawn, and an orientation for the object. You must also consider the measurements of the object. In AutoCAD, the graphics area of the screen operates like a zoom lens on a camera through which you are viewing an imaginary drawing sheet. The imaginary drawing area itself is limitless (although you can limit the area in which points can be specified). The AutoCAD graphics area dimensions are relative to dimensions on the imaginary drawing sheet, depending on the "zoom" factor in effect.

One of the great advantages of computer-aided drafting is that objects can be drawn with real-world dimensions. You can zoom in to view a circuit on a 1-millimeter chip so that it fills the screen or you can zoom out to view a map of the United States so that it fills the screen. In either case, AutoCAD allows you draw the objects at their real size and then use the viewing commands and features to display all or part of the objects. Once completed, the true size drawing can be scaled up or down to fit the final plotted sheet.

The initial AutoCAD startup graphics area depends on the video setup of your computer and monitor. It can range from a full view of a 12-unit-wide by 9-unit-high drawing sheet to a 60-unit-wide by 30-unit-high area or even greater. The startup graphics area relative to the startup coordinate system is explained later in this chapter. Features and commands in AutoCAD permit you to move your view around the drawing area and zoom in for a closer look or zoom out to see a broader area.

Note: Do not assume that because the screen drawing area on the monitor is approximately 12" wide by 9" high, the units of measurement (12 units by 9 units) in the drawing must be inches. As described later, one unit can be whatever distance of measurement you need, perhaps a millimeter, or even a mile.

Startup with an Existing Drawing

You can start the AutoCAD program by choosing a drawing file (with the extension of *.dwg*) from the Windows Explorer window and double-clicking on its icon or filename. The AutoCAD program will be started. This is similar to the way other Windows-based programs are started by double-clicking on one of the types of files that is created and edited by that particular program. When AutoCAD is started in this manner (double-clicking on a *.dwg* file), the initial screen will normally display the drawing that was double-clicked using the view in which it was last saved.

AutoCAD SCREEN

The AutoCAD screen (see Figure 1–2) consists of the following elements: the graphics window, status bar and status bar tray, title bar, toolbars, menu bar, and model/layout tabs. Other elements that appear from time to time are tool palettes, dialog boxes, on-screen prompts, the DesignCenter window, Command window, and graphic geometric values.

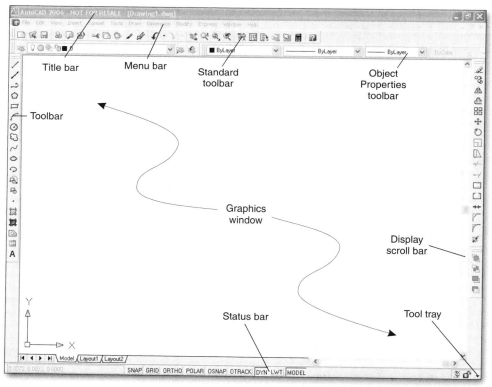

Figure 1–2 *The AutoCAD screen*

GRAPHICS WINDOW

The graphics window is the area on the screen where you can view the objects you create and modify. In this work area, AutoCAD displays the cursor, indicating your current working point. As you move your pointing device (usually a mouse or puck) around on a digitizing tablet, mouse pad, or other suitable surface, the cursor mimics your movements on the screen. Near the cursor is where on-screen prompts and graphic geometric values are displayed.

When AutoCAD prompts you to select a point, the cursor is in the form of crosshairs. When you are required to select an object on the screen, the cursor changes to a small pick box. AutoCAD uses combinations of crosshairs, boxes, dashed rectangles, and arrows in various situations so you can quickly see what type of selection or pick mode is in effect. After objects have been created and are visible on the screen, they become highlighted when the cursor passes over them to indicate that they can be selected for modifying or duplicating.

 Note: It is possible to enter coordinates outside the viewing area for AutoCAD to use for creating objects. As you become more adept at AutoCAD, you may find a need to do this. Until then, working within the viewing area is recommended.

STATUS BAR AND STATUS BAR TRAY

The status bar at the bottom of the screen displays the cursor's coordinates and important information on the status of various modes. On the right end of the status bar is the status bar tray, which contains icons for quick access to the Communications Center, Xref Manager, Locking Toolbar and Windows status, CAD Standards alert, and Digital Signature authenticator, as shown in Figure 1–3.

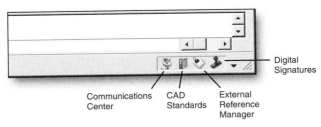

Figure 1–3 *Status bar with tool tray icons*

Communications Center

The Communications Center icon, when selected, causes the Communication Center dialog box to be displayed. A Welcome message appears stating "The Communication Center is your direct connection to the latest software updates, product support announcements and more." From here you specify your country and preferred update frequency, connect to the Internet and download available information, and specify which information channels you wish to view.

External Reference Manager

The Xref manager icon appears when your drawing has an external drawing attached. A message appears when an Xref needs to be reloaded or resolved. Chapter 11 explains using External References.

CAD Standards

The CAD Standards icon appears when there is a standards file associated with your drawing. A message appears when a standards violation occurs. Chapter 13 explains using CAD Standards.

Digital Signatures

The Validate Digital Signatures icon appears when the drawing has a digital signature. Select the icon to validate a digital signature. Chapter 13 explains using Digital Signatures.

Toolbar and Window Locking

The Toolbar and Window Locking icon allows you to lock toolbars and windows to prevent accidentally changing their size or location. You can temporarily override the locking status by holding CTRL while moving a toolbar or window that has been locked.

Controlling Status Bar Display and Tray Settings

The down-arrow at the right end of the status bar tray, when selected, displays a shortcut menu. This menu provides a way to change the appearance of the status bar. Items with a check mark next to them are displayed on the status bar.

Checking CURSOR COORDINATE VALUES causes the coordinate values to be displayed at the cursor location on the left end of the status bar.

Checking any of the button names causes that button to be displayed. These include toggle buttons for snap, grid, ortho, polar tracking, object snap, object snap tracking, dynamic input, lineweight, and model.

Selecting TRAY SETTINGS causes the Tray Settings dialog box to be displayed. From this dialog box you can select **Display icons from services**, which, when cleared, causes the tray to not be displayed. Checking **Display notifications from services** (under which you can select **Display time**, or **Display until closed**) displays notifications from services such as Communications Center, a service to AutoCAD users from Autodesk.

TITLE BAR

The title bar displays the drawing name with the path for the active drawing in the AutoCAD application window.

TOOLBARS

The toolbars contain tools, represented by icons, from which you can invoke commands. Click a toolbar button to invoke a command, and then select options from a dialog box or respond to the on-screen prompts or the prompts on the command line in the Command window. If you position your pointer over a toolbar button and wait a moment, the name of the tool is displayed, as shown in Figure 1–4. This is called the tooltip. In addition to the tooltip, AutoCAD displays a very brief explanation of the function of the command on the status bar.

Figure 1–4 *Toolbar with a ToolTip displayed*

Some of the toolbar buttons have a small triangular symbol in the lower-right corner of the button indicating that there are *flyout* menus underneath that contain subcommands. Figure 1–5 shows the ZOOM command flyout located on the Standard toolbar. The last option utilized from a flyout remains on top to become the default option.

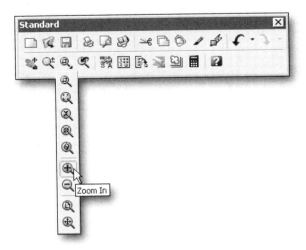

Figure 1–5 *Display of the Zoom flyout located on the Standard toolbar*

You can display multiple toolbars, change their contents, resize them, and dock or float them. A *docked* toolbar attaches to any edge of the graphics window. A *floating* toolbar can lie anywhere on the screen and can be resized. If the AutoCAD window does not fill your monitor screen, you can even locate a floating toolbar outside the window.

Figure 1–6 shows the command icons available on the Standard toolbar; Figure 1–7 shows the commands available on the Properties toolbar. Appendix C lists the available toolbars.

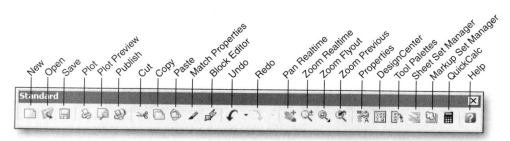

Figure 1–6 *The Standard toolbar*

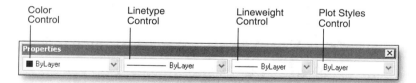

Figure 1–7 *The Properties toolbar*

Selecting the CUSTOMIZE option at the bottom of the shortcut menu that appears when you right-click on a toolbar causes the Customize User Interface dialog to be displayed. The Customize User Interface dialog box allows you to create or modify toolbars. Refer to Chapter 17, "Customizing AutoCAD," for a detailed explanation.

Docking and Undocking a Toolbar

To dock a floating toolbar, position the cursor on the caption, press and hold the pick button on the pointing device. Drag the toolbar to a dock location at the top, bottom, or either side of the graphics window. When the outline of the toolbar appears in the docking area, release the pick button. To undock a toolbar, position the cursor on the left end (for horizontal toolbars) or the top end (for vertical toolbars) of the toolbar and drag and drop it outside the docking regions. To place a toolbar in a docking region without docking it, press CTRL as you drag. By default, the Standard toolbar and the Properties toolbar are docked at the top of the graphics window (see Figure 1–2). Figure 1–8 shows the Standard toolbar and the Properties toolbar docked at the top of the graphics window, and the Draw and Modify toolbars docked on the sides of the graphics window.

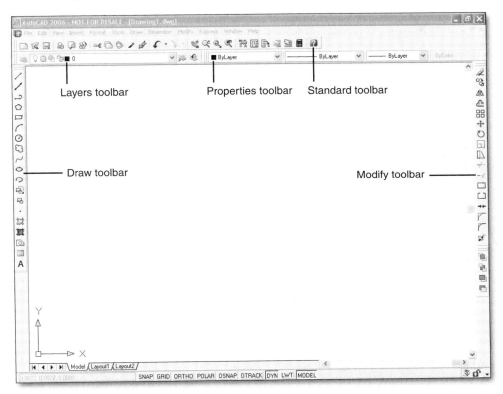

Figure 1–8 *Docking of toolbars in the graphics window*

Resizing a Floating Toolbar

If necessary, you can resize a floating toolbar. To resize a floating toolbar, position the cursor anywhere on the border of the toolbar, and drag it in the direction you want to resize. Figure 1–9 shows different combinations of resizing the Draw toolbar.

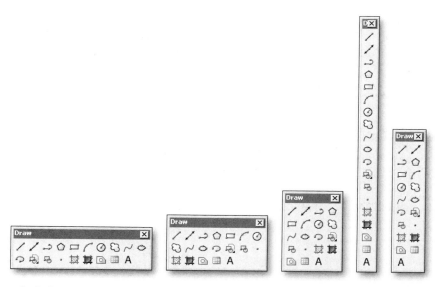

Figure 1–9 *Draw toolbar in different resizing positions*

Closing a Floating Toolbar

To close a floating toolbar, position the cursor on the X located in the upper right corner of the toolbar (see Figure 1–10), and press the pick button on your pointing device. The toolbar will disappear from the graphics window.

Figure 1–10 *Positioning the cursor to close a toolbar*

Opening a Toolbar or Closing a Docked Toolbar

AutoCAD 2006 comes with 30 regular toolbars and 4 "ET" toolbars. To open any of the closed regular toolbars, place the cursor anywhere on any docked or floating toolbar that is already open and press the right button on your pointing device. A shortcut menu appears, listing all the available toolbars (see Figure 1–11) with a check beside each open toolbar. Select the closed (unchecked) toolbar you wish to open. You can also close an open toolbar by selecting it, thereby removing the check mark next to its name. To open an "ET" toolbar, instead of right-clicking on a toolbar, right-click on the blank area behind where toolbars can be placed just off the screen (not on the blank area to the right of the menus). Choose *EXPRESS* from the shortcut menu and then choose *ET:BLOCKS*, *ET:LAYERS*, *ET:STANDARD*, or *ET:TEXT*.

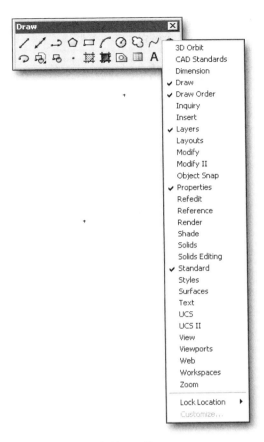

Figure 1–11 *Shortcut menu listing available toolbars*

Note: It is advisable to always keep at least one toolbar visible on the screen. Having a toolbar visible makes it possible to right-click on it to cause the toolbar shortcut menu to appear. This provides easy access for turning other toolbars on and off. If all toolbars have been closed, you must use the TOOLBAR command to display the Customize User Interface dialog box (explained in Chapter 17) to open at least one toolbar.

Locking and Unlocking Toolbars and Windows

When you right-click on a toolbar, it causes a shortcut menu to be displayed. Selecting LOCK LOCATION at the bottom of the shortcut menu causes a flyout menu to appear (see Figure 1–12) with options to lock and unlock windows. You can lock and unlock floating and docked toolbars and windows. When one of the options is in the locked mode, a check will be displayed beside that option. Toolbars and windows that are locked cannot be repositioned or resized. This prevents inadvertently changing a toolbar or window position or size. The ALL option allows you to lock or unlock all floating and docked toolbars and windows in one step.

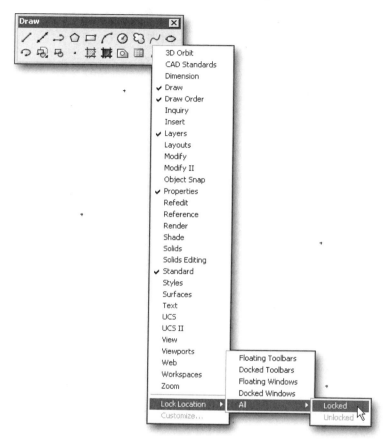

Figure 1-12 *Shortcut menu with Lock Location flyout displayed*

TOOL PALETTES WINDOW

Tool palettes are tabbed areas within the Tool Palettes window that provide an efficient method for organizing, sharing, and placing blocks and hatches. Tool palettes can also contain custom tools provided by third-party developers. Blocks (see Chapter 10) and hatch patterns (see Chapter 9) are managed with tool palettes. You can also create a tool on a tool palette that executes a single AutoCAD command or a string of commands. The Tool Palettes feature allows blocks and hatch patterns of similar usage and type to be grouped in their own tool palette. For example, one tool palette is named **Electrical** and contains, of course, blocks representing electrical symbols as shown in Figure 1-13.

Chapter 1 • *Getting Started* 13

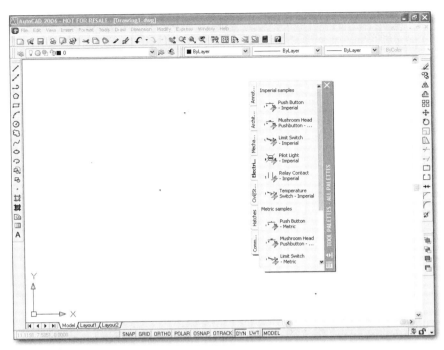

Figure 1–13 *The Tool Palettes window in the floating position*

Figure 1–13 shows the default Tool Palettes window that comes with AutoCAD 2006. Other tool palettes attached to the Tool Palettes window are **Annotation**, **Architectural**, **Mechanical**, **Electrical**, **Civil/Structural**, **Hatches**, and **Command Tools**, all of which contain icons representing blocks, hatch patterns, or commands. The TOOLPALETTES command opens the Tool Palettes window.

The default position for the Tool Palettes window is floating on the right side of the screen. When the Tool Palettes window is undocked, it can be docked by double-clicking in the title bar (which may be on the left or right side of the window) or by placing the cursor over the title bar and dragging the window all the way to the side where you wish to dock it. Its position can be changed by placing the cursor over the double line bar at the top of the window and either double-clicking or dragging the window into the screen area (or across to a docking position on the right side of the screen as shown in Figure 1–14). Double-clicking causes the Tool Palettes window to docked and to float in the drawing area as shown in Figure 1–13.

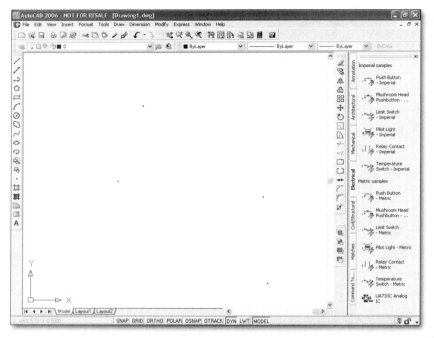

Figure 1–14 *The Tool Palettes window in the docked position with the Electrical tool palette displayed*

To insert a block from a tool palette, simply place the cursor on the block symbol in the tool palette, press the pick button, and drag the symbol into the drawing area. The block will be inserted at the point where the cursor is located when the pick button is released. This procedure is best implemented by using the appropriate Object Snap mode (see Chapter 3 for Object Snap modes).

To use a tool created from a hatch, click a hatch tool and drag it to an object in the drawing.

Once you add a command to a tool palette, you can click the tool to execute the command. For example, clicking a New tool on a tool palette creates a new drawing just as the New button on the Standard toolbar does. You can also create a tool that executes a string of commands or customized commands, such as an AutoLISP® routine, a VBA macro or application, or a script.

MENU BAR

Pull-down menus are available from the menu bar at the top of the screen. To select any of the available commands, move the cursor into the menu bar area and press the pick button on your pointing device, which pops that menu bar onto the screen (see Figure 1–15). To select a command or feature from the list, move the cursor until the desired item is highlighted and press the pick button on the pointing device. If a menu item has an arrow to the right, it has a cascading submenu. To display the submenu, move the pointer over the item and press the pick button. Menu items that include ellipses (...) display dialog boxes. To select one of these, pick that menu item.

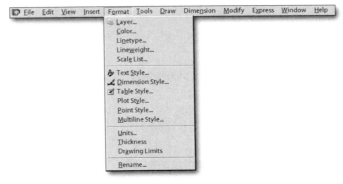

Figure 1–15 *Example of a menu bar*

MODEL TAB AND LAYOUT TAB
AutoCAD allows you to change the drawing environment between model space (for drawing objects) and paper space (for layouts for plotting). You generally create your designs in model space, and then create layouts to plot your drawing in paper space. Refer to Chapter 8 for a detailed explanation on working with and plotting from layouts.

FOCUSING ON THE DESIGN — DYNAMIC INPUT
Dynamic input allows you to "keep your eyes on the road" when working by providing a prompt and input interface where you work, on the screen, at the cursor. How to customize dynamic input is explained in Chapter 3 (in the "Dynamic Input, On-Screen Prompts, and Geometric Values Display" section). Previous versions of AutoCAD emphasized the need to keep a constant vigil on the Command line in the Command window (discussed next). This meant having to look back and forth between the Command window and the point where you were working. Now, with the on-screen (near the cursor) interface, almost everything you need to know about what is going on with the program and your current work is right there, where you are working.

Dynamic Input and Feedback On-Screen
Figure 1–16 shows four stages of the information displayed on the screen during the process of drawing a circle. The first view is how the cursor appears when AutoCAD is ready for you to enter a command. The second view shows the text box that appears if you type in a command name from the keyboard (this step is skipped if you choose

to initiate the command by another method like selecting it from a toolbar). The third view shows the prompt that appears, asking you to specify a center point for the circle. The "or" and the down arrow "" indicate that options (other than just entering the center point) are available and that this is the time to choose one of them. The fourth view shows the graphics feedback that appears while you are being prompted to input the circle radius. This type of feedback varies with the command in effect. The prompt for the second point of a line being drawn might include both the distance and the angle.

Note: Make sure **DYN** located on the status bar is depressed to enable the dynamic input feature.

CURSOR   COMMAND INPUT   COMMAND PROMPT   GRAPHIC FEEDBACK

Figure 1–16 *On-Screen Input, Prompts, and Graphic Feedback*

Locking in Values During Input

Responding to a prompt that is asking for a point or vector normally requires you to enter two values; an *X* coordinate value and a *Y* coordinate value when using rectangular coordinate input or a distance and an angle when using polar coordinate input. If you wish to type in the first value, lock it in, and then use the cursor to specify the second value, you can press TAB after typing the value and AutoCAD displays a lock icon alongside the value. You can then move the cursor to specify the second required point.

COMMAND WINDOW

The Command window is a window in which you enter commands and in which AutoCAD displays prompts and messages. The Command window can be a floating window with a caption and frame. You can move the floating Command window anywhere on the screen and resize its width and height by dragging a side, bottom, or corner of the window. With the introduction of the On-Screen input/prompt/graphics feedback feature in AutoCAD 2006, the Command window has taken a "back seat" and is no longer the primary place to interface with the program. It is now possible to do most of the design/drafting work with the Command window closed. However, if you close the window, you will need to press CTRL+9 or invoke the COMMANDLINE command to open the Command window.

There are two components to the Command window: the single command line where AutoCAD prompts for input and you see your input echoed back, as shown in Figure 1–17, and the command history area, which shows what has transpired in the current drawing session. One display of the single command line remains at the bottom of the screen when the Command window is open.

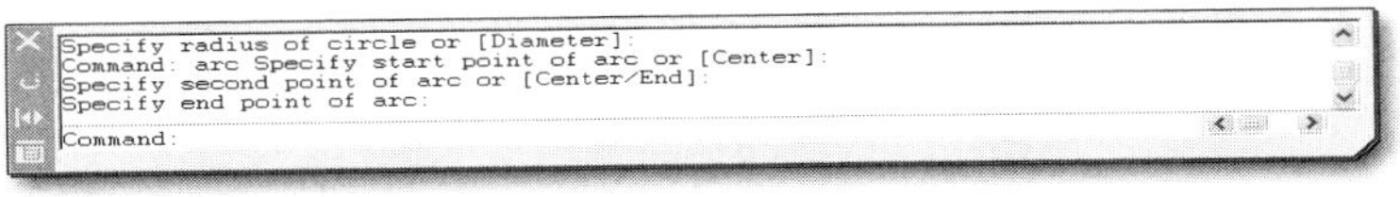

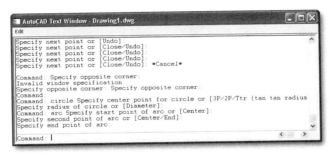

Figure 1–17 *Command window*

The command history area can be enlarged like other windows by picking the top edge and dragging it to a new size. You can also scroll inside the enlarged area to see previous command activity by using the scroll bars (see Figure 1–18).

Figure 1–18 *Command history*

When you press F2, the command history text window switches between being displayed and being hidden. When the text window is displayed, you can scroll through the command history.

When you see "Command:" displayed in the Command window, it signals that AutoCAD is ready to accept a command. After you enter a command name and press ENTER or select a command from one of the menus or toolbars, the prompt area continues to inform you of the type of response(s) that you must furnish, until the command is either completed or terminated. For example, when you pick the LINE command, the prompt displays "Specify first point". After selecting a starting point by appropriate means, you will see "Specify next point or [Undo]", asking for the endpoint of the line. This duplicates what is displayed in the On-Screen prompt.

Each command has its own series of prompts. The prompts that appear when a particular command is used in one situation may differ from the prompts or sequence of prompts when invoked in another situation. You will become familiar with the prompts as you learn to use each command.

When you enter the command name or give any other response by typing from the keyboard, make sure to press ENTER or SPACEBAR. Pressing ENTER sends the input to the program for processing. For example, after you enter **line**, you must press ENTER or SPACEBAR in order for AutoCAD to start the part of the program that lets you draw lines. If you type **lin** and press ENTER or SPACEBAR, you will get an error message, unless someone has customized the program and created a command alias or command named "lin." Likewise, typing **lines** and pressing ENTER or SPACEBAR is not a standard AutoCAD command.

Pressing SPACEBAR has the same function as ENTER except when entering strings of words, letters, or numbers in response to the TEXT and MTEXT commands.

To repeat the previous command, you can press ENTER or SPACEBAR at the "Command:" prompt. A few commands skip some of their normal prompts and assume default settings when repeated in this manner.

Terminating a Command

There are three ways to terminate a command:

- Complete the command sequence and return to the "Command:" prompt.
- Press ESC to terminate the command before it is completed.
- Invoke another command from one of the menus or toolbars, which automatically cancels any command in progress.

Note: All the examples of prompts and responses to commands in this book are based on what is displayed on-screen with dynamic input active.

AUTOCAD COMMANDS AND INPUT METHODS

This section introduces the methods available to initiate or invoke AutoCAD commands.

As much as possible, AutoCAD divides commands into related categories. For example, "Draw" is not a command, but a category of commands used for creating primary objects such as lines, circles, arcs, text (lettering), and other useful objects that are visible on the screen. Categories include **Modify**, **View**, and **Tools**, listing various commands and tools that will help in managing AutoCAD drawing. The commands under **Format** are also referred to as drawing aids and utility commands throughout the book. Learning the program can progress at a better pace if the concepts and commands are mentally grouped into their proper categories. This not only helps you find them when you need them, but also helps you grasp the fundamentals of computer-aided drafting more quickly.

INPUT METHODS

There are several ways to input an AutoCAD command: the keyboard, toolbars, menu bars, the side screen menu, dialog boxes, the shortcut menu, or a digitizing tablet.

Keyboard

To invoke a command from the keyboard, simply type the command name at the On-Screen prompt ("Command:" prompt if you are using the Command window) and then press ENTER or SPACEBAR (ENTER and SPACEBAR are interchangeable except when entering a space in a text string).

To repeat a command you have just used, press ENTER, SPACEBAR, or right-click and choose REPEAT <LAST COMMAND> from the shortcut menu. If you are using the Command window, then you can also repeat a command by using the UP-ARROW and DOWN-ARROW keys to display the commands you previously entered from the keyboard. Use the UP-ARROW key to display the previous line in the command history; use the DOWN-ARROW key to display the next line in the command history. Depending on the buffer size, AutoCAD stores all the information you entered from the keyboard in the current session.

AutoCAD also allows you to use certain commands transparently, which means they can be entered on the command line while you are using another command. Transparent commands are usually commands that change drawing settings or drawing tools, such as GRID, SNAP, and ZOOM. To invoke a command transparently, enter an apostrophe (') before the command name while you are using another command. After the transparent command is completed, the original command resumes.

Toolbars

The toolbars contain tools that represent commands. Click a toolbar button to invoke the command, and then select options from a dialog box or follow the prompts on-screen or on the command line in the Command window.

Menus

The menus are available from the menu bar at the top of the screen. You can invoke almost all of the available commands from the menu bar. You can choose menu options in one of the following ways:

- First, select the menu name to display a list of available commands, and then select the appropriate command.
- Press and hold down ALT and then enter the underlined letter in the menu name. For example, to invoke the LINE command, first hold down ALT then press D (that is, press ALT+D) to open the Draw menu, and then press L.

The default menu file (customized user interface) is *acad.cui*. You can load a different menu file by invoking the MENU command.

Dialog Boxes

Many commands, when invoked, cause a dialog box to appear unless you prefix the command with a hyphen. For example, entering **insert** causes the dialog box to be displayed (see Figure 1–19), and entering **-insert** causes responses to be displayed in the On-Screen prompt area. Dialog boxes display the lists and descriptions of options, long rectangles for receiving your input data, and, in general, are the more convenient and user-friendly method of communicating with the AutoCAD program for that particular command.

The commands listed in the menu bar that include ellipses (...), such as **Plot**... and **Hatch**..., display dialog boxes when selected. AutoCAD dialog boxes have features that are similar to Windows file management dialog boxes.

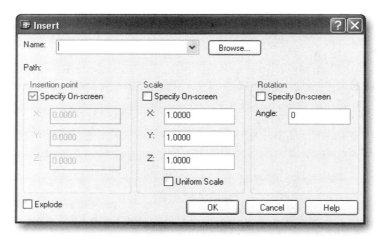

Figure 1–19 *Dialog box invoked from the* INSERT *command*

Cursor Menu

The AutoCAD cursor menu (see Figure 1–20) appears at the location of the cursor when you press the middle button on a three-or-more-button mouse. On a two-button mouse you can invoke this feature by pressing SHIFT and right-clicking. On a two-button mouse, the right button usually causes the shortcut menu to appear. The cursor menu (different from the shortcut menu) includes the handy Object Snap mode options along with the *X,Y,Z* filters. The reason that the Object Snap modes and Tracking are in such ready access will become evident when you learn the significance of these functions.

Figure 1–20 *Cursor menu*

Shortcut Menu

The AutoCAD shortcut menu appears at the location of the cursor when you press the right button (right-click) on the pointing device. The contents of the shortcut menu depend on the situation at hand.

If you right-click in the drawing window when there are no commands in effect, then the shortcut menu will include options to repeat the last command, a section for editing objects such as CUT and COPY, a section with UNDO, PAN, and ZOOM, and a section with QUICK SELECT, FIND, and OPTIONS (see Figure 1–21). Selections that cannot be invoked under the current situation will appear in lighter text than those that can be invoked.

Figure 1–21 *Shortcut menu when no command is in effect*

If you select one or more objects (under the default setup conditions), and no commands are in effect, then right-click, and the shortcut menu will include some of the editing commands, as shown in Figure 1–22.

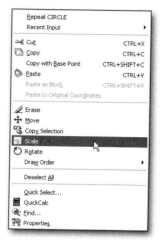

Figure 1–22 *Shortcut menu with one or more objects selected when no command is in effect*

After you have initiated a command and wish to use an option other than the default option, you can invoke the shortcut menu and select the desired option with the mouse. For example, if instead of the default center-radius method of drawing a circle, you wished to use the TTR (tangent-tangent-radius), 2P (two-point), or 3P (three-point) option, you can select one of them from the shortcut menu, as shown in Figure 1–23. You can also select the PAN and ZOOM commands (transparently) or cancel the command. If pressing ENTER is required, that is also available. If you are using dynamic input, then you can also access the available options by pressing the down-arrow key.

Figure 1–23 *Shortcut menu when the CIRCLE command is in effect*

If the Command window is open, right-click anywhere in it and the shortcut menu provides an access to the six most recently used commands (see Figure 1–24).

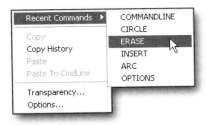

Figure 1–24 *Shortcut menu while the cursor is in the Command window*

Right-click on any of the buttons in the status bar, and the shortcut menu provides toggle options for drawing tools and a means to modify their settings.

Right-click on the **Model** tab or **Layout** tabs in the lower left corner of the drawing area, and the shortcut menu provides display plotting, page setup, and various layout options.

Right-click on any of the open AutoCAD dialog boxes and windows, and the shortcut menu provides context-specific options. Figure 1–25 shows an example for the Layer Properties Manager dialog box with a shortcut menu.

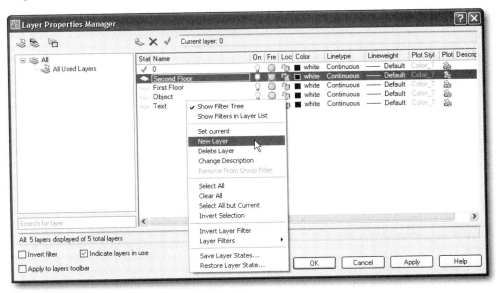

Figure 1–25 *Shortcut menu in the Layer Properties Manager dialog box*

Digitizing Tablet

Another input device in addition to the mouse is the digitizing tablet. It combines the screen cursor control of a mouse with its own printed menu areas for selecting items. However, with the new Heads-up features and customizability in AutoCAD since release 2000, the tablet overlay with command entries is practically obsolete. One powerful feature of the tablet (not related to entering commands) is that it allows you to lay a map or other picture on the tablet and trace over it with the puck (the specific pointing device for a digitizing tablet), thereby transferring the objects to the AutoCAD drawing. The new interfaces with other platforms that allow you to insert a picture of an aerial photograph, for example, make tablet digitizing less of a demand.

Side Screen Menu

The side screen menu provides another, traditional way to enter AutoCAD commands. By default, the side screen menu is turned off. While this book does not refer to the side screen menu, users who are familiar with DOS AutoCAD may be more comfortable using it. To display the side screen menu, choose **Display screen menu** from the **Display** tab of the Options dialog box. AutoCAD displays the side screen menu (see Figure 1–26).

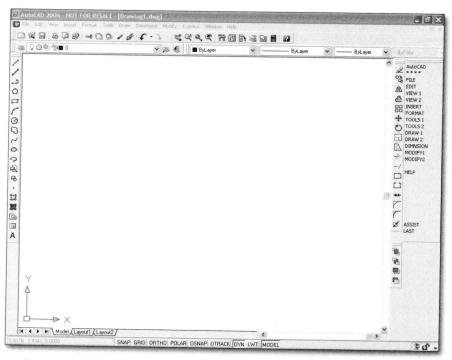

Figure 1–26 *AutoCAD screen window with the side screen menu*

Moving the pointing device to the right will cause the cursor to move into the screen menu area. Moving the cursor up and down in the menu area will cause selectable items to be highlighted. When the desired item is highlighted, you can choose that item by pressing the designated pick button on the pointing device. If the item is a command, either it will be put into action or the menu area will be changed to a list of actions that are options of that command. The screen menu is made up of menus and submenus. At the top of every screen menu is the word *AutoCAD*. When selected, it will return you to what is called the *root menu*. The root menu is the menu that is displayed when you first enter AutoCAD. It lists the primary classifications of commands or functions available.

GETTING HELP

Help is available through either a continuously resident help window, through a traditional Windows-type help interface, or through Online help on the Internet.

INFO PALETTE

Invoking the ASSIST command or choosing the Info Palette from the Help menu provides automatic or on-demand context-sensitive quick help in the form of an Info Palette (see Figure 1–27).

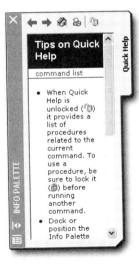

Figure 1–27 *Info Palette*

By dragging a side, bottom, or corner of the Info Palette, you can enlarge it to display more (sometimes all) of the information in the palette.

While the Info Palette is being displayed, if you invoke a command, the Info Palette will display information about the command just invoked. For example, when you invoke the CIRCLE command, Help topics about the CIRCLE command will be displayed (see Figure 1–28). When you select one of the Help topics, the quick help is displayed with information about the topic selected. The procedure that is displayed in the palette can be locked by choosing the Lock from the toolbar located in the Info Palette. If you lock the procedure, when you invoke another command, the procedure will not reflect the information of the active command. Close the Info Palette to terminate the on-demand context-sensitive quick help.

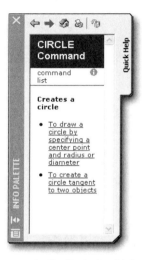

Figure 1–28 *The Info Palette automatically displaying Help topics about the* CIRCLE *command*

TRADITIONAL HELP

The Help window provides a context-sensitive help facility to list commands and what they do. The Help window provides online assistance within AutoCAD. When an invalid command is entered, AutoCAD displays a message to remind you of the availability of the Help facility.

The Help window can be opened transparently while you are in the middle of a command. For example, to use Help transparently, enter **'help** or **'?** in response to any prompt that is not asking for a text string. AutoCAD displays help for the current command. Often the help is general in nature, but sometimes it is specific to the command's current prompt.

As an alternative, press the function key F1 to open the Help window. You can also open the Help window by choosing the Help menu from the menu bar at the top of your screen or choose Help from the Standard toolbar (see Figure 1–29). AutoCAD displays the AutoCAD 2006 Help window (see Figure 1–30).

Figure 1–29 *Invoking the* HELP *command from the Standard toolbar*

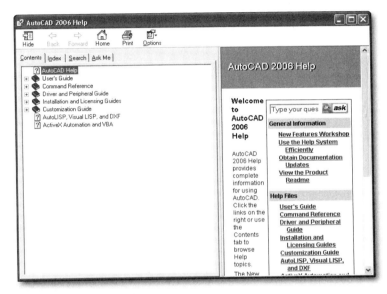

Figure 1–30 *AutoCAD 2006 Help*

Help switches to an independent window, so that when you are through with the help utility, you will need to switch to the AutoCAD program window to continue drawing. You do not have to close the AutoCAD 2006 Help window to work in an AutoCAD drawing session.

> At the top of the AutoCAD Help:User Documentation window is a button that will let you hide the tab (left) side, therefore shrinking the size of the window by choosing **Hide** or, if it has been shrunk, enlarge the window and show the tabs by choosing **Show**. The information (right) side is the instruction or description area where information about the selected subject or command is displayed. The tab side has text boxes and list boxes to aid in getting the help you need on all subjects and commands in AutoCAD.
>
> Choosing **Back** returns you to the previous screen when possible. Choosing **Forward** reverses the action of **Back**.

Choosing **Home** will take you to the home page of the user documentation.

Choosing **Print** sends the contents of the information area to the printer.

The **Contents** tab of the AutoCAD 2006 Help dialog box presents an overview of the available documentation in a list of topics and subtopics. It allows you to browse by selecting and expanding topics.

The **Index** tab displays an alphabetical list of keywords related to the topics listed on the **Contents** tab. You can access information quickly when you already know the name of a feature, command, or operation, or when you know what action you want AutoCAD to perform.

The **Search** tab provides full-text search of all the topics listed on the **Contents** tab. It allows you to perform an exhaustive search for a specific word or phrase. It displays a ranked list of topics that contain the word or words entered in the keyword field.

The **Ask Me** tab allows you to find information using a question phrased in everyday language. It displays a ranked list of topics that correspond to the word or phrase entered in the question field.

NEW FEATURES WORKSHOP

The **New Features Workshop** option of the **Help** menu, when selected, causes the New Features Workshop dialog box to be displayed. New features introduced in AutoCAD 2006 are explained with examples.

DEVELOPER HELP

The **Developer Help** option of the **Help** menu, found under the **Additional Resources** cascading menu, when selected, causes the AutoCAD 2006 Help: Developer Documentation dialog box to be displayed. The topics in this dialog box have to do with subjects important to third-party developers such as customization, AutoLISP, DXF, ActiveX, and VBA applications.

ONLINE RESOURCES

The **Additional Resources** option of the **Help** menu, when selected, displays a submenu with options for **Support Knowledge Base, Online Training Resources, Online Developer Center,** and **Autodesk User Group International**, in addition to **Developer Help**. Each of these options, when selected, launches your Internet browser and links to the Autodesk address associated with the option selected.

BEGINNING A NEW DRAWING

When the first new drawing is started in an AutoCAD drawing session, it is given the temporary name *drawing1.dwg*. It will not be saved with a name of your choice until you use a form of the SAVE command. The second new drawing in a session is given the temporary name *drawing2.dwg* (and so on).

The setting of the system variable STARTUP affects what you see on the screen when an AutoCAD drawing session is begun. It also controls the type of dialog box that is displayed when the NEW command is invoked. Here we discuss using the NEW command when AutoCAD is configured with the system variable STARTUP set to 0 (default setting).

The NEW command allows you to create a new drawing. With the system variable STARTUP set to 0 (default setting), when the AutoCAD program is started, it opens up with a new drawing from "scratch" using the *acad.dwt* template file to put you right into a drawing session with the temporary name *drawing1.dwg*. Therefore, when you started AutoCAD, it automatically invoked the NEW command once. If you invoke the NEW command in this situation, it is really for the second time in the session. AutoCAD prompts for the selection of a template and will utilize the temporary name *drawing2.dwg*.

To create a new drawing, invoke the NEW command from the Standard toolbar, by selecting **QNew** (see Figure 1–31). AutoCAD displays the Select template dialog box (see Figure 1-32).

Figure 1–31 *Invoking the* NEW *command from the Standard toolbar*

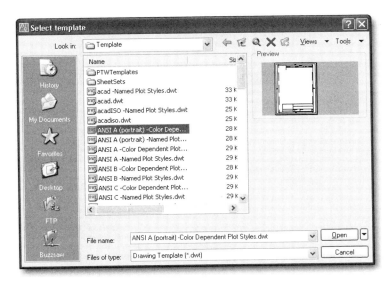

Figure 1–32 *Select template dialog box*

The Select template dialog box operates like a Windows file management dialog box. It contains a **Preview** window that will show a thumbnail sketch (if it is available) of the template file selected.

 Note: The NEW command causes one dialog box to be displayed when the STARTUP system variable is set to 0 and a different dialog box to be displayed when STARTUP is set to 1 (see the explanation in the following section of this chapter)

A drawing template file is a drawing file with selected parameters already preset to meet certain requirements, so that you do not have to go through the process of setting them up each time you wish to begin drawing with those parameters. A template drawing might have the imaginary drawing sheet dimensions preset, or the units of measurement preset, or it could contain objects already drawn. In many cases a blank title block has already been created on a standard sheet size. Or the type of coordinate system that is needed to make the drawing could already be set up with the origin (*X*-, *Y*-, and *Z*-coordinates of 0,0,0) located where needed relative to the edges of the envisioned drawing sheet.

The AutoCAD program files include over 60 templates for drawings of various standard sizes containing a pre-drawn title block conforming to standards such as ANSI, DIN, Gb, ISO, and JIS. You can create templates by making a drawing with the desired preset parameters and pre-drawn objects and then saving the drawing as a template file with the extension of *dwt*.

OPENING AN EXISTING DRAWING

The OPEN command allows you to open an existing drawing. Invoke the OPEN command from the Standard toolbar, by selecting **Open** (see Figure 1–33) and AutoCAD displays the Select File dialog box (see Figure 1–34).

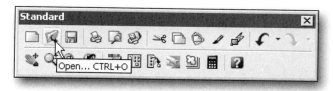

Figure 1–33 *Invoking the OPEN command from the Standard toolbar*

The Select File dialog box is similar to the standard file selection dialog box, except that it includes options for selecting an initial view and for setting **Open Read-Only**, **Partial Open**, and **Partial Open Read-Only** modes. In addition, when you click on the file name, AutoCAD displays a bitmap image in the **Preview** section. And, there is a window on the left side of the dialog box displaying quick access icons to folders on your computer: **Desktop** and **My Documents**; icons for **History** (recently opened drawings) and **Favorites**; and locations on the Internet: **Buzzsaw** and **FTP**. Select the drawing from the appropriate folder and choose **Open** to open the drawing. You can also select multiple drawing files to open at the same time.

Figure 1–34 *Select File dialog box*

The **Select Initial View** check box permits you to specify a view name in the named drawing to be the startup view. If there are named views in the drawing, an M or P beside their name will tell if the view is model or paper space, respectively, as shown in the Initial View dialog.

Buttons to the right of the **Look in** list box are **Back to <the last folder>**, **Up one level**, **Search the Web**, **Delete**, **Create New Folder**, **Views**, and **Tools**. The **Back to <the last folder>**, **Up one level**, **Search the Web**, **Delete**, and **Create New Folder** options are similar to most file handling dialog boxes for reaching the location of the drawing you wish to open.

Choosing **Views** causes a menu to be displayed with the options of **List, Details,** or **Thumbnails** to determine how folders and files are displayed and a **Preview** option that opens a **Preview** area to show a thumbnail sketch of drawing selected.

Choosing **Tools** causes a menu to be displayed with the options of **Find, Locate, Add/Modify FTP Locations, Add current folder to Places,** and **Add to Favorites**.

Choosing **Find** causes the Find dialog box to be displayed (see Figure 1–35). Various drives and folders are searched using search criteria. The Find dialog box combines the usual Windows file/path search of files by name and location in the **Name & Location** tab and by date ranges in the **Date Modified** tab.

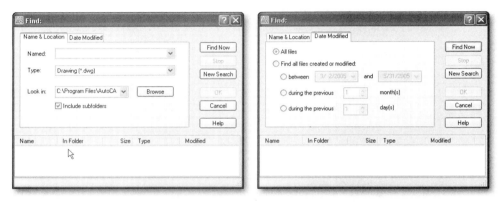

Figure 1–35 *Name and Location tab of the Find dialog box and Date Modified tab of the Find dialog box*

The **Date Modified** tab (see Figure 1–35), lets you specify dates and/or date ranges as criteria to search for drawings.

Note: You can open and edit an AutoCAD Release 12, 13, or 14 drawing. If necessary, you can save the drawing in other formats by using the SAVEAS command. Possible formats include 2006, 2004, and 2000/LT2000 Drawings [*.*dwg*], Drawing Standards [*.*dws*], Template [*.*dwt*], 2006, 2004, 2000/LT2000 DXF [*.*dxf*], and R12/LT 2 DXF [*.*dxf*]. Certain limitations apply when doing this, which are explained in the section about the SAVEAS command later in this chapter.

A drop-down menu is displayed when you select the down arrow to the right of the **Open** button. From this menu you may open a drawing in the **Open Read-Only** option, which permits you to view the drawing but not save it with its current name. You can open a drawing with the **Partial Open** option, which when selected displays the Partial Open dialog box (see Figure 1–36). It displays the drawing views and layers available for specifying what geometry to load into the selected drawing. When working with large drawing files, you can select the minimal amount of geometry you need to load when opening a drawing. You can also select the **Partial Open Read-Only** option and save it under a different name by invoking the SAVEAS command.

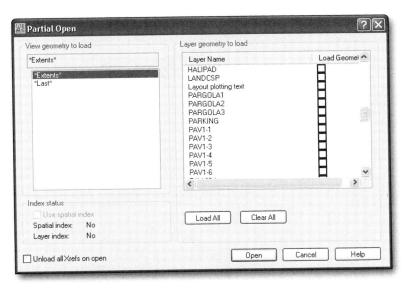

Figure 1–36 *Partial Open dialog box*

 Note: Even though a drawing is partially open, all objects are still loaded. All layers are available, but only the specified layers will have their geometry appear in the drawing when it is opened.

SETTING UP THE DRAWING ENVIRONMENT

This section covers the commands and features used to communicate the physical appearance of objects. A rectangle might represent a very small computer chip on a printed circuit, a building, or the state of Colorado on a map. Whichever object is depicted, the appropriate type of units (metric, architectural, surveyor's) should be utilized. The types of units include both linear and angular measurement. What shape and how much of the drawing area is to be set aside needs to be determined. The UNITS and LIMITS commands are used to accomplish these tasks.

SETTING DRAWING UNITS

The UNITS command lets you change the linear and angular units by means of the Drawing Units dialog box. In addition, it lets you set the display format measurement and precision of your drawing units. You can change any or all of the following:

Unit display format	Angle display precision
Unit display precision	Angle base
Angle display format	Angle direction

Invoke the UNITS command from the **Format** menu and AutoCAD displays the Drawing Units dialog box (see Figure 1–37).

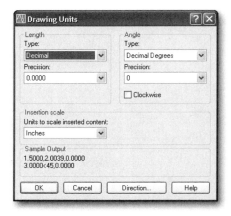

Figure 1–37 *The Drawing Units dialog box*

Setting Linear Units

The **Length** section of the Drawing Units dialog box allows you to change the units of linear measurement. From the **Type** list box select one of the five types of report formats you prefer. For the selected report format, choose precision from the **Precision** list box.

The engineering and architectural units display as feet-and-inches displays. These units assume each drawing unit represents 1 inch. The other units (scientific, decimal, and fractional) make no such assumptions, and they can represent whatever real-world units you like.

Drawing a 150-ft-long object might, however, differ depending on the units chosen. For example, if you use decimal units and decide that 1 unit = 1 foot, then the 150-ft-long object will be 150 units long. If you decide that 1 unit = 1 inch, then the 150-ft-long object will be drawn 150 × 12 = 1,800 units long. In architectural and engineering units modes, the unit automatically equals 1 inch. You may then give the length of the 150-ft-long object as 150' or 1800" or simply 1800.

Setting Angular Units

The **Angle** section of the Drawing Units dialog box allows you to set the drawing's angle measurement. From the **Type** list box select one of the five types of report formats you prefer. For the selected format, choose the precision from the **Precision** list.

Select the direction in which the angles are measured, clockwise or counterclockwise. If the **Clockwise** check box is checked, then the angles will increase in value in the clockwise direction. If it is not checked, the angles will increase in value in the counterclockwise direction. (see Figure 1–38).

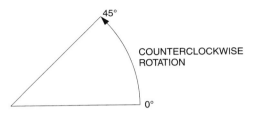

Figure 1–38 *The default, counterclockwise direction of angle measurement*

 Note: The default of 0 degrees being *East* and angle values increasing in the *counterclockwise* direction is used for the angular prompts and responses throughout this book, unless otherwise noted.

Insertion Scale

The units setting that you select from the **Insertion Scale** list box determines the unit of measure used for block insertions from AutoCAD DesignCenter, Tool Palettes, or i-drop. If a block is created in units different from the units specified in the list box, they will be inserted and scaled in the specified units. If you select **Unitless**, the block will be inserted as is, and the scale will not be adjusted to match the specified units.

Setting the Base Angle for Angle Measurement

To set the base angle for angle measurement, choose **Direction** and AutoCAD displays the Direction Control dialog box (see Figure 1–39).

Figure 1–39 *Direction Control subdialog box*

AutoCAD, by default, assumes that 0 degrees is to the right (east, or 3 o'clock) (see Figure 1–40), and that angles increase in the counterclockwise direction.

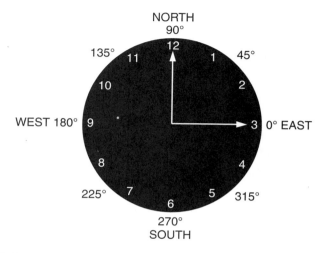

Figure 1–40 *Default angle setting direction*

You can change angle measurement to start at any compass point by selecting one of the five available options. You can also show AutoCAD the direction you want for angle 0 by specifying two points. This can be done by selecting **Other** and choosing **Angle**. AutoCAD prompts you for two points and sets the direction for angle 0. Choose **OK** to close the Direction Control dialog box.

Once you are satisfied with all of the settings in the Drawing Units dialog box, choose **OK** to accept the settings for the current working drawing and close the dialog box.

 Note: When AutoCAD prompts for a distance, displacement, spacing, or coordinates, you can always reply with numbers in integer, decimal, scientific, or fractional format. If the engineering or architectural format is in effect, you can also input feet, inches, or a combination of feet and inches. However, feet-and-inches input format differs slightly from the displayed format because it cannot contain a blank space. For example, a distance of 75.5 inches can be entered in the feet/inches/fractions format as 6'3-1/2". Note the hyphen in the unconventional location between the inches and the fraction and the absence of spaces. Normally, it will be displayed in the status area as 6'-3 1/2.

If you wish, you can use the SETVAR command to set the UNITMODE system variable to 1 (the default UNITMODE setting is 0) to display feet-and-inches output in the accepted format. For example, if you set UNITMODE to 1, AutoCAD displays the fractional value of 45 1/4 as you enter it: 45-1/4. The feet input should be followed by an apostrophe ('). The trailing double quote (") after the inches input is optional.

When the engineering or architectural format is in effect, the drawing unit equals 1 inch, so you can omit the trailing double quote ("). When you enter feet-and-inches values combined, the inch values should immediately follow the apostrophe, without an intervening space. Distance input does not permit spaces because, except when entering text, pressing SPACEBAR functions the same as pressing ENTER.

SETTING DRAWING LIMITS

The LIMITS command allows you to place an imaginary rectangular drawing sheet in the CAD drawing space. But, unlike the limitations of the drawing sheet of the board drafter, you can move or enlarge the CAD electronic sheet (the limits) after you have started your drawing. The LIMITS command does not affect the current display on the screen. The defined area determined by the limits governs the portion of the drawing indicated by the visible grid, if GRID is turned ON (for more on the GRID command, see Chapter 3). The rectangle specified by the limits also determines how much of the drawing is displayed by the ZOOM ALL command (for more on the ZOOM ALL command, see Chapter 3).

The limits are expressed as a pair of *2D* points in the World Coordinate System, a lower left and an upper-right limit. For example, to set limits for an A-size sheet, set lower-left as 0,0 and upper right as 11,8.5 or 12,9; for a B-size sheet, set lower-left as 0,0 and upper-right as 17,11 or 18,12. Many architectural floor plans are drawn at a scale of 1/4" = 1'-0". To set limits to plot on a C-size (22" × 17") paper at 1/4" = 1'-0", the limits are set lower-left as 0,0 and upper-right as 88',68' (4 × 22, 4 × 17).

Invoke the LIMITS command from the **Format** menu and AutoCAD prompts:

> Reset Model space limits:
> Specify lower left corner or ⬇ <current>: *(press ENTER to accept the current setting, specify the lower-left corner, or right-click and select one of the options)*
> Specify upper right corner <current>: *(press ENTER to accept the current setting, or specify the upper-right corner)*

The response you give for the upper-right corner gives the location of the upper-right corner of the imaginary rectangular drawing sheet.

There are two additional options available for the LIMITS command. When AutoCAD prompts for the lower-left corner, you may respond with the ON or OFF option. The ON and OFF options determine whether or not you can specify a point outside the limits when prompted to do so.

When you select the ON option, limits checking is turned on, and you cannot start or end an object outside the limits, nor can you specify displacement points required by the MOVE or COPY command outside the limits. You can, however, specify two points (center and point on circle) that draw a circle, part of which might be outside the limits. The limits check is simply an aid to help you avoid drawing off the imaginary rectangular drawing sheet. Leaving the limits checking ON is a sort of safety net to keep you from inadvertently specifying a point outside the limits. On the other hand, limits checking is a hindrance if you need to specify such a point.

When you select the OFF option (default), AutoCAD disables limits checking, allowing you to draw the objects and specify points outside the limits.

Whenever you change the limits, you will not see any change on the screen unless you use the ALL option of the ZOOM command. The ALL option of the ZOOM command lets you see entire newly set limits on the screen. For example, if your current limits are 12 by 9 (lower-left corner 0,0 and upper-right corner 12,9) and you change the limits to 42 by 36 (lower-left corner 0,0 and upper-right corner 42,36), you still see the 12-by-9 area. You can draw the objects anywhere in the limits 42-by-36 area, but you will see on the screen the objects that are drawn only in the 12-by-9 area. To see the entire limits, invoke the ZOOM command using the ALL option.

When you invoke the ALL option of the ZOOM command directly from the **View** menu. You see the entire limits or current extents (whichever is greater) on the screen. If objects are drawn outside the limits, all objects are displayed. (For a detailed explanation of the ZOOM command, see Chapter 3.)

Whenever you change the limits, you should always invoke ALL option of the ZOOM command to see the entire limits or current extents on the screen.

For example, the following command sequence shows steps to change limits for an existing drawing (see Figures 1–41 and 1–42).

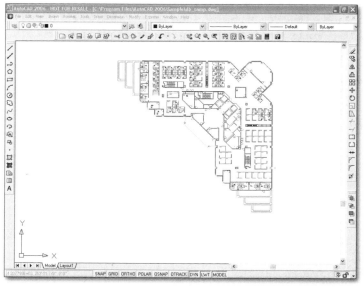

Figure 1–41 *The limits of an existing drawing before being changed by the* LIMITS *command*

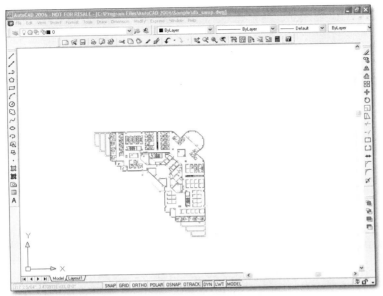

Figure 1–42 *The new limits of the drawing, after being changed by the* LIMITS *command*

    limits (ENTER)
    Reset Model space limits:
    Specify lower left corner or [↓] <current>: **0,0** (ENTER)
    Specify upper right corner <current>: **42,36** (ENTER)
    zoom (ENTER)
    Specify corner of window, enter a scale factor (nX or nXP), or [↓]: **all** (ENTER)

BEGINNING A NEW DRAWING BY USING THE CREATE NEW DRAWING DIALOG BOX (STARTUP SYSTEM VARIABLE SET TO 1)

AutoCAD can be configured to display the Create New Drawing dialog box (see Figure 1–43) instead of the Select Template dialog box whenever the NEW command is invoked to create a new drawing. In order to set up AutoCAD to display the Create a New Drawing dialog box, you must set the system variable named STARTUP to the value of 1. AutoCAD is shipped with the STARTUP system variable set to 0. The following command sequence is used to change the value of the STARTUP system variable from 0 to 1:

    startup (ENTER)
    Enter the new value for STARTUP <0>: **1** (ENTER)

Figure 1–43 *Create New Drawing dialog box*

Whenever you begin a new drawing, whether by means of one of the two available wizards or one of the available templates or by starting from scratch, AutoCAD creates a new drawing called *drawing1.dwg*. You can begin working immediately and save the drawing to a file name later, using the SAVE or SAVEAS command. During any one session of AutoCAD, subsequent new drawings will be named *drawing2.dwg*, *drawing3.dwg* until each is either saved to a name you specify or is terminated without saving.

STARTING A NEW DRAWING BY USING WIZARDS

The wizards let you start and set up the environment in a new drawing with the help of a step-by-step guide. You can choose from two wizards: **Quick Setup** and **Advanced Setup**. The **Quick Setup** wizard allows you to specify the units and area for your new drawing. The **Advanced Setup** wizard allows you to specify the units, angle, angle measure, angle direction, and area for your new drawing.

The initial drawing settings correspond to those in either the template *acad.dwt* (English units) or the template *acadiso.dwt* (metric units), based on the current setting of the MEASUREINIT system variable in the registry. When MEASUREINIT is set to 0, the drawing settings are based on the template *acad.dwt*; when it is set to 1, the drawing settings are based on the template *acadiso.dwt*.

Choose **Use a Wizard** (see Figure 1–44) to list the two available wizards: **Quick Setup** and **Advanced Setup**.

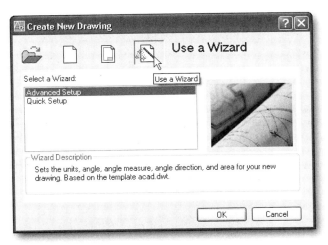

Figure 1–44 *Create New Drawing dialog box with Use a Wizard selection*

Quick Setup

The **QuickSetup** wizard has two pages: the **Units** page (see Figure 1–45) and the **Area** page.

The **Units** page applies to how the linear units of the drawing are entered. They also determine how the linear units are displayed in the status bar. When you select one of the radio buttons, an example of how linear units will be written is displayed to the right of the radio buttons. AutoCAD allows you to choose from several formats for the display and entry of the coordinates and distances.

> Choose **Decimal** to display measurements in decimal notation. Selection of Decimal units allows you to draw in inches, feet (but not feet and inches), millimeters, or whatever units you require, as long it is a single unit of measurement.
>
> Choose **Engineering** to display measurements in feet and decimal inches.
>
> Choose **Architectural** to display measurements in feet and fractional inches.
>
> Choose **Fractional** to display measurements in mixed-number (integer and fractional) notation.
>
> Choose **Scientific** to display measurements in scientific notation (numbers expressed in the form of the product of a decimal number between 0 and 10 and a power of 10).

Once the drawing is complete, you can plot it at whatever scale you like. As mentioned earlier, drawing to real-world size is one of the great advantages of AutoCAD. You can plot a drawing at several different scales, thereby eliminating the need for separate drawings at different scales.

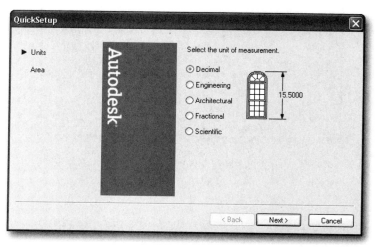

Figure 1-45 *QuickSetup dialog box: Units page*

After specifying the units, choose **Next**. AutoCAD displays the **Area** page (see Figure 1–46).

The **Area** page allows you to set your drawing area's width (left-to-right dimension) and length (bottom-to-top dimension), also known as the *limits*. You may set the limits to accommodate your drawing. For example, if you are drawing a printed circuit board that is 8 inches wide by 6 inches long, you can choose a decimal drawing unit and set the width to 8 and length to 6. It is a common practice to make the limits large enough to allow for space around the item being drawn, say 10 units by 8 units for an object that is 8 units by 6 units. If the drawing exceeds your original plans or the drawing limits become too restrictive, you can change the drawing limits. A detailed description of how to change limits appears earlier in the chapter.

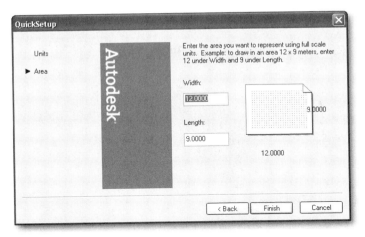

Figure 1-46 *QuickSetup dialog box: Area page*

Choose **Finish** to close the QuickSetup dialog box.

Advanced Setup
The **Advanced Setup** wizard has five pages: **Units, Angle, Angle Measure, Angle Direction**, and **Area**.

The **Units** page and **Area** page settings are the same as for the **Quick Setup** wizard. In the **Units** page, the **Precision** text box lets you set the number of decimal places or the fractional precision of the denominator to which coordinates and linear distances are reported.

In the **Angle** page (see Figure 1-47) you can set the type of units in which angular input and reporting is given. When you select one of the radio buttons, an example of how the angular units will be written is displayed to the right of the radio buttons. You can select the format used for the display and entry of angles. Degrees in decimal form are a common choice. However, you might also select gradient, radians, degrees/minutes/seconds, or surveyor's units. The **Precision** text box lets you set the number of decimal places or the degrees, minutes, seconds, or decimal precision of seconds to which angles are reported. After specifying the angle, choose **Next**. AutoCAD displays the **Angle Measure** page.

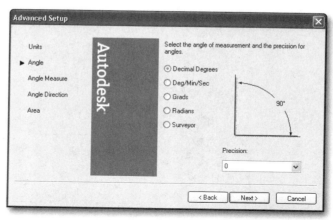

Figure 1-47 *Advanced Setup dialog box: Angle page*

The **Angle Measure** page (see Figure 1-48) lets you set zero degrees for your drawing relative to the universally accepted map compass, where North is up on the drawing and East is 90 degrees clockwise from North. You can select the **East**, **North**, **West**, or **South** radio buttons or the **Other** button, which lets you enter the direction in the text box. After specifying the angle measure, choose **Next**. AutoCAD displays the Angle Direction page.

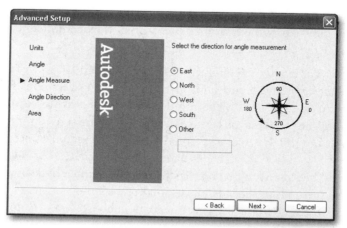

Figure 1-48 *Advanced Setup dialog box: Angle Measure page*

The **Angle Direction** page (see Figure 1-49) lets you set the direction (clockwise or counterclockwise) in which angle values increase. After specifying the angle direction, choose **Next**. AutoCAD displays the **Area** page. Set your drawing area's width (left-to-right dimension) and length (bottom-to-top dimension).

After specifying the area, choose **Finish** to close the Advanced Setup dialog box.

 Note: The wizards let you set up the drawing parameters that most commonly vary from one drawing to the next. It is convenient to have those variables accessible in a single place, and it lessens the chance of forgetting one. You should note, however, that the wizards' **Area** page does not permit you to use a point other than 0,0 as a lower left corner of the limits.

Settings you establish in the wizard can be changed again later by invoking the UNITS and LIMITS commands. Detailed discussions of the UNITS and LIMITS commands are provided later in the chapter.

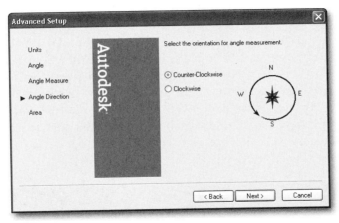

Figure 1–49 *Advanced Setup dialog box: Angle Direction page*

STARTING A NEW DRAWING WITH A TEMPLATE

If AutoCAD has been configured to display the Create New Drawing dialog box, then from it you can select the **Use a Template** option (see Figure 1–50). In the **Select a Template** section AutoCAD lists the available templates. An AutoCAD template is a drawing file with a file extension of *.dwt* instead of *.dwg*. AutoCAD does not actually let you make changes to the template, but lets you begin with a drawing that is exactly like the template, only it is named *drawing*n.*dwg*. This is useful to be able to have the drawing parameters such as the paper size, units of measure, and layers already set up and also have a border and title block already drawn, ready to fill in the blanks.

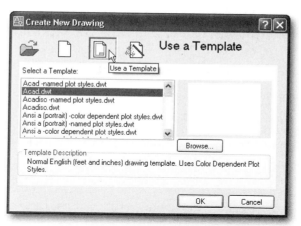

Figure 1–50 *Create New Drawing dialog box with the Use a Template option selection*

The preview graphics box on the right of the **Select a Template** list box lets you see a thumbnail sketch of the template that is highlighted in the **Select a Template** list box. If a description has been saved with the selected template, it will be displayed in the **Template Description** section. To begin your drawing with the identical setup as the selected template, either choose **OK** or double-click the highlighted name of the template in the **Select a Template** list box. You can also choose **Browse** to invoke the Select Template File dialog box to select a template file from a different folder, drive or Web site (see Figure 1–51).

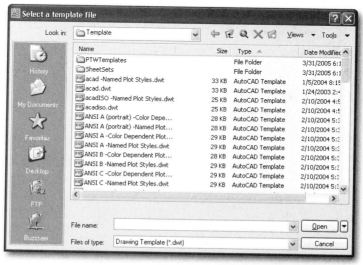

Figure 1–51 *Select a Template File dialog box.*

STARTING A NEW DRAWING FROM SCRATCH

If you choose the **Start from Scratch** option (see Figure 1–52) in the Create New Drawing dialog box, choose from the two options: **Imperial (feet and inches)** or **Metric**.

 Note: When you choose *Start from Scratch*, it is the same as using the Template option with the drawing named *acad.dwt* as the template. The *acad.dwt* drawing comes with drawing environmental variables (called system variables) set "at the factory" so to speak. Any and all of these system variables can be changed. Also, there are no objects drawn in the *acad.dwt* template. It is advisable to NOT replace *acad.dwt* with a drawing having settings of the system variables different from the ones in the original template unless you know exactly how you want the settings of all of the system variables to be each time you choose *Start from Scratch*. Also it is not advisable to have objects already drawn in the new drawing started from scratch.

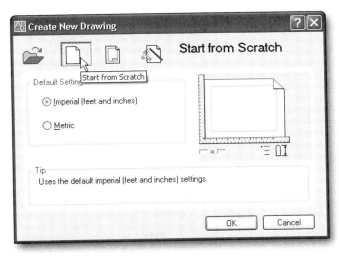

Figure 1–52 *Create New Drawing dialog box with the Start from Scratch option selection*

The **Imperial (feet and inches)** option sets the units to feet and inches settings; the **Metric** option sets the units to metric settings. By default the limits are set to 0,0 for the lower-left corner and 12,9 for the upper-right corner. If necessary, you can change the settings at the beginning of the drawing session or at any time during the drawing session.

 Note: With the STARTUP system variable set to 1, whenever a new AutoCAD session is started, AutoCAD displays the Startup dialog box. It is similar to the Create New Drawing dialog box except it has an additional option called **Open a Drawing**, which allows opening an existing drawing.

WORKING WITH MULTIPLE DRAWINGS

AutoCAD allows you to work on more than one drawing in a single AutoCAD session (Multiple Document). When multiple drawings are open, you can switch between them by selecting the appropriate name of the drawing from the **Window** menu. If the drawing is arranged in tiled or cascaded view, then simply click anywhere in the drawing to make it active. You can also use CTRL+F6 or CTRL+TAB to switch between open drawings. However, you cannot switch between drawings during certain long operations such as regenerating the drawing.

You can copy and paste objects between drawings, and use the Properties palette or DesignCenter to transfer properties from objects in one drawing to objects in another drawing. You can also use AutoCAD object snaps, the copy with base point command, and the paste to original coordinates command for accurate placement, especially when copying objects from one drawing to another.

If you want to turn off the Multiple Document mode, select **Options** from the **Tools** menu and AutoCAD displays the Options dialog box. Select the **System** tab, and under **General Options** check the **Single-drawing compatibility** checkbox (see Figure 1–53). Choose **OK** to save the changes. AutoCAD will allow you to open only one drawing at a time (similar to AutoCAD Release 14).

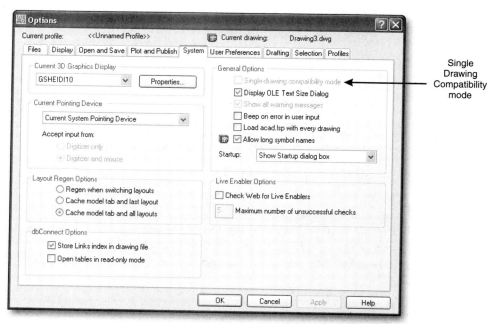

Figure 1–53 *Options dialog box – System tab*

SAVING A DRAWING

While working in AutoCAD, you should save your drawing once every 10 to 15 minutes without exiting AutoCAD. By saving your work periodically, you are protecting your work from possible power failures, editing errors, and other disasters. This can be done automatically by setting the SAVETIME system variable to a specific interval (in minutes). In addition, you can also manually save by using the SAVE, SAVEAS, and QSAVE commands.

The SAVE command saves an unnamed drawing with a file name that you specify. If the drawing is already named, then it works like the SAVEAS command, when the SAVE command is invoked from the On-Screen prompt.

The SAVEAS command saves an unnamed drawing to a new file name or saves the current drawing to a specified new name, makes it the new current drawing while retaining a copy of the drawing from which it is being created in the stage of its last save. If you specify a file name that already exists in the current folder, AutoCAD displays a message warning you that you are about to overwrite another drawing file. If you do not want to overwrite it, specify a different file name. The SAVEAS command also allows you to save in various formats. Possible formats include 2004 and 2000/LT2000 Drawings [*.dwg], Drawing Standards [*.dws], Template [*.dwt], 2004 and 2000/LT2000 DXF [*.dxf], R12/LT 2 DXF [*.dxf].

The QSAVE command saves an unnamed drawing with a file name that you specify. If the drawing is named, AutoCAD saves the drawing without requesting a file name.

Invoke the SAVE command from the standard toolbar (see Figure 1–54) and AutoCAD displays the Save Drawing As dialog box.

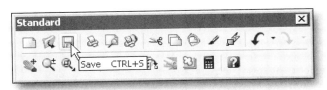

Figure 1–54 Invoke the SAVE command from the Standard toolbar

Select the appropriate folder in which to save the file, and type the name of the file in the **File name** text field. There is a window on the left side of the dialog box displaying quick access icons to folders on your computer: **Desktop** and **My Documents**; icons for **History** (recently opened drawings) and **Favorites**; and locations on the Internet: **Buzzsaw** and **FTP**. This lets you quickly specify where to save the drawing. The file name can contain up to 255 characters including embedded spaces and punctuation. File names cannot include any of the following characters: forward slash (/), backslash (\), greater than sign (>), less than sign (<), asterisk (*), question mark (?), quotation mark ("), pipe symbol (|), colon (:), or semicolon (;). Following are the examples of valid filenames:

> this-is-my-first-drawing
> first_house
> machine part 123

AutoCAD automatically appends *.dwg* as a file extension. If you save it as a template file, then AutoCAD appends *.dwt* as a file extension.

Invoke the SAVEAS command, by selecting **Save As**, from the **File** menu and AutoCAD displays the Save Drawing As dialog box. Name and path restrictions are the same as in the SAVE command described earlier in this section.

If you want to save the current drawing to the given file name, then invoke the SAVE command without changing the file name. AutoCAD saves the drawing without requesting a file name.

If you are working in multiple drawings, the CLOSE command closes the active drawing. Invoke the CLOSE command from the Window menu and AutoCAD closes the active drawing. If you have not saved the drawing since the last change, AutoCAD displays the AutoCAD alert box – Save changes to *Filename.dwg*. If you select **No**, then AutoCAD closes the drawing. If you select **Yes**, then AutoCAD displays the Save Drawing As dialog box. Select the appropriate folder in which to save the file, and type the name of the file in the **File name** text field.

Note: The CLOSE command is not available when AutoCAD is in Single Document mode.

If you are working in multiple drawings, the CLOSEALL command closes all the open drawings. Invoke the CLOSEALL command, by selecting **Close All**, from the **Window** menu and AutoCAD closes all the open drawings. AutoCAD displays a message for any unsaved drawing, in which you can save any changes (since the last SAVE) to the drawing before closing it.

Note: The CLOSEALL command is not available when AutoCAD is in Single Document mode.

EXITING AUTOCAD

The EXIT or QUIT command allows you to exit AutoCAD. The EXIT or QUIT command exits the current drawing if there have been no changes since the drawing was last saved. If the drawing has been modified, AutoCAD displays the Drawing Modification dialog box to prompt you to save or discard the changes before quitting.

Invoke the EXIT command from the **File** menu, by selecting **Exit**. If the drawing has been modified, AutoCAD displays the Drawing Modification dialog box to prompt you to save or discard the changes before exiting.

Invoke the QUIT command by typing **quit**. If the drawing has been modified, AutoCAD displays the Drawing Modification dialog box to prompt you to save or discard the changes before quitting.

REVIEW QUESTIONS

1. If you executed the following commands in order: LINE, CIRCLE, ARC, and ERASE, what would you need to do to re-execute the CIRCLE command?

    a. press PGUP, PGUP, PGUP, ENTER

    b. press DOWN-ARROW, DOWN-ARROW, DOWN-ARROW, ENTER

    c. press UP-ARROW, UP-ARROW, UP-ARROW, ENTER

    d. press PGDN, PGDN, PGDN, ENTER

    e. press LEFT-ARROW, LEFT-ARROW, LEFT-ARROW, ENTER

2. In a dialog box, when there is a set of mutually exclusive options (a list of several from which you must select only one), these are called:

    a. text boxes

    b. check boxes

    c. radio buttons

    d. scroll bars

    e. list boxes

3. What is the extension used by AutoCAD for template drawing files used by the setup wizard?

    a. .dwg   d. .tem

    b. .dwt   e. .wiz

    c. .dwk

4. If you selected the ALL option of the ZOOM command and your drawing shrank to a small portion of the screen, one possible problem might be:

    a. out of computer memory

    b. misplaced drawing object

    c. grid and snap are set incorrectly

    d. this should never happen in AutoCAD

    e. limits are set much larger than current drawing objects.

5. What is an external tablet used to input absolute coordinate addresses to AutoCAD by means of a puck or stylus called?

   a. digitizer

   b. input pad

   c. coordinate tablet

   d. touch screen

6. The menu that can be made to appear at the location of the crosshairs is called:

   a. mouse menu

   b. cross-hair menu

   c. cursor menu

   d. none of the above; there is no such menu

7. In order to save basic setup parameters (such as snap, grid, etc.) for future drawings, you should:

   a. create an AutoCAD Macro

   b. create a prototype drawing template file

   c. create a new configuration file

   d. modify the ACAD.INI file

8. To cancel an AutoCAD command, press

   a. CTRL+A

   b. CTRL+X

   c. ALT+A

   d. ESC

   e. CTRL+ENTER

9. The SAVE command:

   a. saves your work

   b. does not exit AutoCAD

   c. is a valuable feature for periodically storing information to disk

   d. all of the above

Fundamentals I

INTRODUCTION

This chapter introduces some of the basic commands and concepts in AutoCAD that can be used to complete a simple drawing. The project exercise drawings (on the accompanying CD under Exercise Files for Chapter 2) used in this chapter may appear relatively uncomplicated, but for the newcomer to AutoCAD it presents ample challenges. This chapter provides useful lessons in drawing setup, as well as in creating and editing objects. Once you learn how to access and use the basic commands, how to find your way around the screen, and how AutoCAD makes use of coordinate geometry, you can apply these skills to the chapters containing more advanced drawings and projects.

After completing this chapter, you will be able to do the following:

- Construct geometric figures with LINE, RECTANGLE, CIRCLE, and ARC commands
- Use coordinate systems
- Use various object selection methods
- Use the ERASE command

CONSTRUCTING GEOMETRIC FIGURES

AutoCAD gives you an ample variety of drawing elements, called objects. It also provides you with many ways to generate each object in your drawing. You will learn about the properties of these objects as you progress in this text. It is important to keep in mind that the examples in this text of how to generate the various lines, circles, arcs, and other objects are not always the only methods available. You are invited, even challenged, to find other more expedient methods to perform the tasks demonstrated in the lessons. You will progress at a better rate if you make an effort to learn as much as possible, as soon as possible, about the descriptive properties of the individual objects. When you become familiar with how AutoCAD creates, manipulates, and stores the data that describes the objects, you are then able to create drawings more effectively.

DRAWING LINES

A line is a primary drawing object; it may be defined as a path connecting two end points. In AutoCAD, a series of line segments may be drawn by invoking the LINE command and then specifying a sequence of end points. AutoCAD connects the points with a line. The LINE command is one of the few AutoCAD commands that automatically repeats. It uses the ending point of the previous line as the starting point for the next segment, continuing to prompt you for subsequent end points. To terminate this continuation behavior, you must press ENTER or right-click and choose ENTER from the shortcut menu. Even though a sequence of lines may be drawn using a single LINE command, each segment is a separate line object.

You can specify the end points using two-dimensional (X,Y), three-dimensional (X,Y,Z) coordinates, or a combination of the two. If you enter two-dimensional coordinates, AutoCAD uses the current elevation as the Z element of the point (zero is the default).

Invoke the LINE command from the Draw toolbar (see Figure 2–1) and AutoCAD displays the On-Screen prompt: "Specify first point:" (see Figure 2–2).

Figure 2–1 *Invoking the* LINE *command from the Draw toolbar*

Figure 2–2 *On-Screen prompt displayed near cursor*

Where to Start

You can designate the starting point of a line by entering absolute coordinates (see the following section on Coordinate Systems for a detailed explanation) or by specifying a location on the screen using your pointing device. After you specify the first point, AutoCAD prompts:

    Specify next point or ⬇

Where to from Here?

You can specify the end point of a line by entering absolute coordinates, relative coordinates, or by using your pointing device. AutoCAD now displays the prompt shown in Figure 2–3.

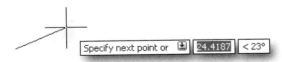

Figure 2–3 *On-Screen prompt asking you to specify next point and displaying current geometric values*

As previously mentioned, to save time, the LINE command remains active and prompts "Specify next point or ⬇" after each point you specify. When you have finished, press ENTER to terminate the LINE command.

If you are placing points with a pointing device instead of specifying coordinates, a rubber-band preview line is displayed between the starting point and the crosshairs. This helps you visualize where the line will be drawn. In Figure 2–4 the dotted lines represent previous cursor positions.

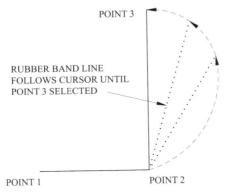

Figure 2–4 *Placing points with a pointing device rather than with keyboard coordinates input*

Continuing from a Previously Drawn Line or Arc

When you invoke the LINE command and respond to the "Specify first point:" prompt by pressing ENTER or the SPACEBAR, AutoCAD will automatically set the start of the new line sequence at the end of the most recently drawn line or arc. This provides a simple method of constructing a tangentially connected line in an arc-line continuation.

The prompt sequence that follows will depend on whether a line or arc was more recently drawn. If a line was more recent, the starting point of the new line will be set at the end point of the previous line, the continuation prompt appears as usual. For instance, continuing with the example shown in Figure 2–4, an additional three line segments can be drawn (see Figure 2–5) using the following sequence:

Specify first point: *(to continue the next line from Point 3, press* ENTER *or the* SPACEBAR*)*
Specify next point or ⬇ *(specify Point 4)*
Specify next point or ⬇ *(specify Point 5)*
Specify next point or ⬇ *(specify Point 6)*
Specify next point or ⬇ (ENTER)

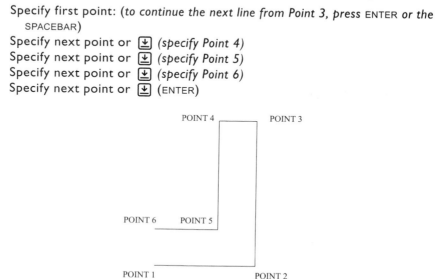

Figure 2–5 *Using the* LINE *command's Continue feature*

If an arc was more recent, its end point and ending direction define the starting point and direction of the new line. In this case, AutoCAD prompts:

Length of line:

After specifying the length of the line to be drawn, AutoCAD continues with the normal "Specify next point or ⬇" prompt.

Line Command Options

Most of the AutoCAD commands have a variety of options. For the LINE command, two options are available: CLOSE and UNDO. You can access the available options from the shortcut menu that appears when you right-click your pointing device, after you invoke the LINE command. You can also access the options from the On-Screen prompt by pressing the DOWN-ARROW (⬇)key. (If you are using the Command Window, you can access the options by typing the key letter(s) of the option name which are shown in capital letters.)

When drawing a sequence of lines to form a polygon, choose CLOSE to automatically join the last and first points. AutoCAD performs two steps when you choose the CLOSE option. The first step closes the polygon, and the second step terminates the LINE command.

The following command sequence shows an example of using the CLOSE option (see Figure 2–6).

> line (ENTER)
> Specify first point: *(specify Point 1)*
> Specify next point or ⤓ *(specify Point 2)*
> Specify next point or ⤓ *(specify Point 3)*
> Specify next point or ⤓ *(specify Point 4)*
> Specify next point or ⤓ *(specify Point 5)*
> Specify next point or ⤓ *(specify Point 6)*
> Specify next point or ⤓ *(choose CLOSE from the shortcut menu)*

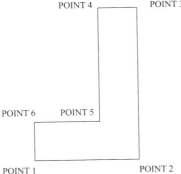

Figure 2–6 *Using the* LINE *command's* CLOSE *option*

Note: Using the CLOSE option causes the LINE command to automatically draw the last line of a sequence from the end point of the most recent line to the starting point of the first line of the sequence. If the series of lines in Figure 2–6 had been drawn in two sequences, one from Point 1 through Point 3 and then continued by invoking the LINE command to draw the second sequence from Point 3 through Point 6, then the CLOSE option would cause a line to be drawn from Point 6 to Point 3, not to Point 1.

Choosing UNDO erases the most recent line without terminating the LINE command. For instance, while drawing a sequence of connected lines, you may wish to erase the most recent line segment and continue from the end point of the previous line segment. You do so by selecting the UNDO option from the shortcut menu. If necessary, you can select UNDO multiple times; this will erase previously drawn line segments one at a time. Once you exit the LINE command, its UNDO option to erase the most recent line segments one at a time, is no longer available.

The following command sequence shows an example using the UNDO option (see Figure 2–7).

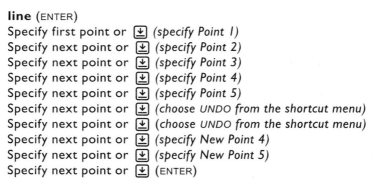

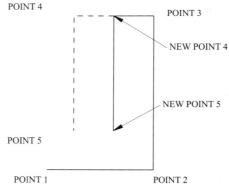

Figure 2–7 *Using the* LINE *command's* UNDO *option*

DRAWING RECTANGLES

The RECTANGLE command creates a closed polyline in a rectangular shape. (Polylines are discussed in detail in chapter 4.) You can specify the length, width, area, and rotation parameters. You can also control the type of corners on the rectangle: fillet, chamfer, or square.

Invoke the RECTANGLE command from the Draw toolbar (see Figure 2–8) and AutoCAD displays the On-Screen prompt shown in Figure 2–9.

Figure 2–8 *Invoking the* RECTANGLE *command from the Draw toolbar*

Figure 2–9 *On-Screen prompt displayed near cursor*

AutoCAD prompts:

>Specify first corner point or ⬇ *(specify first corner point to define the start of the rectangle or right-click for the shortcut menu and choose one of the available options)*
>Specify other corner point or ⬇ *(specify a point to define the opposite corner of the rectangle or right-click for the shortcut menu and choose one of the available options)*

Rectangle Command Options

Choosing CHAMFER sets the chamfer distance for the rectangle to be drawn. Refer to Chapter 4 for a detailed explanation on the usage of the CHAMFER command and its available settings.

Choosing ELEVATION specifies the elevation of the rectangle to be drawn. Refer to Chapter 15 for a detailed explanation on the usage of the Elevation setting.

Choosing FILLET sets the fillet radius for the rectangle to be drawn. Refer to Chapter 4 for a detailed explanation on the usage of the FILLET command and its available settings.

Choosing THICKNESS sets the thickness of the rectangle to be drawn. Refer to Chapter 15 for a detailed explanation on the Thickness setting.

Choosing WIDTH allows you to set the line width for the rectangle to be drawn. The default width is set to 0.0.

Choosing AREA allows you to create a rectangle of a specified area. You also specify either the length or the width value. When you choose the AREA option after specifying the first corner point, AutoCAD prompts:

>Enter area of rectangle in current units <default>: *(specify area of the rectangle)*
>Calculate rectangle dimensions based on *(specify either length or width)*

If you have specified length, AutoCAD prompts:

> Enter rectangle length <default>: *(specify length of the rectangle)*

If you have specified width, AutoCAD prompts:

> Enter rectangle width <default>: *(specify width of the rectangle)*

AutoCAD draws a rectangle using the specified first corner, area, and length or width.

Choosing DIMENSIONS allows you to create a rectangle by specifying length and width values. AutoCAD prompts:

> Specify length for rectangles <default>: *(specify length of the rectangle)*
> Specify width for rectangles <default>: *(specify width of the rectangle)*
> Specify other corner point or ⬇ *(specify a point by moving the cursor to one of the four possible locations for the diagonally opposite corner of the rectangle)*

Choosing ROTATION allows you to create a non-orthogonal rectangle. AutoCAD prompts:

> Specify rotation angle or ⬇ *(specify angle of the base of the rectangle)*
> Specify other corner point or ⬇ *(specify a point by moving the crosshairs to one of the four possible locations for the diagonally opposite corner of the rectangle)*

AutoCAD draws a rectangle using the specified angle for the base and second corner.

DRAWING WIDE LINES

When it is necessary to draw thick lines, the TRACE command may be used instead of the LINE command. Traces are entered just like lines except that the line width is set first. To specify the width, you can specify a distance or select two points and let AutoCAD use the measured distance between them. When you draw using the TRACE command, the previous trace segment is not drawn until the next end point is specified.

Invoke the TRACE command by typing **trace** and pressing ENTER. The TRACE command is one of the few commands that is not available in the AutoCAD default menu. AutoCAD prompts:

> **trace** (ENTER)
> Specify trace width <current>: *(specify the trace width and press ENTER)*
> Specify start point: *(specify the starting point of the trace)*

Specify a series of points to be connected by trace lines. When you have finished specifying a connected series of lines, press ENTER to terminate the TRACE command.

For example, the following command sequence shows the placement of connected lines using the TRACE command (see Figure 2–10).

trace (ENTER)
Specify trace width <current>: **.05** (ENTER)
Specify start point: *(specify Point 1)*
Specify next point: *(specify Point 2)*
Specify next point: *(specify Point 3)*
Specify next point: *(specify Point 4)*
Specify start point: *(specify Point 1)*
Specify next point: (ENTER)

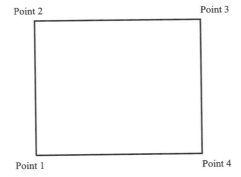

Figure 2–10 *Placing wide connected lines using the* TRACE *command*

COORDINATE SYSTEM

A coordinate system allows you to specify the locations of points in space or on a plane. Anytime you are drawing in AutoCAD, only one drawing plane is active. A drawing plane is an infinite 2D plane on which 2D drawing objects, such as lines and arcs, are drawn. All dimensions and points on the current drawing plane are expressed in numerical terms as a distance from an origin point (0,0), as shown in Figure 2-11. AutoCAD's Coordinate System allows for an infinite number of drawing planes that can be rotated or tilted as needed.

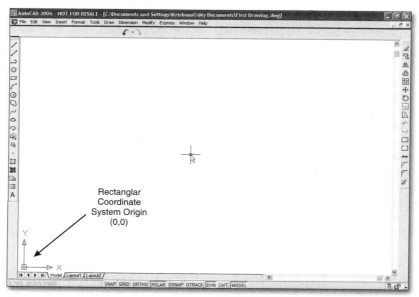

Figure 2–11 AutoCAD Drawing Area showing the Coordinate System icon at the origin (0,0)

What's the Point?

A child's picture puzzle book guides you to create a "connect-the-dots" picture by drawing lines between points that are numbered 1, 2, 3, and so on. These numbers have nothing to do with distances from one dot to another or their relative location in the picture. They specify only the sequence of starting and ending a series of lines. However, if the picture had been generated in AutoCAD, the points would have other numbers associated with them, their coordinates. Because an AutoCAD drawing plane has a built-in coordinate system, whether you elect to use it or not, all points in your drawing have pairs (triplets, when you advance to 3D) of numbers associated with them (see Figure 2–12). These pairs of numbers describe where the point is located as determined by its distances from two axes that intersect at the origin (coordinate 0,0) in a specific coordinate system.

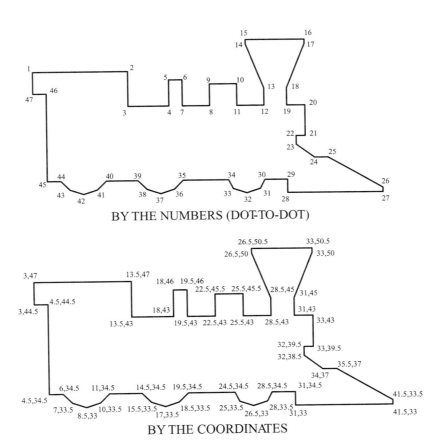

Figure 2–12 *"Connect-the-dots" versus coordinates*

Most board drafters' drawings do not utilize a coordinate system, especially architectural and mechanical drawings, as well as electrical schematics and piping flow diagrams that are not concerned with dimensions and spatial relationships. However, mapping by surveyors and layouts of manufacturing plants often associate objects in their drawings to some basic coordinate system. For instance, when a large petrochemical plant is situated on several hundred acres, those who design the original layout will establish two major axes (usually one east-west and the other north-south). These imaginary lines are perpendicular to each other (see Figure 2–14). Where they cross is called the "origin." If the origin is in the middle of the plant, then depending on which quadrant a point is situated in, one or both coordinates might be negative.

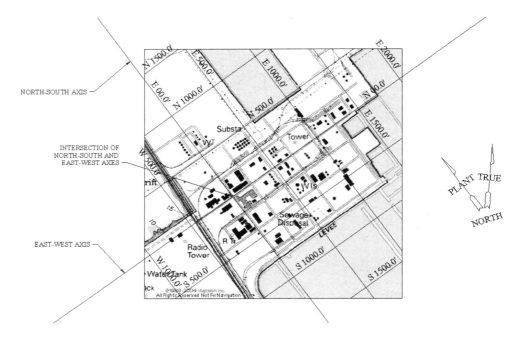

Figure 2–13 *Plant layout with axes*

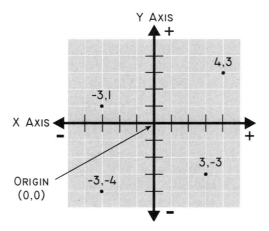

Figure 2–14 *A 2D drawing plane showing the X and Y axes, origin, and points in each quadrant*

AutoCAD's coordinate system *is always present* and sooner or later you will need to interact with it. For instance, if you wish to plot a drawing, you must understand how the coordinate system functions and affects the arrangement and location of objects on the plotted sheet. The fact that every object in AutoCAD has a set of coordinates associated with it offers tremendous advantages. For example, if a grouping of objects are drawn correctly, but not located as they should be, they can be moved en masse to the proper location with relative ease, thanks to the coordinate system.

THREE TYPES OF COORDINATE SYSTEMS

In three-dimensional space, there are three commonly used coordinate systems. Each system uses a set of three numbers to describe the location of a point.

The Spherical Coordinate System

The Spherical coordinate system is used for specifying points on a sphere. It is the basis for navigation on the surface of the earth (latitude and longitude). The first number of this coordinate system is the radius of the sphere (**r** in Figure 2–15). The second number (θ in Figure 2–15) is the angle between a line through a zero point on the equator and a line that goes through where the point is projected onto the equatorial plane of the sphere. This number is given as the longitude in navigational terms. The third number (Φ in Figure 2–15) is the angle between a line through the point and the equatorial plane. This number is given as the latitude in navigational terms. AutoCAD expects Spherical coordinates to be entered as **r** (distance from the origin)<θ (angle from X axis)<Φ (angle from XY plane). In Figure 2–15, the illustration on the right shows how AutoCAD calculates the Spherical Coordinates 10<75<60.

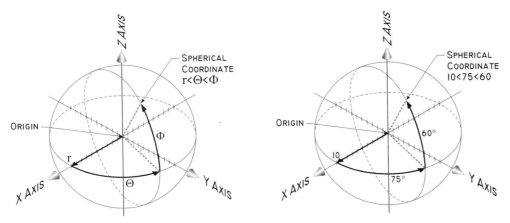

Figure 2–15 *Spherical coordinate system: mathematical model (left); AutoCAD's calculation of point 10<75<60 (right)*

The Cylindrical Coordinate System

The Cylindrical coordinate system is used for specifying a point on a cylinder. It has as its base a horizontal plane that is perpendicular to the centerline of the cylinder. On the base plane there is also a zero base line from the center of the cylinder that extends to the surface of the cylinder. The first number (**r** in Figure 2–16) is the radius of the cylinder. The second number (θ in Figure 2–16) is the angle of rotation from the zero baseline. The third point (**z** in Figure 2–16) is the distance along the Z axis from the base plane. In Figure 2–16, the illustration on the right shows how AutoCAD calculates the Cylindrical coordinate 5<75,9.

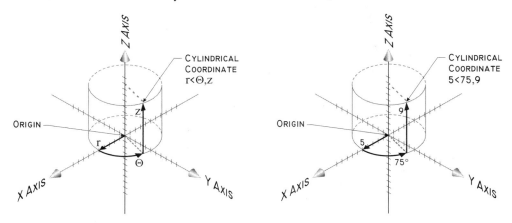

Figure 2–16 *Cylindrical coordinate system: mathematical model (left); AutoCAD's calculation of point 5<75,9 (right)*

The Cartesian Coordinate System – AutoCAD's Default System

AutoCAD comes with a rectangular coordinate system that does not need curved surfaces, circles, arcs, or angles to describe points in 2D and 3D space. It consists of three mutually perpendicular planes. One plane is considered horizontal, which means that the other two are vertical. The three lines created by the intersections of the three pairs of planes are called the axes (see Figure 2–17). Where the three axes intersect is known as the origin, with the coordinates 0,0,0.

AutoCAD adheres to the conventions of the Cartesian coordinate system. From the origin (0,0,0) distances along the X axis (to the right) increase in value. On the Y axis, points above the origin have positive values. On the Z axis, points closer to the viewer than the origin, have a positive value; this provides a sense of depth. These axes define the World Coordinate System (WCS).

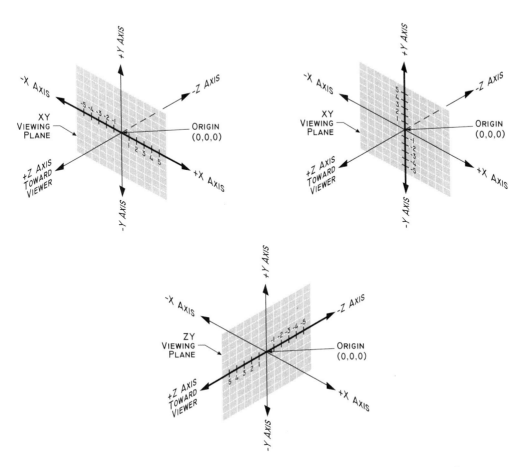

Figure 2-17 *Cartesian coordinate system showing the Origin, XY and ZY viewing planes*

World and User Coordinate Systems

The significance of the WCS is that it is always in your drawing; it cannot be altered. However, an infinite number of other coordinate systems can be established relative to it. These others are called *User Coordinate Systems* (UCS) and are created with the UCS command. Even though the WCS is fixed, you can view it from any angle, side, or rotation without changing to another coordinate system.

AutoCAD provides what is called a coordinate system icon to help you keep your bearings while working in different coordinate systems. The icon shows you the orientation of your current UCS by indicating the positive directions of the X and Y axes. Figure 2–18 shows some examples of coordinate system icons.

Figure 2-18 *Examples of the UCS icon*

Scaling, Plotting and Planning

The board drafter represents real life objects with lines, circles, arcs, etc. that are scaled to fit on a certain size sheet. The AutoCAD drafter does just the opposite by drawing objects at their true sizes and then creates a plot configuration so that everything is reduced to fit on the actual sheet size. He or she then makes the border, title block, and other non-object-associated features fit around the object. The completed combination is scaled to fit the sheet size required for plotting.

A more complicated situation arises when you need to draw objects at different scales on the same drawing. This can be handled easily by using the more advanced features and commands provided by AutoCAD.

Drawing a schematic that is not to scale is a simple task in AutoCAD. Even though the symbols and the distances between them have no relationship to any real-life dimensions, you must still consider factors such as the sheet size, text size, line widths, and other visible characteristics of the drawing in order to make your schematic easily read. The point is that some planning, including scaling and plotting, needs to be applied to all drawings.

METHODS TO SPECIFY POINTS

When AutoCAD prompts for the location of a point, you can use one of several point entry methods, including Spherical, Cylindrical, and Cartesian coordinates. Cartesian coordinates can be entered in the following formats: absolute rectangular, relative rectangular, absolute polar and relative polar.

Absolute Rectangular Coordinates

Entering Absolute rectangular coordinates specifies the location of a point by providing its distances from the intersection of two axes (2D) or from three intersecting axes (3D). Each point's distance is measured along the X, Y, and Z axes, relative to the intersection of the axes. The intersection of the axes is called the origin (2D = 0,0; 3D = 0,0,0; see Figure 2-14). In AutoCAD, by default the origin (0,0) is located at the lower left corner of the Grid display, as shown in Figure 2-19.

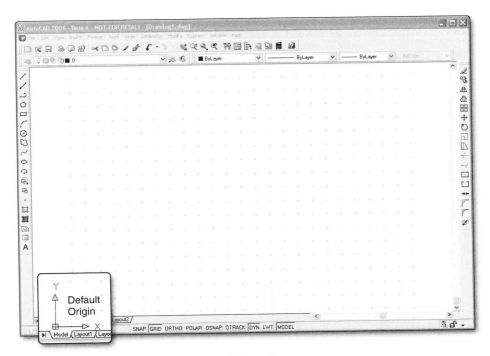

Figure 2–19 *Default location of the AutoCAD origin*

As mentioned earlier, moving along the X axis, toward the right and away from the origin, increases the positive value of X. Movement along the Y axis, above and away from the origin, increases the value of Y. You may specify a point by entering its X and Y coordinates in decimal, architectural, fractional, or scientific notation, separated by commas. AutoCAD automatically assigns the current elevation as the Z coordinate. Unless it has been changed, the default value of Z is zero (0). Three-dimensional drafting involves specifying X,Y, *and* Z coordinates.

Relative Rectangular Coordinates

Entering relative rectangular coordinates are specifies the location of a point relative to the last specified point, rather than the origin. In AutoCAD, to specify relative coordinates, the @ ("at" symbol) must precede your entry. This symbol is entered by holding down the SHIFT key and simultaneously pressing the number 2 key at the top of the keyboard. The following table shows examples of relative rectangular coordinate keyboard input when specifying a point. The first column shows the absolute coordinates of the last point specified (from which the newly specified point is offset). The third column shows the absolute coordinates resulting in from the relative coordinate keyboard input shown in the second column.

Absolute Coordinates of Last Specified Point	Relative Rectangular Coordinates Keyboard Input	Resulting Absolute Coordinates of Point Specified by Keyboard Input
3,4	@2,2	5,6
5,5	@-7,0	-2,5
3.25,8.0	@0,12.5	3.25,20.5

If you are working in a UCS (User Coordinate System) and would like to enter points with reference to the WCS (World Coordinate System), prefix the coordinates with an asterisk (*). For example, to specify a point with an X coordinate of 3.5 and a Y coordinate of 2.57 as referenced to the WCS, regardless of the current UCS, enter:

*3.5,2.57

In the case of relative coordinates, the asterisk will be preceded by the @ symbol. For example:

@*4,5

This represents an offset of 4,5 from the previous point as referenced to the WCS.

Relative Polar Coordinates

Entering polar coordinates specifies the location of a point based on the distance from a fixed point at a given angle. In AutoCAD, a relative polar coordinate point is determined by the distance from the previous point and an angle measured from the X axis. By default the angle is measured in the counterclockwise direction. It is important to remember that for points located using relative polar coordinates, they are positioned relative to the previous point and not the origin (0,0). You specify a point by entering its distance from the previous point and its angle from the X axis, separated by < (not a comma). This symbol is selected by holding the SHIFT key and simultaneously pressing the COMMA (,) key at the bottom of the keyboard. Failure to use the @ symbol will cause the point to be located relative to the origin (0,0). The following table shows examples of relative polar coordinates keyboard input. The first column shows absolute coordinates of the last point specified (from which the newly specified point is offset). The third column shows the absolute coordinates resulting from the relative polar coordinates keyboard input shown in the second column.

Absolute Coordinates of Last Specified Point	Relative Polar Coordinates Keyboard Input	Resulting Absolute Coordinates of Point Specified by Keyboard Input
3,4	@2<0	5,4
5,5	@4<180	1,5
2.00,2.00	@ 1.4142135623<45	3.00,3.00

OVERRIDES FOR ABSOLUTE/RELATIVE COORDINATE ENTRY

Under certain conditions, when you specify a point by typing in the coordinates, you might be specifying relative coordinates when you are intending to type in absolute coordinates. This can occur if you are using the dynamic input (described in chapter 3) and it has not been changed from the default setting of relative coordinates input mode. In the dynamic input (prompt/responses displayed near the cursor) default mode, AutoCAD automatically prefixes the coordinates with the @ (last point) symbol whether you type it in or not for second and subsequent points. For example, if you start a line at 1,1 and specify the next point as 3,3, the line will be drawn to coordinates 4,4 (not 3,3). This is because the coordinate input you specified will be applied as a displacement from the last point (relative) rather than the point at 3,3 (absolute). To change the pointer input to absolute format, set the DYNPICOORDS system variable to 1 (default is set to 0). This prevents the coordinates from being relative (even though you haven't put in the @ symbol). This rule does not apply if you are using the Command window instead of dynamic input. See chapter 3 (Drafting Settings section) for detailed explanation on dynamic input mode settings.

To change the DYNPICOORDS system variable setting, type in **dynpicoords** at On-Screen prompt and AutoCAD prompts:

> Enter new value for DYNPICOORDS <0>: *(specify 1 and press ENTER)*

AutoCAD will now use the input as absolute and not automatically prefix the coordinates entered in dynamic input with the @ symbol.

 Note: The prompts and responses shown in examples throughout this book and drawing exercises assume that you are using the new dynamic input introduced in AutoCAD 2006. Where they call for you to enter coordinates without the @ prefix, they are based on the input being absolute coordinates. Because the default setting (0) of the DYNPICOORDS system variable causes the @ prefix to be added (without your seeing it on the screen), you should change the setting to 1 if you wish to try out the examples on your system. Or, if you do not wish to change the DYNPICOORDS setting to 1, you can use an override by prefixing the relative coordinates with the # symbol. This forces the input to be absolute and only works in dynamic input, not at the Command line in the Command window.

COORDINATES DISPLAY

The Coordinates display (see Figure 2–20) reports the cursor coordinates in the Status bar. It has three settings: Static, Dynamic, and Distance and Angle displays. On most systems, the F6 function key toggles between the three settings. The three settings are as follows:

1. Static display updates only when you specify a point.
2. Dynamic display updates as you move the cursor.

3. Distance and Angle display updates the relative distance (distance<angle) as you move the cursor. This option is available only when you draw lines or other objects that prompt for more than one point.

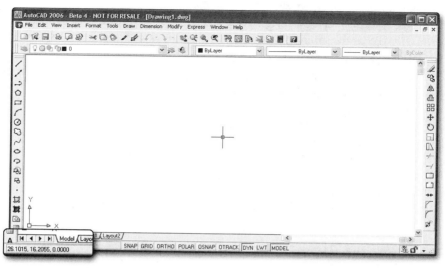

Figure 2–20 *Coordinates display in the Status bar*

DRAWING CIRCLES

The CIRCLE command creates a circle and offers five different methods for drawing circles: Center-Radius (default), Center-Diameter, 2 Point, 3 Point, and Tangent, Tangent, Radius (TTR).

Center-Radius

The Center-Radius method allows you to draw a circle by specifying a center point and a radius. Invoke the CIRCLE command from the Draw toolbar (see Figure 2–21) and AutoCAD prompts:

Figure 2–21 *Invoking the CIRCLE command from the Draw toolbar*

> Specify center point for circle or ⊡ *(specify center point to define the center of the circle)*
> Specify radius of circle or ⊡ *(specify the radius of the circle)*

The following command sequence shows an example (see Figure 2–22):

circle (ENTER)
Specify center point for circle or ⬇ **2,2** (ENTER)
Specify radius of circle or ⬇ **1** (ENTER)

The same circle can be drawn as follows (see Figure 2–23):

circle (ENTER)
Specify center point for circle or ⬇ **2,2** (ENTER)
Specify radius of circle or ⬇ **3,2** (ENTER)

In the last example, AutoCAD used the distance between the center point and the second point given as the value for the radius of the circle.

Figure 2–22 A circle drawn with the CIRCLE command's default method: Center-Radius

Figure 2–23 A circle drawn with the Center-Radius method by specifying the coordinates

Center-Diameter

The Center-Diameter method allows you to draw a circle by specifying a center point and a diameter. Invoke the CIRCLE command from the Draw toolbar (see Figure 2–21) and AutoCAD prompts:

Specify center point for circle or ⬇ *(specify center point to locate the center of the circle)*
Specify radius of circle or ⬇ *(choose DIAMETER from the shortcut menu)*
Specify diameter of circle or ⬇ *(specify the diameter of the circle)*

The following command sequence creates a circle with a diameter of 2 units:

circle (ENTER)
Specify center point for circle or ⬇ **2,2** (ENTER)
Specify radius of circle or ⬇ *(choose DIAMETER from the shortcut menu)*
Specify diameter of circle: **2** (ENTER)

The same circle can be generated as follows (see Figure 2–24):

> **circle** (ENTER)
> Specify center point for circle or ⬇ **2,2** (ENTER)
> Specify radius of circle or ⬇ *(choose DIAMETER from the shortcut menu)*
> Specify diameter of circle: **4,2** (ENTER)

 Note: When you specify a point coordinate as the diameter of a circle, AutoCAD calculates the distance from the center of the circle to the specified point and uses the resulting value for the diameter of the circle.

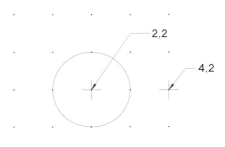

Figure 2–24 *A circle drawn using the Center-Diameter method by specifying coordinates*

Three-Point Circle

The Three-Point circle method allows you to draw a circle by specifying three points on the circumference.

Invoke the CIRCLE command from the Draw toolbar (see Figure 2–21) and AutoCAD prompts:

> Specify center point for circle or ⬇ *(choose 3P from the shortcut menu)*
> Specify first point on circle: *(specify a point or a coordinate)*
> Specify second point on circle: *(specify a point or a coordinate)*
> Specify third point on circle: *(specify a point or a coordinate)*

The following command sequence shows an example (see Figure 2–25).

> **circle** (ENTER)
> Specify center point for circle or ⬇ *(choose 3P from the shortcut menu)*
> Specify first point on circle: **2,1** (ENTER)
> Specify second point on circle: **3,2** (ENTER)
> Specify third point on circle: **2,3** (ENTER)

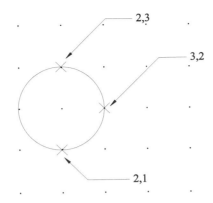

Figure 2–25 *A circle drawn with the Three-Point method*

Two-Point Circle

The Two-Point Circle method allows you to draw a circle by specifying two end points of a diameter.

Invoke the CIRCLE command from the Draw toolbar (see Figure 2–21) and AutoCAD prompts:

> Specify center point for circle or ⬇ *(choose 2P from the shortcut menu)*
> Specify first end point of circle's diameter: *(specify a point or a coordinate)*
> Specify second end point of circle's diameter: *(specify a point or a coordinate)*

The following command sequence shows an example (see Figure 2–26).

> **circle** (ENTER)
> Specify center point for circle or ⬇ *(choose 2P from the shortcut menu)*
> Specify first end point of circle's diameter: **1,2** (ENTER)
> Specify second end point of circle's diameter: **3,2** (ENTER)

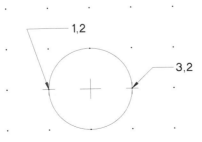

Figure 2–26 *A circle drawn with the Two-Point method*

Tangent, Tangent, Radius (TTR)

The Tangent, Tangent, Radius method allows you to draw a circle tangent to two objects (either lines, arcs, or circles) with a specified radius.

Invoke the CIRCLE command from the Draw toolbar (see Figure 2–21) and AutoCAD prompts:

> Specify center point for circle or ⬇ *(choose TTR from the shortcut menu)*
> Specify point on object for first tangent of circle: *(specify a point on an object for first tangent of circle)*
> Specify point on object for second tangent of circle: *(specify a point on an object for second tangent of circle)*
> Specify radius of circle: *(specify radius)*

When specifying the "tangent-to" objects, it normally does not matter where on the objects you make your selection. However, if more than one circle can be drawn to the specifications given, AutoCAD will draw the one whose tangent point is nearest to the selection made.

Note: Until it is changed, the radius/diameter you specify in any one of the options becomes the default setting for subsequent circles.

DRAWING ARCS

The ARC command offers four different combinations by which you can draw arcs: a combination of three points, a combination of two points and an included angle or starting direction, a combination of two points and a length of chord or radius, and a continuation from line or arc.

Three Points (Start, Point-on-Circumference or Center, and End)

There are three ways to use a Three-Points method of drawing arcs:

- Three-Point
- Start, Center, End (S,C,E)
- Center, Start, End (C,S,E)

Three-Point

The Three-Point (default) method draws an arc using three specified points on the arc's circumference. The first point specifies the start point, the second point specifies a point on the circumference of the arc, and the third point is the arc end point.

Invoke the ARC command from the Draw toolbar (see Figure 2–27), and specify a three-point arc either clockwise or counterclockwise, as shown in the following command sequence:

Specify start point of arc or ⬇ **1,2** (ENTER)
Specify second point of arc or ⬇ **2,1** (ENTER)
Specify end point of arc: **3,2** (ENTER)

Figure 2–27 *Invoking the* ARC *command from the Draw toolbar*

AutoCAD draws an arc based on a start point at 1,2 with a circumference point at 2,1 and an end point at 3,2; as shown in Figure 2–28.

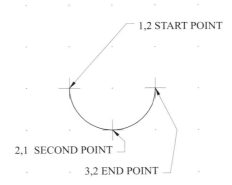

Figure 2–28 *An arc drawn with the* ARC *command's default option: Three Points*

Start, Center, End (S,C,E)

The Start, Center, End method also draws an arc using three specified points. The first point specifies the start point, the second point is the center point, and the third point is the arc end point.

Invoke the ARC command from the Draw toolbar (see Figure 2–27), and specify a three-point arc either clockwise or counterclockwise, as shown in the following command sequence:

Specify start point of arc or ⬇ **1,2** (ENTER)
Specify second point of arc or ⬇ (choose CENTER *from the shortcut menu*)
Specify center point of arc or ⬇ **2,2** (ENTER)
Specify end point of arc: **2,3** (ENTER)

AutoCAD draws an arc based on a starting point at 1,2 with the center point at 2,2 and an end point at 2,3; as shown in Figure 2–29.

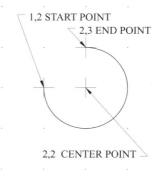

Figure 2-29 *An arc drawn with the Start, Center, End (S,C,E) method*

 Note: Arcs drawn by this method are always drawn counterclockwise from the starting point. The distance between the center point and the starting point determines the radius. Therefore, the point specified in response to "end point" needs only to be on the same radial line as the desired end point.

Center, Start, End (C,S,E)

The Center, Start, End method is similar to the Start, Center, End (S,C,E) method, except the first point selected is the center point of the arc rather than the start point.

Two Points and an Included Angle or Starting Direction

There are four ways to use a Two Points and an Included Angle or Starting Direction method of drawing arcs:

- Start, Center, Angle (S,C,A)
- Center, Start, Angle (C,S,A)
- Start, End, Angle (S,E,A)
- Start, End, Direction (S,E,D)

Start, Center, Angle (S,C,A)

The Start, Center, Angle method draws an arc similar to the Start, Center, End method, but it places the end point on a radial line at the specified angle from the line between the center point and the start point. If you specify a positive number as the value of the included angle, the arc is drawn counterclockwise. When you specify a negative number for the value of the angle, the arc is drawn clockwise.

Invoke the ARC command from the Draw toolbar (see Figure 2–27). Specify the first point as the start point, the second point as the center point of the arc to be drawn, and then specify the included angle, as shown in the following example:

Specify start point of arc or ⤓ **1,2** (ENTER)
Specify second point of arc or ⤓ *(choose CENTER from the shortcut menu)*
Specify center point of arc or ⤓ **2,2** (ENTER)
Specify end point of arc: *(choose ANGLE from the shortcut menu)*
Specify included angle: **270** (ENTER)

AutoCAD draws an arc based on a starting point at 1,2 with the center point at 2,2 and an included angle of 270 degrees, as shown in Figure 2–30.

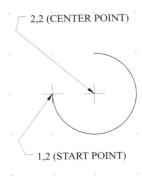

Figure 2–30 *An arc drawn with the Start, Center, Angle (S,C,A) method*

 Note: In the previous example, if a point directly below the specified center were selected in response to the "Included angle:" prompt, AutoCAD would read the angle of the line (270 degrees from zero) as the included angle for the arc.

Center, Start, Angle (C,S,A)

The Center, Start, Angle method is similar to the Start, Center, Angle (S,C,A) method, except the first point selected is the center point of the arc rather than the start point.

Start, End, Angle (S,E,A)

The Start, End, Angle method draws an arc similar to the Start, Center, Angle method, but places the end point on a radial line at the specified angle from the line between the center point and the start point. If you specify a positive angle as the included angle, the arc is drawn counterclockwise. Specify a negative included angle and the arc is drawn clockwise.

Invoke the ARC command from the Draw toolbar (see Figure 2–27). Specify the first point as the start point, the second point as the end point of the arc, and then specify the included angle, as shown in the following example:

Specify start point of arc or ⤓ **3,2** (ENTER)

Specify second point of arc or ⬇ *(choose* END *from the shortcut menu)*
Specify end point of arc: **2,3** (ENTER)
Specify center point of arc or ⬇ *(choose* ANGLE *from the shortcut menu)*
Specify included angle: **90** (ENTER)

AutoCAD draws an arc based on a starting point at 3,2 with an end point at 2,3 and an included angle of 90 degrees, as shown in Figure 2–31.

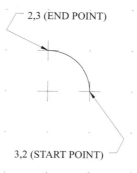

Figure 2–31 *An arc drawn counterclockwise with the Start, End, Angle (S,E,A) method*

The arc shown in Figure 2–32 is drawn with a negative angle using the following sequence:

arc (ENTER)
Specify start point of arc or ⬇ **3,2** (ENTER)
Specify second point of arc or ⬇ *(choose* END *from the shortcut menu)*
Specify end point of arc: **2,3** (ENTER)
Specify center point of arc or ⬇ *(choose* ANGLE *from the shortcut menu)*
Specify included angle: **–270** (ENTER)

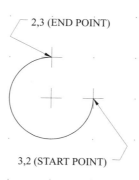

Figure 2–32 *An arc drawn clockwise with the Start, End, Angle (S,E,A) method*

Start, End, Direction (S,E,D)

The Start, End, Direction method allows you to draw an arc between selected points by specifying the direction in which the arc will be drawn from the selected start point. Either the direction can be keyed in or you can select a point on the screen with your pointing device. If you select a point on the screen, AutoCAD uses the angle from the start point to the end point as the starting direction.

Invoke the ARC command from the Draw toolbar (see Figure 2–27). Specify the first point as the start point, the second point as the end point of the arc, and then specify the starting direction, as shown in the following example:

> Specify start point of arc or ⬇ **3,2** (ENTER)
> Specify second point of arc or ⬇ (choose END from the shortcut menu)
> Specify end point of arc: **2,3** (ENTER)
> Specify center point of arc or ⬇ (choose DIRECTION from the shortcut menu)
> Specify tangent direction for the start point of arc: **90** (ENTER)

AutoCAD draws an arc based on a starting point at 3,2 with an end point at 2,3 and direction set to 90 degrees, as shown in Figure 2–33.

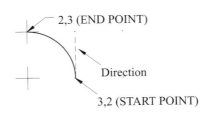

Figure 2–33 *An arc drawn with the Start, End, Direction (S,E,D) method*

Two Points and a Length of Chord or Radius

There are three ways to use a Two Points and a Length of Chord or Radius method of drawing arcs:

- Start, Center, Length of Chord (S,C,L)
- Center, Start, Length of Chord (C,S,L)
- Start, End, Radius (S,E,R)

Start, Center, Length of Chord (S,C,L)

The Start, Center, Length of Chord method uses the specified chord length as the straight-line distance from the start point to the end point. With any chord length (equal to or less than the diameter length), four possible arcs can be drawn: a major arc in either direction and a minor arc in either direction. Therefore, all arcs drawn by this method are counterclockwise from the start point. A positive value as the length of chord will cause AutoCAD to draw the minor arc; a negative value will result in the major arc being drawn. Invoke the ARC command from the Draw toolbar (see Figure 2–27). Specify the first point as the start point, the second point as the center point of the arc to be drawn, and then specify the chord length, as shown in the following example:

Specify start point of arc or ⬇ **1,2** (ENTER)
Specify second point of arc or ⬇ *(choose* CENTER *from the shortcut menu)*
Specify center point of arc: **2,2** (ENTER)
Specify end point of arc or ⬇ *(choose* CHORD LENGTH *from the shortcut menu)*
Specify length of chord: **1.4142** (ENTER)

AutoCAD draws the arc based on a starting point at 1,2 with the center point at 2,2 and a length of chord set to 1.4142, as shown in Figure 2–34.

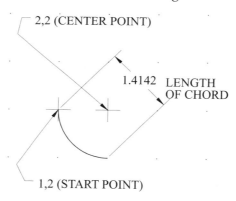

Figure 2–34 *A minor arc drawn with the Start, Center, Length of chord (S,C,L) method.*

The following example shows drawing a major arc.

arc (ENTER)
Specify start point of arc or ⬇ **1,2** (ENTER)
Specify second point of arc or ⬇ *(choose* CENTER *from the shortcut menu)*
Specify center point of arc: **2,2** (ENTER)
Specify end point of arc or ⬇ *(choose* CHORD LENGTH *from the shortcut menu)*
Specify length of chord: **–1.414** (ENTER)

AutoCAD draws an arc based on a starting point at 1,2 with the center point at 2,2 and a length of chord set to –1.414, as shown in Figure 2–35.

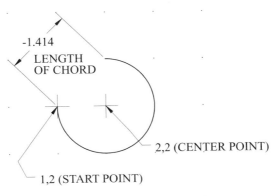

Figure 2–35 *A major arc drawn with the Start, Center, Length of Chord (S,C,L) method*

Center, Start, Length (C,S,L)

The Center, Start, Length method is similar to the Start, Center, Length (S,C,L) method, except the first point selected is the center point of the arc rather than the start point.

Start, End, Radius (S,E,R)

The Start, End, Radius method allows you to specify a radius after selecting the start and end points of the arc. As with the Length of Chord method, four possible arcs can be drawn: a major arc in either direction and a minor arc in either direction. Therefore, all arcs drawn by this method are counterclockwise from the start point. A positive value for the radius causes AutoCAD to draw the minor arc; a negative value results in the major arc.

Invoke the ARC command from the Draw toolbar (see Figure 2–27). Specify the first point as the start point, the second point as the end point, and then specify the radius, as shown in the following example:

    Specify start point of arc or ⬇ **1,2** (ENTER)
    Specify second point of arc or ⬇ *(choose* END *from the shortcut menu)*
    Specify end point of arc: **2,3** (ENTER)
    Specify center point of arc or ⬇ *(choose* RADIUS *from the shortcut menu)*
    Specify radius of arc: **–1** (ENTER)

AutoCAD draws the arc based on a starting point at 1,2 with an end point at 2,3 and a radius of –1, as shown in Figure 2–36.

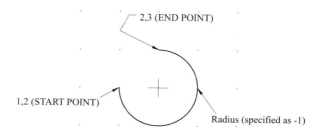

Figure 2–36 A major arc drawn with the Start, End, Radius (S,E,R) method

The following example shows drawing a minor arc.

    arc (ENTER)
    Specify start point of arc or ⬇ **2,3** (ENTER)
    Specify second point of arc or ⬇ (choose END from the shortcut menu)
    Specify end point of arc: **1,2** (ENTER)
    Specify center point of arc or ⬇ (choose RADIUS from the shortcut menu)
    Specify radius of arc: **1** (ENTER)

AutoCAD draws an arc based on a starting point at 2,3 with an end point at 1,2 and a radius of 1 unit, as shown in Figure 2–37.

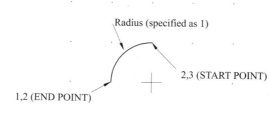

Figure 2–37 A minor arc drawn with the Start, End, Radius (S,E,R) method

Line-Arc and Arc-Arc Continuation

You can use an automatic Start Point, End Point, Starting Direction method to draw an arc by pressing ENTER as a response to the first prompt of the ARC command. After you press ENTER, the only other input needed is to select or specify the end point of the arc. AutoCAD uses the end point of the previous line or arc (whichever was drawn last) as the start point of the new arc. To demonstrate these continuation methods, we must create the first arc. The next sequence creates an arc with the start point at 2,1 with the end point at 3,2 and a radius of 1. This makes the ending direction of the existing arc 90 degrees, as shown in Figure 2–38. Invoke the ARC command from the Draw toolbar (see Figure 2–27), and AutoCAD prompts:

    Specify start point of arc or ⬇ **2,1** (ENTER)
    Specify second point of arc or ⬇ *(choose END from the shortcut menu)*
    Specify end point of arc: **3,2** (ENTER)
    Specify center point of arc or ⬇ *(choose RADIUS from the shortcut menu)*
    Specify radius of arc: **1** (ENTER)

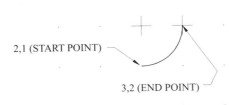

Figure 2–38 *The initial arc drawn with a start point (2,1), an end point (3,2), and a radius of 1*

The next sequence demonstrates the Arc-Arc-Continuation method of drawing an arc from the last-drawn arc (Figure 2–39). The resulting arc is shown in Figure 2–39. Invoke the ARC command from the Draw toolbar (see Figure 2–27), and AutoCAD prompts:

    Specify start point of arc or ⬇ (ENTER)
    Specify end point of arc: **2,3** (ENTER)

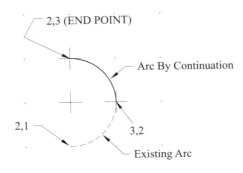

Figure 2–39 *An arc drawn by means of the Arc-Arc Continuation method*

The next example demonstrates the Arc-Arc Continuation method in a clockwise direction. The following sequence draws a initial arc clockwise from the start point at 3,2 to the end point at 2,1 (see Figure 2-40). Invoke the ARC command from the Draw toolbar (see Figure 2–27), AutoCAD prompts:

> Specify start point of arc or ⬇ **3,2** (ENTER)
> Specify second point of arc or ⬇ *(choose* END *from the shortcut menu*
> Specify end point of arc: **2,1** (ENTER)
> Specify center point of arc or ⬇ *(choose* DIRECTION *from the shortcut menu)*
> Specify tangent direction for the start point of arc: **-90** (ENTER)

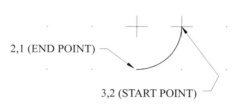

Figure 2–40 *An arc drawn clockwise with start point (3,2) and end point (2,1)*

The next sequence uses the Arc-Arc Continuation method to continue drawing a clockwise arc from the end point of the last-drawn arc (Figure 2–40). The resulting arc is shown in Figure 2–41. Invoke the ARC command from the Draw toolbar (see Figure 2–27), and AutoCAD prompts:

> Specify start point of arc or ⬇ (ENTER)
> Specify end point of arc: **2,3** (ENTER)

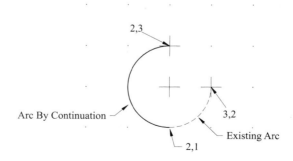

Figure 2–41 *A clockwise arc drawn by means of the Arc-Arc Continuation method*

In the last example, the direction used was 180 degrees. The same arc would have been drawn if the last object drawn was a line starting at 4,1 and ending at 2,1.

 Note: This method uses the last drawn arc or a line. If you draw an arc, draw a line, draw a circle, and then use this continuation method, AutoCAD will use the line as the basis for the start point and direction. This is because the line was the last of the "line-or-arc" objects drawn.

OBJECT SELECTION

Many of AutoCAD's modify and construct commands prompt you to select one or more objects for manipulation. When you select one or more objects, AutoCAD highlights them by displaying them as dashed lines. The group of objects selected for manipulation is called the *selection set*. There are several different ways of selecting objects. These include Window, Last, Crossing, Box, All, Fence, Wpolygon (WP), Cpolygon (CP), Group, Add, Remove, Multiple, Previous, Undo, Auto, and Single.

When a modify and construct command requires a selection set, AutoCAD prompts:

Select objects:

AutoCAD replaces the screen crosshairs with a small box called the object selection target. Using your pointing device (or the keyboard's cursor keys), position the target box so it touches only the desired object or a visible portion of it. The object selection target helps you select the object without having to be very precise. Every time you select an object, the "Select objects" prompt reappears. To indicate your acceptance of the selection set, press ENTER at the "Select objects" prompt.

Sometimes it is difficult to select objects that are close together or lie directly on top of one another. You can cycle through objects for selection at the "Select Objects" prompt by holding down the CTRL key and selecting a point as near as possible to the object. Press the pick button on your pointing device repeatedly until the object you want is highlighted, then press ENTER to select the object.

SELECTION BY WINDOW

The Window method allows you to select all objects that are contained completely within the rectangular area. You define this area by specifying two corner points. At the "Select objects" prompt, use your pointing device to specify the first corner point. This must be above or below and to the left of the objects you want to select. AutoCAD then prompts:

Specify opposite corner: *(specify opposite corner)*

Move your cursor up or down and to the right to create the rectangular area. When the rectangle encompasses the target objects, press the pick button on your pointing device to specify the second corner point (see Figure 2-42).

You can also define a selection window by entering coordinates. Enter **w** in response to the "Select Objects:" prompt, and AutoCAD prompts:

> Specify first corner: *(specify first corner coordinate)*
> Specify opposite corner: *(specify opposite corner coordinate)*

If there is an object that is partially included in the rectangular area, then that object will not be included in the selection set. You can select only objects currently visible on the screen. The selection shown in Figure 2–42 will only include the lines, not the circles, because a portion of each of the circles is outside the rectangular area.

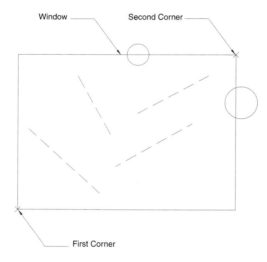

Figure 2–42 *Selecting objects by means of the Window option*

SELECTION BY CROSSING

The Crossing method allows you to select all objects that are contained completely or partially within the rectangular area. You define this area by specifying two corner points. At the "Select objects:" prompt, use your pointing device to specify the first corner point. This must be to the right of the objects you want to select. AutoCAD then prompts:

> Specify opposite corner: *(specify opposite corner)*

Move your cursor to the left to create the rectangular area. When the rectangle crosses a portion of the target objects, press the pick button on your pointing device to specify the second corner point (see Figure 2-43).

You can also define a crossing window for selection of objects, by entering **c** in response to the "Select Objects" prompt, and AutoCAD prompts:

Specify first corner: *(specify first corner coordinate)*
Specify opposite corner: *(specify opposite corner coordinate)*

The selection shown in Figure 2–43 includes all of lines and circles, though parts of the circles are outside the rectangle.

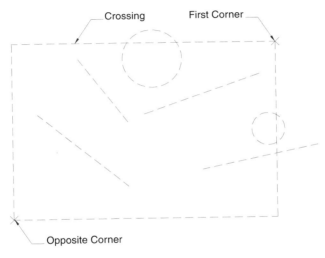

Figure 2–43 *Selecting objects by means of the Crossing option*

PREVIOUS SELECTION

The Previous method enables you to perform several operations on the same object or group of objects. AutoCAD remembers the most recent selection set and allows you to reselect it with the Previous option. For example, if you moved several objects and now wish to copy them elsewhere, you can invoke the COPY command and respond to the "Select objects:" prompt by entering **p** to select the same objects again. (There is a command called SELECT that does nothing but create a selection set; you can then use the Previous option to select this set in subsequent commands.)

LAST SELECTION

The Last option is an easy way to select the most recently created object currently visible. Only one object is selected, no matter how often you use the Last option. The Last option is invoked by entering **l** at the "Select Objects" prompt.

The Wpolygon (WP), Cpolygon (CP), Fence, All, Group, Box, Auto, Undo, Single, Multiple, Add, and Remove methods are explained in Chapter 5.

MODIFYING OBJECTS

AutoCAD not only allows you to draw objects easily, but also allows you to modify the objects you have drawn. Of the many modification commands available, the ERASE command will probably be the one you use most often. Everyone makes mistakes; in AutoCAD it is easy to erase them. Or, you may have created an object to aid in the construction of other objects, if you are finished with it, you may wish to erase it.

ERASING OBJECTS

To erase objects from a drawing, invoke the ERASE command from the Modify toolbar (see Figure 2–44) and AutoCAD prompts:

    Select objects: *(select objects to be erased, and then press the* SPACEBAR *or* ENTER*)*

Figure 2–44 *Invoking the* ERASE *command from the Modify toolbar*

You can use one or more available object selection methods in response to the next "Select objects:" prompt. After selecting the object(s), press ENTER to complete the ERASE command. All the objects that were selected will disappear.

The following command sequence shows an example of erasing individual objects (shown Figure 2–45).

    erase
    Select objects: *(select Line 2, the line is highlighted)*
    Select objects: *(select Line 4, the line is highlighted)*
    Select objects: (ENTER)

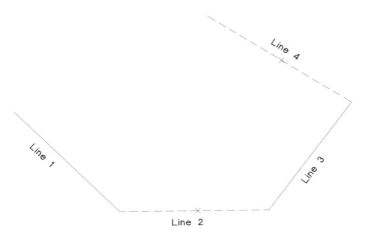

Figure 2–45 *Selection of individual objects to erase*

The following command sequence shows an example of erasing objects using the Window selection method (see Figure 2–46).

erase
Select objects: (*specify a point*)
Specify opposite corner: (*specify a point for the diagonally opposite corner*)
Select objects: (ENTER)

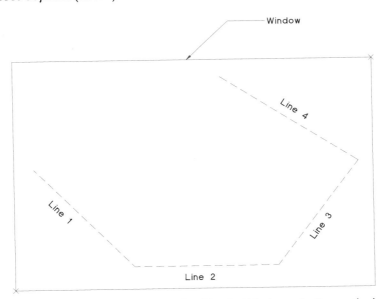

Figure 2–46 *Erasing a group of objects defined by the Window selection method*

GETTING IT BACK

The OOPS command restores objects that have been unintentionally erased. Whenever the ERASE command is used, the last group of objects erased is stored in memory. The OOPS command will restore the objects; it can be used at any time. It only restores the objects erased by the most recent ERASE command. See Chapter 3 on the UNDO command, if you need to step back further than one ERASE command.

To restore objects erased by the last ERASE command, invoke the OOPS command by typing **oops** at the On-Screen prompt. AutoCAD restores the objects.

The following example shows the sequence for using the OOPS command in conjunction with the ERASE command:

    erase (ENTER)
    Select objects: *(specify a point)*
    Other corner: *(specify a point for the diagonally opposite corner)*
    Select objects: (ENTER)
    oops (ENTER) *(the last erased objects are restored)*

Open the Exercise Manual PDF file for Chapter 2 on the accompanying CD for project and discipline specific exercises.

If you have the accompanying Exercise Manual, refer to Chapter 2 for project and discipline specific exercises.

REVIEW QUESTIONS

1. The RECTANGLE command requests what information?

    a. an initial corner, the width, and the height

    b. the coordinates of the four corners of the rectangle

    c. the coordinates of diagonally opposite corners of the rectangle

    d. the coordinates of three adjacent corners of the rectangle

2. When drawing a trace line, after you select the second point:

    a. nothing appears on the screen

    b. you are prompted for the trace width

    c. the segment is drawn and the command terminates

    d. the first segment is drawn and you are prompted for the next point

3. To draw multiple connected line segments, you must invoke the LINE command multiple times.

    a. True

    b. False

4. The file extension .BAK stands for:

    a. backup drawing file

    b. binary file

    c. binary attribute file

    d. drawing file

    e. both b and c

5. The HELP command cannot be used:

    a. while in the LINE command

    b. while in the CIRCLE command option TTR

    c. to list commands

    d. for a text string

6. Points are located by relative rectangular coordinates in relation to:
   a. the last specified point or position
   b. the global origin
   c. the lower left corner of the screen
   d. all of the above

7. Polar coordinates are based on a distance from:
   a. the global origin
   b. the last specified position at a given angle
   c. the center of the display
   d. all of the above

8. To enter a command from the keyboard, simply enter the command name at the On-Screen prompt:
   a. in lowercase letters
   b. in uppercase letters
   c. lowercase, uppercase, or mixed case
   d. commands cannot be entered via the keyboard

9. The "C" option used in the LINE command at the "From Point:" prompt will:
   a. continue the line from the last line or arc that was drawn
   b. close the previous set of line segments
   c. display an error message

10. A rectangle generated by the RECTANGLE command will always have horizontal and vertical sides.
    a. True
    b. False

11. Which of the following coordinates will define a point at the screen default origin?
    a. 000
    b. 00
    c. 0.0,0.0
    d. 112
    e. @0,0

12. Switching between graphics and text screen can be accomplished by:

    a. pressing (ENTER) twice

    b. pressing (CTRL) and (ENTER) at the same time

    c. pressing the (ESC) key

    d. pressing the (F2) function key

    e. both b and c

13. To draw a line a length of 8 feet, 4-5/8 inches in the 12 o'clock direction from the last point selected, enter:

    a. @8'4-5/8<90

    b. 8'-4-5/8<90

    c. @8-45/8<90

    d. 8'-45/8<90

    e. none of the above

14. By default, what direction does a positive number indicate when specifying angles in degrees?

    a. Clockwise

    b. Counterclockwise

    c. Has no impact when specifying angles in degrees

    d. None of the above

15. When erasing objects, if you select a point that is not on any object, AutoCAD will:

    a. terminate the ERASE command

    b. delete the selected objects and continue with the ERASE command

    c. allow you to drag a window to select multiple objects within the area

    d. ignore the selection and continue with the ERASE command

16. Regarding the ARC options, what does "S,C,E" mean?

    a. Start, Center, End

    b. Second, Continue, Extents

    c. Second, Center, End

    d. Start, Continue, End

17. The number of different methods by which a circle can be drawn is:

    a. 1
    b. 3
    c. 4
    d. 7
    e. none of the above

18. When using the ERASE command, AutoCAD deletes each object from the drawing as you select it.

    a. True
    b. False

19. Once an object is erased from a drawing, which of the following commands could restore it to the drawing?

    a. OOPS
    b. RESTORE
    c. REPLACE
    d. CANCEL

20. When drawing a circle with the two-point option, the distance between the two points is equal to:

    a. the circumference
    b. the perimeter
    c. the shortest chord
    d. the radius
    e. the diameter

21. A circle may be created by any of the following options, except:

    a. 2P
    b. 3P
    c. 4P
    d. Cen,Rad
    e. TTR

CHAPTER 3

Fundamentals II

INTRODUCTION

AutoCAD provides various tools to make your drafting and design work easier. The drawing tools described in this chapter will assist you in creating drawings rapidly, while ensuring the highest degree of precision.

After completing this chapter, you will be able to do the following:

- Use and control drafting settings (e.g., GRID, SNAP, ORTHO, POLAR TRACKING, and OBJECT SNAP)
- Use Tracking and Direct Distance
- Use display control commands (e.g., ZOOM, PAN, REDRAW, and REGEN)
- Create tiled viewports
- Use layering techniques
- Use the UNDO and REDO commands

DRAFTING SETTINGS

The SNAP, GRID, ORTHO, POLAR/OBJECT SNAP TRACKING, OBJECT SNAP, DYNAMIC TRACKING, and LINEWEIGHT commands do not create objects. However, they make it possible to create and modify objects more easily and accurately. Each of these Drafting Settings commands can be readily toggled ON when needed and OFF when not. These commands, when turned ON, function according to settings that can be changed easily. When used appropriately, these commands provide the power, speed, and accuracy associated with Computer-Aided Design/Drafting.

SNAP COMMAND

The SNAP command provides an invisible reference grid in the drawing area. When set to ON, the Snap feature forces the cursor to lock onto the nearest point on the specified Snap Grid. With the Snap value set appropriately, you can specify points quickly, letting AutoCAD ensure that they are placed precisely. You can always override the snap spacing by entering absolute or relative coordinate points from the keyboard, or by simply turning the Snap mode OFF. When the Snap mode is OFF, it has no effect on the cursor. When it is ON, you cannot place the cursor on a point that is not on one of the specified Snap Grid locations.

Setting Snap ON and OFF

In AutoCAD, there are five methods for setting the Snap ON and OFF. Of the five explained in this section, the first method (selecting **SNAP** on the status bar) is the easiest and most commonly used with the pointing device. The second method (pressing the function key F9) is the most commonly used from the keyboard. The others, while not quite as convenient as the first two, might be convenient in certain situations.

1. Snap can be toggled ON and OFF by choosing **SNAP** on the status bar.
2. Snap can be toggled ON and OFF by pressing the function key F9.
3. Right-click **SNAP** on the status bar at the bottom of the screen and choose GRID SNAP ON or POLAR SNAP ON from the shortcut menu, or choose OFF to set the Snap OFF.
4. From the Tools menu, choose Drafting Settings. Select the **Snap and Grid** tab, if it is not already selected. Snap can be toggled ON and OFF by selecting **Snap On (F9)** (see Figure 3–1).
5. At the On-Screen prompt, type **snap** and AutoCAD prompts:

Specify snap spacing or ⬇ (choose ON or OFF from the shortcut menu)

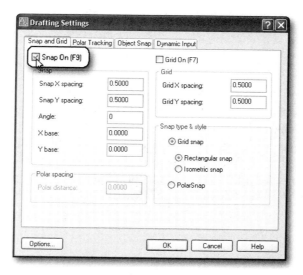

Figure 3–1 Choosing the **Snap On (F9)** check box from the **Snap and Grid** tab of the Drafting Settings dialog box

Changing Snap Spacing

In AutoCAD there are three methods for changing the Snap spacing:

1. Right-click SNAP on the status bar at the bottom of the screen and from the shortcut menu, choose SETTINGS. AutoCAD will display the Drafting Settings dialog box with the **Snap and Grid** tab selected. From the **Snap** section, you can change the settings of the X and Y spacing by entering the desired values in the **Snap X spacing** and **Snap Y spacing** boxes, respectively (see Figure 3–2).

2. From the Tools menu, choose Drafting Settings. Then select the **Snap and Grid** tab. See the explanation in Method 1 for changing settings.

3. At the On-Screen prompt, type **snap** and AutoCAD prompts;

Specify snap spacing or ⤓ (specify distance to be used for both X and Y Snap Spacings)

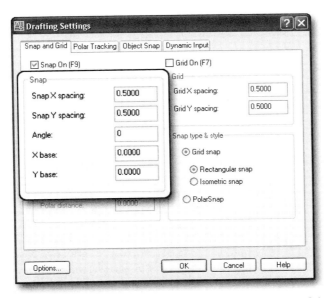

Figure 3–2 Changing the Snap settings from the Snap and Grid section of the Drafting Settings dialog box

Snap Options

There are several snap options available at the On-Screen prompt. In addition to ON or OFF, the following options are available: ASPECT, ROTATE, STYLE, TYPE.

Choose ASPECT to set the Y Snap Spacing different from the X Snap Spacing. AutoCAD prompts for the X and Y values independently. This is handy if the X and Y modular dimensions of your design are of unequal multiples. There are three methods for changing the Snap Aspect setting.

1. From the **Snap and Grid** tab of the Drafting Settings dialog box, you can set the Y spacing different from the X spacing. This is done by entering a value in the **Snap Y spacing** box that is different from the value in the **Snap X spacing** box.

2. From the Tools menu, choose Drafting Settings. Then select the **Snap and Grid** tab. See the explanation in Method 1 for setting the Y spacing different from the X spacing.

3. At the On-Screen prompt, type **snap** and AutoCAD prompts:

Specify snap spacing or ⬇ *(choose ASPECT from the shortcut menu)*
Specify horizontal spacing <current>: *(specify X Snap spacing)*
Specify vertical spacing <current>: *(specify Y Snap spacing)*

For example, the following sequence shows how you can set a horizontal snap spacing of 0.5 and vertical snap spacing of 0.25.

snap (ENTER)
Specify snap spacing or ⬇ *(choose ASPECT from the shortcut menu)*
Specify horizontal spacing <Current>: **0.5** (ENTER)
Specify vertical spacing <Current>: **0.25** (ENTER)

Choose ROTATE to specify an angle to rotate both the visible Grid and the invisible Snap grid. This method is similar to using the more complex User Coordinate System. It permits you to set a Snap grid with an origin (*X* coordinate, *Y* coordinate of 0,0) and an angle of rotation specified with respect to the default origin and Zero-East system of direction. In conjunction with the *X* and *Y* spacing of the Snap grid, the ROTATE option can make it easier to draw certain shapes.

Note: While you can enter the value for the angle of Snap rotation in the Angle text box of the Drafting Settings dialog box, in order to select two points on the screen to specify the angle, use the SNAP command by entering **snap** from the keyboard and then right-click and select ROTATE from the shortcut menu.

The plot plan in Figure 3–3 is an example of where the ROTATE option of the SNAP command can be applied. The property lines are drawn using surveyor's units of angular display. In this example, the architectural units and the surveyor's angular units are selected (both of these can be set in the Drawing Units dialog box). The limits are set up with the lower-left corner at –20',–10', and the upper-right corner at 124',86'. After making sure the ORTHO mode is set to ON, the sequence for drawing the property lines is as follows:

snap (ENTER)
Specify snap spacing or ⊥ (choose ROTATE from the shortcut menu)
Specify base point <0'-0.0",0'-0.0">: (press ENTER to accept the default value)
Specify rotation angle <0>: **4d45'08"** (ENTER)

line (ENTER) (invoking the LINE command)
Specify first point: **0,0** (ENTER) (selecting the first point)
Specify next point or ⊥ **85'** (ENTER) (point selected with the cursor north of the first point)
Specify next point or ⊥ **120'** (ENTER) (point selected with the cursor east of the previous point)
Specify next point or ⊥ **85'** (ENTER) (point selected with the cursor south of the previous point)
Specify next point or ⊥ **c** (choose CLOSE from the shortcut menu to complete the property boundary by closing the rectangle and exiting the LINE command)

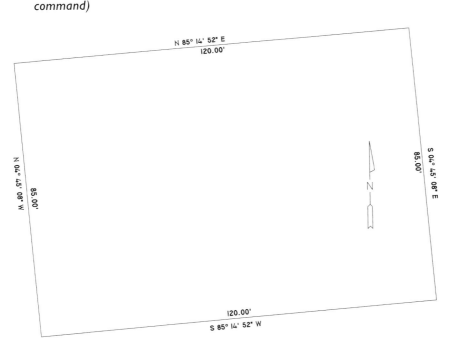

Figure 3–3 *Example drawing where the* ROTATE *option of the* SNAP *command can be used*

Figure 3–4 shows the property boundary drawn.

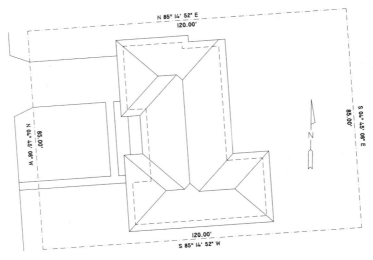

Figure 3–4 *Layout of the property boundary lines*

Choose STYLE to select one of two available formats, Standard or Isometric. Standard refers to the normal rectangular grid (default), and Isometric refers to a Grid and Snap designed for Isometric drafting purposes (see Figure 3–5).

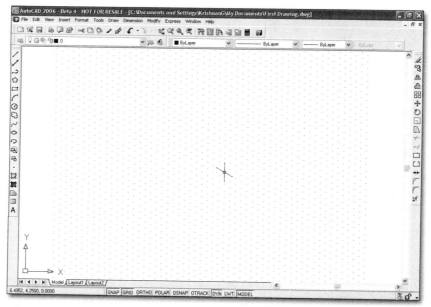

Figure 3–5 *Setting the snap for isometric drafting*

You can switch the Isoplanes between Left (90- and 150-degree angles), Top (30- and 150-degree angles), and Right (30- and 90-degree angles) by pressing CTRL+E (the combination keystrokes of holding down CTRL and then pressing E) or by simply pressing the function key F5.

Choose TYPE to select one of two available Snap types, Polar or Grid. Choosing the Polar type sets the snap to Polar Tracking angles. See the explanations on Polar Tracking later in this chapter. Choosing the Grid type sets the snap spacing equal to the grid spacing.

GRID COMMAND

The GRID command is used to display a visible array of dots. AutoCAD creates a grid that is similar to a sheet of graph paper. You can set the grid display ON and OFF, and can change the dot spacing. The grid is a drawing tool and is not part of the drawing; it is for visual reference and is never plotted. In the World Coordinate System, the grid fills the area defined by the limits of the drawing area.

The grid has several uses within AutoCAD. First, it shows the extent of the drawing limits. For example, if you set the limits to 42 by 36 units and grid spacing is set to 0.5 units, then each row will have 85 dots and each column will have 73 dots. This will give you a better sense of the drawing's size relative to the limits than if it were on a blank background.

Second, using the GRID command with the SNAP command is helpful when you create a design in terms of evenly spaced units. For example, if your design is in multiples of 0.5 units, then you can set grid spacing as 0.5 to facilitate point entry. You could check your drawing visually by comparing the locations of the grid dots and the crosshairs. Figure 3–6 shows a drawing with a grid spacing of 0.5 units, with limits set to 0,0 and 17,11.

Note: While the GRID, SNAP, ORTHO, POLAR TRACKING, and OBJECT SNAP commands do not create objects, they make it possible to create them more easily and accurately. Each of these drafting settings commands, when toggled ON, operates according to the value(s) to which you have set it and, when toggled OFF, has no effect. You are advised to identify and master the two skills involved in using these utilities: one is to learn how to change the settings of the utility commands and the other is to learn how and when to best set them ON and OFF. Changing the settings of these utilities is normally done more easily by means of their associated dialog box(es). You can also set them ON and OFF from a dialog box. But, because these features are normally switched ON and OFF so frequently during a drawing session, special buttons are provided on the status bar at the bottom of the screen for this purpose.

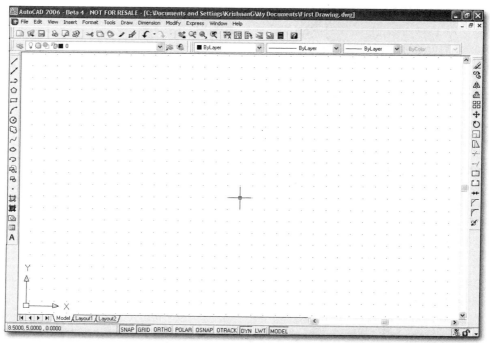

Figure 3–6 A grid spacing of 0.5 units, with limits set to (0,0) and (17,11)

Setting the Grid ON and OFF

In AutoCAD, there are five methods for setting the Grid ON and OFF. Of the five explained in this section, the first method (selecting **GRID** on the status bar) is the easiest and most commonly used with the pointing device. The second method (pressing the function key F7) is the most commonly used from the keyboard. The others, while not quite as convenient as the first two, might be convenient in certain situations.

1. Grid can be toggled ON and OFF by choosing **GRID** on the status bar.
2. Grid can be toggled ON and OFF by pressing the function key F7.
3. Right-click **GRID** on the status bar at the bottom of the screen and choose ON or OFF.
4. From the Tools menu, choose Drafting Settings. Then select the **Snap and Grid** tab if it is not already selected. Grid can be toggled ON and OFF by selecting **Grid On (F7)** (see Figure 3–7).
5. At the On-Screen prompt, type **grid** and AutoCAD prompts:

Specify grid spacing(X) or ⬇ *(choose ON or OFF from the shortcut menu)*

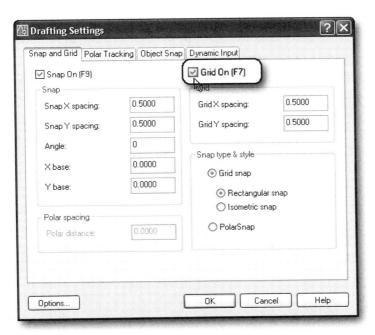

Figure 3–7 *Choosing the **Grid On (F7)** check box from the **Snap and Grid** tab of the Drafting Settings dialog box*

Changing Grid Spacing

In AutoCAD there are three methods for changing the Grid spacing:

1. Right-click **GRID** on the status bar at the bottom of the screen and from the shortcut menu, choose SETTINGS. AutoCAD will display the Drafting Settings dialog box with the **Snap and Grid** tab selected. From the **Grid** section, you can change the settings of the X and Y spacings by entering the desired values in the **Grid X spacing** and **Grid Y spacing** boxes, respectively (see Figure 3–8).

2. From the Tools menu, choose Drafting Settings. Then select the **Snap and Grid** tab. See the explanation in Method 1 for changing settings.

3. At the On-Screen prompt, type **grid** and AutoCAD prompts:

Specify grid spacing(X) or ⊠ *(specify distance to be used for both X and Y Grid Spacings)*

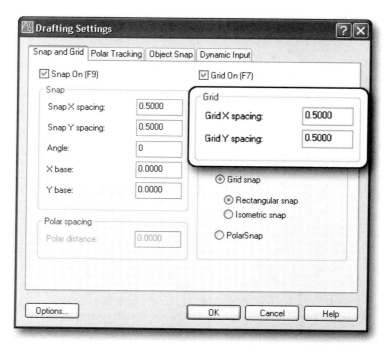

Figure 3-8 *Changing the Grid settings from the Snap and Grid section of the Drafting Settings dialog box*

Grid Options

Choose ASPECT to set the Y Grid Spacing different from the X Grid Spacing. AutoCAD then prompts for the X and Y values independently. This is handy if the X and Y modular dimensions of your design are of unequal multiples. There are three methods for changing the Snap Aspect setting.

1. From the **Snap and Grid** tab of the Drafting Settings dialog box, you can set the Y spacing different from the X spacing. This is done by entering a value in the **Grid Y spacing** box that is different from the value in the **Grid X spacing** box.

2. From the Tools menu, choose Drafting Settings. Then select the **Snap and Grid** tab. See the explanation in Method 1 for setting the Y spacing different from the X spacing.

3. At the On-Screen prompt, type **grid** and AutoCAD prompts:

Specify grid spacing(X) or ⬇ *(choose ASPECT from the shortcut menu)*
Specify the horizontal spacing(X) <current>: *(specify X Grid spacing)*
Specify the vertical spacing(Y) <current>: *(specify Y Grid spacing)*

For example, the following sequence of commands shows how you can set a horizontal grid spacing of 0.5 and vertical grid spacing of 0.25.

> **grid** (ENTER)
> Specify grid spacing(X) or ⬇ (choose ASPECT *from the shortcut menu*)
> Specify the horizontal spacing(X) <current>: **0.5** (ENTER)
> Specify the vertical spacing(Y) <current>: **0.25** (ENTER)

Applying the ASPECT option in this example provides the Grid dot spacing shown in Figure 3–9.

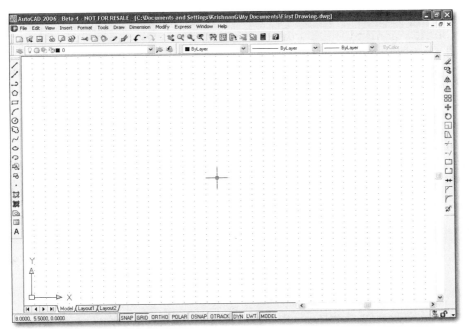

Figure 3–9 *Display after setting the grid aspect to 0.5 for horizontal and 0.25 for vertical spacing*

Choose SNAP from the shortcut menu to set the grid spacing equal to the snap resolution, or make it a multiple of the snap spacing. To specify the grid spacing as a multiple of the snap value, enter **x** after the value. For example, to set up the grid value as three times the current snap value (snap = 0.5 units), enter **3x** for the prompt (shown in the following example), which is the same as setting it to 1.5 units:

> **grid** (ENTER)
> Specify grid spacing(X) or ⬇ **3x** (ENTER)

If the spacing of the visible grid is set too small, AutoCAD will not display the grid. To display the grid, invoke the GRID command again and specify a larger spacing.

 Note: The relationship between the Grid setting and the Snap setting, when established as described in the previous section, is based on the current Snap setting. If the Snap setting is subsequently changed, the Grid setting does not change accordingly. For example, if the Snap setting is 1.00 and you enter **s** in response to the Specify grid spacing(X) or <current>: prompt, the Grid setting becomes 1.00 and remains 1.00 even if the Snap setting is later set to something else. Likewise, if you set the Grid setting to **3x**, it becomes 3.00 and will not change with a subsequent change in the Snap setting.

ORTHO COMMAND

The ORTHO command lets you draw lines and specify point displacements that are parallel to either the X or Y axis. Lines drawn with the Ortho mode set to ON are therefore either parallel or perpendicular to each other. This mode is helpful when you need to draw lines that are exactly horizontal or vertical. Also, when the Snap Style is set to Isometric, it forces lines to be parallel to one of the three isometric axes.

Setting Ortho ON and OFF

In AutoCAD, four methods for setting the ORTHO command ON and OFF are available. Of the four explained in this section, the first method (selecting **ORTHO** on the status bar) is the easiest and most commonly used with the pointing device. The second method (pressing the function key F8) is the most commonly used from the keyboard. The others, while not quite as convenient as the first two, might be convenient in certain situations.

1. Ortho can be toggled ON and OFF by choosing **ORTHO** on the status bar.
2. Ortho can be toggled ON and OFF by pressing the function key F8.
3. Right-click **ORTHO** on the status bar at the bottom of the screen and choose ON or OFF.
4. At the On-Screen prompt, type: **ortho** and AutoCAD prompts:

Enter mode *(choose ON or OFF from the shortcut menu)*

 Note: The Ortho and Polar Tracking modes (explained later in this chapter) cannot both be set to ON at the same time. They can both be set to OFF, or either one can be set to ON.

When the Ortho mode is active, you can draw lines and specify displacements only in the horizontal or vertical directions, regardless of the cursor's on-screen position. The direction in which you draw is determined by the change in the X value of the cursor movement compared to the change in the cursor's distance to the Y axis. AutoCAD allows you to draw horizontally if the distance in the X direction is greater than the distance in the Y direction; conversely, if the change in the Y direction is greater than the change in the X direction, then it forces you to draw vertically. The Ortho mode does not affect keyboard entry of points.

OBJECT SNAP

The Object Snap (or Osnap, for short) feature lets you specify points on existing objects in the drawing. For example, if you need to draw a line from an end point of an existing line, you can apply the Object Snap mode, called ENDpoint. In response to the "Specify first point:" prompt, enter **end** and place the cursor so that it touches the line nearer the desired end point. AutoCAD will lock onto the end point of the existing line when you press the pick button of your pointing device. This end point becomes the starting point of the new line. This feature is similar to the basic SNAP command, which locks to invisible reference grid points.

You can invoke an Object Snap mode whenever AutoCAD prompts for a point. Object snap modes can be invoked while executing an AutoCAD command that prompts for a point, such as the LINE, CIRCLE, MOVE, and COPY commands.

Applying Object Snap Modes

Object Snap modes can be applied in either of two ways:

1. From the Object Snap toolbar, choose **Osnap Settings** (see Figure 3–10), or on the status bar at the bottom of the screen, right-click on **OSNAP** and choose SETTINGS. The Drafting Settings dialog box will be displayed with the **Object Snap** tab selected, as shown in Figure 3–11. An Object Snap mode has been chosen if there is a check in its associated check box. Choose the desired Object Snap mode(s) by placing the cursor over its check box and pressing the pointing device pick button. This makes it possible to use the checked mode(s) any time you are prompted to specify a point. This is referred to as the *Running Osnap* method. Running Osnap can be set to ON and OFF by choosing **OSNAP** on the status bar. You can also toggle the Running Osnap by pressing the function key F3.

Figure 3–10 *Choosing Object Snap Settings from the Object Snap toolbar*

2. When prompted to specify a point, enter the first three letters of the name of the desired Object Snap mode, or select it from the Object Snap toolbar before specifying the point. This is a *one-time-only* Osnap method. This will override any other Running Osnap mode for this one point selection.

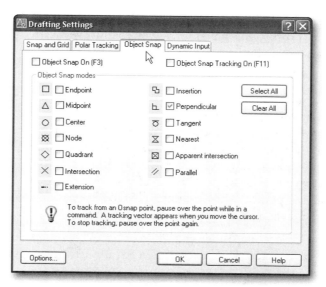

Figure 3–11 *Drafting Settings dialog box displaying the **Object Snap** tab*

To draw a line from an end point of an existing line, invoke the LINE command, and AutoCAD prompts:

>**line** (ENTER)
>Specify first point: **end** *(enter the first three letters of the Osnap mode Endpoint)*
>Specify first point: *(move the cursor near the desired end of a line and press the pick button)*

The first point of the new line is drawn from the end point of the selected object.

Osnap Markers and Tooltips

Whenever one or more Object Snap modes are activated and you move the cursor target box over a snap point, AutoCAD displays a geometric shape (Marker) and Tooltip. By displaying a Marker on the snap points with a Tooltip, you can see the point that will be selected and the Object Snap mode in effect. AutoCAD displays the Marker depending on the Object Snap mode selected. In the **Object Snap** tab of the Drafting Settings dialog box, each Marker is displayed next to the name of its associated Object Snap mode.

From the **Object Snap** tab of the Drafting Settings dialog box, choose **Options** and AutoCAD displays the Options dialog box with the **Drafting** tab selected. This is where the settings for displaying the Markers and Tooltips can be changed by checking or clearing the appropriate check boxes in the **AutoSnap Settings** section (see Figure 3–12).

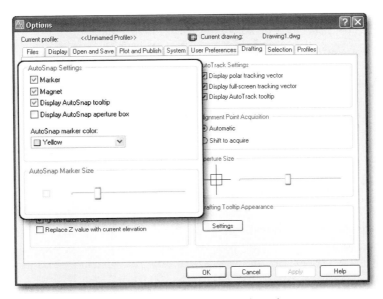

Figure 3–12 *Options dialog box with the Drafting tab selected*

The **AutoSnap Settings** section controls the size, appearance, action of the cursor, and if a tooltip is displayed while Object Snap is in effect.

Choosing **Marker** causes the geometric shape (unique for each Osnap mode) to be displayed when the cursor moves over a snap point.

Choosing **Magnet** causes the cursor to lock onto the snap point when the cursor is near.

Choosing **Display AutoSnap tooltip** causes a tooltip to display the name of the Object Snap mode when AutoCAD is prompting for a point and an Osnap mode is in effect.

Choosing **Display AutoSnap aperture box** causes the square target box to be displayed when AutoCAD is prompting to specify a point.

The **AutoSnap marker color** list box lets you change the color of the Marker.

The **AutoSnap Marker Size** section controls the size of the AutoSnap Marker. To change the size of the AutoSnap Marker, press and hold the pick button of the pointing device while the cursor is over the slide bar and move it to the right to make the Marker larger and to the left to make it smaller. The Image tile shows the current size of the Marker.

After making the necessary changes to the AutoSnap settings, choose **OK** to close the Options dialog box and then choose **OK** again to close the Drafting Settings dialog box.

Object Snap Modes

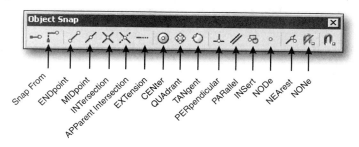

Figure 3-13 *The Object Snap toolbar.*

There are 16 available Object Snap modes: Endpoint, Midpoint, Intersection, Apparent Intersection, Extension, Center, Quadrant, Tangent, Perpendicular, Parallel, Insert, Node, Nearest, None, From, and MTP (snapping between specified points).

 Note: It takes practice! And you must be alert whenever you specify a point near objects, when more than one Running Osnap modes are in effect. For example, if you are trying to specify an end point on a line that is near the circumference of a circle while the Center Osnap mode is active and the cursor pick box touches the circle, AutoCAD may select the center of the circle as the specified point. This kind of error is usually obvious and you have the opportunity to correct it. But if AutoCAD places a point slightly off its intended location due to a Running Osnap mode taking over unintentionally, it could result in an undetected error in your drawing.

Endpoint, Intersection, Midpoint, and Perpendicular

The Endpoint mode allows you to snap to the closest end point of a line, arc, elliptical arc, multiline, polyline segment, spline, region, or ray, or to the closest corner of a trace, solid, or 3D face. As shown in Figure 3-14, LINE B is drawn from the indicated starting point to the end of LINE A by using the Endpoint mode.

The Intersection mode allows you to snap to the intersection of two objects, which can include arcs, circles, ellipses, elliptical arcs, lines, multilines, polylines, rays, splines, or xlines. As shown in Figure 3-14, LINE C is drawn from the indicated starting point to the intersection of the CIRCLE C and LINE A by using the Intersection mode.

The Midpoint mode allows you to snap to the midpoint of a line, arc, elliptical arc, broken ellipse, multiline, polyline segment, xline, solid, or spline. As shown in Figure 3-14, LINE D is drawn from the indicated starting point to the midpoint of LINE A by using the Midpoint mode.

The Perpendicular mode allows you to snap to a point perpendicular to a line, arc, circle, elliptical arc, multiline, polyline, ray, solid, spline, or xline. Deferred Perpendicular snap mode is automatically turned on when more than one perpendicular snap is required by the object being drawn. As shown in Figure 3–14, LINE E is drawn from the indicated starting point to point on and perpendicular to LINE A by using the Perpendicular mode. The location of the cursor may have to be adjusted to ensure that LINE A is selected; otherwise, if CIRCLE C is selected, the result will be LINE F, drawn perpendicular to the circle.

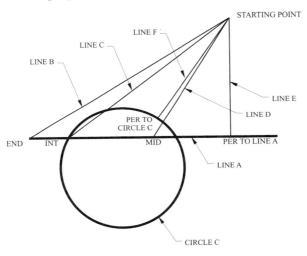

Figure 3–14 *Lines drawn to designated points of another line using the Endpoint, Intersection, Midpoint, and Perpendicular Object Snap modes.*

Quadrant, Tangent, and Center

The Quadrant mode allows you to snap to one of the quadrant points of a circle, arc, ellipse, or elliptical arc. The quadrant points are located at 0°, 90°, 180°, and 270° from the center of the circle or arc. The quadrant points are determined by the zero-degree direction of the current coordinate system. As shown in Figure 3–15, LINE B is drawn from the indicated starting point to one of the quadrants of CIRCLE C by using Quadrant mode.

The Tangent mode allows you to snap to the tangent of an arc, circle, ellipse, or elliptical arc. Deferred Tangent snap mode is automatically turned on when more than one tangent is required by the object being drawn. As shown in Figure 3–15, LINE C is drawn from the indicated starting point a point on and tangent to CIRCLE C by using Tangent mode.

The Center mode allows you to snap to the center of an arc, circle, ellipse, elliptical arc, or line. As shown in Figure 3–15, LINE D is drawn from the indicated starting point to the center of CIRCLE C by using Center mode.

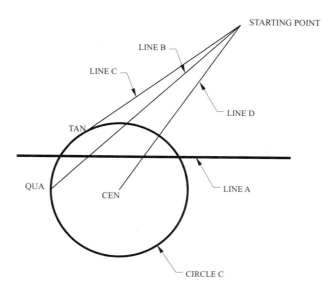

Figure 3–15 *Lines drawn to designated points of a circle using the Quadrant, Tangent, and Center Object Snap modes.*

 Note: The Quadrant point selected is one of four possible points on the circle, which include points at 0°, 90°, 180°, and 270°.

Apparent Intersection, Extension, and Parallel

The Apparent Intersection mode allows you to snap to the apparent intersection of two objects, which can include an arc, circle, ellipse, elliptical arc, line, multiline, polyline, ray, spline, or xline. These objects may or may not actually intersect, but would intersect if either or both objects were extended. In the following example, a rectangle box is drawn from the apparent intersection of two rectangles, as shown in Figure 3–16:

> **rectangle** (ENTER)
> Specify first corner point or ⬇ **app** *(invoke the Apparent Intersection object snap, and select LINE 1 and LINE 4, as shown in Figure 3–16)*
> Specify other corner point or ⬇ **app** *(invoke the Apparent Intersection Object Snap mode, and select LINE 2 and LINE 3, as shown in Figure 3–16)*

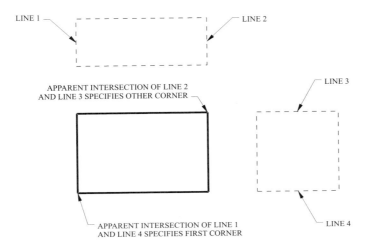

Figure 3-16 *Identifying the lines to draw a rectangle*

The Extension mode causes a temporary extension line to be displayed when you pass the cursor over the end point of objects, so you can draw objects to and from points on the extension line. As shown in Figure 3–17, LINE A was drawn using extensions of LINE B and ARC C.

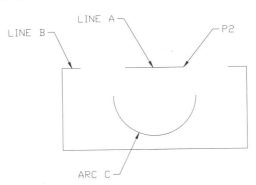

Figure 3-17 *LINE A is drawn using extensions of LINE B and ARC C*

The Parallel mode allows you to draw a line that is parallel to another object. Once the first point of the line has been specified (and the Parallel Object Snap mode is selected), move the cursor over the object to which you wish to make the new line parallel. Then move the cursor near a line from the first point that is parallel to the object selected, and a construction line will appear. While the construction line is visible, specify a point, and the new line will be parallel to the selected object.

Node, Insertion, Nearest, From, and None

The Node mode allows you to snap to a point on an object.

The Insertion mode allows you to snap to the insertion point of a block, text string, attribute, or shape.

The Nearest mode lets you select any object (except text) in response to a prompt for a point, and AutoCAD snaps to the point on that object nearest the cursor.

The From mode locates a point offset from a reference point within a command. At an AutoCAD prompt for locating a point, select From Object snap mode, and then specify a temporary reference or base point from which you can specify an offset to locate the next point. Enter the offset location from this base point as a relative coordinate, or use direct distance entry.

The None mode temporarily overrides any running object snaps that may be in effect.

Snapping Between Specified Points

AutoCAD 2006 includes a new Object Snap mode for snapping to a point that is midway between two specified points. You can enter **mtp** or **m2p** when prompted to specify a point. For example, if you want to start a line at the midpoint between the centers of two circles, you can invoke the Midway Between Two Points Object Snap mode as follows:

> **line** (ENTER)
> Specify first point: **mtp** (ENTER)
> First point of mid: **cen**
> First point of mid: *(specify the center of the first circle)*
> Second point of mid: **cen**
> Second point of mid: *(specify the center of the second circle)*
> Specify next point or *(specify the end point of the line)*
> Specify next point or (ENTER)

Note: The Midway Between Two Points Object Snap mode is not included on the **Object Snap** tab of the Drafting Settings dialog box, nor on the Object Snap toolbar in the initial release of AutoCAD 2006.

DRAWING OBJECTS USING TRACKING

Tracking, or moving through nonselected point(s) to a selected point, could be called a command "enhancer." It can be used whenever a command prompts for a point. If the desired point can best be specified relative to some known point(s), you can "make tracks" to the desired point by invoking the Tracking option and then specifying one or more points relative to previous point(s) "on the way to" the actual point that the command is prompting for. These intermediate tracking points are not necessarily associated with the object being created or modified by the command. The primary significance of tracking points is that they are used to establish a path to the point you wish to specify as the response to the command prompt. Some of the objects in the partial plan shown in Figure 3–18 can be drawn more easily by means of Tracking.

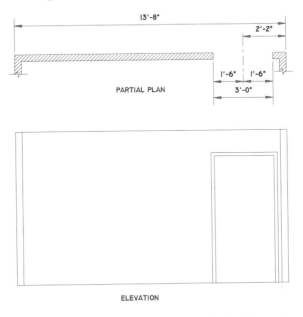

Figure 3–18 *Example of a partial plan to demonstrate the Tracking option*

The following example will use Tracking to draw lines A and B in the Partial Plan in Figure 3–19, leaving the 3'-0" door opening in the correct place. By means of Tracking, we can draw the lines with the given dimension information without having to calculate the missing information.

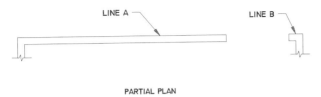

Figure 3–19 *Lines A and B to be drawn with the help of the Tracking option*

In Figure 3–20, line A from TK1/SP1 (tracking point 1 and starting point 1) to EP1 (ending point 1) and line B from SP2 (starting point 2) to TK2/EP2 (tracking point 2 and ending point 2) can be drawn by using the TRACKING command enhancer. To complete this example, change AutoCAD's drawing units to Architectural. Invoke the LINE command and AutoCAD prompts:

> line (ENTER)
> Specify first point: **0,12'** (ENTER) *(specify point TK1/SP1)*
> Specify next point or ⬇ **track** (ENTER) *(invoke the Tracking feature)*
> First tracking point: **0,12'** (ENTER) *(specify point TK1/SP1 again as the first tracking point)*
> Next point (Press enter to end tracking): **@13'8,0** (ENTER) *(locates the second tracking point, TK2, as shown in Figure 3–20)*
> Next point (Press enter to end tracking): **@2'2<180** (ENTER) *(locates the third tracking point, TK3, as shown in Figure 3–20)*
> Next point (Press enter to end tracking): **@1'6<180** (ENTER) *(locates the fourth tracking point, EP1, as shown in Figure 3–20)*
> Next point (Press enter to end tracking): *(press ENTER to exit Tracking; by this you are designating the point to which you have "made tracks" as the response to the prompt that was in effect when you entered Tracking)*
> Specify next point or ⬇ *(press ENTER to exit the LINE command)*

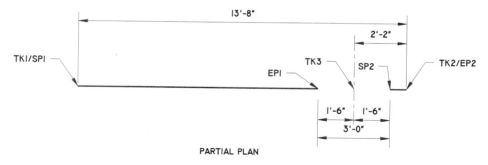

Figure 3–20 *Various points to be drawn with the help of the Tracking option*

Once the Tracking option is invoked, you establish a path to EP1 by specifying the initial tracking point, TK1, and then each subsequent point relative to the previous point, that is, TK2 relative to TK1, TK3 relative to TK2, and EP1 relative to TK3. The first track in this example is specified by the Relative Rectangular method, and the next two are Relative Polar. Also, because the tracking points are all on one horizontal line, you could use the direct distance feature (explained in the next section) by placing the cursor in the correct direction (with ORTHO set to ON) and entering the distance from the keyboard.

 Note: If you knew the coordinates of one of the intermediate tracking points, then it probably should be the initial tracking point. The idea behind Tracking is to establish a point by means of a path from and through other points. Thus, the shortest path is the best. If the coordinates of TK2 were known, or if you could specify it by some other method, it could become the initial tracking point. Keep this in mind as you learn to use Object Snap. You don't necessarily need to know the coordinates if you can use Object Snap to select a point from which a tracking path could be specified.

The line from SP2 to TK2/EP2 can be started and ended in a similar manner as the line from TK1/SP1 to EP1 was started, with some minor modifications. The sequence (which involves using Tracking twice) is as follows:

line (ENTER)
Specify first point: **track** (ENTER) *(invoke the Tracking option)*
First tracking point: **0,12'** (ENTER) *(specify point TK1/SP1 as the first tracking point)*
Next point (Press enter to end tracking): **@13'8,0** (ENTER) *(locates the second tracking point, TK2, as shown in Figure 3–20)*
Next point (Press enter to end tracking): **@2'2<180** (ENTER) *(locates the third tracking point, TK3, as shown in Figure 3–20)*
Next point (Press enter to end tracking): **@1'6<0** (ENTER) *(locates the fourth tracking point, SP2, as shown in Figure 3–20)*
Next point (Press enter to end tracking): *(press ENTER to exit Tracking; AutoCAD establishes point SP2)*
Specify next point or [Undo]: **track** (ENTER) *(invoke the Tracking option again)*
First tracking point: **0,12'** (ENTER) *(specify point TK1/SP1 as the first tracking point)*
Next point (Press enter to end tracking): **@13'8,0** (ENTER) *(locates the second tracking point, TK2/EP2, as shown in Figure 3–20)*
Next point (Press enter to end tracking): *(press ENTER to exit Tracking, AutoCAD establishes point EP2)*
Specify next point: *(press ENTER to terminate the LINE command)*

This example shows an application of the Tracking option in which the points were established in reference to known points.

DRAWING OBJECTS USING DIRECT DISTANCE

The Direct Distance feature for specifying a point relative to another point can be used with a command like LINE as a variation of the Relative Coordinates mode. In the case of the Direct Distance option, the distance is keyed in and the direction is determined by the current location of the cursor. This option is very useful when you know the exact distance, between two points, but specifying the exact angle is not as easy as placing the cursor on a point that forms the exact angle desired.

Figure 3–21 shows a shape that can be drawn more easily by using the Direct Distance option along with setting the Snap and Ortho modes to ON and OFF at the appropriate times. To draw the shape in Figure 3–21, you need to have the end point and perpendicular snap modes enabled. Points A, B, and C are on the Snap grid (X,Y coordinates 3,3 for A, 3,6 for B, and 9,6 for C). By setting the Snap mode to ON with the value set to 1 or perhaps 0.5, the cursor can be placed on the required points to draw lines A-B and B-C. After drawing lines A-B and B-C (with Ortho set to either ON or OFF), the cursor should be placed on point A (with Ortho set to OFF).

If you have exited the LINE command after drawing line B-C, you must invoke the LINE command and specify C as the first point before moving the cursor to A. The rubber-band line indicates that the next line is drawn from C to A, as shown in Figure 3–22. However, you wish to draw a line only 2 units long but in the same direction as a line from C to A. With the cursor placed on A, enter **2** and press ENTER.

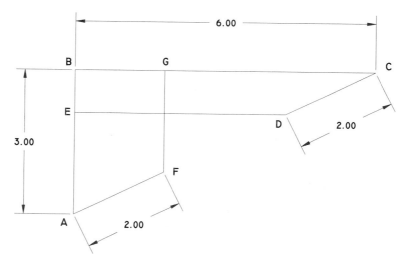

Figure 3–21 *Example of a direct distance application*

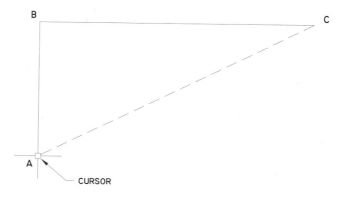

Figure 3–22 *Drawing a line from C to A by means of the Direct Distance option*

AutoCAD draws the line of 2 units, and Point D is established. To draw a line from D to E without exiting the LINE command, first set Ortho and Snap to ON, and then place the cursor on line A-B. The rubber-band line will indicate line D-E, as shown in Figure 3–23. In this case, you do not know the distance, but you do know that the line terminates on line A-B. Therefore, simply press the pick button on your pointing device, and line D-E is drawn. You do not have to specify point D as the starting point of the line. Using the line-line continuation after drawing line C-D does that for you.

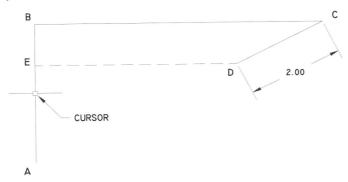

Figure 3–23 *Drawing a line from D to E*

Line A-F can be drawn in a similar manner as line C-D was drawn. Unlike line C-D, line A-F is not a continuation of another line. However, the starting point, A, can be selected with Snap set to ON. After starting point A has been specified, place the cursor on point C and enter **2**. From line A-F, line F-G can be drawn in a similar manner as line D-E was drawn. This example shows an application of the Direct Distance option in which the distance was entered in and the direction was controlled by the cursor.

POLAR TRACKING AND OBJECT SNAP TRACKING

The Polar Tracking feature lets you draw lines and specify point displacements in directions that are multiples of a specified increment angle. You can specify whether the increment angles are measured from the zero point of the current coordinate system or from the angle of a previous object. You can also add up to 10 additional angles to which a line or displacement can be diverted from a base direction. The Object Snap Tracking feature lets you track your cursor from a strategic (Osnap) point on an object in a specified direction, either orthogonal or preset polar angles.

Turning Polar Tracking ON and OFF

In AutoCAD, four primary methods are available to turn Polar Tracking ON and OFF to the specified increment angle:

1. Polar Tracking can be toggled ON and OFF by choosing **POLAR** on the status bar.
2. Polar Tracking can be toggled ON and OFF by pressing the function key F10.
3. Right-click **POLAR** on the status bar at the bottom of the screen and from the shortcut menu, choose SETTINGS. The Drafting Settings dialog box will be displayed with the **Polar Tracking** tab selected. You can toggle the setting by choosing the **Polar Tracking On (F10)** box.
4. Right-click **POLAR** on the status bar at the bottom of the screen and choose ON or OFF.

When the Polar Tracking mode is set to ON, it allows you to specify a displacement or draw in a direction that is a multiple of a specified increment angle. But even with Polar Tracking set to ON, you can still specify a "next point" of a line or displacement with the cursor that is not at one of the multiples of the specified Polar Tracking increment angle. This is different from the ORTHO and SNAP features where you are forced (when specifying a point with the cursor) to accept a point orthogonally or on the Snap grid when the respective mode is set to ON. You will notice, however, that when the Polar Tracking mode is set to ON and you move the cursor near one of the radials (a multiple of the increment angle), the cursor will snap to that radial and the radial will appear as a dotted construction line indicating the direction of the line that will be drawn if you specify that point by pressing the pick button on your pointing device.

Changing Polar Tracking Increment and Additional Angles

In AutoCAD there are two primary methods to change the Polar Tracking increment angle and add additional angles:

1. From the status bar at the bottom of the screen, right-click on **POLAR** and choose SETTINGS. AutoCAD displays the Drafting Settings dialog box with the **Polar Tracking** tab selected. From the **Polar Angle Settings** section, enter the desired increment angle in the **Increment angle** box. This is the base increment angle whose multiples are used by Polar Tracking. Up to 10 additional angles can be added in the text box under the **Additional angles** box. Select **New** and then enter the desired additional angle(s). If there is no check in the **Additional angles** box, the additional angles are not available for use when Polar Tracking is set to ON. To make the additional angle(s) available when the Polar Tracking mode is set to ON, set **Additional angles** box to ON (see Figure 3–24).
2. From the Tools menu, choose Drafting Settings. Then select the **Polar Tracking** tab. See the explanation in Method 1 for changing settings.

Note: The Polar Tracking and the Ortho modes cannot both be set to ON simultaneously. They can both be set to OFF, or either one can be set to ON.

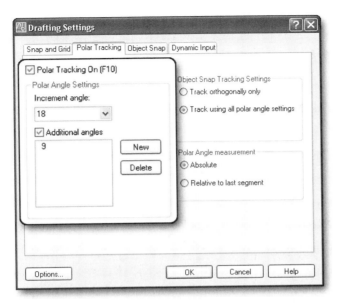

Figure 3–24 *Selecting the **Polar Tracking On (F10)** check box and setting the increment angle to 18° and one additional angle to 9°*

With the 18° increment angle and the Polar Angle measurement set to Absolute, you can draw lines and specify displacements at 18°, 36°, 54°, 72°, 90°, 108°… …324°, 342°, 360° around the compass by just snapping to the construction lines when they appear. The accompanying AutoTrack Tooltip will appear when you have snapped to a multiple of the increment angle, if the **Display AutoTrack Tooltip** box is set to ON on the **Drafting** tab of the Options dialog box. The Tooltip displays the cursor's distance and direction from the first specified point. You can, with the cursor snapped to one of the Polar Tracking angles, use the Direct Distance option and enter a distance from the keyboard and press ENTER. AutoCAD will draw the line or apply the displacement in accordance with the distance entered and the direction set by the cursor. For example, in Figure 3–25, the line from P1 to P2 was drawn 2 units long at 0°. This could have been done with Direct Distance and either ORTHO set to ON or with Polar Tracking. But with P2 to P3 being at 18° and from P3 to P4 being at 36°, it is easier to use Polar Tracking with the settings shown in the example in Figure 3–24. Figure 3–26 shows the three line segments drawn using Polar Tracking with the increment angle set to 18° and the Polar Angle measurement set to **Absolute**. Figure 3–27 shows how you can use the Additional angles setting of 9° to draw a 1-unit-long line at 9° from absolute zero.

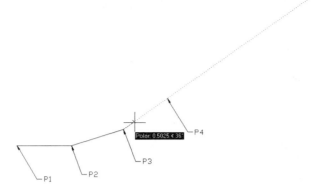

Figure 3–25 *Drawing line segments at multiples of the 18° increment angle*

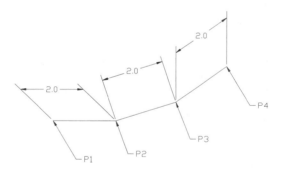

Figure 3–26 *Three line segments drawn using Polar Tracking with the increment angle set to 18° and the Polar Angle measurement set to Absolute*

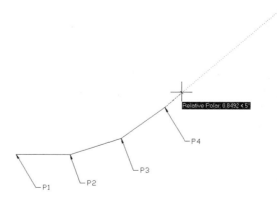

Figure 3–27 *Four line segments drawn using Polar Tracking with the increment angle set to 18° and at one additional angle at 9° from Polar Angle measurement set to Absolute*

After completing the three segments drawn at multiples of 18°, instead of adding the segment at 9°, let's change the increment angle to 5° and the Polar Angle measurement set to **Relative to last segment**. Continuing with three more segments, each 1 unit long, the first increment angle will be measured from the last segment, which is at 36°. The Tooltip displays "Relative Polar (distance) < 5°" but the resulting angle is 41°. Figure 3–28 shows the three additional line segments drawn using Polar Tracking with the increment angle set to 5° and the Polar Angle measurement set to Relative to last segment. The last two segments will actually be at 46° and 51°.

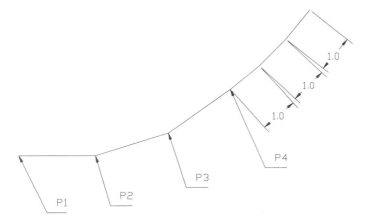

Figure 3–28 *Three line segments drawn using Polar Tracking with the increment angle set to 5° and the Polar Angle measurement set to Relative to last segment*

Polar Snap

The Polar Snap feature can be used to make drawing the 2-unit-long and 1-unit-long segments in the previous examples even easier. On the **Snap and Grid** tab of the Drafting Settings dialog box, choose the **Polar Snap** radio button in the **Snap type & style** section. Then in the **Polar distance** text box of the **Polar spacing** section, enter the desired distance. In the case of the above mentioned example, the 2-unit-long segments at 0°, 18°, and 36°, you can enter 2.0 for the distance. For the 41°, 46°, and 51° segment (additions of 5° each in the **Relative to last segment** mode), you can use a distance of 1.0.

Object Snap Tracking

Object Snap Tracking (OTRACK) helps you draw objects at specific angles or in specific relationships to other objects. When you set OTRACK to ON, temporary alignment paths help you create objects at precise positions and angles. Object Snap Tracking works in conjunction with object snaps and polar angle settings. You must set an object snap before you can track from an Object Snap Point. You can toggle OTRACK on and off with OTRACK on the status bar or the function key F11.

OTRACK includes two tracking options: Object Snap Tracking and Polar Tracking. Use Object Snap Tracking to track along alignment paths that are based on Object Snap points. Acquired points display a small plus sign (+), and you can acquire up to seven tracking points at a time. After you acquire a point, horizontal, vertical, or polar alignment paths relative to the point are displayed as you move the cursor over their drawing paths. For example, you can select a point along a path based on an object end point or an intersection between objects.

The **Object Snap Tracking Settings** section (**Polar Tracking** tab of the Drafting Settings dialog box) lets you select options for Object Snap Tracking. Choose **Track orthogonally only** to display orthogonal (horizontal/vertical) Object Snap Tracking paths for acquired Object Snap points when Object Snap Tracking is set to ON. Choose **Track using all polar angle settings** to permit the cursor to track along any polar angle tracking path for unacquired Object Snap Point when Object Snap Tracking is ON and while specifying points.

This next example shows how to apply Polar Tracking and Object Snap Tracking to draw the vertical line B and the horizontal line C from existing line A and arc D, as shown in Figures 3–29 and 3–30.

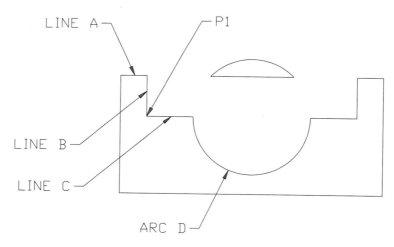

Figure 3–29 *Completed object including lines A, B, and C, and arc D*

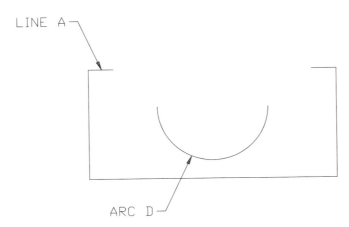

Figure 3–30 *Line A and arc D from which lines B and C will be drawn using Polar Tracking*

From the status bar at the bottom of the screen, right-click on **POLAR** and choose SETTINGS. The Drafting Settings dialog box is displayed with the **Polar Tracking** tab selected. From the **Object Snap Tracking Settings** section choose **Track orthogonal only**. Check to be sure that **OTRACK** is set to ON. Invoke the LINE command:

> **line** (ENTER)
> Specify first point: **end** *(enter the first three letters of the Osnap mode Endpoint, press ENTER and then select the right end of LINE A to establish the start point of LINE B)*
> Specify next point or ⬇ *(move the aperture cursor to the right end of LINE A, keep it there until the Endpoint marker appears, and without pressing a button, move it downward from the line)*

As you move the cursor downward from line A, a dashed construction line will follow as long as the line between the end of line A and the cursor is vertical or horizontal, as shown in Figure 3–31.

Next, move the cursor to the left end of arc D, keep it there until the Endpoint marker appears and then move it away from it toward the left, again without pressing the pick button. A horizontal construction line will follow until the cursor is near the vertical construction line through the end of line A, as shown in Figure 3–32. At this time, with both the vertical and horizontal construction lines being displayed, you can press the pick button on your pointing device and point P1, the end point of line B will be established.

After pressing the pick button to specify point P1 and complete line B, move the cursor to the left end of arc D, invoke the Endpoint Osnap mode, and select arc D. This will complete line C, as shown in Figure 3–33.

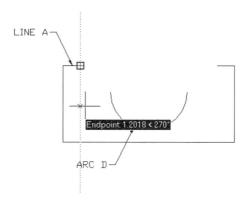

Figure 3–31 *A vertical construction line passes through the cursor and the end of LINE A*

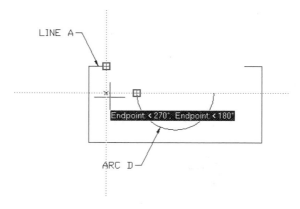

Figure 3–32 *A dashed arc follows the cursor indicating the extension of ARC D to the dashed extension of LINE B, at which point P1 is specified*

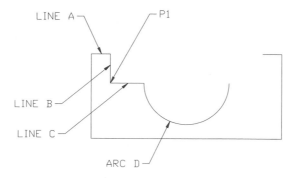

Figure 3–33 *Lines B and C are drawn with point P1 at their intersection*

DYNAMIC INPUT, ON-SCREEN PROMPTS, AND GEOMETRIC VALUES DISPLAY

Dynamic Input provides a command interface near the cursor to help you keep your focus in the drafting area. When Dynamic Input is set to ON, tooltips display information near the cursor and are dynamically updated as the cursor moves. When a command is active, a tooltip provides the field for user entry. The actions required to complete a command are the same as when you work from the command line. The difference is that your attention can stay near the cursor.

Dynamic Input can be toggled ON/OFF by choosing **DYN** on the status bar. Dynamic Input has three components: pointer input, dimensional input, and dynamic prompts. To change the format and visibility of Pointer Input, Dimension Input, and Tooltip appearance (Drafting Prompts), right-click **DYN** on the status bar and choose SETTINGS from the shortcut menu. AutoCAD will display the Drafting Settings dialog box with the **Dynamic Input** tab selected (see Figure 3–34).

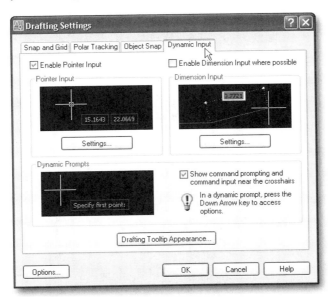

Figure 3–34 *Drafting Settings dialog box displaying the Dynamic Input tab*

Choosing **Enable Pointer Input** causes the cursor to be displayed with coordinate values in a tooltip. When a command prompts for a point, you can enter coordinate values in the tooltip instead of on the command line. The Preview Area shows an example of pointer input.

When **Enable Pointer Input** and **Enable Dimensional Input where possible** are both checked, dimensional input supersedes pointer input when it is available. The **Dynamic Prompts** section allows you to set how AutoCAD displays prompts in a tooltip near the cursor when necessary in order to complete the command. You can enter values in the tooltip instead of on the command line.

Selecting **Show command prompting and command input near the crosshairs** causes prompts to be displayed in a tooltip near the cursor. The Preview Area shows an example of dynamic prompts.

Pointer Input

The **Pointer Input** section controls how AutoCAD displays coordinate values in a tooltip near the cursor.

Choose **Settings** settings in the **Pointer Input** section to display the Point Input Settings dialog box (see Figure 3–35), which allows you to control the coordinate format in the tooltips that are displayed and the visibility of the pointer when pointer input is set to ON.

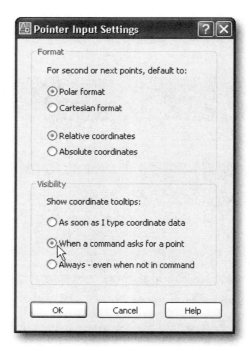

Figure 3–35 *Pointer Input Settings dialog box*

The **Format** section controls coordinate format in the tooltips that are displayed when pointer input is turned on. Selecting **Polar format** (default selection) causes the tooltip to be displayed for the second or next point in polar coordinate format. To switch to a Cartesian format, specify an *X* coordinate then enter a comma (,).

Selecting **Cartesian format** causes the tooltip to be displayed for the second or next point in Cartesian coordinate format. To switch to a polar format, specify the distance and then enter an angle symbol (<).

Selecting **Relative coordinates** (default selection) causes the tooltip to be displayed for the second or next point in relative coordinate format. To switch to a absolute format, enter a pound sign (#). Selecting **Absolute coordinates** causes the tooltip to be displayed for the second or next point in absolute coordinate format. To switch to a relative format, enter an at sign (@).

Note: You can use the direct distance method when pointer input is set to absolute coordinates.

The **Visibility** section controls when pointer input is displayed.

Selecting **As soon as I type coordinate data** causes tooltips to be displayed only when you start to enter coordinate data (when pointer input is turned on). Selecting **When a command asks for a point** (default selection) causes tooltips to be displayed whenever a command prompts for a point (when pointer input is turned on). Selecting **Always–even when not in a command** causes tooltips to always be displayed (when pointer input is turned on). Choose **OK** to exit the Pointer Input settings dialog.

Note: The prompts and responses shown in examples throughout this book and drawing exercises assume that you are using the new dynamic input introduced in AutoCAD 2006. Where they call for you to enter coordinates for second or next points without the @ prefix, they are based on the input being absolute coordinates. Because the default setting of the Pointer Input setting is set to **Relative coordinates** causes the @ prefix to be added (without your seeing it on the screen). You should change the setting of the Pointer Input to **Absolute coordinates** if you wish to try out the examples on your system. Or, if you do not wish to change the setting, you can use an override by prefixing the coordinates with the # symbol. This forces the input to be absolute for second or next points and only works in dynamic input, not at the Command line in the Command window.

Dimension Input

The **Dimension Input** section allows you to set how AutoCAD displays a dimension with tooltips for distance value and angle value when a command prompts for a second point or a distance. The values in the dimension tooltips change as you move the cursor. You can enter values in the tooltip instead of on the command line.

Choosing **Enable Dimension Input where possible** allows dimensional input whenever possible. Dimensional input is not available for some commands that prompt for a second point. Preview Area shows an example of dimensional input.

Choose **Settings** to display the Dimension Input Settings dialog box (see Figure 3–36), which controls which tooltips are displayed during grip stretching when dimensional input is turned on.

Figure 3–36 *Dimension Input Settings dialog box*

Choose one of the three options in the **Visibility** section, which controls which tooltips are displayed during grip stretching, when dimensional input is turned on.

Selecting **Show only 1 dimension input field at a time** causes only the Length Change dimensional input tooltip to be displayed, when you are using grip editing to stretch an object.

Selecting **Show 2 dimension input fields at a time** causes the Length Change and Resulting Dimension input tooltips to be displayed when you are using grip editing to stretch an object.

Selecting **Show the following dimension input fields simultaneously** causes the selected dimensional input tooltips to be displayed when you are using grip editing to stretch an object.

The input fields that can be displayed include: Resulting Dimension, Length Change, Absolute Angle, Angle Change, and Arc Radius.

Selecting **Resulting Dimension** causes a length dimension to be displayed and updated as you move the grip.

Selecting **Length Change** causes the change in length to be displayed as you move the grip.

Selecting **Absolute Angle** causes an angle dimension to be displayed and updated as you move the grip.

Selecting **Angle Change** causes the change in the angle to be displayed as you move the grip.

Selecting **Arc Radius** causes the radius of an arc to be displayed and updated as you move the grip.

Drafting Tooltip Appearance

Choose **Drafting Tooltip Appearance** to open the Tooltip Appearance dialog box (see Figure 3–37), which can be used to control the appearance of tooltips.

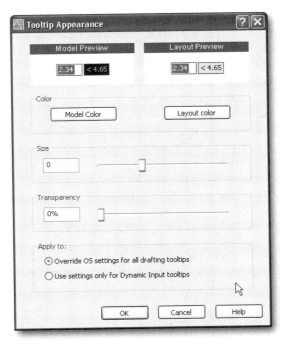

Figure 3–37 *Tooltip Appearance dialog box*

The **Model Preview** and **Layout Preview** show examples of the current tooltip appearance settings in Model space and Layouts, respectively.

Choosing **Model Color** and **Layout Color** causes the Select Color dialog box to be displayed, where you can specify a color for tooltips in model space and in layouts, respectively.

The **Size** slider bar and text box allow you to specify the size for tooltips. The default size is 0. The slider bar is used to make tooltips larger or smaller.

The **Transparency** slider bar and text box allow you to control the transparency of tooltips. The lower the setting, the less transparent the tooltip. A value of 0 sets the tooltip to opaque. The slider bar is used to make tooltips more or less transparent.

The **Apply to** section allows you to specify whether the settings apply to all drafting tooltips or only to Dynamic Input tooltips.

After making changes, choose **OK** to close the Tooltip Appearance dialog box. Choose **OK**, again to close the Drafting Settings dialog box.

DISPLAY CONTROL

There are many ways to view a drawing in AutoCAD. These viewing options vary from on-screen viewing to hard-copy plots. The hard-copy options are discussed in Chapter 8. Using the display commands, you can select the portion of the drawing displayed, establish 3D perspective views, and much more. By enabling you to view the drawing in different ways, AutoCAD provides a way to draw faster. The commands that are explained in this section are utility commands. They make your job easier and help you to draw more accurately.

ZOOM COMMAND

The ZOOM command is like a zoom lens on a camera. You can increase or decrease the viewing area, although the actual size of objects remains constant. As you increase the visible size of objects, you view a smaller area of the drawing in greater detail as though you were closer. As you decrease the visible size of objects, you view a larger area as though you were farther away. This ability provides a close-up view for better accuracy and detail or a distant view to see the whole drawing.

AutoCAD provides smooth transition (rather than instantaneous) zooms and pans so that you can watch as the view changes from the current view to the selected view. This allows you to better keep track of where you are going in relation to where you were. You can change to instantaneous transitions by changing the setting of the VTENABLE system variable. There are various methods by which the ZOOM command can be used.

Zoom Realtime

The Zoom Realtime selection (invoked from the shortcut menu, as shown in Figure 3–38) lets you zoom interactively to a logical extent. The cursor changes to a magnifying glass with a "±" symbol. To zoom in closer, hold the pick button and move the cursor vertically toward the top of the window. To zoom out further, hold the pick button and move the cursor vertically toward the bottom of the window. To discontinue zooming, release the pick button. To exit the Zoom Realtime option, press ENTER, ESC, or from the shortcut menu, select EXIT. In addition, you can perform other operations related to ZOOM and PAN by selecting appropriate commands from the shortcut menu.

Figure 3-38 *Invoking the* ZOOM *(zoom realtime) option from the shortcut menu*

The current drawing window is used to determine the zooming factor. If the cursor is moved by holding the pick button from the bottom of the window to the top of the window vertically, the zoom-in factor would be 200%. Conversely, when holding the pick button from the top of the window and moving vertically to the bottom of the window, the zoom-out factor would be 200%.

When you reach the zoom-out limit, the "-" symbol on the cursor disappears while attempting to zoom-out indicating that you can no longer zoom out. Similarly, when you reach the zoom-in limit, the "+" symbol on the cursor disappears while attempting to zoom in indicating that you can no longer zoom in.

Zoom Window

The Zoom Window selection (invoked from the Zoom toolbar, as shown in Figure 3-39) lets you specify a smaller portion of the drawing and have that portion fill the drawing area. This is done by specifying two diagonally opposite corners of a rectangle, similar to a selection window. The center of the area selected becomes the new display center, and the area inside the window is enlarged to fill the drawing area as completely as possible.

Figure 3-39 *Invoking the Zoom Window option from the Zoom toolbar*

You can enter two opposite corner points to specify an area by means of entering their coordinates or using the pointing device to pick them on the screen. The following command sequence shows an example of using the Zoom Window selection (see Figure 3–40).

AutoCAD prompts:

> Specify first corner: *(specify a point to define the first corner of the window, as shown in Figure 3–40)*
> Specify opposite corner: *(specify a point to define the diagonally opposite corner of the window, as shown in Figure 3–40)*

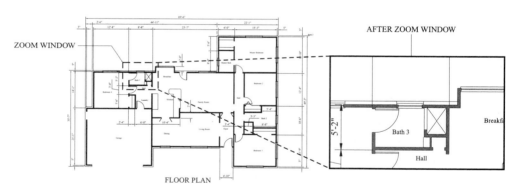

Figure 3-40 *Display of the drawing before and after Zoom Window selection*

Zoom All

The Zoom All selection (invoked from the Zoom toolbar as shown Figure 3–41) lets you see the entire drawing. In a plan view, it zooms to the drawing's limits or current extents, whichever is larger. If objects in the drawing extend outside the drawing limits, the display shows all objects in the drawing.

Figure 3–41 *Invoking the Zoom All option from the Zoom toolbar*

Zoom Extents and Zoom Object

The Zoom Extents selection (invoked from the Zoom toolbar, as shown in Figure 3–42) lets you see the entire drawing on screen. Unlike the Zoom All method, the Extents method uses only the drawing extents of all objects (whether on visible layers or not) and not the drawing limits.

Figure 3–43 illustrates the difference between the Zoom All selection and Zoom Extents selection.

Figure 3–42 *Invoking the Zoom Extents option from the Zoom toolbar*

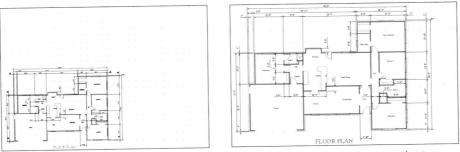

Figure 3–43 *Display of the drawing after Zoom All selection and Zoom Extents selection*

The Zoom Object selection from the Zoom toolbar (see Figure 3–44) zooms to the largest possible display of only selected object(s) instead of all objects.

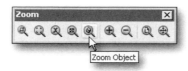

Figure 3–44 *Invoking the Zoom Objects option from the Zoom toolbar*

Zoom Scale

The Zoom Scale selection (invoked from the Zoom toolbar, as shown in Figure 3–45) lets you enter a display scale (or magnification) factor. The scale factor (a numerical value and not expressed in units of measure) is applied to the area covered by the drawing limits. For example, if you enter a scale factor of 3, each object appears three times as large as in the Zoom All view. A scale factor of 1 displays the entire drawing (the full view), which is defined by the established limits. If you enter a value less than 1, AutoCAD decreases the magnification of the full view. For example, if you enter a scale factor of 0.5, each object appears half its size in the full view while the viewing area is twice the size in horizontal and vertical dimensions. When you use this option, the object in the center of the screen remains centered.

Figure 3–45 *Invoking the Zoom Scale option from the Zoom toolbar*

If you enter a number followed by X, the scale is determined relative to the current view. For instance, entering 2X causes each object to be displayed two times its current size on the screen. The scale factor XP option, related to the layout of the drawing, is explained in Chapter 7. Figure 3–46 shows the difference between a full view and a 0.5 zoom.

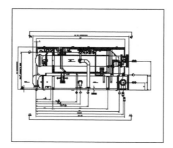 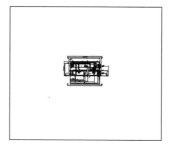

Figure 3–46 *Drawing display showing the difference between a full view and a 0.5 zoom display*

Zoom Center

The Zoom Center selection (invoked from the Zoom toolbar, as shown in Figure 3–47) lets you select a new view by specifying its center point and the magnification value or height of the view in current units. A smaller value for the height increases the magnification; a larger value decreases the magnification. Figure 3–48 illustrates before and after Zoom Center selection.

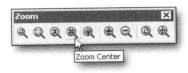

Figure 3–47 *Invoking the Zoom Center option from the Zoom toolbar*

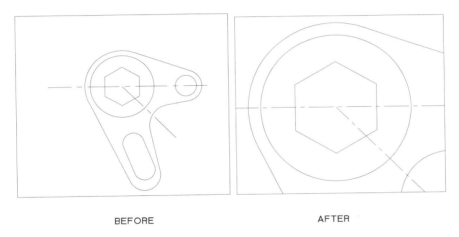

Figure 3–48 *Display before and after Zoom Center selection*

In addition to providing coordinates for a center point, you can also specify the center point by specifying a point on the view window. The height can also be specified in terms of the current view height by specifying the magnification value followed by an X. A response of 3X will make the new view height three times as large as the current height. The fact that the coordinates of the specified center of the new view are at the center of the circle is coincidental. The coordinates you specify can be located anywhere on the drawing.

Zoom Dynamic

The AutoCAD Zoom Dynamic selection (invoked from the Zoom toolbar, as shown in Figure 3–49) provides a quick and easy method to move to another view of the drawing. With Zoom Dynamic, you can see the entire drawing and then simply select the location and size of the next view by means of cursor manipulations. Using Zoom Dynamic is one means by which you can visually select a new display area that is not entirely within the current display. The current viewport is then transformed into a selecting view that displays the drawing extents, as shown in the example in Figure 3–50.

Figure 3–49 *Invoking the Zoom Dynamic option from the Zoom toolbar*

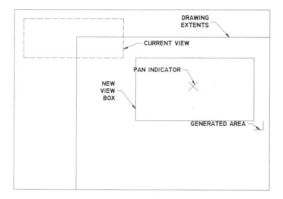

Figure 3–50 *Using the* ZOOM DYNAMIC *command to display the drawing extents*

When the selecting view is displayed, you see the drawing extents marked by a white or black box, the current display marked by a blue or magenta dotted box. A new view box, the same size as the current display, appears. Its location is controlled by the movement of the pointing device. Its size is controlled by a combination of the pick button and cursor movement. When the new view box has an X in the center, the box pans around the drawing in response to cursor movement. After you press the pick button on the pointing device, the X disappears and an arrow appears at the right edge of the box. The new view box is now in Zoom mode. While the arrow is in the box, moving the cursor left decreases the box size; moving the cursor right increases the size.

When the desired size has been chosen, press the pick button again to pan, or press ENTER to accept the view defined by the location and size of the new view box. Pressing ESC cancels the Zoom Dynamic and returns you to the current view.

Zoom Previous

The Zoom Previous selection (invoked from the Standard toolbar, as shown in Figure 3–51) displays the last displayed view. While editing or creating a drawing, you may wish to zoom into a small area, back out to view the larger area, and then zoom into another small area. To do this, AutoCAD saves the coordinates of the current view whenever it is being changed by any of the zoom options or other view commands, so you can return to the previous view by selecting the Previous option, which can restore the previous 10 views.

Figure 3–51 *Invoking the Zoom Previous option from the Standard toolbar*

PAN COMMAND

AutoCAD lets you view a different portion of the drawing in the current view without changing the magnification. You can move your viewing area to see details that are currently off screen. Imagine that you are looking at your drawing through the display window and that you can slide the drawing left, right, up, and down without moving the window.

The Pan Realtime selection (invoked from the Standard toolbar, as shown in Figure 3–52) lets you pan interactively to the logical extent (edge of the drawing space). Once you invoke the command, the cursor changes to a hand cursor. To pan, hold the pick button on your pointing device to lock the cursor to its current location relative to your drawing, and move the cursor in any direction. Graphics within the window are moved in the same direction as the cursor. To discontinue the panning, release the pick button.

When you reach the logical extent of your drawing, a line-bar is displayed adjacent to the hand cursor. The line-bar is displayed at the top, bottom, or left or right side of the drawing, depending upon whether the logical extent is at the top, bottom, or side of the drawing.

Figure 3–52 *Invoking the* PAN REALTIME *command from the Standard toolbar*

To exit Pan Realtime, press ESC or ENTER. You can also exit by selecting EXIT from the shortcut menu that is displayed when you right-click with your pointing device. In addition, you can perform other operations related to zoom and pan by selecting the appropriate commands from the shortcut menu.

Choosing one of the Left, Right, Up, or Down options from the View > Pan flyout menu causes AutoCAD to pan the view left, right, up, or down accordingly.

AERIAL VIEW

The DSVIEWER command is used to activate the Aerial View window, which provides a quick method of visually panning and zooming. By default, AutoCAD displays the Aerial View window, with the entire drawing displayed in the window, as shown in Figure 3–53. You can select any portion of the drawing in the Aerial View window by visually panning and zooming, this behavior is similar to the Zoom Dynamic option; AutoCAD displays the selected portion in the view window (current viewport).

Figure 3–53 *The Aerial View window*

Invoke the DSVIEWER command from the View menu and AutoCAD displays the Aerial View window. Two option menus are provided to pan and zoom visually.

The **View** menu in the Aerial View window has three options. The zoom in option causes the view to appear closer, enlarging the details of objects, but displaying a smaller area. The zoom out causes the view to appear farther away, decreasing the size of objects, but displaying a larger area. The global option causes the entire drawing to be viewable in the Aerial View window. You can also select the three options from the toolbar provided in the Aerial View window.

The **Options** menu in the Aerial View window has three options. **Auto Viewport** causes the active viewport to be displayed in model space. **Dynamic Update** toggles whether the view is updated or not in response to editing. The **Realtime Zoom** controls whether or not the AutoCAD window updates in real time when you zoom using the Aerial View.

CONTROLLING DISPLAY WITH INTELLIMOUSE

AutoCAD also allows you to control the display of the drawing with the small wheel provided with the IntelliMouse (two-button mouse). You can use the wheel to zoom and pan in your drawing any time without using any AutoCAD commands. By rotating the wheel forward you can zoom in and backwards you can zoom out. When double-clicking the wheel button, AutoCAD displays the drawing to the extent of the view window. To pan the display of the drawing, press the wheel button and drag the mouse. By default, each increment in the wheel rotation changes the zoom level by 10 percent. The ZOOMFACTOR system variable controls the incremental change, whether forward or backward. The higher the setting, the smaller the change.

AutoCAD also allows you to display the Object Snap shortcut menu when you click the wheel button. To do so, set the MBUTTONPAN system variable to 0. By default, the MBUTTONPAN is set to 1.

REDRAW COMMAND

The REDRAW command is used to refresh the on-screen image. You can use this command whenever you see an incomplete image of your drawing. You can use the REDRAW command to remove the blip marks, temporary markers created when in blipmode, from the screen. A redraw is considered a screen refresh as opposed to database regeneration. You can invoke the REDRAW command from the View menu. AutoCAD does not provide any options for the REDRAW command.

REGEN COMMAND

The REGEN command is used to regenerate the drawing's on-screen data. In general, you should use the REGEN command if the image presented by REDRAW does not correctly reflect your drawing. REGEN goes through the drawing's entire database and projects the most up-to-date information on the screen; this command will give you the most accurate image possible. Because of the manner in which it functions, a REGEN takes significantly longer than a REDRAW. There are certain AutoCAD commands for which REGEN takes place automatically, unless REGENAUTO is set to OFF. You can invoke the REGEN command from the View menu. AutoCAD does not provide any options for the REGEN command.

SETTING MULTIPLE VIEWPORTS

The ability to divide the display into two or more separate viewports is one of the most useful features of AutoCAD. Multiple viewports divide your drawing screen into regions, permitting several different areas for drawing instead of just one. It is like having multiple zoom lens cameras, with each camera being used to look at a different portion of the drawing.

Each viewport maintains a display of the current drawing independent of the other viewports. You can simultaneously display a viewport showing the entire drawing, and another viewport showing a close-up of part of the drawing in greater detail. A view in one viewport can be from a different point of view than those in other viewports. You can begin drawing (or modifying) an object in one viewport and complete it in another viewport. For example, three viewports could be used in a 2D drawing, two of them to zoom in on two separate parts of the drawing, showing two widely separated features in great detail on the screen simultaneously, and the third to show the entire drawing, as in Figure 3–54. In a 3D drawing, four viewports could be used to display simultaneously four views of a wireframe model: top, front, right side, and isometric, as in Figure 3–55.

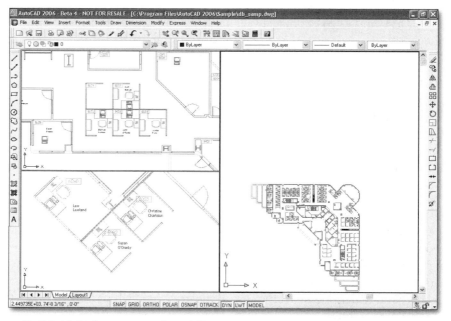

Figure 3–54 *Multiple viewports show different parts of the same 2D drawing*

Chapter 3 • *Fundamentals II* 147

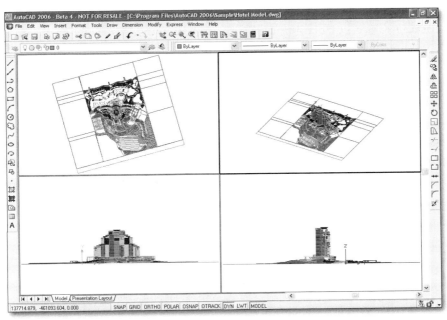

Figure 3–55 *Using viewports to show four views simultaneously for a 3D wireframe model*

AutoCAD allows you to divide the graphics area of your display screen into multiple, non-overlapping (tiled) viewports, as in Figures 3–54 and 3–55, when you are in model space and the system variable TILEMODE set to 1. The maximum number of active tiled viewports that you can have is set by the system variable MAXACTVP, and the default is 64. In addition you can also create multiple overlapping (floating) viewports when the system variable TILEMODE is set to 0. For a detailed explanation on how to create floating viewports, refer to Chapter 8.

You can work in only one viewport at a time. It is considered the current viewport. A viewport is set to current by moving the cursor into it with your pointing device and then pressing the pick button. You can even switch viewports in midcommand (except during some of the display commands). For example, to draw a line using two viewports, you must start the line in the current viewport, make another viewport current by clicking in it, and then specify the end point of the line in the second viewport. When a viewport is current, its border will be thicker than the other viewport borders. The cursor is active for specifying points or selecting objects only in the current viewport; when you move your pointing device outside the current viewport, the cursor appears as an arrow pointer.

Display commands like ZOOM and PAN and drawing tools like GRID, SNAP, ORTHO, and UCS modes are set independently in each viewport. The most important thing to remember is that the images shown in multiple viewports are all of the same drawing. An object added to or modified in one viewport will affect its image in the other viewports. You are not making copies of your drawing, just viewing its image in different viewports. When you are working in tiled viewports, visibility of the layers is controlled globally in all the viewports. If you turn off a layer, AutoCAD turns it off in all viewports.

CREATING TILED VIEWPORTS

AutoCAD allows you to display tiled viewports in various configurations. Display of the viewports depends on the number and size of the views you need to see. By default, whenever you start a drawing, AutoCAD displays a single viewport that fills the entire drawing area. To create multiple viewports, choose Display Viewports Dialog from the Layouts toolbar (see Figure 3–56). AutoCAD then displays a Viewports dialog box similar to Figure 3–57.

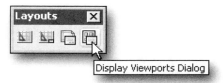

Figure 3–56 *Invoking the* VPORTS *command from the Layouts toolbar*

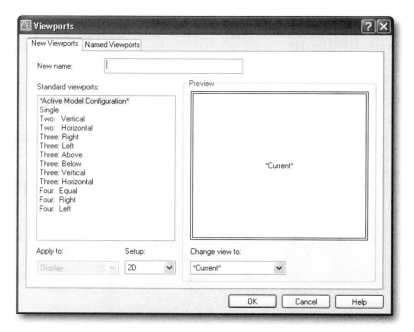

Figure 3–57 *Viewports dialog box*

Choose the name of the configuration you want to use from the **Standard viewports** list. AutoCAD displays the corresponding configuration in the **Preview** window. If necessary, you can save the selected configuration by providing a name in the **New name** box. Select the Display option from the **Apply to** menu and select 2D from the **Setup** menu for 2D viewport setup or select 3D for 3D viewport setup. Choose **OK** to create the selected viewport configuration.

If you need additional viewports other than the standard configurations, you can subdivide a selected viewport. First, select the viewport you want to subdivide. Next, select the number of viewports in which to divide the active viewport. This selection is made from the **Viewports** flyout under the **View** menu. When the cursor is moved into the active viewport, choose a configuration from the shortcut menu. For more configuration options, open the Viewports dialog box and select a viewport, and then choose a view from the **Change view to** drop down list. The preview image will update to reflect the change.

The **Named Viewports** tab lists all saved viewport configurations. To save the current viewport configuration, enter a name in the **New name** field and select **OK**. At any time you can restore one of the saved viewport configurations.

CREATING AND MODIFYING LAYERS

AutoCAD offers a means of grouping objects in layers in a manner similar to the manual drafter's separating complex drawings into simpler ones on individual transparent sheets, superimposed in a single stack. In AutoCAD you can draw only on the current layer. However, AutoCAD permits you to transfer selected objects from one layer to another (neither of which needs to be the current layer) with commands called CHANGE, CHPROP, PROPERTIES, and several others. (Let's see the manual drafter try that!)

A common application of the layer feature is to use one layer for construction (or layout) lines and the others for drawing objects. You can create geometric constructions with objects, such as lines, circles, and arcs. These generate intersections, end points, centers, points of tangency, midpoints, and other useful data that might take the manual drafter considerable time to calculate with a calculator or to hand-measure on the board. From these you can create other objects using intersections or other data generated from the layout. Then the layout layer can be turned off (making it no longer visible) or set the layout not to plot. The layer is not lost, but can be recalled (set to ON) for viewing later as required. The same drawing limits, coordinate system, and zoom factors apply to all layers in a drawing.

To draw an object on a particular layer, first make sure that the layer is set as the "current layer." There is one, and only one, current layer. Whatever you draw will be placed on the current layer. To draw an object on a particular layer, that layer must first have been created; if it is not the current layer, you must make it the current layer.

You can always move, copy, or rotate any object, whether it is on the current layer or not. When you copy an object that is not on the current layer, the copy is placed on the layer that the original object is on. This is also true with the mirror or array of an object or group of objects.

A layer can be visible (ON) or invisible (OFF). Only visible layers are displayed or plotted. If necessary, AutoCAD allows you to set a visible layer not to plot. The layer(s) that are visible and set not to plot, will not be plotted. Invisible layers are still part of the drawing; they are just not displayed or plotted. You can turn layers on and off at will, in any combination. It is possible to turn off the current layer. If this happens and you draw an object, it will not appear on the screen; it will be placed on the current layer and will appear on the screen when that layer is set to ON (provided you are viewing the area in which the object was drawn). This is not a common occurrence, but it can cause concern to both the novice and the more experienced operator who has not faced the problem before. Do not turn OFF the current layer; the results can be very confusing. When the TILEMODE system variable is set to OFF, you can make specified layers visible only in certain viewports.

Each layer in a drawing has an associated name, color, lineweight, and linetype. The name of a layer may be up to 255 characters long. It may contain letters, digits, and the special characters: dollar ($), hyphen (-), underscore (_), and spaces. Always give a descriptive name appropriate to your application, such as "floor-plan" or "plumbing". The first several characters of the current layer's name are displayed in the layer list box located on the Layers toolbar (see Figure 3–58). You can change the name of a layer any time you wish, and you can delete unused layers except layer 0.

Figure 3–58 *The current layer name is displayed in the list box located on the Layers toolbar*

AutoCAD allows you to assign a color to a layer in a drawing. See the description later in this section of how to assign colors to layers from the range and types of colors available. If necessary, you can assign the same color to more than one layer.

Similar to assigning color, you can also assign a specific lineweight to a layer. Lineweights add width to your objects, both on screen and on paper. Using lineweights, you can create heavy and thin lines to show varying object thicknesses in details. For example, by assigning varying lineweights to different layers, you can differentiate between new, existing, and demolition construction.

AutoCAD allows you to assign a lineweight to a specific layer or an object in either inches or millimeters, with millimeters being the default. If necessary, you can change the default setting by invoking the LINEWEIGHT command. A lineweight value of 0 is displayed as one pixel in model space and plots at the thinnest lineweight available on the specified plotting device. Appendix G lists all the available lineweights as well as associated industry standards.

Lineweights are displayed differently in model space than in a paper space layout (refer to Chapter 8 for a detailed explanation on layouts). In model space, lineweights are displayed in relation to pixels. In a paper space layout, lineweights are displayed in the exact plotting width. You can recognize that an object has a thick or thin lineweight in model space but the lineweight does not represent an object's real-world width. A lineweight of 0 will always be displayed on screen with the minimum display width of one pixel. All other lineweights are displayed using a pixel width in proportion to its real-world unit value. A lineweight displayed in model space does not change with the zoom factor. For example, a lineweight value that is represented by a width of four pixels is always displayed using four pixels regardless of how close you zoom in to your drawing.

If necessary, you can change the display scale of object lineweights in model space to appear thicker or thinner. Changing the display scale does not affect the lineweight plotting value. However, AutoCAD regeneration time increases with lineweights that are represented as more than one pixel wide. By default, all lineweight values that are less than, or equal to, 0.01 in. or 0.25 mm are displayed one pixel wide and do not slow down performance in AutoCAD. If you want to optimize AutoCAD performance, when working in the Model Space, set the lineweight display scale to the minimum value. The LINEWEIGHT command allows you to change the display scale of lineweights.

In model space, AutoCAD allows you to turn ON and OFF the display of lineweights by toggling **LWT** on the status bar. With **LWT** set to OFF, AutoCAD displays all the objects with a lineweight of 0 and reduces regeneration time. When exporting drawings to other applications or cutting and copying objects to the clipboard, objects retain lineweight information.

A linetype is a repeating pattern of dashes, dots, and blank spaces. AutoCAD adds the capability of creating custom linetypes. The assigned linetype is normally used to draw all objects on the layer unless you set the current linetype to another linetype.

The following are some of the linetypes that are provided in AutoCAD in a library file called *acad.lin*:

Border	Dashdot	Dot
Center	Dashed	Hidden
Continuous	Divide	Phantom

See Appendix G for examples of each of these linetypes. Linetypes are another means of conveying visual information. You can assign the same linetype to any number of layers. In some drafting disciplines, conventions have been established giving specific meanings to particular dash-dot patterns. If a line is too short to hold even one dash-dot sequence, AutoCAD draws a continuous line between the end points. When you are working on large drawings, you may not see the gap between dash-dot patterns in a linetype, unless the scaling for the linetype is set for a large value. This can be done by means of the LTSCALE command. This command is discussed in more detail later in this chapter.

Every drawing will have a layer named 0 (zero). By default, layer 0 is set to ON and assigned the color white, the default lineweight, and the linetype of continuous. Layer 0 cannot be renamed or deleted.

If you need additional layers, you must create them. By default, each new layer is assigned the properties of the layer directly above it in the layer table. If necessary, you can always reassign the color, lineweight, and linetype of a newly created layer.

CREATING AND MANAGING LAYERS WITH THE LAYER PROPERTIES MANAGER DIALOG BOX

The Layer Properties Manager dialog box can be used to set up and control layers. Choose Layer Properties Manager from the Layers toolbar (see Figure 3–59) to open the Layer Properties Manager (Figure 3–60).

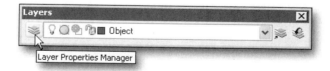

Figure 3–59 *Selecting the Layer Properties Manager on the Layers toolbar*

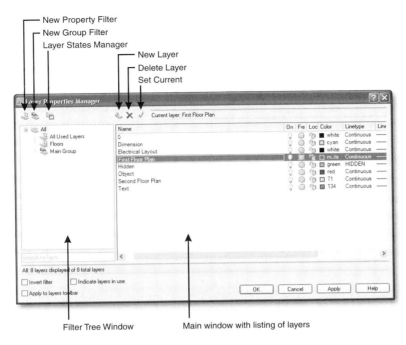

Figure 3–60 *Layer Properties Manager dialog box with Filter Tree displayed*

The main window lists each layer, along with the status of its associated properties. The Filter Tree window, as shown in Figure 3–60, lists the available filters. A layer filter limits the display of layers in the Layer Properties Manager and in the Layer control on the Layers toolbar. In a large drawing, you can use layer filters to display only the layers on which you need to work.

Creating New Layers

To create a new layer, choose **New Layer** from the Layer Properties Manager dialog box or choose NEW LAYER from the shortcut menu, available in the main window. AutoCAD then creates a new layer by assigning the name "Layer1," and the new layer inherits the properties of the currently selected layer in the layer list (color, on or off state, and so on). When first listed in the layer list box, the name "Layer1" is highlighted and ready to be edited. Just enter the desired name for the new layer. If you accept the name Layer1, you can still change it later. To rename a layer, click anywhere on the line where the layer is listed, and the whole line is highlighted. Then click again on the layer name, and the entire name is highlighted—you can type a new name and press ENTER. If you click again on the name, the highlighting is removed and you can then change part of the name.

Note: The layer name cannot contain wild-card characters (such as * and ?). You cannot duplicate existing layer names.

Making a Layer Current

To make a layer current, first choose the layer name from the layer list box in the Layer Properties Manager dialog box, and then choose **Set Current** located at the top of the dialog box, or double-click the icon corresponding to the layer name located under the **Status** column (first column from the left), or choose SET CURRENT from the shortcut menu.

Visibility of Layers

When you turn a layer OFF, the objects on that layer are not displayed in the drawing area and they are not plotted. The objects are still in the drawing, but they are not visible on the screen. And they are still calculated during regeneration of the drawing, even though they are invisible.

To change the setting for the visibility of selected layer(s), select the icon corresponding to the layer name located under the **On** column (third column from the left) in the Layer Properties Manager dialog box. The icon is a toggle for setting layers ON or OFF. It is possible to turn OFF the current layer. If this happens and you draw an object, it will not appear on the screen; it will be placed on the current layer and will appear on the screen when that layer is set to ON.

Freezing and Thawing Layers

In addition to turning the layers OFF, you can freeze layers. The layers that are frozen will not be visible in the view window, nor will they be plotted. In this respect, frozen layers are similar to layers that are OFF. However, layers that are simply turned OFF still go through screen regeneration each time the system regenerates your drawing, whereas the layers that are frozen are not considered during screen regeneration. If you want to see the frozen layer later, you simply thaw it, and automatic regeneration of the drawing area takes place.

To change the setting for the visibility of a layer by freezing or thawing, select the icon corresponding to the layer name located under the **Freeze** column (fourth column from the left). The icon is a toggle for the freezing or thawing of layers. You cannot freeze the current layer.

Locking and Unlocking Layers

Objects on locked layers are visible in the view window but cannot be modified by means of the modifying commands. However, it is still possible to draw on a locked layer by making it the current layer, and using any of the inquiry commands or Object Snap modes on them.

To lock or unlock a layer, select the icon corresponding to the layer name located under the **Lock** column (fifth column from the left). The icon is a toggle for the locking or unlocking of layers.

Changing the Color of Layers

By default, AutoCAD assigns the color of the currently selected layer in the layer list to the newly created layer. To change the assigned color, choose the icon under the **Color** column (sixth column from the left) corresponding to the layer name. AutoCAD displays the Select Color dialog box (see Figure 3–61) with the **Index Color** tab displayed, which allows you to change the color of the selected layer(s). You can select one of 256 colors. Use the cursor to select the color you want, or enter its name or number in the **Color** box. Choose **OK** to accept the color selection. There are two additional tabs available in the Select Color dialog box: **True Color** (see Figure 3–62) and **Color Books** (see Figure 3–63). The **True Color** tab specifies color settings using true colors (24-bit color) with the Hue, Saturation, and Luminance (HSL) color model or the Red, Green, and Blue (RGB) color model. Over 16 million colors are available when using true color functionality. The **Color Books** tab specifies colors using third-party color books (such as PANTONE®) or user-defined color books. Once a color book is selected, the **Color Books** tab displays the name of the selected color book. **True Color** and **Color Books** make it easier to match colors in your drawing with colors of actual materials.

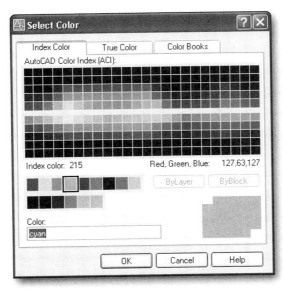

Figure 3–61 *Select Color dialog box with Index Color tab displayed*

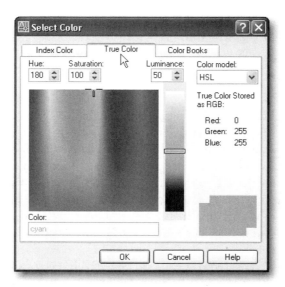

Figure 3-62 *Select Color dialog box with True Color tab displayed*

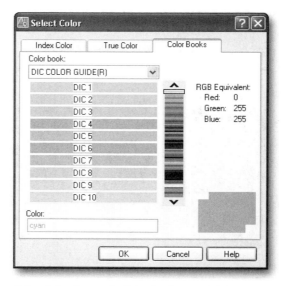

Figure 3-63 *Select Color dialog box with Color Books tab displayed*

Changing the Linetype of Layers

By default, AutoCAD assigns the linetype of the currently selected layer in the layer list to the newly created layer. To change the assigned linetype, choose the linetype name corresponding to the layer name located under the **Linetype** column (seventh column from the left). AutoCAD displays the Select Linetype dialog box, as shown in Figure 3–64, which allows you to change the linetype of the selected layer(s). Select the appropriate linetype from the list box, and choose **OK** to accept the linetype selection.

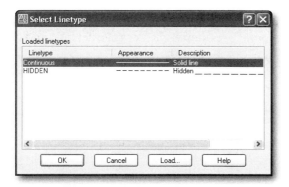

Figure 3–64 *Select Linetype dialog box*

Loading Linetypes

In the Select Linetype dialog box, AutoCAD lists only the linetypes that are loaded in the current drawing. To load additional linetypes in the current drawing, choose **Load**. AutoCAD displays the Load or Reload Linetype dialog box similar to Figure 3–65. AutoCAD lists the available linetypes from *acad.lin*, the default linetype file. Select all the linetypes that need to be loaded, and choose **OK** to load them into the current drawing. If necessary, you can change the default linetype file *acad.lin* to another file by choosing **File** and selecting the desired linetype file.

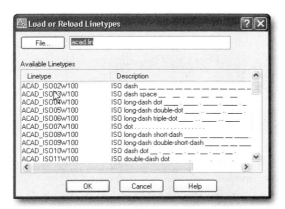

Figure 3–65 *Load or Reload Linetypes dialog box*

Changing the Lineweight of Layers

By default, AutoCAD assigns the lineweight of the currently selected layer in the layer list to the newly created layer. To change the assigned lineweight, choose the lineweight name corresponding to the layer name located under the **Lineweight** column (eighth column from the left). AutoCAD displays the Lineweight dialog box, similar to Figure 3–66, which allows you to change the lineweight of the selected layer(s). Select the appropriate lineweight from the list box, and choose **OK** to accept the lineweight selection.

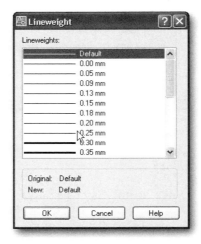

Figure 3–66 *Lineweight dialog box*

Assigning Plot Style to Layers

Plot styles are a collection of property settings (such as color, linetype, and lineweight) that can be assigned to a layer or to individual objects. These property settings are contained in a named plot style table. (For a detailed explanation on creating plot styles, refer to Chapter 8.) When applied, the plot style can affect the appearance of the plotted drawing. By default, AutoCAD assigns the plot style Normal to a newly created layer if the Default plot style behavior is set to Named plot styles. To change the assigned plot style, choose the plot style name corresponding to the layer name located under the **Plot Style** column (ninth column from the left) as shown in Figure 3–67. AutoCAD displays the Select Plot Style dialog box, which allows you to change the Plot Style of the selected layer(s). Select the appropriate plot style from the list box, and choose **OK** to accept the plot style selection. A layer that is assigned a Normal plot style assumes the properties that have already been assigned to that layer. You can create new plot styles, name them, and assign them to individual layers. You cannot change the plot style if the Default plot style behavior is set to Color Dependent plot styles.

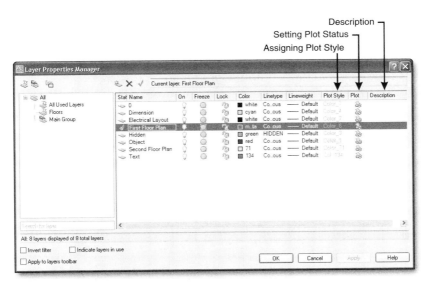

Figure 3-67 *Layer Properties Manager dialog box*

Setting Plot Status for Layers
AutoCAD allows you to turn plotting ON or OFF for visible layers. For example, if a layer contains construction lines that need not be plotted, you can specify that the layer is not plotted. If you turn OFF plotting for a layer, the layer is displayed but is not plotted. At plot time, you do not have to turn OFF the layer before you plot the drawing.

To change the setting for the plotting of layers, select the icon corresponding to the layer name located under the **Plot** column (tenth column from the left) as shown in Figure 3-67. The icon is a toggle for plotting or not plotting layers.

Adding Description to Layers
In addition to assigning a layer name, AutoCAD allows you to add a description to individual layers. Click the appropriate layer name and add the description in the **Description** column (eleventh column from the left) as shown in Figure 3-67. You may edit a layer's description by selecting it and choosing CHANGE DESCRIPTION from the shortcut menu. The description is limited in length to 255 characters.

Saving Layer Properties
At any time during a drawing session, the collective status of all layer properties settings is known as the Layer State. This state can be saved and given a name by which it can be recalled later, thus having every setting of selected properties of every layer revert to what they were when that particular layer state was named and saved. Layer states are saved in files with the extension of *.las*.

To create a new Layer State, in the Layer Properties Manager dialog box choose **Layer States Manager**, which causes the Layer States Manager dialog box to be displayed, as shown in Figure 3–68. From this dialog box, choose **New**, which causes the New Layer State to Save dialog box to be displayed. In the **New Layer State** and **Description** text boxes, enter the name and descriptions respectively and then choose **OK**. This will return you to the Layer States Manager dialog box, and the new Layer State will be saved based on settings of selected properties. Once the desired properties have been selected, choose **Close**. For example, you may wish to save and name a layer state based only on the visibility of the layers. Therefore you would select **On/Off** and when the named layer state is restored, all of the layers will revert to the visibility status they had when the layer state was created. Other properties would not be changed.

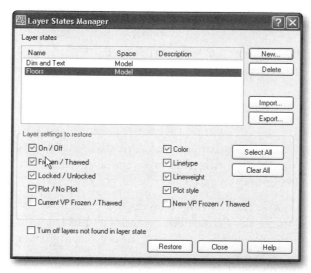

Figure 3–68 *The Layer States Manager dialog box*

A saved Layer State can be restored. To restore a saved layer state, select its name in the Layer States Manager dialog box and then choose **Restore**.

To edit a saved layer state, the procedure is the same as for creating a new Layer State, except that you enter the name of the existing Layer State you wish to be changed in the New Layer State to Save dialog box. AutoCAD will prompt "There is already a layer state named <*layerstatename*> Overwrite existing state?" Choose **Yes**.

To delete a named and saved layer state, in the Layer States Manager dialog box choose **Delete** when the layer state you wish to delete is highlighted in the **Layer states** list.

To import one or more layer states into your drawing, in the Layer States Manager dialog box, choose **Import**. This causes the Import Layer State dialog box to be displayed. This dialog box is similar to other Windows file search and management dialog boxes. From this dialog box you can select a saved layer state to import.

To export one or more layer states from your drawing, in the Layer States Manager dialog box choose **Export**. This causes the Export Layer State dialog box to be displayed. This dialog box is similar to other Windows file search and management dialog boxes. From this dialog box you can select a saved layer state to export.

Setting Filters for Listing Layers

AutoCAD allows you to create two kinds of filters: Layer Property and Layer Group. Layer Property filters are the ones that include layers that have names or other properties in common. For example, you can define a filter that includes all layers that are blue and whose names include the letters "floor." Layer Group filters are the ones that include the layers that are included into the filter when you define it, regardless of their names or properties.

The tree view in the Layer Properties Manager displays default layer filters and any named filters that you create and save in the current drawing. The icon next to a layer filter indicates the type of filter, as shown in Figure 3–69.

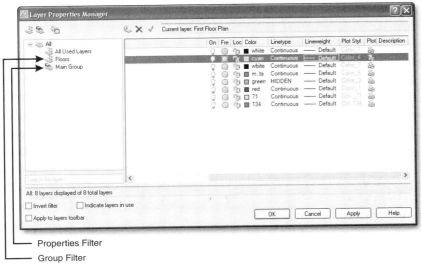

Figure 3–69 *Layer Properties Manager with listing of filters*

By default, AutoCAD creates three filters in a newly created drawing. The **All** filter selection displays all the layers in the current drawing. The **All Used Layers** filter selection displays all the layers on which objects in the current drawing are drawn.

And the **Xref** filter selection displays all the layers being referenced from other drawings, if any Xrefs are attached.

Once you have named and defined a layer filter, you can select it in the tree view to display the layers in the list view. You can also apply the filter to the Layers toolbar, so that the Layer control displays only the layers in the current filter. When you select a filter in the tree view and right-click, options on the shortcut menu can be used to DELETE, RENAME, or MODIFY filters.

Choosing **New Property Filter** (or choosing NEW PROPERTIES FILTER from the shortcut menu) causes the Layer Filter Properties dialog box to be displayed, as shown in Figure 3–70.

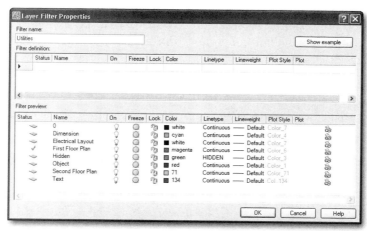

Figure 3–70 *Layer Filter Properties dialog box*

Specify the name of the filter in the **Filter name** box. In the **Filter definition** section, you can use one or more properties to define the filter. For example, you can define a filter that displays all layers that are either green or red and in use. To include more than one color, linetype, or lineweight, duplicate the filter on the next line and select a different setting. Choose **Show example** to display the examples (from the help file) of layer property filter definitions.

The **Status** column can be set to display the In Use icon or the Not In Use icon. In the **Name** column you can use wild-card characters to filter layer names. For example, enter **floor*** to include all layers that start with "floor" in the name. Set any of the one or more corresponding properties to define the filter. AutoCAD displays the results of the filter as you define in the **Filter preview** section of the dialog box. The filter preview shows which layers will be displayed in the layer list in the Layer Properties Manager when you select this filter. To rename or delete a property filter, first select the property filter in the tree view and then from the shortcut menu, choose RENAME or DELETE, respectively. Choose **OK** to save the newly created or modified filter definition.

Choose **New Group Filter** (or choose NEW GROUP FILTER from the shortcut menu) to create a new layer group filter. A new layer group filter named GROUP FILTER1 is created in the tree view. Rename the filter to an appropriate name. In the tree view, choose **All** or one of the other layer property filters to display layers in the list view. In the list view, select the layers you want to add to the filter, and drag them to the newly created Group filter name in the tree view. To rename or delete a group filter in the tree view, first select the group filter and then from the shortcut menu, choose RENAME or DELETE, respectively.

Inverting a filter is another method for listing selected layers. When the **Invert filter** box on the Layer Properties Manager is OFF (not checked), then the layers that are shown in the main window are those that have all the matching characteristic(s) of the specified filter. For example, if you were to create and name a set of filters that specify the color yellow and the linetype dashed, then the list would include only layers whose color is yellow and linetype is dashed. If you set the **Invert filter** box to ON, then all layers would be listed *except* those with both the color yellow and the linetype dashed.

Apply to Layers Toolbar
The **Apply to layers toolbar** option controls the display of layers in the list of layers on the Layers toolbar by applying the current layer filter. When **Apply to layers toolbar** is not checked, then the layers that are filtered (or Invert filtered) are shown in the main window of the Layer Properties Manager dialog box only. When **Apply to layers toolbar** is checked, then the list in the text box for layers in the Layers toolbar will also be filtered (or Invert filtered, depending on the status of the **Invert filter** check box).

Changing the Appearance of the Layer Properties Manager Dialog Box
If necessary, you can drag the widths of the column headings to see additional characters of the layer name, full legend for each symbol and color name, or number in the list box. You can sort the order in which layers are displayed in the list box by choosing the column headings, which causes AutoCAD to list the layers in descending order (Z to A, then numbers). Choosing the column heading again causes AutoCAD to list the layers in ascending order (numbers, A to Z). Choosing the **Status** column headers lists the layers by the property in the list.

Applying and Closing the Layer Properties Manager
After making the necessary changes, choose **Apply** to apply changes that have been made to layers and filters but not close the dialog box. Choose **OK** to accept the

changes and close the dialog box. To discard the changes, choose **Cancel**, which closes the dialog box without making any changes to the layer properties.

Changing Layer Status from the Layers Toolbar

You can toggle on or off, freeze or thaw, lock or unlock, or plot or not plot, in addition to making a layer current in the Layer list box provided on the Layers toolbar. Select the appropriate icon next to the layer name you wish to toggle, as shown in Figure 3–71.

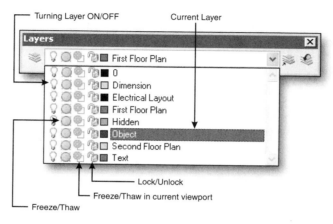

Figure 3–71 *Layer list box on the Layers toolbar*

MAKING AN OBJECT'S LAYER CURRENT

AutoCAD allows you to select an object in the drawing and make its layer the current layer. To do so, choose Make Object's Layer Current (see Figure 3–72) from the Layers toolbar. AutoCAD prompts:

> Select object whose layer will become current: (*select the object in the drawing to make its layer current*)

AutoCAD makes the selected object's layer current.

UNDOING LAYER SETTINGS

AutoCAD allows you to undo the last change or set of changes made to layer settings. To do so, choose Layer Previous (see Figure 3–73) from the Layers toolbar.

Figure 3–72 *Choosing the Make Object's Layer Current icon from the Layer toolbar*

Figure 3–73 *Invoking the* LAYERP *command from the Layers toolbar*

This command undoes changes you have made to layer settings such as color or linetype. The LAYERP command restores the original properties, but not the original name if the layer is renamed, and it also does not restore if you delete or purge a layer.

SETTING THE LINETYPE SCALE FACTOR

The linetype scale factor allows you to change the relative lengths of dashes and spaces between dashes and dots linetypes per drawing unit. The definition of the linetype instructs AutoCAD on how many units long to make dashes and the spaces between dashes and dots. As long as the linetype scale is set to 1.0, the displayed length of dashes and spaces coincides with the definition of the linetype. The LTSCALE command allows you to set the linetype scale factor.

Invoke the LTSCALE command by typing **ltscale** from the On-Screen prompt and AutoCAD prompts:

Enter new linetype scale factor <current>: *(specify the scale factor)*

Changing the linetype scale affects all linetypes in the drawing. If you want dashes that have been defined as 0.5 units long in the dashed linetype to be displayed as 10 units long, you set the linetype scale factor to 20. This also makes the dashes that were defined as 1.25 units long in the center linetype display as 25 units long, and the short dashes (defined as 0.25 units long) display as 5 units long. Note that the 1.25-unit-long dash in the center linetype is 2.5 times longer than the 0.5-unit-long dash in the dashed linetype. This ratio will always remain the same, no matter what the setting of LTSCALE. So if you wish to have some other ratio of dash and space lengths between different linetypes, you will have to change the definition of one of the linetypes in the *acad.lin* file.

Remember that linetypes are for visual effect. The actual lengths of dashes and spaces are bound more to how they should look on the final plotted sheet than to distances or sizes of any objects on the drawing. An object plotted full size can probably use an LTSCALE setting of 1.0. A 50'-long object plotted on an 18" x 24" sheet might be plotted at a 1/4" = 1'-0" scale factor. This would equate to 1 = 48. An LTSCALE setting of 48 would make dashes and spaces plot to the same lengths as the full-size plot with a setting of 1.0. Changing the linetype scale factor causes the drawing to regenerate.

WILD CARDS AND NAMED OBJECTS

AutoCAD lets you use a variety of wild cards for specifying selected groups of named objects when responding to prompts during commands. By placing one or more of these wild cards in the string (your response), you can specify a group that includes (or excludes) all of the objects with certain combinations or patterns of characters.

The types of objects associated with a drawing that are referred to by name include blocks, layers, linetypes, text styles, dimension styles, named User Coordinate Systems, named views, shapes, plot styles and named viewport configurations.

The wild-card characters include the two commonly used in DOS (* and ?), as well as eight more that come from the UNIX operating system. Here is a table listing the wild cards and their uses.

Wild Card	Use
# (pound)	Matches any numeric digit
@ (at)	Matches any alphanumeric character
. (period)	Matches any character except alphabetic
* (asterisk)	Matches any string. It can be used anywhere in the search pattern: the beginning, middle, or end of the string.
? (question mark)	Matches any single character
~ (tilde)	Matches anything but the pattern
[...]	Matches any one of the characters enclosed
[~...]	Matches any character not enclosed
[-] (hyphen)	Specifies single-character range
` (reverse quote)	Reads next character in string literally

The following table shows some examples of wild-card patterns.

Pattern	Will match or include . . .	But not . . .
ABC	Only ABC	
~ABC	Anything but ABC	
?BC	ABC through ZBC	AB, BC, ABCD, XXBC
A?C	AAC through AZC	AC, ABCD, AXXC, ABCX
AB?	ABA through ABZ	AB, ABCE, XAB
A*	Anything starting with A	XAAA
A*C	Anything starting with A and ending with C	XA, ABCDE
*AB	Anything ending with AB	ABCX, ABX
AB	AB anywhere in string	AXXXB
~*AB*	All strings without AB	AB, ABX, XAB, XABX
[AB]C	AC or BC	ABC, XAC
[A-K]D	AD, BD, through KD	ABC, AKC, KD

U, UNDO, AND REDO COMMANDS

The UNDO command undoes the effects of the previous command or group of commands, depending on the option employed. The U command reverses the most recent operation, and the REDO command is a one-time reversal of the effects of the previous U and UNDO commands.

U COMMAND

The U command undoes the effects of the previous command. Pressing ENTER after using the U command undoes the next-previous command, and continues stepping back with each repetition until it reaches the state of the drawing at the beginning of the current editing session.

When an operation cannot be undone, AutoCAD performs no action. An operation external to the current drawing, such as plotting or writing to a file, cannot be undone. You can invoke the UNDO command by typing **u** at the On-Screen prompt.

AutoCAD reverses the most recent operation. For example, if the previous command sequences drew a circle and then copied it, two U commands would undo the two previous commands in sequence, as follows:

u (ENTER)
COPY
(ENTER)
CIRCLE

Using the U command after commands that involve transparent commands or subcommands causes the entire sequence to be undone. For example, when you set a dimension variable and then perform a dimension command, a subsequent U command nullifies the dimension drawn and the change in the setting of the dimension variable.

UNDO COMMAND

The UNDO command permits you to select a specified number or marked group of prior commands for undoing. You can invoke the UNDO command by typing **undo** at the On-Screen prompt (or from the Standard toolbar, as shown in Figure 3–74).

Figure 3–74 *Invoking the* UNDO *command from the Standard toolbar*

After invoking UNDO AutoCAD prompts:

> Enter the number of operations to undo or ↓ *(specify number of undo operations to undo or select one of the available options from the shortcut menu)*

The default number of undo operations is 1. You can specify a higher number of undo operations. If you select a number higher than the available number of undo operations, AutoCAD will undo all of the operations available.

Choose CONTROL from the shortcut menu to set the available undo options. By limiting the number of undo options, you can free up memory and disk space that is otherwise used to save undo operation information. AutoCAD prompts as follows when the CONTROL option is selected:

> Enter an UNDO control option *(select one of the four available options)*

Selecting ALL (the default setting of the UNDO command) enables all Undo options.

Selecting NONE disables the U and UNDO commands but not the CONTROL option of the UNDO command that re-enables the various options.

Selecting ONE sets the UNDO commands functionality to that of the U command.

Selecting COMBINE controls whether multiple, consecutive zoom and pan commands are combined as a single operation for UNDO and REDO (discussed later) operations.

Choose MARK to set a mark in the undo information.

Choose BACK to undo all the operations back to the last mark.

If you are at a point in the editing session where you would like to experiment, but would still like the option of undoing the experiment, you can mark that point. An example of the use of the MARK and BACK options is as follows:

>   **line** (ENTER) *(draw a line)*
>   **circle** (ENTER) *(draw a circle)*
>   **undo** (ENTER)
>   Enter number of operations to undo or ⬇ **m** (ENTER)
>   **text** (ENTER) *(enter text)*
>   **arc** (ENTER) *(draw an arc)*
>   **undo** (ENTER)
>   Enter number of operations to undo or ⬇ **b** (ENTER)

The BACK option returns you to the state of the drawing that has the line and the circle. Following this UNDO BACK with a U removes the circle. Another U removes the line.

Using the BACK option when no Mark has been established will prompt:

>   This will undo everything. OK? <Y>

Responding Y undoes everything done since the current editing session was begun or since the last SAVE command.

Note: The default is Y; think twice before pressing ENTER in response to the "This will undo everything" prompt.

Choose BEGIN from the shortcut menu to group a sequence of actions into a set. All subsequent actions become part of this group until you choose the END option. Entering **undo begin** while a group is already active ends the current group and begins a new one. UNDO and U treat grouped actions as a single action. The BEGIN and END options are normally intended for use in strings of menu commands, where a menu pick involves several operations.

Choose AUTO from the shortcut menu to group the actions of a single command, making them reversible by a single U command. When the AUTO option is on, starting a command groups all actions until you exit that command.

The effects of the following commands cannot be undone: AREA, ATTEXT, DBLIST, DELAY, DIST, DXFOUT, GRAPHSCR, HELP, HIDE, ID, LIST, MSLIDE, PLOT, QUIT, REDRAW, REDRAWALL, REGENALL, RESUME, SAVE, SHADE, SHELL, STATUS, and TEXTSCR.

When the down arrow to the right of the UNDO command icon on the Standard menu is selected, a drop-down menu is displayed with a list of commands that can be undone by using the UNDO command. They are listed from the most recent to the earliest. You can select any command on the list and AutoCAD will undo all of the commands back to and including the command selected.

REDO COMMAND

The REDO command permits reversal of prior U or UNDO commands. All UNDO commands can be reversed with the REDO command. In order to function, the REDO command must be used following the U or UNDO command and prior to any other action. You can invoke the REDO command from the Standard toolbar (see Figure 3–75).

Figure 3–75 *Invoking the REDO command from the Standard toolbar*

The REDO command does not have any options. When the down arrow to the right of the REDO command icon on the Standard menu is selected, a drop-down menu is displayed with a list of commands that can be redone by using the REDO command. They are listed from the most recent to the earliest.

Open the Exercise Manual PDF file for Chapter 3 on the accompanying CD for project- and discipline-specific exercises.

 If you have the accompanying Exercise Manual, refer to Chapter 3 for project- and discipline-specific exercises.

REVIEW QUESTIONS

1. The invisible grid, which the crosshairs lock onto, is called:
    a. SNAP
    b. GRID
    c. ORTHO
    d. cursor lock

2. To globally change the sizes of the dashes for all dashed lines, you should adjust:
    a. line scale
    b. LTSCALE
    c. SCALE
    d. layer scale

3. To reverse the effect of the last 11 commands, you could:
    a. use the UNDO command
    b. use the U command multiple times
    c. either A or B
    d. it is not possible because AutoCAD only retains the last 10 commands

4. The following are AutoCAD tools available, except:
    a. GRID
    b. SNAP
    c. ORTHO
    d. TSNAP
    e. OSNAP

5. If you just used the U command, what command would restore the drawing to the state before the U command?
    a. RESTORE
    b. U
    c. REDO
    d. OOPS

6. Which of the following cannot be modified in the Drafting Settings dialog box?
    a. Snap
    b. Grid
    c. Ortho
    d. Limits
    e. All of the above

7. If the spacing of the visible grid is set too small, AutoCAD responds as follows:

   a. does not accept the command

   b. produces a "Grid too dense to display" message

   c. produces a display that is distorted

   d. automatically adjusts the size of the grid so it will display

   e. displays the grid anyway

8. The smallest number that can be displayed in the denominator when setting units to architectural units is:

   a. 8

   b. 16

   c. 64

   d. 128

   e. none of the above

9. Which of the following is not a valid option of the Layer Properties Manager dialog box?

   a. Close

   b. Lock

   c. On

   d. Freeze

   e. Color

10. After having drawn a 3-point circle, you want to begin a line at the exact center of the circle. What tool in AutoCAD would you use?

    a. Snap

    b. Object Snap

    c. Entity Snap

    d. Geometric Calculator

11. How many previous zooms are available with the previous option of the ZOOM command?

    a. 4

    b. 6

    c. 8

    d. 10

12. In general, a REDRAW is quicker than a REGEN.

    a. True

    b. False

13. When a layer is ON and THAWed:

    a. the objects on that layer are visible on the monitor

    b. the objects on that layer are not visible on the monitor

    c. the objects on that layer are ignored by a REGEN

    d. the drawing REDRAW time is reduced

    e. the objects on that layer cannot be selected

14. To ensure that the entire limits of the drawing are visible on the display, you should perform a ZOOM-

    a. ALL

    b. PREVIOUS

    c. EXTENTS

    d. LIMITS

15. A layer where objects may not be edited or deleted, but are still visible on the screen and may be OSNAPed to, is considered:

    a. Frozen   d. unSet

    b. Locked   e. fiXed

    c. On

16. Which Osnap option allows you to select the closest end point of a line, arc, or polyline segment?

    a. ENDpoint   d. INSertion point

    b. MIDpoint   e. PERpendicular

    c. CENter

17. Which Osnap option allows you to select the point in the exact center of a line?

    a. ENDpoint

    b. MIDpoint

    c. CENter

    d. INSertion point

    e. PERpendicular

18. When drawing a line, which Osnap option allows you to select the point on a line or polyline segment where the angle formed with the line is a 90-degree angle?

   a. ENDpoint

   b. MIDpoint

   c. CENter

   d. INSertion point

   e. PERpendicular

19. Which Osnap option allows you to select the location where two lines, arcs, or polyline segments cross each other?

   a. NODe

   b. QUADrant

   c. TANgent

   d. NEArest

   e. INTersection

20. Which Osnap option allows you to select the location where a "point" has been established?

   a. NODe

   b. QUADrant

   c. TANgent

   d. NEArest

   e. INTersection

21. Which Osnap option allows you to select a point on a circle that is 0, 90, 180, or 270 degrees from the circle's center?

   a. NODe

   b. QUADrant

   c. TANgent

   d. NEArest

   e. INTersection

22. Which Osnap option allows you to select a point on any object, except text, that is closest to the cursor's position?

   a. NODe

   b. QUADrant

   c. TANgent

   d. NEArest

   e. INTersection

23. Which Osnap option allows you to select the location where two lines or arcs may or may not cross each other in 3D space?

   a. QUADrant

   b. APParent intersection

   c. NEArest

   d. INTersection

CHAPTER 4

Fundamentals III

INTRODUCTION

After completing this chapter, you will be able to do the following:

- Draw construction lines using the XLINE and RAY commands
- Construct geometric figures with polygons, ellipses, and polylines
- Create single line text and multi-line text using appropriate styles and sizes, to annotate drawings
- Use the construct commands: COPY, ARRAY, OFFSET, MIRROR, FILLET, and CHAMFER
- Use the modify commands: MOVE, TRIM, BREAK, and EXTEND

DRAWING CONSTRUCTION LINES

AutoCAD provides powerful tools for drawing lines (XLINE and RAY commands) that extend infinitely in one or both directions. These lines have no effect, however, on the ZOOM EXTENTS command. They can be moved, copied, and rotated like any other objects. If necessary, you can trim the lines, break them anywhere with the BREAK command, draw an arc between two non-parallel construction lines with the FILLET command, and draw a chamfer between two non-parallel construction lines with the CHAMFER command.

The construction lines can be used as reference lines for creating other objects. To keep the construction lines from being plotted, draw them on a separate layer and set that layer to a non-printing status.

XLINE COMMAND

The XLINE command allows you to draw lines that extend infinitely in both directions from the point selected when being created.

Invoke the XLINE command by selecting Construction Line from the Draw toolbar (see Figure 4–1) and AutoCAD prompts:

> Specify a point or ⬇ *(specify a point or right-click and choose one of the options)*

Figure 4–1 *Invoking the XLINE command by selecting Construction Line from the Draw toolbar*

When you specify a point, AutoCAD prompts:

> Specify through point: *(specify a point through which the construction line should pass)*

AutoCAD draws a line that passes through the two points and extends infinitely in both directions. AutoCAD continues to prompt for additional through points. AutoCAD uses these to draw construction lines through each of them and the first point specified. To terminate the command, press ENTER or the SPACEBAR.

Xline Command Options

Choosing HOR (Horizontal) from the shortcut menu allows you to draw a construction line through a specified point that is parallel to the *X* axis of the current UCS.

Choosing VER (Vertical) from the shortcut menu allows you to draw a construction line through a specified point that is parallel to the *Y* axis of the current UCS.

Choosing ANG (Angle) from the shortcut menu allows you to draw a construction line at a specified angle. AutoCAD prompts:

> Enter angle of xline (0) or ⬇ *(specify an angle at which to place the construction line)*
> Specify through point: *(specify a point through which the construction line should pass)*

AutoCAD draws the construction line through the specified point, using the specified angle.

Choosing REFERENCE from the shortcut menu allows you to draw a construction line at a specific angle from a selected reference line. The angle is measured counterclockwise from the reference line.

Choosing BISECT from the shortcut menu allows you to draw a construction line through the first point bisecting the angle determined by the second and third points, with the first point being the vertex. AutoCAD prompts:

> Specify angle vertex point: *(specify a point for the vertex of an angle to be bisected and through which the construction line will be drawn)*
> Specify angle start point: *(specify a point to determine one boundary line of an angle)*
> Specify angle endpoint: *(specify a point to determine second boundary line of angle)*

The construction line lies in the plane determined by the three points.

Choosing OFFSET from the shortcut menu allows you to draw a construction line parallel to and at the specified distance from the line object selected and on the side selected. AutoCAD prompts:

> Specify offset distance or ⬇ *(specify an offset distance or right-click and choose one of the available options)*
> Select a line object: *(select a line, pline, ray, or xline)*
> Specify side to offset: *(specify a point to draw a construction line parallel to the selected object)*
> Select a line object: *(press ENTER to terminate this command sequence)*

Choosing THROUGH allows you to specify a point through which an offset construction line is drawn.

RAY COMMAND

The RAY command allows you to draw lines that extend infinitely in one direction from a specified point.

Invoke the RAY command from the Draw menu and AutoCAD prompts:

> Specify start point: *(specify the start point to draw the ray)*
> Specify through point: *(specify a point through which you want the ray to pass)*
> Specify through point: *(specify a point to draw additional rays or press ENTER to terminate the command sequence)*

A ray is drawn starting at the first point and extending infinitely in one direction through the second point. AutoCAD continues to prompt for through points until you terminate the command sequence.

DRAWING POLYGONS

The POLYGON command creates an equilateral (edges with equal length) closed polyline. It offers three different methods for drawing 2D polygons: Inscribed in Circle, Circumscribed about Circle, and Edge. The number of sides can vary from 3 (which forms an equilateral triangle) to 1024. Invoke the POLYGON command from the Draw toolbar (see Figure 4–2) and AutoCAD prompts:

> Enter number of sides<current>: **6** (ENTER)
> Specify center of polygon or ⬇ **3,3** (ENTER)

Figure 4–2 *Invoking the* POLYGON *command from the Draw toolbar*

After you specify the number of sides and the center of the polygon, AutoCAD prompts for one of two options, INSCRIBED IN CIRCLE or CIRCUMSCRIBED ABOUT CIRCLE.

POLYGON OPTIONS

Choosing INSCRIBED IN CIRCLE draws the polygon of equal length for all sides inscribed inside an imaginary circle having the same diameter as the distance across opposite polygon corners (for an even number of sides), as shown in the following example:

> Enter an option *(choose* INSCRIBED IN CIRCLE*)*
> Specify radius of circle: **2** (ENTER)

AutoCAD draws a polygon (see Figure 4–3) with six sides, centered at 3,3; whose edge vertices are 2 units from the center of the polygon.

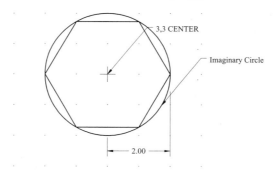

Figure 4–3 *Polygon drawn with six sides using the* INSCRIBED IN CIRCLE *option*

 Note: Specifying the radius with a specific value draws the bottom edge of the polygon at the current snap rotation angle. If instead, you specify the radius with your pointing device or by means of coordinates, AutoCAD places one apex of the polygon on the specified point, which determines the rotation and size of the polygon.

Choosing CIRCUMSCRIBED ABOUT CIRCLE draws a polygon circumscribed around the outside of an imaginary circle having the same diameter as the distance across the opposite polygon sides (for an even number of sides) as shown in the following example:

    polygon (ENTER)
    Enter number of sides <current>: **8** (ENTER)
    Specify center of polygon or ⬇ **3,3** (ENTER)
    Enter an option (*choose* CIRCUMSCRIBED ABOUT CIRCLE)
    Specify radius of circle: **2** (ENTER)

AutoCAD draws a polygon (see Figure 4–4) with eight sides, centered at 3,3, whose edge midpoints are 2 units from the center of the polygon.

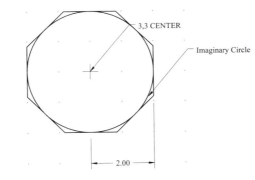

Figure 4–4 *Polygon drawn with eight sides using the* CIRCUMSCRIBED ABOUT CIRCLE *option*

 Note: Specifying the radius draws the bottom edge of the polygon at the current snap rotation angle. If instead, you specify the radius with your pointing device or by means of coordinates, AutoCAD places the midpoint of one edge of the polygon at the specified point, which determines the rotation and size of the polygon.

Choosing EDGE allows you to draw a polygon by specifying the end points of the first edge, as shown in the following example:

    polygon (ENTER)
    Enter number of sides <current>: **7** (ENTER)
    Specify center of polygon or ⬇ (*choose* EDGE *from the shortcut menu*)
    Specify first endpoint of edge: **1,1** (ENTER)
    Specify second endpoint of edge: **3,1** (ENTER)

AutoCAD draws a polygon (see Figure 4–5) with seven sides for the specified end points of one of the sides of the polygon.

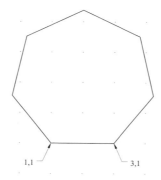

Figure 4–5 *Polygon drawn with seven sides using the EDGE option*

DRAWING ELLIPSES

AutoCAD allows you to draw an ellipse or an elliptical arc with the ELLIPSE command. Invoke the ELLIPSE command from the Draw toolbar (see Figure 4–6), and AutoCAD prompts:

> Specify axis endpoint of ellipse: *(specify axis end point of the ellipse to be drawn or right-click and choose one of the options)*

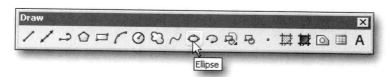

Figure 4–6 *Invoking the ELLIPSE command from the Draw toolbar*

DRAWING AN ELLIPSE BY SPECIFYING AXIS END POINTS

This option allows you to draw an ellipse by specifying the end points of the axes. AutoCAD prompts for two end points of the first axis. The first axis can define either the major or the minor axis of the ellipse. Then AutoCAD prompts for an end point of the second axis as the distance from the midpoint of the first axis to the specified point.

For example, the following command sequence shows steps in drawing an ellipse by defining axis end points (see Figure 4–7).

> **ellipse** (ENTER)
> Specify axis endpoint of ellipse or ⬇ **1,1** (ENTER)
> Specify other endpoint of axis: **5,1** (ENTER)
> Specify distance to other axis or ⬇ **3,2** (ENTER)

AutoCAD draws an ellipse whose major axis is 4.0 units long in a horizontal direction and whose minor axis is 2.0 units long in a vertical direction (see Figure 4–7).

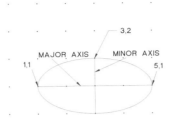

Figure 4–7 *An ellipse drawn by specifying the major and minor axes*

Ellipse Command Options

Choosing ROTATION from the shortcut menu, after specifying two axis end points, allows you to draw an ellipse by specifying a rotation angle. The rotation angle defines the major-axis-to-minor-axis ratio of the ellipse by rotating a circle about the first axis. The greater the rotation angle value, the greater the ratio of major to minor axes. AutoCAD draws a circle if you set the rotation angle to 0 degrees.

For example, the following command sequence shows steps in drawing an ellipse by specifying the rotation angle.

> **ellipse** (ENTER)
> Specify axis endpoint of ellipse or ⊥ **3,–1** (ENTER)
> Specify other endpoint of axis: **3,3** (ENTER)
> Specify distance to other axis or ⊥ (*choose* ROTATION *from the shortcut menu*)
> Specify rotation around major axis: **30** (ENTER)

See Figure 4–8 for examples of ellipses with various rotation angles.

Figure 4–8 *Ellipses drawn with different rotation angles*

Choosing CENTER from the shortcut menu before specifying two axis end points allows you to draw an ellipse by defining the center point first and then the axis end points. First, AutoCAD prompts for the ellipse center point. Then, AutoCAD prompts for an end point of an axis as the distance from the center of the ellipse to the specified point. The first axis can define either the major or the minor axis of the ellipse. Next,

AutoCAD prompts for an end point of the second axis as the distance from the center of the ellipse to the specified point.

For example, the following command sequence shows steps in drawing an ellipse by defining the ellipse center point.

> **ellipse** (ENTER)
> Specify axis endpoint of ellipse or ⬇ *(choose CENTER from the shortcut menu)*
> Specify center of ellipse: **3,1** (ENTER)
> Specify endpoint of axis: **1,1** (ENTER)
> Specify distance to other axis or ⬇ **3,2** (ENTER)

AutoCAD draws an ellipse similar to the first example (see Figure 4–7), with a major axis 4.0 units long in a horizontal direction and a minor axis 2.0 units long in a vertical direction.

Choosing ARC from the shortcut menu, before specifying two axis end points, allows you to draw an elliptical arc. After you specify the major and minor axis end points, AutoCAD prompts for the start and end angle points for the elliptical arc to be drawn. Instead of specifying the start angle or the end angle, you can toggle to the PARAMETER option, which prompts for the Start parameter and End parameter point locations. AutoCAD creates the elliptical arc using the following parametric vector equation:

$$p(u) = c + a^x + \cos(u) + b^x \sin(u)$$

where c is the center of the ellipse, and a and b are its major and minor axes, respectively. Instead of specifying the end angle, you can specify the included angle of the elliptical arc to be drawn.

The ELLIPSE command, with the ARC option, is accessible directly from the Draw toolbar or you can select the ARC option while in the ELLIPSE command.

For example, the following command sequence shows steps in drawing an elliptical arc.

> **ellipse** (ENTER)
> Specify axis endpoint of ellipse or ⬇ *(choose ARC from the shortcut menu)*
> Specify axis endpoint of elliptical arc or ⬇ **1,1** (ENTER)
> Specify other endpoint of axis: **5,1** (ENTER)
> Specify distance to other axis or ⬇ **3,2** (ENTER)
> Specify start angle or ⬇ **4,1** (ENTER)
> Specify end angle or ⬇ **-2,1** (ENTER)

AutoCAD draws an elliptical arc with the start angle at 3,2 and the ending angle at 1,1.

ISOMETRIC CIRCLES (OR ISOCIRCLES)

By definition, Isometric Planes (*iso* meaning "same" and *metric* meaning "measure") are all viewed at the same angle of rotation (see Figure 4–9). The angle is approximately 54.7356 degrees. AutoCAD uses this angle of rotation automatically when you wish to represent circles in one of the isoplanes by drawing ellipses with the ISOCIRCLE option.

Normally, a circle 1 unit in diameter viewed in one of the isoplanes, will project a short axis dimension of 0.577350 units. Its diameter parallel to an isoaxis will project a dimension of 0.816497 units. A line drawn in isometric mode that is parallel to one of the three main axes will also project a dimension of 0.816497 units. We would like these lines and circle diameters to project a dimension of exactly 1.0 unit. Therefore, you automatically increase the entire projection by a fudge factor of 1.22474 (the reciprocal of 0.816497) in order to use true dimensioning parallel to one of the isometric axes.

This means that isocircles, 1 unit in diameter ,will be measured along one of their isometric diameters, rather than along their long axis. This facilitates using true lengths as the lengths of distances projected from lines parallel to one of the isometric axes. So, a 1-unit-diameter isocircle will project a long axis that is 1.224744871 units and a short axis that is 0.707107 (0.577350 x 1.22474) units. These "fudge" factors are built into AutoCAD isocircles.

The *ISOMETRIC CIRCLE* option is available as one of the options of the ELLIPSE command when you are in the Isometric Snap mode. To enter Isometric Snap mode, right-click on **SNAP** (found on the status bar). In the **Snap type & style** section, select **Isometric snap**.

> **ellipse** (ENTER)
> Specify axis endpoint of ellipse or ⬇ *(choose ISOCIRCLE from the shortcut menu)*
> Specify center of isocircle: *(select the center of the isometric circle)*
> Specify radius of isocircle or ⬇ *(specify the radius or right-click and choose one of the options)*

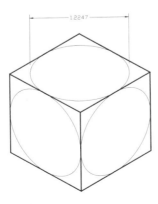

Figure 4-9 *Ellipses drawn using the ISOCIRCLE option of the ELLIPSE command*

Choosing DIAMETER causes AutoCAD to prompt:

Circle diameter: *(specify the desired diameter)*

 Note: The ISOCIRCLE and DIAMETER options will work only when you are in the Isometric Snap mode.

DRAWING POLYLINES

The "poly" in polyline refers to a single object with multiple connected straight-line and/or arc segments. The polyline is drawn by invoking the PLINE command and then selecting a series of points. In this respect, PLINE functions much like the LINE command. However, when completed, the segments act like a single object when operated on by modify commands. You specify the end points using only 2D (X,Y) coordinates.

The versatile PLINE command also draws lines and arcs of different widths, linetypes, tapered lines, and a filled circle. The area and perimeter of a 2D polyline can be calculated.

By default, polylines are drawn as optimized polylines. An optimized polyline provides most of the functionality of 2D polylines but with much improved performance and reduced drawing file size. The vertices are stored as an array of information on one object. When you use the PEDIT command to edit the polyline into a spline fitting or a curve fitting, the polyline loses its optimization feature and vertices are stored as separate entities, but it still behaves as a single object when operated on by modify commands.

Invoke the PLINE command from the Draw toolbar (see Figure 4–10) and AutoCAD prompts:

Specify start point: *(specify the start point of the polyline)*
Specify next point or ⤓ *(specify next point or right-click and choose one of the options)*

Figure 4–10 *Invoking the* PLINE *command from the Draw toolbar*

You can specify the end of the line by means of absolute coordinates, relative coordinates, or by using your pointing device to specify the end of the line on the screen. After you do, AutoCAD repeats the prompt:

Specify next point or ⤓

Having drawn a connected series of lines, you can press ENTER to terminate the PLINE command. The resulting figure is recognized by AutoCAD modify commands as a single object.

The following command sequence presents an example of connected lines drawn by means of the PLINE command (see Figure 4–11).

> pline
> Specify start point: **2,2** (ENTER)
> Specify next point or ⬇ **4,2** (ENTER)
> Specify next point or ⬇ **5,1** (ENTER)
> Specify next point or ⬇ **7,1** (ENTER)
> Specify next point or ⬇ **8,2** (ENTER)
> Specify next point or ⬇ **10,2** (ENTER)
> Specify next point or ⬇ **10,4** (ENTER)
> Specify next point or ⬇ **9,5** (ENTER)
> Specify next point or ⬇ **8,5** (ENTER)
> Specify next point or ⬇ **7,4** (ENTER)
> Specify next point or ⬇ **5,4** (ENTER)
> Specify next point or ⬇ **4,5** (ENTER)
> Specify next point or ⬇ **3,5** (ENTER)
> Specify next point or ⬇ **2,4** (ENTER)
> Specify next point or ⬇ *(choose CLOSE from the shortcut menu)*

Figure 4–11 *Example of connected line segments drawn by means of the PLINE command*

Polyline Command Options

Choosing CLOSE or UNDO from the shortcut menu works similarly to the corresponding options in the LINE command.

Choosing WIDTH from the shortcut menu allows you to specify a starting and an ending width for a wide segment. When you select this option, AutoCAD prompts:

> Specify starting width <current>: *(specify starting width)*
> Specify ending width <current>: *(specify ending width)*

You can specify a width by entering a value at the prompt or by selecting points on the screen. When you specify points on the screen, AutoCAD uses the distance from the starting point of the polyline to the point selected as the starting width. You can accept the default value for the starting width by pressing ENTER, or enter a new value. The starting width is the default for the ending width. If necessary, you can change the ending width to another value, which will result in a tapered segment or an arrow. The ending width, in turn, is the uniform width for all subsequent segments until you change the width again.

The following command sequence presents an example of connected lines with tapered width (see Figure 4–12) drawn by means of the PLINE command.

Specify start point: **2,2** (ENTER)
Specify next point or ⬇ *(choose WIDTH from the shortcut menu)*
Specify starting width <current>: **0** (ENTER)
Specify ending width <current>: **.25** (ENTER)
Specify next point or ⬇ **2,2.5** (ENTER)
Specify next point or ⬇ **2,3** (ENTER)
Specify next point or ⬇ *(choose WIDTH from the shortcut menu)*
Specify starting width <0.2500>: (ENTER)
Specify ending width <0.2500>: **0** (ENTER)
Specify next point or ⬇ **2,3.5** (ENTER)
Specify next point or ⬇ *(choose WIDTH from the shortcut menu)*
Specify starting width <0.0000>: (ENTER)
Specify ending width <0.0000>: **.25** (ENTER)
Specify next point or ⬇ **2.5,3.5** (ENTER)
Specify next point or ⬇ **3,3.5** (ENTER)
Specify next point or ⬇ *(choose WIDTH from the shortcut menu)*
Specify starting width <0.2500>: (ENTER)
Specify ending width <0.2500>: **0** (ENTER)
Specify next point or ⬇ **3.5,3.5** (ENTER)
Specify next point: *(choose WIDTH from the shortcut menu)*
Specify starting width <0.0000>: (ENTER)
Specify ending width <0.0000>: **.25** (ENTER)
Specify next point or ⬇ **3.5,3** (ENTER)
Specify next point or ⬇ **3.5,2.5** (ENTER)
Specify next point: *(choose WIDTH from the shortcut menu)*
Specify starting width <0.2500>: (ENTER)
Specify ending width <0.2500>: **0** (ENTER)
Specify next point or ⬇ **3.5,2** (ENTER)
Specify next point or ⬇ *(choose WIDTH from the shortcut menu)*
Specify starting width <0.0000>: (ENTER)
Specify ending width <0.0000>: **.25** (ENTER)
Specify next point or ⬇ **3,2** (ENTER)

```
Specify next point or ⬇ 2.5,2 (ENTER)
Specify next point ⬇ (choose WIDTH from the shortcut menu)
Specify starting width <0.2500>: (ENTER)
Specify ending width <0.2500>: 0 (ENTER)
Specify next point or ⬇ (choose CLOSE from the shortcut menu)
```

Figure 4–12 *Example of connected line segments with tapered width, drawn by means of the PLINE command*

Choosing HALFWIDTH from the shortcut menu operates similar to the WIDTH option, including the prompts, except it lets you specify the width from the center of a wide polyline to one of its edges. In other words, you specify half of the total width. For example, it is easier to input 1.021756 as the half width than to figure out the total width by doubling. You can specify a half width by selecting points on the screen in the same manner used to specify the full width.

Choosing ARC allows you to draw a polyline arc. When you select the ARC option, AutoCAD displays:

```
Specify endpoint of arc or ⬇ (specify end point of arc or right-click and
    choose one of the available options)
```

If you respond with a point, it is interpreted as the end point of the arc. The end point of the previous segment is the starting point of the arc, and the starting direction of the new arc will be the ending direction of the previous segment (whether the previous segment is a line or an arc). This resembles the ARC command's Start, End, Direction (S,E,D) option, but requires only the end points to be specified or selected on the screen.

Polyline Arc Options

The CLOSE, WIDTH, HALFWIDTH, and UNDO options on the shortcut menu operate similar to the corresponding options for the straight-line segments described earlier.

Choosing ANGLE lets you specify the included angle by prompting:

> Specify included angle: *(specify an angle)*

The arc is drawn counterclockwise if the value is positive, clockwise if it is negative. After the angle is specified, AutoCAD prompts for the end point of the arc.

Choosing CENTER lets you override with the location of the center of the arc and AutoCAD prompts:

> Specify center point: *(specify center point)*

When you provide the center point of the arc, AutoCAD prompts for additional information:

> Specify endpoint of arc or ⬇ *(specify end point of arc or right-click and choose one of the options)*

If you respond with a point, it is interpreted as the end point of the arc.

Choosing ANGLE or LENGTH allows you to specify the arc's included angle or chord length.

Choosing DIRECTION lets you override the direction of the last segment, and AutoCAD prompts:

> Specify the tangent direction for the start point of arc: *(specify the direction)*

If you respond with a point, AutoCAD interprets the starting point and the direction from this point and then prompts for the end point for the arc.

Choosing LINE reverts to drawing straight-line segments.

Choosing RADIUS allows you to specify the radius by prompting:

> Specify radius of arc: *(specify the radius of the arc)*

After the radius is specified, you are prompted for the end point of the arc.

Choosing SECOND PT causes AutoCAD to use the three-point method of drawing an arc by prompting:

> Specify second point of arc: *(specify second point)*

If you respond with a point, it is interpreted as the second point and then you are prompted for the end point of the arc. This resembles the ARC command's Three-point option.

Choosing LENGTH continues the polyline in the same direction as the last segment for a specified distance.

DRAWING TEXT

You have learned how to draw some geometric shapes that make up your design. Now it is time to learn how to annotate your design. When you draw on paper, adding descriptions of the design components and the necessary shop and fabrication notes is a tedious, time-consuming process. AutoCAD provides several text commands and tools (including a spell checker) that greatly reduce the tedium of text placement and the time it takes.

Text is used to label the various components of your drawing and to create the necessary shop or field notes needed for fabrication and construction of your design. AutoCAD includes a large number of text fonts. Text can be stretched, compressed, obliqued, mirrored, or drawn in a vertical column by applying a style. Each text string can be sized, rotated, and justified to meet your drawing needs. You should be aware that AutoCAD considers a text string (all the characters that comprise the line of characters using a single TEXT command) as one object.

AutoCAD has a toolbar just for text-related commands. Included are Multiline Text, Single Line Text, Edit, Find and replace, Text Style, Scale, Justify, and Convert distance between spaces (see Figure 4–13).

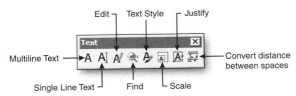

Figure 4–13 *Text toolbar*

CREATING A SINGLE LINE OF TEXT

The TEXT command allows you to create several lines of text in the current style. If necessary, you can change the current style. To modify a style or to create a new style, refer to the section Creating and Modifying Text Styles, later in this chapter.

Invoke the TEXT command for a single line of text from the Text toolbar (see Figure 4–13) and AutoCAD prompts:

> Specify start point of text or ⊡ *(specify start point of text or right-click and choose one of the available options)*

The start point indicates the lower left corner of the text. If necessary, you can change the location of the justification point. You can specify the starting point in absolute coordinates or by using your pointing device. After you specify the starting point, AutoCAD prompts:

> Specify height <current>: *(specify the text height)*

This allows you to set the text height. You can accept the current text height by pressing ENTER. You can set a new text height by using your pointing device, or entering the appropriate text height. Next, AutoCAD prompts:

>   Specify rotation angle of text <current>: *(specify the rotation angle)*

This allows you to place the text at any angle in reference to 0 degrees (default). A rotation angle of 0 degrees causes the text to be placed horizontally at the specified start point.

A cursor appears on screen at the starting point you have selected. After you enter the first line of text and press ENTER, you will notice the cursor drop down to the next line, anticipating that you wish to enter another line of text. If this is the case, enter the next line of text string; when you are through with typing text strings, press ENTER twice to terminate the command sequence.

If you are in the TEXT command and notice a mistake (or simply want to change a character or word), you can use the arrow keys on your keyboard to position the cursor and make the necessary changes.

The following command sequence shows placement of left-justified text by providing the starting point of the text (see Figure 4–14).

>   **text** (ENTER)
>   Specify start point of text or 🔽 *(specify start point of text as shown in Figure 4–14)*
>   Specify height <current>: **0.25** (ENTER)
>   Specify rotation angle of text <current>: (ENTER)
>   *(at the screen text cursor, type in the desired text)* **Sample Text Left Justified** *(press ENTER to accept first line of text)*
>   *(enter the next line of text or press ENTER to accept single line as entered)*

Sample Text Left Justified
— Start Point

Figure 4–14 *Using the* TEXT *command to place left-justified text by specifying a start point*

Single-Line Text Options

When AutoCAD prompts for the start point, you can right-click for options to change justification and style. In addition to the standard justification options, you can choose to apply the *ALIGN* and *FIT* options.

Choosing *JUSTIFY* allows you to place text in one of the 14 available justification points. When you select this option, AutoCAD prompts:

>   Enter an option *(choose one of the options listed)*

Options available from the On-Screen prompt include: ALIGN, FIT, CENTER, MIDDLE, RIGHT, TL, TC, TR, ML, MC, MR, BL, BC, and BR.

Choosing CENTER allows you to select the center point for the baseline of the text. Baseline refers to the line along which the bases of the capital letters lie. Letters with descenders, such as g, q, and y, dip below the baseline. After providing the center point, enter the text height and rotation angle.

For example, the following command sequence shows placement of center-justified text, by providing the center point of the text (see Figure 4–15).

    text (ENTER)
    Specify start point of text or ⬇ *(choose JUSTIFY from the shortcut menu)*
    Enter an option *(choose CENTER)*
    Specify center point of text: *(specify center point)*
    Specify height <current>: **.25** (ENTER)
    Specify rotation angle of text:<current>: **0** (ENTER)
    Enter text: **Sample Text Center Justified** (ENTER)

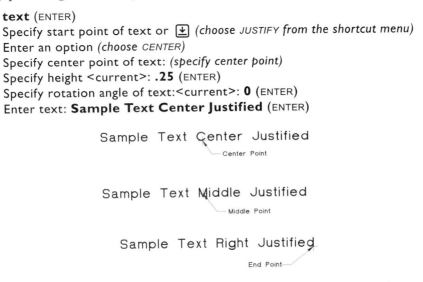

Figure 4–15 *Using the* TEXT *command to place text by specifying a center point (center justified), a middle point (middle justified), or an end point (right justified)*

Choosing MIDDLE allows you to center the text both horizontally and vertically at a given point. After providing the middle point, enter the text height and rotation angle.

For example, the following command sequence shows placement of middle-justified text by providing the middle point of the text (see Figure 4–15).

    text (ENTER)
    Specify start point of text or ⬇ *(choose JUSTIFY from the shortcut menu)*
    Enter an option *(choose MIDDLE)*
    Specify middle point of text: *(specify middle point)*
    Specify height <current>: **.25** (ENTER)
    Specify rotation angle of text: **0** (ENTER)
    Enter text: **Sample Text Middle Justified** (ENTER)

Choosing RIGHT allows you to place the text in reference to its lower right corner (right justified). Here, you provide the point where the text will end. After providing the right point, enter the text height and rotation angle.

For example, the following command sequence shows placement of right-justified text (see Figure 4–15).

>text (ENTER)
Specify start point of text or ⬇ *(choose JUSTIFY from the shortcut menu)*
Enter an option *(choose RIGHT)*
Specify right endpoint of text baseline: *(specify right end point point)*
Specify height <current>: .25 (ENTER)
Specify rotation angle of text: 0 (ENTER)
Enter text: **Sample Text Right Justified** (ENTER)

Other options are combinations of the previously mentioned options:

TL	top left
TC	top center
TR	top right
ML	middle left
MC	middle center
MR	middle right
BL	bottom left
BC	bottom center
BR	bottom right

Choosing ALIGN allows you to place the text by designating the end points of the baseline. AutoCAD computes the text height and orientation so that the text just fits proportionately between two points. The overall character size adjusts in proportion to the height. The height and width of the character will be the same.

For example, the following command sequence shows placement of text using the Align option (see Figure 4–16).

>text (ENTER)
Specify start point of text or ⬇ *(choose JUSTIFY from the shortcut menu)*
Enter an option *(choose ALIGN)*
Specify first endpoint of text baseline: *(specify the first point)*
Specify second endpoint of text baseline: *(specify the second point)*
Enter text: **Sample Text Aligned** (ENTER)

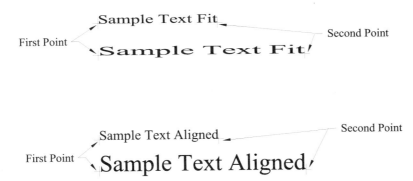

Figure 4–16 Using the ALIGN and FIT options of the TEXT command to place text

Choosing FIT is similar to the ALIGN option, but in the case of the Fit option AutoCAD uses the current text height and adjusts only the text's width, expanding or contracting it to fit between the points you specify.

For example, the following command sequence shows placement of text using the Fit option (see Figure 4–16).

> **text** (ENTER)
> Specify start point of text or ⬇: *(choose JUSTIFY from the shortcut menu)*
> Enter an option *(choose FIT)*
> Specify first endpoint of text baseline: *(specify the first point)*
> Specify second endpoint of text baseline: *(specify the second point)*
> Specify height <current>: **0.25** (ENTER)
> Enter text: **Sample Text Fit** (ENTER)

Choosing STYLE allows you to select one of the available styles in the current drawing. To modify a style or to create a new style, refer to the section "Creating and Modifying Text Styles," later in this chapter.

CREATING MULTILINE TEXT

The MTEXT command draws text by "processing" the words in paragraph form; the width of the paragraph is determined by the user-specified rectangular boundary. It is an easy way to have your text automatically formatted as a paragraph, with left, right, or center justification as a group.

Each multiline text object is a single object, regardless of the number of lines it contains. If one character on one line is selected for editing, all characters on all lines are included in the selection set. The text boundary remains part of the object's framework, although it is not plotted or printed.

Multiline text features include indents and tabs, making it easier to correctly align your text for tables and numbered, lettered, or bulleted lists. At the top of the user-specified rectangle is a ruler (see Figure 4–17) similar to those in word processors. Characters in the Multiline Text Editor can be selected individually or in a group for applying formatting styles such as bold, underline, and italics.

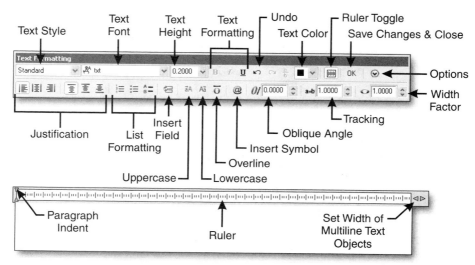

Figure 4–17 *Text Formatting toolbar, Options toolbar, ruler, and multiline text input (editing) area*

Invoke the MTEXT command by selecting Multiline Text from the Text toolbar (see Figure 4–18) and AutoCAD prompts:

Specify first corner: *(specify the first corner of the rectangular boundary)*
Specify opposite corner or ⬇: *(specify the opposite corner of the rectangular boundary, or right-click and choose one of the options)*

Figure 4–18 *Invoking the* MTEXT *command from the Text toolbar*

When you drag the cursor after specifying the first corner of the rectangular boundary (referred to as the bounding box), AutoCAD displays an arrow within the rectangle to indicate the direction of the paragraph's text flow. After you specify the opposite corner of the bounding box, AutoCAD displays the Text Formatting toolbar and input (editing) area, shown in Figure 4–17.

Before you specify the opposite corner of the bounding box, you can choose one of the available property settings available from the shortcut menu, which include HEIGHT, JUSTIFY, LINE SPACING, ROTATION, STYLE, and WIDTH.

Choosing HEIGHT sets the character height in drawing units. Once the text height is specified, AutoCAD returns to the previous prompt to specify the opposite corner of the rectangular boundary.

Choosing JUSTIFY allows you to place text in one of the nine available justification points, similar to Single Line Text explained earlier.

Choosing LINE SPACING sets the spacing between two lines of text. Two options are available: AT LEAST and EXACTLY. The AT LEAST option adjusts lines of text automatically based on the height of the largest character in the line. The EXACTLY option forces the line spacing to be the same for all lines of text in the multiline text object.

Choosing ROTATION sets the rotation angle for new or selected text, in the current unit of angle measurement.

Choosing STYLE allows you to select one of the available styles for the new text.

Choosing WIDTH sets the paragraph width for new text. Once the width is specified, AutoCAD returns to the previous prompt to specify the opposite corner of the rectangular boundary.

After you specify the opposite corner of the bounding box, AutoCAD displays the Text Formatting toolbar and input (editing) area. Enter the text in the text input (editing) bounding box. If you right-click in the bounding box, AutoCAD displays a shortcut menu with options that include UNDO, REDO, CUT, COPY, PASTE, INDENTS AND TABS, JUSTIFICATION, FIND AND REPLACE, SELECT ALL, CHANGE CASE, AUTOCAPS, REMOVE FORMATTING, COMBINE PARAGRAPHS, SYMBOL, IMPORT TEXT, and HELP.

To edit all or part of the text in the bounding box, highlight the text to be edited. The text can be highlighted in the following three ways: holding down the pick button while dragging across the selected text, double-clicking to select an entire word, or triple-clicking to select an entire line of text.

Text Formatting Toolbar

The options provided on the Text Formatting toolbar (see Figure 4–17) include character formatting: **Style, Font, Text Height, Bold, Italic, Underline, Undo, Redo, Stack, Color, Ruler, OK,** and **Options**, a button to display an abbreviated Text Editor shortcut menu.

Choosing **Style** allows you to apply an existing style to new text or selected text. If you apply a new style to an existing text object, AutoCAD overrides character formatting such as font, height, bold, and italic attributes. Styles that have backwards or upside-down effects are not applied.

Choosing **Font** allows you to specify a font for new text or changes the font of selected text. All of the available TrueType fonts and SHX fonts are listed in the drop-down list box.

Choosing **Text Height** sets the character height in drawing units. The default value for the height is based on the current style. (A detailed discussion is provided later in the chapter on creating or modifying a text style.) If the current style is set to 0, then the value of the height is based on the value stored in the TEXTSIZE system variable. Each multiline text object can contain a text string of varying text size. When you highlight the text string in the dialog box, AutoCAD displays the selected text height in the list box. If necessary, you can specify a new height in addition to those listed.

Choosing **Bold** allows you to turn bold formatting ON and OFF for new text or selected text and is available only for the characters that belong to the TrueType font.

Choosing **Italic** allows you to turn italic formatting ON and OFF for new text or selected text and is available only for the characters that belong to the TrueType font.

Choosing **Underline** allows you to turn underlining ON and OFF for new text or selected text.

Choosing **Undo** undoes the last edit action in the Multiline Text Editor dialog box that includes changes in the content of the text string or formatting.

Choosing **Redo** cancels the previous Undo.

Choosing **Stack** allows you to place one part of a selected group of text over the remaining part. Before you choose Stack, the selected text must contain a forward slash (/) to separate the top part (to the left of the /) from the bottom part (to the right of the /). The slash will cause the horizontal bar to be drawn between the upper and lower parts, necessary for fractions with center justification. Instead of a slash (/), you can use the caret (^) symbol. In this case, AutoCAD will not draw a horizontal bar between the upper and lower parts with left justification, which is useful for placing tolerance values.

Choosing **Color** sets the color for new text or changes it for the selected text. You can assign the color as Bylayer, Byblock, or one of the available colors.

Choosing **Ruler** causes the ruler at the top of the bounding box to be displayed or hidden.

Choosing **OK** accepts the text in the boundary box and terminates the MLINE command.

Choosing **Options**, at the right end of the Text Formatting toolbar, causes an abbreviated Text Editor shortcut menu to be displayed that includes some of the buttons listed in the Text Editor shortcut menu described below under Text Editor shortcut menu.

Options Toolbar

The Options toolbar attached to the bottom of the Text Formatting toolbar includes buttons for quick access to some of the options that are also available in the Text Editor shortcut menu. Options included on the Options toolbar are **Left, Center,** and **Right** justification, **Top, Middle,** and **Bottom** justification, **Numbering** (lists), **Bullets** (lists), and **Uppercase Letters** (lists), **Insert Field, Uppercase, Lowercase, Overline, Symbol, Oblique Angle, Tracking,** and **Width Factor**.

Choosing **Left, Center,** or **Right** sets the horizontal justification for new text or selected text.

Choosing **Top, Middle,** or **Bottom** sets the vertical justification for new text or selected text.

Choosing **Numbering** causes AutoCAD to establish a numbered list beginning with the number 1 for new text or selected text.

Choosing **Bullets** causes AutoCAD to establish a bulleted list for new text or selected text.

Note: Numbered, lettered, and bulleted lists can be created in the multiline text object. AutoCAD, with the Auto-list option, automatically converts text lines in the specified type of list and increments the numbers and letters appropriately. If one line in the list is removed or added, the numbers or letters of the lines that follow are adjusted accordingly. If you enter a special character, such as an asterisk or hyphen, it will be automatically repeated in subsequent bulleted lines.

You can access the NUMBERED, LETTERED, and BULLETED lists from the BULLETS AND LISTS flyout in the Text Editor shortcut menu, which has the options RESTART, CONTINUE, and AUTO-LIST.

Choosing RESTART causes the current numbered or lettered line in a list to start over with a "1" or "A" respectively.

Choosing CONTINUE causes the current numbered or lettered line in a list to start with the next number or letter. This is necessary if creation of the lines needs to be continued from where it was interrupted.

Choosing AUTO-LIST when you have started a line with a letter or number followed by a period automatically converts it to a numbered or lettered list designation. For example, if you start a line with "5." or "C." with the Auto-list option on, the next line will begin with "6." or "D." respectively.

Choosing UPPERCASE LETTERS causes AutoCAD to establish an alphabetized list using uppercase letters beginning with the letter A for new text or selected text. For a list using lowercase letters, right-click and choose BULLETS AND LISTS, LETTERED and then choose LOWERCASE from the shortcut menu flyout.

Choosing INSERT FIELD causes the Field dialog box (see Figure 4–19) to be displayed, from which you can select one of the available fields to be inserted in the text at the cursor.

Figure 4–19 *The Field dialog box shown with Field category "All" and Field "Author" selected*

The **Field category** list box allows you to filter the fields that are displayed in the **Field names** list box into the following categories: All, Date & Time, Document, Linked, Objects, Other, Plot, and SheetSet. The **Field names** list box is visible for all categories, but the other text boxes vary with the category. The **Field names** list box allows you to select from the list of available fields according to the filter selected in the **Field category** list box.

Choose All from the **Field category** list box to list all the available fields in the **Field names** list box.

Choose Date & Time from the **Field category** list box to list fields related to the current drawing: CreateDate, Date, PlotDate, and SaveDate. The text boxes that appear when the Date & Time category is selected are **Date format** and **Examples**. The Date format text box displays the format of the example that is selected in the **Examples** text box. For example, if you choose 1/12/05 in the **Examples** text box, then M/d/yy is displayed in the **Date format** text box.

Choose Document from the **Field category** list box to list fields related to working in a document: Author, Comments, Filename, Filesize, HyperlinkBase, Keywords, LastSavedBy, Subject, and Title. The top right text box that appears when the

Document category is selected corresponds to the field selected in the Field names list box. The items listed in the **Format** list box are (none), Uppercase, Lowercase, First capital, and Title case, except the Filesize item, which has Bytes, Kilobytes, and Megabytes listed.

Choose Linked from the **Field category** list box to list fields related to text display for the hyperlinks. The top right text box that appears when the Linked category is selected is **Text to display,** which allows to you enter a text string that will be inserted in the multiline text box and a link to the specified hyperlink when selected. Choosing **Hyperlink** causes the Insert Hyperlink dialog box to be displayed. Creation of Hyperlinks is explained in Chapter 14.

Choose Objects from the **Field category** list box to list fields related to named objects: BlockPlaceholder, Formula, NamedObject, and Object.

Choose Other from the **Field category** list box to list fields related to SystemVariable and DieselEexpression.

Choose Plot from the **Field category** list box to list fields related to plot-related variables: DeviceName, Login, PageSetupName, PaperSize, PlotDate, PlotOrientation, PlotScale, and PlotStyleTable.

Choose SheetSet from the **Field category** list box to list fields related to sheet set variables: CurrentSheetCategory, CurrentSheetCustom, CurrentSheetDescription, CurrentSheetIssuePurpose, CurrentSheetNumber, CurrentSheetNumberAndTitle, CurrentSheetRevisionDate, CurrentSheetRevisionNumber, CurrentSheetSet, CurrentSheetSetCustom, CurrentSheetSetDescription, CurrentSheetSetProjectMilestone, CurrentSheetSetProjectName, CurrentSheetSetProjectNumber, CurrentSheetSetProjectPhase, CurrentSheetSubSet, CurrentSheetTitle, SheetSet, SheetSetPlaceholder, SheetView.

Choosing UPPERCASE causes new text or selected text to be drawn in uppercase letters.

Choosing LOWERCASE causes new text or selected text to be drawn in lowercase letters.

Choosing OVERLINE causes new text or selected text to be drawn with an overscore line over the letters.

Choosing SYMBOL causes a shortcut menu to be displayed from which symbols for DEGREES, PLUS/MINUS, and DIAMETER can be selected along with the NON-BREAKING SPACE. Also on the shortcut menu is the OTHER option, which causes the Character Map dialog box to be displayed. The Character Map dialog box operates in the same manner as the similar dialog box in other Windows-based word processors and text-handling programs. In addition to the options provided in the Multiline Text Editor dialog box for drawing special characters, you can draw them by means of the control characters. The control characters for a symbol begin with a double percent sign (%%).

The next character you enter represents the symbol. The control sequences defined by AutoCAD are presented in the following table.

Control Character Sequences for Drawing Special Characters and Symbols

Special Character or Symbol	Control Character Sequence	Text String	Example of Control Character Sequence
° (degree symbol)	%%d	104.5°F	104.5%%dF
± (plus/minus tolerance symbol)	%%p	34.5±3	34.5%%p3
Ø (diameter symbol)	%%c	56.06Ø	56.06%%c
% (single percent sign; necessary only when it must precede another control sequence)	%%%	34.67%±1.5	34.67%%%%%P1.5
Special coded symbols (where nnn stands for a three-digit code)	%%nnn	@	%%064

Choosing OBLIQUE ANGLE allows you to set the obliquing angle of the text. If it is set to 0 (zero) degrees, the text is drawn upright (in AutoCAD, 90 degrees). A positive value slants the top of the characters toward the right, or in the clockwise direction. A negative value slants the characters in the counterclockwise direction.

Choosing TRACKING allows you to decrease or increase the space between the selected characters.

Choosing WIDTH FACTOR allows you to set the character width relative to text height. If it is set to more than 1.0, the text widens; if set to less than 1.0, it narrows.

Ruler

At the top of the bounding box is a ruler similar to those in word processors. The ruler allows you to set indents and tabs and adjust the width of the bounding box. To set indents, use the arrows on the left of the ruler. To set the indent of the first line of a paragraph, drag the top arrow to the desired point. To set the indent of the rest of the lines in the paragraph, drag the bottom arrow to the desired point. To set a tab, specify a point on the ruler. To change the width of the bounding box, drag the double arrow on the right end of the ruler to the desired width. You can also right-click in the ruler and select INDENTS AND TABS or SET MTEXT WIDTH from the shortcut menu. These menu items open dialog boxes in which numeric values can be entered for setting indents, tabs, and bounding-box width.

Text Editor Shortcut Menu

In addition to the options available on the toolbars, you can also access additional options from the shortcut menu available within the Text Editor (see Figure 4–20).

Figure 4–20 *Text Editor shortcut menu*

UNDO, REDO, CUT, COPY, and PASTE operate in the same manner as similar commands in other Windows-based word processors and text-handling programs.

Choosing LEARN ABOUT MTEXT causes the New Features Workshop dialog box to be displayed from which you can access help and information about the MTEXT command.

Choosing SHOW TOOLBAR, SHOW OPTIONS, or SHOW RULER turns the corresponding toolbars and Ruler ON and OFF.

Choosing OPAQUE BACKGROUND allows you to turn the opaque background ON and OFF inside the bounding box.

Choosing IMPORT TEXT causes the Select File dialog box to be displayed, from which a file of ASCII or RTF format can be imported. The file imported is limited to 32K.

Choosing FIND AND REPLACE causes the Find and Replace dialog box to be displayed (see Figure 4–21), which includes the **Find what** and **Replace with** text boxes, and the **Match Case** and **Match whole word only** check boxes. These are used to search for specified text strings and replace them with new text.

Enter the text string to search for in the **Find what** text box, and then choose the **Find Next** button to start the search. AutoCAD highlights the appropriate text string in the bounding box. To continue the search, choose the **Find Next** button again.

Figure 4–21 *Find and Replace dialog box*

In the **Replace with** text box, type the text string that you want as a replacement for the text string in the **Find what** text box. Then choose **Replace** to replace the highlighted text with the text in the **Replace with** text box. If you choose **Replace All**, all instances of the specified text will be replaced.

Selecting **Match Case** causes AutoCAD to find text only if the case of all characters in the text object matches the text characters in the **Find what** text box. When it is not selected, AutoCAD finds a match for the specified text string regardless of the case of the characters.

Selecting **Match whole word only** causes AutoCAD to find text only if the text string is a single word. If the text is part of another text string, it is ignored. When it is not selected, AutoCAD finds a match for the specified text string whether it is a single word or part of another word.

The SELECT ALL option, when selected, causes all the text in the Multiline text object to be selected and highlighted.

The CHANGE CASE option, when selected, causes a shortcut menu to be displayed with options for UPPERCASE and LOWERCASE. Selecting UPPERCASE causes any selected and highlighted text to be uppercase. Selecting LOWERCASE causes any selected and highlighted text to be lowercase.

The AUTOCAPS option, when selected, operates in a manner similar to pressing the CAPS LOCK key, toggling the CAPS LOCK key on and off.

The REMOVE FORMATTING option, when selected, removes any bold, italics, or underline formatting from selected text.

The COMBINE PARAGRAPHS option, when selected, combines selected paragraphs into a single paragraph and replaces each paragraph return with a space.

The CHARACTER SET option, when selected, displays a shortcut menu from which you can select from the following character sets: Central Europe, Cyrillic, Hebrew, Arabic, Baltic, Greek, Turkish, Vietnamese, Japanese, Korean, CHINESE_GB2312, CHINESE_BIG5, Western, and Thai.

EDITING TEXT

Several methods are available to edit existing text objects. You can choose Edit from the Text toolbar, and the Text Formatting toolbars will be displayed when you select an Mtext object; or you can double-click on an Mtext object, and the Text Formatting toolbars will be displayed for that Mtext object. You can select a text object while the Properties palette is displayed and change the values of the properties in the appropriate text boxes. From the Text toolbar you can initiate a global Find and Replace, or change the Text Style, Scale, Justification, and Convert distances between spaces (Model and Paper). From the Tools menu you can invoke the spell checking feature. While most of the features used to edit text have been explained in the preceding section on creating text, the following features apply primarily to editing text.

The DDEDIT command allows you to edit text and attributes in place. An attribute is informational text associated with a block. See Chapter 10 for a detailed discussion of blocks and attributes.

Invoke the DDEDIT command by choosing Edit from the Text toolbar (see Figure 4–22) and AutoCAD prompts:

Select an annotation object or ⊥: *(select the text or attribute definition)*

Figure 4–22 *Invoking the* DDEDIT *command from the Text toolbar*

If you select a text string created by means of a TEXT command, AutoCAD highlights the selected text with a contrasting background. Make the necessary changes in the text string and press ENTER to accept the changes.

If instead, you select text created by means of the MTEXT command, AutoCAD displays the Multiline Text Formatting toolbar shown in Figure 4–23. Make the necessary changes in the text string, and select **OK** to accept the changes.

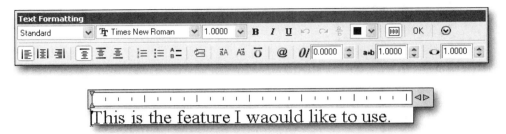

Figure 4–23 *Multiline Text Formatting toolbar, Options toolbar, Ruler, and Bounding box*

AutoCAD continues to prompt you to select a new text string to edit, or you can enter U to undo the last change made to the text. To terminate the command sequence, press ENTER.

FINDING AND REPLACING TEXT

The FIND command is used to find a string of specified text and replace it with another string of specified text. Invoke the FIND command from the Text toolbar (see Figure 4–24) and AutoCAD displays the Find and Replace dialog box, similar to Figure 4–25.

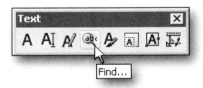

Figure 4–24 *Invoking the FIND command from the Text toolbar*

In the **Find text string** text box, enter the string that you wish to replace. In the **Replace with** text box, enter the new text string. Choose **Find** and an instance of the string to be replaced is displayed in the **Context** area of the **Search results** section of the dialog box. Select **Replace** to replace this instance with the string in the **Replace with** text box. To skip over this instance without replacing it, select **Find**. To replace all instances of the string entered in the **Find text string** text box, select **Replace All**.

Figure 4–25 *Find and Replace dialog box*

In the **Search in** text box, you can direct AutoCAD to search either in the Entire drawing, or the Current selection, for the string to be replaced. The **Select objects** button returns you to the graphics screen to select text objects. Then AutoCAD searches for the string to be replaced. Selecting **Options** causes the Find and Replace Options dialog box to appear, similar to Figure 4–26.

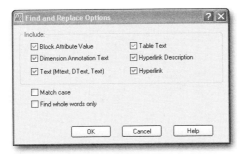

Figure 4–26 *Find and Replace Options dialog box*

The **Include** section of the Find and Replace Options dialog box lets you filter the type of text to be included in the Find search. Optional categories for text to be included are: Block Attribute Value, Dimension Annotation Text, Text (Mtext, Dtext, Text), Table Text, Hyperlink Description, and Hyperlink. The **Match case** check box, when checked, causes AutoCAD to include text strings whose case matches that of the specified text string to search. The **Find whole words only** check box, when checked, causes AutoCAD to include only whole words that match the specified text string.

JUSTIFYING TEXT

The JUSTIFYTEXT command lets you change the justification point of a text string without having to change its location.

Invoke the JUSTIFYTEXT command by selecting Justify from the Text toolbar (see Figure 4–27) and AutoCAD prompts:

> Select objects: *(select a text object)*
> Select objects: (ENTER)
> Enter a justification option *(select one of the available options to change the justification)*

Figure 4–27 *Invoking the* JUSTIFYTEXT *command from the Text toolbar*

CHANGING TEXT FROM ONE SPACE TO ANOTHER

The SPACETRANS command converts distances between model space units and paper space units. By using SPACETRANS transparently, you can provide commands with distance entries relative to another space. For example, if you wish to create a text object in model space that matches the height of text in a layout, enter the following from model space (this command cannot be executed in the model tab):

> **text**
> Specify start point of text or ⬇ **6.5,2.75**
> Specify height <0.20>: **'spacetrans**
> Specify paper space distance <1.000>: **1/2**
> Specify rotation angle of text <0>: (ENTER)

When the command is complete, a text object is created in model space with a height of 0.875, which appears as 1/2 when viewed from a layout.

SPELL-CHECKING

The SPELL command is used to correct the spelling of text objects created with the TEXT or MTEXT command in addition to attribute values in Blocks.

Invoke the SPELL command from the Tools menu and AutoCAD prompts:

> Select objects: *(select one or more text strings, and press* ENTER *to terminate object selection)*

AutoCAD displays the Check Spelling dialog box (see Figure 4–28), only when it finds a dubious word in the selected text objects.

AutoCAD displays the name of the current dictionary in the top of the Check Spelling dialog box. If necessary, you can change to a different dictionary by choosing **Change Dictionaries** and selecting the appropriate dictionary from the Change Dictionaries dialog box.

AutoCAD displays each misspelled word in the **Current Word** section and lists the suggested alternate spellings in the **Suggestions** list box. Click **Change** to replace the current word with the selected suggested word, or click **Change All** to replace all instances of the current word. Alternatively, click **Ignore** to skip the current word, or click **Ignore All** to ignore all subsequent entries of the current word.

Choosing **Add** allows you to include the current word (up to 63 characters) in the current or custom dictionary.

Choosing **Lookup** allows you to check the spelling of the word in the **Suggestions** box.

Figure 4–28 *Check Spelling dialog box*

After completion of the spelling check, AutoCAD displays an AutoCAD message informing you that the spelling check is complete.

LINE SPACING

Line spacing in an Mtext object can be changed by selecting the Mtext object while the Properties palette is displayed. Line spacing can be changed by entering a new value for either of the two options: Line Space Factor, which sets the line spacing to a multiple of single-line spacing (single spacing is 1.66 times the height of the text character) and Line Space Distance, which sets the line spacing to an absolute value measured in drawing units. You can also change the line spacing by choosing one of the two options for Line Space Style: At Least (default setting), which adjusts lines of text automatically based on the height of the largest character in the line (more space is added between lines of text with taller characters), and Exactly, which forces the line spacing to be the same size for all lines of text regardless of format differences, such as font or text height.

CONTROLLING THE DISPLAY OF TEXT

The QTEXT command is a utility command for TEXT and MTEXT that is designed to reduce the redraw and regeneration times of a drawing. Regeneration time becomes a significant factor if the drawing contains a great amount of text and attribute information and/or if a fancy text font is used. Using QTEXT, the text is replaced with rectangular boxes of a height corresponding to the text height. These boxes are regenerated in a fraction of the time required for the actual text.

If a drawing contains many text and attribute items, it is advisable to set QTEXT to ON. However, before plotting the final drawing, or inspection of text details, the QTEXT command is set to OFF and is followed by the REGEN command.

Invoke the QTEXT command: **qtext** (ENTER) and AutoCAD prompts:

Enter mode *(select one of the options)*

CREATING AND MODIFYING TEXT STYLES

The *STYLE* option of the TEXT and MTEXT commands (in conjunction with the STYLE command) lets you determine how text characters and symbols appear, other than the usual adjusting of height, slant, and angle of rotation. To specify a text style from the *STYLE* option of the TEXT and MTEXT commands, it must have been defined by using the STYLE command. In other words, the STYLE command creates a new style or modifies an existing style. The *STYLE* option under the TEXT or MTEXT command allows you to choose a specific style from the styles available.

There are three things to consider in creating a new style with the STYLE command. First, you must name the newly defined style. Style names may contain up to 255 characters, numbers, and special characters ($, –, and _). Names like "title block," "notes," and "bill of materials" can remind you of the purpose for which the particular style was designed.

Second, you may apply a particular font to a style. The font that AutoCAD uses as a default is called TXT. It has blocky looking characters, which are economical to store in memory. But the TXT.SHX font, made up entirely of straight-line (non-curved) segments, is not considered as attractive or readable. Other fonts offer many variations in characters, including those for foreign languages. All fonts are stored for use in files of their font name with an extension of .SHX. The most effective way to get a distinctive appearance in text strings is to use a specially designed font. You can also use TrueType and Type 2 PostScript fonts. See Appendix F for a list of fonts that come with AutoCAD. If necessary, you can buy additional fonts from third-party vendors. AutoCAD can also read hundreds of PostScript fonts available in the marketplace.

The third consideration of the STYLE command is in how AutoCAD treats the general physical properties of the characters, regardless of the font that is selected. These properties are the height, width-to-height ratio, obliquing angle, backwards, upside-down, and orientation (horizontal/vertical) options.

Invoke the STYLE command from the Text toolbar and AutoCAD displays the Text Style dialog box (see Figure 4–29).

Choose **New** to create a new style. AutoCAD displays the New Text Style dialog box (see Figure 4–30). Enter an appropriate name for the text style and choose **OK** to create the new style.

To rename an existing style, first select the style from the **Style Name** list box in the New Text Style dialog box, and then choose **Rename**. AutoCAD displays the Rename Text Style dialog box. Make the necessary changes in the name of the style, and choose **OK** to rename the text style.

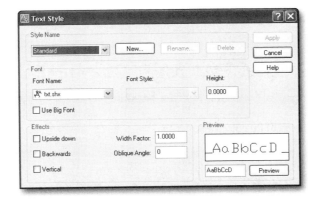

Figure 4–29 *Text Style dialog box*

Figure 4–30 *New Text Style dialog box*

To delete an existing style, first select the style from the **Style Name** list box in the Text Style dialog box, and then choose **Delete**. AutoCAD displays the AutoCAD Alert dialog box to confirm the deletion of the selected style. Click **Yes** to confirm the deletion or **No** to cancel the deletion of the selected style.

To assign a font to the selected text style, select the appropriate font from the **Font Name** list box. Similarly, select a font style from the **Font Style** list box. The font style specifies font character formatting, such as italic, bold, or regular.

The **Height** text box sets the text height to the value you enter. If you set the height to 0 (zero), then when you use this style in the TEXT or MTEXT command, you are given an opportunity to change the text height with each occurrence of the command. If you set it to any other value, then that value will be used for this style and you will not be allowed to change the text height.

Selecting **Backwards** or **Upside Down** controls whether the text is drawn right to left (with the characters backward) or upside down (left to right), respectively (see Figure 4–31).

Figure 4–31 *Examples of Backward and Upside-down text*

Selecting **Vertical** controls the display of the characters aligned vertically. The Vertical option is available only if the selected font supports dual orientation. See Figure 4–32 for an example of vertically oriented text.

```
T        E        V        T
H        X        E        E
I        A        R        X
S        M        T        T
         P        I
I        L        C
S        E        A
                  L
A        O
N        F
```

Figure 4–32 *Example of vertically oriented text*

The **Width Factor** text box sets the character width relative to text height. If it is set to more than 1.0, the text widens; if set to less than 1.0, it narrows.

The **Oblique Angle** text box sets the obliquing angle of the text. If it is set to 0 (zero) degrees, the text is drawn upright (in AutoCAD, 90 degrees). A positive value slants the top of the characters toward the right, or in the clockwise direction. A negative value slants the characters in the counterclockwise direction. See Figure 4–33 for examples of oblique angle settings applied to a text string.

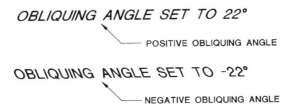

Figure 4–33 *Example of oblique angle settings applied to a text string*

The **Preview** section of the Text Style dialog box displays sample text that changes dynamically as you change fonts and modify the effects. To change the sample text, enter characters in the box below the larger preview image.

After making the necessary changes in the Text Style dialog box, choose **Apply** to apply the changes. Choose **Close** to close the Text Style dialog box.

CREATING AND MODIFYING TABLES

AutoCAD's TABLE command makes it easy to create tables that contain text in the row-and-column format customarily found on drawings for listing revisions, finish schedules, specifications, and other structured textual information. A combination of table characteristics such as row and column sizes, border lineweights, text alignments and associated text styles, and colors can be saved in a Table Style with a specified name to be recalled and applied to tables when required.

INSERTING TABLES

Figures 4–34 and 4–35 show typical examples of tabular information included in drawings. After you invoke the TABLE command from the Draw toolbar (see Figure 4–36), AutoCAD displays the Insert Table dialog box (see Figure 4–37).

ROOM FINISH SCHEDULE				
ROOM NO.	ROOM DESCRIP	WALLS	FLOOR	CEILING
100	ENTRY	PLASTER	CARPET	SUSP. ACOUS.
101	HALL	GYPSUM	CARPET	GYPSUM
102	RECEPTION	PANELLING	VINYL TILE	GYPSUM
103	OFFICE	GYPSUM	CARPET	GYPSUM

Figure 4–34 *Example of Room Finish Schedule drawn as table from top down*

3	CLOSET DIM	T.A.S.	01/23/04
2	WDW DETAIL	G.V.K.	01/12/04
1	DOOR 101	A.B.C.	01/08/04
REVISION NO.	DESCRIPTION	BY	DATE
REVISIONS			

Figure 4–35 *Example of Revision History drawn as table from bottom up*

Figure 4–36 *Invoking the TABLE command from the Draw toolbar*

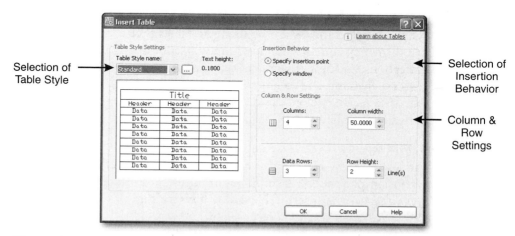

Figure 4–37 *The Insert Table dialog box*

The **Table Style name** text box in the **Table Style Settings** section of the Insert Table dialog box enables you to select the table style to be applied. Refer to Chapter 13 for information about creating and modifying table styles.

The **Insertion Behavior** section of the Insert Table dialog box allows you to specify which characteristics of the table control insertion and determines how selections can be made in the **Column & Row Settings** section.

Choosing **Specify insertion point** creates a table with reference to the location of the upper left corner of the table (default table style) and based on the number of the columns, column width, number of data rows, and row height specified in the **Column & Row Settings** section of the dialog box. If the table style calls for the direction of the table to read from the bottom up, the insertion point is the lower left corner of the table.

Choosing **Specify window** creates a table based on the constraints set in the **Column & Row Settings** section of the table. You can specify the number of columns with the column width automatically set, or you can specify the column width with the number of columns automatically set. You can specify the number of **Data Rows** with the **Row Height** automatic, or you can specify the number of lines for the Row Height and the number of **Data Rows** automatic.

After setting up the Table Style and Column/Row configuration, choose **OK** to close the Insert Table dialog box. AutoCAD then prompts you to "Specify first corner:" and shows a phantom image of the potential table attached to the cursor, following its movement.

Once you specify the insertion point, if you have chosen **Specify insertion point** in the **Insertion Behavior** section, AutoCAD draws the table, shows the text-entry cursor in the heading at the justification specified, and displays the Text Formatting toolbar. If you have chosen **Specify window** in the **Insertion Behavior** section, AutoCAD prompts you to "Specify second corner:", allowing you to drag the cursor to determine the number of columns and rows desired. Then AutoCAD displays the Text Formatting dialog box.

While the Text Formatting toolbar is displayed, text can be input into each cell by typing or pasting from the Clipboard. You can move from a cell to the cell below it by pressing the ENTER key, to the cell above by pressing SHIFT+ENTER, to the cell on the right by pressing the TAB key, or to the cell on the left by pressing SHIFT+TAB.

EDITING TEXT IN A CELL

You can edit text by selecting the cell in which the text has to be changed and choosing EDIT CELL TEXT from the shortcut menu. AutoCAD displays the Text Formatting toolbar; make the necessary changes and choose **OK** to close the text editor.

Additional options are available from the shortcut menu when text is selected in a cell. The options include the following:

Choosing CELL ALIGNMENT allows you to change the alignment of the selected text in a cell.

Choosing CELL BORDERS sets the properties of the borders of table cells.

Choosing MATCH CELL applies the properties of a selected table cell to other table cells.

Choosing INSERT BLOCK allows you to insert a block or drawing that is stored locally or in a network.

Choosing INSERT COLUMNS allows you to insert a column right or left of the selected cell.

Choosing DELETE COLUMNS deletes the selected column.

Choosing INSERT ROWS allows you to insert a row above or below the selected cell.

Choosing DELETE ROWS deletes the selected row(s).

Choosing DELETE CELL CONTENTS deletes the text objects in the selected cell(s).

MODIFYING TABLES

Similar to modifying the individual cells, AutoCAD allows you to modify the table. First select the table, and then select the available options from the shortcut menu. The options include the following:

Choosing SIZE COLUMNS EQUALLY resizes the columns to equal width.

Choosing SIZE ROWS EQUALLY resizes the rows to equal width.

Choosing REMOVE ALL PROPERTY OVERRIDES removes any property overrides applied to the selected table.

Choosing EXPORT allows you to export all the text objects in the table to a text file in *.csv* format.

CREATING AND MODIFYING TABLE STYLES

The TABLESTYLE command is used to create a new table style or modify an existing one. The appearance of the table is controlled by its table style. The table style can specify a different justification and appearance for the text and gridlines for the title, column heads, and data. The Insert Table dialog box has a style menu from which you can select available table styles.

To create a new table style, invoke the TABLESTYLE command by selecting Table Style Manager from the Styles toolbar (see Figure 4-38). AutoCAD then displays the Table Style dialog box, as shown in Figure 4-39.

Figure 4–38 *Invoking the* TABLESTYLE *command from the Styles toolbar*

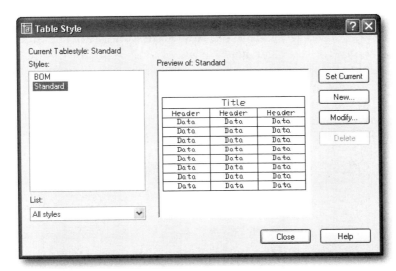

Figure 4–39 *Table Style dialog box*

The **Current Tablestyle** heading shows the name of the table style to which the next drawn table will confirm.

The **Styles** section displays the name(s) of the current style(s) and any other style(s) available depending on the option chosen in the **List** selection box.

From the **List** selection box you can choose to list all of the styles available in the drawing or only those in use.

The **Preview of** window shows how the table will appear using the current table style.

Choosing **Set Current** causes the table style chosen in the **Styles** section to become the current table style. To delete a table style, select the style in the **Styles** section and choose **Delete**. The Standard style cannot be deleted.

To create a new table style, choose **New**. AutoCAD displays the Create New Table Style dialog box, similar to Figure 4-40. You can create a new table style based on an existing table style whose settings are the default for the new table style.

Figure 4–40 *Create New Table Style dialog box*

Specify the name of the new style in the **New Style Name** box. The **Start With** list box lets you choose the existing table style whose settings are the default for the new table style. Choose **Continue** to display the New Table Style dialog box, similar to Figure 4-41, in which you define the new style properties. There are three tabs to choose from: **Data**, **Column Heads**, and **Title**. Options on each tab set the appearance of the data cells, the column heads, or the table title. The dialog box initially displays the properties of the table style that you selected as the **Start With** style.

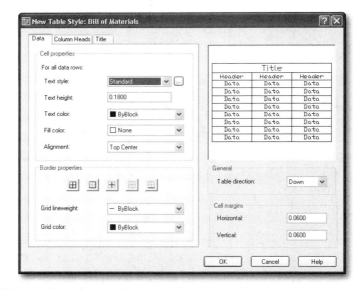

Figure 4–41 *New Table Style dialog box with the Data tab selected*

Depending on which tab is active, the **Cell Properties** section allows you to set the Text style, Text height, Text color, Fill color, and Alignment for the text in the applicable cell.

The **Border properties** section allows you to set the **Grid lineweight** and **Grid color** for the cells.

The **General** section allows you specify whether the table has the title at the top (the Down option) or the bottom (the Up option) and determines the direction in which additional rows are added.

The **Cell margins** section allows you to specify the size of the horizontal and vertical margins for the applicable cells.

The preview window displays how a table conforming to the selected properties will appear. After setting up the table style and cell and border properties, choose **OK** to create a new table style and close the New Table Style dialog box.

To modify an existing table style, first select the style to modify from the Styles list and choose **Modify**. AutoCAD displays the Modify Table Style dialog box. The content of the Modify Table Style dialog boxes is identical to the New Table Style dialog box, however you are modifying an existing table style rather than creating a new one.

CREATING OBJECTS FROM EXISTING OBJECTS

AutoCAD not only allows you to draw objects easily, but also allows you to create additional objects from existing objects. This section discusses six important commands that will make your job easier: COPY, ARRAY, OFFSET, MIRROR, FILLET, and CHAMFER.

COPYING OBJECTS

The COPY command places copies of the selected objects at the specified displacement, leaving the original objects intact. The copies are oriented and scaled the same as the original. If necessary, you can make multiple copies of selected objects. Each resulting copy is completely independent of the original and can be edited and manipulated like any other object.

Invoke the COPY command from the Modify toolbar (see Figure 4–42) and AutoCAD prompts:

>Select objects: *(select the objects and press* ENTER *to complete the selection)*
>Specify base point or ⊥ *(specify base point or right-click and choose one of the options)*
>Specify second point or <use first point as displacement>: *(specify a point for displacement, or press* ENTER *to use first point as displacement)*

Figure 4–42 *Invoking the* COPY *command from the Modify toolbar*

If you specify two data points, AutoCAD computes the displacement and places a copy accordingly. If you press ENTER in response to the second point of displacement, AutoCAD considers the point provided as the second point of a displacement vector with the origin (0,0,0) as the first point, indicating how far to copy the objects and in what direction.

The following command sequence shows an example of copying a group of objects (see Figure 4–43), by placing two data points:

>**copy** (ENTER)
>Select objects: *(select objects with Crossing selection option as shown in Figure 4–43)*
>Select objects: (ENTER)
>Specify base point or ⬇ *(specify the base point as shown in Figure 4–43)*
>Specify second point or <use first point as displacement>: *(specify the second point as shown in Figure 4–43)*
>Specify second point or ⬇ *(specify additional points to copy the selected objects and press ENTER to complete the command sequence)*

AutoCAD copies the selected object(s), placing them at a new location displaced from the location of the original objects at a direction and distance determined by the base point/second displacement point vector. By default, AutoCAD allows you to make multiple copies. To terminate the command sequence, press ESC or choose *EXIT* from the shortcut menu. To erase the most recent copy of the select objects, choose *UNDO* from the shortcut menu.

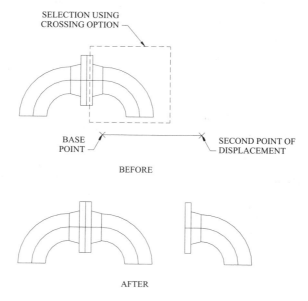

Figure 4–43 *Using the* COPY *command*

Note: The first and second points of displacement do not have to be on or near the object. For example, you can enter 1,1 and 3,4 as the first and second points respectively, causing the objects to be moved or copied 2 units in the X direction and 3 units in the Y direction. To have the same result, you could have entered 0,0 at the first point and 2,3 at the second prompt. Or if you were moving or copying the object 24" to the right, you could select a point on the screen for the base point and then enter **@24<0** for the displacement relative to the point specified. You can simplify specifying the move or copy displacement vector if you know how far in the X direction and how far in the Y direction you wish to move or copy the selected objects. To do this, enter the coordinates for the second point of displacement at the first prompt (Specify base point of displacement). Then you press ENTER in response to the second prompt (Specify second point of displacement or <use first point as displacement>). AutoCAD uses the origin (0,0,0) as the first point. This is a stroke-saving procedure. For the example at the beginning of this note, you could have entered **2,3** at the first prompt and pressed ENTER at the second prompt. It is as if you had entered **0,0,0** at the first prompt and **2,3** at the second prompt. If you wish to move the object in the direction opposite to that of the example, you could enter **-2,-3** at the first prompt and press ENTER at the second prompt.

CREATING A PATTERN OF COPIES

The ARRAY command is used to make multiple copies of selected objects in either rectangular or polar arrays (patterns). In the rectangular array, you can specify the number of rows, the number of columns, and the spacing between rows and columns (row and column spacing may differ). The whole rectangular array can be rotated at a selected angle. In the polar array, you can specify the angular intervals, the number of copies, the angle that the group covers, and whether or not the objects maintain their orientation as they are arrayed.

Invoke the ARRAY command from the Modify toolbar (see Figure 4–44) and AutoCAD displays the Array dialog box as shown in Figure 4–45.

Figure 4–44 *Invoking the* ARRAY *command from the Modify toolbar*

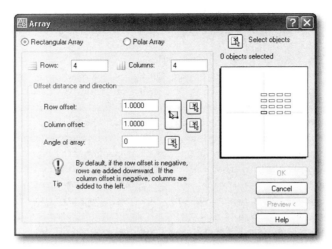

Figure 4–45 *Array dialog box*

In the Array dialog box, you can select the type of array (rectangular or polar). As mentioned earlier, for the rectangular type of array you can select the number and spacings of rows and columns of the array and the angle of the array and for the polar type of array you can select the center of the array, the angle through which objects are arrayed, the number of objects in the array and whether the arrayed objects are rotated as they are arrayed or they keep the orientation of the original object when arrayed.

Rectangular Array

In the Array dialog box, select **Rectangular Array** (see Figure 4–45). Enter the number of rows in the **Rows** text box and the number of columns in the **Columns** text box. At least one of them must be greater than one.

 Note: The numbers for rows and columns you enter include the original object. For an array of objects that will be four rows high and three columns wide the numbers four and three must be entered in the *Rows* and *Columns* text boxes. A positive number for the column and row spacing causes the elements to array toward the right and upward, respectively. Negative numbers for the column and row spacing cause the elements to array toward the left and downward, respectively.

The **Offset distance and direction** section of the Array dialog box allows you to set the value for the spacing between objects in rows and the value for the spacing between objects in columns.

The **Row offset** text box lets you set the value for the spacing between objects in rows. Or you can select the **Pick Row Offset** button, to the far right of the **Row offset** text box (with a single arrow and crossmark icon). This will allow you to specify the row offset (spacing) by selecting two points on the screen with the cursor. The two points can be in any direction. AutoCAD will use the distance between them for the row offset.

The **Column offset** text box lets you set the value for the spacing between objects in columns. Or you can select the **Pick Column Offset** button to the far right of the **Column offset** text box (with a single arrow and crossmark icon). This will allow you to specify the column offset (spacing) by selecting two points on the screen with the cursor. The two points can be in any direction. AutoCAD will use the distance between them for the row column.

 Note: The offset you specify determines the distance between corresponding points of adjacent objects and not the space between adjacent objects. For example, if you arrayed 3" diameter circles with 2" offsets, the adjacent circles would overlap 1".

You can also specify the row and column offsets in one maneuver by first selecting the large **Pick Both Offsets** button just to the right of the **Row offset** and **Column offset** text boxes. Then select two points on the screen with the cursor that specify the opposite corners of a rectangle called a unit cell. AutoCAD uses the width of the unit cell as the horizontal distance(s) between columns and the height as the vertical distance(s) between rows.

The array can be rotated by entering the angle in the **Angle of array** text box. Or you can select the **Pick Angle of Array** button to the right of the **Angle of array** text box (with a single arrow and crossmark icon). This will allow you to specify the rotation angle by selecting two points on the screen with the cursor. The two points can be in anywhere on the screen. AutoCAD will use the direction between them (measured from the first point selected to the second) for the angle that the array will be rotated.

The array preview window on the right of the dialog box shows the rows and columns selected. Choose **Select objects** to return to the drawing screen in order to select the objects to be arrayed. Once one or more objects are selected, the number of objects is shown above the array sample window and you can preview the resulting array on the screen by choosing **Preview**.

Once the settings have been satisfactorily specified, choose **OK**.

Polar Array

In the Array dialog box, select **Polar Array** (see Figure 4–46). Next to **Center point**, enter the *X* coordinate in the **X** text box and the *Y* coordinate in the **Y** text box for the center of the polar array. Or you can choose the button with the arrow and crossmark, which will return you to the drawing screen to specify the center point with your cursor.

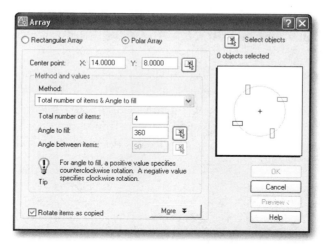

Figure 4–46 *Array dialog box—Polar array selection*

The **Method and values** section provides three methods to create polar arrays: **Total number of items & Angle to fill**, **Total number of items & Angle between items**, and **Angle to fill & Angle between items**. The method you select will determine which two of the three text boxes will be active. The boxes are the **Total number of items** text box, the **Angle to fill** text box, and the **Angle between items** text box.

The **Total number of items** text box lets you specify the number of items in the array. The total number of items includes the original item, just as in the rectangular array.

The **Angle to fill** text box lets you specify the total angle between the original item and the last item in the array. Note that a positive angle causes items to array in a counterclockwise direction. A negative angle causes them to array in a clockwise direction.

The **Angle between items** text box lets you specify the angle between individual items in the array.

The array preview window on the right of the dialog box shows a representative example of the number of items selected, the total angle of the array, whether the items are rotated as they are arrayed, and the angle between items. Choose **Select objects** to return to the drawing screen in order to select the objects to be arrayed. Once one or more objects are selected, the number of objects is shown above the array sample window and you can preview the resulting array on the screen by choosing **Preview**.

Choose **Rotate items as copied** to cause the items to each be rotated from the original item's orientation the same angle that the particular item is arrayed.

Choose **More** to display additional options in the dialog box.

The **Object base point** section lets you specify a new reference (base) point relative to the selected objects that will remain at a constant distance from the center point of the array as the objects are arrayed. AutoCAD uses the distance from the array's center point to a base point on the last object selected. The point used is determined by the type of object, as shown in the following table.

Type of Object	Base Point
Arc, circle, ellipse	Center point
Polygon, rectangle	First corner
Donut, line, polyline, 3D polyline, ray, spline	Starting point
Block, paragraph text, single-line text	Insertion point
Construction lines	Midpoint
Region	Grip point

Selecting **Set to Object's Default** causes AutoCAD to use the default base point of the object to position the arrayed object.

The **Base Point** text boxes allow you to set new *X* and *Y* base point coordinates. Or you can choose **Pick Base Point** to temporarily close the dialog box and specify a point with your cursor. After you specify a point, the Array dialog box is redisplayed.

Once the settings have been satisfactorily specified, choose **OK** .

The following example of values entered in their respective text boxes draws a rectangular array with four rows and six columns (see Figure 4–47).

Rows:	4
Columns:	6
Row offset:	1
Column offset:	1.5
Angle of array:	0

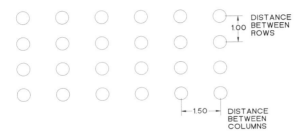

Figure 4–47 *Using the* ARRAY *command to place a rectangular array*

The following example of values entered in their respective text boxes draws a polar array with eight items and a 360° angle to fill (see Figure 4–48).

Center point:	(specify the center point as shown in Figure 4–48)
Method	**Total number of items & angle to fill**
Total number of items:	**8**
Angle to fill:	**360**
Rotate items as copied	(checked to rotate, cleared for non-rotated)

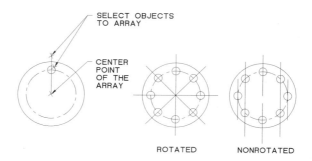

Figure 4–48 *Using the* ARRAY *command to place rotated and non-rotated polar arrays*

CREATING PARALLEL LINES, PARALLEL CURVES, AND CONCENTRIC CIRCLES

The OFFSET command creates parallel lines, parallel curves, and concentric circles relative to existing objects, (see Figure 4–49). Special precautions must be taken when using the OFFSET command to prevent unpredictable results from occurring when using the command on arbitrary curve/line combinations in polylines.

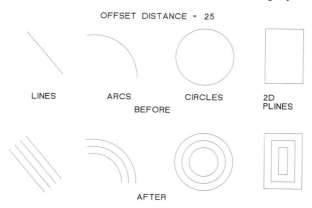

Figure 4–49 *Examples created using the* OFFSET *command*

Invoke the OFFSET command from the Modify toolbar (see Figure 4–50) and AutoCAD prompts:

> Specify offset distance or ⤓ *(specify offset distance, or right-click and choose one of the options)*
> Select object to offset: *(select an object to offset)*
> Specify point on side to offset or ⤓ *(specify a point to one side of the object to offset)*
> Select object to offset or ⤓ *(continue selecting additional objects for offset, and specify the side of the object to offset, or press ENTER to terminate the command sequence)*

Figure 4–50 *Invoking the* OFFSET *command from the Modify toolbar*

Offset Command Options

Choosing THROUGH from the shortcut menu causes AutoCAD to prompt for a through point and an object is created passing through the specified point.

Choosing ERASE from the shortcut menu causes AutoCAD to erase the source object after it is offset.

Choosing LAYER from the shortcut menu lets you determine whether offset objects are created on the current layer or on the layer of the source object.

Valid Objects to Offset

Valid objects include the line, spline curve, arc, circle, and 2D polyline. If you try to select another type of object, such as text, you will not be able to select the object. The object selected for offsetting must be in a plane parallel to the current coordinate system.

Offsetting Miters and Tangencies

The OFFSET command affects single objects in a manner different from a polyline made up of the same objects. Polylines whose arcs join lines and other arcs in a tangent manner are affected differently than polylines with nontangent connecting points. For example, in Figure 4–51 the seven lines are separate objects. When you specify a side to offset as shown, there are gaps and overlaps at the ends of the newly created lines.

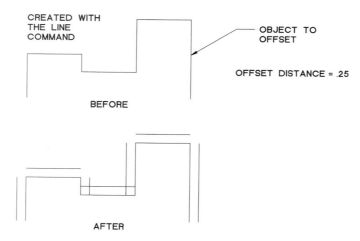

Figure 4–51 *Using the* OFFSET *command with single objects*

In Figure 4–52, the lines have been joined together (see PEDIT in Chapter 5) as a single polyline. See how the OFFSET command affects the corners where the new polyline segments join.

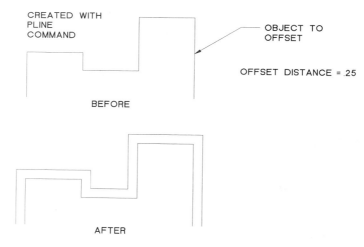

Figure 4–52 *Using the* OFFSET *command with polylines*

 Note: The results of offsetting polylines with arc segments that connect other arc segments and/or line segments in dissimilar (non tangent) directions might be unpredictable. Examples of offsetting such polylines are shown in Figure 4–53.

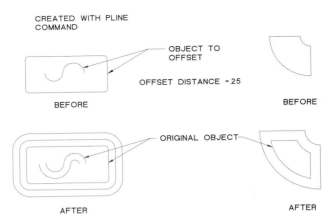

Figure 4–53 *Using the* OFFSET *command with non-tangent arc and/or line segments*

If you are not satisfied with the resulting polyline configuration, you can use the PEDIT command to edit it. Or, you can explode the polyline and edit the individual segments.

CREATING A MIRROR COPY OF OBJECTS

The MIRROR command creates a copy of selected objects in reverse, that is, mirrored about a specified line. Invoke the MIRROR command from the Modify toolbar (see Figure 4–54) and AutoCAD prompts:

> Select objects: *(select the objects and then press* ENTER *to complete the selection)*
> Specify first point of mirror line: *(specify a point to define the first point of the mirror line)*
> Specify second point of mirror line: *(specify a point to define the second point of the mirror line)*
> Erase source objects? ⏎ *(enter* **y**, *for yes to delete the original objects, or* **n**, *not to delete the original objects, that is, to retain them)*

Figure 4–54 *Invoking the* MIRROR *command from the Modify toolbar*

The first and second points of the mirror line become the end points of an imaginary line about which the selected objects will be mirrored.

The following command sequence shows an example of mirroring a group of selected objects by means of the Window option, as shown in Figure 4–55:

mirror
Select objects: *(specify Point 1 to place one corner of a window)*
Specify opposite corner: *(specify Point 2 to place the opposite corner of the window)*
Select objects: (ENTER)
Specify first point of mirror line: *(specify Point 3, as shown in Figure 4–55)*
Specify second point of mirror line: *(specify Point 4, as shown in Figure 4–55)*
Erase source objects? ⬇: (ENTER)

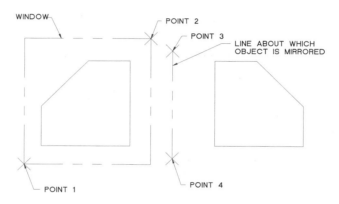

Figure 4–55 *Mirroring a group of objects selected by means of the Window option*

Text is mirrored relative to other objects within the selection group. But text will or will not retain its original orientation, depending on the setting of the system variable called MIRRTEXT. If the value of MIRRTEXT is set to 1, then text items in the selected group will have their orientations and location mirrored. That is, if their characters were normal and they read from left-to-right in the original group, in the mirrored copy they will read from right-to-left and the characters will be backwards. If MIRRTEXT is set to 0 (zero), then the text strings in the group will have their locations mirrored, but the individual text strings will retain their normal, left-to-right, character appearance. The MIRRTEXT system variable, like other system variables, can be changed by using the SETVAR command or by typing MIRRTEXT at the On-Screen prompt, as follows:

mirrtext (ENTER)
Enter New Value for MIRRTEXT <1>: **0** (ENTER)

This setting causes mirrored text to retain its readability. Figures 4–56 and 4–57 show the result of the MIRROR command when the MIRRTEXT variable is set to 1 and 0, respectively.

Figure 4–56 *The* MIRROR *command with the* MIRRTEXT *variable set to 1*

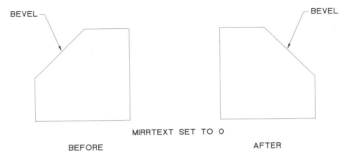

Figure 4–57 *The* MIRROR *command with the* MIRRTEXT *variable set to 0*

CREATING A FILLET BETWEEN TWO OBJECTS

The FILLET command fillets (rounds) the intersecting ends of two arcs, circles, lines, elliptical arcs, polylines, rays, xlines, or splines with an arc of a specified radius.

If the TRIMMODE system variable is set to 1 (default), then the FILLET command trims the intersecting lines to the end points of the fillet arc. If TRIMMODE is set to 0 (zero), then the FILLET command leaves the intersecting lines at the end points of the fillet arc.

Invoke the FILLET command from the Modify toolbar (see Figure 4–58) and AutoCAD prompts:

> Select first object or ⬇ *(select one of the two objects to fillet, or right-click and choose one of the options)*

Figure 4–58 *Invoking the* FILLET *command from the Modify toolbar*

By default, AutoCAD prompts you to select an object. When you select an object to fillet, then AutoCAD prompts:

Select second object or shift-select to apply corner: *(select the second object to fillet or hold down* SHIFT *and select an object to create a sharp corner)*

AutoCAD joins the two objects with an arc having the specified radius. If the objects selected to be filleted are on the same layer, AutoCAD creates the fillet arc on the same layer. If not, AutoCAD creates the fillet arc on the current layer.

AutoCAD allows you to draw a fillet between parallel lines, xlines, and rays. The first selected object must be a line or ray, but the second object can be a line, xline, or ray. The diameter of the fillet arc is always equal to the distance between the lines. The current fillet radius is ignored and remains unchanged.

Fillet Command Options

When AutoCAD is prompting for the first point, you can right-click for options on the shortcut menu including RADIUS, POLYLINE, TRIM, and MULTIPLE.

Selecting RADIUS allows you to change the current fillet radius. The following command sequence sets the fillet radius to 0.25 and draws the fillet between two lines (see Figure 4–59).

fillet (ENTER)
Select first object: *(choose* RADIUS *from the shortcut menu)*
Specify fillet radius <default>: **0.25** (ENTER)
Select first object or ⬇ *(select one of the lines, as shown in Figure 4–59)*
Select second object or shift-select to apply corner: *(select the other line to fillet, as shown in Figure 4–59)*

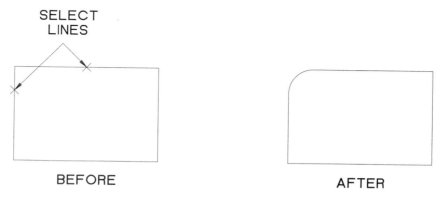

Figure 4–59 *Fillet drawn with a radius of 0.25*

If you select lines or arcs, AutoCAD extends these lines or arcs until they intersect or trims them at the intersection, keeping the selected segments if they cross. The following command sequence sets the fillet radius to 0 and draws the fillet between two lines (see Figure 4–60).

> **fillet** (ENTER)
> Select first object: *(choose RADIUS from the shortcut menu)*
> Specify fillet radius: **0** (ENTER)
> Select first object or ⬇ *(select the first object, as shown in Figure 4–60)*
> Select second object or shift-select to apply corner: *(select the second object to fillet as shown in Figure 4–60)*

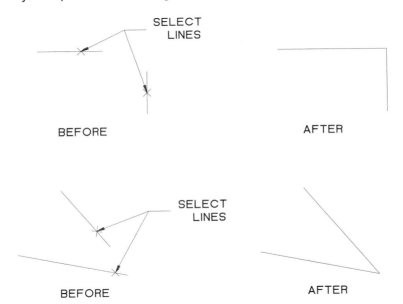

Figure 4–60 *Fillet drawn with a radius of 0*

Choosing POLYLINE causes AutoCAD to draw fillet arcs at each vertex of a selected 2D polyline where two line segments meet. The following command sequence sets the fillet radius to 0.5 and draws the fillet at each vertex of a 2D polyline (see Figure 4–61):

> **fillet** (ENTER)
> Select first object: *(choose RADIUS from the shortcut menu)*
> Specify fillet radius <default>: **0.5** (ENTER)
> Select first object or ⬇: *(choose POLYLINE from the shortcut menu)*
> Select 2D polyline: *(select the polyline as shown in Figure 4–61)*

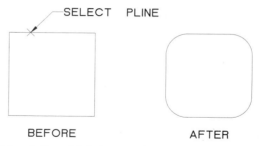

Figure 4–61 *Fillet with a radius of 0.5 drawn to a polyline*

Choosing TRIM controls whether or not AutoCAD trims the selected edges to the fillet arc end points. This option is similar to setting the TRIMMODE system variable from 1 to 0 or 0 to 1, as explained earlier.

Selecting MULTIPLE allows you to specify multiple pairs of objects to be filleted without exiting the FILLET command. To exit the command, select ENTER or CANCEL from the shortcut menu, press ESC, or invoke another command.

CREATING A CHAMFER BETWEEN TWO OBJECTS

The CHAMFER command allows you to draw an angled corner between two lines. The size of the chamfer is determined by the settings of the first and the second chamfer distances. If it is to be a 45-degree chamfer for perpendicular lines, then the two distances are set to the same value.

If the TRIMMODE system variable is set to 1 (default), then the CHAMFER command trims the intersecting lines to the end points of the chamfer line. If TRIMMODE is set to 0 (zero), then the CHAMFER command leaves the intersecting lines at the end points of the chamfer line.

Invoke the CHAMFER command from the Modify toolbar (see Figure 4–62) and AutoCAD prompts:

> Select first line or ⬇ *(select one of the two lines to chamfer, or right-click and choose one of the options)*

Figure 4–62 *Invoking the CHAMFER command from the Modify toolbar*

By default, AutoCAD prompts you to select the first line to chamfer. If you select a line to chamfer, then AutoCAD prompts:

Select second line or shift-select to apply corner: *(select the second line to chamfer or hold down* SHIFT *and select an object to create a sharp corner)*

AutoCAD draws a chamfer to the selected lines. If the selected lines to be chamfered are on the same layer, AutoCAD creates the chamfer on the same layer. If not, AutoCAD creates the chamfer on the current layer.

Chamfer Command Options

When AutoCAD prompts for the first point, you can right-click for options on the shortcut menu including DISTANCE, ANGLE, METHOD, POLYLINE, TRIM, and MULTIPLE.

Choosing DISTANCE allows you to set the first and second chamfer distances. The following command sequence sets the first chamfer and second chamfer distance to 0.5 and 1.0, respectively, and draws the chamfer between two lines (see Figure 4–63).

chamfer (ENTER)
Select first line or ⬇: *(choose* DISTANCE *from the shortcut menu)*
Specify first chamfer distance <default>: **0.5** (ENTER)
Specify second chamfer distance <0.5000>: **1.0** (ENTER)
Select first line or ⬇: *(select the first line, as shown in Figure 4–63)*
Select second line or shift-select to apply corner: *(select the second line, as shown in Figure 4–63)*

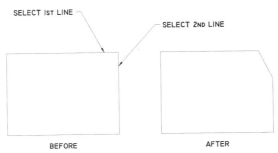

Figure 4–63 *Chamfer drawn with distances of 0.5 and 1.0*

Choosing POLYLINE causes AutoCAD to draw chamfers at each vertex of a selected 2D polyline where two line segments meet. The following command sequence sets the chamfer distances to 0.5 and draws the chamfer at each vertex of a 2D polyline (see Figure 4–64).

chamfer (ENTER)
Select first line or ⬇: *(choose* DISTANCE *from the shortcut menu)*
Specify first chamfer distance <default>: **0.5** (ENTER)
Specify second chamfer distance <0.5000>: **0.5** (ENTER)
Select first line or ⬇: *(choose* POLYLINE *from the shortcut menu)*
Select 2D polyline: *(select the polyline, as shown in Figure 4–64)*

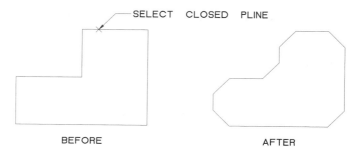

Figure 4–64 *Chamfer with distances of 0.5 drawn on a polyline*

Choosing ANGLE is similar to the DISTANCE option, but instead of prompting for the first and second chamfer distances, AutoCAD prompts for the first chamfer distance and an angle from the first line.

Choosing METHOD controls whether AutoCAD uses two distances or a distance and an angle to create the chamfer line.

Choosing TRIM controls whether or not AutoCAD trims the selected edges to the chamfer line end points. This option is similar to setting the TRIMMODE system variable from 1 to 0 or from 0 to 1, as explained earlier.

 Note: The CHAMFER command set to zero distance operates the same way the FILLET command operates when set to zero radius.

Selecting MULTIPLE allows you to specify multiple pairs of objects to be chamfered without exiting the CHAMFER command. To exit the command, select ENTER or CANCEL from the shortcut menu, press ESC, or invoke another command.

MODIFYING OBJECTS

In this section, four additional MODIFY commands are explained: MOVE, TRIM, BREAK, and EXTEND. (The ERASE command was explained in Chapter 2.)

MOVING OBJECTS

The MOVE command lets you move one or more objects from their present location to a new one without changing orientation or size.

Invoke the MOVE command from the Modify toolbar (see Figure 4–65) and AutoCAD prompts:

  Select objects: *(select the objects, press* ENTER *when done)*
  Specify base point or ⬇ *(specify a point)*
  Specify second point or <use first point as displacement>: *(specify a point*
   for displacement, or press ENTER*)*

Chapter 4 • *Fundamentals III* 237

Figure 4–65 *Invoking the* MOVE *command from the Modify toolbar*

You can use one or more object selection methods to select the objects. If you specify two data points, AutoCAD computes the displacement and moves the selected objects accordingly. If you specify the points on the screen, AutoCAD assists you in visualizing the displacement by drawing a rubber-band line from the first point to the second point, as you move the crosshairs. If you press ENTER at the prompt for the second point of displacement, AutoCAD interprets the base point as relative X,Y,Z displacement.

The following command sequence shows an example of moving a group of objects, selected by means of the Window option, by relative displacement, (see Figure 4–66).

move (ENTER)
Select objects: *(specify Point 1 to place one corner of a window)*
Specify opposite corner: *(specify Point 2 to place the opposite corner of the window)*
Select objects: (ENTER)
Specify base point or ⬇ **2,3** (ENTER)
Specify second point or <use first point as displacement>: (ENTER)

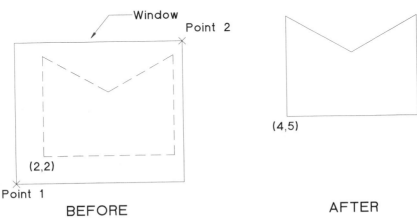

Figure 4–66 *Using the Window option of the* MOVE *command to move a group of objects by means of vector displacement*

The following command sequence shows an example of moving a group of objects selected by the Window option and moving the objects by specifying two data points, (see Figure 4–67).

> **move** (ENTER)
> Select objects: *(pick Point 1 to place one corner for a window)*
> Specify opposite corner: *(pick Point 2 to place the opposite corner of the window)*
> Select objects: (ENTER)
> Specify base point or *(specify Base Point)*
> Specify second point or <use first point as displacement>: *(specify Second Point)*

Note: See the note following Figure 4–43 concerning specifying displacements.

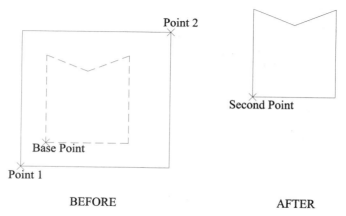

Figure 4–67 *Using the Window option of the* MOVE *command to move a group of objects by specifying two data points*

TRIMMING OBJECTS

The TRIM command is used to trim a portion of the selected object(s) that is drawn past a cutting edge or from an implied intersection defined by other objects. Objects that can be trimmed include lines, arcs, elliptical arcs, circles, 2D and 3D polylines, xlines, rays, and splines. Valid cutting edge objects include lines, arcs, circles, ellipses, 2D and 3D polylines, floating viewports, xlines, rays, regions, splines, and text.

Invoke the TRIM command from the Modify toolbar (see Figure 4–68) and AutoCAD prompts:

> Select objects or <select all>: *(select the objects and press* ENTER *to terminate the selection of the cutting edges)*
> Select object to trim or shift-select to extend or : *(select object(s) to trim or right-click and choose one of the options)*

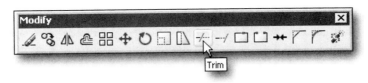

Figure 4–68 *Invoking the* TRIM *command from the Modify toolbar*

The TRIM command initially prompts you to "Select objects or <select all>:". After selecting one or more cutting edges to establish where to trim, press ENTER. You are then prompted to "Select object to trim or shift-select to extend or ⬇:". Select one or more objects to trim and then press ENTER to terminate the command. If you press ENTER in response to "Select objects or <select all>" prompt without selecting any objects, by default AutoCAD selects all the objects. You can switch to the EXTEND command by holding the SHIFT key while selecting the objects which will then be extended to the specified cutting edge instead of being trimmed. See the section on the EXTEND command in this chapter.

 Note: Don't forget to press ENTER after selecting the cutting edge(s). Otherwise, the program will not respond as expected. In fact, TRIM continues to expect more cutting edges until you terminate the cutting edge selection mode.

Trim Command Options

When AutoCAD prompts you to select objects to trim, you can right-click for options on the shortcut menu including EDGE, UNDO, and PROJECT.

Selecting EDGE determines whether objects that extend past a selected cutting edge or to an implied intersection are trimmed. AutoCAD prompts you as follows when the EDGE option is selected:

Enter an implied edge extension mode: *(select one of the two options)*

Selecting EXTEND from the EDGE shortcut causes the object's cutting edge to extend along its natural path to intersect an object in 3D (implied intersection).

Selecting NO EXTEND from the EDGE shortcut specifies that the object is to be trimmed only at a cutting edge that intersects it in 3D space.

Choosing UNDO reverses the most recent change made by the TRIM command.

Selecting PROJECT specifies the projection mode and coordinate system AutoCAD uses when trimming objects. By default, it is set to the current UCS.

Figure 4–69 shows examples of using the TRIM command.

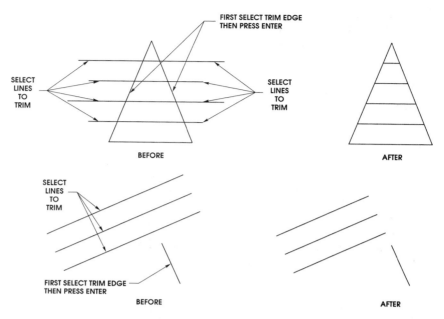

Figure 4–69 *Examples of using the* TRIM *command*

ERASING PARTS OF OBJECTS

The BREAK command is used to remove parts of objects, to make a circle into an arc or to split an object in two parts, and it can be used on lines, xlines, rays, arcs, circles, ellipses, splines, donuts, traces, and 2D and 3D polylines.

Invoke the BREAK command from the Modify toolbar (see Figure 4–70) and AutoCAD prompts:

> Select object: *(select an object)*
> Specify second break point or ⬇: *(specify the second break point or right-click and choose one of the options)*

Figure 4–70 *Invoking the* BREAK *command from the Modify toolbar*

See Figure 4–71 for examples of applications of the BREAK command.

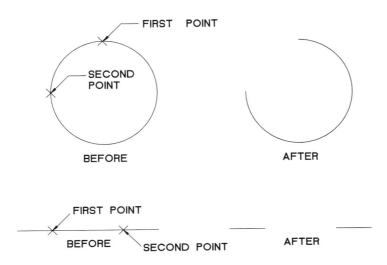

Figure 4–71 *Examples of applications of the* BREAK *command*

AutoCAD erases the portion of the object between the first point (the point where the object was selected) and second point. If the second point is not on the object, then AutoCAD selects the nearest point on the object. If you need to erase an object to one end of a line, arc, or polyline, then specify the second point beyond the end to be removed.

If instead of specifying the second point, you choose FIRST POINT from the shortcut menu, AutoCAD prompts for the first point and then for the second point.

An object can be split into two parts without removing any portion of the object by selecting the same point as the first and second points. You can do so by entering @ to specify the second point.

If you select a circle, then AutoCAD converts it to an arc by erasing a piece, moving counterclockwise from the first point to the second point. For a closed polyline, the part is removed between two selected points, moving in direction from the first to the last vertex. And in the case of 2D polylines and traces with width, the BREAK command will produce square ends at the break points.

EXTENDING OBJECTS TO MEET ANOTHER OBJECT

The EXTEND command is used to change one or both end points of selected lines, arcs, elliptical arcs, open 2D and 3D polylines, and rays to extend to lines, arcs, elliptical arcs, circles, ellipses, 2D and 3D polylines, rays, xlines, regions, splines, text string, or floating viewports.

Invoke the EXTEND command from the Modify toolbar (see Figure 4–72) and AutoCAD prompts:

> Select objects or <select all>: *(select the objects and then press ENTER to complete the selection)*
> Select object to extend or shift-select to trim or ⬇ *(select the object(s) to extend and press ENTER to terminate the selection process, or right-click and choose one of the options)*

Figure 4–72 *Invoking the EXTEND command from the Modify toolbar*

The EXTEND command initially prompts you to "Select objects or <select all>:" After selecting one or more boundary edges, press ENTER to terminate the selection process. Then AutoCAD prompts you to "Select object to extend or shift-select to trim or ⬇". Select one or more objects to extend to the selected boundary edges. After selecting the required objects to extend, press ENTER to complete the selection process. You can switch to the TRIM command by holding the SHIFT key while selecting the objects which will be trimmed to the specified cutting edge instead of being extended. See the section on the TRIM command in this chapter.

The EXTEND and TRIM commands are very similar in this method of selecting. With EXTEND you are prompted to select the boundary edge to extend to; with TRIM you are prompted to select a cutting edge. If you press ENTER in response to "Select objects or <select all>:" prompt without selecting any objects, by default AutoCAD selects all the objects.

Extend Command Options

Selecting *EDGE* determines whether objects are extended past a selected boundary or to an implied edge. AutoCAD prompts as follows when the *EDGE* option is selected:

> Enter an implied edge extension mode *(select one of the two options)*

Select *EXTEND* from the *EDGE* shortcut to extend the boundary object along its natural path to intersect another object in 3D space (implied edge).

Select *NO EXTEND* from the *EDGE* shortcut to specify that the object is to extend only to a boundary object that actually intersects it in 3D space.

Choosing *UNDO* reverses the most recent change made by the EXTEND command.

Select PROJECT to specify the projection and coordinate system that AutoCAD uses when trimming objects. By default, it is set to the current UCS.

Figure 4–73 shows examples of the use of the EXTEND command.

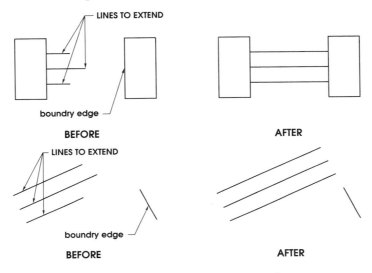

Figure 4–73 *Examples of applications of the* EXTEND *command*

Open the Exercise Manual PDF file for Chapter 4 on the accompanying CD for project and discipline specific exercises.

 If you have the accompanying Exercise Manual, refer to Chapter 7 for project and discipline specific exercises.

REVIEW QUESTIONS

1. In order to draw two rays with different starting points, you must use the RAY command twice.

    a. True   b. False

2. When you place xlines on a drawing, they:

    a. may affect the limits of the drawing

    b. may affect the extents of the drawing

    c. always appear as construction lines on layer 0

    d. can be constructed as offsets to an existing line

    e. none of the above

3. Filleting two non-parallel, non-intersecting line segments with a zero radius will:

    a. return an error message

    b. have no effect

    c. create a sharp corner

    d. convert the lines to rays

4. Objects can be trimmed at the points where they intersect existing objects.

    a. True   b. False

5. The default justification for text is:

    a. TL   d. BR

    b. BL   e. None of the above

    c. MC

6. Which command allows you to change the location of the objects and allows a duplicate to remain intact?

    a. CHANGE   c. COPY

    b. MOVE   d. MIRROR

7. The maximum number of sides accepted by the POLYGON command is:

   a. 8

   b. 32

   c. 128

   d. 1024

   e. infinite (limited by computer memory, but VERY large)

8. Portions of objects can be erased or removed by using the command:

   a. ERASE

   b. REMOVE

   c. BREAK

   d. EDIT

   e. PARERASE

9. Ellipses are drawn by specifying:

   a. The major and minor axes

   b. The major axis and a rotation angle

   c. Any three points on the ellipse

   d. Any of the above

   e. Both a and b

10. Which of the following commands can be used to place text on a drawing?

    a. TEXT

    b. CTEXT

    c. MTEXT

    d. Both a and c

    e. None of the above

11. Two lines are drawn. Then the first line is erased and the second line is moved. Executing the OOPS command at the On-Screen prompt will:

    a. restore the erased line

    b. execute the LINE command automatically

    c. replace the second line at its original position

    d. restore both lines to their original positions

    e. none of the above

12. To move multiple objects, which selection option would be more efficient?

    a. OBJECTS

    b. LAST

    c. WINDOW

    d. ADD

    e. UNDO

13. While you're using the TEXT command, AutoCAD will display the text you are typing:
    a. in the command prompt area
    b. in the drawing screen area
    c. both a and b
    d. neither a or b

14. To insert the diameter symbol into a string of text, you should enter:
    a. %%d
    b. %%c
    c. %%phi
    d. %%dia

15. It is possible to force all text on a drawing to display as an open rectangle in order to speed up the redisplay of the drawing by using the:
    a. TEXT command
    b. RTEXT command
    c. QTEXT command
    d. DTEXT command

16. In order to select a different font for use in the TEXT command, a text style must be created.
    a. True
    b. False

17. Which of the following are not valid options when creating a text style?
    a. width factor
    b. upside down
    c. vertical
    d. backwards
    e. none of the above (all answers are valid)

18. To locate text such that it is centered exactly within a circle, a reasonable justification would be:
    a. Center
    b. Middle
    c. Full
    d. Both A and B
    e. none, simply zoom in and approximate it

19. The MOVE command allows you to:
    a. move objects to new locations on the screen
    b. dynamically drag objects on the screen
    c. move only the objects that are on the current layer
    d. move an object from one layer to another
    e. both a and b

20. To create a rectangular array of objects, you must specify:
    a. the number of items and the distance between them
    b. the number of rows, the number of items, and the unit cell size
    c. the number of rows, the number of columns, and the unit cell size
    d. none of the above

21. A polyline:
    a. can have width
    b. can be exploded
    c. is one object
    d. all of the above

22. Polylines are:
    a. made up of line and arc segments, each of which is treated as an individual object
    b. are connected sequences of lines and arcs
    c. both a and b
    d. none of the above

23. To create an arc that is concentric with an existing arc, you could use what command?
    a. ARRAY
    b. COPY
    c. OFFSET
    d. MIRROR

24. The following are all options of the PLINE command except:
    a. UNDO
    b. HALFWIDTH
    c. ARC
    d. LTYPE
    e. WIDTH

Directions: Answer the following questions based on Figure QX4–1.

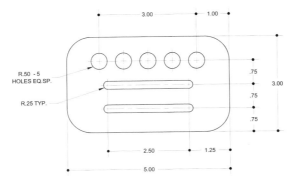

Figure QX4–1 *Mechanical part*

25. If the RECTANGLE command were used to create the object outline with the first corner at 0,0 what "*specify other corner point:*" coordinate location input would be required to complete the rectangle shape?

   a. @3.0,3.0   c. 3.0,5.0
   b. 5.0,3.0   d. @3.0,5.0

26. Which command would be the easiest and quickest to create rounded corners on the object?

   a. ARC   c. FILLET
   b. ROUND   d. CHAMFER

27. Creating the five holes equally spaced can be best achieved with which command?

   a. MIRROR   c. COPY
   b. OFFSET   d. ARRAY

24. Which ARRAY type would the five equally spaced holes be created with?

   a. POLAR   c. COPY
   b. RECTANGULAR   d. ARRAY

25. If the first corner of the object outline is at 0,0, what coordinate location input is required to position the center of the circle on the left?

   a. @1.25,2.5
   b. -2.25,1.25
   c. @1.00,2.25
   d. 1.25,2.25

26. What is the *distance between cells* (holes) value required when using the ARRAY command to locate the five holes?

   a. 3.00   c. 0.50
   b. 0.75   d. 0.25

27. If the top horizontal line of the long slot is drawn with the LINE command, which command can easily establish the location of the remaining horizontal lines for the slots?

   a. XLINE
   b. MIRROR
   c. OFFSET
   d. MOVE

CHAPTER 5

Fundamentals IV

INTRODUCTION

After completing this chapter, you will be able to do the following:

- Construct geometric figures with the DONUT, SOLID, and POINT commands
- Create freehand line segments with the SKETCH command
- Use advanced object selection methods and modes to modify objects
- Use the modify commands: LENGTHEN, STRETCH, ROTATE, SCALE, PEDIT, JOIN, and MATCHPROP

CONSTRUCTING GEOMETRIC FIGURES

DRAWING SOLID-FILLED CIRCLES

The DONUT (or DOUGHNUT) command lets you draw solid-filled circles and rings by specifying the outer and inner diameters of the filled area. The fill display depends on the setting of the FILLMODE system variable.

Invoke the DONUT command from the Draw menu and AutoCAD prompts:

> Specify inside diameter of donut <current>: *(specify a distance, or press ENTER to accept the current setting)*
> Specify outside diameter of donut <current>: *(specify a distance, or press ENTER to accept the current setting)*
> Specify center of donut or <exit>: *(specify a point to draw the donut)*

You may specify the inside and outside diameters of the donut by using your pointing device to specify two points at the appropriate distance apart for either or both diameters, and AutoCAD will use the measured distance for the diameter(s). Or you may enter the distances from the keyboard.

You can select the center point by entering its coordinates or by selecting it with your pointing device. After you specify the center point, AutoCAD prompts for the center of the next donut and continues prompting for subsequent center points. To terminate the command, press ENTER.

Note: Be sure the FILLMODE system variable is set to ON (a value of 1). Invoke the FILLMODE comand by typing **fill** at the On-Screen prompt and press ENTER. If FILLMODE is set to OFF (value of 0), then the PLINE, TRACE, DONUT, and SOLID commands display only the outline of the shapes. With FILLMODE set to ON (value of 1), the shapes you create with these commands appear solid. If FILLMODE is reset to ON after it has been set to OFF, you must use the REGEN command in order for the screen to display as filled any unfilled shapes created by these commands. Switching between ON and OFF affects only the appearance of shapes created with the PLINE, TRACE, DONUT, and SOLID commands. Solids can be selected or identified by specifying the outlines only. The interior of a solid area is not recognized as an object when specified with the pointing device.

For example, the following command sequence shows placement of a solid-filled circle (see Figure 5–1) by use of the DONUT command.

```
donut (ENTER)
Specify inside diameter of donut <0.5000>: 0 (ENTER)
Specify outside diameter of donut <1.0000>:1 (ENTER)
Specify center of donut or <exit>: 3,2 (ENTER)
Specify center of donut or <exit>: (ENTER)
```

Figure 5–1 *Using the* DONUT *command to place a solid-filled circle and a filled donut*

The following command sequence shows placement of a filled donut (see Figure 5–1) by use of the DONUT command.

```
donut (ENTER)
Specify inside diameter of donut <0.0000>: 0.5 (ENTER)
Specify outside diameter of donut <1.0000>: 1 (ENTER)
Specify center of donut or <exit>: 6,2 (ENTER)
Specify center of donut or <exit>: (ENTER)
```

DRAWING SOLID-FILLED POLYGONS

The SOLID command creates a solid-filled, straight-sided polygon whose outline is determined by points you specify. Two important factors should be kept in mind when using the SOLID command: (1) the points must be selected in a specified order or else the four corners generate a bow-tie instead of a rectangle; and (2) the polygon generated has straight sides. (Closer study reveals that even filled donuts and PLINE-generated curved areas are actually straight-sided, just as arcs and circles generate as straight-line segments of small enough length to appear smooth.)

Invoke the SOLID command by selecting 2D Solid from the Surfaces toolbar (see Figure 5–2) and AutoCAD prompts:

> Specify first point: *(specify a first point)*
> Specify second point: *(specify a second point)*
> Specify third point: *(specify a third point diagonally opposite the second point)*
> Specify fourth point or <exit>: *(specify a fourth point, or press ENTER to exit)*

Figure 5–2 *Invoking the* SOLID *command from the Surfaces toolbar*

When you specify a fourth point, AutoCAD draws a quadrilateral area. If instead you press ENTER, AutoCAD creates a filled triangle.

For example, the following command sequence shows how to draw a quadrilateral area (see Figure 5–3).

> **solid** (ENTER)
> Specify first point: *(specify Point 1)*
> Specify second point: *(specify Point 2)*
> Specify third point: *(specify Point 3)*
> Specify fourth point or <exit>: *(specify Point 4)*
> Specify third point: *(press ENTER to exit)*

Figure 5–3 *The point-specification order for creating a quadrilateral area with the* SOLID *command*

To create the solid shape, the odd-numbered picks must be on one side and the even-numbered picks on the other side. If not, you get a bow-tie effect (see Figure 5–4).

Figure 5–4 *Results of using the* SOLID *command when odd/even points are not specified correctly*

You can use the SOLID command to create an arrowhead or triangle (see Figure 5–5). Polygon shapes can be created with the SOLID command by keeping the odd picks along one side and the even picks along the other side of the object (see Figure 5–6).

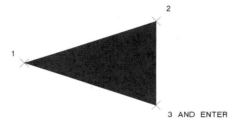

Figure 5–5 *Using the* SOLID *command to create a solid triangular shape*

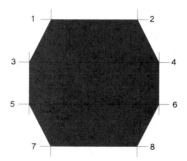

Figure 5–6 *Using the* SOLID *command to create a polygonal shape*

DRAWING POINT OBJECTS

The POINT command draws points on the drawing, and these points are drawn on the plotted drawing sheet with a single "pen down." You can enter such points to be used as reference points for object snapping when necessary. When the drawing is finished, simply erase them from the drawing or freeze their layer. Points are entered by specifying 2D or 3D coordinates or with the pointing device.

Invoke the POINT command from the Draw toolbar (see Figure 5–7) and AutoCAD prompts:

Point: (*specify a point*)

Figure 5–7 *Invoking the* POINT *command from the Draw toolbar*

Options available if you select Point from the flyout of the Draw menu include POINT, MULTIPLE POINT, DIVIDE, and MEASURE. The POINT option draws a single point object. The MULTIPLE POINT option continues to prompt for points until you terminate the command by pressing ESC. The DIVIDE and MEASURE options are described in Chapter 10.

When you draw the point, it appears on the display as a blip (+) if the BLIPMODE system variable is set to ON (default is OFF). After a REDRAW command, it appears as a dot (.). You can make the point appear as a +, x, 0, or any of the available symbols by changing the PDMODE system variable. This can be done by entering PDMODE at the On-Screen prompt and entering the appropriate value. You can also change the PDMODE value by using the icon menu (see Figure 5–8) invoked by typing DDPTYPE at the On-Screen prompt and pressing ENTER. The default value of PDMODE is zero, which means the point appears as a dot. If PDMODE is changed, all previous points drawn are replaced with the current setting.

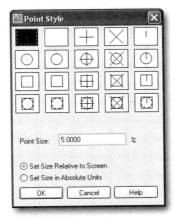

Figure 5–8 *The Point Style icon menu lets you select the shape and size of the point object*

The size that the point appears on screen depends on the value the PDSIZE system variable. If necessary, you can change the size via the PDSIZE command. The default for PDSIZE is zero (one pixel in size). Any positive value larger than this will increase the size of the point accordingly.

DRAWING SKETCH LINE SEGMENTS

The SKETCH command creates a series of freehand line segments. It is useful for freehand drawings, contour mapping, and signatures. The sketched lines are not added to the drawing until they are recorded.

Invoke the SKETCH command from the On-Screen prompt by typing **sketch.** *You will not see an On-Screen prompt.* Before you can begin sketching, you must tell AutoCAD the length it should use for each line segment drawn by entering the segment length. You may also respond by specifying two points, either keyed in or specified on the screen, causing AutoCAD to use the distance between the points as the record increment distance. Once a record increment is specified, AutoCAD provides the following list of options (although they are not shown at the On-Screen prompt):

    Pen   eXit   Quit   Record   Erase   Connect

Note: You will not see an On-Screen prompt but you will be able to see the command prompts if it is invoked from Command window.

You can use any of the available options while you are in the SKETCH command. They are accessible either as single-key entries or as a mouse/puck button, provided your mouse/puck has the number of buttons corresponding to the option. The following table shows the optional subcommands, and their key, button number, and function. Normal button functions are not available while in the SKETCH mode.

Command Character	Pointer Button	Function
P	Pick	Raise/lower pen
.(period)	1	Line to point
R	2	Record lines
X, SPACEBAR, or ENTER	3	Record lines and exit
Q, or ESC	4	Discard lines and exit
E	5	Erase
C	6	Connect

An imaginary pen follows the cursor movement. When the pen is down, AutoCAD sketches a connected segment whenever the cursor moves the specified increment distance from the previously sketched segment. When the pen is up, the pen follows the cursor movement without drawing.

The pen is raised (up) and lowered (down) either by pressing the pick button on the mouse/puck or by pressing P on the keyboard. When you invoke a PEN UP, the current location of the pen will be the end point of the last segment drawn, which will be shorter than a standard increment length.

A PEN UP does not take you out of the SKETCH mode. Nor does a PEN UP permanently record the lines drawn during the current SKETCH session.

While the pen is up, you cause AutoCAD to draw a straight line from the last segment to the current cursor location and return to the PEN UP status by entering . (period) from the keyboard. This is convenient for long, straight lines that might occur in the middle of irregular shapes.

Lines being displayed while the cursor is moved (with the pen down) are temporary. They will appear green (or red if the current color for that layer or object is green) until they are permanently recorded. These temporary segments are subject to being modified with special Sketch options until you press **R** to record the latest lines. These may include several groups of connected lines drawn during PEN DOWN sequences separated by PEN UPs.

Prior to any group(s) of connected lines being recorded with the Record option, you may use the E (Erase) option to remove any or all of the lines from the last segment back to the first. The pen is automatically set to UP, and you may then use the cursor to remove segments, starting from the last segment. Press **P** or the pick button to accept the selected segments to delete. To abort the deletion of the selected segments and return to the SKETCH mode, press **E** again (or any other option).

Whenever disconnect occurs (PEN UP or ERASE), you can reconnect and continue sketching from the point of the last disconnect by pressing **C** as long as you have not exited the SKETCH command.

At this prompt you can move the cursor near the end of the last segment. When you are within a specified increment length, sketching begins, connected to that last end point. This option is meaningless if invoked during PEN DOWN. The Connect option can be canceled by pressing **C** a second time.

The X option exits SKETCH mode after recording all temporary lines. This can also be accomplished by pressing either ENTER or the SPACEBAR. The Q option exits SKETCH mode without recording any temporary lines. It is the same as pressing ESC.

OBJECT SELECTION

As mentioned earlier, all the modify commands initially prompt you to select objects. In most modify commands, the prompt allows you to select any number of objects. In some of the modify commands, however, AutoCAD limits your selection to only one object, for instance, the BREAK, DIVIDE, and MEASURE commands. In the case of the FILLET and CHAMFER commands, AutoCAD requires you to select two objects. And, whereas in the DIST and ID commands AutoCAD requires you to select a point, in the AREA command AutoCAD permits selection of either a series of points or an object.

AutoCAD has various options to control the appearance of objects during selection preview. You can make the changes to the appearance of the selection preview from the Visual Effect Settings dialog box (see Figure 5–9), which can be opened from the Selection tab of the Options dialog box.

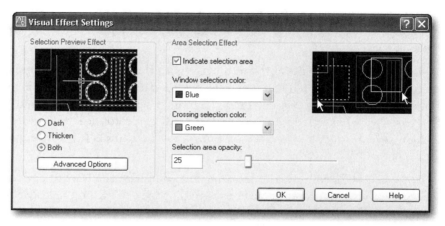

Figure 5–9 *Visual Effect Settings dialog box*

The preview area displays the effects of the current settings.

Selecting **Dash** causes dashed lines to be displayed when the pick box cursor rolls over an object. This selection previewing indicates that the object would be selected if you clicked.

Selecting **Thicken** causes thickened lines to be displayed when the pick box cursor rolls over an object. This selection previewing indicates that the object would be selected if you clicked.

Selecting **Both** (default setting) causes thickened, dashed lines to be displayed when the pick box cursor rolls over an object. This selection previewing indicates that the object would be selected if you clicked.

Choosing **Advanced Options** causes AutoCAD to open the Advanced Preview Options dialog box (see Figure 5–10) from which you can select object types to exclude from selection previewing.

Figure 5–10 *Advanced Preview Options dialog box*

Selecting **Exclude Objects on locked layers** (default set to ON) causes selection previewing to be excluded for objects on locked layers.

Select appropriate object types from the **Exclude** section to exclude selection previewing. The available object types include Xrefs, Tables, Groups, Multiline Text, and Hatches. By default Xrefs and Tables are set to ON.

The **Area Selection Effect** section of the Visual Effect Settings dialog box controls the appearance of selection areas during selection preview. The preview area displays the effect of the current settings. Select **Indicate selection area** to indicate the selection area with the selected background color. Choose background colors for Window and Crossing selection area from the **Window selection color** and **Crossing selection color** drop-down lists respectively. The slider bar for **Selection area opacity** controls the degree of transparency background for window selection areas.

Choose **OK** to save and close the Visual Effect Settings dialog box.

Compared to the basic object selection options of Pick, Window, Crossing, Previous, and Last, the options covered in this section give you more flexibility and greater ease of use when you are prompted to select objects for use by the modify commands. The options that are explained in this section include *WPOLYGON (WP)*, *CPOLYGON (CP)*, *FENCE*, *ALL*, *MULTIPLE*, *BOX*, *AUTO*, *UNDO*, *ADD*, *REMOVE*, and *SINGLE*.

WPOLYGON (WP) SELECTION

The *WPOLYGON* option is similar to the *WINDOW* option, but it allows you to define a polygon-shaped window rather than a rectangular area. You define the selection area as you specify the points about the objects you want to select. The polygon can be of any shape, but may not intersect itself. The polygon is formed as you select the points and includes rubber-band lines to the graphics cursor indicating the selection area. When the selected points define the desired polygon, press ENTER. Only those objects that are totally inside the polygon shape are selected.

To select the WPOLYGON option, type **WP** and press ENTER at the "Select objects:" prompt. The UNDO option lets you undo the most recent polygon pick point.

CPOLYGON (CP) SELECTION

The CPOLYGON option is similar to the WPOLYGON option, but it selects all objects within or crossing the polygon boundary. If there is an object that is partially inside the polygon area, then the whole object is included in the selection set.

To select the CPOLYGON option, type **CP** and press ENTER at the "Select objects:" prompt. The UNDO option lets you undo the most recent polygon pick point.

FENCE (F) SELECTION

The FENCE option is similar to the CPOLYGON option, except you do not close the last vector of the polygon shape. The selection fence selects only those objects it crosses or intersects. Unlike WPOLYGON and CPOLYGON, the fence can cross over and intersect itself.

To select the FENCE option, type **F** and press ENTER at the "Select objects:" prompt.

ALL SELECTION

The ALL option selects all the objects in the drawing, including objects on frozen or locked layers. After selecting all the objects, you may use the REMOVE (R) option to remove some of the objects from the selection set.

The ALL option must be spelled out in full (All) and not applied as an abbreviation as you may do with the other options.

MULTIPLE SELECTIONS

The MULTIPLE option helps you overcome the limitations of the PICK, WINDOW, and CROSSING options. The PICK option is time-consuming for use in selecting many objects. AutoCAD does a complete scan of the screen each time a point is picked. By using the MULTIPLE modifier option, you can pick many points without delay and when you press ENTER, AutoCAD applies all of the points during one scan.

Selecting one or more objects from a crowded group of objects is sometimes difficult with the POINT option. It is often impossible with the WINDOW option. For example, if two objects are very close together and you wish to point to select them both, AutoCAD normally selects only one no matter how many times you select a point that touches them both. By using the MULTIPLE option, AutoCAD excludes an object from being selected once it has been included in the selection set. As an alternative, use the CROSSING option to cover both objects. If this is not feasible, then the MULTIPLE modifier may be the best choice.

BOX SELECTION

The *BOX* option is usually employed in a menu macro to give the user an option of using both the *WINDOW* and *CROSSING* methods, depending on how and where the picks are made on the screen.

When *BOX* is invoked as a response to a prompt to select an object, the options are applied as follows:

If the picks are made left to right (the first point is to the left of the second), then the two points become diagonally opposite corners of a rectangle that is used as a *WINDOW* option. That is, all visible objects totally within the rectangle are part of that selection.

If the picks are right to left, then the selection rectangle becomes the *CROSSING* option. That is, all visible objects that are within or partially within the rectangle are part of that selection.

AUTO SELECTION

The *AUTO* option is actually a triple option. It includes the *POINT* option with the two *BOX* options. If the target box touches an object, then that object is selected as you would in using the *POINT* option. If the target box does not touch an object, then the selection becomes either a *WINDOW* or *CROSSING* option, depending on where the second point is picked in relation to the first.

Note: The *AUTO* option is the default option when you are prompted to "Select objects:".

UNDO SELECTION

The *UNDO* option allows you to remove the last item(s) selected from the selection set without aborting the "Select objects:" prompt and then to continue adding to the selection set. It should be noted that if the last option to the selection process includes more than one object, the *UNDO* option will remove all the objects from the selection set that were selected by that last option.

ADD SELECTION

The *ADD* option lets you switch back from the *REMOVE* mode in order to continue adding objects to the selection set by however many options you wish to use.

REMOVE SELECTION

The *REMOVE* option lets you remove objects from the selection set. The "Select objects:" prompt always starts in the *ADD* mode. The *REMOVE* mode is a switch from the *ADD* mode, not a standard option. Once invoked, the objects selected by whatever and however many options you use will be removed from the selection set. It will be in effect until reversed by the *ADD* option.

SINGLE SELECTION

The *SINGLE* option causes the object selection to terminate and the command in progress to proceed after you use only one object selection option. It does not matter if one object is selected or a group is selected (only one selection opportunity is given). If no object is selected and the point selected cannot be the first point of a *WINDOW* or *CROSSING* rectangle, AutoCAD will not abort the command in progress; however, once there is a successful selection, the command proceeds.

OBJECT SELECTION MODES

AutoCAD provides six selection modes that will enhance object selection. You can toggle on/off one or more object selection modes from the Selection tab of the Options dialog box. By having the appropriate selection mode set to ON, you have various methods that give you more flexibility and greater ease of use in the selection of the objects.

To make changes to the selection modes settings, open the Options dialog box by selecting Options from the Tools menu. Choose the **Selection** tab (see Figure 5–11):

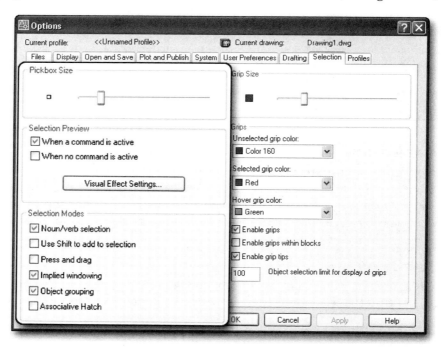

Figure 5–11 *Options dialog box with Selection tab selected*

You can toggle any one or more combinations of the settings provided under the **Selection Modes**. **Noun/verb selection**, **Implied windowing**, and **Object grouping** are the defaults.

In the **Pickbox Size** section, you can adjust the size of the pickbox using the **Pickbox Size** slider bar provided in the dialog box. In the **Selection Modes** section, you can customize how objects will be selected for modification or for use by AutoCAD commands.

Selecting **Noun/verb selection** allows the traditional verb-noun command syntax to be reversed for most modifying commands. When **Noun/verb selection** is checked, AutoCAD allows you to select objects when there is no command in progress and then invoke the appropriate modifying command you want to use on the selection set. For example, instead of invoking the COPY command followed by selecting the objects to be copied, with **Noun/verb selection** checked, you can select the objects first and then invoke the COPY command, and AutoCAD skips the object selection prompt. When **Noun/verb selection** is checked, the cursor changes to resemble a Running-Osnap cursor. Whenever you want to use the **Noun/verb selection** feature, create a selection set first. Subsequent modify commands you invoke execute by using the objects in the current selection set without prompting for object selection. To clear the current selection set, press the ESC key. This clears the selection set, so any subsequent editing command will once again prompt for object selection.

Note: Another way to turn ON/OFF **Noun/verb selection** is to use the PICKFIRST system variable. TRIM, EXTEND, BREAK, CHAMFER, and FILLET are the commands not supported by the **Noun/Verb selection** feature.

Selecting **Use Shift to add to selection** controls how you add objects to an existing selection set. When **Use Shift to add to selection** is checked, it activates an additive selection mode in which the SHIFT key must be held down while adding more objects to the selection set. For example, if you first pick an object, it is highlighted. If you pick another object, it is highlighted and the first object is no longer highlighted. The only way you can add objects to the selection set is to select objects by holding down the SHIFT key. Similarly, the way to remove objects from the selection set is to select the objects by holding down the SHIFT key. When **Use Shift to add to selection** is not checked (default), objects are added to the selection set by just picking them individually or by using one of the selection options; AutoCAD adds the objects to the selection set.

Note: Another way to select objects with this method is to set the PICKADD system variable to 0.

Selecting **Press and drag** controls the manner by which you draw the selection window with your pointing device. When **Press and drag** is checked, you can create a selection window by holding down the pick button and dragging the cursor diagonally while you create the window. When **Press and drag** is not checked (default), you need to use two separate picks of the pointing device to create the selection window. In other words, you need to pick once to define one corner of the selection window, and pick a second time to define its diagonal corner.

Note: Another way to control how selection windows are drawn is to set the PICKDRAG system variable to 0 to draw the selection window using two points and to 1 to draw the selection window using dragging.

Selecting **Implied windowing** allows you to create a selection window automatically when the "Select objects:" prompt appears. When **Implied windowing** is checked (default), it works like the box option, explained earlier. If **Implied windowing** is not checked, you can create a selection window by using the WINDOW or CROSSING selection set methods.

Note: Another way to control the **Implied windowing** option is to set the PICKAUTO system variable to 0 to turn PICKAUTO OFF and to 1 to turn PICKAUTO ON.

Selecting **Object grouping** controls the automatic group selection. If **Object grouping** is checked, then selecting an object that is a member of a group selects the whole group. Refer to Chapter 6 for a detailed description of how to create groups.

Selecting **Associative Hatch** controls which objects will be selected when you select an associative hatch. If **Associative Hatch** is checked, then selecting an associative hatch also selects the boundary objects. Refer to Chapter 9 for a detailed description of hatching.

MODIFYING OBJECTS

In this section, seven modify commands are described: LENGTHEN, STRETCH, ROTATE, SCALE, PEDIT, JOIN, and MATCHPROP. Additional modify commands were explained in Chapters 2 and 4.

LENGTHENING OBJECTS

The LENGTHEN command is used to increase or decrease the length of line objects or the included angle of an arc.

Invoke the LENGTHEN command from the Modify menu and AutoCAD prompts:

    Select an object or ⬇ *(select an object or right-click and choose one of the options)*

Lengthen Command Options

After invoking the LENGTHEN command, you can right-click and choose one of the options, which include DELTA, PERCENT, TOTAL, and DYNAMIC.

Choosing DELTA allows you to change the length or, where applicable, the included angle from the end point of the selected object closest to the pick point. A positive value results in an increase in extension; a negative value results in a trim. When you select the DELTA option, AutoCAD prompts:

> Enter delta length or ⤓ (specify positive or negative value)
> Select an object to change or ⤓ (select an object, and its length is changed on the end nearest the selection point)
> Select an object to change or ⤓ (select additional objects; when done, press ENTER to exit the command sequence)

Instead of specifying the delta length, if you select the ANGLE option, AutoCAD prompts:

> Enter delta angle or ⤓ (specify positive or negative angle)
> Select an object to change or ⤓ (select an object, and its included angle is changed on the end nearest the selection point)
> Select an object to change or ⤓ (select additional objects; when done, press ENTER to exit the command sequence)

Choosing PERCENT allows you to set the length of an object by a specified percentage of its total length. It will increase the length/angle for values greater than 100 and decrease them for values less than 100. For example, a 12-unit-long line will be changed to 15 units by using a value of 125. A 12-unit-long line will be changed to 9 units by using a value of 75. When you select the PERCENT option, AutoCAD prompts:

> Enter percentage length: (specify positive nonzero value and press ENTER)
> Select an object to change or ⤓ (select an object, and its length is changed on the end nearest the selection point)
> Select an object to change or ⤓ (select additional objects; when done, press ENTER to exit the command sequence)

Choosing TOTAL allows you to change the length/angle of an object to the value specified. When you select the TOTAL option, AutoCAD prompts:

> Specify total length or ⤓ (specify distance, right-click and choose ANGLE, and then specify an angle for change)
> Select an object to change or ⤓ (select an object, and its length is changed to the specified distance on the end nearest the selection point)
> Select an object to change or ⤓ (select additional objects; when done, press ENTER to exit the command sequence)

Choosing DYNAMIC allows you to change the length/angle of an object in response to the cursor's final location, relative to the end point nearest to where the object is selected. When you select the DYNAMIC option, AutoCAD prompts:

Select an object to change or ⤓ *(select an object to change the endpoint)*
Specify new endpoint: *(specify new endpoint for the selected object)*
Select an object to change or ⤓ *(select additional objects; when done, press* ENTER *to exit the command sequence).*

Choosing UNDO reverses the most recent change made by the LENGTHEN command.

STRETCHING OBJECTS

The STRETCH command allows you to stretch the shape of an object without affecting other crucial parts that remain unchanged. A common example is to stretch a square into a rectangle. The length is changed while the width remains the same.

AutoCAD stretches lines, polyline segments, rays, arcs, elliptical arcs, and splines that cross the selection window. The STRETCH command moves the end points that lie inside the window, leaving those outside the window unchanged. If the entire object is inside the window, then the STRETCH command operates like the MOVE command.

It is possible to make multiple selection sets using the crossing method, and AutoCAD will modify all of the selection sets with a single STRETCH command operation. If you use the pick method of selecting objects, the STRETCH command operates like the MOVE command.

Invoke the STRETCH command from the Modify toolbar (see Figure 5–12) and AutoCAD prompts:

Select objects: *(select the first corner of a crossing window or polygon)*
Specify opposite corner: *(select the opposite corner of the crossing window or polygon)*
Select objects: *(select additional objects; when done, press* ENTER *to exit the command sequence)*
Specify base point or ⤓: *(specify a base point or press* ENTER*)*
Specify second point <or use first point as displacement>: *(specify the second point of displacement or press* ENTER *to use first point as displacement)*

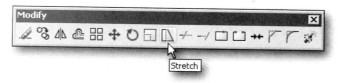

Figure 5–12 *Invoking the* STRETCH *command from the Modify toolbar*

If you provide the base point and second point of displacement, AutoCAD stretches the selected objects the vector distance from the base point to the second point. If you press ENTER at the prompt for the second point of displacement, then AutoCAD considers the first point as the *X,Y* displacement value.

Figure 5–13 shows some examples of using the STRETCH command.

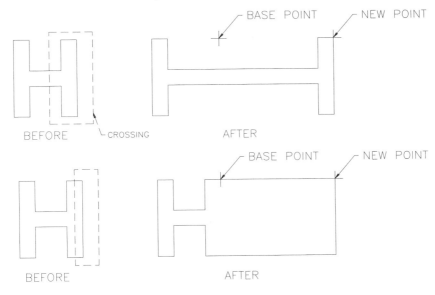

Figure 5–13 *Examples of using the* STRETCH *command*

ROTATING OBJECTS

The ROTATE command changes the orientation of existing objects by rotating them about a specified point, labeled as the base point. Design changes often require that an object, feature, or view be rotated. By default, a positive angle rotates the object in the counterclockwise direction, and a negative angle rotates in the clockwise direction.

Invoke the ROTATE command from the Modify toolbar (see Figure 5–14) and AutoCAD prompts:

> Select objects: *(select the objects to rotate and press* ENTER *to complete the selection)*
> Specify base point: *(specify a base point about which selected objects are to be rotated)*
> Specify rotation angle or ⊡ *(specify a positive or negative rotation angle, or choose* REFERENCE *from the shortcut menu)*

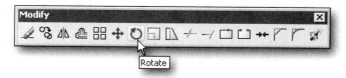

Figure 5–14 *Invoking the* ROTATE *command from the Modify toolbar*

The base point can be anywhere in the drawing. If the base point selected is on the selected object itself, then the selected base point becomes an anchor point for rotation. The rotation angle determines how far an object rotates around the base point.

The following command sequence shows an example of rotating a group of objects selected by the WINDOW option (see Figure 5–15).

>   **rotate** (ENTER)
>   Select objects: *(select the objects to rotate as shown in Figure 5–15 and press* ENTER *to complete the selection)*
>   Specify base point: *(pick the base point as shown in Figure 5–15)*
>   Specify rotation angle or ⬇ **45** (ENTER)

AutoCAD rotates the selected objects by 45 degrees (see Figure 5–15).

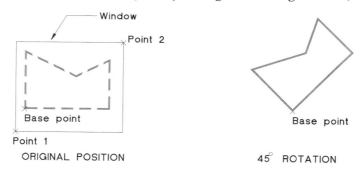

Figure 5–15 *Rotating a group of objects by means of the* ROTATE *command*

Rotate Command Options

Choosing REFERENCE allows you to rotate objects from a specified angle to a new, absolute angle. Specify the current orientation as referenced by the angle, or provide AutoCAD the angle by selecting two end points of a line to be rotated, and specifying the desired rotation angle. AutoCAD automatically calculates the rotation angle and rotates the object appropriately. This method of rotation is very useful when you want to straighten an object or align it with other features in a drawing.

The following command sequence shows an example of rotating a group of objects, selected by the WINDOW option, in reference to the current orientation (see Figure 5–16).

rotate
Select objects: *(select the objects to rotate, as shown in Figure 5–16)*
Specify base point: *(pick the base point, as shown in Figure 5–16)*
Specify rotation angle or ⬇ *(choose REFERENCE from the shortcut menu)*
Specify the reference angle <current>: **60** (ENTER)
Specify the new angle or ⬇ **30** (ENTER)

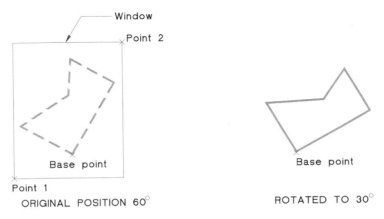

Figure 5–16 *Rotating a group of objects by means of the* ROTATE *command in reference to the current orientation*

Choosing COPY creates a copy of the selected objects for rotation.

SCALING OBJECTS

The SCALE command lets you change the size of selected objects or the complete drawing. Objects are made larger or smaller; the same scale factor is applied to the X, Y, and Z directions. To enlarge an object, specify a scale factor greater than 1. For example, a scale factor of 3 makes the selected objects three times larger. To reduce the size of an object, use a scale factor between 0 and 1. Do not specify a negative scale factor. For example, a scale factor of 0.75 would reduce the selected objects to three-quarters of their current size.

Invoke the SCALE command from the Modify toolbar (see Figure 5–17) and AutoCAD prompts:

Select objects: *(select the objects to scale and press* ENTER *to complete the selection)*
Specify base point: *(specify a base point about which selected objects are to be scaled)*

Specify scale factor or 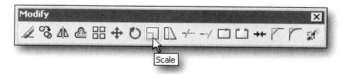: *(specify a scale factor, or right-click and choose* REFERENCE*)*

Figure 5–17 *Invoking the* SCALE *command from the Modify toolbar*

The base point can be anywhere in the drawing. If the base point selected is on the selected object itself, then the selected base point becomes an anchor point for scaling. The scale factor multiplies the dimensions of the selected objects by the specified scale.

The following command sequence shows an example of enlarging a group of objects by a scale factor of 3 (see Figure 5–18).

> scale
> Select objects: *(select the objects to enlarge, as shown in Figure 5–18, and press* ENTER *to complete the selection)*
> Specify base point: *(pick the base point, as shown in Figure 5–18)*
> Specify scale factor or ⬇ 3 (ENTER)

AutoCAD enlarges the selected objects by a factor of 3 (see Figure 5–18).

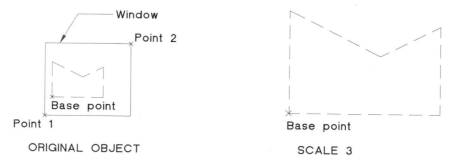

Figure 5–18 *Enlarging a group of objects by means of the* SCALE *command*

Scale Command Options

Choosing REFERENCE causes the selected objects to be scaled based on a reference length and a specified new length. Specify the current dimension as a reference length, or select two end points of a line to be scaled, and specify the desired new length. AutoCAD will automatically calculate the scale factor and enlarge or shrink the object appropriately.

The following command sequence shows an example of using the SCALE command to enlarge a group of objects selected by means of the WINDOW option in reference to the current dimension (see Figure 5–19).

scale
Select objects: *(select the objects to enlarge, as shown in Figure 5–19)*
Specify base point:*(pick the base point, as shown in Figure 5–19)*
Specify scale factor or ⬇ *(choose REFERENCE from the shortcut menu)*
Specify reference length <1>:**3.8** (ENTER)
Specify new length or ⬇: **4.8** (ENTER)

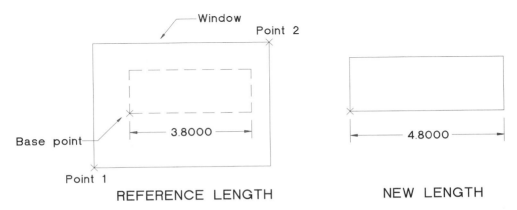

Figure 5–19 *Enlarging a group of objects by means of the* SCALE *command in reference to the current dimension*

Choosing COPY creates a copy of the selected objects for scaling.

MODIFYING POLYLINES

The PEDIT command allows you to modify polylines. In addition to using such modify commands as MOVE, COPY, BREAK, TRIM, and EXTEND, you can use the PEDIT command to modify polylines. The PEDIT command has special editing features for dealing with the unique properties of polylines and is perhaps the most complex AutoCAD command, with several multi-option submenus totaling some 70 command options.

Invoke the PEDIT command from the Modify II toolbar (see Figure 5–20) and AutoCAD prompts:

Select polyline or ⬇: *(select polyline, line, or arc, or select MULTIPLE from the shortcut menu)*

If you select a line or an arc instead of a polyline, you are prompted as follows:

Do you want it to turn into one? <Y>

Figure 5–20 *Invoking the* EDIT POLYLINE *command from the Modify II toolbar*

Responding Y or pressing ENTER turns the selected line or arc into a single-segment polyline that can then be edited. Normally this is done in order to use the JOIN option to add other connected segments that, if not polylines, will also be transformed into polylines. It should be emphasized, at this time, that in order to join segments together into a polyline, their end points must coincide. This occurs during line-line, line-arc, arc-line, and arc-arc continuation operations, or by using the Endpoint Object Snap mode.

The second prompt does not appear if the first segment selected is already a polyline. It may even be a multi-segment polyline. If you select the MULTIPLE option, then AutoCAD allows you to select more than one polyline to modify. After the object selection process, you will be returned to the multi-option prompt as follows:

Enter an option *(select one of the available options)*

PEDIT Command Options

Choosing CLOSE performs in a manner similar to the CLOSE option of the LINE command. If, however, the last segment was a polyline arc, then the next segment will be similar to the arc-arc continuation, using the direction of the last polyarc as the starting direction and drawing another polyarc, with the first point of the first segment as the ending point of the closing polyarc.

Figures 5–21 and 5–22 show some examples of the application of the CLOSE option.

Figure 5–21 *Using the* PEDIT *command* CLOSE *option with polylines*

Figure 5–22 *Using the* PEDIT *command* CLOSE *option with polyarcs*

Choosing OPEN deletes the segment that was drawn with the CLOSE option. If the polyline had been closed by drawing the last segment to the first point of the first segment without using the CLOSE option, then the OPEN option will not have a visible effect.

Choosing JOIN takes selected lines, arcs, or polylines and combines them with a previously selected polyline into a single polyline if all segments are connected at sequential and coincidental end points.

Choosing WIDTH permits uniform or varying widths to be specified for polyline segments.

Choosing EDIT VERTEX allows you to edit the vertices of the polyline. A vertex is the point where two segments join. The visible vertices are marked with an X to indicate which one is to be modified. You can modify vertices of polylines in several ways. When you select the EDIT VERTEX option, AutoCAD prompts you with additional suboptions:

Enter a vertex editing option

Choose NEXT or PREVIOUS when you wish to move the mark to the next or previous vertex, whether or not you have modified the marked vertex.

Choosing BREAK establishes the marked vertex as one vertex for the BREAK option and then prompts:

Enter an option

The choices of the BREAK option permit you to step to another vertex for the second break point, or to initialize the break, or to exit the option. If two vertices are selected, you may use the GO option to have the segment(s) between the vertices removed. If you select the end points of a polyline, this option will not work. If you select the GO option immediately after the BREAK option, the polyline will be divided into two separate polylines. Or, if it is a closed polyline, it will be opened at that point.

Choosing INSERT allows you to specify a point and have the segment between the marked vertex and the next vertex become two segments meeting at the specified point. The selected point does not have to be on the polyline segment.

For example, the following command sequence shows the application of the INSERT option (see Figure 5–23).

```
pedit (ENTER)
Select polyline or ⬇: (select the polyline as shown in Figure 5–23)
Enter an option (choose EDIT VERTEX)
Enter a vertex editing option (choose INSERT)
Specify location for new vertex: (specify a new vertex)
```

Figure 5–23 *Using the* PEDIT *command* INSERT *option*

Choosing MOVE allows you to specify a point and have the marked vertex be relocated to the selected point.

For example, the following command sequence shows the application of the MOVE option (see Figure 5–24).

```
pedit (ENTER)
Select polyline or ⬇ (select the polyline as shown in Figure 5–24)
Enter an option (choose EDIT VERTEX)
Enter a vertex editing option (choose MOVE)
Specify new location for marked vertex: (specify the new location)
```

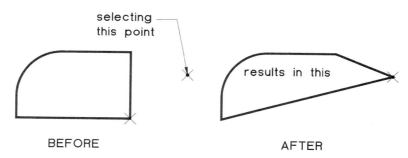

Figure 5–24 *Using the* PEDIT *command* MOVE *option*

Choosing REGEN regenerates the polyline without having to cancel the PEDIT command.

Choosing STRAIGHTEN establishes the marked vertex as one vertex for the STRAIGHTEN option and then prompts:

Enter an option

These choices of the STRAIGHTEN option permit you either to step first to another vertex for the second point or to exit the option. When the two vertices are selected, you may use the GO option to have the segment(s) between the vertices replaced with a single straight-line segment.

For example, the following command sequence shows the application of the STRAIGHTEN option (see Figure 5–25).

    pedit (ENTER)
    Select polyline or ⬇: *(select the polyline as shown in Figure 5–25)*
    Enter an option *(choose EDIT VERTEX)*
    Enter a vertex editing option *(choose STRAIGHTEN)*
    Enter an option *(choose NEXT)*
    Enter an option *(choose GO)*
    Enter a vertex editing option *(choose EXIT)*

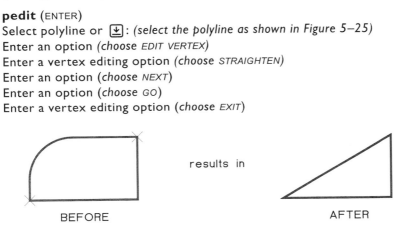

Figure 5–25 *Using the* PEDIT *command* STRAIGHTEN *option*

Choosing TANGENT permits you to assign to the marked vertex a tangent direction that can be used for the curve fitting option. The prompt is as follows:

    Specify direction of vertex tangent:

You can either specify the direction with a point or type the coordinates at the keyboard.

Choosing WIDTH permits you to specify the starting and ending widths of the segment between the marked vertex and the next vertex. The prompt is as follows:

    Specify new width for all segments:

For example, the following command sequence shows the application of the WIDTH option (see Figure 5–26).

    pedit (ENTER)
    Select polyline or ⬇ *(select a polyline)*
    Enter an option *(select EDIT VERTEX)*
    Enter a vertex editing option *(choose EDIT VERTEX)*
    Enter an option *(choose WIDTH)*
    Specify new width for all segments: **0.25** (ENTER)

Choosing EXIT exits from the VERTEX EDITING option and returns to the PEDIT multi-option prompt.

Figure 5–26 *Using the* PEDIT *command* WIDTH *option*

Choosing FIT draws a smooth curve through the vertices, using any specified tangents.

Choosing SPLINE provides several ways to draw a curve based on the polyline being edited. These include Quadratic B-spline and Cubic B-spline curves.

Choosing DECURVE returns the polyline to the way it was drawn originally. See Figure 5–27 for differences between FIT, SPLINE, and DECURVE.

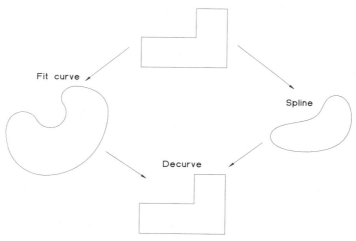

Figure 5–27 *Comparing* PEDIT *command* FIT CURVE, SPLINE CURVE, *and* DECURVE *options*

Choosing LTYPE GEN controls the display of linetype at vertices. When it is set to ON, AutoCAD generates the linetype in a continuous pattern through the vertices of the polyline. And when it is set to OFF, AutoCAD generates the linetype starting and ending with a dash at each vertex. The LTYPE GEN option does not apply to polylines with tapered segments.

Choosing UNDO reverses the latest PEDIT operation.

JOINING SIMILAR OBJECTS

The JOIN command allows you to create one line, arc, elliptical arc, polyline, or spline from multiple like objects. Invoke the JOIN command from the Modify toolbar (see Figure 5–28) and AutoCAD prompts:

> Select source object: (select a line, polyline, arc, elliptical arc, or spline)

Figure 5–28 *Invoking the JOIN command from the Modify toolbar*

If the source object selected is a line, AutoCAD prompts:

> Select lines to join to source: (select one or more lines and press ENTER)

The line objects must be collinear (lying on the same infinite line), and can have gaps between them.

If the source object selected is a polyline, AutoCAD prompts:

> Select objects to join to source: (select one or more objects and press ENTER)

The objects can be lines, polylines, or arcs. The objects cannot have gaps between them, and must lie on the same plane parallel to the UCS *XY* plane.

If the source object selected is an arc, AutoCAD prompts:

> Select arcs to join to source or ⊠ (select one or more arcs and press ENTER)

To convert the arc into a circle, select CLOSE from the shortcut menu. The arc objects must lie on the same imaginary circle, and can have gaps between them.

If the source object selected is an elliptical arc, AutoCAD prompts:

> Select elliptical arcs to join to source or ⊠ (select one or more elliptical arcs and press ENTER)

To convert the elliptical arc into an ellipse, select CLOSE from the shortcut menu. The elliptical arcs must lie on the same imaginary ellipse, and can have gaps between them.

If the source object selected is a spline, AutoCAD prompts:

> Select splines to join to source: (select one or more splines and press ENTER)

The spline objects must lie in the same plane, and must be contiguous (lying end-to-end).

MATCHING PROPERTIES

The MATCHPROP command allows you to copy selected properties from one object to one or more other objects located in the current drawing or any other drawing currently open. Properties that can be copied include color, layer, linetype, linetype scale, lineweight, thickness, plot style, and in some cases, dimension, text, polyline, viewport, and hatch.

Invoke the MATCHPROP command by selecting Match Properties from the Standard toolbar (see Figure 5–29) and AutoCAD prompts:

> Select source object: *(select the object whose properties you want to copy)*
> Select destination object(s) or *(select the destination objects and press* ENTER *to terminate the selection, or choose* SETTINGS *from the shortcut menu to display the Property Settings dialog box)*

Figure 5–29 *Invoking the Match Properties command from the Standard toolbar*

Selecting the SETTINGS option displays the Property Settings dialog box (see Figure 5–30). Use the SETTINGS option to control which object properties are copied. By default, all object properties in the Property Settings dialog box are set to ON for copying.

Figure 5–30 *Property Settings dialog box*

Choosing **Color** changes the color of the destination object to that of the source object.

Choosing **Layer** changes the layer of the destination object to that of the source object.

Choosing **Linetype** changes the linetype of the destination object to that of the source object. Available for all objects except attributes, hatches, multiline text, points, and viewports.

Choosing **Linetype Scale** changes the linetype scale factor of the destination object to that of the source object. Available for all objects except attributes, hatches, multiline text, points, and viewports.

Choosing **Lineweight** changes the lineweight of the destination object to that of the source object.

Choosing **Thickness** changes the thickness of the destination object to that of the source object. Available only for arcs, attributes, circles, lines, points, 2D polylines, regions, text, and traces.

Choosing **PlotStyle** changes the plot style of the destination object to that of the source object. If you are working in color-dependent plot style mode (PSTYLEPOLICY is set to 1), this option is unavailable.

Choosing **Dimension** changes the dimension style of the destination object to that of the source object. Available only for dimension, leader, and tolerance objects.

Choosing **Polyline** changes the width and linetype generation properties of the destination polyline to those of the source polyline. The fit/smooth property and the elevation of the source polyline are not transferred to the destination polyline. If the source polyline has variable width, the width property is not transferred to the destination polyline.

Choosing **Text** changes the text style of the destination object to that of the source object.

Choosing **Viewport** changes the following properties of the destination paper space viewport to match those of the source viewport: on/off, display locking, standard or custom scale, shade plot, snap, grid, and UCS icon visibility and location. The settings for clipping and for UCS per viewport and the freeze/thaw state of the layer are not transferred to the destination object.

Choosing **Hatch** changes the hatch pattern of the destination object to that of the source object.

Choosing **Table** changes the table style of the destination object to that of the source object.

After making the necessary changes, choose **OK** to close the Property Settings dialog box. AutoCAD continues with the "Select destination object(s):" prompt. Press ENTER to complete the object selection.

Open the Exercise Manual PDF file for Chapter 5 on the accompanying CD for project- and discipline-specific exercises.

 If you have the accompanying Exercise Manual, refer to Chapter 5 for project- and discipline-specific exercises.

REVIEW QUESTIONS

1. The command that allows you to draw freehand lines is:

   a. SKETCH

   b. DRAW

   c. PLINE

   d. FREE

2. To create a six-sided area that would select all the objects completely within it, you should respond to the "Select Objects:" prompt with:

   a. WP

   b. CP

   c. W

   d. C

3. The MATCHPROP command does not allow you to modify an object's:

   a. Linetype

   b. Fillmode

   c. Color

   d. Layer

4. By default, most selection set prompts default to which option:

   a. ALL

   b. AUTO

   c. BOX

   d. WINDOW

   e. CROSSING

5. The MULTIPLE option selection sets allows you to:

    a. select multiple objects that lie on top of each other

    b. scan the database only once to find multiple objects

    c. use the WINDOW or CROSSING options

    d. both A and B

    e. both B and C

6. In regard to using the SOLID command, which of the following statements is true?

    a. The order of point selection is unimportant

    b. FILL must be turned ON in order to use the SOLID command

    c. The points must be selected on existing objects

    d. The points must be selected in a clockwise order

    e. None of the above

7. What command is commonly used to create a filled rectangle?

    a. FILL-ON

    b. PLINE

    c. RECTANGLE

    d. LINE

    e. SOLID

8. To turn a series of line segments into a polyline, you would use which option of the PEDIT command?

    a. JOIN

    b. FIT CURVE

    c. SPLINE

    d. CONNECT

    e. Line segments cannot be connected.

9. Which command allows you to change the size of an object, where the X and Y scale factors are changed equally?

    a. ROTATE

    d. MODIFY

    b. SCALE

    e. MAGNIFY

    c. SHRINK

10. The BOX option for creating selection sets is most useful for:

    a. employed in a menu macro

    b. creating rectangular selection areas

    c. extending a selection into 3D

    d. Nothing—it is not a valid option.

11. Using the SCALE command, what number would you enter to enlarge an object by 50%?

    a. 0.5

    b. 50

    c. 3

    d. 1.5

12. To avoid changing the location of an object when using the SCALE command:

    a. the reference length should be less than the limits

    b. the scale factor should be less than one

    c. the base point should be on the object

    d. the base point should be at the origin

13. The ROTATE command is used to rotate objects around:

    a. any specified point

    b. the point -1,-1 only

    c. the origin

    d. is only usable in 3D drawings

    e. none of the above

14. The REMOVE option for forming selection sets deletes objects from the drawing in much the same manner as the ERASE command.

    a. True

    b. False

15. Which of the following is not supported by the noun/verb feature?

    a. COPY

    b. MOVE

    c. TRIM

    d. ROTATE

    e. ERASE

16. When shift to add selection is set to ON, you may add objects to the selection set by:

    a. picking the objects

    b. windowing the objects

    c. using SHIFT to add

    d. using CTRL to add

Directions: Use Figure QX5–1 to answer the following questions.

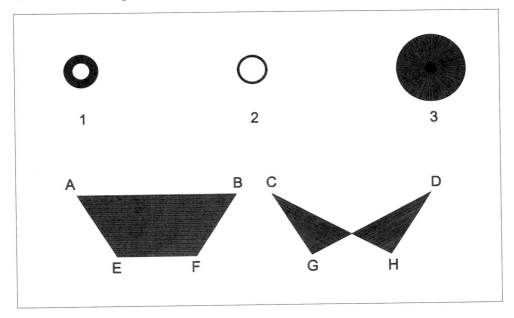

Figure QX5–1

17. Which of the figures were drawn with the DONUT command?

    a. 1

    b. 2

    c. 3

    d. all of the above

    e. 1, 2, and 3

18. Which of the donut objects has the smallest inside diameter?

    a. 1

    b. 2

    c. 3

19. Which of the options below is the correct sequence to create the solid object on the left?

   a. A, B, F, E
   b. A, E, F, B
   c. A, B, E, F
   d. F, E, A, B
   e. B, F, A, E

20. Which of the options below is the correct sequence to create the solid object on the right?

   a. H, G, C, D
   b. G, H, D, C
   c. G, D, H, C
   d. D, C, H, G
   e. C, G, H, D

21. If the space between the two solid shapes were to be filled, what sequence of points would completely fill this space?

   a. G, B, F, C
   b. B, G, F, C
   c. C, B, F, G
   d. F, G, C, B
   e. B, F, C, G

CHAPTER 6

Fundamentals V

INTRODUCTION

After completing this chapter, you will be able to do the following:

- Construct geometric figures by means of the MULTILINE (MLINE), SPLINE, WIPEOUT and REVCLOUD commands
- Modify multilines and splines
- Create or modify multiline styles
- Use grips to modify objects, group objects, and filter certain types of objects by Quick Select for modification
- Use inquiry commands
- Change the settings of system variables

MULTILINES

The MLINE command allows you to draw multiple parallel line segments called multilines. You can modify the intersection of two or more multilines or cut gaps in them with the MLEDIT command. AutoCAD also allows you to create new multiline styles or edit existing styles composed of up to 16 lines, called elements, with the MLSTYLE command.

DRAWING MULTIPLE PARALLEL LINES

Multilines are multiple parallel line segments, similar to polyline segments that have been offset one or more times. Examples of multilines are shown in Figure 6–1.

Figure 6–1 Examples of multilines

The properties of each element of a multiline are determined by the current style when the multiline is drawn. The properties of multilines controlled by the style include whether to display the line at the joints (miters) and ends and, to close the ends, and rather to connect inner and/or outer elements. In addition, the style controls element properties such as color, linetype, and offset distance between two parallel lines.

Invoke the MLINE command by choosing **Multiline** from the Draw menu and AutoCAD prompts:

> Specify start point or ⬇ (specify a point or right click and select one of the options)

Specify the starting point of the multiline, known as its origin. Once you specify the origin for a multiline, AutoCAD prompts:

> Specify next point:

When you respond by selecting a point, the first multiline segment is drawn according to the current style. You are then prompted:

> Specify next point or ⬇

If you specify a point, the next segment is drawn. After two segments have been drawn, the shortcut menu will include the CLOSE option. Choosing CLOSE causes the next segment to join the origin of the multiline, fillets all elements, and exits the command.

Choosing UNDO after any segment is drawn (and the MLINE command has not been terminated) causes the last segment to be erased, and you are prompted again for the next point.

Multiline command options

After invoking the MLINE command, you can right-click and choose one of the options, which include JUSTIFICATION, SCALE, and STYLE.

Choosing JUSTIFICATION determines the relationship between the elements of the multiline and the line you specify by way of the placement of the points. The justification is set by selecting one of the three available suboptions.

Enter justification type

Choosing TOP causes the element with the most positive offset value to be drawn aligned with the specified points. All other elements will be below or to the right of the element with the most positive offset value. In other words, if the line is drawn left-to-right, the element with the most positive offset value will be above all other elements.

Choosing ZERO causes the elements to center between the specified points. Elements with positive offsets will be to the right or above, and those with negative offsets to the left or below, the selected points.

Choosing BOTTOM causes the element with the most negative offset value to be drawn aligned with the specified points. All other elements will be to the left or above the element with the most negative offset value. In other words, if the line is drawn left-to-right, and the element with the most negative value will be below all other elements.

Figure 6–2 shows the location for various justifications.

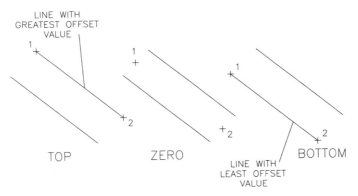

Figure 6–2 *Location of various justifications*

Choosing SCALE determines the value used for offsetting elements relative to the values assigned in the style. For instance, if the scale is changed to 3.0, elements that are assigned 0.5 and −1.5 will be drawn with offsets of 1.5 and −4.5, respectively. If a negative value is given for the scale, then the signs of the values assigned to them in the style will be inverted (positive to negative and negative to positive). The value can be entered in decimal form or as a fraction. Also, a 0 (zero) scale value produces a single line.

Choosing STYLE allows you to choose the current multiline style from the available styles. A detailed explanation is provided later in this chapter for creating or modifying multiline styles.

EDITING MULTIPLE PARALLEL LINES

The MLEDIT command allows you to modify the intersections of two or more multilines or cut gaps in the lines of one multiline. The tools available depend on the type of intersection to be modified (cross, tee, or vertex) and if one or more elements need to be cut or welded.

Invoke **Multiline** from the Object flyout of the Modify menu and AutoCAD displays the Multilines Edit Tools dialog box, as shown in Figure 6–3.

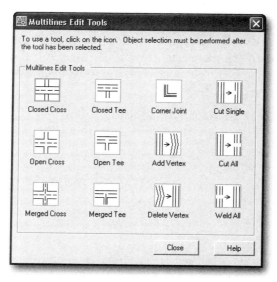

Figure 6–3 *Multilines Edit Tools dialog box*

To choose one of the available tools, select the appropriate image tile. AutoCAD then prompts for the required information.

The first column in the Multilines Edit Tools dialog box contains tools for multilines that cross, the second for multilines that form a tee, the third for corner joints and vertices, and the fourth for multilines to be cut or welded.

Choose **Closed Cross** (see Figure 6–3) to cut all lines that make up the first multiline you select at the point where it crosses the second multiline (see Figure 6–4)

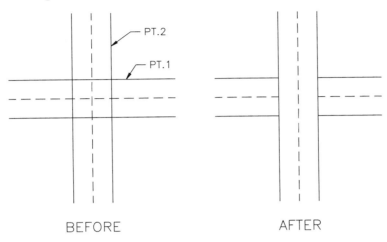

Figure 6–4 *An example of a closed cross*

Choose **Open Cross** (see Figure 6–3) to cut all lines that make up the first multiline you select and cuts only the outside line of the second multiline (see Figure 6–5).

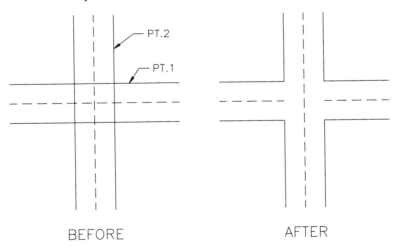

Figure 6–5 *An example of an open cross*

Choose **Merged Cross** (see Figure 6–3) to cut all lines that make up the intersecting multiline, you select, except the centerlines (see Figure 6–6).

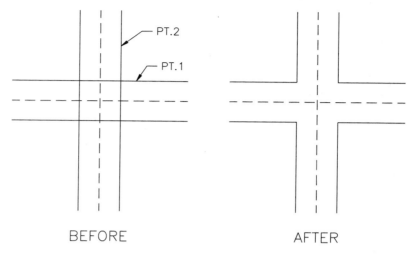

Figure 6–6 *An example of a merged cross*

Choose **Closed Tee** (see Figure 6–3) to extend or shorten the first multiline you select to its intersection with the second multiline (see Figure 6–7).

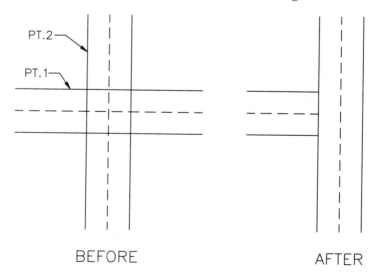

Figure 6–7 *An example of a closed tee*

Choose **Open Tee** (see Figure 6–3) to create an open-tee intersection between two multilines (see Figure 6–8). The first multiline is trimmed or extended to its intersection with the second multiline.

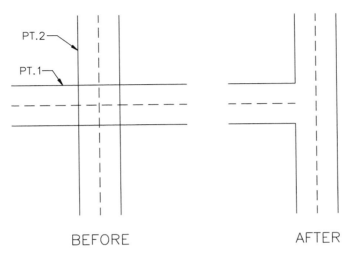

Figure 6–8 *An example of an open tee*

Choose **Merged Tee** (see Figure 6–3) to create a merged-tee intersection between two multilines (see Figure 6–9). The multiline selected first is trimmed or extended to its intersection with the second multiline.

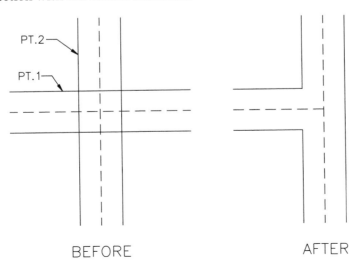

Figure 6–9 *An example of a merged tee*

Choose **Corner Joint** (see Figure 6–3) to lengthen or shorten each of the two multilines you select as necessary to create a clean intersection forming a corner (see Figure 6–10).

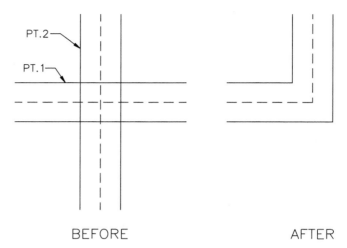

Figure 6–10 *An example of a corner joint*

Choose **Add Vertex** (see Figure 6–3) to add a vertex to a multiline (see Figure 6–11).

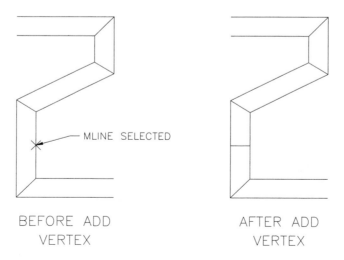

Figure 6–11 *An example of adding a vertex*

Choose **Delete Vertex** (see Figure 6–3) to delete a vertex from a multiline (see Figure 6–12).

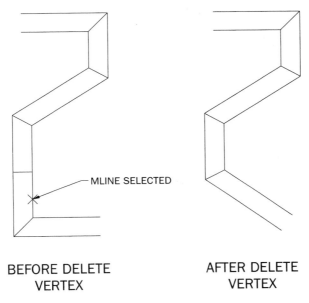

Figure 6–12 *An example of deleting a vertex*

Choose **Cut Single** (see Figure 6–3) to cut a selected element of a multiline between two cut points (see Figure 6–13).

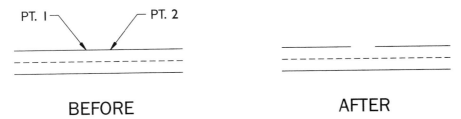

Figure 6–13 *An example of removing a selected element of a multiline between two cut points*

Choose **Cut All** (see Figure 6–3) to remove a portion of a multiline between two selected cut points (see Figure 6–14).

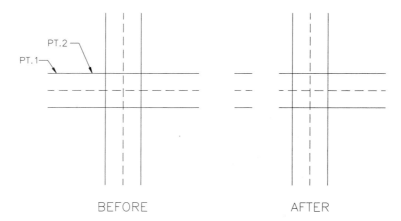

Figure 6–14 *An example of removing a portion of a multiline between two cut points*

Choose **Weld All** (see Figure 6–3) to rejoin multiline segments that have been cut (see Figure 6–15).

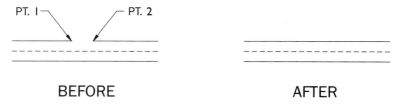

Figure 6–15 *An example of rejoining multiline segments that have been cut*

AutoCAD prompts to select additional multilines to continue with the selected multiline tool. Choosing UNDO undoes the edited intersection and AutoCAD continues with the "Select first mline or ⬇" prompt. To terminate the command sequence, press ENTER.

AutoCAD allows you to trim and extend Multilines in the same manner as other objects such as lines, arcs and polylines.

CREATING AND MODIFYING MULTILINE STYLES

The MLSTYLE command is used to create a new Multiline Style or edit an existing one. You can define a multiline style consisting of up to 16 lines, called elements. The style controls the number of elements and the properties of each element. In addition, you can specify the background color and the end caps of each multiline.

Invoke the **Multiline Style** from the Format menu and AutoCAD displays the Multiline Style dialog box (see Figure 6–16).

Figure 6–16 *Multiline Style dialog box*

Next to the label **Current Multiline Style** is the name of the multiline style that current. The current multiline style is the style that will be used for all new multilines.

The **Styles** list box allows you to choose from the available multiline styles loaded in the current drawing.

The **Description** area displays a description (if available) of the selected multiline style.

The **Preview of** area displays the name and an image of the selected multiline style.

Set Current is used to make the selected style current.

Selecting **New** causes the Create New Multiline Style dialog box to be displayed (see Figure 6–17).

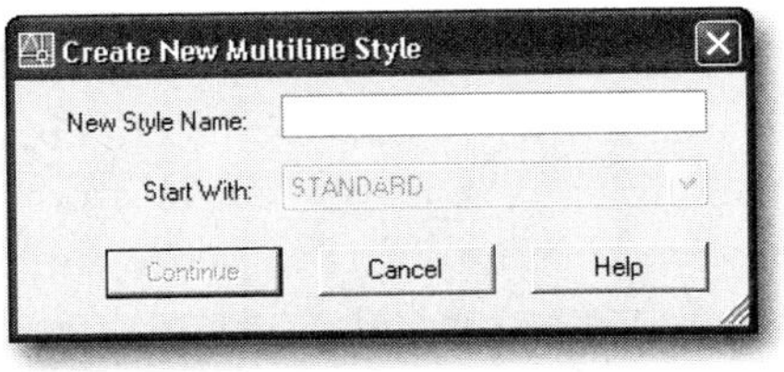

Figure 6–17 *Create New Multiline Style dialog box*

To create a new Multiline Style, enter a style name in the **New Style Name** text box. The **Start With** list box lets you choose an existing multiline style to use as a basis for the new style. After selecting the base style and naming the new one, choose **Continue** and AutoCAD displays the New Multiline Style dialog box (see Figure 6–18).

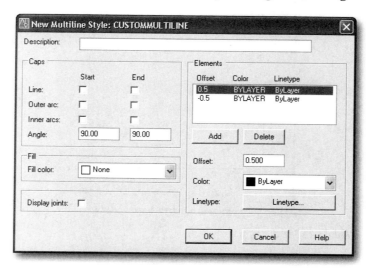

Figure 6–18 *New Multiline Style dialog box*

The **Description** text box allows you to add a description of up to 255 characters, including spaces. The description entered here will be displayed in the **Description** text box when this style is selected in the Multiline Styles dialog box.

The **Elements** section sets element properties, such as the offset, color, and linetype, of new and existing multiline elements. AutoCAD lists all the elements in the current multiline style. Choose **Add** to add a new element to the multiline style. Each new element in the style is defined by its offset, its color (default is set to ByLayer), and its linetype (default is set to ByLayer). The **Offset**, **Color**, and **Linetype** fields allow you to specify an elements offset distance from the middle of the multiline, color, and linetype. Elements are always displayed in descending order of their offsets. To delete the selected element, choose **Delete**.

The **Caps** section allows you to specify the appearance of multiline start and end caps. The **Line** check boxes control the display of the start and end caps, by adding a straight line to the start or end of a multiline, as shown in Figure 6–19. The **Outer arc** check boxes control the display of the start and end caps by connecting the ends of the outermost elements with a semicircular arc, as shown in Figure 6–20. The **Inner arcs** check boxes control the display of the start and end caps by connecting the ends of the innermost elements with a semicircular arc, as shown in Figure 6–21. For a multiline with an odd number of elements, the center element is not connected. For an even number of elements, connected elements are paired with elements that are the same number from each edge. For example, the second element from the outer-left will be connected to the second element from the outer-right, the third to the third, and so forth. The **Angle** edit field sets the angle of endcaps. Figure 6–22 shows the display of the end caps with an angular cap.

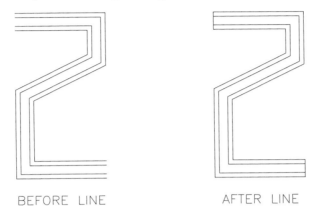

Figure 6–19 *Display of the line for start and end caps*

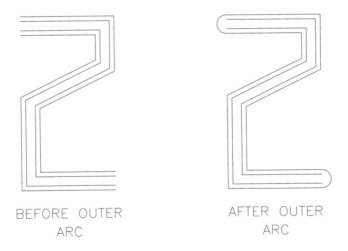

Figure 6–20 *Display of the outer arc for start and end caps*

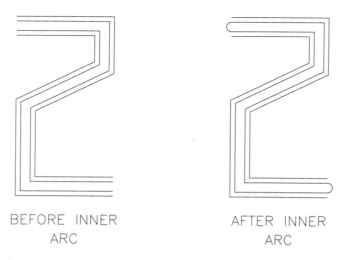

Figure 6–21 *Display of the inner arc for start and end caps*

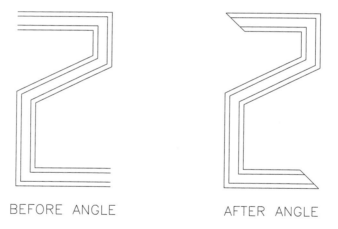

Figure 6–22 *Display of the end caps with an angular cap*

The **Fill** section contains a **Fill color** list box which allows you to include a specified background color for the multiline style. Choosing **Select Color** displays the Select Color dialog box from which you can specify a non-standard color for the background fill.

Selecting **Display joints** causes a line (miter) to be drawn at the vertices of each multiline segment (see Figure 6–23).

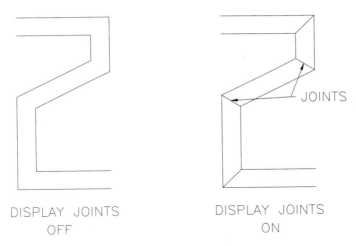

Figure 6–23 *Display of the joints*

Choose **OK** to create the selected multiline style.

Choose **Modify** to change the element properties of an existing multiline style. AutoCAD displays the Modify Multiline Style dialog box which is similar to the Create New Multiline Style dialog box (see Figure 6–17).

Choose **Rename** to change the name of the selected multiline style.

Choose **Delete** to delete the selected multiline style.

Choose **Load** to load multiline style from a specified *.mln* file.

Choose **Save** to save multiline style to a multiline library (*.mln*) file. If you specify an *.mln* file that already exists, the new style definition is added to the file and existing definitions are not erased. The default file name is *acad.mln*.

Choose **OK** to close the New Multiline Style dialog box to create your new multiline style.

SPLINE CURVES

The SPLINE command is used to draw a curve through specified points. The curve drawn is a nonuniform rational B-spline (NURBS). This type of curve has irregularly varying radii, and is used to draw objects such as topographical contour lines.

The spline curve is drawn through a series of two or more points, with options either to specify end tangents or to use CLOSE to join the last segment to the first. Another option, FIT TOLERANCE, lets you specify a tolerance, which determines how close to the selected points the curve is drawn.

Invoke **Spline** (SPLINE) from the Draw toolbar (see Figure 6–24) and AutoCAD prompts:

Specify first point: *(specify a point or choose Object from the shortcut menu)*

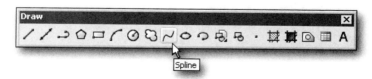

Figure 6–24 *Invoking Spline from the Draw toolbar*

The default option lets you specify the point from which the spline starts. After you enter the first point, a preview line appears. You will then be prompted:

Specify next point:

When you respond by specifying a point, the spline segments are displayed as a preview spline, curving from the first point, through the second point, and ending at the cursor. You are then prompted:

Specify next point or ⤓

If you specify a point, the next segment is added to the spline. This will occur with each additional point specified until you use the CLOSE option or by pressing ENTER.

You are then prompted for a start tangent determining point as follows:

Specify start tangent:

If you specify a point (for tangency), its direction from the start point determines the start tangent. If you press ENTER, the direction from the first point to the second point determines the tangency. After the start tangency is established, you are prompted:

Specify end tangent:

If you specify a point (for tangency), its direction from the endpoint determines the end tangent. If you press ENTER, the direction from the last point to the previous point determines the tangency.

The CLOSE option uses the original starting point of the first spline segment as the endpoint of the last segment and terminates segment placing sequence. You are then prompted:

Specify tangent:

You can specify a point to determine the tangency at the connection of the first and last segments. If you press ENTER, AutoCAD calculates the tangency and draws the spline accordingly. You can also use the perpendicular or tangent object snap modes to cause the tangency of the spline to be perpendicular or tangent to a selected object.

The FIT TOLERANCE option lets you vary how the spline is drawn relative to the selected points. You are then prompted:

> Specify fit tolerance <current>:

Entering 0 (zero) causes the spline to pass through the specified points. A positive value causes the spline to pass within the specified value of the points.

Instead of specifying the first point, if you choose the OBJECT option, AutoCAD prompts to select an object. AutoCAD converts selected 2D or 3D quadratic or cubic spline-fit polylines to equivalent splines and (depending on the setting of the DELOBJ system variable) deletes the polylines.

EDITING SPLINE CURVES

Splines created by means of the SPLINE command have numerous characteristics that can be changed with the SPLINEDIT command. These include quantity and location of fit points, end characteristics such as opened or closed, tangency, and tolerance of the spline (how near the spline is drawn to fit points).

SPLINEDIT operates on control points (which are different than fit points) of the selected spline. Features include adding control points and changing the weight of individual control points, which determines how close the spline is drawn to the individual control points.

Invoke **Edit Spline** (SPLINEDIT) from the Modify II toolbar (see Figure 6–25) and AutoCAD prompts:

> Select Spline: *(select a spline curve)*
> Enter an option *(choose one of the available options from the shortcut menu)*

Figure 6–25 *Invoking Edit Spline from the Modify II toolbar*

Control points appear in the grip color, and, if the spline has fit data, fit points also appear in the grip color. If you select a spline whose fit data is deleted, then the FIT DATA option is not available. A spline can lose its fit data if you use the PURGE option while editing fit data, refining the spline, moving its control vertices, fitting the spline to a tolerance, or opening or closing the spline.

SPLINEDIT command options

The OPEN option will replace CLOSE if you select a closed spline, and vice versa.

Choosing FIT DATA allows you to edit the spline by providing the following suboptions: ADD, OPEN, DELETE, MOVE, PURGE, TANGENTS, TOLERANCE, and EXIT.

> Choosing ADD allows you to add fit points to the selected spline. AutoCAD prompts:
>
>> Specify control point <exit>: *(select a fit point)*
>
> After you select one of the fit points, AutoCAD highlights it and you are prompted for the next point:
>
>> Specify new point or ⬇ *(specify a point)*
>> Specify new point or ⬇ *(specify another point or press ENTER)*
>
> Selecting a point places a new fit point between the highlighted ones.
>
> Choosing CLOSE closes an open spline smoothly with a segment or smooths a spline with coincidental starting and ending points.
>
> Choosing OPEN opens a closed spline, disconnecting it and changing the starting and ending points.
>
> Choosing DELETE deletes a selected fit point.
>
> Choosing MOVE moves fit options to a new location by prompting:
>
>> Specify new location or ⬇ *(choose one of the available options to navigate through fit point)*
>
> Choosing PURGE deletes fit data for the selected spline.
>
> Choosing TANGENTS edits the start and end tangents of a spline.
>
> Choosing TOLERANCE refits the spline to the existing points with new tolerance value.
>
> Choosing EXIT exits the FIT DATA option and returns to the main prompt.

Choosing CLOSE causes the spline to be joined smoothly at its start point.

Choosing OPEN opens a closed spline. Previously open splines with coincidental starting and ending points will lose their tangency. Others will be restored to a previous state.

Choosing MOVE VERTEX relocates a spline's control vertices by providing the following suboptions: NEXT, PREVIOUS, SELECT POINT, and EXIT.

Choosing REFINE allows you to fine-tune a spline definition by providing the following suboptions: ADD CONTROL POINT, ELEVATE ORDER, WEIGHT, and EXIT.

> Choosing ADD CONTROL POINT increases the number of control points that control a portion of a spline.
>
> Choosing ELEVATE ORDER increases the order of the spline. You can increase the current order of a spline up to 26 (the default is 4), causing an increase in the number of control points.
>
> Choosing WEIGHT changes the weight at various spline control points by providing the following sub-options: NEXT, PREVIOUS, SELECT POINT, and EXIT.
>
> The default weight value for a control point is 1.0. Increasing it causes the spline to be drawn closer to the selected point. A negative or zero value is not valid.
>
> From the REFINE menu, the EXIT suboption returns you to the main prompt.

Choosing REVERSE reverses the direction of the spline. Reversing the spline does not delete the fit data.

Choosing UNDO undoes the effects of the last subcommand.

Choosing EXIT terminates the SPLINEDIT command.

WIPEOUT

The WIPEOUT command allows you to create an area on the screen that obscures previously drawn objects within its boundary. These objects can be displayed with or without a visible boundary (called a frame).

Invoke **Wipeout** (WIPEOUT) from the Draw menu and AutoCAD prompts:

> Specify start point or ⬇ (specify the start point of the wipeout area or right-click and select one of the options)
> Specify next point: (specify the second point)
> Specify next point or ⬇ (specify the third point or right-click and select one of the options)

After drawing a connected series of lines, press ENTER to terminate the WIPEOUT command. The polygonal boundary of the wipeout object is determined from the series of specified points. The shortcut menu displayed when you right-click at the prompt for the first point includes the FRAMES and POLYLINE options.

Selecting FRAMES determines whether the edges of all wipeout objects are displayed or hidden. Choose ON to display all wipeout frames and choose OFF to suppress the display of all wipeout frames. When the frames are turned off the wipeout object cannot be selected on the screen with the pointing device.

Selecting POLYLINE allows you to select a polyline, which determines the polygonal boundary of the wipeout area.

After selecting POLYLINE AutoCAD prompts:

Select a closed polyline: *(Use an object selection method to select a closed polyline)*
Erase polyline? *(Enter y or n)*

Enter **y** to erase the polyline that was used to create the wipeout object. Enter **n** to retain the polyline. The shortcut menu displayed when you right-click at the prompt for the third or subsequent point includes CLOSE and UNDO options. These operate in the same way as in the LINE and POLYLINE commands. Also available are the ENTER and CANCEL options.

 Note: The wipeout object created by the WIPEOUT command will cover existing objects. Objects drawn on top of the wipeout object will not be hidden. If one or more of the objects being covered is modified, by the MOVE command for example, then it will no longer be obscured. Likewise, if the wipeout area is modified, it will then cover all objects that overlap it. You can use the DRAWORDER command to change the effect of the wipeout object on other objects.

REVISION CLOUD

The REVCLOUD command allows you to draw a connected series of arcs encircling objects in a drawing to signify an area on the drawing that has been revised (see Figure 6–26).

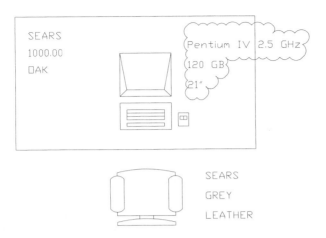

Figure 6–26 *An example of using the* REVCLOUD *command*

Invoke **Revision Cloud** (REVCLOUD) from the Draw toolbar (see Figure 6–27) and AutoCAD prompts:

> Specify start point or ⊙ *(specify the start point of the revision cloud)*
> Guide crosshairs along cloud path... *(move the cursor crosshairs along the path of the desired revision cloud)*

Figure 6–27 *Invoking Revision Cloud from the Draw toolbar*

When the crosshairs cursor approaches the starting point, the cloud is automatically closed without requiring any additional action or input.

At the "Specify start point or ⊙" prompt right-click for the shortcut menu which includes the options ARC LENGTH, OBJECT, and STYLE.

Choose ARC LENGTH to specify the size range of the arcs when prompted for minimum and maximum arc lengths. These values are then multiplied by the value of the dimension variable DIMSCALE to compensate for drawings with different scale factors.

Choose OBJECT to select a closed shape (polyline, rectangle, circle, etc.) from which AutoCAD creates a revision cloud. You are then prompted:

> Reverse direction *(Choose* YES *or* NO*)*

Choose YES to cause the arcs to be redrawn on the opposite side of the line or arc defining the cloud. Choose NO to cause the revision cloud to remain as drawn. To reverse the bulges of an existing revision cloud use the OBJECT option to select it and then choose the YES option at the "Reverse direction" prompt.

Choose STYLE to select between NORMAL and CALLIGRAPHY. NORMAL causes AutoCAD to draw the revision cloud in the current linetype and lineweight. CALLIGRAPHY causes the revision cloud to be drawn in bulges of tapered line width, producing an artistic effect. To change the style of an existing revision cloud, invoke the REVCLOUD command, select STYLE from the shortcut menu, change the style, then select OBJECT from the shortcut menu and select the revision cloud.

EDITING WITH GRIPS

Grips allow you to edit AutoCAD drawings in an entirely different way than using the traditional AutoCAD modify commands. With grips you can move, stretch, rotate, copy, scale, and mirror selected objects without invoking one of the regular AutoCAD modify commands. When you select an object with grips, small squares appear at strategic points on the object enabling you to edit the selected objects.

To change the grip options, choose **Options** from the Tools menu and AutoCAD displays the Options dialog box (see Figure 6–28).

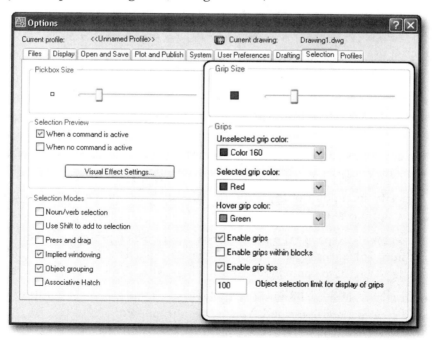

Figure 6–28 *Options dialog box*

Choose the **Selection** tab, and AutoCAD displays various options related to Grips (see Figure 6–28).

Selecting **Enable grips** controls the display of the grips. If it is set to on, the grips display is enabled; if it is set off, the grips display is disabled. You can also enable the grips feature by setting the GRIPS system variable to 1.

Selecting **Enable grips within blocks** controls the display of grips on objects within blocks. If it is enabled, the grips are displayed on all objects within the block; if it is disabled, a grip is displayed only on the insertion point of the block.

Selecting **Enable grip tips** causes a grip-specific tip to be displayed when the cursor hovers over a grip on an object that supports grip tips. Standard AutoCAD objects are not affected by this option.

The **Unselected grip color** list box allows you to change the color of the unselected grips.

The **Selected grip color** list box allows you to change the color of the selected grips.

The **Hover grip color** list box allows you to change the color of the grips in hover mode.

The **Object selection limit for display of grips** text box allows you to specify a number that limits the display of grips when the initial selection set includes more than the specified number of objects. The valid range is 1 to 32,767. The default setting is 100.

The **Grip Size** slider bar allows you to change the size of the grips. To adjust the size of the grips, move the slider box left or right. As you move the slider, the size is illustrated to the right of the slider.

After making changes, choose **OK** to accept the changes and close the Options dialog box.

AutoCAD gives you a visual cue when grips are enabled by displaying a pick box at the intersection of the crosshairs, even when you have not invoked any of the AutoCAD command (see Figure 6–29).

Figure 6–29 *The pick box displayed at the intersection of the crosshairs*

Note: The pick box is also displayed on the crosshairs when the PICKFIRST (Noun/Verb selection) system variable is set to ON. Grips appear on the endpoints and midpoint of lines and arcs, on the vertices and endpoints of polylines, on quadrants and the center of circles, on dimensions, text, solids, 3dfaces, 3dmeshes, and viewports, and on the insertion point of a block. Figure 6–30 shows location of the grips on some of the commonly used objects.

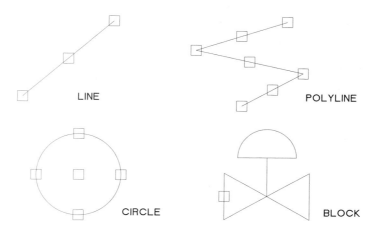

Figure 6–30 *Locations of grips on commonly used objects*

USING GRIPS

This section explains how to utilize grips to modify your drawing. Learning to use grips speeds up the editing of your drawing while at the same time maintaining the accuracy of your work.

When you move your cursor over a grip, it automatically snaps to the grip point. This allows you to specify exact locations in the drawing without having to use grid, snap, ortho, object snap, or coordinate entry tools.

To clear a selection set, press the ESC key. The grips on the selected objects will disappear. When you invoke a non-modifying AutoCAD command, such as LINE or CIRCLE, AutoCAD clears the selection set.

To edit an object using grips, select the object and then select a grip (place the cursor over the grip and press the pick button) to act as the base point for the editing operation. You can use multiple grips to keep the shape of the object intact. Hold down SHIFT as you select the grips. Selecting a grip starts the Grip modes, which includes *STRETCH*, *MOVE*, *ROTATE*, *SCALE*, and *MIRROR*. You can cycle through the grip modes by pressing the SPACEBAR, pressing ENTER, entering a keyboard shortcut, or selecting from the right-click shortcut menu (see Figure 6–31). To cancel a grip mode, enter **x** (for the mode's EXIT option);. You can also use a combination of the current grip mode and a copy operation on the selection set.

Figure 6–31 *Shortcut menu displaying the grip modes*

Stretch

STRETCH works like the STRETCH command. It allows you to stretch the shape of an object without affecting other parts. When you are in the stretch mode, the following prompt appears:

Specify stretch point or ⊡

STRETCH (the default) refers to the stretch displacement point. As you move the cursor, you see that the shape of the object is stretched dynamically from the base point. You can specify the new point with the cursor or by entering coordinates.

If necessary, you can change the base point to be a point other than the base grip, by selecting BASE POINT from the shortcut menu. Then specify the new base point with the cursor, or enter the coordinates. You can also select multiple base points by holding the SHIFT key and choosing additional grips.

To make multiple copies while stretching objects, choose COPY from the shortcut menu. Then specify destination copy points with the cursor, or enter their coordinates.

Move

MOVE works like the MOVE command. It allows you to move one or more objects from their present location to a new one without changing their orientation or size. In addition, you can make copies of the selected objects at the specified displacement, leaving the original objects intact. To invoke the move mode, cycle through the modes by pressing ENTER until it takes you to the move mode, or choose MOVE from the shortcut menu.

At the "Specify move point or ⊻" prompt enter a displacement point. As you move the cursor, AutoCAD moves all the objects in the current selection set to a new point relative to the base point. You can specify the new point with the cursor or by entering the coordinates.

If necessary, you can change the base point to be a point other than the base grip, by selecting BASE POINT from the shortcut menu. Then specify the new base point with the cursor, or enter the coordinates. You can also select multiple base points by holding the SHIFT key and choosing additional grips.

To make multiple copies while stretching objects, choose COPY from the shortcut menu. Then specify destination copy points with the cursor, or enter their coordinates.

Rotate Mode

ROTATE works like the ROTATE command. It allows you to change the orientation of objects by rotating them about a specified base point. In addition, you can make copies of the selected objects and at the same time rotate them about a specified base point.

To invoke the ROTATE mode, cycle through the modes by pressing ENTER until it takes you to the rotate mode, or choose ROTATE from the shortcut menu.

At the "Specify rotation angle or ⊻" prompt enter the rotation angle to which objects are to be rotated. As you move the cursor, AutoCAD allows you to specify the rotation angle, relative to the base point, with your cursor. You can specify the new orientation with the cursor or from the keyboard. If you specify an angle by entering a value from the keyboard, this is taken as the degree that the objects should be rotated based on their current orientation. A positive value rotates counterclockwise, and a negative value rotates clockwise. Similar to the ROTATE command, you can use the REFERENCE option to specify the desired new rotation.

If necessary, you can change the base point to be a point other than the base grip, by selecting BASE POINT from the shortcut menu. Then specify the new base point with the cursor, or enter the coordinates. You can also select multiple base points by holding the SHIFT key and choosing additional grips.

To make multiple copies while rotating objects, choose COPY from the shortcut menu. Then specify destination copy points with the cursor, or enter their coordinates.

Scale Mode

SCALE works like the SCALE command. It allows you to change the size of objects. In addition, you can make copies of the selected. To invoke the scale mode, cycle through the modes by pressing ENTER until it takes you to the scale mode, or choose SCALE from the shortcut menu.

At the "Specify scale factor or ⬇" prompt enter the scale factor by which objects should be scaled. As you move the cursor, AutoCAD allows you to specify the scale factor, in relation to the base point, by moving your cursor closer to or further from the base point. You can specify the new scale factor with the cursor or from the keyboard. If you specify the scale factor by entering a value from the keyboard, this is taken as a relative scale factor by which all dimensions of the objects in the current selection set are to be multiplied. To enlarge an object, enter a scale factor greater than 1; to reduce an object, use a scale factor between 0 and 1. Similar to the SCALE command, you can use the REFERENCE option to specify the current length and the desired new length.

If necessary, you can change the base point to be a point other than the base grip, by selecting BASE POINT from the shortcut menu. Then specify the new base point with the cursor, or enter the coordinates. You can also select multiple base points by holding the SHIFT key and choosing additional grips.

To make multiple copies while scaling objects, choose COPY from the shortcut menu. Then specify destination copy points with the cursor, or enter their coordinates.

Mirror Mode

MIRROR works like the MIRROR command. It allows you to make mirror images of existing objects. To invoke the mirror mode, cycle through the modes by pressing ENTER until it takes you to the mirror mode, or choose MIRROR from the shortcut menu.

Two points are required in AutoCAD to define a line about which the selected objects are mirrored. AutoCAD considers the base grip point as the first point; the second point is the one you specify or enter in response to the "Specify second point or ⬇" prompt.

If necessary, you can change the base point to be a point other than the base grip, by selecting BASE POINT from the shortcut menu. Then specify the new base point with the cursor, or enter the coordinates. You can also select multiple base points by holding the SHIFT key and choosing additional grips.

To make multiple copies while mirroring objects, choose COPY from the shortcut menu. Then specify destination copy points with the cursor, or enter their coordinates.

SELECTING OBJECTS BY QUICK SELECT

The QSELECT command is used to create selection sets based on filters that select objects that have similar characteristics or properties. For example, you can create a selection set of all lines that are equal to or less than 2.5 units long. Or you can create a selection set of all objects that are not text objects on a specific layer. The combinations of possible filters are almost limitless.

To create a filtered selection set, invoke **Quick Select** (QSELECT) from the Tools menu and AutoCAD displays the Quick Select dialog box (see Figure 6–32).

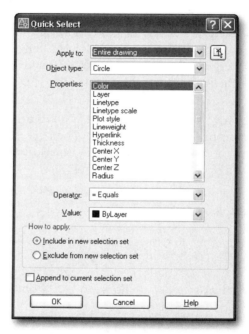

Figure 6–32 *Quick Select dialog box*

The **Apply to** list box lets you select whether to apply the specified filters to the Current selection or to the Entire drawing. If there is a current selection, then Current selection is the default, otherwise, Entire drawing is the default. If the **Append to current selection set** check box is checked, then Current selection is not an option. If you wish to create a selection set, choose the **Select Objects** icon (located next to the **Apply to** list box). **Select Objects** is available only when **Append to current selection set** is not checked.

Choosing **Select Objects** returns you to the drawing screen so you can select objects. After selecting the objects to be included in the selection set, press ENTER.

The **Object type** drop down list lets you select whether to include certain types of objects or multiple objects.

The **Properties** list box lists the properties that can be used for filters for objects specified in the **Object type** drop down list box.

The **Operator** drop down list box lets you apply logical operators to values. These include Equals, Not Equal, Greater than, Less than, and Select All. Greater than and Less than apply primarily to numeric values, and Select All applies to text strings.

The **Value** field is dynamic, based on the **Properties** selection and lets you specify a value to which the operator applies. If you specify the Greater than operator and the value of 1.0 for the length of lines, then lines with lengths greater than 1.0 will be filtered to be either included in or excluded from the selection set, depending on which radio button in the **How to apply** section has been chosen.

The **How to apply** section lets you specify whether the filtered objects will be included in or excluded from the selection set.

The **Append to current selection set** check box lets you specify whether the filtered selection set replaces the current selection or is appended to it.

Once you have all the required selection criteria set, choose **OK** to close the dialog box. AutoCAD displays grips on the selected objects. You can proceed with the appropriate modification for the selected objects.

SELECTION SET BY FILTER TOOL

The FILTER command displays the Object Selection Filters dialog box which lets you create filter lists that you can apply to a selection set. This is another method of selecting objects. With the FILTER command, you can select objects based on object properties, such as: location, object type, color, linetype, layer, block name, text style, and thickness. For example, you could use the FILTER command to select all the blue lines and arcs with a radius of 2.0 units. You can even name filter lists and save them. The selection set created by the FILTER command can be reselected by using PREVIOUS option at the next "Select object" prompt.

Invoke the FILTER command by typing **filter** at the On-Screen prompt and AutoCAD displays the Object Selection Filters dialog box (see Figure 6–33).

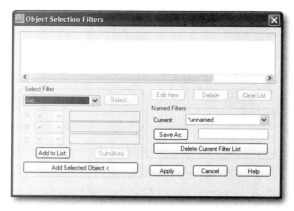

Figure 6-33 *Object Selection Filters dialog box*

The filter list box at the top of the dialog box displays the filters being used. The first time you use the FILTER command, the list box is empty.

The **Select Filter** section lets you add filters to the filters list box based on object properties. Your selection filters may be based on an object or multiple objects compared by boolean operators, both options are found in the **Select Filter** drop down list box. The boolean operators must be paired and balanced correctly in the filter list. For example, each Begin OR operator must have a matching End OR operator. If you select more than one filter object, AutoCAD by default uses an AND as a grouping operator between each filter.

Choosing **Select** displays a dialog box that lists all items of the specified type within the drawing. From the list, you can select as many items as you want to filter. This process is more accurate than typing the specific filter parameters.

Choosing **Add to List** adds the filter in the **Select Filter** section to the filter list box.

Choosing **Add Selected Object** allows you to select an object from the drawing and add it to the filter list box.

Choosing **Substitute** replaces the selected filter with the one in the **Select Filter** section.

Choosing **Edit Item** allows you to edit a selected filter from the filter list box by moving the filter into the **Select Filter** section for editing. Selecting the filter from the filter list box and then choosing the **Edit Item** button. When finished making changes in the **Select Filter** section, choose the **Substitute** button. The edited filter replaces the selected filter.

Choosing **Delete** deletes the selected filter in the filter list box.

Choosing **Clear List** deleted all of the filters from the filter list box.

To save the filter list, name the filter list in the **Save As** field, and choose **Save As** to save the list.

The **Current** list box displays saved filter lists by name. Select a list to make it current.

Choosing **Delete Current Filter List** deletes the named filter lists from the filter file.

Choosing **Apply** closes the dialog box; AutoCAD displays the "Select Objects" prompt. AutoCAD applies the filter to the objects you select.

CHANGING PROPERTIES OF SELECTED OBJECTS

The PROPERTIES command is used to manage and change the properties of selected objects by means of the Properties palette.

To change the objects in a selection set, invoke **Properties** (PROPERTIES) from the Standard toolbar (see Figure 6–34) and AutoCAD displays the Properties palette (see Figure 6–35).

Figure 6–34 *Invoking Properties from the Standard toolbar*

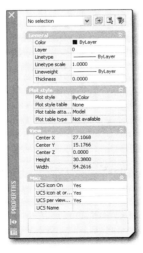

Figure 6–35 *Properties palette*

The default position for the Properties palette is floating on the left side of the screen. The Properties palette can be docked by double-clicking in the title bar or by placing the cursor over the title bar and dragging the window all the way to the side you wish to dock it. When docked, in can be undocked by placing the cursor over the double line bar at the top of the window and either double-clicking or dragging the window into the screen area (or across to a docking position on the right side of the screen). Double-clicking causes the Properties palette to become undocked and to float in the drawing area.

When the Properties palette is displayed and object, say a circle, is selected (make sure the PICKFIRST and/or GRIPS system variable is set to on), then the Properties palette lists all the properties of that circle. Not only are properties like color, linetype and layer listed, but the center's X, Y, and Z coordinates, radius, diameter, circumference, and area also listed. However, if two circles are selected, only the values of properties that are the same for both circles are listed. Properties that are not the same, but are common properties, show as *VARIES*. If you enter a value for an uncommon value, the new value will be applied to both circles. For example, if the centers of the two circles are different and you enter X and Y coordinates in the **Center X** and **Center Y** edit fields respectively, both circles will be moved to have their centers coincide with the specified coordinates. If different types of objects are in the selection set, then the properties listed in the Properties palette will include only those common to the selected objects such as color, layer and linetype. Whenever you change the properties in the Properties palette, AutoCAD reflects the changes immediately in the drawing window.

You can also open the Properties palette by double-clicking on an object when the PICKFIRST system variable is set to on. You can also open the Quick Select dialog box to select objects by choosing the **Quick Select** icon, located top right side of the Properties palette. If you need to select objects by the traditional method (by window and/or crossing), then choose the **Select Objects** icon located on the top right side of the Properties palette. In addition, you can change PICKADD system variable value by choosing the toggle value of the PICKADD system variable located on the top right side of the Properties palette. The PICKADD system variable setting controls how you add objects to an existing selection set.

You can also change properties of the selected objects by invoking the DDCHPROP or CHANGE commands.

The DDCHPROP command also displays the Properties palette, as shown in Figure 6–35.

The CHANGE command also allows you to change properties for selected objects such as their color, lineweight, linetype, lineweight, layer, elevation, thickness, and plotstyle. In addition, the CHANGE command lets you modify some of the characteristics of lines, circles, text, and blocks. Invoke the CHANGE command by typing **change** at the On-Screen prompt and AutoCAD prompts:

> Select objects: *(select objects and press* ENTER *to complete the selection)*
> Specify change point or ⬇ *(specify change point for the selected objects, or choose properties from the shortcut menu to change one of the properties of the selected object)*

CHANGE allows you to modify some of the characteristics of lines, circles, text, and blocks. If you select one or more lines, the closest endpoint(s) are moved to the new change point. If you select a circle, change point allows you to change the radius of the circle. If you selected more than one circle, AutoCAD moves on to the next circle and repeats the prompt. If the selected object is a text string, then AutoCAD allows you to change one or more parameters such as text height, text style, text rotation angle and text string. And if the selected object is a block, then specifying a new location repositions the block.

Instead of changing a point, you can change the properties of selected objects by right-clicking and choosing the PROPERTIES option from the shortcut menu. Properties that can be changed include color, lineweight, linetype, lineweight, layer, elevation, thickness, and plotstyle.

GROUPING OBJECTS

The GROUP command adds flexibility when modifying a group of objects. It allows you to name a selection set. Naming a selection set combines two powerful AutoCAD drawing features. One is being able to modify a group of unrelated objects as a group. It is similar to using PREVIOUS to select the last selection set when prompted to "Select object" for a modify command. The advantage of using GROUP instead of PREVIOUS is that you are not restricted to only the last selection set. The other feature combined in the GROUP command is that of giving a name to a selected group of objects for recall later by the name of the group. This is similar to the BLOCK command. The advantage of using GROUP instead of BLOCK is that the GROUP command's "selectable" switch can be set to off, allowing for modification of an individual member without losing its "membership" in the group. Also, named groups, like blocks, are saved with the drawing.

A named group can be selected for modification as a group only when its "selectable" switch is set to on. Figure 6–36 shows the result of trimming an object with the selectable switch set to on and to off. Modifying objects (such as with the MOVE or COPY command) that belong to a group can be accomplished by two methods. One is to select one of its members. The other method is to select the GROUP option by typing **g** at the "Select Objects" prompt. AutoCAD prompts for the group's name.

Enter the group name and press ENTER or the SPACEBAR. AutoCAD highlights the objects that belong to the specified group.

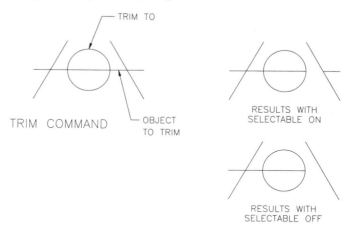

Figure 6–36 *Trimming an object with the selectable switch set to ON and to OFF*

To create a new group or edit an existing group, invoke the GROUP command by typing **group** at the On-Screen prompt and AutoCAD displays the **Object Grouping** dialog box (see Figure 6–37).

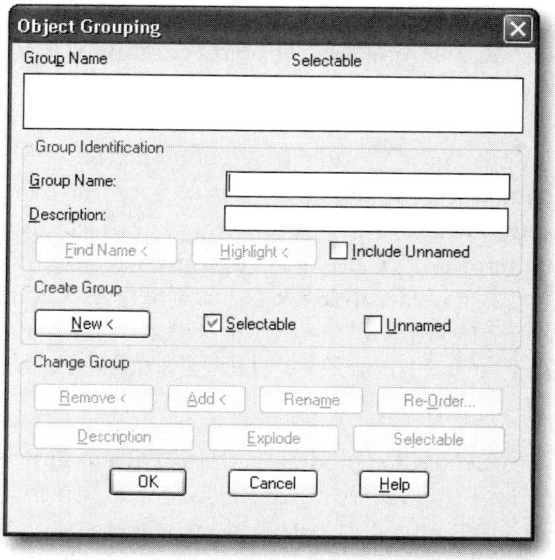

Figure 6–37 *Object Grouping dialog box*

The **Group Name** list box lists the names of existing groups defined in the current drawing. The **Selectable** column indicates whether a group is selectable. If it is listed as selectable, then selecting a single group member selects all the members except those on locked layers. If it is listed as unselectable, then selecting a single group member selects only that object.

The **Group Identification** section is where AutoCAD displays the group name and its description, when a group is selected in the **Group Name** list.

Choosing **Find Name** lists the groups to which an object belongs. AutoCAD prompts for the selection of an object and displays the **Group Member List** dialog box, which lists the group or groups to which the selected object belongs.

Choosing **Highlight** lets you see the members of the group selected from the **Group Name** list box.

Choosing **Include Unnamed** controls the listing of the unnamed groups in the **Group Name** list box.

The **Create Group** section is used for creating a new group. In addition, you can set whether or not it is initially selectable and if the group will have a name.

To create a new group, enter the group name and description in the **Group Name** and **Description** text boxes. Group names can include letters, numbers, and the special characters $ and _. To create an unnamed group, check **Unnamed**. AutoCAD assigns a default name, *A*n, to unnamed groups. The *n* represents a number that increases with each new group.

To create a selectable group, check **Selectable** and then choose **New**. AutoCAD prompts for the selection of objects. Select all the objects to be included in the new group, and press ENTER to complete the selection.

The **Change Group** section is for making changes to individual members of a group or to the group itself. The buttons are disabled until a group name is selected in the **Group Name** list box.

Remove lets you remove objects from the selected group. To remove objects from the selected group, choose **Remove** and AutoCAD prompts:

> Remove objects: *(select objects that are to be removed from the selected group and press* ENTER*)*

AutoCAD then redisplays the Object Grouping dialog box.

Choosing **Add** allows you to add objects to the selected group. To add objects to the selected group, choose **Add** and AutoCAD prompts:

Select objects: *(select objects that are to be added to the selected group and press* ENTER*)*

AutoCAD then redisplays the Object Grouping dialog box.

Choosing **Rename** allows you to change the name of the selected group to the name entered in the **Group Name** text box in the **Group Identification** section.

Re-Order allows you to change the numerical order of objects within the selected group. Initially, the objects are numbered in the order in which they were selected. Reordering is useful when creating tool paths. Choose **Re-Order** and AutoCAD displays the Order Group dialog box shown in Figure 6–38.

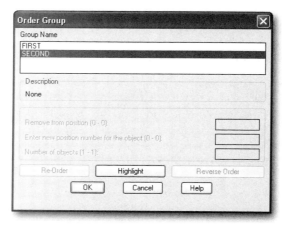

Figure 6–38 *Order Group dialog box*

The **Group Name** list box gives the names of the groups defined in the current drawing. Members of a group are numbered sequentially starting with 0 (zero).

Remove from position (0 – 0) identifies the position number of an object.

Enter new position number for the object (0-0) identifies the new position number of the object.

Number of objects (1 – 1) identifies the number/range of objects to reorder.

Re-Order and **Reverse Order** allow you to change the numerical order of objects as specified and reverses the order of all members, respectively.

Highlight allows AutoCAD to display the members of the selected group in the graphics area.

Choosing **Description** allows you to change the description of the selected group.

Choosing **Explode** in the **Change Group** section of the Object Grouping dialog box deletes the selected group from the current drawing. Thus, the group no longer exists as a group. The members remain in the drawing and in any other group(s) of which they are members.

Choosing **Selectable** controls whether the selected group is selectable.

After making changes in the Object Grouping dialog box, choose **OK** to accept the changes and close the dialog box.

INFORMATION ABOUT OBJECTS

AutoCAD provides several commands for displaying information about the objects in the drawing. These commands do not create anything, nor do they modify or have any effect on the drawing or objects therein. The only effect is that on single-screen systems, the screen switches to the AutoCAD Text window (not to be confused with the TEXT command) and the information requested by the particular inquiry command is then displayed on the screen. If you are new to AutoCAD, it is helpful to know you can switch between the text and graphics screens so you can continue with your drawing. On most systems this is accomplished with the F2 function key. You can also change back and forth between graphic and text screens with the GRAPHSCR and TEXTSCR commands. The inquiry commands include LIST, AREA, ID, DBLIST, and DIST.

LIST COMMAND

The LIST command displays information about individual objects stored by AutoCAD in the drawing database. The location, layer, object type, and space (model or paper) of the selected object as well as the color, lineweight, and linetype, if not set to BYLAYER or BYBLOCK is listed. In addition, the distance of the main axes between the endpoints of a line, that is, the delta-*X*, delta-*Y*, and delta-*Z*. Also, the area and circumference of a circle or the area of a closed polyline. And the insertion point, height, angle of rotation, style, font, obliquing angle, width factor, and actual character string of a selected text object. The object handle, reported in hexadecimal, is also listed

Invoke **List** (LIST) from the Inquiry toolbar (see Figure 6–39) and AutoCAD prompts:

    Select objects: *(select the objects and press* ENTER *to terminate object selection)*

AutoCAD lists the information about the selected objects in the AutoCAD Text Window.

Figure 6-39 *Invoking List from the Inquiry toolbar*

DBLIST COMMAND

The DBLIST command lists the data about all of the objects in the drawing. It can take a long time to scroll through all the data in a large drawing. DBLIST can, like other commands, be terminated by pressing the ESC key.

Invoke the DBLIST command by typing **dblist** at the On-Screen prompt and AutoCAD lists the information about all the objects in the drawing. To view the results in AutoCAD Text Window press F2

AREA COMMAND

The AREA command is used to report the area (in square units) and perimeter of a selected closed geometric figure, such as a circle, polygon, or polyline. You may also specify a series of points that AutoCAD considers to be a closed polygon.

Invoke **Area** (AREA) command from the Inquiry toolbar (see Figure 6-40) and AutoCAD prompts:

> Specify first corner point or ⬇: *(specify a point or select one of the available options from the shortcut menu)*

Figure 6-40 *Invoking Area from the Inquiry toolbar*

The default behavior calculates the area when you select the vertices of the objects. If you want to know the area of a specific object such as a circle, polygon, or closed polyline, select *OBJECT* from the shortcut menu.

The following command sequence is an example of finding the area of a polygon using the *OBJECT* option (see Figure 6-41).

> **area**
> Specify first corner point or ⬇: *(select OBJECT from the shortcut menu)*
> Select objects: *(select an object, as shown in Figure 6-41)*
> Area = 12.21, Perimeter = 13.79

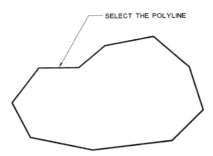

Figure 6–41 *Finding the area of a polygon using the* AREA *command* OBJECT *option*

The ADD option allows you to add selected objects to form a total area; then you can use the SUBTRACT option to remove selected objects from the running total.

The following example demonstrates the application of the ADD and SUBTRACT options. In this example, the area is determined for the closed shape after subtracting the area of the four circles shown in Figure 6–42. When using the ADD and SUBTRACT options of the AREA command, the results are not shown at the On-Screen prompt. Therefore, you should press F2 to toggle to AutoCAD Text Window to view the results after each addition or subtraction. When using the add and subtract options of the area command, the results are not shown at the On-Screen prompt. You should press F2, to toggle between the drawing area and the AutoCAD Text Window, to see the results.

>**area** (ENTER)
>Specify first corner point or ⊻ *(select Add, from the shortcut menu)*
>Specify first corner point or ⊻ *(select Object, from the shortcut menu)*
>(ADD mode) Select objects: *(select the polyline, as shown in Figure 6–42)*
>
>**Result:** Area = 12.9096, Perimeter = 15.1486
>**Result:** Total area = 12.9096
>
>(ADD mode) Select objects: (ENTER)
>Specify first corner point or ⊻ *(select Subtract, from the shortcut menu)*
>Specify first corner point or ⊻ *(select Object, from the shortcut menu)*
>(SUBTRACT mode) Select objects: *(select circle A, as shown in Figure 6–42)*
>
>**Result:** Area = 0.7125, Circumference = 2.9992
>**Result:** Total area = 12.1971
>
>(SUBTRACT mode) Select objects: *(select circle B, as shown in Figure 6–42)*
>
>**Result:** Area = 0.5452, Circumference = 2.6175
>**Result:** Total area = 11.6179
>
>(SUBTRACT mode) Select objects: *(select circle C, as shown in Figure 6–42)*

Result: Area = 0.7125, Circumference = 2.9922
Result: Total area = 10.9394

(SUBTRACT mode) Select objects: *(select circle D, as shown in Figure 6–42)*

Result: Area = 0.5452, Circumference = 2.6175
Result: Total area = 10.3942

(SUBTRACT mode) Select objects: (ENTER)
Specify first corner point or ⬇ (ENTER)

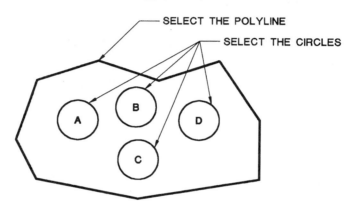

Figure 6–42 *Using the ADD and SUBTRACT options of the AREA command*

ID COMMAND

The ID command is used to obtain the coordinates of a selected point. If you do not use an Object Snap mode to select a point that is not in the current construction plane, AutoCAD assigns the current elevation as the Z coordinate of the point selected.

Invoke **Locate Point** (ID) from the Inquiry toolbar (see Figure 6–43) and AutoCAD prompts:

Specify point: *(select a point)*

Figure 6–43 *Invoking Locate Point from the Inquiry toolbar*

AutoCAD then displays the information about the selected point.

If the BLIPMODE system variable is set to on, a blip appears on the screen at the specified point, provided it is in the viewing area.

DIST COMMAND

The DIST command prints out the distance, in the current units, between two points, either selected on the screen or keyed in from the keyboard. Included in the information are the horizontal and vertical distances (delta-X and delta-Y, respectively) between the points and the angles in and from the XY plane.

Invoke **Distance** (DIST) from the Inquiry toolbar (see Figure 6–44) and AutoCAD prompts:

>Specify first point: *(specify the first point to measure from)*
>Specify second point: *(specify the endpoint to measure to)*

Figure 6–44 *Invoking Distance from the Inquiry toolbar*

AutoCAD displays the distance between two selected points.

SYSTEM VARIABLES

AutoCAD stores the settings (or values) for its operating environment and some of its commands in system variables. Each system variable has an associated type: integer (for switching or for numerical value), real, point, or text string. Unless they are read-only, you can examine and change these variables at the On-Screen prompt by typing the name of the system variable, or you can change them by means of the SETVAR command.

Integers (for switching) system variables that have limited nonnumerical settings and can be switched by setting them to the appropriate integer value. For example, the snap can be either on or off. The purpose of the SNAPMODE system variable is to turn the snap on or off by using the AutoCAD SETVAR command or the AutoLISP (setvar) function.

Turning the snap on or off is demonstrated in the following example by changing the value of its SNAPMODE system variable. First, its current value is set to "0", which is off. Invoke the SETVAR command from by typing **setvar** at the On-Screen prompt and AutoCAD prompts:

>Enter variable name or ⬇: **snapmode** (ENTER)
>Enter new value for SNAPMODE <0>: 1 (ENTER)

For any system variable whose status is associated with an integer, the method of changing the status is just like the preceding example. In the case of SNAPMODE, "0" turns it off and "1" turns it on. In a similar manner, you can use SNAPISOPAIR to switch one from one isoplane to another by setting the system variable to one of three integers: 0 is the left isoplane, 1 is the top, and 2 is the right isoplane.

It should be noted that the settings for the Osnap system variable named OSMODE are members of the binomial sequence. The integers are 1, 2, 4, . . . , 512, 1024, 2048, 4096, 8192. See Table 6–1 for the meaning of OSMODE values. While the settings are switches, they are more than just on and off. There may be several Object Snap modes active at one time. It is important to note that the value of an integer (switching) has nothing to do with its numerical value.

Table 6–1 *Values for the* OSMODE *System Variable*

NONe	0
ENDpoint	1
MIDpoint	2
CENter	4
NODe	8
QUAdrant	16
INTersection	32
INSertion	64
PERpendicular	128
TANgent	256
NEArest	512
QUIck	1024
APP int	2048
EXTension	4096
PARallel	8192

Integers (for Numerical Value) System variables such as APERTURE and AUPREC are changed by using an integer whose value is applied numerically in some way to the setting, rather than just as a switch. For instance, the size of the aperture (the target box that appears for selecting Osnap points) is set in pixels according to the integer value entered in the SETVAR command. For example, setting the value of APERTURE to 9 should render a target box that is three times larger than setting it to 3.

AUPREC is the variable that sets the precision of the angular units in decimal places. The value of the setting is the number of decimal places; therefore, it is considered a numerical integer setting.

Real System variables that have a real number for a setting, such as VIEWSIZE, are called real.

Point (*X* Coordinate, *Y* Coordinate) LIMMIN, LIMMAX, and VIEWCTR are examples of system variables whose settings are points in the form of the *X* coordinate and *Y* coordinate.

Point (Distance, Distance) Some system variables, whose type is point, are primarily for setting spaces rather than a particular point in the coordinate system. For instance, the SNAPUNIT system variable, though called a point type, uses its *X* and *Y* distances from (0,0) to establish the snap *X* and *Y* resolution, respectively.

String These variables have names like CLAYER, for the current layer name, and DWGNAME, for the drawing name.

Open the Exercise Manual PDF file for Chapter 6 on the accompanying CD for project and discipline specific exercises.

 If you have the accompanying Exercise Manual, refer to Chapter 6 for project and discipline specific exercises.

REVIEW QUESTIONS

1. The elements of a multiline can have different colors.

    a. True

    b. False

2. A SPLINE object:

    a. does not actually pass through the control points

    b. is always shown as a continuous line

    c. requires you to specify tangent information for each control point

    d. requires you to specify tangent information for the first and last points only

3. Once objects are grouped together, they can be ungrouped by:

    a. using the EXPLODE command

    b. using the UNGROUP command

    c. using the GROUP command

    d. they cannot be ungrouped

4. The options dialog box allows you to change all of the following, except:

    a. grip size

    b. grip color

    c. toggle grips system variable ON/OFF

    d. specify the location of the grips

5. Grips do not allow you to _____ an object.

   a. trim

   b. erase

   c. mirror

   d. move

   e. stretch

6. Multiline styles can be saved to an external file, thus allowing their use in multiple drawings.

   a. True

   b. False

7. The ID command will:

   a. display the serial number of the AutoCAD program

   b. display the X, Y, and Z coordinates of a selected point

   c. allow you to password protect a drawing

   d. none of the above

8. A quick way to find the length of a line is to use what command?

   a. LIST

   b. DISTANCE

   c. DIMENSION

   d. LENGTH

   e. COORDINATES

9. To obtain a full listing of all the objects contained in the current drawing, you should use:

   a. LIST

   b. DBLIST

   c. LISTALL

   d. DBDUMP

   e. DUMP

10. You can change the setting of a system variable by using the command:

    a. SYSVAR

    b. SETVAR

    c. VARSET

    d. VARSYS

    e. none of the above

11. Which of the following is not a valid type of system variable?

    a. real

    b. integer

    c. point

    d. string

    e. double

Directions: Use the following X and Y locations to construct figure Qx6–1. Your completed drawing will be used to answer the remaining questions.

Point	X value	Y value
A	1.5000	1.0000
B	2.1717	1.0000
C	2.7074	1.7500
D	4.5000	1.7500
E	4.8200	1.0000
F	5.5000	1.0000
G	4.1838	4.0711
H	3.5071	3.8100
J	3.7974	3.3861
K	3.6104	3.3139
L	2.8125	2.2500
M	4.2843	2.2500

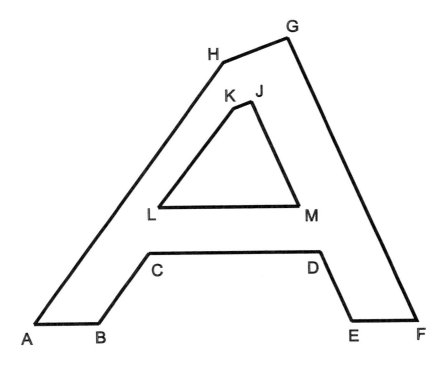

Figure Qx6–1

12. Using the appropriate AutoCAD command what is the length of Line AH?

    a. 3.5432

    b. 3.4523

    c. 3.4532

13. Using the appropriate AutoCAD command what is the angle of Line AH?

    a. 324 degrees

    b. 234 degrees

    c. 432 degrees

14. Using the appropriate AutoCAD command what is the length of Line CD?
    a. 1.7962
    b. 1.7926
    c. 1.7629

15. Using the appropriate AutoCAD command what is the length of Line FG?
    a. 3.3143
    b. 3.3314
    c. 3.3413

16. Using the appropriate AutoCAD command what is the angle of Line FG?
    a. 131 degrees
    b. 311 degrees
    c. 113 degrees

17. Using the appropriate AutoCAD command what is the length of Line GH?
    a. 0.7253
    b. 0.7532
    c. 0.7325

18. Using the appropriate AutoCAD command what is the angle of Line GH?
    a. 210 degrees
    b. 102 degrees
    c. 201 degrees

19. What is the AREA of shape JKLM?
    a. .8833
    b. .9067
    c. .8838

20. What is the perimeter of shape JKLM?

    a. 4.1281

    b. 4.2382

    c. 4.1182

21. What is the linear distance from point E to point J?

    a. 2.2589

    b. 2.5960

    c. 2.5958

CHAPTER 7

Dimensioning

INTRODUCTION

AutoCAD provides a full range of dimensioning commands and utilities. These enable the drafter to comply with the conventions of most disciplines, including architectural, civil, electrical, mechanical engineering, and many others.

After completing this chapter, you will be able to do the following:

- Draw linear dimensioning
- Draw aligned dimensioning
- Draw angular dimensioning
- Draw diameter and radius dimensioning for arcs and circles
- Draw arc length dimensioning
- Draw leaders with annotation and geometric tolerance
- Draw ordinate dimensioning
- Draw baseline and continue dimensioning
- Use Quick dimensioning
- Disassociate and reassociate dimensioning
- Draw center marks for circles and arcs
- Edit dimension text
- Create and modify dimension styles

Figure 7–1 is a three-dimensional drawing of a lever arm. This "pictorial" rendering gives a viewer a good idea of what the object looks like, especially if the viewer has trouble understanding engineering drawings. Figure 7–2 is a typical engineering drawing of the same object. Because of the symmetry of the object, two views are sufficient. But in order to communicate the size and location information necessary to make the object, the drawing in Figure 7–3 has the vital data that the manufacturer needs. It's possible that other information, such as material and finish specifications, and tolerances could also be included. This section covers how to use AutoCAD to produce dimensions and notes for this type of drawing, as well as drawings for other disciplines such as architecture, civil engineering, and surveying.

Figure 7–1 *Pictorial view of lever arm*

AutoCAD provides commands to draw the full range of dimension types: Linear, Angular, Arc Length, Diameter and Radius, and Ordinate. Each type includes primary and secondary commands. For example, Linear dimensioning includes Horizontal, Vertical, Angle, and Rotated options. Figure 7–3 has both horizontal and vertical formats of the Linear dimension type. There are also other general utility, editing, and style-related commands and subcommands that assist you in drawing the correct dimensions quickly and accurately.

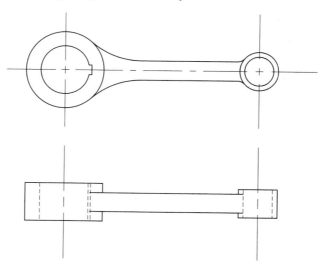

Figure 7–2 *Orthogonal views of lever arm*

Chapter 7 • *Dimensioning* **335**

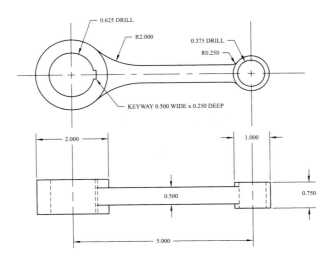

Figure 7–3 *Dimensioned drawing of lever arm*

AutoCAD makes drawing dimensions easy. For example, the width of the rectangle shown in Figure 7–4 can be dimensioned by selecting the two endpoints of the top corners (using an OSNAP mode such as ENDPOINT or INTERSECTION) and then specifying a point to determine the location of the dimension line. AutoCAD provides a preview image of the dimension to indicate how it will look while you move the cursor to specify the location of the dimension line.

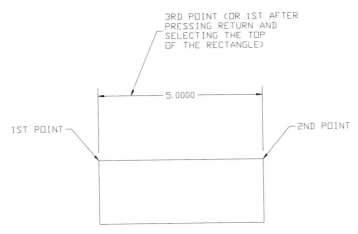

Figure 7–4 *An example of linear dimensioning*

The Linear dimensioning options include HORIZONTAL, VERTICAL, ANGLE, and ROTATED. Angular dimensioning is covered later, under the Angular Dimensioning section.

Diameter, radius, arc length, and ordinate dimensioning are also covered later, under their respective sections.

Approximately 60 dimensioning system variables are available. Most of these have names that begin with "DIM." They are used for such purposes as determining the size of the gap between the extension line and the point specified on an object or whether one or both of the extension lines will be drawn or suppressed. It is combinations of these variable settings that can be named and saved as Dimension Styles and later recalled for applying where needed. See Appendix C for the listing of available Dimensioning System Variables.

Dimension variables change when their associated settings are changed in the Modify Dimension Style dialog box. For example, the value of the DIMEXO (extension line offset) dimensioning system variable is established by the number in the **Offset from origin** text box in the Modify Dimension Style dialog box. See the section later in this chapter on "Dimension Styles."

Dimension utilities include Override, Center, Leader, Baseline, Initial Length, Dimscale, Oblique, Linetype, Continue, and Feature Control Frames for adding tolerancing information.

The dimension text editing command options include HOME, NEW, OBLIQUE, EDIT, and ROTATE.

DIMENSION TERMINOLOGY

Following are the terms for the different parts that make up dimensions in AutoCAD.

Figure 7–5 shows the different components of a typical dimension.

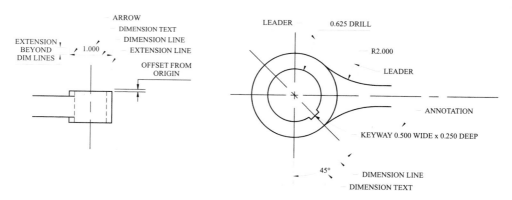

Figure 7–5 *Components of a typical dimension*

DIMENSION LINE

The dimension line is offset from the measured feature. The dimension line (sometimes drawn as two segments outside of the extension lines if a single line with its related text will not fit between the extension lines) indicates the direction and length of the measured distance. It is normally offset for visual clarity. If the dimension is measured between the parallel lines of one or two objects, then it may not be offset, but is drawn on the object or between the two objects. Dimension lines are usually terminated with markers such as arrows or ticks (short slanted lines). Angular dimension lines become arcs whose centers are at the vertex of the angle.

ARROWHEAD

The arrowhead is a mark at the end of a dimension line to indicate its termination. Shapes other than arrows are used in some styles.

EXTENSION LINE

When a dimension line is offset from the measured feature, the extension lines (sometimes referred to as witness lines) indicate such offset. Unless you have invoked the OBLIQUE option, the extension lines will be perpendicular to the direction of the measurement.

DIMENSION TEXT

The dimension text consists of numbers, words, characters, and symbols used to indicate the measured value and type of dimension. Unless the text settings have been altered in the Standard dimension style, the number/symbol format is decimal. This conforms to the same linear and angular units as the default of the drawing. The text style conforms to that of the Current text style.

LEADER

The leader is a radial line used to point from the dimension text to the circle or arc whose diameter or radius is being dimensioned. A leader can also be used for general annotation.

CENTER MARK

The center mark is made up of lines or a series of lines that cross in the center of a circle for the purpose of marking its center.

ASSOCIATIVE, NON-ASSOCIATIVE & EXPLODED DIMENSIONS

Dimensions in AutoCAD can be drawn as associative, non-associative, or exploded, depending on the setting of the DIMASSOC dimensioning system variable. The variable setting for associative is 2 (default), non-associative is 1, and exploded is 0.

ASSOCIATIVE

An associative dimension becomes associated with an object by selecting points on the object (using an Object Snap mode) when prompted to do so during a dimensioning command. If the object is subsequently modified, in a manner that changes the location of one or both of the selected points, the associated dimension is automatically updated to correctly indicate the new distance or angle. Associative dimensioning does not support multilines. The associativity between a dimension and a block is lost when the block is redefined.

An associative dimension drawn with the DIMASSOC dimensioning system variable set to 2 will have all of its separate parts become members of a single object. Therefore, if any one of its members is selected for modifying, all members are highlighted and subject to being modified. This is similar to the manner in which member objects of a block reference are treated. If you have dimensioned the width of a rectangle with an associative dimension and then specified one end of the rectangle to stretch, the dimension will also be stretched, and the dimension text will be changed to correspond to the new measurement. (In addition to the customary visible parts, AutoCAD draws point objects at the ends, where the measurement actually occurs on the object.)

The associativity of associative dimensions created in AutoCAD 2006 is maintained when saved to a previous release and then reopened in the current release. However, if you modify dimensioned objects (forming new objects) using an earlier version of AutoCAD, the dimension associations change when the drawing is loaded into the current release. For example, if a circle that was dimensioned is broken so that portions of the circle are removed, two arc objects result and the associated dimension applies to only one of the arc objects.

Figure 7–6 shows an example of a dimensioned object that has been revised. If the second extension line origin for the dimension on the right, is moved horizontally with the STRETCH command, the associative dimension and the dimension text will reflect the new location.

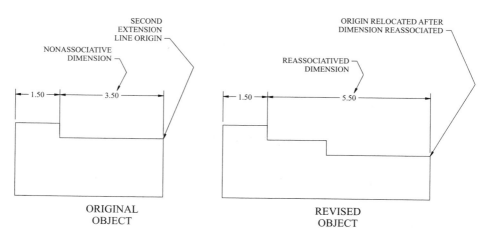

Figure 7–6 *Non-associative dimension that has been reassociated*

NON-ASSOCIATIVE

Non-associative dimensions are drawn while the DIMASSOC dimensioning system variable is set to 1. The separate parts of the non-associative dimension are, like those of the associative dimensions, considered members of a single object when any of them are selected for modification. However, if the object that was dimensioned is selected for modification, without selecting the dimension, the dimension itself will remain unchanged.

EXPLODED

Exploded dimensions are drawn while the DIMASSOC dimensioning system variable is set to 0, and the members are drawn as separate objects. If one of the components of the dimension is selected for modification, that component will be the only one modified.

Note: An associative dimension can be converted to an exploded dimension with the EXPLODE command. Once the dimension is exploded, you cannot recombine the separate parts back into the associative dimension from which they were exploded (except by means of the UNDO command, if feasible). Also, when you explode an associative dimension, the measurement-determining points (nodes) remain in the drawing as point objects.

DIMENSIONING COMMANDS

The dimensioning commands are accessible from the Dimension toolbar. You can also access them from the Dimension menu or by entering their names directly from the keyboard at the On-Screen prompt. Dimension types include Linear, Aligned, Ordinate, Radius, Diameter, Arc Length, Angular, Baseline, Continue, Quick, Qleader, Leader (with Tolerances), Cross Marks (for circles and arcs), and Oblique.

LINEAR DIMENSIONING

Invoke **Linear** (DIMLINEAR) from the Dimension toolbar (see Figure 7–7) and AutoCAD prompts:

   Specify first extension line origin or <select object>: *(specify a point)*

Figure 7–7 *Invoking the* DIMLINEAR *command by selecting Linear from the Dimension toolbar*

AutoCAD uses the first point specified as the origin for the first extension line. This point can be the endpoint of a line, the intersection of objects, the center point of a circle, the insertion point of a text object, or a point on the object itself.

AutoCAD provides a gap between the object and the extension line that is equal to the value of the DIMEXO dimensioning system variable. After you specify the origin, AutoCAD prompts:

> Specify second extension line origin: *(specify a point at which the second extension line should start)*

Dynamic Horizontal and Vertical Dimensioning

Dynamic horizontal and vertical dimensioning is the default mode after you have specified the second point, in response to the DIMLINEAR command. If you specify two points on the same horizontal line, moving the cursor above or below the line causes a preview image of the dimension to appear. AutoCAD assumes you wish to draw a horizontal dimension. Likewise, AutoCAD assumes a vertical dimension if the specified points are on the same vertical line.

Note: Horizontal and vertical formats of Linear dimensions refer to their orientation in the drawing, not their orientation on the object or in real space. Horizontal dimensions are aligned with the X axis (orthogonal left to right) in the drawing coordinate system and Vertical dimensions are aligned with the Y axis (orthogonal up and down).

Dynamically switching between horizontal and vertical dimensioning is more applicable when the two points specified are not on the same horizontal or vertical line. That is, they can be considered diagonally opposite corners of an imaginary rectangle with both width and height. After the two points are specified, you are prompted to specify the location of the dimension line. You are also shown a preview image of the dimension on the screen, as you move the cursor. If you position the cursor above the top line, or below the bottom line of the rectangle, then the dimension will be horizontal. If the cursor moves to the right of the right side, or to the left of the left side of the rectangle, then the dimension will be vertical. If the cursor is dragged to one of the outside quadrants or inside of the rectangle, it will maintain the orientation in effect before the cursor was moved.

After two points have been specified, AutoCAD prompts:

> Specify dimension line location or ⬇ *(specify location for dimension line or select one of the options from the shortcut menu)*

Drawing the Dimension Line

After specifying the two points (or using the shortcut menu to change the dimension text or type), specify a point through which the dimension line is to be drawn. When AutoCAD draws the dimension line, if there is enough room between the extension lines, the dimension text will be centered in line with or above this line. If the dimension text is to be in line with the dimension line, then the dimension line will be broken to allow room for the text. However, if the dimension line, arrows, and text do not fit between the extension lines, they are drawn outside and the text will be drawn near the second extension line.

Changing Dimension Text

The MTEXT and TEXT options allow you to change the measured dimension text. The ANGLE option allows you to change the rotation angle of the dimension text. After responding appropriately for TEXT or ANGLE, AutoCAD repeats the prompts for the dimension line location.

Changing Dimension Type

The HORIZONTAL option forces a horizontal dimension to be drawn (even when dynamic dimensioning calls for a vertical dimension). Likewise, the VERTICAL option forces a vertical dimension to be drawn (even when dynamic dimensioning calls for a horizontal dimension).

Rotated Dimension

The ROTATED option allows you to draw the dimension at a specified angle. Figure 7–8 shows a situation where ROTATED dimensions are useful. Here is a case where the desired angle of dimensioning is the angle created by a line from point A to point B. The dimension from point A to point C can be properly oriented by choosing the ROTATED option, after specifying points A and C. Then points A and B are specified to determine the angle.

Choose ROTATED from the shortcut menu, and then AutoCAD prompts:

> Specify angle of dimension line <0>: *(specify the dimension line angle or specify points A and B to determine the angle)*

The points specified do not have to be parallel to the dimensioned object.

The following command sequence is an example of drawing linear dimensioning for a horizontal line by providing data points for the first and second line origins, respectively, as shown in Figure 7–9.

> **dimlinear** (ENTER)
> Specify first extension line origin or <select object>: *(specify the origin of the first extension line)*
> Specify second extension line origin: *(specify the origin of the second extension line)*
> Specify dimension line location or ⬇ *(specify the location for the dimension line*

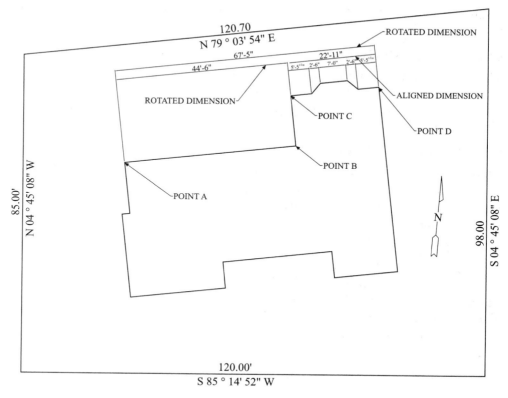

Figure 7–8 Example of using rotated dimensioning

The following command sequence is an example of drawing a linear dimension for a vertical line by providing data points for the first and second line origins, respectively, as shown in Figure 7–10.

> **dimlinear** (ENTER)
> Specify first extension line origin or <select object>: *(specify the origin of the first extension line)*
> Specify second extension line origin: *(specify the origin of the second extension line)*
> Specify dimension line location or *(specify the location for the dimension line)*

> **Note:** You can dynamically switch dimensions between parallel and perpendicular, to the angle between the endpoints. This is similar to dynamically switching between horizontal and vertical dimensioning for normal linear dimensioning.

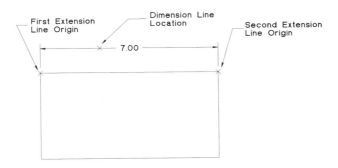

Figure 7-9 *Drawing a linear dimension for a horizontal line*

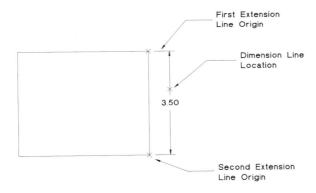

Figure 7-10 *Drawing linear dimensioning for a vertical line*

LINEAR DIMENSIONING BY SELECTING AN OBJECT

If the dimension you wish to draw is between endpoints of a line, arc, or circle diameter, AutoCAD allows you to bypass specifying those endpoints separately. This speeds up drawing the dimension, especially if you have to invoke the endpoint object snap modes. When you have invoked the DIMLINEAR command and you press ENTER in response to the "Specify first extension line origin or <select object>:" prompt, AutoCAD prompts:

> Select object to dimension: *(select line, polyline segment, arc, or circle object)*
> Specify dimension line location or ⬇ *(specify location for dimension line or select one of the options from the shortcut menu)*

If you select a line object, AutoCAD automatically uses the line's endpoints as the first and second points to determine the distance to measure. You will be prompted to specify the location of the dimension line. If you right-click and choose HORIZONTAL, a horizontal dimension is drawn accordingly. In a similar manner, if you choose VERTICAL, a vertical dimension is drawn. If you do not choose the type of dimension to be drawn, then a horizontal dimension is drawn if the specified point for the dimension line is above or below the line object, and a vertical dimension is drawn if the specified point for the dimension line is to the right or left of the line object. If you right-click and choose ROTATED from the shortcut menu, AutoCAD will use the endpoints of the line for the distance reference points and then use the next two points specified to determine the direction of the dimension and measure the distance between the two endpoints of the line object in that direction. The dimension line passes through the last point specified.

The following command sequence shows an example of drawing a linear dimension applied to a single object (see Figure 7–11).

dimlinear (ENTER)
Specify first extension line origin or <select object>: (ENTER)
Select object to dimension: *(select Line A)*
Specify dimension line location or ⏎ *(specify the location for the dimension line location)*

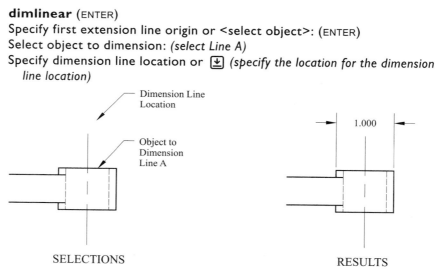

Figure 7–11 *Drawing a linear dimension for a single object*

If the object is a circle, AutoCAD automatically uses the diameter of the circle as the distance to dimension. If you are chose the HORIZONTAL option, a horizontal dimension is drawn using the endpoints of the horizontal diameter. Likewise, with the VERTICAL option the endpoints of the vertical diameter is used. If you are chose the ROTATED option, AutoCAD uses the first two points specified to determine the direction of the dimension and to measure the distance between the two endpoints of a diameter in that direction. The dimension line passes through the last point specified.

If the object is an arc, AutoCAD automatically uses the arc's endpoints as the distance to dimension. You will be prompted to specify the location of the dimension line. If you are chose the HORIZONTAL option, a horizontal dimension is drawn accordingly. If you are chose the VERTICAL option, a vertical dimension is drawn accordingly If you have not yet specified the orientation of the dimension to be drawn, then a horizontal dimension will be drawn if the specified point for the dimension line is above or below the arc, and a vertical dimension is drawn if the specified point for the dimension line is to the right or left of the arc. If you are chose the ALIGNED option, AutoCAD uses the endpoints of the selected arc as the first and second points and the direction from one endpoint to the other endpoint as the direction to measure. If you are chose the ROTATED option, AutoCAD uses the first two points specified to determine the dimension's direction and the distance to dimension. The dimension line passes through the last point specified.

ALIGNED DIMENSIONING

Invoke **Aligned** (DIMALIGNED) from the Dimension toolbar (see Figure 7–12) and AutoCAD prompts:

> Specify first extension line origin or <select object>: *(specify the origin of the first extension line)*
> Specify second extension line origin: *(specify the origin of the second extension line)*

Figure 7–12 *Invoking Aligned from the Dimension toolbar*

After two points have been specified, AutoCAD prompts:

> Specify dimension line location or ⬇ *(specify the location for the dimension line or select one of the options from the shortcut menu)*

Specify a point where the dimension line is to be drawn or choose one of the available options. The MTEXT, TEXT, and ANGLE options are the same as in Linear dimensioning, explained earlier in this section.

The DIMALIGNED command creates an aligned linear dimension using the three familiar points on the object (see Figure 7–13). To draw the proper dimension: "first extension line origin," "second extension line origin," and "dimension line location" as prompted in the following command sequence:

> **dimaligned**
> Specify first extension line origin or <select object>: *(specify Point A)*
> Specify second extension line origin: *(specify Point B)*
> Specify dimension line location or ⬇ *(specify Point C)*

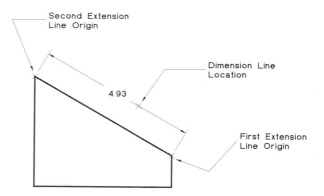

Figure 7–13 *Drawing aligned dimensioning for a line drawn at an angle*

ORDINATE DIMENSIONING

AutoCAD uses the mutually perpendicular X and Y axes of the World Coordinate System or current User Coordinate System as the reference lines from which to base the X or Y coordinate displayed in an ordinate dimension (sometimes referred to as a datum dimension). In the following examples, Figure 7–15 is valid when the base of the rectangle lies on the X axis, giving it a Y value of 0.0000 and Figure 7–16 is valid when the left side of the rectangle lies on the Y axis, giving it an X value of 0.0000. If this is not where the objects are located in the drawing and you still wished to have their values to be 0.0000, a new coordinate system would have to be created (even if temporarily for drawing the dimensions). See Chapter 15 for a detailed discussion on creating a User Coordinate System.

Invoke **Ordinate** (DIMORDINATE) from the Dimension toolbar (see Figure 7–14) and AutoCAD prompts:

Specify feature location: *(specify a point)*

Figure 7–14 *Invoking Ordinate from the Dimension toolbar*

Although the default prompt is "Specify feature location:", AutoCAD is actually looking for a point that is significant in locating a feature point on an object, such as the endpoint or an intersection where planes meet, or the center of a circle representing a hole or shaft. Therefore, an object snap mode, such as endpoint, intersection, quadrant, or center, will need to be invoked when responding to the "Specify feature location:"

prompt. Specifying a point determines the origin of a single orthogonal leader that will point to the feature when the dimension is drawn. AutoCAD prompts:

Specify leader endpoint or ⬇ (specify a point or select one of the options from the shortcut menu)

If the Ortho mode is on, the leader will be a single horizontal line for a Ydatum ordinate dimension (see Figure 7–15) or a single vertical line for an Xdatum ordinate dimension (see Figure 7–16).

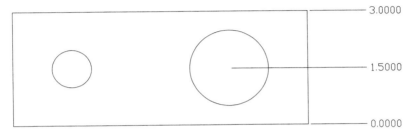

Figure 7–15 *Ydatum dimension*

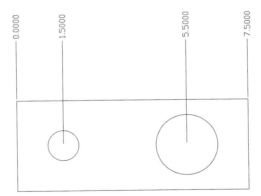

Figure 7–16 *Xdatum dimension*

If the Ortho mode is off, the leader will be a three-part line consisting of orthogonal lines on each end and joined by a diagonal line in the middle. It may be necessary to use the non-orthogonal leader if the text has to be offset to avoid interfering with other objects in the drawing. The type of dimension drawn (Ydatum or Xdatum) depends on the direction of the the second endpoint from the fist endpoint. A preview image of the dimension is displayed during specification of the "leader endpoint".

You can right-click and choose XDATUM or YDATUM from the shortcut menu, and AutoCAD then draws an Xdatum dimension or Ydatum dimension, respectively, regardless of the location of the "leader endpoint" point relative to the "feature location" point.

RADIUS DIMENSIONING

Radius dimensioning allows you to create radius dimensions (see Figure 7–17) for circles and arcs. The type of dimensions that AutoCAD utilizes depends on the Dimensioning System Variable settings (see the section Dimension Styles, later in this chapter, for how to change the Dimensioning System Variable settings to draw appropriate radius dimensioning).

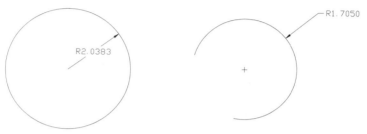

Figure 7–17 *Radius dimensioning of a circle and an arc*

Invoke **Radius** (DIMRADIUS) from the Dimension toolbar (see Figure 7–18) and AutoCAD prompts:

> Select arc or circle: *(select an arc or a circle to dimension)*
> Specify dimension line location or ⊥ *(specify the location for the dimension leader line or select one of the options from the shortcut menu)*

Figure 7–18 *Invoking Radius from the Dimension toolbar*

The MTEXT, TEXT, and ANGLE options are the same as those in linear dimensioning, explained earlier in this section. Dimension text for radius dimensioning is preceded by the letter R.

The following command sequence shows an example of drawing radius dimensioning for a circle (see Figure 7–19).

dimradius
Select arc or circle: *(select the circle object)*
Specify dimension line location or ⬇ *(specify a point to draw the dimension leader line or select one of the options from the shortcut menu)*

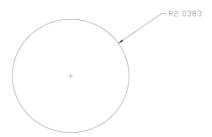

Figure 7–19 *Radius dimensioning of a circle*

DIAMETER DIMENSIONING

Diameter dimensioning allows you to create diameter dimensions (see Figure 7–20) for circles and arcs. The type of dimensions that AutoCAD utilizes depends on the Dimensioning System Variable settings (see the section on Dimension Styles, later in this chapter, for how to change the Dimensioning System Variable settings to draw an appropriate diameter dimensioning).

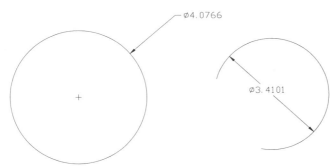

Figure 7–20 *Diameter dimensioning of a circle and an arc*

Invoke **Diameter** (DIMDIAMETER) from the Dimension toolbar (see Figure 7–21) and AutoCAD prompts:

Select arc or circle: *(select an arc or a circle to dimension)*
Specify dimension line location or ⬇ *(specify the location for the dimension line or select one of the options from the shortcut menu)*

Figure 7–21 *Invoking Diameter from the Dimension toolbar*

The MTEXT, TEXT, and ANGLE options are the same as those in linear dimensioning, explained earlier in this section.

Specifying a point determines the location of the diameter dimension. The following command sequence shows an example of drawing diameter dimensioning for a circle (see Figure 7–22).

    dimdiameter
    Select arc or circle: *(select an arc or a circle to dimension)*
    Specify dimension line location or ⤓ *(specify a point to draw the dimension)*

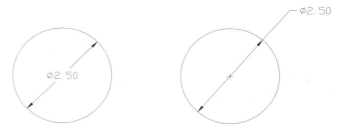

Figure 7–22 *Diameter dimensioning of a circle*

ARC LENGTH DIMENSIONING

Arc Length allows you to create length dimensions (see Figure 7–23) for an arc.

Figure 7–23 *Arc Length dimensioning*

Invoke **Arc Length** (DIMARC) from the Dimension toolbar (see Figure 7–24) and AutoCAD prompts:

Select arc or polyline arc segment: *(select an arc or a circle to dimension)*
Specify arc length dimension location, or ⬇ *(specify the location for the dimension line or select one of the options from the shortcut menu)*

Figure 7–24 *Invoking Arc Length from the Dimension toolbar*

The MTEXT, TEXT, and ANGLE options are the same as those in linear dimensioning, explained earlier in this section.

Selecting PARTIAL allows you to specify a point other than the second endpoint of the arc, and AutoCAD will dimension the length of the portion of the arc between the first endpoint of the arc and the point specified (see Figure 7–25).

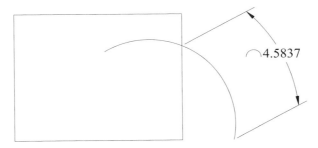

Figure 7–25 *Partial Arc Length dimensioning of a arc*

ANGULAR DIMENSIONING

Angular allows you to draw angular dimensions using various methods: three points (Vertex/Point/Point), between two nonparallel lines, on an arc (between the two endpoints of the arc, with the center as the vertex), and on a circle (between two points on the circle, with the center as the vertex).

Invoke **Angular** (DIMANGULAR) from the Dimension toolbar (see Figure 7–26) and AutoCAD prompts:

Select arc, circle, line, or <specify vertex>: *(select an object or press* ENTER *to specify a vertex)*

Figure 7–26 *Invoking Angular from the Dimension toolbar*

The default method of angular dimensioning is to select an object. If the object selected is an arc (see Figure 7–27) AutoCAD automatically uses the arc's center as the vertex and its endpoints for the first angle endpoint and second angle endpoint to determine the three points of a Vertex/Endpoint/Endpoint angular dimension. AutoCAD prompts:

> Specify dimension arc line location or ⬇ *(specify the location for the dimension arc line or right-click and select one of the options from the shortcut menu)*

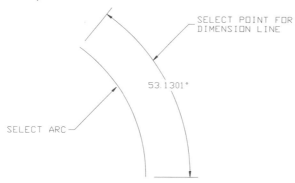

Figure 7–27 *Angular dimensioning of an arc*

The MTEXT, TEXT, and ANGLE options are the same as those in linear dimensioning, explained earlier in this section. Specify a point for the location of the dimension and AutoCAD automatically draws radial extension lines.

If the object selected is a circle (see Figure 7–28) AutoCAD automatically uses the circle's center as the vertex and the point at which you selected the circle as the endpoint for the first angle endpoint, and prompts:

> Specify second angle endpoint: *(specify a point to determine the second angle endpoint)*

Specify a point, and AutoCAD makes this point the second angle endpoint to use along with the previous two points as the three points of a Vertex/Endpoint/Endpoint angular dimension. Note that the last point does not have to be on the circle; however, it does determine the origin for the second extension line. Then, AutoCAD prompts:

> Specify dimension arc line location or ⬇ *(specify the location for the dimension arc line or select one of the options from the shortcut menu)*

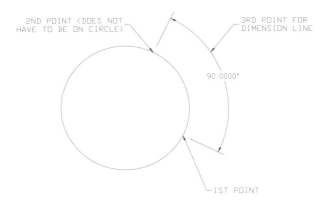

Figure 7–28 *Angular dimensioning of a circle*

The MTEXT, TEXT, and ANGLE options are the same as those in linear dimensioning, explained earlier in this section.

When you specify a point for the location of the dimension arc line, AutoCAD automatically draws radial extension lines and draws either a minor or a major angular dimension, depending on whether the point used to specify the location of the dimension arc line is in the minor or major projected sector.

If the object selected is a line (see Figure 7–29) then AutoCAD prompts:

> Select second line: *(select a line object)*

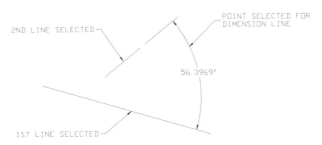

Figure 7–29 *Angular dimensioning of a line*

Select another line, and AutoCAD uses the apparent intersection of the two lines as the vertex for drawing a Vertex/Vector/Vector angular dimension. You are then prompted to specify the location of the dimension arc, which will always be less than 180 degrees.

If the dimension arc is beyond the end of either line, AutoCAD adds the necessary radial extension lines. Then AutoCAD prompts:

Specify dimension arc line location or ⬇ (specify the location for the dimension arc line or select one of the options from the shortcut menu)

The MTEXT, TEXT, and ANGLE options are the same as those in linear dimensioning, explained earlier in this section.

If you specify a point for the location of the dimension arc, AutoCAD automatically draws extension lines and draws the dimension text.

If, instead of selecting an arc, a circle, or two lines for angular dimensioning, you press ENTER, AutoCAD allows you to do three-point angular dimensioning. The following command sequence shows an example of creating an angular dimension by providing three data points (see Figure 7–30).

dimangular (ENTER)
Select arc, circle, line, or <specify vertex>: (ENTER)
Specify angle vertex: *(specify point 1)*
Specify first angle endpoint: *(specify point 2)*
Specify second angle endpoint: *(specify point 3)*
Specify dimension arc line location or ⬇ *(specify the dimension arc line location or select one of the options from the shortcut menu)*

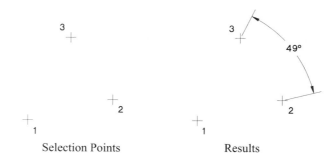

Figure 7–30 *Using three points for angular dimensioning*

BASELINE DIMENSIONING

Baseline dimensioning (sometimes referred to as parallel dimensioning) is used to draw dimensions to multiple points from a single datum baseline (see Figure 7–31). The first extension line origin of the initial dimension (it can be a linear, angular, or ordinate dimension) establishes the base from which the subsequent baseline dimensions are drawn. That is, all of the dimensions in the series of baseline dimensions will share a common first extension line origin. AutoCAD automatically draws a dimension line/arc beyond the initial (or previous baseline) dimension line/arc. The location of the new dimension line/arc is an offset distance established by the DIMDLI (for dimension line increment) dimensioning variable.

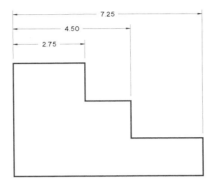

Figure 7–31 *Baseline dimensioning*

Invoke **Baseline** (DIMBASELINE) command from the Dimension toolbar (see Figure 7–32) and AutoCAD prompts:

> Specify a second extension line origin or ⤓ *(specify a point or select one of the options from the shortcut menu)*

Figure 7–32 *Invoking Baseline from the Dimension toolbar*

After you specify a point for the second extension line origin, AutoCAD will use the first extension line origin of the previous linear, angular, or ordinate dimension as the first extension line origin for the new dimension, and the prompt is repeated. To exit the command, press the ESC key, right-click and choose CANCEL from the shortcut menu.

For the DIMBASELINE command to be valid, there must be an existing linear, angular, or ordinate dimension. If the previous dimension was not a linear, angular, or ordinate dimension, or if you press ENTER without providing the second extension line origin, AutoCAD prompts:

> Select base dimension: *(select a dimension object)*

You may select the base dimension, with the baseline being the extension line nearest to where you select the dimension.

The following command sequence shows an example of drawing baseline dimensioning for a circular object to an existing angular dimension (see Figure 7–33).

    dimbaseline (ENTER)
    Second extension line origin or ⬇ *(specify a point)*
    Second extension line origin or ⬇ *(specify a point)*
    Second extension line origin or ⬇ *(press ESC or right-click and choose ENTER from the shortcut menu)*

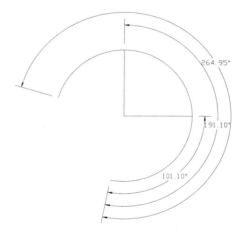

Figure 7–33 *Baseline dimensioning of a circular object*

CONTINUE DIMENSIONING

Continue, shown in Figure 7–34, is used for drawing a string of dimensions, each of whose second extension line origin coincides with the next dimension's first extension line origin.

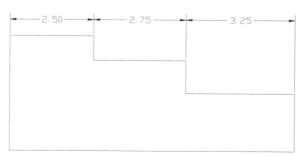

Figure 7–34 *Continue dimensioning*

Invoke **Continue** (DIMCONTINUE) from the Dimension toolbar (see Figure 7–35) and AutoCAD prompts:

> Specify a second extension line origin or ⤓: *(specify a point or select one of the options from the shortcut menu)*

Figure 7–35 *Invoking Continue from the Dimension toolbar*

After you specify a point for the second extension line origin, AutoCAD will use the second extension line origin of the previous linear, angular, or ordinate dimension as the first extension line origin for the new dimension, and the prompt is repeated. To exit the command, right-click and select ENTER from the shortcut menu.

For the DIMCONTINUE command to be valid, there must be an existing linear, angular, or ordinate dimension. If the previous dimension was not a linear, angular, or ordinate dimension, or if you press ENTER without providing the second extension line origin, AutoCAD prompts:

> Select continued dimension: *(select a dimension object)*

You may select the continued dimension, with the coincidental extension line origin being the one nearest to where the existing dimension is selected.

The following command sequence shows an example of drawing continue dimensioning for a linear object to an existing linear dimension (see Figure 7–36).

> **dimcontinue** (ENTER)
> Specify a second extension line origin or ⤓ *(specify the right end of the 3.60-unit line)*
> Specify a second extension line origin or ⤓ *(specify the right end of the right 1.80-unit line)*
> Specify a second extension line origin or ⤓ *(press ESC or right-click and choose ENTER from the shortcut menu)*

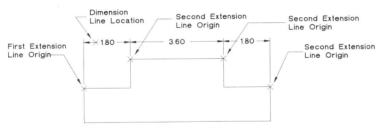

Figure 7–36 *Continue dimensioning of a linear object*

QUICK DIMENSIONING

Quick Dimension is used to draw a string of dimensions between all of the end and center points of the selected object(s).

Invoke **Quick Dimension** (QDIM) from the Dimension toolbar (see Figure 7–37) and AutoCAD prompts:

> Select geometry to dimension: *(select one or more objects and then press* ENTER*)*
> Specify dimension line position, or ⊥ *(specify location for dimension line or select one of the options from the shortcut menu)*

Figure 7–37 *Invoking Quick Dimension from the Dimension toolbar*

If you specify a location for a dimension line, AutoCAD will draw continuous dimensioning between all end or center points of the objects selected horizontally or vertically, depending on where the dimension line location is specified (see Figure 7–38 for an example). You can right-click and select one of the options that will draw the type of dimension chosen.

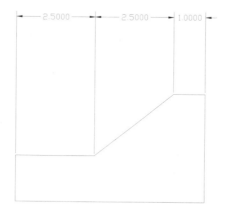

Figure 7–38 *An example of Quick Dimensioning*

QLEADER AND LEADER

The Quick Leader or QLEADER, the default version of the LEADER command, is used to create leaders and leader annotation. It assumes that you wish to enter multiline text that is top-left justified. The QLEADER command minimizes the steps required to draw text with the line and arrowhead pointing to an object (or a feature on an object) for annotations and callouts used to describe them.

Invoke Quick Leader from the Dimension toolbar (see Figure 7-39) and AutoCAD prompts:

> Specify first leader point, or ⬇ *(specify a point for the arrowhead end of the leader line or select one of the options from the shortcut menu)*
> Specify next point: *(specify another point for the end of the first leader segment opposite the arrowhead)*
> Specify text width <0.0000>: *(specify a point to determine the maximum width of the multiline text)*
> Enter first line of annotation text <Mtext>: *(enter the first line of annotation text or right-click for the Multiline Text Editor dialog box)*

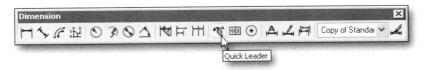

Figure 7–39 *Invoking Quick Leader from the Dimension toolbar*

Note that the text width specified causes the text to wrap to the next line when the text on the current line exceeds the specified width. If no width is specified, then the entire text is drawn on one line.

If you choose SETTINGS from the shortcut menu instead of specifying first leader point, AutoCAD displays the Leader Settings dialog box, shown in Figure 7–40.

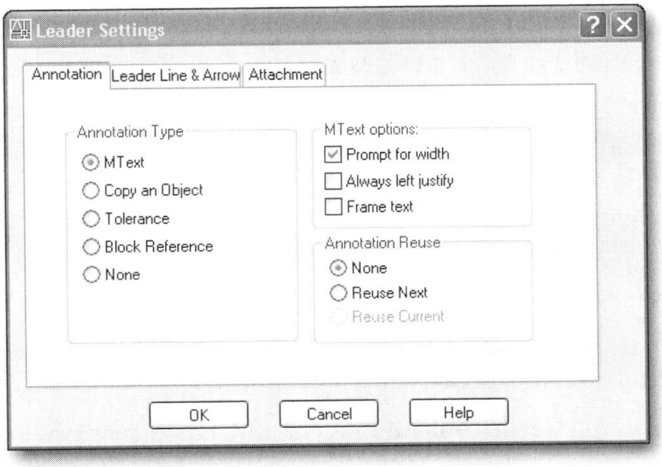

Figure 7–40 *Leader Settings dialog box with the Annotation tab chosen*

The **Annotation** tab (see Figure 7–40) allows you to set the **Annotation Type**, select Mtext options, and determine how annotations are reused.

In the **Annotation Type** section there are five radio buttons that let you choose the type of annotation that will be used. The options are as follows:

> Choosing **Mtext** causes AutoCAD to use the MTEXT command when the annotation is called for (default setting).

> Choosing **Copy an Object** causes AutoCAD to prompt you to "Select an object to copy:" when the annotation is called for. The text object you select will become the text for leader annotation.

> Choosing **Tolerance** causes the Tolerance option to be used (as described in the following section) when the annotation is called for.

> Choosing **Block Reference** causes AutoCAD to use the INSERT command (for inserting a block) when the annotation is called for.

> Choosing **None** causes AutoCAD to draw the leader and exit the LEADER command without prompting for or placing text or other object.

In the **MText options** section there are three check boxes that let you specify how MText will be formatted. The **MText options** section is active for use only if **Mtext** has been selected in the **Annotation Type** section.

> Selecting **Prompt for width** causes AutoCAD to prompt you for the width of the MText box when the annotation is called for.

> Selecting **Always left justify** causes the annotation text to be left justified whether the leader is to the left or the right of the annotation. When this box is checked, **Prompt for width** becomes disabled.

> Selecting **Frame text** causes AutoCAD to draw a rectangle around the annotation text.

In the **Annotation Reuse** section there are three radio buttons that let you specify whether or not AutoCAD reuses the next or the current annotation text for subsequent leader annotation text.

> Choosing **None** causes AutoCAD to not reuse annotation text and to prompt you for annotation text to be used.

> Choosing **Reuse Next** causes AutoCAD to reuse the next annotation text used for subsequent annotation text.

> Choosing **Reuse Current** causes AutoCAD to reuse the current annotation text for subsequent annotation text.

The Leader Line & Arrow Tab

The **Leader Line & Arrow** tab (see Figure 7–41) allows you to modify the geometry of the lines and arrow heads that make up the leader, along with specifying the number of points permitted (which limits the number of leader segments that may be drawn) and the angle constraints of the first and second leader lines.

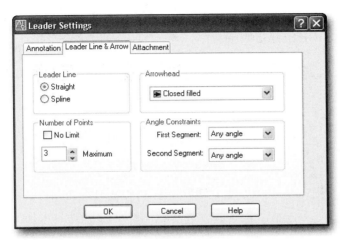

Figure 7–41 *Leader Settings dialog box with Leader Line & Arrow tab chosen*

In the **Leader Line** section, choosing **Straight** causes leaders to be drawn with straight segments. Choosing **Spline** causes leaders to be drawn with curved segments using the same input as in the SPLINE command.

In the **Number of Points** section, selecting **No Limit** lets you put in any number of leader segments. If **No Limit** is not checked, the number of points you can specify to draw leader segments is limited to the number (between 2 and 999) in the Maximum text box. The number of segments will be one less than the number of points specified.

In the **Arrowhead** section, from the list box you can select one of the many available arrowhead shapes. Select **None** to have no arrowhead or select **User Arrow** and name a user-defined block to be used as an arrowhead.

In the **Angle Constraints** section, the **First Segment** list box lets you specify if the first segment can be drawn at any angle or will be constrained to one of the following angles: 90, 45, 30, or 15 degrees. The **Second Segment** list box lets you specify if the first segment can be drawn at any angle or will be constrained to one of the following angles: 90, 45, 30, or 15 degrees.

Attachment Tab

The **Attachment** tab (see Figure 7–42) allows you to specify which part of Multi-line text will be lined up with the annotation end of the leader. The attachment can be specified to line up one way when the text is to the left of the leader and a different way when it is to the right of the leader. An option in the **Multi-line Text Attachment** section is specified by selecting its radio button. The five options are as follows: **Top of top line, Middle of top line, Middle of multi-line text, Middle of bottom line,** and **Bottom of bottom line**.

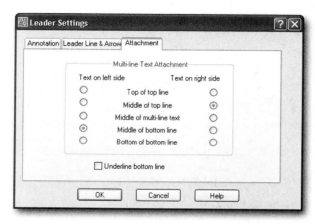

Figure 7–42 *Leader Settings dialog box with the Attachment tab chosen*

The **Attachment** tab has a check box that, when checked, causes the leader to terminate at the bottom line of text and continue as an underline of the text. This overrides any of the first four attachment options.

TOLERANCES

Tolerance symbols and text can be included with the dimensions that you draw in AutoCAD. AutoCAD provides a special set of subcommands for the two major methods of specifying tolerances. One set is for lateral tolerances; the other is for geometric tolerances.

Lateral tolerances are the traditional tolerances. Though they are often easier to apply, they do not always satisfy tolerancing in all directions (circularly and cylindrically), and are more subject to misinterpretations, especially in the international community.

Geometric tolerance values are the maximum allowable distances that a form or its position may vary from the stated dimension(s).

 Note: Lateral tolerance symbols and text are accessible from the Dimension Style dialog box. Geometric tolerance symbols and text are accessible from the Dimension toolbar.

Lateral Tolerance

Lateral tolerancing draws the traditional symbols and text for Limit, Plus or Minus (unilateral and bilateral), Single Limit, and Angular tolerance dimensioning. Lateral tolerances will appear with the dimension text in accordance with how they are set up in the **Tolerances** tab of the Modify Dimension Style dialog box.

Lateral tolerance is the range from the smallest to the greatest that a dimension is allowed to deviate and still be acceptable. For example, if a dimension is called out as 2.50 ± 0.05, then the tolerance is 0.1 and the feature being dimensioned may be anywhere between 2.45 and 2.55. This is the symmetrical plus-or-minus convention. If the dimension is called out as $2.50^{+0.10}_{-0.00}$, then the feature may be 0.1 greater than 2.50, but may not be smaller than 2.50. This is referred to as unilateral. The Limits tolerance dimension may also be shown as $^{2.55}_{2.45}$.

To set the tolerance method, first open the Dimension Style Manager dialog box by typing dimstyle (for a discussion of the other features of the Dimension Style manager see the Dimension Style Manager Dialog Box section). Choose **Modify,** and AutoCAD displays the Modify Dimension Style dialog box. Select the **Tolerances** tab and select the **Tolerance** method from the **Method** option menu. Figure 7–43 shows all the options available in the **Tolerance Format** section and the **Alternate Unit Tolerance** section of the **Tolerances** tab. Selecting a method (other than **None**) in the **Method** text box applies only to lateral tolerancing.

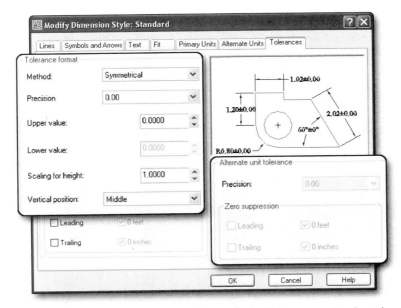

Figure 7–43 *Modify Dimension Style dialog box with the Tolerances tab selected.*

Choosing the **Method** list box in the **Tolerance Format** section of the Tolerances tab displays the following options: **None, Symmetrical, Deviation, Limits,** and **Basic**. See the following table for examples of the various options.

Examples of Lateral Tolerance Methods

Tolerance Method	Description	Example
Symmetrical	Only the Upper Value box is usable. Only one value is required.	1.00 ± 0.05
Deviation	Both Upper Value and Lower Value boxes are active. A 0.00 value may be entered in either box, indicating that the variation is allowed in only one direction.	$1.00^{+0.07}_{-0.03}$
Limits	Both Upper and Lower Value boxes are active. In this case, there is no base dimension. Using an Upper value of 0.0500 and a Lower value of −0.0250 will cause annotation of a 1.000 basic dimension to be $^{0.0500}_{0.0250}$.	1.070 0.930
Basic	The dimension value is drawn in a box, indicating that it is the base value from which a general tolerance is allowed. The general tolerance is usually given in other notes or specifications on the drawing or other documents.	1.00

Note: To change how your tolerance will appear in relation to the dimension, the format is set in the **Primary Units** tab of the Modify Dimension Style dialog box.

In the **Tolerance Format** and **Alternate Unit Tolerance** sections, the **Precision** text box lets you determine how many decimals will be shown in decimal units. This value is recorded in DIMDEC. The **Upper value** and **Lower value** text boxes allow you to preset tolerance values. The **Scaling for height** text box lets you set the height of the tolerance value text. The **Vertical position** selection box determines if the tolerance value will be at the top, middle, or bottom of the text space.

In the **Alternate Unit Tolerance** section, the **Zero suppression** subsection lets you specify whether or not zeros are displayed, as follows:

Selecting the **Leading** check box causes zeros ahead of the decimal point to be suppressed. *Example:* .700 is displayed instead of 0.700.

Selecting the **Trailing** check box causes zeros behind the decimal point to be suppressed. *Example:* 7 or 7.25 is displayed instead of 7.000 or 7.250, respectively.

Selecting the **0 Feet** check box causes zeros representing feet to be suppressed if the dimension text represents inches and/or fractions only. *Example:* 7" or 7 1/4" is displayed instead of 0'-7" or 0'-7 1/4".

Selecting the **0 Inches** check box causes zeros representing inches to be suppressed if the dimension text represents feet only. *Example:* 7' is displayed instead of 7'-0".

Geometric Tolerance

Geometric tolerancing draws a Feature Control Frame for use in describing standard tolerances according to the geometric tolerance conventions. Geometric tolerancing is applied to forms, profiles, orientations, locations, and runouts. Forms include squares, polygons, planes, cylinders, and cones.

Invoke **Tolerance** (TOLERANCE) from the Dimension toolbar (see Figure 7–44) and AutoCAD displays the Geometric Tolerance dialog box (see Figure 7–45).

Figure 7–44 *Invoking the Tolerance from the Dimension toolbar*

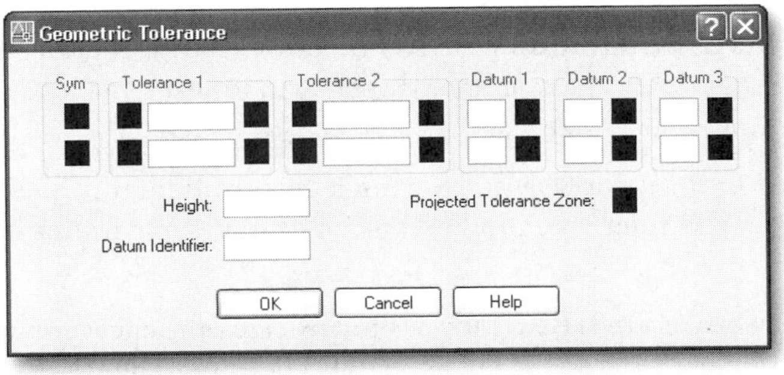

Figure 7–45 *Geometric Tolerance dialog box*

The conventional method of expressing a geometric tolerance for a single dimensioned feature is in a Feature Control Frame, which includes all necessary tolerance information for a particular dimension. A Feature Control Frame has the Geometric Characteristic Symbol box and a Tolerance Value box. Datum reference and or material condition datum boxes may be added where needed. The Feature Control Frame is shown in Figure 7–46. An explanation of the Characteristic Symbols is given in

Figure 7–47. The supplementary material conditions of datum symbols are shown in Figure 7–48.

Once the symbol button located in the **Sym** column in the Geometric Tolerance dialog box is chosen, AutoCAD displays the Symbol dialog box (see Figure 7–49). Select one of the available symbols and specify appropriate tolerance values in a Tolerance field.

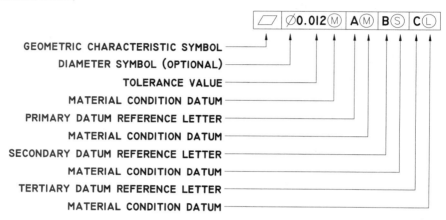

Figure 7–46 *Feature control frame*

Maximum Materials Condition (MMC) means a feature contains the maximum material permitted within the tolerance dimension; for example, the minimum hole size or the maximum shaft size.

Least Material Condition (LMC) means a feature contains the least material permitted within the tolerance dimension; for example maximum hole size or the minimum shaft size.

Regardless of Feature Size (RFS) applies to any size of the feature within its tolerance. This is more restrictive. For example, RFS does not allow the tolerance of the center-to-center dimension of a pair of pegs fitting into a pair of holes greater leeway if the peg diameters are smaller or the holes are bigger, whereas MMC does.

The diameter symbol, Ø, is used in lieu of the abbreviation DIA.

CHARACTERISTIC SYMBOLS		
FORM	⌗	FLATNESS
	—	STRAIGHTNESS
	○	ROUNDNESS
	⌭	CYLINDRICITY
	⌒	LINE PROFILE
	⌓	SURFACE PROFILE
	∠	ANGULARITY
	∥	PAREALLELISM
	⊥	PERPENDICULARITY
LOCATION	◎	CONCENTRICITY
	⌖	POSITION
	═	SYMMETRY
RUNOUT	↗	CIRCULAR RUNOUT
	↗↗	TOTAL RUNOUT

Figure 7–47 *Geometric characteristics symbols*

MATERIAL CONDITIONS OF DATUM SYMBOLS	
Ⓜ	MAXIMUM MATERIAL CONDITIONS (MMC)
Ⓢ	REGARDLESS OF FEATURE SIZE
⌀	DIAMETER

Figure 7–48 *Material conditions of datum symbols*

Figure 7–49 *Symbol dialog box*

Geometric tolerancing is becoming widely accepted. It is highly recommended that you study the latest drafting texts concerning the significance of the various symbols, so that you will be able to apply geometric tolerancing properly.

DRAWING CROSS MARKS FOR ARCS OR CIRCLES

The DIMCENTER command is used to draw the cross marks that indicate the center of an arc or circle (see Figure 7–50).

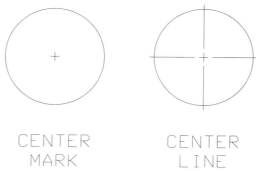

Figure 7–50 *Circles with center cross marks*

Invoke **Center Mark** (DIMCENTER) from the Dimension toolbar (see Figure 7–51) and AutoCAD prompts:

    Select arc or circle:

Figure 7–51 *Invoking Center mark from the Dimension toolbar*

Select an arc or circle, and AutoCAD draws the cross marks in accordance with the setting of the DIMCEN dimensioning variable.

OBLIQUE DIMENSIONING

The OBLIQUE command allows you to slant the extension lines of a linear dimension to a specified angle. The dimension line will follow the extension lines, retaining its original direction. This is useful for having the dimension stay clear of other dimensions or objects in your drawing. It is also a conventional method of dimensioning isometric drawings.

Invoke the OBLIQUE command from the Dimension menu and AutoCAD prompts:

    Select objects: *(select the dimension(s) for obliquing, press* ENTER*)*

Enter obliquing angle (press ENTER for none): *(specify an angle or press ENTER)*

 Note: The OBLIQUE command is actually the DIMEDIT command in which AutoCAD automatically chooses the OBLIQUE option for you. All you have to do is select a dimension.

The extension lines of the selected dimension(s) will be slanted at the specified angle.

EDITING DIMENSIONS AND DIMENSION TEXT

AutoCAD allows you to edit dimensions with modification commands and grip editing options. Also, AutoCAD provides two additional Modify commands specifically designed to work on dimension text objects, DIMEDIT and DIMTEDIT.

DIMEDIT COMMAND

The options that are available with the DIMEDIT command allow you to replace the dimension text with new text, rotate the existing text, move the text to a new location, and if necessary, restore the text back to its home position (which is the position defined by the current style). In addition, these options allow you to change the angle of the extension lines (normally perpendicular) relative to the direction of the dimension line (by means of the OBLIQUE option).

Invoke Dimension Edit (DIMEDIT) from the Dimension toolbar (see Figure 7–52) and AutoCAD prompts:

Enter type of dimension editing *(press ENTER for the HOME option or select one of the options)*

Figure 7–52 *Invoking Dimension Edit from the Dimension toolbar*

Dimension Editing Options

Choosing HOME returns the dimension text to its default position. AutoCAD prompts:

Select objects: *(select the dimension objects and press ENTER)*

Choosing NEW allows you to change the original dimension text to the new text. AutoCAD opens the mtext editor. Enter the new text and choose **OK** and AutoCAD then prompts:

>Select objects: *(select the dimension objects for which the existing text will be replaced by the new text)*

Choosing ROTATE allows you to change the angle of the dimension text. AutoCAD prompts:

>Specify angle for dimension text: *(specify the rotation angle for text)*
>Select objects: *(select the dimension objects for which the dimension text has to be rotated)*

Choosing OBLIQUE adjusts the obliquing angle of the extension lines for linear dimensions. This is useful to keep the dimension parts from interfering with other objects in the drawing. Also, it is an easy method by which to generate the slanted dimensions used in isometric drawings. AutoCAD prompts:

>Select objects: *(select the dimension objects, press* ENTER*)*
>Enter obliquing angle (press ENTER for none): *(specify the angle or press* ENTER*)*

DIMTEDIT COMMAND

The DIMTEDIT command is used to change the location of dimension text (with the LEFT, RIGHT, CENTER, and HOME options) along the dimension line and its angle (with the ROTATE option).

Invoke Dimension Text Edit (DIMTEDIT) command from the Dimension toolbar (see Figure 7–53) and AutoCAD prompts:

>Select dimension: *(select the dimension object to modify)*

Figure 7–53 *Invoking Dimension Text Edit from the Dimension toolbar*

A preview image of the dimension selected is displayed near the cursor. You will be prompted:

>Specify new location for dimension text or ⬇ *(specify a new location for the dimension text or right-click for the shortcut menu and select one of the available options)*

By default, AutoCAD allows you to position the dimension text with the cursor, and the dimension updates dynamically as it is moved.

Dimension Text Editing Options

Choosing CENTER will cause the text to be drawn at the center of the dimension line.

Choosing LEFT will cause the text to be drawn toward the left extension line.

Choosing RIGHT will cause the text to be drawn toward the right extension line.

Choosing HOME returns the dimension text to its default position.

Choosing ANGLE changes the angle of the dimension text. AutoCAD prompts:

> Specify angle for dimension text: *(specify the angle)*

The angle specified becomes the new angle for the dimension text.

EDITING DIMENSIONS WITH GRIPS

If Grips are enabled, you can select an associative dimension object and its grips will be displayed at strategic points. The grips will be located at the object ends of the extension lines, the intersections of the dimension and extension lines, and at the insertion point of the dimension text. In addition to the normal grip editing of the dimension objects as a group (ROTATE, MOVE, COPY, etc.), each grip can be selected for editing the dimension as follows: moving the object end grip of an extension line will move that specified point, making the value change accordingly. Horizontal and vertical dimensions remain horizontal and vertical. Aligned dimensions follow the alignment of the relocated point. Moving the grip at the intersection of the dimension line and one of the extension lines causes the dimension line to be nearer to or farther from the object dimensioned. Moving the grip at the insertion point of the text does the same as the intersection grip and also permits you to move the text back and forth along the dimension line.

DIMDISASSOCIATE COMMAND

The DIMDISASSOCIATE command removes associativity from selected associative dimensions. Invoke the DIMDISASSOCIATE command by typing at the On-Screen prompt. AutoCAD prompts:

> Select objects: *(select the dimension objects and press* ENTER *to complete the selection)*

AutoCAD removes the associativity of the selected dimensions.

DIMREASSOCIATE COMMAND

The DIMREASSOCIATE command associates selected dimensions to geometric objects. When prompted, select the dimension to be reassociated to an object and then step through the point selection process (using Object Snap mode) when each extension line point is marked with an X. If the X is in a box, then that point is already associated with a point on an object. You may skip this point by pressing ENTER. Invoke the DIMREASSOCIATE command by typing at the On-Screen prompt. AutoCAD prompts:

Select objects: *(select the dimension objects and press* ENTER *to complete the selection)*

Follow through the prompts by selecting the points to reassociate the dimensions.

DIMREGEN COMMAND

The DIMREGEN command updates the locations of all associative dimensions. This is sometimes necessary after panning or zooming with a wheel mouse, after opening a drawing that was modified in an earlier version of AutoCAD, or after opening a drawing that contains external references that have been modified. Invoke the DIMREGEN command by typing at the On-Screen prompt.

EDITING DIMENSIONS USING THE SHORTCUT MENU

Dimensions already drawn can be edited easily by selecting them when Noun/Verb selection (PICKFIRST system variable) is enabled and choosing the appropriate available options from the shortcut menu (see Figure 7–54).

The DIM TEXT POSITION flyout includes options for ABOVE DIM LINE, CENTERED, HOME TEXT, MOVE TEXT ALONE, MOVE WITH LEADER, and MOVE WITH DIM LINE. Choosing one of these options causes the text in the selected dimension to conform to the position option chosen.

The FLIP ARROW (AIDIMFLIPARROW) command lets you reverse the position of a selected arrow with relation to the extension line.

The *PRECISION* flyout includes a range of *0.0* to *0.000000* for Decimal dimensions and a range of *0'-0* to *0'-0 1/128* for Architectural and Fractional dimensions. Choosing one of the decimal or fractional options causes the distances or angles in the selected dimension to conform to the level of precision chosen.

Figure 7–54 *Shortcut menu showing the dimension editing section*

 Note: If the dimensions selected include styles with both decimal and fractional units, the PRECISION option is not available in the shortcut menu.

The DIM STYLE flyout includes an option for SAVE AS NEW STYLE, and a list of available dimstyles (dimension styles) to choose from. If you have made some changes to settings of individual variables in a particular dimension, you can choose SAVE AS NEW STYLE to create and name a new, unique dimstyle conforming to the selected dimension's variable settings.

DIMENSION STYLES

Each time a dimension is drawn, it conforms to the settings of the Dimensioning System Variables in effect at the time. The entire set of Dimensioning System Variable settings can be saved in their respective states as a dimension style. Named dimstyles be recalled for application to a dimension later in the drawing session or in a subsequent session. Some Dimensioning System Variables affect every dimension. For example, every time a dimension is drawn, the DIMSCALE setting determines the relative size of the dimension. But DIMDLI, the variable that determines the offsets for Baseline Dimensions, comes into effect only when a Baseline Dimension is drawn. However, when a dimension style is created and named, all Dimensioning System Variable settings are recorded in that dimension style. The associativity of dimensions is not controlled by dimstyles. The DIMASSOC dimensioning variable settings are saved separately from the dimension styles. As stated at the beginning of this chapter: "Dimension variables change when their associated settings are changed in the Modify Dimension Style dialog box. For example, the value of the DIMEXO (extension line offset) dimensioning system variable is established by the number in the **Offset from origin** text box in the Modify Dimension Style dialog box."

DIMENSION STYLE MANAGER DIALOG BOX

AutoCAD provides a comprehensive set of dialog boxes accessible through the Dimension Style Manager dialog box for creating new Dimension Styles and managing existing ones. In turn, these dialog boxes compile and store Dimensioning System Variable settings. Creating Dimension Styles through use of the DIMSTYLE command's dialog boxes allows you to make the desired changes to the appearance of dimensions without having to search for or memorize the names of the Dimensioning System Variables in order to change the settings directly.

Invoke **Dimension Style** (DIMSTYLE) command from the Dimension toolbar (see Figure 7–55) and AutoCAD displays the Dimension Style Manager, shown in Figure 7–56.

Figure 7–55 *Invoking Dimension Style from the Dimension toolbar*

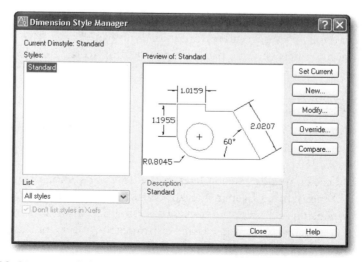

Figure 7–56 *Dimension Style Manager dialog box*

The **Current Dimstyle** label shows the name of the dimension style to which the next drawn dimension will conform. The name of the current dimstyle is recorded as the value of the DIMSTYLE dimensioning system variable.

The **Styles** list box displays the name(s) of the current style and any other style(s) available depending on the option chosen in the **List** selection box. If you choose one of the styles in the list, it is highlighted and will be the dimstyle acted upon when one of the buttons (**Set Current, New, Modify, Override**, or **Compare**) is chosen. It will also be the dimstyle whose appearance is shown in the **Preview of** viewing area.

The **List** text box allows you to list all of the styles available in the drawing displayed or only those in use by choosing the appropriate option from this selection box. Because the Standard style cannot be deleted and is always available.

The area below the **Preview of** heading shows how dimensions will appear when drawn using the current dimstyle.

The **Description** section shows the difference(s) between the dimstyle chosen in the Styles section and the current dimstyle. For example, if the current dimstyle is Standard and the dimstyle chosen in the Styles section is named Harnessing, then the **Description** section might say "Standard + Angle format = 1, Fraction format = 1, Length units = 4" to indicate that the angle format is in Degrees Minutes Seconds, the fraction format is in diagonal, and the length units are architectural, these settings are different from the corresponding settings for the Standard dimstyle.

Choosing **Set Current** causes the dimstyle chosen in the **Styles** section to become the current dimstyle.

Choosing **New** causes the Create New Dimension Style dialog box to appear (see Figure 7–57). The Create New Dimension Style dialog box allows you to create and name a new dimstyle.

Figure 7–57 *Create New Dimension Style dialog box*

The **New Style Name** text box is where you enter the name of the new dimstyle you wish to create.

The **Start With** list box lets you choose an existing dimstyle that you would like to start with in creating your new dimstyle. The dimensioning system variables will be the same as those in the **Start With** list box until you change their settings. In many cases you may wish to change only a few Dimensioning System Variable settings.

The **Use For** list box lets you choose the type(s) of dimensions for which to apply the new Dimstyle you will be creating.

Choosing **Continue** causes the New Dimension Style dialog box to be displayed, from which the dimension variables of the style specified in the **Start With** text box can be modified and then saved with the new name specified. The New Dimension Style dialog box is similar to the Modify Dimension Style dialog box described in the next section.

Choosing **Modify** in the Dimension Style Manager dialog box causes the Modify Dimension Style dialog box to appear (see Figure 7–58). There are seven tabs to choose from: **Lines, Symbols and Arrows, Text, Fit, Primary Units, Alternate Units,** and **Tolerances**.

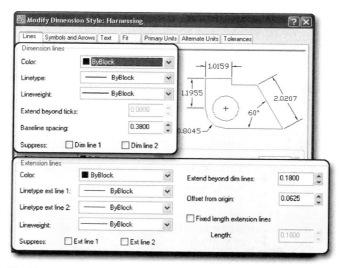

Figure 7–58 *Modify Dimension Style dialog box with Lines tab chosen*

The **Lines** tab allows you to modify the geometry of the various elements that make up the dimensions, such as dimension lines and extension lines. A preview showing how dimensions will appear when drawn using the settings, as you change them.

The **Dimension Line** and **Extension Line** sections allow you to control the color, linetype, and lineweight of the dimension and extension lines.

The **Color** list boxes let you choose how the color of dimension lines and extension lines will be determined. **ByLayer** causes the color of the line to match that of its layer. **ByBlock** causes the color of the line to match that of its block reference if it is part of a block reference. You can also choose one of the standard colors or an Other color for the line color to match. The value is recorded in the DIMCLRD system variable.

The **Linetype** list boxes let you choose how the linetype of Dimension and extension lines will be determined. **ByLayer** causes the linetype of the line to match that of its layer. **ByBlock** causes the linetype of the line to match that of its block reference if it is part of a block reference. You can also choose one of the standard linetypes or another linetype to load through the Select Linetype dialog box.

The **Lineweight** list boxes let you choose how the lineweight of Dimension and extension lines will be determined. **ByLayer** causes the lineweight of the line to match that of its layer. **ByBlock** causes the lineweight of the line to match that of its block reference if it is part of a block reference. You can also choose one of the standard lineweights or enter a value. The value is recorded in the DIMLWD system variable.

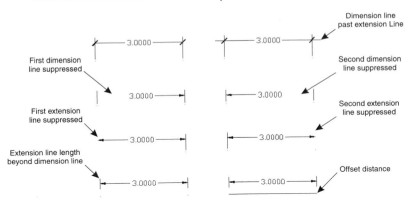

Figure 7–59 *Various examples with Dimension and Extension Lines settings*

In the **Dimension Line** section only:

> The **Extend beyond ticks** text box lets you enter the distance that dimension lines will be drawn past extension lines (see Figure 7–59) if the arrowhead type is a tick (small diagonal line). The value is recorded in the DIMDLE system variable.
>
> The **Baseline spacing** text box lets you enter the distance that AutoCAD uses between dimension lines when they are drawn using the BASELINE DIMENSION command. The value is recorded in the DIMDLI system variable.

Selecting **Suppress** let you determine if one or more of the dimension lines is suppressed when the dimension is drawn (see Figure 7–59). The value is recorded in the DIMSD1 and DIMSD2 system variables.

In the **Extension Line** section only:

The **Extend beyond dim lines** text box lets you enter the distance that extension lines will be drawn past dimension lines (see Figure 7–59).

The **Offset from origin** text box lets you enter the distance that AutoCAD uses between the extension line and the origin point specified when drawing an extension line (see Figure 7–59). The value is recorded in the DIMEXO system variable. Selecting **Fixed length extension lines** causes dimension extension lines to be drawn to the length specified in the **Length** text box.

Selecting **Suppress** lets you determine if one or more of the extension lines is suppressed when the dimension is drawn (see Figure 7–59). The value is recorded in the DIMSE1 and DIMSE2 system variables.

The **Symbols and Arrows** tab lets you specify the appearance, location, and display of symbols and arrowheads (see Figure 7–60).

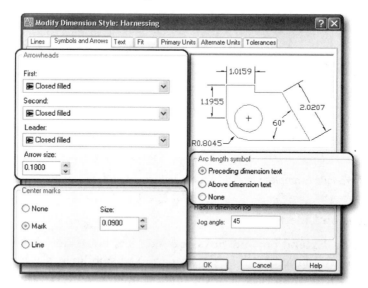

Figure 7–60 *Modify Dimension Style dialog box with Symbols and Arrows tab chosen*

The **Arrowheads** section controls the appearance of the arrowheads.

The **First, Second,** and **Leader** list boxes let you determine whether and how arrowheads are drawn at the termini of the extension lines or the start point of a Leader. Unless you specify a different type for the second line terminus, it will be the same type as the first line terminus. The types of arrowheads are recorded in DIMBLK; if both the first and second are the same, or in DIMBLK1 and DIMBLK2 if they are different.

The **Arrow size** text box lets you enter the distance that AutoCAD uses for the length of an arrowhead when drawn in a dimension or leader. The value is recorded in the DIMASZ system variable.

Some of the arrowhead types included in the standard library are **None, Closed, Dot, Closed filled, Oblique, Open, Origin Indication,** and **Right angle**. Figure 7–61 presents an example of each.

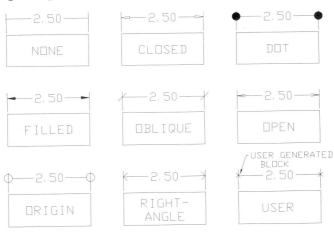

Figure 7–61 *Arrowhead types*

Choosing **User Arrow** lets you use a saved block definition by entering its name. Figure 7–62 shows the Select Custom Arrow Block dialog box. The block definition should be created as though it were drawn for the right end of a horizontal dimension line, with the insertion point at the intersection of the extension line and the dimension line.

Figure 7–62 *Select Custom Arrow dialog box*

The **Center Marks** section controls the size and type of center marks drawn with the DIMCENTER command.

    Selecting **None** suppresses the center mark (none will be drawn).

    Selecting **Mark** causes cross marks to be drawn according the size specified in the **Size** text box.

    Selecting **Line** causes cross lines to be drawn extending outside the circle in addition to the cross marks. Figure 7–63 presents an example of each.

Figure 7–63 *Circles with various center marks*

The **Arc length symbol** section controls the appearance of the symbol used for the DIMARC command.

    Selecting **Preceding dimension text** causes the arc length symbol to be drawn before the dimension text.

    Selecting **Above dimension text** causes the arc length symbol to be drawn above the dimension text.

    Selecting **None** suppresses the arc length symbol (none will be drawn).

The **Radius dimension jog** section controls the appearance of the line used to dimension long radius arcs whose center is located out of the drawing area.

    The **Jog Angle** lets you specify the angle that is drawn in the shortened dimension line for the long radius arc. Figure 7–64 shows an example of a line with a jog in it representing a long radius arc being dimensioned.

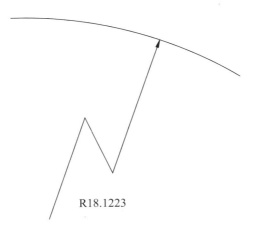

Figure 7–64 *A long radius arc with jog*

The **Text** tab lets you modify the appearance, location, and alignment of dimension text that is included when a dimension is drawn. There is a preview showing how dimensions will appear when drawn using the settings as you change them (see Figure 7–65)

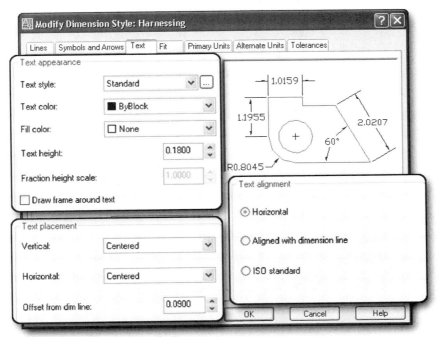

Figure 7–65 *Modify Dimension Style dialog box with Text tab chosen*

The **Text Appearance** section controls the appearance of dimension text.

The **Text style** list box lets you choose the text style to which the dimension text will conform. The value is recorded in the DIMTXSTY system variable.

Note: Do not confuse Text Style with Dimension Style. Dimensions are drawn in accordance with the current Dimstyle, which has as part of its configuration a Text Style to which Dimension Text will conform.

The **Text color** list box lets you choose how the color of dimension text will be determined. **ByLayer** causes the color of the text to match that of its layer. **ByBlock** causes the color of the text to match that of its block reference if it is part of a block reference. You can also choose one of the standard colors for the text color to match. The value is recorded in the DIMCLRT system variable.

The **Fill color** list box lets you choose how the color for the text background in dimensions will be determined.

The **Text height** text box lets you enter the distance that AutoCAD uses for text height when drawing dimension text as part of the dimension. The value is recorded in the DIMTXT system variable.

The **Fraction height scale** text box lets you enter a scale that determines the distance that AutoCAD uses for text height when drawing dimension text for fractions as part of the dimension text. This scale is the ratio of the text height for normal dimension text to the height of the fraction text. The value is recorded in the DIMTFAC system variable.

Selecting **Draw frame around text** causes AutoCAD to draw the dimension text inside a rectangular frame (see Figure 7–66). The value is recorded in the DIMGAP system variable as a negative value.

The **Text Placement** section controls how text is placed with relation to the dimension line.

The **Vertical** list box lets you choose how the dimension text will be drawn in relation to the dimension line. The options include **Centered**, **Above**, **Outside**, and **JIS**. The value is recorded in the DIMTAD system variable. Figure 7–66 shows an example of each.

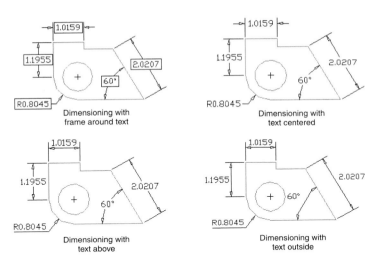

Figure 7–66 *Dimensioning Text with various Text appearance selections*

The **Horizontal** list box lets you choose how the dimension text will be drawn in relation to the extension lines. The options include **Centered**, **At Ext Line 1**, **At Ext Line 2**, **Over Ext Line 1**, and **Over Ext Line 2**. The value is recorded in the DIMJUST system variable. Figure 7–67 shows an example of each.

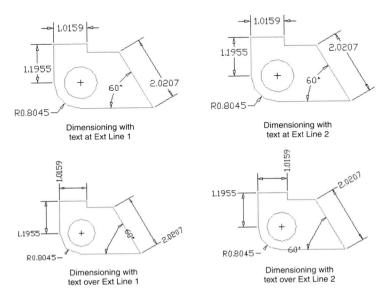

Figure 7–67 *Dimensioning Text with various Text placement (horizontal) selection*

The **Offset from dim line** text box lets you enter the distance from the dimension line that AutoCAD uses when drawing dimension text as part of the dimension. The value is recorded in the DIMGAP system variable.

The **Text Alignment** section controls the alignment of the dimension text. The value is recorded in the DIMTIH and DIMTOH system variables.

Selecting **Horizontal** causes dimension text in non-horizontal dimension lines to be drawn horizontally.

Selecting **Aligned with dimension line** causes dimension text in non-horizontal dimension lines to be drawn aligned with the dimension line.

Selecting **ISO standard** causes dimension text to comply with ISO standards.

The **Fit** tab lets you determine the arrangement of the various elements of dimensions. There is a preview showing how dimensions will appear using the settings as you change them (see Figure 7–68).

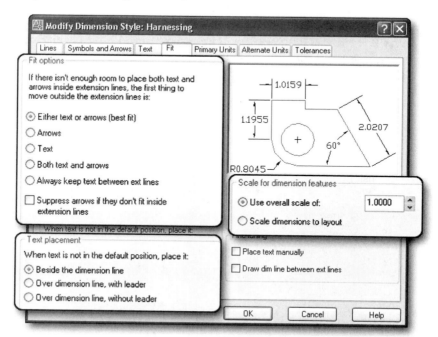

Figure 7–68 *Modify Dimension Style dialog box with Fit tab chosen*

The **Fit Options** section lets you choose which of the text or arrowheads will be drawn between the extension lines. The value is recorded in the DIMATFIT, DIMTIX, and DIMSOXD system variables.

Selecting **Either text or the arrows (best fit)** causes the text to be drawn outside and the arrows inside the extension lines if there is room for the arrows only.

Selecting **Arrows** causes the text to be drawn outside and the text inside the extension lines if there is room for the arrows only.

Selecting **Text** causes the arrows to be drawn outside and the text inside the extension lines if there is room for the text only.

Selecting **Both text and arrows** causes both the text and arrows to be drawn outside if the dimension lines are forced outside (see Figure 7–69) and both to be drawn inside if space is available.

Selecting **Always keep text between ext lines** causes text to always be drawn between the extension lines (see Figure 7–69). The value is recorded in the DIMTIX system variable.

Selecting **Suppress arrows if they don't fit inside the extension lines** causes AutoCAD to not draw the arrowheads between extension lines if they do not fit (see Figure 7–69). The value is recorded in the DIMSOXD system variable.

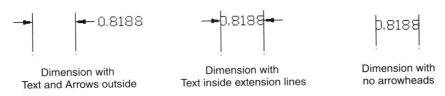

Figure 7–69 *Dimensioning examples with various Fit options*

The **Text Placement** section lets you choose how the text will be placed when text is not in the default position. The value is recorded in the DIMTMOVE system variable.

Selecting **Beside the dimension line** causes text to be drawn beside the dimension lines (see Figure 7–70).

Selecting **Over the dimension line, with leader** draws a leader line connecting the text to the dimension line (see Figure 7–70). The leader line is omitted when text is too close to the dimension line. Selecting **Over**

the dimension line, without leader draws the text without a leader line connecting the text to the dimension line (see Figure 7–70).

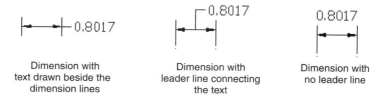

Figure 7–70 *Dimensioning example with various Text placement options*

The **Scale for Dimension Features** section sets the overall dimension scale value or the paper space scaling. This lets you increase or decrease the size of all of the dimensioning components uniformly by the factor entered.

Selecting **Use overall scale of** activates the text box that lets you set a scale for all dimension style settings that specify size, distance, or spacing, including text and arrowhead sizes. This scale does not change dimension measurement values. A scale factor greater than 0 (zero) causes text sizes, arrowhead sizes, and other scaled distances to plot at their face values. The value is recorded in the DIMSCALE system variable. Selecting **Scale dimensions to layout** (paperspace) determines a scale factor based on the scaling between the current model space viewport and paper space. When you work in paper space, but not in a model space viewport, or when TILEMODE is set to 1, the default scale factor of 1.0 is used or the DIMSCALE system variable.

The **Fine Tuning** section lets you place dimension elements manually.

Selecting **Place text manually** allows you to dynamically specify where dimension text is placed. The value is recorded in the DIMUPT system variable.

Selecting **Draw dim line between ext lines** causes a dimension line to be drawn between the extension lines regardless of the distance between extension lines. The value is recorded in the DIMTOFL system variable.

The **Primary Units** tab lets you to determine the appearance and format of numerical values of distances and angles. There is a preview showing how dimensions will appear when drawn using the settings as you change them (see Figure 7–71).

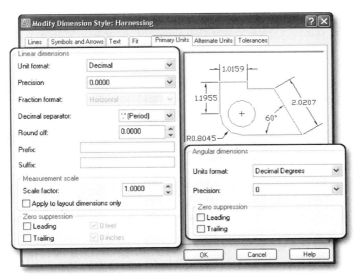

Figure 7–71 *Modify Dimension Style dialog box with Primary Units tab chosen*

The **Linear Dimensions** section controls the type, format, and precision of the primary units used in the dimension text for calling out numeric values of distances and angles.

The **Unit format** list box lets you choose the format of the units that AutoCAD uses when drawing dimension text as part of the dimension. The options include **Scientific, Decimal, Engineering, Architectural, Fractional**, and **Windows Desktop**. The value is recorded in the DIMLUNIT system variable. Figure 7–72 shows examples of these options.

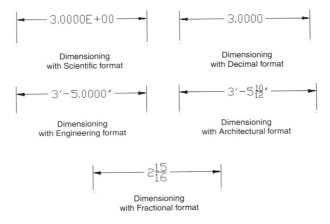

Figure 7–72 *Dimensioning example with various dimensioning unit formats*

The **Precision** list box lets you determine how many decimal places the text will be shown if the format is in Scientific, Decimal, or Windows Desktop units. It shows how many decimal places the inch measurments will have, if the format is in Engineering units. It shows the precision fractions will have in Architectural or Fractional units. The value is recorded in the DIMDEC system variable.

The **Fractional format** text box shows whether the fractions will be shown Horizontal, Diagonal, or Not Stacked. The value is recorded in the DIMFRAC system variable.

The **Decimal separator** list box lets you choose whether the decimal will be separated by a Period, Comma, or a Space if the format is in Scientific, Decimal, or Windows Desktop units. The value is recorded in the DIMDSEP system variable.

The **Round off** text box lets you enter the value to which distances will be rounded off. For example, a value of 0.5 causes dimensions to be rounded to the nearest 0.5 units.

The **Prefix** text box allows you to include a prefix in the dimension text (see Figure 7–73). The prefix text will override any default prefixes, such as those used in radius (R) dimensioning. The value is recorded in the DIMPOST system variable.

The **Suffix** text box allows you to include a suffix in the dimension text (see Figure 7–73). If you specify tolerances, AutoCAD includes the suffix in the tolerances, as well as in the main dimension.

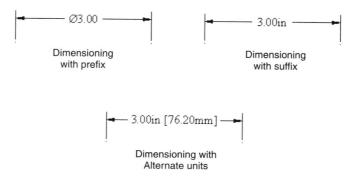

Figure 7–73 *Dimensioning example with prefix, suffix, and alternate units*

The **Measurement Scale** section controls the scale of the specified dimension values.

> The **Scale factor** text box lets you enter the number AutoCAD uses as a ratio of the true dimension distances, or the linear scale factor for linear measured distances of a dimension without affecting the components, angles, or tolerance values. For example, you are drawing with the intention of plotting at quarter-size scale, or 3" = 1'-0". You scale up a detail by a factor of 4 so that it will plot to full scale. If you wish to dimension it after it has been enlarged, you can set the measurment scale factor in this section of the Primary Units dialog box to .25 so that dimensioned distances will represent the dimensions of the object before it was scaled up. This method keeps the components at the same size as dimensions created without a scale change. The value is recorded in DIMLFAC. Adjusting the scale factor is useful when you wish to create different parts of the drawing at different scales but have the dimension elements uniform in size.
>
> Selecting **Apply to layout dimensions only** causes the ratio to be applied only to layout dimensions.

The **Zero Suppression** section controls whether or not leading and trailing zeros are displayed. The value is recorded in the DIMZIN system variable.

> Selecting **Leading** causes zeros ahead of the decimal point to be suppressed. For example: .700 is displayed instead of 0.700.
>
> Selecting **Trailing** causes zeros behind the decimal point to be suppressed. For example: 7 or 7.25 is displayed instead of 7.000 or 7.250.
>
> Selecting **0 Feet** causes zeros representing feet to be suppressed if the dimension text represents inches and/or fractions only. For example: 7" or 7 1/4" is displayed instead of 0'-7" or 0'-7 1/4".
>
> Selecting **0 Inches** causes zeros representing inches to be suppressed if the dimension text represents feet only. For example: 7' is displayed instead of 7'-0".

The **Angular Dimensions** section controls the format and precision of angular units. The value is recorded in the DIMAUNIT system variable.

> The **Units format** list box lets you choose the format of the units for the angular dimension text. Options include **Decimal Degrees, Degrees Minutes Seconds, Grads,** and **Radians**.

The **Precision** text box lets you set the number of decimal places for angular dimension units corresponding to the specified angular format. The value is recorded in the DIMADEC system variable.

The **Zero Suppression** section controls whether or not leading or trailing zeros are displayed. The value is recorded in the DIMAZIN system variable.

Selecting **Leading** causes zeros ahead of the decimal point to be suppressed.

Selecting **Trailing** causes zeros behind the decimal point to be suppressed.

The **Alternate Units** tab (see Figure 7–74) is similar to the **Primary Units** tab, except for the **Display alternate units** check box, which lets you enable (DIMALT set to 1) or disable (DIMALT set to 0) alternate units, and there is no **Angular Dimensions** section. See Figure 7–73 for an example of dimensioning with alternate units. There is a **Placement** section with **After primary value** and **Before primary value**, which let you determine where to place the Alternate units.

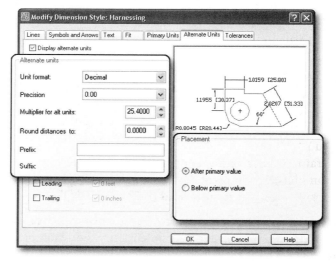

Figure 7–74 *Modify Dimension Style dialog box with Alternate Units tab chosen*

The **Tolerances** tab (see Figure 7–75) allows you to set the tolerance method and appropriate settings. Detailed explanation is provided in the "Tolerances" section, earlier in this chapter.

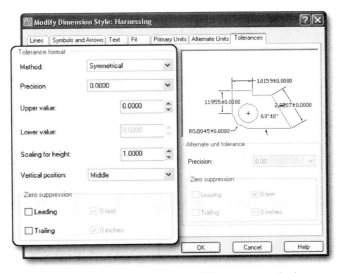

Figure 7–75 Modify Dimension Style dialog box with Tolerances tab chosen

Once the necessary changes are made for the appropriate settings, choose **OK** to close the Modify Dimension Style dialog box and then choose **Close** to close the Dimension Style Manager dialog box.

OVERRIDING THE DIMENSION FEATURE

The DIMOVERRIDE command allows you to change one of the features in a dimension without having to change its dimension style or create a new dimension style. For example, you may wish to have one leader with the text centered at the ending horizontal line, rather than over the ending horizontal line in the manner that the current dimension style may call for. By invoking the DIMOVERRIDE command, you can respond to the prompt with the name of the dimensioning system variable (DIMTAD in this case) and set the value to 0 rather than 1. Then you can select the dimension leader you wish to have overridden.

Invoke the DIMOVERRIDE command by typing **dimoverride** and AutoCAD prompts:

> Enter dimension variable name to override or ⊡ *(specify the dimensioning system variable name or choose* CLEAR *from the shortcut menu to clear overrides)*

If you specify a dimensioning system variable, you are prompted:

> Enter new value for dimension variable <current>: *(specify the new value and press* ENTER*)*

Enter dimension variable name to override: *(specify another dimension system variable or press* ENTER*)*

If you press ENTER, you are prompted:

Select objects: *(select the dimension objects)*

The dimensions selected will have the specified Dimensioning System Variable settings overridden in accordance with the value specified.

If you choose CLEAR, you are prompted:

Select objects: *(select the dimension to clear the overrides)*

The dimensions selected will have the Dimensioning System Variable setting overrides cleared.

UPDATING DIMENSIONS

Dimension Update permits you to make selected existing dimension(s) conform to the settings of the current dimension style.

Invoke Dimension Update from the Dimension toolbar (see Figure 7–76) and AutoCAD prompts:

Select objects: *(select any dimension(s) whose settings you wish to have updated to conform to the current dimension style).*

Figure 7–76 *Invoking Dimension Update from the Dimension toolbar*

 Note: Dimension Update is actually the DIMSTYLE command in which AutoCAD automatically chooses the APPLY option for you. All you have to do is select a dimension.

REVIEW QUESTIONS

 Open the Exercise Manual PDF file for Chapter 7 on the accompanying CD for project- and discipline-specific exercises.

 If you have the accompanying Exercise Manual, refer to Chapter 7 for project- and discipline-specific exercises.

1. Dimension types available in AutoCAD include:
   a. Linear
   b. Angular
   c. Diameter
   d. Radius
   e. All of the above

2. The associative dimension drawn with the DIMASSOC variable set to ON has all of its separate parts drawn as separate objects.
   a. True
   b. False

3. Linear (DIMLINEAR) allows you to draw horizontal, vertical, and aligned dimensions.
   a. True
   b. False

4. To draw a linear dimension you must (1) specify the first extension line origin, (2) locate the dimension line, and then (3) specify the second extension line.
   a. True
   b. False

5. Angular (DIMANGULAR) allows you to draw angular dimensions between two parallel lines.
   a. True
   b. False

6. By default, the dimension text for a radius dimension is preceded by:
   a. Radius
   b. Rad
   c. R

7. Using the PROPERTIES (modify properties) command will allow you to override the Dimensioning System Variable settings for a single dimension, without modifying the base dimension style.
   a. True
   b. False

8. Baseline (DIMBASELINE) is used to draw dimensions from a single datum baseline.

   a. True

   b. False

9. Center Mark (DIMCENTER) allows you to draw center cross marks or centerlines in a circle.

   a. True

   b. False

10. You must explode a dimension before you can use the DIMTEDIT command to edit the dimension text.

    a. True

    b. False

11. The suppress option in the **Lines** tab of the Modify Dimension Style dialog box allows you to suppress only one extension line at a time.

    a. True

    b. False

12. The **Arrowhead** section in **Symbols and Arrows** tab of the Modify Dimension Style dialog box allows you to change the size and style of your arrowheads.

    a. True

    b. False

13. You must use the UNITS command to determine how many decimal places will be displayed in the dimension text.

    a. True

    b. False

Directions: For each of the following questions, select which of the dimensioning tabs the option can be found on in the following list.

   a. Symbols and Arrows tab

   b. Text tab

   c. Fit tab

   d. Primary Units tab

   e. Tolerances

   f. cannot be found in the dimensioning dialog boxes

14. Arrowhead size
15. Associative/Non-associative setting
16. Text height
17. Arrowhead style
18. Text location
19. Suppressing of extension lines
20. Linear tolerance settings
21. Number of decimal places for a linear dimension

Directions: Answer the following questions based on Figure QX7–1.

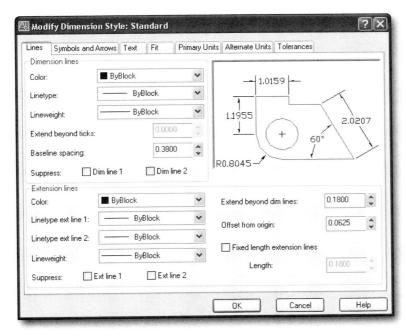

Figure QX7–1 *Modify Dimension Style dialog box with Lines and Arrows tab selected*

22. Which Dimension Variable must be changed to modify the size of Arrowheads?

   a. DIMARO
   b. DIMASZ
   c. DIMTSZ
   d. DIMDLI

23. Which Dimension Variable must be changed to modify Arrowhead types?

    a. DIMASZ   c. DIMBLK
    b. DIMTSZ   d. DIMEXE

24. Which Dimension Variable must be modified to change the color of dimension lines?

    a. DIMCLRE   c. DIMCLRL
    b. DIMCLRT   d. DIMCLRD

25. Which Dimension Variable is used to hide the first extension line?

    a. DIMEXL   c. DIMSE1
    b. DIMOFF   d. DIMEL1

26. Which Dimension Variable establishes the distance from the object to the start of the extension line?

    a. DIMEXO   c. DIMGAP
    b. DIMSEL   d. DIMSPC

27. Which Dimension Variable establishes the distance from the first dimension line to the second dimension line when using Baseline Dimensions?

    a. DIMD12   c. DIMDLI
    b. DIMSPC2   d. DIMDSPC

28. Which Dimension Variable is used to modify tick terminators?

    a. DIMARC   c. DIMTIC
    b. DIMART   d. DIMTSZ

29. Which Dimension Variable is used to hide the second dimension line?

    a. DIMSDL   c. DIM2OFF
    b. DIMHD2   d. DIMSD2

30. When one desires for the extension lines to be drawn past the dimension lines which Dimension Variable is used to perform this function?

    a. DIMEXE   c. DIMEXT
    b. DIMELD   d. DIMDLE

31. Which dimension variable is used to have a second arrowhead that is different from the first?

    a. DIMARW2   c. DIMBLK2
    b. DIMABLK   d. DIMARO2

CHAPTER 8

Plotting and Layouts

INTRODUCTION

One task has not changed much in the transition from board drafting to CAD: obtaining a hard copy. The term "hard copy" describes a tangible reproduction of a screen image. The hard copy is usually a reproducible medium from which prints are made and can take many forms, including slides, videotape, prints, or plots. This chapter describes the most commonly used processes for getting a hard copy: plotting/printing.

In manual drafting, if you need your drawing to be done in two different scales, you physically draw the drawing for two different scales. In CAD, with minor modifications, you plot or print the same drawing in different scale factors on different sizes of paper. In AutoCAD, you can even compose your drawing in Paper Space with limits that equal the sheet size and plot it at 1:1 scale.

After completing this chapter, you will be able to do the following:

- Plan the plotted sheet
- Plot from model space
- Set up a layout
- Create and modify a layout
- Create floating viewports
- Scale views relative to paper space
- Control the visibility of layers within viewports
- Plot from Layout (WYSIWYG)
- Create and modify plot style tables
- Change the Plot Style Property for an Object or Layer
- Configure plotters

PLANNING THE PLOTTED SHEET

Planning ahead is still required in laying out the objects to be drawn on the final sheet. The objects drawn on the plotted sheet must be arranged. At least in CAD, with its true-size capability, an object can be started without first laying out a plotted sheet.

But eventually, limits, or at least a displayed area, must be determined. For schematics, diagrams, and graphs, plotted scale is of little concern. But for architectural, civil, and mechanical drawings, plotting to a conventional scale is a professionally accepted practice that should not be abandoned just because it can be circumvented.

When setting up the drawing limits, you must take the plotted sheet into consideration to get the entire view of the object(s) on the sheet. So, even with all the power of the CAD system, some thought must still be given to the concept of scale, which is the ratio of true size to the size plotted. In other words, before you start drawing, you should have an idea at what scale the final drawing will be plotted or printed on a given size of paper.

The limits should correspond to some factor of the plotted sheet. If the objects will fit on a 24" x 18" sheet at full size with room for a border, title block, bill of materials, dimensioning, and general notes, then set up your limits to (0,0) (lower left corner) and (24,18) (upper right corner). This can be plotted or printed at 1:1 scale, that is, one object unit equals one plotted unit.

Plot scales can be expressed in several formats. Each of the following five plot scales is exactly the same; only the display formats differ.

    1/4" = 1'-0"
    1" = 4'
    1 = 48
    1:48
    1/48

A plot scale of 1:48 means that a line 48 units long in AutoCAD will plot with a length of 1 unit. The units can be any measurement system, including inches, feet, millimeters, nautical miles, chains, angstroms, and light-years, but, by default, plotting units in AutoCAD are inches.

There are four variables that control the relationship between the size of objects in the AutoCAD drawing and their sizes on a sheet of paper produced by an AutoCAD plot:

- Size of the object in AutoCAD. For simplification it will be referred to as ACAD_size.
- Size of the object on the plot. For simplification it will be referred to as ACAD_plot.
- Maximum available plot area for a given sheet of paper. For simplification it will be referred to as ACAD_max_plot.
- Plot scale. For simplification it will be referred as to ACAD_scale.

The relationship between the variables can be described by the following three algebraic formulas:

$$\text{ACAD\_scale} = \text{ACAD\_plot} / \text{ACAD\_size}$$
$$\text{ACAD\_plot} = \text{ACAD\_size} \times \text{ACAD\_scale}$$
$$\text{ACAD\_size} = \text{ACAD\_plot} / \text{ACAD\_scale}$$

EXAMPLE OF COMPUTING PLOT SCALE, PLOT SIZE, AND LIMITS

An architectural elevation of a building 48' wide and 24' high must be plotted on a 36" x 24" sheet. First, you determine the plotter's maximum available plot area for the given sheet size. This depends on the model of plotter you use.

In the case of an HP plotter, the available area for 36" x 24" is 33.5" x 21.5". Next, you determine the area needed for the title block, general notes, and other items, such as an area for revision notes and a list of reference drawings. For the given example, let's say that an area of 27" x 16" is available for the drawing.

The objective is to arrive at one of the standard architectural scales in the form of x in. = 1 ft. The usual range is from 1/16" = 1'-0" for plans of large structures to 3" = 1'-0" for small details. To determine the plot scale, substitute these values for the appropriate variables in the formula:

$$\text{ACAD\_scale} = \text{ACAD\_plot}/\text{ACAD\_size}$$
$$\text{ACAD\_scale} = 27"/48' \text{ for } X \text{ axis}$$
$$= 0.5625"/1'\text{-}0" \text{ or } 0.5625"=1'\text{-}0"$$

The closest standard architectural scale that can be used in the given situation is 1/2" = 1'-0" (0.5" = 1'-0", 1/24 or 1:24).

To determine the size of the object on the plot, substitute these values for the appropriate variables in the formula:

$$\text{ACAD\_plot} = \text{ACAD\_size} \times \text{ACAD\_scale}$$
$$\text{ACAD\_plot} = 48' \times (0.5"/1') \text{ for } X \text{ axis}$$
$$= 24" \text{ (less than the 27" maximum allowable space on the paper)}$$
$$\text{ACAD\_plot} = 24' \times (0.5"/1') \text{ for } Y \text{ axis}$$
$$= 12" \text{ (less than the 16" maximum allowable space on the paper)}$$

If instead of 1/2" = 1'-0" scale, you wish to use a scale of 3/4" = 1'-0", then the size of the object on the plot will be 48' x (0.75"/1') = 36" for the X axis. This is more than the available space on the given paper, so the drawing will not fit on the given paper size. You must select a larger paper size.

Once the plot scale is determined and you have verified that the drawing fits on the given paper size, you can then determine the drawing limits for the plotted sheet size of 33.5" x 21.5".

To determine the limits for the X and Y axes, substitute the appropriate values in the formula:

$$\text{ACAD\_limits } (X \text{ axis}) = \text{ACAD\_max\_plot}/\text{ACAD\_scale}$$
$$= 33.5"/(0.5"/1'\text{-}0")$$
$$= 67'$$
$$\text{ACAD\_limits } (Y \text{ axis}) = 21.5"/(0.5"/1'\text{-}0")$$
$$= 43'$$

Appropriate limits settings in AutoCAD for a 36" x 24" sheet with a maximum available plot area of 33.5" x 21.5" at a plot scale of 0.5" = 1'-0" would be:

lower left corner: 0,0
upper right corner: 67',43'

Another consideration in setting up a drawing for user convenience is to have the (0,0) coordinates at some point other than the lower left corner of the drawing sheet. Many objects have a reference point from which other parts of the object are dimensioned. Being able to set that reference point to (0,0) is very helpful. In many cases, the location of (0,0) is optional. In other cases, the coordinates should coincide with real coordinates, such as those on an industrial plant area block. In still other cases, only one set of coordinates might be a governing factor.

In this example, the 48' wide x 24' high front elevation of the building is to be plotted on a 36" x 24" sheet at a scale of 1/2" = 1'-0". It has been determined that (0,0) should be at the lower left corner of the front elevation view, as shown in Figure 8–1.

Centering the view on the sheet requires a few minutes of layout time. Several approaches allow the drafter to arrive at the location of (0,0) relative to the lower left corner of the plotted sheet or limits. Having computed the limits to be 67' wide x 43' high, the half-width and half-height (dimensions from the center) of the sheet are 33.5' and 21.5' to scale, respectively. Subtracting the half-width of the building from the half-width of the limits will set the X coordinate of the lower left corner at

−9.5' (from the equation, 24' − 33.5'). The same is done for the *Y* coordinate −9.5' (12' − 21.5'). Therefore, the lower left corner of the limits is at (−9.5',−9.5').

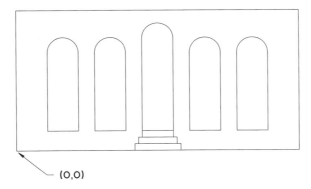

Figure 8–1 *Setting the reference point to the origin (0,0) in a location other than the lower left corner*

Appropriate limits settings in AutoCAD for a 36" x 24" sheet with a maximum available plot area of 33.5" x 21.5" by centering the view at a plot scale of 0.5" = 1'-0" (see Figure 8–2) are:

> lower left corner: −9.5',−9.5'
> upper right corner: 57.5',33.5'

 Note: The absolute *X* coordinates, when added (57.5' + 9.5'), equal 67', which is the width of the limits, and the absolute *Y* coordinates (33.5' + 9.5') equal 43', which is the height of the limits.

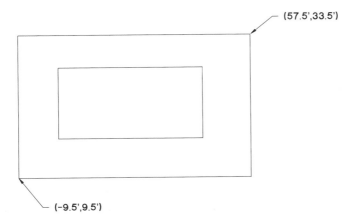

Figure 8–2 *Setting the limits to the maximum available plot area by centering the view*

SETTING FOR LTSCALE

As explained in Chapter 3, the LTSCALE system variable provides a method of adjusting the linetypes to a meaningful scale for the drawing. This sets the length of dashes and spaces in linetypes. When the value of LTSCALE is set to the reciprocal of the plot scale, the linetypes provided with AutoCAD plot out on paper at the sizes they are defined in *acad.lin*.

$$\text{LTSCALE} = 1/\text{ACAD\_scale}$$

SETTING FOR DIMSCALE

As explained in Chapter 7, AutoCAD provides a set of dimensioning variables that control the way dimensions are drawn. The DIMSCALE dimension variable is applied globally to all dimensions that are applied to sizes or distances, as an overall scaling factor. The default DIMSCALE value is set to 1. When DIMSCALE is set to the reciprocal of the plot scale, it applies globally to all dimension variables for the plot scale factor.

$$\text{DIMSCALE} = 1/\text{ACAD\_scale}$$

If necessary, you can set individual dimensioning variables to the size that you want the dimension to appear on the paper by substituting the appropriate values in the following formula:

$$\text{size of the plotted dimvars\_value} = \text{dimvars\_value} \times \text{ACAD\_scale} \times \text{DIMSCALE}$$

As an example, here's how to determine the arrow size DIMASZ for a plot scale of 1/2" = 1'-0", DIMSCALE set to 1, and default DIMASZ of 0.18":

$$\text{size of the plotted arrow} = 0.18" \times 24 \times 1$$
$$= 4.32"$$

SCALING ANNOTATIONS AND SYMBOLS

How can you determine the size at which text and symbols (blocks) will plot? As mentioned earlier, you almost always draw objects in their actual size, that is, to real-world dimensions. Even in the case of text and blocks, you place them at the real-world dimensions. In the previous example, the architectural elevation of a building 48' x 24' is drawn to actual size and plotted to a scale of 1/2" = 1'-0'. Let's say you wanted your text to plot at 1/4" high. If you were to create your text and annotations at 1/4", they would be so small relative to the elevation drawing itself that you could not read the characters.

Before you begin placing the text, you need to know the scale at which you will eventually plot the drawing. In the previous example of an architectural elevation, the plot scale is 1/2" = 1'-0" and you want the text to plot 1/4" high. You need to find a relationship between 1/4" on the paper and the size of the text for the real-world dimensions in the drawing. If 1/2" on the paper equals 12" in the model, then 1/4"-high text on the paper equals 6", so text and annotations should be drawn at 6" high in the drawing to plot at 1/4" high at this scale of 1/2" = 1'-0". Similarly, you can calculate the various text sizes for a given plot scale. The following table shows the model text size needed to achieve a specific plotted text height at some common scales.

Text Size Corresponding to Specific Plotted Text Height at Various Scales

SCALE	FACTOR	PLOTTED TEXT HEIGHT								
		1/16"	3/32"	1/8"	3/16"	1/4"	5/16"	3/8"	1/2"	5/8"
1/16" = 1'-0"	192	12"	18"	24"	36"	48"	60"	66"	96"	120"
1/8" = 1'-0"	96	6"	9"	12"	18"	24"	30"	36"	48"	60"
3/16" = 1'-0"	64	4"	6"	8"	12"	16"	20"	24"	32"	40"
1/4" = 1'-0"	48	3"	4.5"	6"	9"	12"	15"	18"	24"	30"
3/8" = 1'-0"	32	2"	3"	4"	6"	8"	10"	12"	16"	20"
1/2" = 1'-0"	24	1.5"	2.25"	3"	4.5"	6"	7.5"	9"	12"	15"
3/4" = 1'-0"	16	1"	1.5"	2"	3"	4"	5"	6"	8"	10"
1" = 1'-0"	12	0.75"	1.13"	1.5"	2.25"	3"	3.75"	4.5"	6"	7.5"
1 1/2" = 1'-0"	8	0.5"	.75"	1"	1.5"	2"	2.5"	3"	4"	5"
3" = 1'-0"	4	0.25"	.375"	0.5"	0.75"	1"	1.25"	1.5"	2"	2.5"
1" = 10'	120	7.5"	11.25"	15"	22.5"	30"	37.5"	45"	60"	75"
1" = 20'	240	15"	22.5"	30"	45"	60"	75"	90"	120"	150"
1" = 30'	360	22.5"	33.75"	45"	67.5"	90"	112.5"	135"	180"	225"
1" = 40'	480	30"	45"	60"	90"	120"	150"	180"	240"	300"
1" = 50'	600	37.5"	56.25"	75"	112.5"	150"	187.5"	225"	300"	375"
1" = 60'	720	45"	67.5"	90"	135"	180"	225"	270"	360"	450"
1" = 70'	840	52.5"	78.75"	105"	157.5"	210"	262.5"	315"	420"	525"
1" = 80'	960	60"	90"	120"	180"	240"	300"	360"	480"	600"
1" = 90'	1080	67.5"	101.25"	135"	202.5"	270"	337.5"	405"	540"	675"
1" = 100'	1200	75"	112.5"	150"	225"	300"	375"	450"	600"	750"

WORKING IN MODEL SPACE AND PAPER SPACE

One of the useful features of AutoCAD is the option to work on your drawing in two different environments, model space and paper space. AutoCAD allows you to plot a drawing from model space as well as paper space, also referred to as Layout. Most of the drafting and design work is created in the *3D* environment of model space, even though your objects may have been drawn only in a *2D* plane. You can plot the drawing from model space to any scale by specifying the plot scale in the plot dialog box.

Paper space is a 2D environment (like paper) used for arranging various views (floating viewports) of what was drawn in model space. It represents the paper on which you arrange the drawing prior to plotting. With AutoCAD, single or multiple paper space environments (layouts) can be easily set up and manipulated. Each layout represents an individual plot output sheet, or an individual sheet in a drawing project. You can apply different scales and specify different visibility for layers in each floating viewport. After arranging the views and scaling appropriately, you can plot the drawing from layout at 1:1 (full scale), allowing each separate viewport to display the selected parts of model space at the scale determined by the factor you specify. Paper space allows you to plot the drawing in WYSIWYG (What You See Is What You Get) mode.

You can switch between model space and paper space by selecting the appropriate tab provided at the bottom of the drawing window. Model space can be accessed from the Model tab. Selecting one of the available Layout tabs will access paper space. You can also switch between model space and paper space by changing the value of the system variable TILEMODE. When TILEMODE is set to 1 (default), you will be working in model space; when it is set to 0, you will be working in paper space.

When you are working in a layout, you can access model space by double-clicking inside one of the floating viewports. If you double-click anywhere outside the viewport, AutoCAD will switch to paper space. You must have at least one floating viewport in paper space to see objects drawn in model space. You can also maximize the active floating viewport to fill the screen and switch to model space for editing.

By default, AutoCAD creates two layout tabs called Layout1 and Layout2. You can rename these and, if necessary, create additional layouts. A detailed explanation is provided later in the chapter for creating and modifying layouts.

PLOTTING FROM MODEL SPACE

To plot/print the current drawing from the model space, invoke the PLOT command from the Standard toolbar (see Figure 8–3) and AutoCAD displays the Plot dialog box (see Figure 8–4).

Figure 8–3 *Invoking the* PLOT *command from the Standard toolbar*

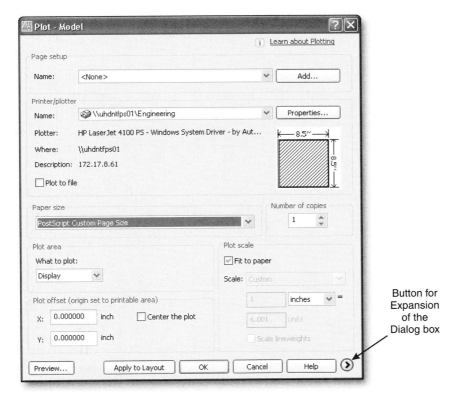

Figure 8–4 *Plot dialog box, not expanded*

PLOT SETTINGS

The Plot dialog box is almost identical to the Page Setup dialog box, except for the title and the **Name** text box in the **Page setup** section. The Plot dialog box can be expanded or contracted in size by choosing the arrow at the lower right corner of the dialog box. The first seven sections described in the following section on dialog box details are displayed when the dialog box is contracted. The last four are displayed only when the dialog box is expanded (see Figure 8–5).

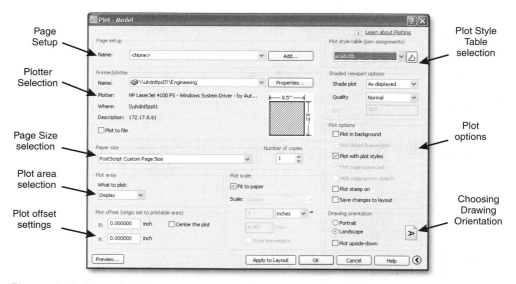

Figure 8–5 *Expanded Plot dialog box with additional options*

The **Page setup** section displays a list of any named and saved page setups in the drawing. You can base the current setup on a named page setup saved in the drawing, or you can create a new named page setup based on the current settings in the Plot dialog box by choosing **Add**.

The **Printer/plotter** section displays the currently configured plotting device.

> The **Name** list box lists the system printers/plotters or PC3 files that are available to select for plotting.
>
> **Plotter** lists the currently selected plotter or plotter assigned in the currently selected page setup.
>
> **Where** gives the port to which the selected plotter is connected, or its network location.
>
> Choosing **Properties** allows access to the Plotter Configuration of the currently configured plotting device. Refer to the section "Configuring Plotters" later in this chapter for a detailed explanation on plotter configuration.
>
> **Description** lists information about the output device currently selected.
>
> Selecting **Plot to file** causes plots to be output to a file rather than to a plotter or printer. If **Plot to file** is selected when you choose **OK** in the Plot dialog box, the Browse for Plot File dialog box is displayed. Specify the file name to save the plot file and press **Save**.

The **Paper size** section allows you to select the paper size to plot. Select the paper size to plot from the **Paper size** box for the selected plotting device. Actual paper sizes are indicated by the width (X axis direction) and height (Y axis direction). A default paper size is set for the plotting device when you create a plotter configuration file with the Add-a-Plotter Wizard.

The **Number of copies** text box lets you specify the number of copies to plot. This option is not available when you plot to file.

The **Plot area** section allows you to select the area to be plotted from the **What to plot** box.

Selecting **Display** causes the area displayed on the screen to be plotted.

Selecting **Limits** causes everything within the drawing limits to be plotted to the specified scale when the PLOT command is invoked from the **Model** tab.

Selecting **Layout** causes everything within the area of the paper size specified by the layout to be plotted when the PLOT command is invoked from paper space in a layout tab.

Selecting **Extents** causes the portion of the current space of the drawing that contains objects to be plotted. All visible geometry in the current space is plotted.

Selecting **View** plots a named view that was previously saved with the VIEW command. You can select a named view from the list. If there are no saved views in the drawing, this option is unavailable.

Selecting **Window** allows you to specify a window on the screen and plot the area covered by the window. The lower left corner of the window becomes the origin of the plot.

The **Plot scale** section controls the plot area.

Selecting **Fit to paper** causes the plot to fit within the selected paper size. To specify the exact scale for the plot, clear the **Fit to paper** check box and select the plot scale from the **Scale** list box. You can create a custom scale by entering the number of inches (or millimeters) equal to the number of drawing units in the appropriate text fields. When you plot from a layout, the default setting is 1:1 (full scale).

Selecting **Scale lineweights** causes AutoCAD to plot lineweights in proportion to the plot scale. Otherwise AutoCAD will plot the objects with assigned lineweights. Lineweights normally specify the linewidth of printed objects and are plotted with the linewidth size regardless of the plot scale.

The **Plot offset** section specifies an offset of the plotting area from the lower left corner of the paper. In a layout, the lower left corner of a specified plot area is positioned at the lower left margin of the plotting area. You can offset the origin by entering a positive or negative value. Select **Center the plot** to automatically center the plot on the paper.

The following sections of the Plot dialog box (see Figure 8–5) are displayed only when the dialog box is expanded by choosing the arrow at the lower right corner of the dialog box.

The **Plot style table** section allows you to set the plot style table, edit a plot style table, or create a new plot style table. Plot style tables are settings that give you control over how objects in your drawing are plotted into hardcopy plots. By modifying an object's plot style, you override that object's color, linetype, and lineweight. You can also specify end, join, and fill styles, as well as output effects such as dithering, grayscale, pen assignment, and screening. You can use plot styles if you need to plot the same drawing in different ways. This section displays the plot style table that is assigned to the current Model tab or Layout tab and provides a list of the currently available plot style tables. If you select **New** from the **Plot style table** list, the Add Color-Dependent Plot Style Table Wizard is displayed, which you can use to create a new plot style table. Choose **Edit** to modify the currently assigned plot style table. Refer to the section "Creating a Plot Style Table" later in this chapter for a detailed explanation about creating and modifying a plot style table.

The **Shaded viewport options** section specifies how shaded and rendered viewports are plotted and determines their resolution level with corresponding dpi.

The **Shade plot** text box displays how views are plotted and it is specified through the Properties dialog box for the selected viewport.

The Quality text box specifies the resolution at which shaded and rendered viewports are plotted. Selecting Draft sets rendered and shaded model space views to plot as wireframe. Selecting Preview sets rendered and shaded model space views to plot at a maximum of 150 dpi. Selecting Normal sets rendered and shaded model space views to plot at a maximum of 300 dpi. Selecting Presentation sets rendered and shaded model space views to plot at the current device resolution, to a maximum of 600 dpi. DPI values may vary with different computer/plotter configurations. Selecting Maximum sets rendered and shaded model space views to plot at the current device resolution with no maximum. Selecting Custom sets rendered and shaded model space views to plot at the resolution setting you specify in the DPI box, up to the current device resolution.

The **Plot options** section specifies the following options for lineweights, plot styles, and the current plot style table:

> Selecting **Plot in background** causes the plot to be processed in the background while you perform other tasks on the computer.
>
> Selecting **Plot object lineweights** plots the objects with assigned lineweights. Otherwise AutoCAD will plot with the default lineweight.
>
> Selecting **Plot with plot styles** plots using the object plot styles that are assigned to the geometry, as defined by the selected plot style table.
>
> Selecting **Plot paperspace last** plots model space geometry before paper space objects are plotted.
>
> Selecting **Hide paperspace objects** plots layouts (paper space) with hidden lines removed from objects.
>
> Selecting **Plot stamp on** includes a plot stamp on the plotted sheet. Refer to the section "Plot Stamp Settings" for a detailed explanation about modifying plot stamp settings.
>
> Select **Save changes to layout** to save the changes that you make in the Plot dialog box to the layout.

The **Drawing orientation** section specifies the orientation of the drawing on the paper for plotters that support landscape or portrait orientation.

> Selecting **Portrait** orients and plots the drawing so that the short edge of the paper represents the top of the page.
>
> Selecting **Landscape** orients and plots the drawing so that the long edge of the paper represents the top of the page.
>
> Selecting **Plot upside-down** orients and plots the drawing upside down.

Choosing **Preview** displays the drawing on the screen as it would appear when plotted. AutoCAD temporarily hides the plotting dialog boxes, draws an outline of the paper size, and displays the drawing as it would appear, using all the current settings, when it is plotted (see Figure 8–6).

The cursor changes to a magnifying glass with plus and minus signs. Holding the pick button and dragging the cursor toward the top of the screen enlarges the preview image. Dragging it toward the bottom of the screen reduces the preview image. Right-click, and AutoCAD displays a shortcut menu offering additional preview options: PAN, ZOOM, ZOOM WINDOW, ZOOM ORIGINAL, PLOT, and EXIT. To end the full preview, choose EXIT from the shortcut menu or press ENTER. AutoCAD returns to the Plot dialog box.

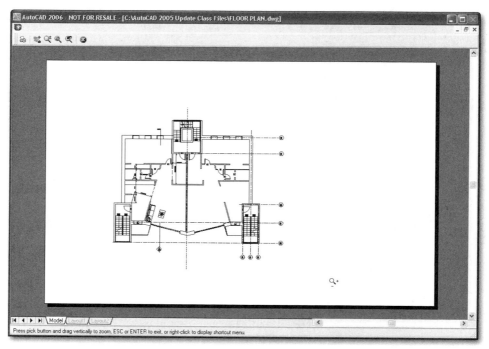

Figure 8–6 *Plot preview*

 Note: You can also access a full preview by invoking the Plot Preview command from the File menu.

After making the necessary changes in the plot settings, choose **OK**. AutoCAD starts plotting and reports its progress as it converts the drawing to the plotter's graphics language by displaying the number of vectors processed.

If something goes wrong or if you want to stop immediately, choose **Cancel** at any time. AutoCAD cancels the plotting.

Plot Stamp Settings

The Plot Stamp dialog box (see Figure 8–7) allows you to specify the information for the plot stamp that can be placed on a specified corner of each drawing and, if necessary, logs the information to a file. Choose **Plot Stamp Settings** (appears only when the **Plot stamp on** check box is set to ON) from the Plot dialog box to open the Plot Stamp dialog box.

Plot stamp information includes drawing name, layout name, date and time, login name, plot device name, paper size, plot scale, and user-defined fields, if any. Once you check the **Plot stamp on** check box in the **Plot options** section of the Plot dialog box, it remains active with whatever settings have been most recently entered until you specifically clear the check box. AutoCAD creates a plot stamp at the time the drawing is being plotted, and it is not saved with the drawing. Before you plot the drawing, you can preview the position of the plot stamp (not the contents) in the Plot Stamp dialog box. The plot stamp can be set to plot at one of the four drawing corners and can print up to two lines.

 Note: Plot stamp information is plotted with pen number 7 or the highest numbered available pen if the plotter doesn't hold seven pens. If you are using a non-pen (raster) device, color 7 is always used for plot stamping.

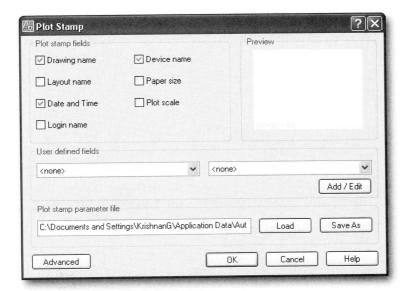

Figure 8–7 *Plot Stamp dialog box*

The **Plot stamp fields** section specifies the drawing information you want applied to the plot stamp. There are seven items that you can include in the plot stamp by selecting them by their field name as follows: **Drawing name, Layout name, Date and Time, Login name, Device name, Paper size,** and **Plot scale**.

The **Preview** section of the Plot Stamp dialog box provides a visual display of the plot stamp location based on the location and rotation values specified in the Advanced Options dialog box.

 Note: The preview displayed in the **Preview** area of the Plot Stamp dialog box is a visual display of the plot stamp location, not a preview of the plot stamp contents.

The **User defined fields** section provides text that can optionally be plotted. You can choose one or both user-defined fields for the plot stamp information. If the user-defined value is set to <none>, no user-defined information is plotted. To add, edit, or delete user-defined fields, choose **Add/Edit**. AutoCAD displays the User Defined Fields dialog box (see Figure 8–8).

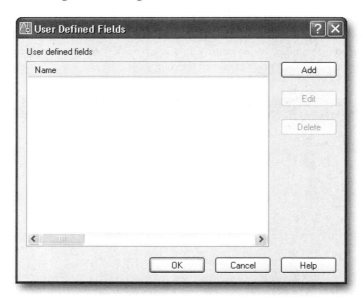

Figure 8–8 *User Defined Fields dialog box*

Choose **Add** to add an editable user-defined field to the bottom of the list, choose **Edit** to edit the selected user-defined field, and choose **Delete** to delete the selected user-defined field. Choose **OK** to save the changes and close the User Defined Fields dialog box. Choose **Cancel** to discard the changes and close the User Defined Fields dialog box.

The **Plot stamp parameter file** section displays the name of the file in which the plot stamp settings are stored. If necessary, you can save the current plot stamp settings to a new file by choosing **Save As** and providing an appropriate file name. AutoCAD stores plot stamp information in a file with a *.pss* extension. If you need to load a different parameter file, choose **Load**. AutoCAD displays a standard file selection dialog box, in which you can specify the location of the parameter file you want to use.

To set the location, text properties, and units of the plot stamp, choose **Advanced**. AutoCAD displays the Advanced Options dialog box (see Figure 8–9).

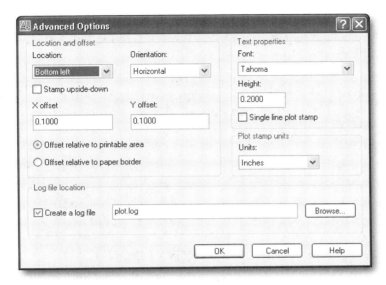

Figure 8–9 *Advanced Options (Plot Stamp) dialog box*

Location allows you to select the area where you want to place the plot stamp. The options include Top Left, Bottom Left (default), Bottom Right, and Top Right. The location is relative to the image orientation of the drawing on the page.

Orientation allows you to select the rotation of the plot stamp in relation to the specified page. The options include Horizontal and Vertical for each location.

Stamp upside-down controls whether to rotate the plot stamp upside down.

X offset and **Y offset** determine the offset distance calculated from either the corner of the paper or the corner of the printable area, depending on which setting you specify. Select one of the two options: **Offset relative to printable area** or **Offset relative to paper border** to set the reference point from which to measure the offset distance.

The **Text properties** section determines the font, height, and number of lines you want to apply to the plot stamp text. **Font** specifies the font you want to apply to the text used for the plot stamp information. **Height** specifies the text height you want to apply to the plot stamp information. **Single line plot stamp** controls whether to place the plot stamp information in a single line of text or not. The plot stamp information can consist of up to two lines of text, but the placement and offset values you specify must accommodate text wrapping and text height. If the plot stamp contains text that is longer than the printable area, the plot stamp text will be truncated. If **Single line plot stamp** is set to OFF, the plot stamp text is wrapped after the third field.

The **Plot stamp units** section allows you to specify the units used to measure X offset, Y offset, and height. From the **Units** box you can select one of the available units: inches, millimeters, or pixels.

Log file location specifies the name of the file to which the plot stamp information is saved instead of, or in addition to, stamping the current plot. The default log file name is *plot.log*, and it is located in the AutoCAD folder. Choose **Browse** to specify a different file name and path. After the initial plot log file is created, the plot stamp information in each succeeding plotted drawing is added to this file. Each drawing's plot stamp information is a single line of text. If necessary the log file can be placed on a network drive and shared by multiple users.

Choose **OK** to save the changes and close the Advanced Options dialog box. Choose **OK** to save the changes and close the Plot Stamp dialog box.

PLOTTING FROM PAPER SPACE (WYSIWYG)

As mentioned earlier, AutoCAD allows you to plot a drawing from paper space as well as model space. The paper space environment allows you to set up multiple layouts. You can have as many layouts as you like, and each one can be set up for a different type of output. A layout is used to compose or lay out your model drawing for plotting. A layout may consist of a title block, one or more viewports, and annotations. As you create a layout, you can design floating viewport configurations to display different details in your drawing. You can apply different scales to each viewport within the layout, and specify different visibility for layers in the viewport. Layouts are accessible by choosing the Layout tab at the bottom of the drawing area. The purpose of using a paper space layout with viewports is to make it easy to produce plotted sheets with AutoCAD, rearranging and scaling objects that are drawn in model space and adding non-object elements in paper space.

Note: Reference to viewports in layouts means floating viewports. The distinction between floating viewports and model space viewports is explained in the section on viewports in this chapter.

The combination of paper space and viewports is a special and powerful application for producing the most commonly used communication tool in the architect/engineer's repertoire, a set of paper drawings, traditionally referred to as "blueprints."

AutoCAD allows you to:

1. Create or import a page setup, applying it to a layout that is the exact width and height of the desired plotted paper sheet.
2. Create (or attach, import, or otherwise position) a border, title block, symbols, tables, and any other annotation or AutoCAD objects in full size (12" = 1'-0") on the sheet.

3. Open windows of specified sizes and locations (referred to as viewports) at specified scales on the sheet for viewing desired parts of objects drawn in model space.
4. Plot the Layout to 1:1 scale.

When configured, the layout will be like a separate drawing sheet that represents the final printed sheet with a border, title block, and note information, making it easy to arrange objects to their desired scale on the sheet (in their viewports) and a simple task to plot.

PLANNING TO PLOT

Figure 8–10 shows a model space drawing of a residential elevation and floor plan without a border, title block, or any notation except dimensions in the plan view. In Figure 8–11, by using the Layout feature, the views are juxtaposed for conventional arrangement and a border, title block, view titles, and text tables are added in paper space. The two figures illustrate the application of model space for creating the design objects and paper space for arranging the design objects on a plot-friendly sheet with border, title block, view titles, and annotation, accomplished in a single drawing file.

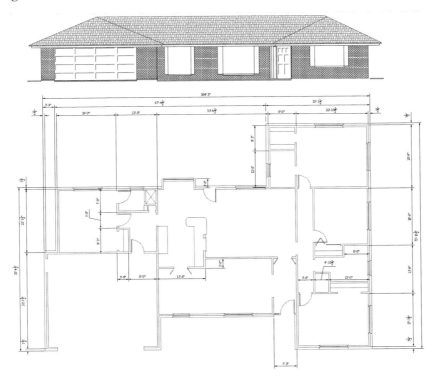

Figure 8–10 *Architectural Floor Plan and Elevation drawn in model space*

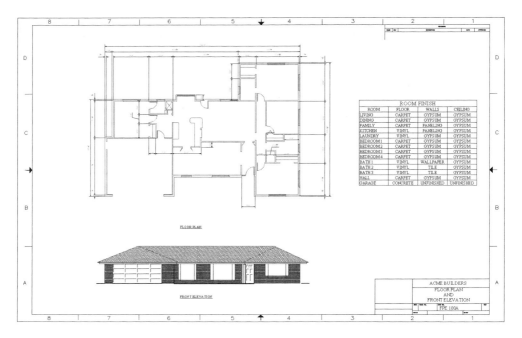

Figure 8–11 *Architectural Floor Plan and Elevation in layout format configured for plotting*

The earlier part of this chapter gave an example of setting up a drawing for plotting from model space (see Figures 8–1 and 8–2). This section describes procedures for setting up a drawing for plotting/printing from layouts. The problem with plotting from model space is that the objects must be located and sized in the coordinate system according to how they will be arranged for plotting. Even with AutoCAD's true-size capability, plotting from model space requires parts that need to be plotted at different scales be drawn at different scales. This is not necessary when the Layout/Paper Space feature is utilized.

Setting up the drawing for plotting from layouts is not as restrictive as it is from model space. Some thought, however, must still be given to the concept of scale, which is the ratio of true size to the size plotted. In other words, before you start drawing, you should have an idea of which parts of the drawing must be plotted at what scale. The example earlier in this chapter showed how to determine the appropriate scale to use so that objects will fit on the desired sheet size. That example can be used as a guide for plotting to scale in layouts, except that each view of the objects drawn in model space can be treated as a separate drawing on the paper space sheet, like pictures on the page of a scrapbook.

SETTING UP LAYOUTS

By default, a new drawing (using the *acad.dwt* file as a template) starts with two layouts, named **Layout1** and **Layout2**. (A drawing started with another template might have only one layout, for example, the *ANSI B -Named Plot Styles.dwt* that is configured with an ANSI B Title Block on an 11 x 17 sheet.) However, each of the two default layouts in the drawing started with *acad.dwt* represents a sheet in landscape mode 11 units by 8.5 units with a dashed rectangle 10.5 units by 8 units outlining the expected printable area.

Figure 8–12 shows how views of a design object drawn in model space will appear with the **Model** tab selected (in the top view) and with the **Layout1** tab selected (in the bottom view). **Layout1** has the basic elements of a layout: a paper space drawing sheet with specific width and height, and a viewport through which you can view design objects that are drawn in model space. If you erase the viewport, model space objects would not be visible.

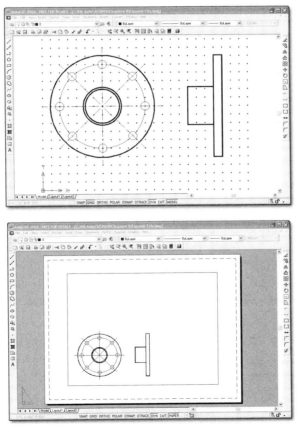

Figure 8–12 *Two views of object in Model space (11 x 8.5 limits) and in Layout1*

Figure 8–13 shows the basic parts of the default layout with only one viewport and nothing drawn in paper space.

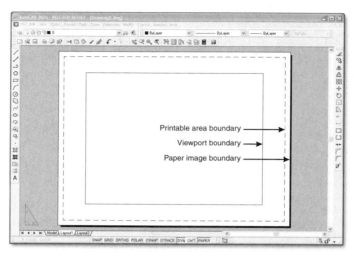

Figure 8–13 *Parts of a Layout tab*

To repeat, by default, every initialized layout has an unnamed page setup associated with it. If necessary, you can change the layout's page setup through the Page Setup Manager (explained in detail in "Reconfiguring the Layout with Page Setup" later in this chapter). You can modify the settings of a page setup at any time.

Layout(s) by Way of a Template

The most common procedure for creating a new drawing is to use a template file that has already been configured for the application/discipline in effect. The template will contain one or more layouts, each created at the desired sheet size and normally having a border, title block, revision history table, or other non-object elements drawn in paper space on it. Non-object elements are things like borders, title blocks, dimensions, call-outs, etc. versus object elements that represent real objects like walls, pipes, switches, streets, etc. A template drawing can also have layers, system variables, and styles for text, dimensions, and other features configured that conform to the standards of the proposed set of drawings.

The model space objects shown in Figure 8–12 will be used as an example of how to use a layout contained in an existing template drawing file (ANSI-A Color Dependent Plot Style template) to produce the desired plot. When the new drawing is created, the starting view will be of the layout named ANSI A Title Block, as shown in Figure 8–14a. After you switch to model space (selecting **Model** in the Status bar) and draw the objects shown in Figure 8–12, when you switch back to the ANSI A Title Block Layout tab, the objects will appear in the one viewport as shown in Figure 8–14b, and you can plot the drawing from paper space at 1:1 scale. The outline of the single viewport in this layout is not very distinguishable because it coincides with the inside lines of the border/title block. If you double-click inside the viewport, you will be switched to model space in the viewport while still in the layout. (This is different from switching to model space by using the **Model** tab.) When you do this, the outline of the viewport becomes a heavy line and is more visible, as shown in Figure 8–14b.

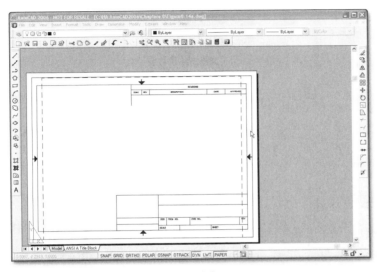

Figure 8–14a *Newly created ANSI-A A Title Block Layout*

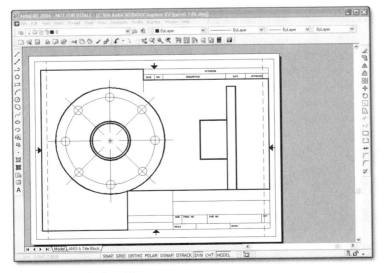

Figure 8–14b *Layout with the objects*

Creating a New Layout

To create a new layout, choose New Layout from the Layouts toolbar (see Figure 8–15). Up to 255 layouts can be created in a single drawing. AutoCAD prompts:

> Enter name of new layout <Layout#>: *(specify a layout name or press* ENTER *to accept the default name)*

Figure 8–15 *Choosing New Layout from the Layouts toolbar*

Layout names must be unique. Layout names can be up to 255 characters long and are not case sensitive. Only the first 31 characters are displayed on the tab.

AutoCAD establishes a new layout, adding its tab at the bottom of the screen. By default, every initialized layout has an unnamed page setup associated with it. Once you create a layout, you can change the settings for the layout's page setup with the help of the Page Setup Manager dialog box (described later in this chapter), which includes the plot device settings and other settings that affect the appearance and format of the output. The settings you specify in the page setup are stored in the drawing file with the layout.

 Note: If you want the Page Setup Manager to be displayed each time you begin a new drawing layout, select the **Show Page Setup Manager for new layouts** option on the **Display** tab in the Options dialog box. If you don't want a viewport to be automatically created for each new layout, clear the **Create Viewport in new layouts** option on the **Display** tab in the Options dialog box.

Figure 8–16 *Choosing Layout from Template from the Layouts toolbar*

Choosing Layout from Template from the Layouts toolbar (see Figure 8–16) creates a new layout tab based on an existing layout in a template (*.dwt*), drawing (*.dwg*), or drawing interchange (*.dxf*) file. AutoCAD displays a standard file selection dialog box to select a file. Once you select a file, AutoCAD displays the Insert Layout(s) dialog box (see Figure 8–17), which displays the layouts saved in the selected file. After you select a layout, the layout and all objects from the specified template or drawing file are inserted into the current drawing.

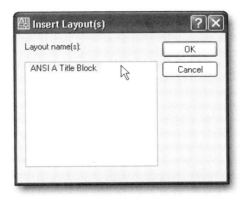

Figure 8–17 *Insert Layout(s) dialog box*

Additional options are available by invoking the LAYOUT command. COPY creates a new layout by copying an existing layout, DELETE deletes an existing layout, Rename renames an existing layout, and SAVEAS saves a layout as a drawing template (*.dwt*) file without saving any unreferenced symbol table and block definition information. You can access all of the available options from the shortcut menu that appears when you right-click the name of the layout tab.

WORKING WITH FLOATING VIEWPORTS

As mentioned earlier, in a layout, you can create multiple, overlapping, contiguous, or separated floating viewports, as shown in Figure 8–18. To repeat, viewports are configurable windows from paper space (a layout) with a view into the model space design that you can move and resize. You can use any of the standard AutoCAD modify commands, such as MOVE, COPY, STRETCH, SCALE, and ERASE, to manipulate the floating viewports. For example, you can use the MOVE command to grab one viewport and move it around the screen without affecting other viewports. A viewport can be of any size and can be located anywhere in the layout. You must have at least one floating viewport to view the objects drawn in model space.

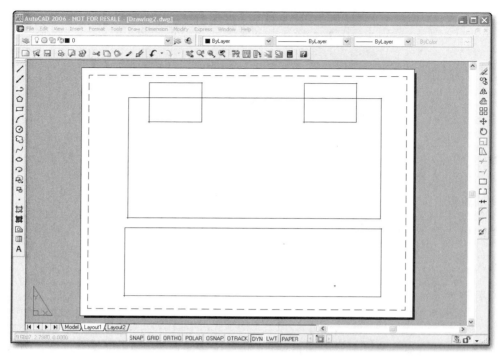

Figure 8–18 *Floating viewports in a layout*

You can create or manipulate floating viewports only while in paper space. While working in a layout, you can switch back and forth between model space and paper space. When you make a floating viewport current in a layout by double-clicking inside of it, you are then working in model space in a floating viewport. Any modification to objects in the drawing in model space is reflected in all paper space viewports in which those objects are visible, as well as tiled viewports. When you double-click outside of a floating viewport, AutoCAD switches the mode to paper space. In paper space you can add annotations or other graphical objects, such as a title block. You can even dimension model space objects while in paper space. Objects you add in paper space do not change the model or other layouts.

 Note: While working in paper space you can use Object Snaps to snap to points on objects in model space, but not vice versa.

Figure 8–19 shows the model that is shown in Figure 8–14b in the desired plot-ready form. It has two floating viewports with dimensions and annotation added in paper space. The topics that follow explain the steps required to get from Figure 8–14b to Figure 8–19.

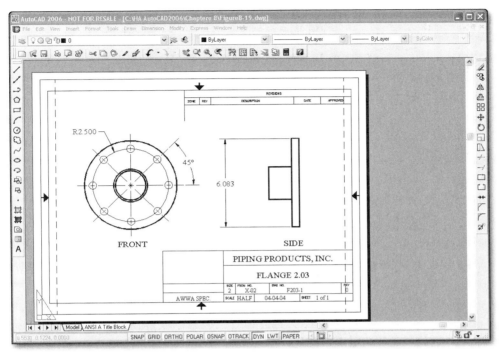

Figure 8–19 *Completed plot-ready drawing*

Almost all of the plotted sheets in a set of project drawings are the same size and share the same border/title block configuration, with repeated title block information such as the name of the project, client, and architect/engineer. But the content, arrangement, and scale of the design objects are naturally different from one drawing sheet to the next. This is where each layout needs to have its own individualized group of floating viewports. So, when a new drawing is created using a template file, the first order of business is to get rid of any unusable floating viewports in the layout that came with the template drawing.

The example in Figure 8–14b shows the model space design in a single floating viewport. It has been decided that the front view and side view are too close together to allow for dimensioning. Rather than relocate the views in model space, we will create two viewports, one for each view of the object, and separate them on the layout to allow for the dimensioning. The existing viewport needs to be erased. To accomplish this, the existing viewport must be selected. Because it coincides with lines in the border/title block, it may be difficult to select by the pick method. So, the selection Window option of the object selection function can be employed. Figure 8–20 shows where the window needs to be in order to select the viewport only, without including the border/title block. When the ERASE command is invoked, the drawing will appear as it is in Figure 8–14a. The model space design objects will not be visible.

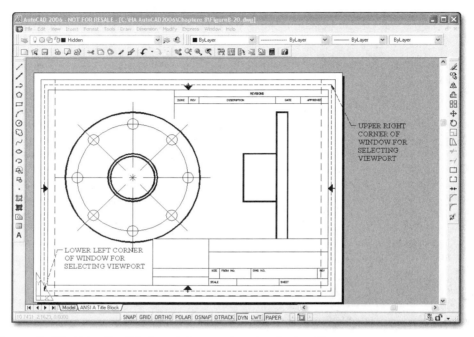

Figure 8–20 *Using the Window option to select the viewport for erasure*

Creating Floating Viewports

You can create a single floating viewport that fits the entire layout or create multiple floating viewports in the layout. Once you create a viewport, you can change its size and properties, and move it as needed.

> **Note:** It is important to create layout viewports on their own separate layer. When you are ready to plot, you can turn off the layer and plot the layout without plotting the boundaries of the layout viewports. Turning off the layer on which a viewport has been created turns off the border of that viewport only. The window to model space is not affected and model space objects remain visible.

To create a single rectangular floating viewport, choose Single Viewport from the Viewports toolbar (see Figure 8–21) and AutoCAD prompts:

Specify corner of viewport or *(specify the first corner of viewport)*
Specify opposite corner: *(specify the opposite corner to create the floating viewport)*

Figure 8–21 *Choosing Single Viewport from the Viewports toolbar*

The viewport will be created on the active layer and if necessary, you can move, copy, rotate, scale, or stretch the viewport or place it on a different layer, just like modifying any other object.

To create an irregularly shaped floating viewport, choose Polygonal Viewport from the Viewports toolbar (see Figure 8–22) and AutoCAD prompts:

> Specify start point: *(specify first point to create an irregularly shaped floating viewport)*
> Specify next point or ⬇: *(specify next point or select one of the options from the shortcut menu)*

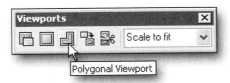

Figure 8–22 *Choosing Polygonal Viewport from the Viewports toolbar*

> The ARC option adds arc segments to the polygonal viewport.
>
> The LENGTH option draws a line segment of a specified length at the same angle as the previous segment. If the previous segment is an arc, AutoCAD draws the new line segment tangent to that arc segment.
>
> The UNDO option removes the most recent line or arc segment added to the polygonal viewport.
>
> The CLOSE option closes the polygon to create the polygonal viewport.

To create a floating viewport from a closed polyline, ellipse, spline, region, or circle, choose Convert Object to Viewport from the Viewports toolbar (see Figure 8–23) and AutoCAD prompts:

> Select object to clip viewport: *(Select an object)*

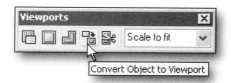

Figure 8–23 *Choosing Convert Object to Viewport from the Viewports toolbar*

The polyline you specify must be closed and contain at least three vertices. It can be self-intersecting, and it can contain an arc as well as line segments.

Choose Display Viewports from the Viewports toolbar (see Figure 8–24) to create multiple floating viewports. This opens the Viewports dialog box (see Figure 8–25). AutoCAD lists standard viewport configurations. Choose the name of the configuration you want to use from the **Standard viewports** list. AutoCAD displays how the corresponding configuration will look in the **Preview** window. The **Setup** list specifies either a 2D or a 3D setup. When you select 2D, the new viewport configuration is initially created with the current view in all of the viewports. When you select 3D, a set of standard orthogonal 3D views is applied to the viewports in the configuration. The **Preview** section displays a preview of the viewport configuration you select and the default views assigned to each individual viewport in the configuration. Choosing **Change view to** lets you replace the view in the selected viewport with the view you select from the list. You can choose a named view, or if you have selected 3D setup, you can select from the list of standard views. Use the **Preview** area to see the choices. After choosing the viewport configuration and setting corresponding values, choose **OK** to close the dialog box; AutoCAD prompts:

> Specify first corner or ⬇: *(specify the first corner to define selected viewport configuration or right-click and select FIT to create the selected viewport configuration to fit the paper size)*
> Specify opposite corner: *(specify opposite corner to define selected viewport configuration)*

Figure 8–24 *Choosing Display Viewports Dialog from the Viewports toolbar*

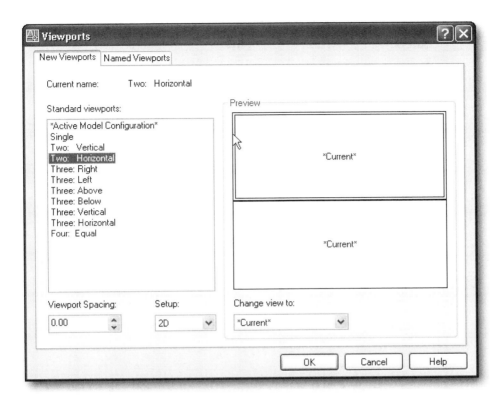

Figure 8–25 *Viewports dialog box*

Using the Viewports dialog box (see Figure 8–25) for creating new viewports will be the method used in the example of transforming the layout shown in Figure 8–14b to the plot-friendly layout of Figure 8–19. Select the points shown in Figure 8–26a when prompted to specify the corners to determine the rectangle outlining the two vertical viewports. The model space design objects will be displayed the same in both viewports (see Figure 8–26b).

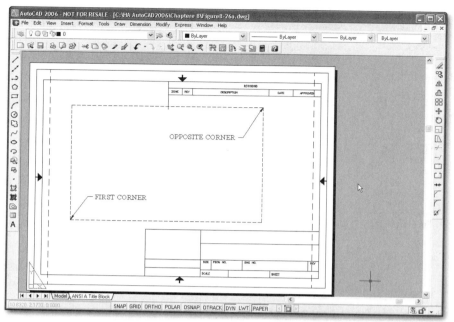

Figure 8–26a *Selection of points to create two vertical viewports*

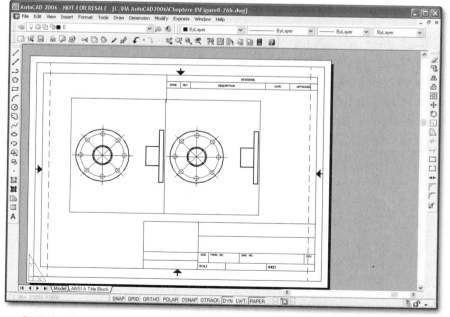

Figure 8–26b *Two vertical viewports with the objects*

Modifying Floating Viewports

As mentioned earlier, once you create a viewport, you can change its size and properties, and reposition it as needed. If you want to change the shape or size of a layout viewport, you can use its grips to edit the vertices just as you edit any object with grips.

By using grips and/or the STRETCH and MOVE commands, the left viewport is widened and the right viewport is moved to the right and made narrower, as shown in Figure 8–27. The model space design objects are visible and will be arranged properly within each viewport in later steps.

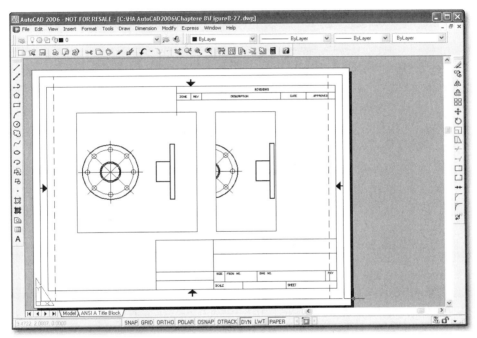

Figure 8–27 *Resizing and rearranging the two viewports*

You can also redefine the boundary of a layout viewport by using the VPCLIP command and maximize the viewport by using the VPMAX command. When you right-click on an active viewport, you can control whether the objects display in a viewport (by changing the DISPLAY VIEWPORT OBJECTS setting), and control the setting of the locking feature which prevents the zoom scale factor in the selected viewport from being changed when working in model space.

The VPCLIP command (see Figure 8–28) allows you to clip a floating viewport to a user-drawn boundary. AutoCAD reshapes the viewport border to conform to a user-drawn boundary. To clip a viewport, you can select an existing closed object, or specify the points of a new boundary. AutoCAD prompts:

Select viewport to clip: *(select viewport to clip)*
Select clipping object or ⬇: *(select clipping object or select one of the options from the shortcut menu)*

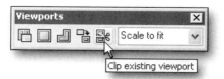

Figure 8–28 *Choosing Clip existing viewport from the Viewports toolbar*

If you select an object for clipping, AutoCAD converts the object to a clipping boundary. Objects that are valid as clipping boundaries include closed polylines, circles, ellipses, closed splines, and regions.

> The POLYGONAL option allows you to draw line segments or arc segments by specifying points to create a polygonal clipping boundary.
>
> The DELETE option deletes the clipping boundary of a selected viewport. This option is available only if the selected viewport has already been clipped. If you clip a viewport that has been previously clipped, the original clipping boundary is deleted and the new clipping boundary is applied.

Choosing MAXIMIZE VIEWPORT (see Figure 8–29) when a viewport is selected causes the selected viewport in the current layout to fill the screen drawing area, making the entire drawing area accessible for viewing and editing. The size of the area displayed depends on the zoom factor in effect. When the viewport has been maximized, choosing the MINIMIZE VIEWPORT option in the shortcut menu returns the display to the previous layout state. You can also maximize or minimize viewports from the button located on the Status bar.

Selecting DISPLAY VIEWPORT OBJECTS when a viewport is selected controls the display of objects in the selected viewport. When Off is selected, objects in the selected viewport are not visible, and the viewport cannot be selected when switching viewports in model space in the current layout. When On (default) is selected, AutoCAD turns on a viewport, making it active and making its objects visible.

Selecting DISPLAY LOCKED when a viewport is selected prevents or enables, respectively, the zoom scale factor in the selected viewport from being changed when working in model space

Figure 8–29 *Choosing Maximize Viewport from the shortcut menu*

SCALING VIEWS RELATIVE TO PAPER SPACE

AutoCAD allows you to scale viewport objects relative to paper space, which establishes a consistent scale for each displayed view. To accurately scale the plotted drawing, you must scale each viewport relative to paper space. Usually the layout is plotted at a 1:1 ratio. The ratio is determined by dividing the paper space units by the model space units. The scale factor of model space design objects in a viewport can be set with the XP option of the ZOOM command while model space is active in that viewport. For example, entering **1/24xp** or **0.04167xp** (1/24 = 0.01467) in response to the ZOOM command prompt will display an image to a scale of 1/2" = 1' 0", which is the same as 1:24 or 1/24. You can also change the plot scale of the viewport using the Viewport Scale Control on the Viewports toolbar (see Figure 8–30).

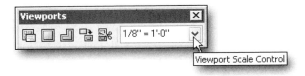

Figure 8–30 *Viewport Scale Control on the Viewports toolbar*

In the case of the viewports in the example drawing shown in Figure 8–27, the views need to be set to half scale. In this case, first double-click in one of the viewports to make it active in model space and then enter a scale factor in the Viewport Scale Control box of the Viewports toolbar of 0.5 or choose 1:2 from the drop-down box. AutoCAD displays 6" = 1', and the objects are rescaled in the selected viewport. Repeat the procedure for the second viewport and the result will appear as shown in Figure 8–31.

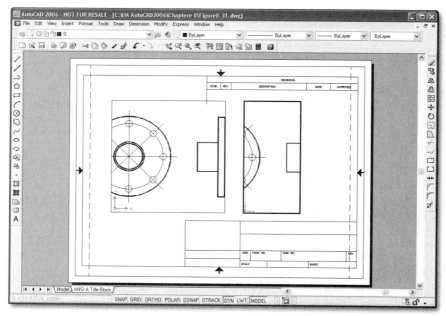

Figure 8–31 *Objects in viewports rescaled*

CENTERING MODEL SPACE OBJECTS IN A VIEWPORT

Next, in order to center the front view of the object in the left viewport, respond to the zoom Center option with the coordinates **3.5,5.0**, which is the center of the circle. Repeat the procedure in the right viewport using the coordinates **9.5,5.0**. Specifying the same Y coordinate for centering the model spaces in both viewports ensures that the objects will line up horizontally. Some practice (and some trial and error) is needed to size the viewports and center the model space design objects to ensure that the desired object views (and only the desired ones) are visible in the appropriate viewports. The result will appear as shown in Figure 8–32.

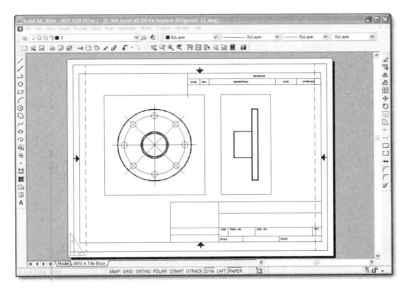

Figure 8–32 *Objects in viewports centered*

HIDING VIEWPORT BORDERS

After the viewports are scaled and the objects are centered, double-click outside the viewports to return to paper space. Turn the layer named Viewports off (initially two viewports were created on layer Viewports) and the result will appear as shown in Figure 8–33.

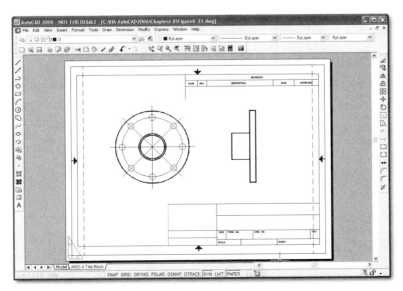

Figure 8–33 *Result after the Viewports layer is set to OFF*

While in paper space, the drawing name and other information can be entered in the title block as appropriate. The views can be named, and the objects can also be dimensioned, as shown in Figure 8–34.

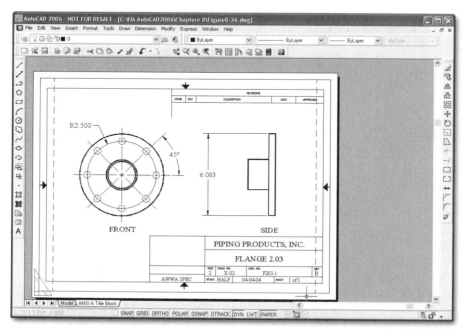

Figure 8–34 *Completed layout*

CONTROLLING THE VISIBILITY OF LAYERS WITHIN VIEWPORTS

The Layer Properties Manager dialog box (see Figure 8-35) controls the visibility of layers in a single viewport or in a set of viewports. This enables you to select a viewport and freeze a layer in it while still allowing the contents of that layer to appear in another viewport. Figure 8–36 shows two viewports containing the same view of the drawing, but in one viewport the layer containing the dimensioning is set to ON, and in the other the dimensioning layer is set to OFF.

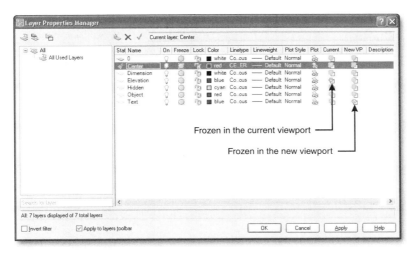

Figure 8–35 *Layer Properties Manager dialog box*

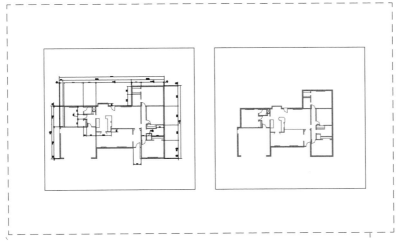

Figure 8–36 *One viewport with* DIMLAYER *ON and the other with* DIMLAYER *OFF*

The **Current VP Freeze** column available only from a layout tab (eleventh column from the left, as shown in Figure 8–36) freezes selected layers in the current layout viewport. You can freeze or thaw layers in the current viewport without affecting layer visibility in other viewports. **Current VP Freeze** is an override to the Thaw setting in the drawing. In other words, you can freeze a layer in the current viewport if it is thawed in the drawing, but you can't thaw a layer in the current viewport if it is frozen or Off in the drawing. A layer is not visible when it is set to Off or Frozen in the drawing.

The **New VP Freeze** column available only from a layout tab (twelfth column from the left, as shown in Figure 8–36) freezes selected layers in new layout viewports. For example, freezing the Text layer in all new viewports restricts the display of text on that layer in any newly created layout viewports, but does not affect the Text layer in existing viewports. If you later create a viewport that requires text, you can override the default setting by changing the current viewport setting.

In Figure 8–36, the layers Dimension, Elevation, and Hidden are frozen in the current viewport, and Object and Text are frozen in all the new viewports.

PLOTTING FROM LAYOUT

Invoke the PLOT command to plot the drawing from the selected Layout. Before you plot the drawing from the layout, make sure you complete the following tasks:

>Create a model drawing.

>Create or activate a layout.

>Open the Page Setup dialog box and set settings such as plotting device (if necessary configure plotting device), paper size, plot area, plot scale, and drawing orientation.

>If necessary, insert a title block or attach a title block as a reference file.

>Create and position floating viewport(s) in the layout.

>Annotate or create geometry in the layout as needed.

>Set the view scale of the floating viewport(s).

>Plot the layout.

To plot the current drawing from the layout, invoke the PLOT command from the Standard toolbar (see Figure 8–37).

Figure 8–37 *Invoking the* PLOT *command from the Standard toolbar*

AutoCAD displays the Plot dialog box as shown in Figure 8–38. This dialog box is similar to one that is displayed when you plot from model space except that by default

Layout is selected in the **What to plot** list box in the **Plot area** section. For a detailed explanation, refer to the "Plotting from Model Space" section.

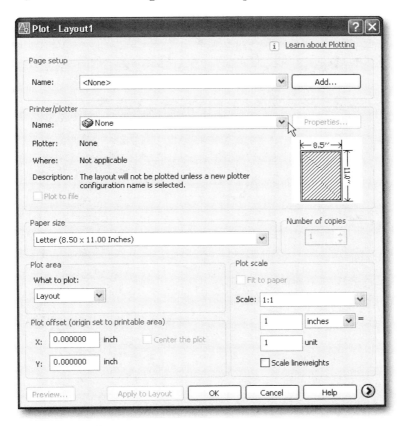

Figure 8–38 *Plot dialog box*

After making the necessary changes in the plot settings, choose **OK**. AutoCAD starts plotting of the current layout.

RECONFIGURING THE LAYOUT WITH PAGE SETUP

As mentioned earlier, by default, every initialized layout has an unnamed page setup associated with it. You can modify the settings for the layout's page setup with the help of the Page Setup Manager dialog box. Choose Page Setup Manager from the Layouts toolbar (see Figure 8–39) and AutoCAD displays the Page Setup Manager dialog box (see Figure 8–40).

Figure 8–39 *Choosing Page Setup Manager from the Layouts toolbar*

Figure 8–40 *Page Setup Manager dialog box*

AutoCAD displays the current layout name in the **Current layout** box. In the **Page setups** section of the Page Setup Manager dialog box, **Current page setup** displays the name of the page setup that is applied to the current layout. If the name is displayed as <None>, an unnamed page setup is assigned to the current layout. The **Page setups** section lists the page setups that are available to apply to the current layout. If the Page Setup Manager is opened from a layout, the current page setup is selected by default. The list includes the named page setups and layouts that are available in the drawing. Layouts that have a named page setup applied to them are enclosed in asterisks, with the named page setup in parentheses; for example, *ANSI A Title Block (portrait)*. You can double-click a page setup or a layout name that has an unnamed page setup associated in this list to set it as the current page setup for the current layout.

Figure 8–40 lists three layouts: *ANSI A Title Block (portrait)*, *Layout1*, and *Layout2*. and one page setup: ANSI A Title Block (portrait). **Current layout** is listed as Layout2. Layout1 and Layout2 have unnamed page setups assigned to them.

CHANGING THE CURRENT PAGE SETUP

To change the page setup for the current layout, first select the named page setup or a layout that has an unnamed page setup associated and choose **Set Current** to set the selected page setup as the current page setup.

MODIFYING THE PAGE SETUP

To modify the page setup assigned to the current layout, choose **Modify**. AutoCAD displays the Page Setup dialog box as shown in Figure 8–41.

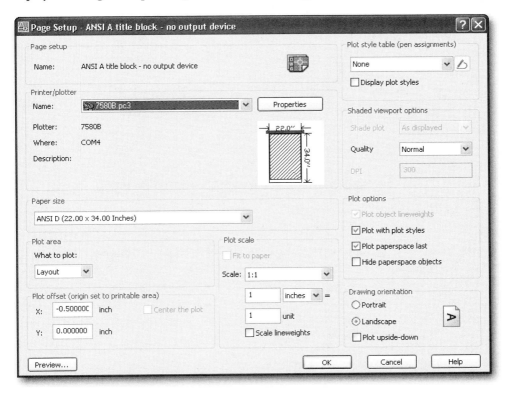

Figure 8–41 *Page Setup dialog box*

The Page Setup dialog box specifies page layout and plotting device settings. The Page Setup dialog box is similar to the expanded Plot dialog box (see Figure 8–5) except it cannot be reduced in size and it does not contain some of the features that apply primarily to plotting. Items that do not appear are **Add** in the **Page setup** section, and **Plot in background**, **Plot stamp on**, and **Save changes to layout** in the **Plot options** section. Also, **Apply to layout** at the bottom of the Plot dialog box is not on the Page Setup dialog box. All of the other options and features are explained in the "Plot Settings" section, earlier in this chapter.

To create a new page setup that can be assigned to any of the layouts, choose **New** in the Page Setup Manager dialog box. AutoCAD displays the New Page Setup dialog box as shown in Figure 8–42.

Figure 8–42 *New Page Setup dialog box*

Specify the name of the new page setup in the **New page setup name** box. Select one of the available page setups to use as a starting point for the new page setup. <None> specifies that no page setup is used as a starting point. <Default output device> specifies the default output device. Choose **OK** to close the dialog box, and AutoCAD displays the Page Setup dialog box with the settings of the selected page setup, which you can modify as necessary.

To import a page setup from a drawing template or drawing file, choose **Import**. AutoCAD displays the Select Page Setup From File dialog box (a standard file selection dialog box), in which you can select a drawing format (*.dwg*), or drawing template (*.dwt*) file from which to import one or more page setups. After selecting the appropriate file, choose **Open**; AutoCAD displays the Import Page Setups dialog box. Choose one of the available page setups to import to the current drawing and choose **OK** to close the dialog box.

In the **Selected page setup details** section of the Page Setup Manager dialog box (see Figure 8–40), AutoCAD displays information relative to the selected page setup. **Device name** displays the name of the plot device, **Plotter** displays the type of plot device, **Plot size** displays the plot size and orientation, **Where** displays the physical location of the output device, and **Description** displays descriptive text about the output device.

The **Display when creating a new layout** box specifies that the Page Setup Manager dialog box is displayed when a new layout tab is selected or a new layout is created.

After making necessary changes in the Page Setup Manager dialog box, choose **Close** to close the dialog box.

CREATING A LAYOUT BY LAYOUT WIZARD

Once you have mastered the concepts of layouts and viewports, you can capitalize on the time-saving features in the Layout Wizard for creating new layouts. The Layout Wizard lets you create a new layout (paper space) for plotting. Each wizard page instructs you to specify different layout and plot settings for the new layout you are creating. Once the layout is created using the wizard, you can modify layout settings using the Page Setup dialog box.

Open the Layout Wizard by invoking Create Layout from the Wizards flyout under the Tools menu. AutoCAD displays the Begin page of the Layout Wizard, as shown at the top left in Figure 8–43.

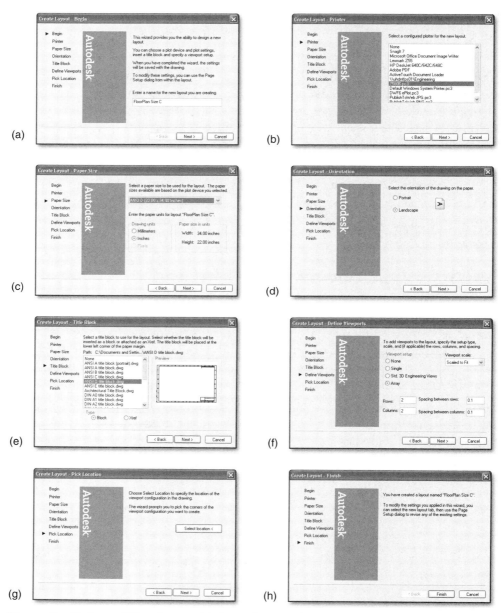

Figure 8–43 Layout Wizard pages: (a) Create Layout – Begin page, (b) Create Layout – Printer page, (c) Create Layout – Paper Size page, (d) Create Layout – Orientation page, (e) Create Layout – Title Block page, (f) Create Layout – Define Viewports page, (g) Create Layout – Pick Location page, (h) Create Layout – Finish page.

Specify the name of the layout in the **Enter a name for the new layout you are creating** text field. Choose **Next**. AutoCAD displays the Create Layout – Printer page of the Layout Wizard, as shown in Figure 8–43(b).

Select a configured plotter for the new layout from the list box. If you do not see the name of the plotter to which you want to plot, refer to the section on "Configuring Plotters" to configure a plotter. Once you have selected the plotter configuration, choose **Next**. AutoCAD displays the **Create Layout – Paper Size** page of the Layout Wizard, as shown in Figure 8–43(c).

Select a paper size to be used for the layout from the list box. The paper sizes available are based on the plot device you selected. Select drawing units from one of the two radio buttons located in the **Drawing units** section. Choose **Next**, and AutoCAD displays the **Create Layout – Orientation** page of the Layout Wizard, as shown in Figure 8–43(d).

Select the orientation of the drawing on the paper from one of the two radio buttons: **Portrait** or **Landscape**. Choose **Next**, and AutoCAD displays the **Create Layout – Title Block** page of the Layout Wizard, as shown in Figure 8–43(e).

Select a title block from the list box to use for the layout. Select whether the title block will be inserted as a block or attached as an external reference. Choose **Next**, and AutoCAD displays the **Create Layout – Define Viewports** page of the Layout Wizard, as shown in Figure 8–43(f). Choose one of the following four available options in the Viewport setup section.

- **None**—if you do not need any floating viewports.
- **Single**—to create one floating viewport.
- **Std. 3D Engineering Views**—to create four viewports with top left set for top view, top right for isometric view, bottom left for front view, and bottom right for right side view. If necessary, you can specify the distance between the viewports in the **Spacing Between rows** and **Spacing between columns** text fields.
- **Array**—to create an array of viewports. Specify the number of viewports in rows and columns in the **Rows** and **Columns** text fields. If necessary, you can specify the distance between the viewports in the **Spacing between rows** and **Spacing between columns** text fields.

Choose **Next**, and AutoCAD displays the **Create Layout – Pick Location** page of the Layout Wizard, as shown in Figure 8–43(g).

Choose **Select location** to specify the location of the viewport configuration in the drawing. The wizard prompts you to specify the corners of the viewport configuration you want to create. After you specify the location, AutoCAD displays the **Create Layout – Finish** page of the Layout Wizard, as shown in Figure 8–43(h).

Choose **Finish** to create the layout. If necessary, you can make any changes to the newly created layout by using the Page Setup dialog box.

MAKING THINGS LOOK RIGHT FOR PLOTTING

Once you have mastered the concepts of layouts and viewports, you can utilize the following features that will enhance the look of your plots.

SETTING PAPER SPACE LINETYPE SCALING

Linetype dash lengths and the space lengths between dots or dashes are based on the drawing units of the model or paper space in which the objects were created. They can be scaled globally by setting the value of the system variable LTSCALE, as explained in Chapter 3. If you want to display objects in viewports at different scales in layout, the linetype objects would be scaled to model space rather than paper space by default. However, by setting paper space linetype scaling (system variable PSLTSCALE) to 1 (default), dash and space lengths are based on paper space drawing units, including the linetype objects that are drawn in model space. For example, a single linetype definition with a dash length of 0.30, displayed in several viewports with different zoom factors, would be displayed in paper space with dashes of length 0.30, regardless of the scale of the viewpoint in which it is being displayed (PSLTSCALE set to 1).

Note: When you change the PSLTSCALE value to 1, the linetype objects in the viewport are not automatically regenerated. Use the REGEN or REGENALL command to update the linetypes in the viewports.

DIMENSIONING IN MODEL SPACE AND PAPER SPACE

Dimensioning can be done in both model space and paper space. There are no restrictions placed on the dimensioning commands by the current mode. For dimensioning in model space, the DIMSCALE factor should be set to 0.0. This causes AutoCAD to compute a scale factor based on the scaling between paper space and the current model space viewport.

Figure 8–44 shows dimensions and view labels that have been drawn in paper space of model space objects.

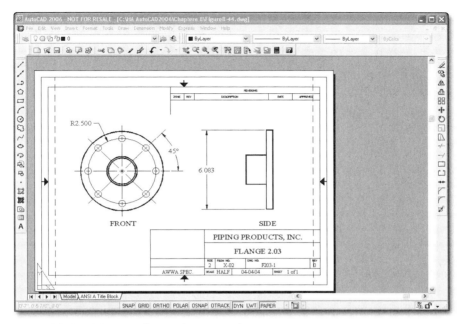

Figure 8–44 *Model space objects dimensioned in paper space*

CREATING A PLOT STYLE TABLE

Plot style tables are settings that give you control over how objects in your drawing are plotted into hard-copy plots. By modifying an object's plot style, you can override that object's color, linetype, and lineweight. You can also specify end, join, and fill styles, as well as output effects such as dithering, gray scale, pen assignment, and screening. You can use plot styles if you need to plot the same drawing in different ways.

By default, every object and layer has a plot style property. The actual characteristics of plot styles are defined in plot style tables that you can attach to a Model tab and layouts within drawings. If you assign a plot style to an object, and then detach or delete the plot style table that defines the plot style, the plot style will not have any effect on the object.

AutoCAD provides two plot style modes: color-dependent and named.

The color-dependent plot styles are based on object color. There are 255 color-dependent plot styles. You cannot add, delete, or rename color-dependent plot styles. You can control the way all objects of the same color plot in color-dependent mode by adjusting the plot style that corresponds to that object color. Color-dependent plot style tables are stored in files with the extension *.ctb*.

Named plot styles work independently of an object's properties. You can assign any plot style to any object regardless of that object's color. Named plot style tables are stored in files with the extension .*stb*.

By default, all the plot style table files are saved in the path that is listed in the **Files** section of the Options dialog box.

The default plot style mode is set in the Plot Style Table Settings dialog box (see Figure 8–45) which can be opened by choosing **Plot Style Table settings** in the **Plot and Publish** tab of the Options dialog box.

Figure 8–45 *Plot Style Table Settings dialog box*

Every time you start a new drawing in AutoCAD, the plot style mode that is set in the Options dialog box is applied. Whenever you change the mode, it is applied only for the new drawings or an open drawing that has not yet been saved in AutoCAD.

The CONVERTPSTYLES command converts a currently open drawing from color-dependent plot styles to named plot styles, or from named plot styles to color-dependent plot styles, depending on which plot style method the drawing is currently using.

CREATING A NEW PLOT STYLE TABLE

AutoCAD allows you to create a named plot style table to utilize all the flexibility of named plot styles, or a color-dependent plot style table to work in a color-based mode. The Add Plot Style Table Wizard allows you to create a plot style from scratch, modify an existing plot style table, import style properties from an *acadr14.cfg* file, or import style properties from an existing *.pcp* or *.pc2* file. After you invoke the Add Plot Style Table option from the Wizards flyout of the Tools menu, AutoCAD displays the introductory text of the Add Plot Style Table Wizard, as shown at top left in Figure 8–46.

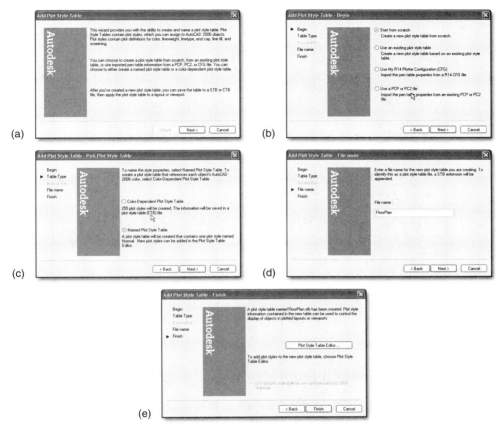

Figure 8–46 *Add Plot Style Table Wizard pages: (a) Add Plot Style Table – Introductory page, (b) Add Plot Style Table – Begin page, (c) Add Plot Style Table – Pick Plot Style Table page, (d) Add Plot Style Table – File name page, (e) Add Plot Style Table – Finish Page.*

Choose **Next**, and AutoCAD displays the **Add Plot Style Table – Begin** page, as shown in Figure 8–46(b). The following four options are available:

Selecting **Start from scratch** allows you to create a new plot style from scratch.

Selecting **Use an existing plot style table** creates a new plot style using an existing plot style table.

Selecting **Use My R14 Plotter Configuration (CFG)** creates a new plot style table using the pen assignments stored in the *acadr14.cfg* file. Select this option if you do not have an equivalent *.pcp* or *.pc2* file.

Selecting **Use a PCP or PC2 file** creates a new plot style table using pen assignments stored in a *.pcp* or *.pc2* file.

 Note: If you selected the *Use an existing plot style table* option to create a new plot style, AutoCAD displays the *Add Plot Style Table – Browse File Name* page. Specify the plot style table file name from which to create a new plot style name.

To create a new pen table, select **Start from scratch** and choose **Next**. AutoCAD displays the **Add Plot Style Table – Pick Plot Style Table** page, as shown in Figure 8–46(c). Select one of the following options:

Select **Color-Dependent Plot Style Table** to create a plot style table with 255 plot styles.

Select **Named Plot Style Table** to create a named plot style table.

Choose **Next**, and AutoCAD displays the **Add Plot Style Table – File name** page, as shown in Figure 8–46(d). Specify the file name in the **File name** box. By default the new style table is saved in the path that is listed in the **Files** section of the Options dialog box.

Choose **Next**, and AutoCAD displays the **Add Plot Style Table – Finish** page, as shown in Figure 8–46(e).

Set the **Use this plot style table for new and pre-AutoCAD 2006 drawings** to ON to attach this plot style table to all new drawings and pre-AutoCAD 2006 drawings by default.

Choose **Finish** to create the plot style table and close the wizard.

MODIFYING A PLOT STYLE TABLE

AutoCAD allows you to add, delete, copy, paste, and modify plot styles in a plot style table by using the Plot Style Table Editor. You can open more than one instance of the Plot Style Table Editor at a time and copy and paste plot styles between the tables. Open the Plot Style Table Editor using any of the following methods:

- Choose the **Plot Style Table Editor** button from the Finish screen in the Add Plot Style Table Wizard.
- Open the Plot Style Manager (File menu), right-click a CTB or STB file, and then choose OPEN from the shortcut menu.
- In the **Plot style table (pen assignment)** section of the Plot dialog box or Page Setup dialog box, select the plot style table you want to edit from the **Plot Style Table** list, and then choose the **Edit** button.
- On the **Plot and Publish** tab of the Options dialog box, choose the **Add or Edit Plot Style Tables Settings** button.

Figure 8–47 shows an example of the Plot Style Table Editor for a named plot style table, and Figure 8–48 shows an example of the Plot Style Table Editor for a color-dependent plot style table.

Figure 8–47 *Plot Style Table Editor dialog box for a named Plot Style table: General tab selection, Table View tab selection, Form View tab selection*

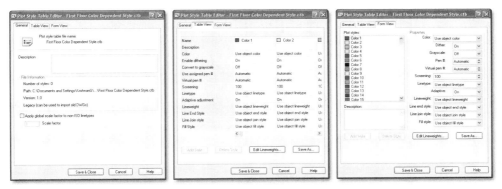

Figure 8–48 *Plot Style Table Editor dialog box for a Color-Dependent Style table: General tab selection, Table View tab selection, Form View tab selection*

Following are the three tabs available in the Plot Style Table Editor:

- **General**—Displays the name of the plot style table, description (if any), location of the file, and version number (see Figure 8–47[a] and Figure 8–48[a]). You can modify the description, and apply scaling to non-ISO lines and to fill patterns.

- **Table View**—Lists entire plot styles in the plot style table and their settings in tabular form (see Figure 8–47[b] and Figure 8–48[b]). The styles are displayed in columns from left to right. The setting names of each row appear at the left of the tab. By default, in the case of a named plot style table,

AutoCAD sets up a style named Normal and represents an object's default properties. You cannot modify or delete the Normal style. In the case of a color-dependent plot style table, AutoCAD lists all the 255 color styles in tabular form. In general this is convenient if you have a small number of plot styles to view them in the tabular form.

- **Form View**—The plot style names are listed under the **Plot styles** list box and the settings for the selected plot style are displayed at the right side of the dialog box (see Figure 8–47[c] and Figure 8–48[c]).

To create a new plot style, choose **Add Style** from the Plot Style Table Editor. AutoCAD adds a new style, and you can change the name to a descriptive name if necessary (cannot exceed 255 characters). You cannot duplicate names within the same plot style table.

Note: You cannot add or change the name for a plot style in the color-dependent style table.

To delete a pen style, click the gray area above the plot style name in the Table View (the entire column will be highlighted) and choose **Delete Style**. In the Form View, select the style name from the **Plot styles** list box and choose **Delete Style**.

Note: You cannot delete a plot style from a color-dependent style table.

The settings on the **Form View** tab include:

Description field allows you to specify a description for plot styles and modify an existing description for a plot style if necessary. The description cannot exceed 255 characters.

The **Color** list box allows you to assign a plot style color. If you assign a color from one of the available colors, then AutoCAD overrides the object's color at plot time. By default, all of the plot styles are set to Use object color.

Set **Dither** to ON for the plotter to approximate colors with dot patterns, giving the impression of plotting with more colors than the number of inks available in the plotter. If you set **Dither** to OFF, then AutoCAD maps colors to the nearest color, which limits the range of colors used for plotting. The most common reason for turning off dithering is to avoid false linetyping from dithering of thin vectors and to make dim colors more visible. If the plotter does not support dithering, the dithering setting is ignored. The default setting is set to ON.

Set **Grayscale** to ON for AutoCAD to convert the object's colors to gray scale if the plotter supports gray scale. If you set **Grayscale** to OFF, AutoCAD uses the RGB values for the object's colors. The default setting is set to OFF.

The **Pen #** setting in the Plot Style Table Editor specifies which pen to use for each plot style. You can specify a pen to use in the plot style by selecting from a range of pen numbers from 1 to 32. By using the BACKSPACE or DELETE keys, you can set the field to read Automatic. AutoCAD uses the information you provided under Physical Pen Configuration in the Plotter Configuration Editor to select the pen closest in color to the object you are plotting. The default is set to Automatic.

Specify a virtual pen number in the **Virtual pen #** edit field for plotters that do not use pens but can simulate the performance of a pen plotter by using virtual pens. The default is set to Automatic to specify that AutoCAD should make the virtual pen assignment from the AutoCAD Color Index. You can specify a virtual pen number between 1 and 255. The virtual pen number setting in a plot style is used only by plotters without pens and only if they are configured for virtual pens. If this is the case, all the other style settings are ignored and only the virtual pen is used. The default is set to Automatic.

Note: If a plotter without pens is not configured for virtual pens, then both the virtual and the physical pen settings in the plot style are ignored and all the other settings are used.

The **Screening** text field sets a color intensity setting that determines the amount of ink AutoCAD places on the paper while plotting. The valid range is 0 through 100. Selecting 0 reduces the color to white. Selecting 100 (default) displays the color at its full intensity. The default is set to 100%.

The **Linetype** list box allows you to assign a plot style linetype. If you assign a linetype from one of the available linetypes, then AutoCAD overrides the object's linetype at plot time. By default, all the plot styles are set to Use object linetype.

The **Adaptive** toggle adjusts the scale of the linetype to complete the linetype pattern. Set **Adaptive**: to ON if it is more important to have complete linetype patterns than correct linetype scaling. Set **Adaptive** to OFF if linetype scale is important. The default is set to ON.

The **Lineweight** list box allows you to assign a plot style lineweight. If you assign a lineweight from one of the available lineweights, then AutoCAD overrides the object's lineweight at plot time. By default, all the plot styles are set to Use object lineweight.

The **Line end style** list box allows you to assign a line end style. The line end style options include: Butt, Square, Round, and Diamond. If you assign a line end style from one of the available line end styles, then AutoCAD overrides the object's line end style at plot time. By default, all the plot styles are set to Use object end style.

The **Line join style** list box allows you to assign a line end style. The line join style options include: Miter, Bevel, Round, and Diamond. If you assign line join style from one of the available line join styles, then AutoCAD overrides the object's line join style at plot time. By default, all the plot styles are set to Use object line join style.

The **Fill style** list box allows you to assign a fill style. The fill style options include: Solid, Checkerboard, Crosshatch, Diamonds, Horizontal Bars, Slant Left, Slant Right, Square Dots, and Vertical Bars. The fill style applies only to solids, plines, donuts, and 3D faces. If you assign a fill style from one of the available fill styles, then AutoCAD overrides the object's fill style at plot time. By default, all the plot styles are set to Use object fill style.

AutoCAD allows you to edit the available lineweights by choosing **Edit Lineweights**. You cannot add or delete lineweights from the list.

To save the changes and close the Plot Style Table Editor, choose **Save & Close**. To save the changes to another plot style table, choose **Save As**. AutoCAD displays the Save As dialog box. Specify the file name in the **File name** text field and choose **Save** to save and close the Save As dialog box.

CHANGING PLOT STYLE PROPERTY FOR AN OBJECT OR LAYER

As mentioned earlier, every object that is created in AutoCAD has a plot style property in addition to color, linetype, and lineweight. Similarly, every layer has a color, linetype, and lineweight, in addition to a plot style property. The default settings for plot styles for objects and layers are set in the Plot Style Table Settings dialog box (see Figure 8–49), which can be opened from the Options dialog box.

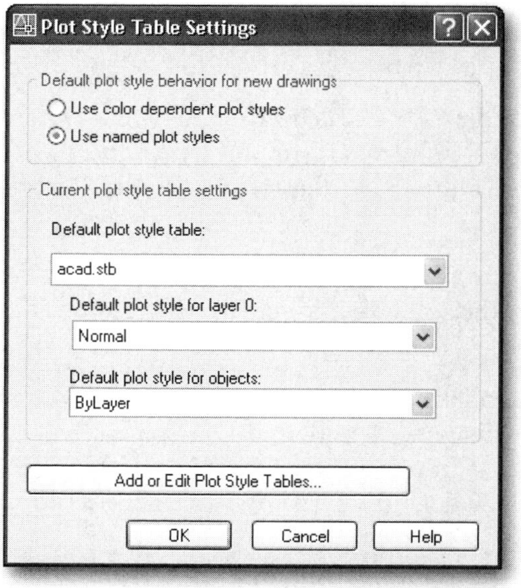

Figure 8–49 *Plot Style Table Settings dialog box*

The default plot style for objects can be any of the following:

- **Normal**—Uses the object's default properties.
- **ByLayer**—Uses the properties of the layer that contains the object.
- **ByBlock**—Uses the properties of the block that contains the object.
- **Named plot style**—Uses the properties of the specific named plot style defined in the plot style table.

The default plot setting for an object is ByLayer, and the initial plot style setting for a layer is Normal. When the object is plotted, it retains its original properties.

If you are working in a Named plot style mode, you can change the plot style for an object or layer at any time. If you are working in a color-dependent plot style mode, you cannot change the plot style for objects or layers. By default, they are set to ByColor.

To change the plot style for one or more objects, first select the objects (system variable PICKFIRST set to ON), and select the plot style from the Plot style control list box on the Properties toolbar, as shown in Figure 8–50. If the plot style does not list the one you want to select, then choose **Other**, and AutoCAD displays the Current Plot Styles dialog box, as shown in Figure 8–51.

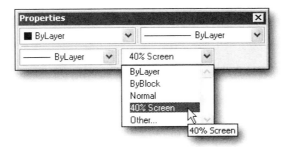

Figure 8–50 *Properties toolbar – Plot Style control*

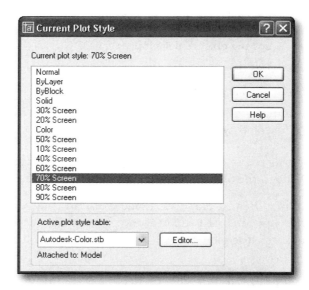

Figure 8–51 *Current Plot Style dialog box*

Select the plot style you want to apply to the selected object(s) from the **Current plot styles** list box. If you need to select a plot style from a different plot style table, then select the plot style table from the **Active plot style table** list box. AutoCAD lists all the available plot styles in the **Current plot styles** list box and selects the one you want to apply to the select object(s). Choose **OK** to close the dialog box. You can also change the plot style of the selected object(s) from the Properties dialog box.

To change the plot style for a layer, open the Layer Properties Manager dialog box. Select the layer you want to change and select a plot style for the selected layer similar to changing color or linetype.

CONFIGURING PLOTTERS

Autodesk® Plotter Manager allows you to configure a local or network non-system plotter. In addition, you can also configure a Windows system printer with non-default settings. AutoCAD stores information about the media and plotting device in configured plot (PC3) files. The PC3 files are stored in the path that is listed in the **Files** section of the Options dialog box. Plot configurations are therefore portable and can be shared in an office or on a project. If you calibrate a plotter, the calibration information is stored in a plot model parameter (PMP) file that you can attach to any PC3 files you create for the calibrated plotter.

AutoCAD allows you to configure plotters for many devices and store multiple configurations for a single device. You can create several PC3 files with different output options for the same plotter. After you create a PC3 file, it's available in the list of plotter configuration names in the **Printer/plotter** section of the Plot dialog box.

Open the Autodesk® Plotter Manager from the File menu and AutoCAD displays the Plotters window explorer (see Figure 8–52) listing all the plotters configured.

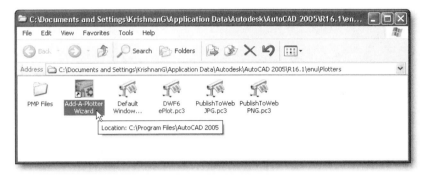

Figure 8–52 *Plotters window explorer (Windows XP version)*

Double-click the Add-A-Plotter Wizard, and AutoCAD displays the Add Plotter – Introduction Page, as shown in Figure 8–53.

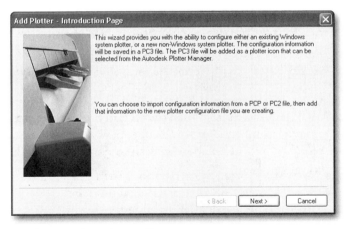

Figure 8–53 *Add Plotter – Introduction Page*

Choose **Next**, and AutoCAD displays the Add Plotter – Begin page, as shown in Figure 8–54.

Figure 8–54 *Add Plotter – Begin page*

Choose one of the three radio buttons:

- **My Computer**—To configure a local non-system plotter.
- **Network Plotter Server**—To configure a plotter that is on the network.
- **System Printer**—To configure a Windows system printer. If you want to connect to a printer that is not in the list, you must first add the printer using the Windows Add Printer wizard in the Control Panel.

If you select the **My Computer** option, then the wizard prompts you to select a plotter manufacturer and model number, identify the port to which the plotter is connected, specify a unique plotter name, and choose **Finish** to close the wizard.

If you select the **Network Plotter Server** option, then the wizard prompts you to identify the network server, select the plotter manufacturer and model number, specify a unique plotter name, and choose **Finish** to close the wizard.

If you select the **System Printer** option, then the wizard prompts you to select one of the printers configured in the Windows operating system, specify a unique plotter name, and choose **Finish** to close the wizard.

AutoCAD saves the configuration file with PC3 file format with a unique given name in the path that is listed in the Files section of the options dialog box.

If necessary, you can edit the PC3 file using the Plotter Configuration Editor. The Plotter Configuration Editor provides options for modifying a plotter's port connections and output settings including media, graphics, physical pen configuration, custom properties, initialization strings, calibration, and user-defined paper sizes. You can drag these options from one PC3 file to another.

You can open the Plotter Configuration Editor using one of the following methods:

- From the File menu, choose **Page Setup**. Choose **Properties**.
- From the File menu, choose **Plot**. Choose **Properties**.
- Double-click a PC3 file from Windows Explorer and right-click the file and choose **Open**.
- Choose **Edit Plotter Configuration...** on the Add Plotter - Finish page in the Add-A-Plotter Wizard.

Figure 8–55 shows the Plotter Configuration Editor for an HP7580B plotter.

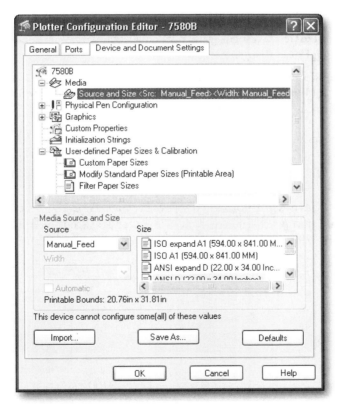

Figure 8–55 *Plotter Configuration Editor for an HP7580B plotter*

The Plotter Configuration Editor contains three tabs:

- **General** tab—Contains basic information about the configured plotter.
- **Ports** tab—Contains information about the communication between the plotting device and your computer.
- **Device and Document Settings** tab—Contains plotting options.

In the **Device and Document Settings** tab, you can change many of the settings in the configured plot (PC3) file. Following are the six areas in which you make the changes:

- **Media**—Specifies a paper source, size, type, and destination.
- **Physical Pen Configuration**—Specifies settings for pen plotters.
- **Graphics**—Specifies settings for printing vector graphics, raster graphics, and TrueType® fonts.
- **Custom Properties**—Displays settings related to the device driver.
- **Initialization Strings**—Sets pre-initialization, post-initialization, and termination printer strings.
- **User-defined Paper Sizes & Calibration**—Attaches a plot model parameter (PMP) file to the PC3 file, calibrates the plotter, and adds, deletes, or revises custom or standard paper sizes.

The areas correspond to the categories of settings in the PC3 file you're editing. Double-click any of the six categories to view and change the specific settings. When you change a setting, your changes appear in angle brackets (<>) next to the setting name unless there is too much information to display. To save the changes to another PC3 file, choose **Save As**. AutoCAD displays the Save As dialog box. Specify the file name in the **File name** text field and choose **Save**. To save the changes to the PC3 file and close the Plotter Configuration Editor, choose the **OK** button.

Open the Exercise Manual PDF file for Chapter 8 on the accompanying CD for project- and discipline-specific exercises.

If you have the accompanying Exercise Manual, refer to Chapter 8 for project- and discipline-specific exercises.

REVIEW QUESTIONS

1. If you were to plot a drawing at a scale of 1"=60', what should you set LTSCALE to?

   a. 60

   b. 1/60

   c. 720

   d. 1/720

2. If you want to plot a drawing requiring multiple pens and you are using a single pen plotter, AutoCAD will:

   a. not plot the drawing at all

   b. pause when necessary to allow you to change pens

   c. invoke an error message

   d. plot all of the drawing using the single pen

   e. none of the above

3. A drawing created at a scale of 1:1 and plotted to "Scaled to Fit" is plotted:

   a. at a scale of 1:1

   b. to fit the specified paper size

   c. at the prototype scale

   d. none of the above

4. To plot a full-scale drawing at a scale of 1/4"=1', use a plot scale of:

   a. 0.25=12

   b. 0.25=1

   c. 48=1

   d. 12=0.25

   e. 24=1

5. What is the file extension assigned to all files created when plotting to a file?

   a. DWG

   b. DRW

   c. DRK

   d. PLO

   e. PLT

6. When plotting, pen numbers are assigned to:

   a. colors

   b. layers

   c. thickness

   d. linetypes

   e. none of the above

You need to draw three orthographic views of an airplane whose dimensions are: wingspan of 102 feet, a total length of 118 feet, and a height of 39 feet. The drawing has to be plotted on a standard 12" x 9" sheet of paper. No dimensions will be added, so you will need only 1" between the views. Answer the following five questions using the information from this drawing:

7. What would be a reasonable scale for the paper plot?

   a. 1=5'

   b. 1=15'

   c. 1=25'

   d. 1=40'

8. What would be a reasonable setting of LTSCALE?

   a. 1

   b. 5

   c. 60

   d. 25

   e. 300

   f. 480

9. If you were plotting from paper space, what ZOOM scale factor would you use?

   a. 1/5X

   b. 1/25X

   c. 1/60X

   d. 1/300X

   e. 1/5XP

   f. 1/25XP

   g. 1/60XP

   h. 1/300XP

10. When inserting your border in paper space, what scale factor should you use?
    a. 1
    b. 5
    c. 25
    d. 60
    e. 300

11. Which of the following options will the plot preview give you?
    a. seeing what portion of your drawing will be plotted
    b. seeing the plotted size of your drawing
    c. seeing the plotted drawing relative to the page size
    d. seeing rulers around the edge of the plotted page for size comparison
    e. none of the above

12. Which of the following determine the relationship between the size of the objects in a drawing and their sizes on a plotted copy?
    a. Size of the object in the AutoCAD drawing
    b. Size of the object on the plot
    c. Maximum available plot area
    d. Plot scale
    e. All of the above

13. AutoCAD permits plotting in which of the following environment modes?
    a. Model space
    b. Paper space
    c. layout
    d. all of the above

14. Which TILEMODE system variable setting corresponds to model space?
    a. 0
    b. 1
    c. either A or B
    d. none of the above

15. When starting a new drawing, how many default plotting layouts does AutoCAD create?
    a. 0
    b. 1
    c. 2
    d. unlimited

16. Within multiple floating viewports you can establish various scale and layer visibility settings for each individual viewport.

    a. True

    b. False

17. Paper sizes are indicated by X axis direction (drawing length) and Y axis direction (drawing width)

    a. True

    b. False

18. Floating viewports, like lines, arcs, and text, can be manipulated using AutoCAD commands such as MOVE, COPY, STRETCH, SCALE, or ERASE.

    a. True

    b. False

19. While in paper space, both the floating viewports and the 3D model can be modified or edited.

    a. True

    b. False

20. Which of the following can be converted into a viewport?

    a. ellipse

    b. splines

    c. circles

    d. all of the above

21. Which of the following commands allows for the control of layer visibility in a specific viewport?

    a. VPORTS

    b. VPLAYER

    c. VIEWLAYER

    d. LAYERVIS

CHAPTER 9

Hatching, Gradients, and Boundaries

INTRODUCTION

After completing this chapter, you will be able to do the following:

- Create hatch and gradient patterns using the HATCH command
- Modify hatch patterns via the HATCHEDIT command
- Control the visibility of hatch and gradient patterns

WHAT IS HATCHING?

Drafters and designers use repeating patterns, called hatching, to fill regions in a drawing for various purposes (see Figure 9–1). In a cutaway (cross-sectional) view, hatch patterns help the viewer differentiate between components of an assembly and indicate the material of each. In surface views, hatch patterns depict material and add to the readability of the view. In general, hatch patterns greatly help the drafter/designer achieve his or her purpose, that is, communicating information. Because drawing hatch patterns is a repetitive task, it is an ideal application of computer-aided drafting.

You can use patterns that are supplied in an AutoCAD support file called *acad.pat*, patterns in files available from third-party custom developers, or you can create your own custom hatch patterns. See Appendix E for the list of patterns supplied with *acad.pat*.

AutoCAD allows you to fill objects with a solid color in addition to a hatch pattern. AutoCAD creates an associative hatch, which updates when its boundaries are modified, or a nonassociative hatch, which is independent of its boundaries. Before AutoCAD draws the hatch pattern, it allows you to preview the hatching and to adjust the definition if necessary.

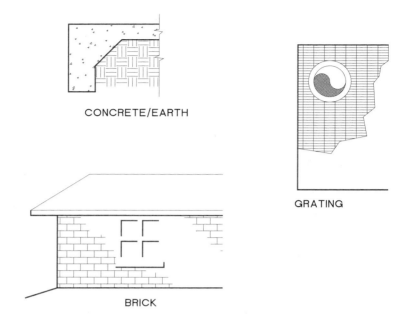

Figure 9–1 *Examples of hatch patterns*

Hatch patterns are considered as separate drawing objects. The hatch pattern behaves as one object; if necessary, you can separate it into individual objects with the EXPLODE command. Once it is separated into individual objects, the hatch pattern will no longer be associated with the boundary object.

Hatch patterns are stored with the drawing, so they can be updated, even if the pattern file containing the hatch is not available. You can control the display of the hatch pattern with the FILLMODE system variable. If FILLMODE is set to OFF, then the patterns are not displayed, and regeneration calculates only the hatch boundaries. By default, FILLMODE is set to ON.

The hatch pattern is drawn with respect to the current coordinate system, current elevation, current layer, color, linetype and current snap origin.

WHAT IS GRADIENT FILL?

A gradient fill is a solid hatch fill that gives the blended-color effect of a surface with light on it. You can use gradient fills to suggest a curved surface in two-dimensional drawings. The color in a gradient fill makes a smooth transition from light to dark, or from dark to light, and back. You may select a predefined pattern (for example, linear, spherical, or radial sweep) and specify an angle for the pattern. In a two-color gradient fill, the transition is both from light to dark and from the first color to the second.

Gradient fills are applied to objects in the same way solid fills are and can be associated with their boundaries. An associated fill is automatically updated when the boundary changes. The Hatch and Gradient dialog box allows you to modify to settings for both hatch and gradient patterns.

 Note: You cannot use plot styles to control the plotted color of gradient fills.

DEFINING THE HATCH OR GRADIENT BOUNDARY

A region of the drawing may be filled with a hatch pattern or gradient fill, if it is enclosed by a boundary of connecting lines, circles, or arc objects. Overlapping boundary objects can be considered as terminating at their intersections with other boundary objects. There must not be any gaps between boundary objects, however. Figure 9–2 illustrates variations of objects and the potential boundaries that might be established from them.

Note in Figure 9–2 how the enclosed regions are defined by their respective boundaries. A boundary might include all or part of one or more objects. In addition to lines, circles, and arcs, boundary objects can include 2D and 3D polylines, 3D faces, and viewports. Boundary objects should be parallel to the current UCS. You can also hatch Block References that have been inserted with unequal X and Y scale factors.

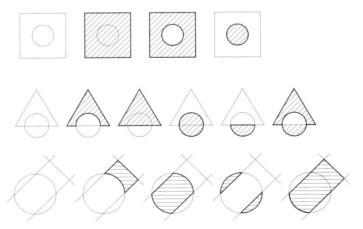

Figure 9–2 *Allowed hatching boundaries made from different objects*

SELECTING OBJECTS VERSUS PICKING A POINT

AutoCAD provides two methods of determining the area to be filled with a hatch or gradient pattern. Using the select object method (see Figure 9–3), you select the four lines via the Window option. These four lines compose the hatching boundary. These four objects are valid boundary segments if they connect at their endpoints and do not overlap. Instead of using the Window option of the object selection process, you can select the four lines individually. This may be desirable if there were unwanted objects within a window used to select them.

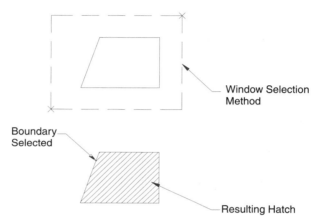

Figure 9–3 *Using the select objects method to define the boundary*

In Figure 9–4, using the pick points method, you select a point in the region enclosed by the four lines. Then, AutoCAD creates a polyline with vertices that coincide with the intersections of the lines. There is also an option that allows you to retain or discard the boundary when the hatching is complete.

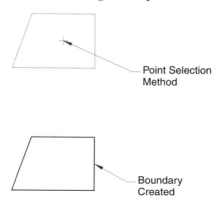

Figure 9–4 *Using the pick points method to create the boundary*

If the four lines shown in Figure 9–3, for example, had been segments of a closed polyline, then you could have selected that polyline by picking it with the cursor. Otherwise, all objects enclosing the region to be hatched must be selected and those objects must be connected at their endpoints. For example, to use the select objects method for the region in Figure 9–5, you would need to draw three lines (from 1 to 2, 2 to 3, and 3 to 4) and an arc from 4 to 1, select the four objects or connect them into a polyline, and select them (or it) to be the boundary.

The select points method permits you to select a point in the region and have AutoCAD automatically create the needed polyline boundary. This ease of use and automation of the select points method almost eliminates the need for the select objects method except for rare, specialized applications.

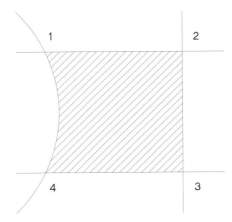

Figure 9–5 *A Region bounded by three lines and an arc*

The dialog box used by the HATCH command provides a variety of easy-to-select options, including a means to preview the hatching before completing the command. This saves time. Consider the variety of effects possible, such as areas to be hatched, angle, spacing between segments in a pattern, and even the pattern selected. The Preview option lets you make necessary changes without having to start over.

HATCH AND GRADIENT FILL WITH THE HATCH COMMAND

Invoke the HATCH command from the Draw toolbar by selecting **Hatch** (see Figure 9–6) and AutoCAD displays the Hatch and Gradient dialog box (see Figure 9–7). The Hatch and Gradient dialog box can be expanded or contracted by choosing the arrow at the lower-right corner of the dialog box.

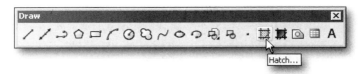

Figure 9–6 *Invoking the* HATCH *command from the Draw toolbar*

Figure 9–7 *Hatch and Gradient dialog box (expanded) with Hatch tab selected*

The majority of the Hatch and Gradient dialog box applies to both hatch and gradient patterns such as the sections on **boundaries**, **Option**s, **Islands**, **Boundary retention**, **Boundary set**, **Gap tolerances**, and **Inherit options**. The last five sections listed are displayed only in the expanded dialog box. There are separate **Hatch** and **Gradient** tabs that have controls and options that apply to hatch patterns and gradients respectively.

HATCH—RELATED SETTINGS

The **Hatch** tab of the Hatch and Gradient dialog box enables you to specify the type of hatch pattern to be applied and the angle, scale, and origin of the pattern.

The **Type and pattern** section controls the type of pattern drawn.

> The **Type** text box lets you select the type of pattern from **Predefined**, **User defined**, or **Custom** hatch patterns.
>
> When the **Predefined** pattern type is chosen, the **Pattern** list box lets you select a pattern from those defined in the *acad.pat* file. Or, you can select one of the available patterns by choosing the button located at the

right of the **Pattern** text box or clicking in the **Swatch** sample box. This causes the Hatch Pattern Palette to be displayed (see Figure 9–8). There are four tabs from which to select predefined patterns: **ANSI, ISO, Other Predefined**, and **Custom**. Each tab displays icons representing a selected pattern group. To select one of the patterns, choose it and then choose **OK** or double-click on the icon. An example of the selected pattern is displayed in the **Swatch** box in the **Hatch** tab of the dialog box. To create a solid fill in an enclosed area, select the **Solid** pattern (located in the Other Predefined section of the Hatch Pattern Palette). The solid fill is drawn with the current color settings, and all pattern properties are disabled, such as scale, angle, and spacing. The selected pattern becomes the value of the HPNAME system variable.

The **User-defined** pattern type allows you to define a simple pattern using the current linetype on the fly. You can specify a simple pattern of parallel lines or two groups of parallel lines (crossing at 90 degrees) at the spacing and angle desired. Specify the angle and spacing for the user-defined pattern in the **Angle** and **Spacing** text boxes in the **Angle and scale** section. To draw a second set of lines at 90 degrees to the original lines, select **Double**.

When the **Custom** pattern type is chosen, the **Custom Pattern** text box lets you select a custom pattern from a *.pat* file other than the *acad.pat* file. Or you can select one of the available patterns by choosing the image tile located at the right end of the **Custom Pattern** text box.

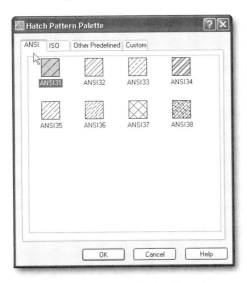

Figure 9–8 *Hatch Pattern Palette with the ANSI tab selected*

The **Angle and scale** section controls the angle and relative size of the pattern drawn.

The **Angle** text box, lets you determine the angle at which the lines in the hatch pattern are drawn relative to the angle at which they were defined. The default angle is set to 0 degrees. The angle 0 (zero) corresponds to the positive X axis of the current UCS. The **Scale** text box lets you change the scale of the pattern. This affects the spacing between the lines of the pattern relative to the spaces at which they were defined. The default scale is set to 1. These settings can be changed to suit the desired appearance, as shown in Figure 9–9.

Figure 9–9 *Hatch pattern with different scale and angle values*

The angle and scale specified in the **Angle** and **Scale** text boxes refer to the selected hatch pattern in the **Pattern** list box and shown in the **Swatch** sample box.

Choosing **Double** causes a user-defined hatch pattern to be drawn as defined in the **Angle, Scale,** and **Spacing** text boxes in the **Angle and scale** section and then repeated at 90 degrees to the original definition. This can only be used when a user-defined hatch pattern has been selected.

Choosing **Relative to paper space** allows you to scale the hatch pattern relative to the units in Paper Space. This can only be used in layout mode.

The **Spacing** text box allows you to specify the spacing between lines of a user-defined hatch pattern. This can only be used when a user-defined hatch pattern has been selected.

The **ISO pen width** list box allows you to specify ISO-related pattern scaling based on the selected pen width. The option is available only if a predefined ISO hatch pattern is selected.

The **Hatch origin** section controls the origin of the pattern drawn. Changing the origin causes the lines in the hatch pattern to be offset by the distance between 0,0 and the specified origin. This is sometimes necessary when you wish to offset the whole pattern of lines for visual effects. For example, if you wish to use the same hatch pattern in adjacent boundaries but do not want their lines to coincide, you can use different origins for the two patterns. Or, if a hatch pattern, such as those for brick and masonry, needs to begin at a certain point, you can specify that point as the origin.

Selecting **Use current origin** causes the hatch pattern to use the current setting of the hatch origin which is stored in the HPORIGIN system variable. By default it is set to 0,0 in the current UCS. If it has been specified as another point, then selecting **Use current origin** uses the current setting.

Choosing **Specified origin** lets you specify a different origin than the current one by utilizing one of the available methods. Choosing **Click to set new origin** lets you specify the new origin on the screen with the pointing device or by entering the new coordinates. Choosing **Default to boundary extents** lets you specify one of the four corners of the rectangular extents of the boundary as the new origin. Choosing **Store as default origin** lets you store the newly specified origin in the HPORIGIN system variable. The **origin preview** sample box shows the newly specified location of the origin.

GRADIENT—RELATED SETTINGS

The **Gradient** tab of the Hatch and Gradient dialog box (see Figure 9–10) lets you specify the type of gradient pattern to be applied and the color, orientation, and angle of the gradient.

The **Color** section controls the color of the gradient drawn.

The **One color** radio specifies a fill that uses a smooth transition between darker shades and lighter tints of one color. When **One color** is selected (see Figure 9–10), AutoCAD displays a color swatch with browse button and a **Shade and Tint** slider. The color swatch specifies the color for the gradient fill, and the **Shade and Tint** slider specifies the tint (the selected color mixed with white) or shade (the selected color mixed with black) of a color to be used for a gradient fill of one color.

Selecting **Two color** specifies a fill that uses a smooth transition between two colors. When **Two color** is selected, AutoCAD displays a color swatch with a browse button for **Color 1** and for **Color 2**. The Color swatch specifies the color for the gradient fill.

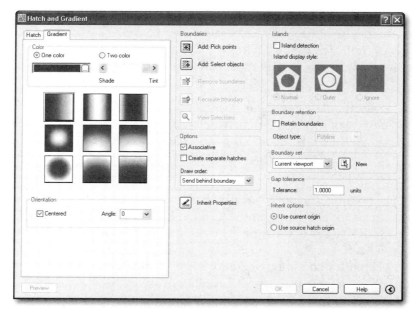

Figure 9–10 *Hatch and Gradient dialog box with Gradient tab selected*

The **Shade and Tint** slider lets you specify the tint (amount of white mixed in) of the selected color or the shade (the amount of black mixed in) of the selected color for a one-color gradient fill.

The **Orientation** section controls the base location and angle of the gradient drawn.

Choosing **Centered** causes the gradient to be drawn with the gradation from the center of the boundary outward. When it is not selected, the gradient fill is shifted up and to the left, creating the illusion of a light source to the left of the object.

The **Angle** text box lets you specify the angle (relative to 0) that the gradation is drawn. Valid values are 0 through 360 degrees and the specified angle is relative to the current UCS. This option is independent of the angle specified for hatch patterns.

Settings for Both Hatching and Gradient Fill Patterns

The options and controls that are applicable to both hatching and gradient fill patterns include: **Boundaries, Options, Islands, Boundary retention, Boundary set, Gap tolerances,** and **Inherit** options. The **Boundaries** section provides two methods by which you can select the objects that determine the boundary for drawing hatch patterns: **Add Pick points** and **Add Select objects**.

Choosing **Add Pick points** lets you determine a boundary from existing objects that form an enclosed area around the specified point. If you set **Island Detection** (available in the **Islands** section) to ON, objects that enclose areas within the outermost boundary are detected as islands. How HATCH detects objects using this option depends on which island detection method (**Normal**, **Outer**, or **Ignore**) you select in the **More Options** area of the dialog box. If it is set to OFF, then AutoCAD draws an imaginary line from the selected point to the nearest object and traces the boundary in the counterclockwise direction. If it cannot trace a closed boundary, AutoCAD will return to the drawing without hatching the object.

For example, in Figure 9–11, point A is valid and point B is not when the **Island Detection** is not chosen. The object nearest to point A is the line that is part of a potential boundary (the square) of which point A is inside and AutoCAD considers the square as the hatch boundary. Conversely, point B is nearest a line that is part of a potential boundary (the triangle) of which point B is outside and AutoCAD displays an error message with the selected point as outside the boundary.

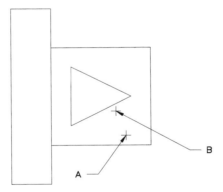

Figure 9–11 *Selecting points for hatching*

To define a hatch or gradient boundary from existing objects, choose **Add Pick points** and the dialog box closes temporarily, and you are prompted to pick a point as follows:

Pick internal point or ⬇ *(specify a point within the area to be hatched)*
Pick internal point or ⬇ *(specify a point, enter* **u** *to undo the selection, or press* ENTER *to end point specification)*

AutoCAD redisplays the dialog box after choosing the internal points. See Figure 9–12 for an example of hatching by specifying a point inside a boundary. While picking internal points, you can right-click in the drawing area at any time to display a shortcut menu that contains several options (see Figure 9–13). You can undo the last selection or all selections, change the boundary selection method, change the island detection style, or preview the hatch or gradient fill.

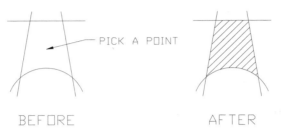

Figure 9–12 *Hatching by specifying a point*

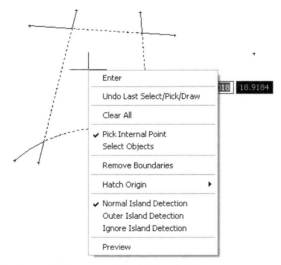

Figure 9–13 *Shortcut menu listing several hatching options*

Choosing **Add Select objects** lets you select specific objects for hatching that determine a boundary to form an enclosed area. The dialog box closes temporarily, and you are prompted to select objects as follows:

Select objects or ⬇ *(select the object(s) by one of the standard methods, and press* ENTER *to terminate object selection)*

When you use the **Add Select objects** option, HATCH does not detect interior objects automatically. You must select the objects within the selected boundary to hatch or fill those objects according to the current island detection style. While selecting objects, you can right-click at any time in the drawing area to display a shortcut menu. You can undo the last selection or all selections, change the selection method, change the island detection style, or preview the hatch or gradient fill.

The **Remove boundaries** and **Recreate boundary** options are not available during creation of a new hatch. They are used on existing hatches and fills with internal boundaries. See their explanation in the section on the Hatch Edit dialog box.

Choosing **View Selections** causes AutoCAD to highlight the defined boundary set.

Caution must be observed when hatching over dimensioning. Dimensions are not affected by hatching as long as the DIMASSOC dimension variable (short for "associative dimensioning") is set to ON when the hatching is created and the dimension has not been exploded. The DIMASSOC system variable toggles between associative and nonassociative dimensioning. If the dimensions are drawn with DIMASSOC set to OFF (or exploded into individual objects), then the lines (dimension and extension) have an unpredictable (and undesirable) effect on the hatching pattern. Therefore, selecting in this case should be done by specifying the individual objects on the screen.

Blocks are hatched as though they are separate objects. Note, however, that when you select a block, all objects that make up the block are selected as part of the group to be considered for hatching.

If the selected items include text objects, shapes, or attribute objects, AutoCAD does not hatch through these items if identified in the selection process. AutoCAD leaves an unhatched area around the text objects so they can be clearly viewed, as shown in Figure 9–14. Using the **Ignore** (Island display) style will negate this feature so that the hatching is not interrupted when passing through the text, shape, and attribute objects.

Figure 9–14 *Hatching in an area where there is text*

Note: When you select objects individually (after selecting the **Select Objects** button in the Hatch and Gradient dialog box), AutoCAD no longer automatically creates a closed border. Therefore, any objects selected that will be part of the desired border must be either connected at their endpoints or a closed polyline.

When a filled solid or trace with width is selected in a group to be hatched, AutoCAD does not hatch inside that solid or trace. However, the hatching stops right at the outline of the filled object, leaving no clear space around the object as it does around text objects, shapes, and attributes.

The **Options** section of the Hatch and Gradient dialog box controls several commonly used hatch or fill options.

Associative controls whether the hatch or fill is associative or nonassociative. Choosing **Associative** causes the hatch pattern elements to be associated with the objects that make up the boundary. For example, if the object is stretched, the hatch pattern expands to fill the new size. Figure 9–15 shows examples of associative and nonassociative hatch patterns.

Figure 9–15 *Examples of associative and nonassociative hatch pattern, when an object is stretched*

Create Separate Hatches controls whether a single hatch object or multiple hatch objects are created when several separate closed boundaries are specified.

The **Draw Order** text box lets you assign draw order to a hatch or fill. You can place a hatch or fill behind all other objects, in front of all other objects, behind the hatch boundary, or in front of the hatch boundary.

Inherit Properties allows you to apply the hatch pattern settings, such as pattern type, pattern angle, and pattern scale, from an existing pattern to another area to be hatched. The dialog box closes temporarily, and you are prompted to select hatch objects as follows:

Select hatch object: *(select a hatch pattern)*
Select objects or ⬇ *(selects a closed area to hatch)*
Select objects or ⬇ *(select a closed area to hatch or press* ENTER *to complete the selection)*

The **Islands** section specifies the method used to hatch or fill objects within the outermost boundary. If no internal boundaries exist, specifying an island detection style has no effect.

> **Island detection** lets you control whether internal closed boundaries, called islands, are detected. Choose one of the three Island display styles: **Normal**, **Outer**, and **Ignore**.
>
> Choosing **Normal** hatches or fills between alternate areas, starting with the outermost area.
>
> Choosing **Outer** hatches or fills only the outermost area and leaves the internal structure blank. Choosing **Ignore** hatches or fills the entire area enclosed by the outermost boundary, regardless of how you select the object, as long as its outermost objects comprise a closed polygon and are joined at their endpoints.
>
> For example, in Figure 9–16, specifying the point shown in the upper-right image in response to the **Add Pick points** option results in hatching for the **Normal** style, as shown in the upper-right; **Outer** style, as shown in the lower-left; and **Ignore** style, as shown in the lower-right.

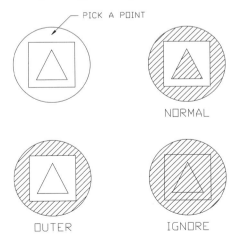

Figure 9–16 *Examples of hatching by specifying a point for Normal, Outer, and Ignore styles*

The **Boundary retention** section specifies whether to retain boundaries as objects and allows you to select the object type AutoCAD applies to those objects.

Choosing **Retain boundaries** specifies whether the area to be hatched is to be retained. If it is to be retained, then the adjacent selection box lets you select whether the area is to be determined by a Polyline boundary or by a Region. The **Object type** list box lets you specify whether the object is a polyline or a region.

The **Boundary set** section allows you to select a set of objects (called a boundary set) that AutoCAD analyzes when defining a boundary from a specified point. The selected boundary set applies only when you use the **Add Pick points** selection to create a boundary to draw hatch patterns. By default, when you use **Add Pick points** to define a boundary, AutoCAD analyzes all objects visible in the current viewport.

The **Boundary set** list box lets you select whether the boundary set will be selected from the Current Viewport or Existing Set. Selecting **Current Viewport** causes the boundary set to be defined from everything in the current viewport extents. Selecting this option discards any current boundary set. Selecting **Existing Set** causes the boundary set to be defined from the objects that you selected with **New**. If you have not created a boundary set with **New**, the **Existing Set** option is not available. Choosing **New** causes AutoCAD to clear the dialog box and return you to the drawing area to select objects from which a new boundary set will be defined. AutoCAD creates a new boundary set from those objects selected that are hatchable; existing boundary sets are abandoned. If hatchable objects are selected, they remain as a boundary set until you define a new one or exit the HATCH command. Defining a boundary set will be helpful when you are working on a drawing that has too many objects to analyze to create a boundary for hatching.

It is important to distinguish between a boundary set and a boundary. A boundary set is the group of objects from which AutoCAD creates a boundary. As explained earlier, a boundary set is defined by selecting objects in a manner similar to selecting objects for some modify commands. The objects (or parts of them) in the group are used in the subsequent boundary. A boundary is created by AutoCAD after it has analyzed the objects (the boundary set) you have selected. It is the boundary that determines where the hatching begins and ends. The boundary consists of line/arc segments, which can be considered to be a closed polygon with segments that connect at their endpoints. If objects in the boundary set overlap, then in creating the boundary

AutoCAD uses only the parts of objects in the boundary set that lie between intersections with other objects in the boundary set.

The **Gap tolerance** section treats a set of objects that almost encloses an area as a closed hatch boundary.

The **Gap tolerance** text box lets you specify a value, in drawing units, from 0 to 5000 to set the maximum size of gaps that can be ignored when the objects serve as a hatch boundary. Any gaps equal to or smaller than the value you specify are ignored, and the boundary is treated as closed. The default value, 0, specifies that the objects enclose the area, with no gaps.

The **Inherit options** section of the Hatch and Gradient dialog box lets you determine the origin the hatch pattern uses.

Choosing **Use current origin** causes the hatch pattern to use the origin stored in the HPORIGIN system variable.

Choosing **Use source hatch origin** causes the hatch pattern to use the origin of the hatch pattern from which properties are to be inherited.

Choose **Preview** to see an example of the hatch pattern to be drawn as specified for the selected objects, AutoCAD displays the currently defined boundaries with the current hatch settings. After previewing the hatch, press ESC to return to the dialog box. If necessary, make any changes to the settings, choose **OK** to apply the hatch pattern, or choose **Cancel** to disregard the selection.

HATCH PATTERNS USING TOOL PALETTES

You can also hatch a closed shape by dragging a hatch pattern from a tool palette (see Figure 9–17).

Tool palettes are tabbed areas within the Tool Palettes window, and hatches that reside on a tool palette are called tools. Several tool properties including scale, rotation, and layer can be set for each tool individually. To change the tool properties, right-click a tool and select PROPERTIES from the shortcut menu. Then you can change the tool's properties in the Tool Properties dialog box as shown in Figure 9–18. The Tool Properties dialog box has two categories of properties—the Pattern properties category, which controls object-specific properties such as scale, rotation, and angle, and the General properties category, which overrides the current drawing property settings such as layer, color, and linetype.

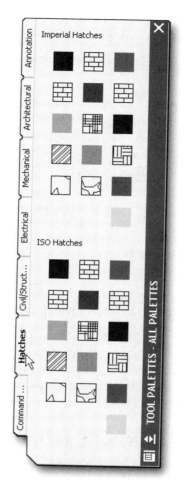

Figure 9–17 *Tool Palettes window with Hatches tab selected*

You can place hatches that you use often on a tool palette by dragging hatch patterns from the DesignCenter by opening the *acad.pat* file. The *acad.pat* file can be found on the following path: *\ACAD2006\Bin\acadFeui\Program Files\Root\UserDataCache\ Support\acad.pat*. When this file is selected in the DesignCenter window, available patterns are displayed in the Content pane. See Chapter 12 for detailed explanation on using DesignCenter. 1

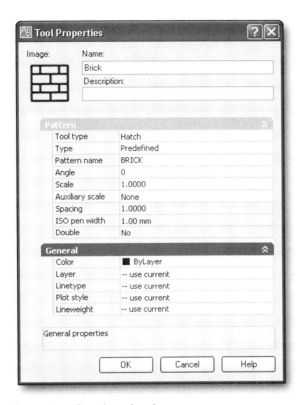

Figure 9–18 *Tool Properties of a selected tool*

EDITING HATCHES AND GRADIENTS

The HATCHEDIT command allows you to modify hatch patterns and gradient fills or choose a new pattern for an existing hatch. In addition, it allows you to change the pattern style of an existing pattern.

Invoke the HATCHEDIT command from the Modify II toolbar by selecting **Edit Hatch** (see Figure 9–19) and AutoCAD prompts:

Select hatch object: *(select an associative hatch object)*

Figure 9–19 *Invoking the* HATCHEDIT *command from the Modify II toolbar*

AutoCAD displays the Hatch Edit dialog box (see Figure 9–20).

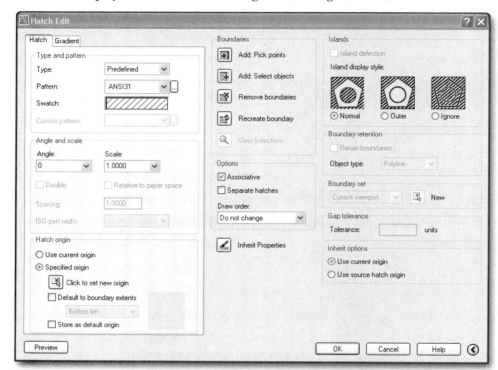

Figure 9–20 *Hatch Edit dialog box*

The Hatch Edit dialog box is similar to the Hatch and Gradient dialog box. After the hatch object is selected, the dialog box will be displayed with the **Hatch** tab selected. To edit a gradient fill, invoke the HATCHEDIT command and when prompted to select a hatch object, select a gradient fill and the dialog box will be displayed with the **Gradient** tab selected. The hatch pattern or gradient fill selected can be modified in the same manner as specifying a new pattern or fill.

If a hatch or gradient is created using the normal island display style, you can cause inner objects that are used as boundaries to no longer be effective with the **Remove boundaries** option. This does not physically remove the objects. It only removes their being used as boundaries. The **Recreate boundary** option can be applied only to boundaries that have been removed with the **Remove boundary** option.

> Choosing **Remove boundaries** causes AutoCAD to prompt you to select a boundary set to be removed from the defined boundary set. You cannot remove the outermost boundary.

Choosing **Recreate boundary** creates a polyline or region around the selected hatch or fill, with the option to associate the hatch object with it. Choosing **Recreate Boundary** causes the dialog box to close temporarily, and AutoCAD prompts:

Select objects: *(select the object(s) by one of the standard methods, and press* ENTER)
Enter type of boundary object *(Enter* ***r*** *to create a region or* ***p*** *to create a polyline)*
Reassociate hatch with new boundary? ⬇ *(Enter* ***y*** *or* ***n****)*

You can also display and change the current properties for hatch or fill objects with the **Properties** palette. Open the **Properties** palette and view and change the settings for all properties of the selected hatch or fill objects (see Figure 9–21). The **Properties** palette lists the current settings for properties of the selected objects. You can modify any property that can be changed by specifying a new value. The **Properties** palette also enables you to view the area of a hatch (see Figure 9–21). If you select multiple hatch objects, you can now view their cumulative area.

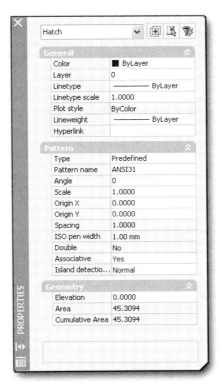

Figure 9–21 *Properties palette*

CONTROLLING THE VISIBILITY OF HATCH PATTERNS

The FILL command controls the visibility of hatch patterns in addition to the filling of multilines, traces, solids, and wide polylines. Invoke the FILL command by typing **fill** and pressing ENTER at the On-Screen prompt. AutoCAD prompts:

> Enter mode *(choose ON to display the hatch pattern and OFF to turn off the display of hatch pattern)*

You have to invoke the REGEN command after changing the setting of FILL to see the effect.

Open the Exercise Manual PDF file for Chapter 9 on the accompanying CD for project- and discipline-specific exercises.

If you have the accompanying Exercise Manual, refer to Chapter 9 for project- and discipline-specific exercises.

REVIEW QUESTIONS

1. AutoCAD will ignore text within a crosshatching boundary.

    a. True   b. False

2. The HATCH command allows you to create an associative hatch pattern that updates when its boundaries are modified.

    a. True   b. False

3. By default, hatch patterns are drawn at a 45-degree angle.

    a. True   b. False

4. All of the following may be used as boundaries of the HATCH command, except:

    a. ARC   d. CIRCLE

    b. LINE   e. PLINE

    c. BLOCK

5. The following are all valid AutoCAD commands except:

    a. ANGLE   d. ELLIPSE

    b. POLYGON   e. MULTIPLE

    c. HATCH

6. When using the HATCH command with a named hatch pattern, you can change

    a. the color and scale of the pattern

    b. the angle and scale of the pattern

    c. the angle and linetype of the pattern

    d. the color and linetype of the pattern

    e. the color and angle of the pattern

7. Boundary hatch patterns inserted with an asterisk "*" preceding the name of the pattern will:

    a. exclude inside objects   d. be inserted on layer 0

    b. ignore inside objects   e. none of the above

    c. be inserted as individual objects

8. The AutoCAD hatch feature:

   a. provides a selection of numerous hatch patterns

   b. allows you to change the color and linetype

   c. hatches over the top of text when the text is contained inside the boundary

   d. all of the above

9. The HATCH command will allow you to create a polyline around the area being hatched and to retain that polyline upon completion of the command.

   a. True   b. False

10. Hatch patterns created with the HATCH command can be nonassociative or associative.

    a. True   b. False

11. The HATCH command will place a hatch pattern over any text contained within the hatch boundary.

    a. True   b. False

12. If you insert a hatch pattern with an asterisk (*) preceding the name, the pattern will be inserted as individual objects.

    a. True   b. False

13. The _____ command automatically defines the nearest boundary surrounding a point you have specified.

    a. HATCH   c. BOUNDARY

    b. BHATCH   d. PTHATCH

14. The Hatch and Gradient and Hatch Edit dialog boxes look the same.

    a. True   b. False

15. What default file does AutoCAD use to load hatch patterns from?

    a. ACAD.mnu   c. ACAD.dwg

    b. ACAD.pat   d. ACAD.hat

16. You can terminate the HATCH command before applying the hatch pattern by pressing ESC.

    a. True   b. False

17. AutoCAD allows the properties of a nonassociative hatch pattern to be inherited.

    a. True   b. False

CHAPTER 10

Block References and Attributes

INTRODUCTION

The AutoCAD BLOCK command feature is a powerful design/drafting tool. The BLOCK command enables a designer to create an object from one or more objects, save it under a user-specified name, and later place it back into the drawing. When block references are inserted in the drawing, they can be scaled up or down in both or either of the X and Y axes. They can also be rotated as they are inserted in the drawing. Block references can best be compared with their manual drafting counterpart, the template. Even though an inserted block reference can be created from more than one object, the block reference acts as a single unit when operated on by certain modify commands, like MOVE, COPY, ERASE, ROTATE, ARRAY, and MIRROR. You can export a block reference to become a drawing file outside the current drawing and create a symbol library from which block references are inserted into other drawings. Like the plastic template, block references greatly reduce repetitious work.

AutoCAD 2006 introduced the Dynamic Block feature (explained in this chapter), which uses parametric data technology to make the use of blocks much more powerful. After you have mastered the concepts for creating and using the BLOCK command, you can apply the power of the dynamic blocks to provide drawings of objects and parts that have similar geometry with more flexibility, ease, and accuracy.

The BLOCK command can save time because you don't have to draw the same object(s) more than once. Block references save computer storage because the computer only needs to store the object descriptions once. When inserting block references, you can change the scale and/or proportions of the original object(s).

After completing this chapter, you will be able to do the following:

- Create and insert block references in a drawing
- Convert individual block references into drawing files
- Define attributes, edit attributes, and control the display of attributes
- Use the DIVIDE and MEASURE commands

CREATING BLOCKS

When you invoke the BLOCK command to create a block, AutoCAD refers to this as defining the block. The resulting definition is stored in the drawing database. The same block can be inserted as a block reference as many times as needed.

Blocks may comprise one or more objects. The first step in creating blocks is to create a block definition. In order to do this, the objects that make up the block must be visible on the screen. That is, the objects that will make up the block definition must have already been drawn so you can select them when prompted to do so during the BLOCK command.

The layer the objects comprising the block are on is very important. Objects that are on layer 0 when the block is created will assume the color, linetype, and lineweight of the layer on which the block reference is inserted. Objects on any layer other than 0 when included in the block definition will retain the characteristics of that layer, even when the block reference is inserted on a different layer. See Figure 10–1 for an example.

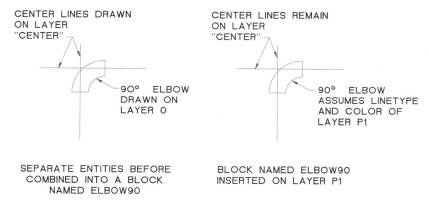

Figure 10–1 *Example of inserting block references drawn in different layers*

You should be careful when invoking the PROPERTIES command to change the color, linetype, or lineweight of elements of a block reference. It is best to keep the color, linetype, and lineweight of block references and the objects that comprise them in the BYLAYER state.

Examples of some common uses of blocks in various disciplines are shown in Figure 10–2.

Chapter 10 • *Block References and Attributes* 489

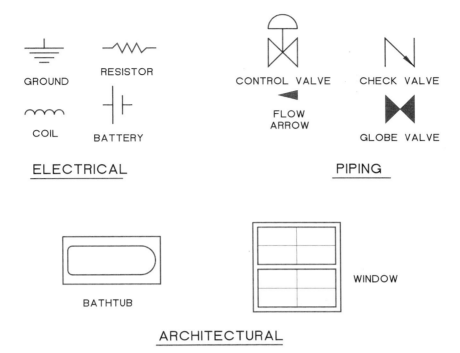

Figure 10-2 *Examples of common uses of blocks in various disciplines*

CREATING A BLOCK DEFINITION

The BLOCK command creates a block definition for selected objects.

Invoke the BLOCK command from the Draw toolbar (see Figure 10-3) and AutoCAD displays the Block Definition dialog box (see Figure 10-4).

Figure 10-3 *Invoking the* BLOCK *command from the Draw toolbar*

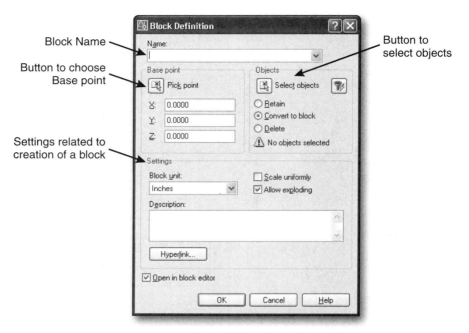

Figure 10–4 Block Definition dialog box

The **Name** text box lets you specify the block name. The block name can be up to 255 characters long and may contain letters, numbers, and any special character not used by Microsoft® Windows® and AutoCAD for other purposes, if the system variable EXTNAMES is set to 1. DIRECT, LIGHT, AVE_RENDER, RM_SDB, SH_SPOT, and OVERHEAD cannot be used as block names. To list the block names in the current drawing, click the down arrow to the right of the **Name** text box. AutoCAD lists the blocks in the current drawing.

The **Base point** section lets you specify the insertion point for the block. The insertion point specified during the creation of the block becomes the basepoint for future insertions of this block as a block reference. It is also the point about which the block reference can be rotated or scaled during insertion. When determining where to locate the base insertion point, it is important to consider what will be on the drawing before you insert the block reference. Therefore, you must anticipate this pre-insertion state of the drawing. It is sometimes more advantageous for the insertion point to be somewhere off the object than on it.

Choose **Pick point** to specify the basepoint on the screen and AutoCAD prompts:

Specify insertion point: *(specify the insertion point)*

The **X, Y,** and **Z** text boxes let you specify the X, Y, and Z coordinates of the insertion point. Once you have specified the insertion point, the Block Definition dialog box reappears.

The **Objects** section lets you select objects to be included in the block and determine how the objects will be treated once the block is created.

Choose **Select Objects** to select objects to include in the block definition and AutoCAD prompts:

Select objects: *(select objects using one of the AutoCAD object selection methods, and press* ENTER *to complete object selection)*

Choosing **Quick Select** lets you use the QSELECT command (explained in Chapter 6) to select the object to include in the block definition.

Choose **Retain** for the objects selected to be included in the block definition to remain in place as separate objects.

Choose **Convert to Block** for the block definition created from the selected objects to become a block reference inserted into the drawing at the location where the block definition was created.

Choose **Delete** for the objects selected to be included in the block definition to be deleted from the drawing after the block definition is created.

The **Settings** section controls units, scale, explode options, description, and hyperlink options.

The **Block unit** text box lets you specify the insertion units for the block reference.

Choosing **Scale Uniformly** prevents the block from being inserted with different X and Y scale factors.

Choosing **Allow exploding** determines that an inserted reference of the block can be exploded.

The **Description** text box can be used to enter a description of the block if desired.

Choosing **Hyperlink** causes AutoCAD to display the Insert Hyperlink dialog box (see Figure 10–5).

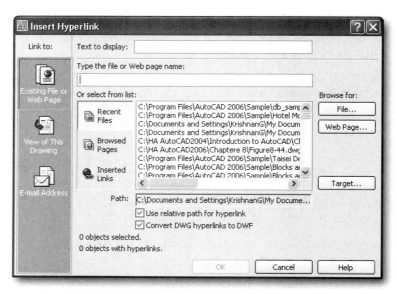

Figure 10–5 *Insert Hyperlink dialog box*

Hyperlinks are created in AutoCAD drawings as pointers to associated files. Hyperlinks can launch a word processing program and open a specific file and even point to a named location in a file. Hyperlinks can activate your Web browser and load a specified HTML page. You can specify a view in AutoCAD or a bookmark in a word processing file. Hyperlinks can be attached to a graphical object in an AutoCAD drawing.

Hyperlinks can be either absolute or relative. Absolute hyperlinks have the full path to a file location stored in them. Relative hyperlinks have only a partial path to a file location stored in them, relative to a default URL or directory you have specified by setting the HYPERLINKBASE system variable.

Hyperlinks can point to locally stored files, files on a network drive, or files on the Internet. Cursor feedback is automatically provided to indicate when the crosshairs are over a graphical object that has an attached hyperlink. You can then select the object and use the Hyperlink shortcut menu to open the file associated with the hyperlink. This hyperlink cursor and shortcut menu display can be turned off in the Options dialog box.

When a hyperlink to an AutoCAD drawing that has a named view is opened, that view is restored. This also applies to a hyperlink created with a named layout. AutoCAD opens that drawing in that layout.

If a hyperlink is created that points to an AutoCAD drawing template (DWT) file, AutoCAD will create a new drawing file based on the template. This prevents overwriting the original template.

For additional information on hyperlinks and how to create and edit them, refer to Chapter 14.

Choose **OK** to create the block definition with the given name. If the given name is the same as an existing block in the current drawing, AutoCAD displays a warning (see Figure 10–6).

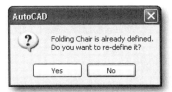

Figure 10–6 *Warning dialog box regarding block definition*

To redefine the block, choose **Yes** in the Warning dialog box. The block with that same name is then redefined. Once the drawing is regenerated, any insertion of this block reference already inserted in the drawing is redefined to the new block definition with this name.

Choose **No** in the Warning dialog box to cancel the block definition. Then, to create a new block definition, specify a different block name in the **Name** text box of the Block Definition dialog box and choose **OK**.

If you create a block without selecting objects, AutoCAD displays a warning that nothing has been selected and provides an opportunity to select objects before the named block is created.

AutoCAD creates the block from the selected objects that make up the definition from the screen using the specified name.

INSERTING BLOCK REFERENCES

You can insert previously defined blocks into the current drawing by invoking the INSERT command. If there is no block definition with the specified name in the current drawing, AutoCAD searches the drives and folders on the path for a drawing of that name and inserts it instead.

 Note: If blocks were created and stored in a template drawing, and you make your new drawing equal to the template, those blocks will be in the new drawing ready to insert. Any drawing inserted into the current drawing will bring with it all of its block definitions, whether they have been inserted or are only stored as definitions.

Invoke the INSERT command from the Draw toolbar (see Figure 10–7) and AutoCAD displays the Insert dialog box (see Figure 10–8).

Figure 10–7 *Invoking the* INSERT *command from the Draw toolbar*

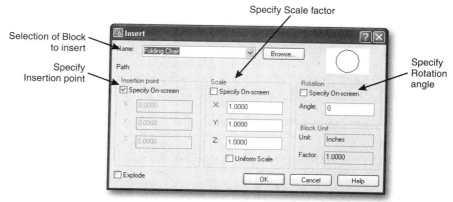

Figure 10–8 *Insert dialog box*

Specify a block name in the **Name** text box. Or choose the down arrow to display a list of blocks defined in the current drawing and select the block you wish to insert.

The **Insertion point** section of the Insert dialog box allows you to specify the insertion point for inserting a copy of the block definition.

> The **X**, **Y**, and **Z** text boxes let you specify the X, Y, and Z coordinates of the insertion point by entering them from the keyboard when **Specify On-screen** is not checked. Set **Specify On-screen** to ON if you prefer to specify the insertion point on screen with your pointing device.

The **Scale** section of the Insert dialog box allows you to specify the scale for the inserted block. The default scale factor is set to 1 (Full scale). You can specify a scale factor between 0 and 1 to insert the block reference smaller than the original size of the block and specify more than 1 to increase the size from the original size.

> The **X**, **Y**, and **Z** text boxes let you specify the X, Y, and Z scales by entering them from the keyboard. If necessary, you can specify different X and Y scale factors for the block reference. If you specify a negative scale factor, then AutoCAD inserts a mirror image of the block about

the insertion point. As a matter of fact, if –1 were used for both *X* and *Y* scale factors, it would "double-mirror" the object, the equivalent of rotating it 180 degrees.

Set **Specify On-screen** to ON if you prefer to specify the scale factor on screen with your pointing device.

Select **Uniform Scale** and then you can enter a value only in the **X** text box. The *Y* and *Z* scales will be the same as that entered for the *X* scale.

The **Rotation** section of the Insert dialog box allows you to specify the rotation angle for the inserted block.

Set **Specify On-screen** to ON if you prefer to specify the rotation angle on screen with your pointing device.

The **Angle** text box lets you specify the angle to rotate the block reference by entering the value from the keyboard.

The **Block Unit** section displays the units (inches, millimeters, etc.) that were used when the selected block was created and the unit scale factor, calculated based on the block units value and the drawing units.

Set **Explode** to ON to insert the block reference as a set of individual objects rather than as a single unit.

To specify a drawing file to insert as a block definition, enter the drawing file name in the **Name** text box. Or choose **Browse** to display a standard file dialog box and select the appropriate drawing file.

 Note: The name of the last block reference inserted during the current drawing session is remembered by AutoCAD. The name becomes the default for subsequent use of the INSERT command.

Choose **OK** to insert the selected block.

NESTED BLOCKS

Blocks can contain other blocks. That is, when you are using the BLOCK command to combine objects into a single object, one or more of the selected objects can themselves be blocks. And the blocks selected can have blocks nested within them. There is no limitation to the depth of nesting. You may not, however, use the name of any of the nested blocks as the name of the block being defined. This would mean that you were trying to redefine a block, using its old definition in the new.

Any objects within blocks (as nested blocks) that were on layer 0 when made into a block will assume the color, linetype, and lineweight of the layer on which the block reference is inserted. If an object (originally on layer 0 when included in a block definition) is in a block reference that has been inserted on a layer other than layer 0, it will retain the color, linetype, and lineweight of the layer it was on when its block was included in a higher-level block. For example, you draw a circle on layer 0 and include it in a block named Z1. Then, you insert Z1 on layer R, whose color is red. The circle would then assume the color of layer R (in this case it will be red). Create another block called Y3 by including the block Z1. If you insert block reference Y3 on a layer whose color is blue, the block reference Y3 will retain the current color of layer R (in this case it will be red) instead of taking up the color of blue.

EXPLODE COMMAND

The EXPLODE command causes block references, hatch patterns, and associative dimensioning to be turned into the separate objects from which they were created. It also causes polylines/polyarcs and multilines to separate into individual simple line and arc objects. The EXPLODE command causes 3D polygon meshes to become 3Dfaces, and 3D polyface meshes to become 3Dfaces and simple line and point objects. When an object is exploded, the new, separate objects are created in the space (model or paper) of the exploded objects.

Invoke the EXPLODE command from the Modify toolbar (see Figure 10–9) and AutoCAD prompts:

> Select objects: *(select objects to explode, and press* ENTER *to complete object selection)*

Figure 10–9 *Invoking the* EXPLODE *command from the Modify toolbar*

You can use one or more object selection methods. The object selected must be eligible for exploding, or an error message will appear. An eligible object may or may not change its appearance when exploded.

POSSIBLE CHANGES CAUSED BY THE EXPLODE COMMAND

A polyline segment having width will revert to a zero-width line and/or arc. Tangent information associated with individual segments is lost. If the polyline segments have width or tangent information, the EXPLODE command will be followed by the message:

> Exploding this polyline has lost (width/tangent) information.
> The undo command will restore it.

Individual elements within blocks that were on layer 0 when the block was created (and whose color was BYLAYER) but were inserted on a layer with a color different than that of layer 0 will revert to the color of layer 0.

Attributes are special text objects that, when included in a block definition, take on the values (names and numbers) specified at the time the block reference is inserted. The power and usage of attributes are discussed later in this chapter. To understand the effect of the EXPLODE command on block references that include attributes, it is sufficient to know that the fundamental object from which an attribute is created is called an attribute definition. It is displayed in the form of an attribute tag before it is included in the block.

An attribute within a block will revert to the attribute definition when the block reference is exploded and will be represented on the screen by its tag. The value of the attribute specified at the time of insertion is lost. The group will revert to those elements created by the ATTDEF command prior to combining them into a block via the BLOCK command.

In brief, an attribute definition is turned into an attribute when the block in which it is a part is inserted; conversely, an attribute is turned back into an attribute definition when the block reference is exploded.

EXPLODING BLOCK REFERENCES WITH NESTED ELEMENTS

Block references containing other blocks and/or polylines are separated for one level only. That is, the highest-level block reference will be exploded, but any nested blocks or polylines will remain block references or polylines. They in turn can be exploded when they come to the highest level.

Viewport objects in a block definition cannot be turned on after being exploded unless they were inserted in paper space.

Block references with equal X, Y, and Z scales explode into their component objects. Block references with unequal X, Y, and Z scales (nonuniformly scaled block references) might explode into unexpected objects.

 Note: Block references inserted via the MINSERT command or external references and their dependent blocks cannot be exploded.

BASE COMMAND

The BASE command allows you to establish a base insertion point for the whole drawing in the same manner that you specify a base insertion point when using the BLOCK command to combine elements into a block. The purpose of establishing this basepoint is primarily so that the drawing can be inserted into another drawing by way of the INSERT command and having the specified basepoint coincide with the specified insertion point. The default basepoint is the origin (0,0,0). You can specify a 2D point, and AutoCAD will use the current elevation as the base Z coordinate. Or you can specify the full 3D point.

Invoke the BASE command from the Draw menu and AutoCAD prompts:

> Enter base point: *(specify a point, or press* ENTER *to accept the default)*

ATTRIBUTES

Attributes can be used for automatic annotation during insertion of a block reference. Attributes are special text objects that can be included in a block definition and must be defined beforehand and then selected when you are creating a block definition.

Attributes have two primary purposes:

> The first use of attributes is to permit annotation during insertion of the block reference to which the attributes are attached. Depending on how you define an attribute, it either appears automatically with a preset (constant) text string or it prompts you (or other users) for a string to be written as the block reference is inserted. This feature permits you to insert each block reference with a string of preset text or with its own unique string.

> The second (perhaps the more important) purpose of attaching attributes to a block reference is to have extractable data about each block reference stored in the drawing database file. Then, when the drawing is complete (or even before), you can invoke the ATTEXT (short for "attribute extract") command to have attribute data extracted from the drawing and written to a file in a form that database-handling programs can use. You can have as many attributes attached to a block reference as you wish. As just mentioned, the text string that makes up an attribute can be either constant or user-specified at the time of insertion.

A DEFINITION WITHIN A DEFINITION

When creating a block, you select objects to be included. Objects such as lines, circles, and arcs are drawn by means of their respective commands. Normal text is drawn with the TEXT command or the MTEXT command.

As with drawing objects, attributes must be drawn before they can be included in the block. This is complicated, and it requires additional steps to place them in the drawing; AutoCAD calls this procedure defining the attribute. Therefore, an attribute definition is simply the result of defining an attribute by means of the ATTDEF command. The attribute definition is the object that is selected during the BLOCK command. Later, when the block reference is inserted, the attributes that are attached to it and the manner in which they become a part of the drawing are a result of how you created (defined) the attribute definition.

VISIBILITY AND PLOTTING

If an attribute is to be used only to store information, then you can, as part of the definition of the attribute, specify whether or not it will be visible. If you plan to use an attribute with a block as a note, label, or callout, you should consider the effect of scaling (whether equal or unequal X/Y factors) on the text that will be displayed. The scaling factor(s) on the attribute will be the same as on the block reference. Therefore, be sure that it will result in the size and proportions desired. You should also be aware of the effect of rotation on visible attribute text. Attribute text that is defined as horizontal in a block will be displayed vertically when that block reference is inserted with a 90-degree angle of rotation.

Like any other object in the drawing, the attribute must be visible on the screen (or would be if the plotted view were the current display) for that object to be eligible for plotting.

ATTRIBUTE COMPONENTS

Four components associated with attributes should be understood before attempting a definition: tag, value, prompt, and default. The purpose of each is described in the following sections.

Tag

An attribute definition has a tag, just as a layer or a linetype has a name. The tag is the identifier of the attribute definition and is displayed where this attribute definition is located, depicting text size, style, and angle of rotation. The tag cannot contain spaces. Two attribute definitions with the same tag should not be included in the same block. Tags appear in the block definition only, not after the block reference is inserted. However, if you explode a block reference, the attribute value (described herein) changes back into the tag.

If multiple attributes are used in one block, each must have a unique tag in that block. This restriction is similar to each layer, linetype, block, and other named object having a unique name within one drawing. An attribute's tag is its identifier at the time that attribute is being defined, before it is combined with other objects and attributes by the BLOCK command.

Value

The value of an attribute is the actual string of text that appears (if the visibility mode is set to ON) when the block reference (of which it is a part) is inserted. Whether visible or not, the value is tied directly to the attribute, which, in turn, associates it with the block reference. It is this value that is written to the database file. It might be a door or window size or, in a piping drawing, the flange rating, weight, or cost of a valve or fitting. In an architectural drawing the value might represent the manufacturer, size, color, cost, or other pertinent information attached to a block representing a desk.

 Note: When an extraction of attribute data is performed, it is the value of an attribute that is written to a file, but it is the tag that directs the extraction operation to that value. This will be described in detail in the later section on "Extracting Attributes."

Prompt

The prompt is what you see when inserting a block reference with an attribute whose value is not constant or preset. During the definition of an attribute, you can specify a string of characters that will appear in the prompt area during the insertion of the block reference to prompt you to enter the appropriate value. What the prompt says to you during insertion is what you told it to say when you defined the attribute.

Default

You can specify a default value for the attribute during the definition procedure. Then, during insertion of the block reference, it will appear behind the prompt in brackets, i.e., <default>. It will automatically become the value if you press ENTER in response to the prompt for a value.

ATTRIBUTE COMMANDS

The four primary commands to manage Attributes are:

        ATTDEF—attribute definition

        ATTDISP—attribute display

        ATTEDIT—attribute edit

        ATTEXT—attribute extract

As explained earlier, the ATTDEF command is used to create an attribute definition. The attribute definition is the object that is selected during the BLOCK command.

The ATTDISP command controls the visibility of the attributes.

The ATTEDIT command provides a variety of ways to edit without exploding the block reference.

The ATTEXT command allows you to extract the data from the drawing and have it written to a file in a form that database-handling programs can use, as shown in the following table.

DOORS						
SIZE	**THKNS**	**CORE**	**FINISH**	**LOCKSET**	**HINGES**	**INSET**
3070	1.750	SOLID	PAINT	PASSAGE	4 x 4	3/4
3070	1.750	SOLID	VARNISH	KEYED	4 x 4	20 x 20
2868	1.375	HOLLOW	PAINT	PRIVACY	3 x 3	3/4

ROOM FINISHES					
NAME	**WALL**	**CEILING**	**FLOOR**	**BASE**	**REMARKS**
LIVING	GYPSUM	GYPSUM	CARPET	NONE	PAINT
FAMILY	PANEL	ACOUSTICAL	TILE	STAIN	STAIN
BATH	PAPER	GYPSUM	TILE	COVE	4'_CERAMIC_TILE
GARAGE	GYPSUM	GYPSUM	CONCRETE	NONE	TAPE_FLOAT_ONLY

You can include an attribute in the WDW block to record the size of the window. A suggested procedure would be to zoom in near the insertion point and create an attribute definition with a tag that reads WDW-SIZE, as shown in Figure 10–10.

If, during the insertion of the WDW block reference, you respond to the prompt for the SIZE attribute with 2054 for a 2'-0"–wide x 5'-4"–high window, the resulting block reference object would be as shown in Figure 10–11, with the normally invisible attribute value shown here for illustration purposes. Figure 10–12 shows the result of the attribute being inserted with unequal scale factors.

Figure 10–10 *Create the attribute definition before defining the block*

Figure 10–11 *The attribute value (2054) visible with the inserted block*

Even though the value displayed is distorted, the string is not affected when extracted to a database file for a bill of materials.

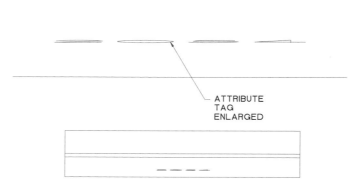

Figure 10–12 *The attribute in a block reference with exaggerated, unequal scale factors*

To solve the distortion and rotation problem, if you wish to have an attribute displayed for rotational purposes, you can create a block that contains attributes only or only one attribute. Then it can be inserted at the desired location and rotated for readability to produce the results shown in Figure 10–13.

Attribute definitions would be created as shown in Figure 10–13 and inserted as shown in Figure 10–14.

Figure 10–13 *Attributes created as separate blocks*

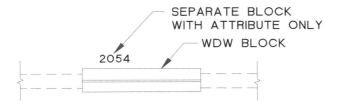

Figure 10–14 *Attribute inserted separately for the window*

The insertion points selected would correspond to the midpoint of the outside line that would result from the insertion of the WDW block reference. The SIZE attributes could be defined into blocks called WDWSIZE and inserted separately with each WDW block reference, thereby providing both annotation and data extraction.

 Note: The main caution in having an attribute block reference separate from the symbol (WDW) block reference is in editing. Erasing, copying, and moving the symbol block reference without the attribute block reference could mean that the data extraction results in the wrong quantity.

There is another solution to the problem of a visible attribute not being located or rotated properly in the inserted block reference. If the attribute is not constant, you can edit it independently after the block reference has been inserted by way of the ATTEDIT command. It permits changing an attribute's height, position, angle, value, and other properties. The ATTEDIT command is covered in detail in the later section on "Editing Attributes," but it should be noted here that the height editing option applies to the X and Y scales of the text. Therefore, for text in the definition of a block that was inserted with unequal X and Y scale factors, you will not be able to edit its proportions back to equal X and Y scale factors.

Creating an Attribute Definition
The ATTDEF command allows you to create an attribute definition through a dialog box.

Invoke the DEFINE ATTRIBUTES (ATTDEF) command from the flyout of the Block option on the Draw menu and AutoCAD displays the Attribute Definition dialog box (see Figure 10–15).

Set one or more of the available modes to ON in the **Mode** section of the Attribute Definition dialog box.

Selecting **Invisible** causes the attribute value *not* to be displayed when the block reference is completed. Even if visible, the value will not appear until the insertion is completed. Attributes needed only for data extraction should be invisible, to quicken regeneration and to avoid cluttering your drawing. You can use the ATTDISP command to override the Invisible mode setting. Setting the Invisible mode to Y (yes, or ON) does not affect the visibility of the tag in the attribute definition.

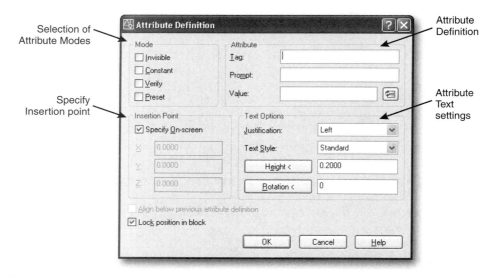

Figure 10–15 *Attribute Definition dialog box*

Selecting **Constant** requires that you must enter the value of the attribute while defining it. That value will be used for that attribute every time the block reference to which it is attached is inserted. There will be no prompt for the value during insertion, and you cannot change the value.

Selecting **Verify** means that you will be able to verify its value when the block reference is inserted. For example, if a block reference with three (non-constant value) attributes is inserted, once you have completed all prompt/value sequences that have displayed the original defaults, you will be prompted again, with the latest values as new defaults, giving you a second chance to be sure the values are correct before the INSERT command is completed. Even if you press ENTER to accept an original default value, it also appears as the second-chance default. If, however, you make a change during the verify sequence, you will *not* get a third chance, that is, a second verify sequence.

Selecting **Preset** causes the attribute to automatically take the value of the default that was specified at the time of defining the attribute. During a normal insertion of the block reference, you will not be prompted for the value. You must be careful to specify a default during the ATTDEF command or the attribute value will be blank. A block consisting only of attributes whose defaults were blank when Preset modes were set to ON could be inserted, but it would not display anything and cannot be purged from the drawing. The only adverse effect would be that of adding to the space taken in memory. One way to get rid of a non-displayable block reference like this is to use a visible entity to create a block with the same name, thereby redefining it to something that can be edited, that is, erased and subsequently purged.

You can duplicate an attribute definition with the COPY command, and use it for more than one block. Or you can explode a block reference and retain one or more of its attribute definitions for use in subsequent blocks.

The **Attribute** section of the Attribute Definition dialog box allows you to set attribute data. Enter the attribute's tag, prompt, and default value in the text boxes.

The attribute's tag identifies each occurrence of an attribute in the drawing. The tag can contain any characters except spaces. AutoCAD changes lowercase letters to uppercase.

The attribute's prompt appears when you insert a block reference containing the attribute definition. If you do not specify the prompt, AutoCAD uses the attribute tag as the prompt. If you turn on the Constant mode, the Prompt field is disabled.

The default Value specifies the default attribute value. This is optional, except if you turn on the Constant mode, for which the default value needs to be specified.

The **Insertion Point** section of the dialog box allows you to specify a coordinate location for the attribute in the drawing, either by choosing **Pick Point** to specify the location on the screen or by entering coordinates in the text boxes provided.

The **Text Options** section of the dialog box allows you to set the justification, text style, height, and rotation of the attribute text.

Selecting **Align below previous attribute definition** allows you to place the attribute tag directly below the previously defined attribute. If you haven't previously defined an attribute definition, this option is unavailable.

Selecting **Lock Position in Block** locks the position of the attribute within the block reference.

Choose **OK** to define the attribute definition.

After you close the Attribute Definition dialog box, the attribute tag appears in the drawing. Repeat the procedure to create additional attribute definitions.

Inserting a Block Reference with Attributes

Blocks with attributes may be inserted in a manner similar to that for inserting regular block references. If there are any non-constant attributes, you will be prompted to enter the value for each. You may set the system variable called ATTREQ to 0 (zero), thereby suppressing the prompts for attribute values. In this case the values will either be blank or be set to the default values if they exist. You can later use either the DDATTE or ATTEDIT command to establish or change values.

Controlling the Display of Attributes

The ATTDISP command controls the visibility of attributes. Attributes will normally be visible if the Invisible mode is set to N (normal) when they are defined.

Invoke the ATTDISP command options by selecting one of the available options from the flyout menu of Display from the View menu.

> Choosing ON makes all attributes visible;
>
> Choosing OFF makes all attributes invisible.
>
> Choosing Normal displays the attributes as you created them.

The ATTMODE system variable is affected by the ATTDISP setting. If REGENAUTO is set to ON, changing the ATTDISP setting causes drawing regeneration.

Editing Attribute Values

Unlike other objects in an inserted block reference, attributes can be edited independently of the block reference. The EATTEDIT command allows you to change the value of attributes in blocks that have been inserted. This permits you to insert a block reference with generic attributes; that is, the default values can be used in anticipation of changing them to the desired values later. Or you can copy an existing block reference that may need only one or two attributes changed to make it correct for its new location. And, of course, there is always the chance that either an error was made in entering the value or design changes necessitate subsequent changes. The ATTEDIT command can be used to change only the value of the attributes. In order to change other characteristics of attributes such as text size and font and visibility, the Block Attribute Manager, described later in this chapter, must be used.

Editing an attribute is accomplished by invoking the EATTEDIT command. The EATTEDIT command edits individual, non-constant attribute values associated with a specific block reference.

Invoke the EDIT ATTRIBUTE (EATTEDIT) command from the Modify II toolbar (see Figure 10–16) and AutoCAD prompts:

Select block reference: *(select the block reference)*

Figure 10–16 *Invoking the* EDIT ATTRIBUTE *command from the Modify II toolbar*

AutoCAD displays the Enhanced Attribute Editor dialog box (see Figure 10–17). Selecting objects that are not block references or block references that contain no attributes will cause an error message to appear.

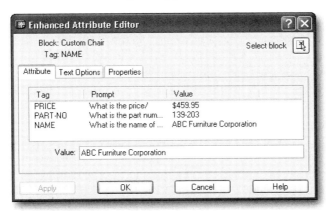

Figure 10–17 *Enhanced Attribute Editor dialog box*

The dialog box lists all the attributes defined with values for the selected block reference. **Block** and **Tag** display the tag identification of the selected attribute and name of the block whose attributes you are editing respectively. Using the pointing device, you can select values to be changed in the dialog box. In addition, you can also change the text attributes and object properties of the selected attribute tag.

Choose **Apply** to update the drawing with the attribute changes you have made, and leave the Enhanced Attribute Editor open. Choose **Select Block** to temporarily close the dialog box while you select a block for modifying its attribute definitions. If you modify attributes of a block and then select a new block before you save the attribute changes you made, you are prompted to save the changes before selecting another block. After making necessary changes, choose **OK** to accept the changes and close the Enhanced Attribute Editor dialog box.

Extracting Attributes

Extracting data from a drawing is one of the most innovative features in CAD. Paper copies of drawings have long been used to communicate more than just how objects look. In addition to dimensions, drawings tell builders or fabricators what materials to use, quantities of objects to make, manufacturers' names and models of parts in an assembly, coordinate locations of objects in a general area, and what types of finishes to apply to surfaces. But, until computers came into the picture (or pictures came into the computer), extracting data from manual drawings involved making lists (usually by hand) while studying the drawing, often checking off the data with a marker.

The AutoCAD attribute feature and the EATTEXT command combine to allow complete, fast, and accurate extraction of (1) data consciously put in for the purpose of extraction, (2) data used during the drawing process, and (3) data that AutoCAD maintains about all objects (block references in this case).

The CAD drawing in Figure 10–18 shows a piping control set. The 17 valves and fittings are a fraction of those that might be on a large drawing. Each symbol is a block with attributes attached to it. Values that have been assigned to each attribute tag record the type (TEE, ELL, REDUCER, FLANGE, GATE VALVE, or CONTROL VALVE), size (3", 4", or 6"), rating (STD or 150#), weld (length of weld), and many other vital bits of specific data. Keeping track of hundreds of valves, fittings, and even cut lengths of pipe is a time-consuming task subject to omissions and errors if done manually, even if the drawing is plotted from CAD. Just as important as extracting data from the original drawing is the need to update a list of data when the drawing is changed. Few drawings, if any, remain unchanged. The AutoCAD attribute feature makes the job fast, thorough, and accurate. Examples of some of the block references with attribute definition are shown in Figures 10–19 through 10–21.

These are three examples of block references as defined and inserted, each attribute having been given a value during the INSERT command. Remember, the value that will be extracted is in accordance with how the template specifies the tags. Figure 10–22 shows the three block references inserted, with corresponding attribute values in tabular form. By using the ATTEXT command, you can write a complete listing of all valves and fittings in the drawing to a file, as shown in the following table.

Chapter 10 • *Block References and Attributes* 509

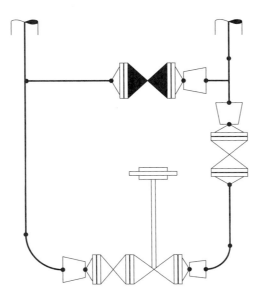

Figure 10-18 *Piping control set*

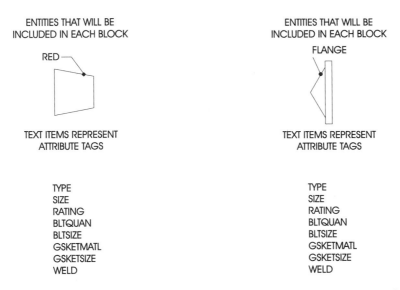

Figure 10-19 *Reducer with attribute definition*

Figure 10-20 *Flange with attribute definition*

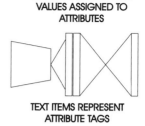

	ENTITIES THAT WILL BE INCLUDED IN EACH BLOCK	BLOCKS INSERTED WITH VALUES ASSIGNED TO ATTRIBUTES	
TYPE		REDUCER	RFWN FLG	GT VLV
SIZE		6X4	4"	4"
RATING		STD	150#	150#
BLTQUAN		-	8	8
BLTSIZE		-	5/8"x3"	5/8"x3"
GSKETMATL		-	COMP	COMP
GSKETSIZE		-	1/8"	1/8"
WELD		34.95	14.14	-

Figure 10–21 *Valve with attribute definition*

Figure 10–22 *Inserted block references with corresponding attribute values*

A List of Inventory as Specified by the ATTEXT Command and Tags

TYPE	SIZE	RATING	BLTQUAN	BLTSIZE	GSKETMATL	GSKETSIZE	WELD
TEE	6"	STD	4	5/8"X3"	COMP	1/8"	62.44
ELL	6"	STD	4	5/8"X3"	COMP	1/8"	41.63
ELL	4"	STD	4	5/8"X3"	COMP	1/8"	14.14
RED	6X4	STD	8	5/8"X3"	COMP	1/8"	34.95
RED	6X4	STD	8	5/8"X3"	COMP	1/8"	34.95
RED	6X3	STD	4	5/8"X3"	COMP	1/8"	31.81
RED	4X3	STD	8	5/8"X3"	COMP	1/8"	25.13
FLG	4"	150#	8	5/8"X3"	COMP	1/8"	14.14
FLG	4"	150#	8	5/8"X3"	COMP	1/8"	14.14
FLG	4"	150#	8	5/8"X3"	COMP	1/8"	14.14
FLG	4"	150#	8	5/8"X3"	COMP	1/8"	14.14
FLG	3"	150#	4	5/8"X3"	COMP	1/8"	11.00
FLG	3"	150#	4	5/8"X3"	COMP	1/8"	11.00
GVL	4"	150#	16	5/8"X3"	COMP	1/8"	11.00
GVL	4"	150#	16	5/8"X3"	COMP	1/8"	11.00
GVL	3"	150#	8	5/8"X3"	COMP	1/8"	11.00
CVL	3"	150#	8	5/8"X3"	COMP	1/8"	11.00

The headings above each column in this table are for your information only. These will not be written to the extract file by the EATTEXT command. They signify the tags whose corresponding values will be extracted.

When operated on by a database program, this file can be used to sort valves and fittings by type, size, or other value. The scope of this book is too limited to cover database applications. But generating a file similar to this inventory table that a database program can use is the important linkage between computer drafting and computer management of data for inventory control, material takeoff, flow analysis, cost, maintenance, and many other applications. The CAD drafter can apply the ATTEXT feature to perform this task.

The EATTEXT command uses the Attribute Extraction Wizard to facilitate selecting objects and/or drawings from which to extract attribute data from inserted block references and then lets you place the data in tabular format for use on the drawing and/or write the data to a file and in a format for use by a program such as Microsoft Excel or Access or other file for handling the resulting database.

Extracted data can be manipulated by a database application program. The telephone directory, a database, is an alphabetical listing of names, each followed by a first name (or initial), an address, and a phone number. A listing of pipe, valves, and fittings in a piping system can be a database if each item has essential data associated with it, such as its size, flange rating, weight, material of manufacture, product that it handles, cost, and location within the system, among many others.

The two elementary terms used in a database are the record and the field. A record is like a single listing in the phone book, made up of a name and its associated first name, address, and phone number. The name Jones with its data is one record. Another Jones with a different first name (or initials) is another record. Another Jones with the same first name or initials at a different address is still another record. Each listing is a record. The types of data that may be in a record come under the heading of a field. Name is a field. All the names in the list come under the name field. Address is a field. Phone number is a field. And even though some names may have first names and some may not, first name is a field. It is possible to take the telephone directory that is listed alphabetically by name, feed it into a computer database program, and generate the same list in numerical order by phone number. You can generate a partial list of all the Joneses sorted alphabetically by the first name. Or you can generate a list of everyone who lives on Elm Street. The primary purpose of the EATTEXT command is to generate the main list that includes all of the desired objects to which the database program manipulations can be applied.

Invoke the ATTRIBUTE EXTRACTION (EATTEXT) command from the Modify II menu and AutoCAD displays the Attribute Extraction Wizard (see Figure 10–23 top left).

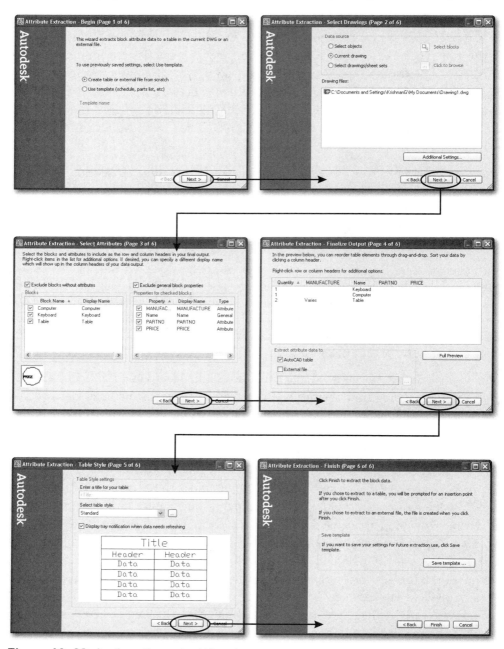

Figure 10–23 *Attribute Extraction Wizard pages*

The **Attributes Extraction - Begin (Page 1 of 6)** page lets you extract block attribute data to a table in the current drawing or to an external file.

>Choose **Create table or external file from scratch** to use the settings you specify as you proceed through the wizard to create a table or save data in an external file.

>Choose **Use template (schedule, parts list, etc)** to use settings previously saved in an attribute extraction template (*.blk*) file. As you move through the wizard, each page is already filled in with the settings in the template file. If necessary, you can change these settings and on the Finish page choose whether to save the new settings. You can specify an attribute extraction template (*.blk*) file in the Template name file or click the [...] button to select the file in a standard file selection dialog box.

Choose **Next** and AutoCAD displays the **Attribute Extraction – Select Drawings (Page 2 of 6)** page (see Figure 10–23 top right). This page lets you select a drawing file from which to extract information from Block Attributes or lets you specify blocks in the current drawing.

The **Data source** section lets you select objects in the current drawing, the current drawing itself, or drawings/sheet sets to search for blocks from which to extract attribute data.

>Choose **Select Objects** to select specific blocks and Xrefs in the current drawing for extracting attribute data. The name of the current drawing is listed in the Drawing Files section. Choose **Select blocks** to select blocks in the current drawing, and AutoCAD temporarily returns to the drawing display so that you can select blocks for extracting attributes.

>Choose **Current Drawing** to select all blocks and xrefs from the current drawing for extracting attribute data. The name of the current drawing is listed in the Drawing Files section.

>Choose **Select drawings/sheet sets** to select drawing/sheet sets from which to select blocks from which to extract attribute data. Choose **Browse** to the right of **Select drawings/sheet sets** and AutoCAD displays the Select Files dialog box to select the drawings and sheet sets.

>Choose **Additional options** to set options for nested and external reference blocks and options that specify which blocks to count.

After selecting the desired objects or drawing(s), choose **Next** and AutoCAD displays the **Attribute Extraction – Select Attributes (Page 3 of 6) page** (see Figure 10–23 second row left). This page allows you to select which blocks and attributes in the selected blocks will have their data extracted to the specified table and/or file.

Selecting **Exclude blocks without attributes** causes only blocks with attributes to be included in the list box in the **Blocks** section.

Selecting **Exclude general block properties** causes only user-defined attributes to be included in the list box in the **Properties for checked blocks** section.

The **Blocks** section list box lists the blocks contained in the drawing(s) specified.

The **Properties for checked blocks** section list box lists the attributes of the block(s) checked in the drawing(s) listed in the list box in the **Blocks** section.

Choose **Next**, and AutoCAD displays the **Attribute Extraction – Finalize Output (Page 4 of 6)** page (see Figure 10–23 second row right). This page lets you rearrange table elements and sort data for the output table. To rearrange the data, select one of the available options from the shortcut menu that appears when you right-click on a header. The options include: Sort Ascending, Sort Descending, Hide Column, Show All Columns, Filter Rows, Reset Filter, Reset All Filters, and Copy To Clipboard.

The **Extract attribute data to** section lets you determine where to extract the data.

Choose **AutoCAD table** to extract the information to a table and AutoCAD displays the Table Style page when you choose **Next**.

Choose **External** to save data to be output to the file named in the text box. Click the [...] button to select the file in a standard file selection dialog box. You can save the data in comma-separated file format (CSV), tab-separated file format (TXT), or Microsoft Excel or Access format.

Choose **Full Preview** to display a full preview of the final output in the text window. The preview is for viewing only. Make any changes to the display on the Finalize Output page. You can also right-click in the list and click **Copy to Clipboard** and then paste the output into a file.

Choose **Next** and AutoCAD displays the **Attribute Extraction - Table Style (Page 5 of 6)** page (see Figure 10–23 third row left). This page is displayed only if **AutoCAD Table** is selected on the Finalize Output page. The **Table Style (Page 5 of 6)** page of the dialog box with the **Table Style settings** section lets you determine the style of table to which the data will be extracted.

The **Enter a title for your table** text box lets you specify the name for the table.

The **Select table style** list box lets you select the style for the table from the available styles.

Choosing **Display tray notification when data needs refreshing** causes AutoCAD to display a message when the data needs to be refreshed.

The sample window shows an example of the table style with its components.

Choose **Next**, and AutoCAD displays the **Attribute Extraction – Finish (Page 6 of 6)** page (see Figure 10–23 third row right), which finishes the process of extracting attributes.

The **Finish (Page 6 of 6)** page lets you save the attribute settings to a template. Choose **Save template** to display the Save As dialog box (a standard file selection dialog box), where you can save the attribute extraction settings to a template file with a *.blk* file extension. The next time you use the Attribute Extraction wizard, you can select a template file on the **Begin** page to fill in each page of the wizard in advance.

Choose **Finish** to close the wizard and AutoCAD returns to the drawing screen, prompting you to specify the location of the table if a table has been created during the attribute extraction process.

REDEFINING A BLOCK AND ITS ASSOCIATED ATTRIBUTES

The ATTREDEF command allows you to redefine a block reference and updates associated attributes. Invoke the ATTREDEF command by typing **attredef** at the On-Screen prompt and AutoCAD prompts:

```
Enter name of the block you wish to redefine: (specify the block name
    to redefine)
Select objects for new Block…
Select objects: (select objects for the block to redefine and press ENTER)
Specify insertion base point of new Block: (specify the insertion basepoint
    of the new block)
```

New attributes assigned to existing block references are given their default values. Old attributes in the new block definition retain their old values. Old attributes not included in the new block definition are deleted from the old block references.

BLOCK ATTRIBUTE MANAGER

The BATTMAN command provides a means of managing blocks that contain attributes.

Invoke the BATTMAN command from the Modify menu by selecting Block Attribute Manager from the flyout of the Attribute option of the Object selection. AutoCAD displays the Block Attribute Manager dialog box, similar to Figure 10–24.

Figure 10–24 *Block Attribute Manager dialog box*

The BATTMAN command lets you edit the attribute definitions in blocks, change the order in which you are prompted for attribute values when inserting a block, and remove attributes from blocks. AutoCAD displays attributes of the selected block in the attribute list of the Block Attribute Manager dialog box. By default, the Tag, Prompt, Default, and Modes attribute properties are shown in the attribute list. You can specify which attribute properties you want displayed in the list by choosing Settings. The number of instances of the selected block is shown in a description below the attribute list.

The **Block** list box lists all block definitions in the current drawing that have attributes. Select the block whose attributes you want to modify.

Instead of selecting a block from the **Block** list box, choose **Select Block** to use your pointing device to select a block from the drawing area. When you choose **Select Block**, the dialog box closes until you select a block from the drawing or cancel by pressing ESC. If you modify attributes of a block and then select a new block before you save the attribute changes you made, you are prompted to save the changes before selecting another block.

Choosing **Sync** causes AutoCAD to update all instances of the selected block with the attribute properties currently defined.

Selecting **Move Up** causes the selected attribute tag to move up earlier in the prompt sequence when the block contains multiple attributes.

Selecting **Move Down** causes the selected attribute tag to move down in the prompt sequence when the block contains multiple attributes.

Selecting **Edit** causes AutoCAD to display the Edit Attribute dialog box (see Figure 10–25) where you can modify selected attribute properties. There are three tabs from which to choose in the Edit Attribute dialog box: the Attribute tab, the Text Options tab, and the Properties tab.

Figure 10–25 *Edit Attribute dialog box with the Attribute tab selected*

The Attribute tab controls the modes and data tag, prompt, and default of the selected attribute. The **Mode** section of the Attribute tab has check boxes for setting the Invisible, Constant, Verify, and Preset modes. The Data section has text boxes in which you can set the tag name, prompt, and default. Selecting **Auto preview changes** causes changes to attributes to be immediately visible.

The **Text Options** tab lets you modify attribute text (see Figure 10–26). It has text boxes to set the Text Style, Justification, Height, Rotation, Width Factor, and Oblique Angle of the Attribute text. There are check boxes for displaying the text Backwards or Upside down.

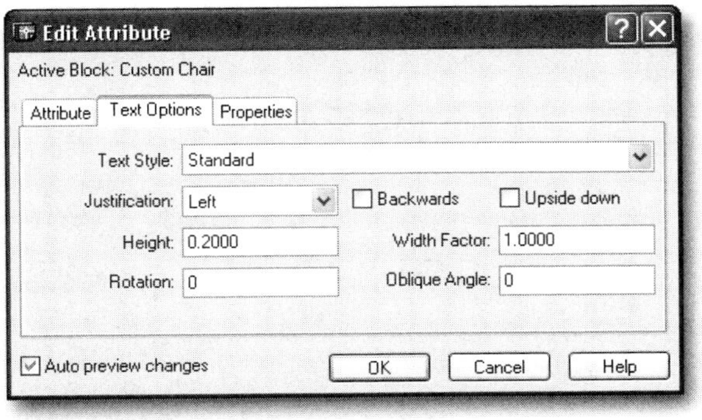

Figure 10–26 *Edit Attribute dialog box with the Text Options tab selected*

The **Properties** tab (see Figure 10–27) has text boxes to set the layer that the attribute is on and the color, lineweight, and linetype for the selected attribute. If the drawing uses plot styles, you can assign a plot style to the attribute.

After making the necessary changes to the Edit Attribute dialog box, choose **OK** to save the changes and return to the Block Attribute Manager dialog box.

Figure 10–27 *Edit Attribute dialog box with the Properties tab selected*

Choosing **Remove** in the Block Attribute Manager dialog box removes the selected attribute from the block definition. **Remove** is not available for blocks with only one attribute.

Choosing **Settings** causes AutoCAD to display the Settings dialog box (see Figure 10–28) where you can customize how attribute information is listed in the Block Attribute Manager.

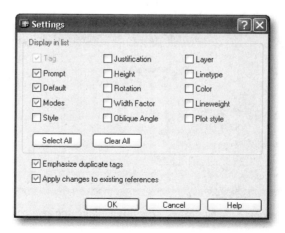

Figure 10–28 *Settings dialog box*

The **Display in List** section specifies the properties to be displayed in the attribute list. Only the selected properties are displayed in the list. The Tag property is always selected.

Choose **Select All** to select all properties' check boxes to be checked.

Choose **Clear All** causes all properties' check boxes to be cleared.

Selecting **Emphasize duplicate tags** causes duplicate attribute tags to be displayed in red type in the attribute list. If this option is cleared, duplicate tags are not emphasized in the attribute list.

Selecting **Apply changes to existing references** updates all instances of the block with the new attribute definitions. If this option is cleared, AutoCAD updates only new instances of the block with the new attribute definitions. You can choose **Sync** in the Block Attribute Manager to apply changes immediately to existing block instances. This temporarily overrides the **Apply Changes to Existing References** option.

Choose **OK** to close the Settings dialog box and return to the Block Attribute Manager.

Choosing **Apply** in the Block Attribute Manager dialog box updates the drawing with the attribute changes you have made and leaves the Block Attribute Manager open. Choose **OK** to close the Attribute Manager dialog box.

DIVIDING OBJECTS

The DIVIDE command causes AutoCAD to divide an object into equal-length segments, placing markers at the dividing points. Objects eligible for application of the DIVIDE command are the line, arc, circle, ellipse, spline, and polyline. Selecting an object other than one of these will cause an error message to appear, and the command will terminate.

Invoke the DIVIDE command from the Draw menu and AutoCAD prompts:

> Select the object to divide: *(select a line, arc, circle, ellipse, spline, or polyline)*
> Enter the number of segments or ⬇: *(specify the number of segments or select the BLOCK option from the shortcut menu)*

You may respond with an integer from 2 to 32767, causing points to be placed along the selected object at equal distances but not actually separating the object. The Object Snap NODe can snap at the divided points. Logically, there will be one less point placed than the number entered, except in the case of a circle. The circle will have the first point placed at the angle from the center of the current snap rotation angle. A closed polyline will have the first point placed at the first point drawn in the polyline. The total length of the polyline will be divided into the number of segments entered without regard to the length of the individual segments that make up the polyline. An example of a closed polyline is shown in Figure 10–29.

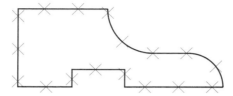

Figure 10–29 *The* DIVIDE *command as used with a closed polyline*

 Note: It is advisable to set the PDSIZE and PDMODE system variables to values that will cause the points to be visible.

The **Block** option allows a named block reference to be placed at the dividing points instead of a point. The sequence of prompts is as follows:

> **divide**
> Select object to divide: *(select a line, arc, circle, ellipse, spline, or polyline)*
> Enter the number of segments or ⬇: *(choose* BLOCK *from the shortcut menu)*
> Enter name of block to insert: *(enter the name of the block)*
> Align block with object? or ⬇: *(press* ENTER *to align the block reference with the object, or enter* **n**, *for not to align with the block reference)*
> Enter the number of segments: *(specify the number of segments)*

If you respond with No or N to the "Align block with object?" prompt, all of the block references inserted will have a zero angle of rotation. If you default to Yes, the angle of rotation of each inserted block reference will correspond to the direction of the linear part of the object at its point of insertion or to the direction of a line tangent to a circular part of an object at the point of insertion.

MEASURING OBJECTS

The MEASURE command causes AutoCAD to divide an object into specified-length segments, placing markers at the measured points. Objects eligible for application of the MEASURE command are the line, arc, circle, ellipse, spline, and polyline. Selecting an object other than one of these will cause an error message to appear and the command will terminate.

Invoke the MEASURE command from the Draw menu and AutoCAD prompts:

> Select object to measure: *(select a line, arc, circle, ellipse, spline, or polyline)*
> Specify length of segment or ⬇: *(specify the length of the segment, or select the* BLOCK *option from the shortcut menu)*

If you reply with a distance, or show AutoCAD a distance by specifying two points, the object is measured into segments of the specified length, beginning with the closest endpoint from the selected point on the object. The **Block** option allows a named block reference to be placed at the measured point instead of a point. The sequence of prompts is as follows:

> **measure**
> Select object to measure: *(select a line, arc, circle, ellipse, spline, or polyline)*
> Specify length of segment or ⬇: *(choose BLOCK from the shortcut menu)*
> Enter name of block to insert: *(enter the name of the block)*
> Align block with object? or ⬇: *(press ENTER to align the block reference with the object, or enter **n**, for not to align with the block reference)*
> Specify length of segment: *(specify the length of segments)*

DYNAMIC BLOCKS

A dynamic block has flexibility and intelligence. Individual objects or groups of objects within a dynamic block reference can easily be changed in a drawing while you work. You can manipulate the geometry in a dynamic block reference through custom grips or custom properties. This allows you to adjust the block in-place as necessary rather than searching for another block to insert or having to redefine the existing one.

For example, after inserting a block in a drawing representing a door, you might need to change the size of the door while you're editing the drawing. If the block is dynamic and defined to have an adjustable size, you can change the size of the door simply by dragging the custom grip or by specifying a different size in the Properties palette. You might also need to change the open angle of the door. The door block might also contain an alignment grip, which allows you to align the door block reference easily to other geometry in the drawing.

Figure 10–30 shows three insertions of a block representing a simplified instrument panel. The block on the left is shown with the geometry as it was drawn when the block was defined with the ON/OFF switch in the center. The middle and right insertions (called block references) are references of the same block with the ON/OFF switch moved to the left and right respectively. All three references are insertions of the same block definition. The dynamic capability of the block allows the user to change the size, shape, location, or orientation of preselected geometry in the reference (after it has been inserted). In Figure 10–30, all three references are drawn from a single block definition.

Figure 10–30 *Three block references of an instrument panel with the ON/OFF switch in the center (as defined), on the left, and on the right*

THREE STEPS TO CREATE A SIMPLE DYNAMIC BLOCK

The first step to create a dynamic block is to create the geometry as shown on the left in the example in Figure 10–30. This can be accomplished either in the normal AutoCAD drawing area or in the Block Editor drawing area. If the objects have been drawn in the normal drawing area, then invoke the MAKE BLOCK command and this causes the Block Definition dialog box to be displayed. While in the dialog box, you can name the block, specify the insertion point, and select the geometry to be included in the definition. Select the **Open in block editor** check box and choose **OK**. AutoCAD then displays the Edit Block Definition dialog box from which you can select the block in the list of block names and choose **OK**. You can go through the process of opening the Block Editor without selecting objects and then create the geometry in the Block Editor drawing area. You can also double-click an insertion (reference) of a nondynamic block. AutoCAD displays the Edit Block Definition dialog box with the name of the block selected in the **Block to create or edit** text box. Choose **OK** to select the block. After you use either method (arriving at the Edit Block Definition dialog box when first creating the block or double-clicking an existing block), AutoCAD opens the Block Editor with the Authoring palette displayed. The Authoring palette has tabs from which you can add parameters, actions, and parameter sets to your block definition to make it dynamic.

The second step to create a dynamic block is to add a parameter. Parameters are used to establish points, distances, and angles on or near objects in the block so that actions can be applied to those objects, making that block dynamic. Parameters also allow you to control visibility of objects in the reference and make use of the properties of the geometry in the reference. Parameters include the Point, Linear, Polar, XY, Rotation, Alignment, Flip, Visibility, Lookup, and Base Point parameters. For this example, **Point Parameter** is selected from the **Parameters** tab of the Block Authoring palette and AutoCAD prompts:

> Specify parameter location or ⬇: *(specify the center of the switch geometry)*

The point parameter is located on the center of the switch geometry (see Figure 10–31).

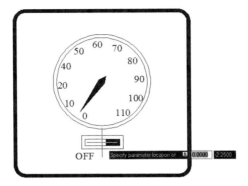

Figure 10–31 Adding a point parameter while defining the dynamic block in the Block Editor.

AutoCAD prompts for the label location. You can locate the label for the point parameter ("Position" in this case) as shown in Figure 10–32. The label name "Position" is the default for a point parameter and can be changed if desired. The grip is displayed where you specified the location of the parameter with a leader to the label. An Alert icon (with the exclamation point) is displayed because there is no action associated with the parameter yet.

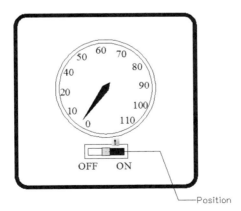

Figure 10–32 The point parameter (labeled as Position) with Alert icon indicating that no action has been added

The third step consists of adding an action, selecting objects, and closing the Block Editor. Actions are added to parameters and then associated with selected objects in the block. These objects are the selection set that will be affected by the subsequent selecting of the appropriate grip in a reference of the block and initiation of the action. Actions include Move, Scale, Stretch, Polar Stretch, Rotate, Flip, Array, and Lookup. Choose **Move Action** from the **Actions** tab (see Figure 10–33) and AutoCAD prompts:

Select parameter: *(select the Position label or other part of the parameter)*

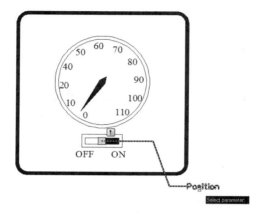

Figure 10–33 *A Move action is added by selecting the point parameter when prompted*

After you select the point parameter labeled Position (see Figure 10–34), AutoCAD prompts:

Select objects: *(in this case use the Window method to select the geometry that makes up the switch and the OFF and ON text)*

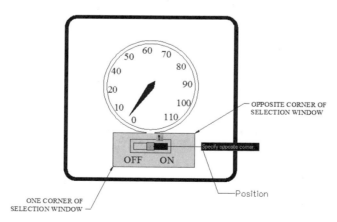

Figure 10–34 *Objects (geometry of Switch) are selected for being associated with the Move action*

After you select the objects to be associated with the Move action (see Figure 10–35), AutoCAD prompts:

> Specify action location or ⬇: *(specify the symbol for the Move action near the label)*

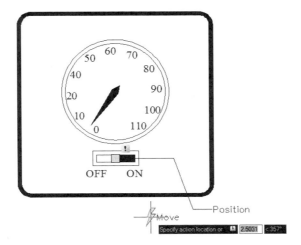

Figure 10–35 *Specifying the location of the Move action*

 Note: The location of the Action symbol does not affect how the action operates. It is for the purpose of identifying the type of action associated with the parameter only when you are in the Block Editor. When you use the window or crossing method to select the objects for adding them to the selection set or to define a stretch frame (explained later), the label, leader, grip, and other Block Editor symbols are not affected. The symbols, leaders, labels, and icons do not appear in a block reference. Only the grips appear when the block reference is selected for modifying.

After the parameter is added, the action added, the objects in the selection set (associated with the action) selected, and the action symbol located, you can exit the Block Editor by right-clicking and choosing CLOSE BLOCK EDITOR from the shortcut menu.

USING THE DYNAMICS OF A DYNAMIC BLOCK

When a dynamic block is inserted and the reference is selected for modification (see Figure 10–34), the geometry is highlighted and a grip will appear at the insertion point of the block as it usually does and a grip will also appear where you located the point parameter on the switch geometry. This grip is square for the Move action. You can select this grip and move the switch geometry and OFF and ON text with the cursor or enter new coordinates from the keyboard. The illustration on the left of Figure 10–36 shows where grips are located. The middle illustration shows the switch being relocated by selecting the grip and initiating the Move action. The illustration on the right shows the result of applying the Move action in the dynamic block reference.

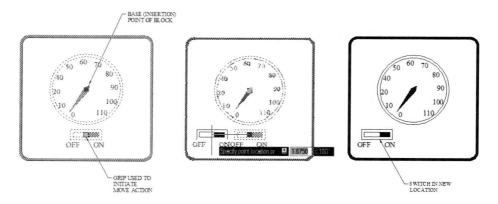

Figure 10–36 *The Move Action grip appears when the block reference is selected for modifying*

The Move action added to a point parameter in the block definition is the simplest form of using the dynamic block feature. Other, more powerful action/parameter combinations make dynamic blocks one of the most useful tools for the application of the computer to design/drafting. For example, the block in the example above can also have a Stretch action added to a linear parameter to allow the shape and/or size of the panel outline to be changed from the one in Figure 10–36 to the one shown in Figure 10–37 and geometry (second dial) that was hidden in the block definition is made visible.

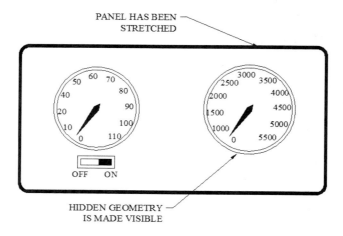

Figure 10–37 *The Dynamic Block after using the Stretch action and making hidden geometry visible*

Commonly used parameter/action combinations can be quickly added by using the Parameter Sets available on the Parameter Sets tab of the Authoring palette. These include Point Move, Linear Move, Linear Stretch, Linear Array, Linear Move Pair, Linear Stretch Pair, Polar Move, Polar Stretch, Polar Array, and Polar Move Pair. Once you have learned how to use the actions and parameters individually, you can use these combinations to speed up dynamic block creation.

Note: Good planning is necessary in creating dynamic blocks that are flexible and useful. And a thorough understanding of how to apply the concepts of parameters, actions, and grips is necessary also. You are advised to begin with a simple geometric shape, perhaps a rectangle, add a parameter and action, save it in the Block Editor, insert it and then see how your parameter/action operates. Like many of the very powerful features of AutoCAD, applying the Dynamic Block feature requires study and practice. But once mastered, the savings in time and the flexibility are worth the effort.

THE BLOCK EDITOR—MAKING A BLOCK DYNAMIC

The Block Editor is an authoring window dedicated especially to creating and editing dynamic blocks. You can use the Block Editor to make a dynamic block out of an existing nondynamic block, or while creating a new block. You can open the Block Editor by selecting **Open in block editor** in the Block Definition dialog box (see Figure 10–38) before choosing **OK** to create a new block.

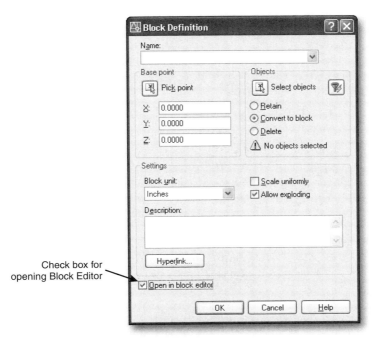

Figure 10–38 *Block Definition dialog box with the Open in Block Editor check box checked*

You can also open the Block Editor by double-clicking on an inserted block reference on the screen. AutoCAD displays the Edit Block Definition dialog box (see Figure 10–39) from which you can select a block from the **Block to create or edit** list and choose **OK**. AutoCAD then opens the Block Editor and the selected block is displayed on the editing screen and the Block Editor toolbar is displayed across the top of the drawing area.

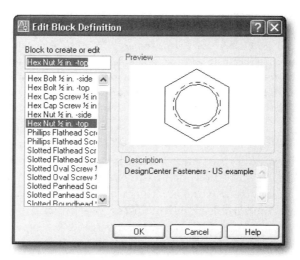

Figure 10–39 *Edit Block Definition dialog box*

AutoCAD displays the Block Editor (see Figure 10–40 with the Parameter tab selected in the Authoring palette). The example shown in Figure 10–40 is for creating a dynamic block named RFWN-FLANGE (for Raised-Face-Weld-Neck Flange).

The Block Editor provides a special Authoring palette. This palette provides quick access to block authoring tools. Figure 10–41 shows the three tabs that are available on the Authoring palette: Parameters, Actions, and Parameter Sets. In addition to the Block Authoring palettes, the Block Editor provides a drawing area in which you can draw and edit geometry as you would in the program's main drawing area. You can specify the background color for the Block Editor drawing area.

 Note: You can use most commands in the Block Editor. When you enter a command that is not allowed in the Block Editor, a message is displayed on the command line. It is recommended that the Command window be left open.

In the Block Editor, a special toolbar is displayed above the drawing area. The toolbar shows the name of the block definition currently being edited and provides several tools (see Figure 10–42).

Chapter 10 • *Block References and Attributes* 529

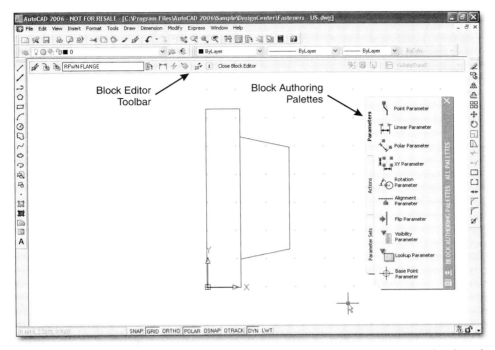

Figure 10–40 *Block Editor displaying the Authoring palette with the Parameters tab selected*

 Note: The objects that you see on the screen of the Block Editor are all part of the geometry of the particular block being edited. Any objects you add will become part of the block when you accept the changes as you exit the Block Editor. The Block Editor window, unlike the Help window, the DesignCenter window, the Properties window, and other palettes, must be closed before you can create and modify objects in the normal AutoCAD drawing area that are not associated with the selected block.

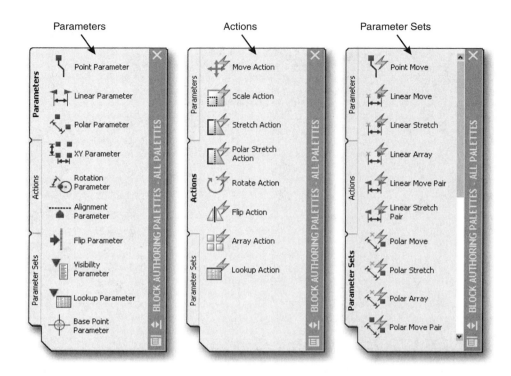

Figure 10–41 *Three tabs of the Block Editor Authoring palette: Parameters, Actions, and Parameter Sets*

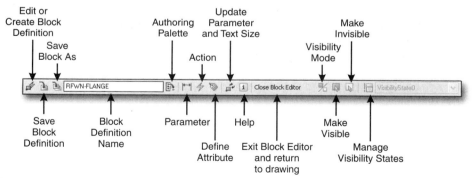

Figure 10–42 *Block Editor toolbar*

When the Block Editor is open for editing an existing named dynamic block or creating a dynamic block from the geometry that has just been defined as a named block, there are numerous special commands that are used for this purpose and can only be used while the Block Editor is open.

PARAMETERS

As mentioned earlier, after the initial planning is done and the geometry has been created, the first step in making a block dynamic is to add parameter(s). Parameters include Point, Linear, Polar, XY, Rotation, Alignment, Flip, Visibility, Lookup, and Base Point. The type of parameter used depends on the type of action desired. Some parameters can have different types of actions added to them. Some are limited to only one type of action. The following table shows which actions can be associated with which parameters. Note that the Alignment, Visibility, and Base Point parameters do not have actions associated with them, while the Rotation, Flip, and Lookup parameters can have only one action associated with them.

Actions	Parameters									
	Point	Linear	Polar	Xy	Rotation	Alignment	Flip	Visibility	Lookup	Base Pt
MOVE	X	X	X	X						
SCALE		X	X	X						
STRETCH	X	X	X	X						
POL. STR			X							
ROTATE					X					
FLIP							X			
ARRAY										
LOOKUP									X	

The parameter, in combination with the associated action, determines the location and appearance of the grips that appear when a reference of the block is selected for modifying.

Move, Scale, and Stretch actions can be associated with at least three parameters. Some of the reasons for associating an action with a particular parameter will become evident as you learn how the action operates with each parameter. For example, a Move action associated with a point or polar parameter allows the user to move the selection set of objects in any direction. But, if the Move action is associated with a linear parameter, it restricts the movement to the direction (even if diagonal) that the linear parameter was drawn.

It is a recommended practice to coordinate parameter geometry with the geometry of the objects in the selection set that are associated with the action added to the parameter. For example, if a rectangle in the block definition is 8 units wide and you plan to allow the user to stretch the rectangle to 10, 11, or 12 units in the block reference, then a linear parameter might be added from one end of the rectangle to the other, making its distance 8 units (see Figure 10–43). For this example the linear parameter label has been changed to "8 UNITS DISTANCE" for illustrative purposes. When a Stretch action is added to this linear parameter, the user will be able to resize the rectangle with a parameter whose base distance coincides with the original width of the rectangle. This is a logical approach to creating a dynamic block.

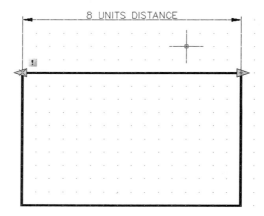

Figure 10–43 *Linear parameter added to rectangle*

It is possible to draw the linear parameter with its endpoints located otherwise at a different distance and still enable the user to stretch the rectangle to the desired dimension, but unless there is a specific reason to do so, it would not be good planning. The power of dynamic blocks can best be mastered when you understand the concept of how the parameters are manipulated by the actions that manipulate the objects in the selection set associated with the action.

Point Parameter

Figure 10–44 shows a dynamic block on the left with a point parameter near the switch geometry and text. In this example, a Move action has been added to the point parameter and associated with the objects that make up the switch and text. The view on the right shows how the switch and text were moved in the block reference by using the Move action.

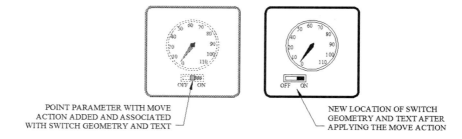

Figure 10–44 *Example of using the Move action with the point parameter*

The point parameter establishes a point by setting user-defined X and Y properties for the block reference. A Move or Stretch action can be associated with a point parameter. Invoke the point parameter from the Authoring palette and AutoCAD prompts:

> Specify parameter location or ⊠: *(specify a location or right-click and select an option)*
> Specify label location: *(specify a location for the label)*

Options for the point parameter include Name, Label, Chain, Description, and Palette.

Choosing NAME lets you specify the parameter name that is displayed in the Properties palette when you select the parameter in the Block Editor. When the parameter has been selected, AutoCAD prompts:

> Enter parameter name <default>: *(enter a name for the parameter or press* ENTER *to use the default name)*

Choosing LABEL lets you specify the parameter label, which identifies the custom property name added to the block. The label is displayed in the Properties palette as a Custom property when you select the block reference in a drawing. In the Block Editor, the parameter label is displayed next to the parameter. AutoCAD prompts:

> Enter position property label <default>: *(enter a label for the position property or press* ENTER *to use the default label)*

Choosing CHAIN lets you specify the Chain Actions property for the parameter. The point parameter may be included in the selection set of an action that is associated with a different parameter. When that parameter is edited in a block reference, its associated action may trigger a change in the values of other parameters included in the action's selection set.

Setting the Chain Actions property to **Yes** triggers any actions associated with the point parameter, just as if you had edited the parameter in the block reference through a grip or custom property.

Setting the Chain Actions property to **No** means that the point parameter's associated actions are not triggered by the changes to the other parameter. AutoCAD prompts:

> Evaluate associated actions when parameter is edited by another action? [Yes/No] <No>: *(enter* **y** *or press* ENTER*)*

Choosing DESCRIPTION lets you specify the description for the custom property name (parameter label). This description is displayed in the Properties palette when you select the parameter in the Block Editor. In a drawing, when you select the custom property name (parameter label) for the block reference in the Properties palette, the description is displayed at the bottom of the Properties palette. AutoCAD prompts:

> Enter property description: *(enter a description for the parameter)*

Choosing PALETTE lets you specify whether or not the parameter label is displayed in the Properties palette when the block reference is selected in a drawing.

> Display property in Properties palette? [Yes/No] <Yes>: *(enter **n** or press* ENTER*)*

Note: You can also specify and edit these properties in the Properties palette at a later time, after you've added the parameter to the block definition.

A grip is displayed where you specified the location of the parameter with a leader to the label. An Alert icon (with the exclamation point) is displayed because there is no action associated with the parameter yet. Refer to the section on Using Actions for a detailed explanation on adding actions.

Linear Parameter

Figure 10–45 shows a dynamic block on the left with a linear parameter added that is the width of the instrument panel. In this example, a Stretch action has been added to the linear parameter and associated with the objects that make up the outline of the right end of the panel. The view in the middle shows the grip being moved to change the distance of the linear parameter and in turn stretch the width of the panel outline. The view on the right shows how the panel outline in the block reference was changed by using the Stretch action.

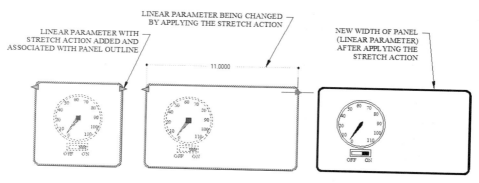

Figure 10–45 *Example of using the Stretch action with the linear parameter*

The linear parameter establishes a vector by setting a user-defined distance property for the block reference. A Move, Scale, Stretch, or Array action can be associated with a linear parameter. After you invoke the linear parameter from the Authoring palette, AutoCAD prompts:

> Specify start point or ⬇: *(specify a start point for the parameter or right-click and select an option)*

Specify endpoint: *(specify an endpoint for the parameter)*
Specify label location: *(specify a location for the label)*

Options for the linear parameter include Name, Label, Chain, Description, Base, Palette, and Value set.

Choosing NAME, LABEL, CHAIN, DESCRIPTION or PALETTE operates the same as described for the point parameter above.

Choosing BASE lets you specify the Base Location property for the parameter. AutoCAD prompts:

Enter base location [Startpoint/Midpoint]: *(specify an option)*

> Choosing STARTPOINT specifies that the start point of the parameter remains fixed when the endpoint of the parameter is edited in the block reference.
>
> Choosing MIDPOINT specifies a midpoint base location for the parameter. It is indicated by an X in the block definition. When you edit the linear parameter in the block reference, the midpoint of the parameter remains fixed, and the start point and endpoint of the parameter move simultaneously equal distances from the midpoint. For example, if you move the grip on the endpoint two units away from the midpoint, the start point simultaneously moves two units in the opposite direction.

Choosing VALUE SET lets you specify a value set for the parameter that limits the available values for the parameter in a block reference to the values specified in the set. AutoCAD prompts:

Enter distance value set type [None/List/Increment] <None>: *(specify a value set type or select one of the available option)*

> Choosing LIST lets you specify a list of available values for the parameter in a block reference. AutoCAD prompts:

Enter list of distance values (separated by commas): *(specify a list of values separated by commas)*

> Choosing INCREMENT lets you specify a value increment and minimum and maximum values for the parameter in the block reference. AutoCAD prompts:

Enter distance increment: *(specify an increment value for the parameter)*
Enter minimum distance: *(specify a minimum distance value for the parameter)*
Enter maximum distance: *(specify a maximum distance value for the parameter)*

Refer to the section on Using Actions for a detailed explanation on adding actions.

Polar Parameter

Figure 10–46 shows a dynamic block on the left with a polar parameter added that is the length and angle of the crane boom. In this example, a Polar Stretch action has been added to the polar parameter and associated with the objects that make up the outline of the boom. The view in the middle shows the grip being selected for modifying the geometry. The view on the right shows the grip being moved to change the distance and angle of the polar parameter and in turn stretching and relocating the end of the boom by using the Polar Stretch action.

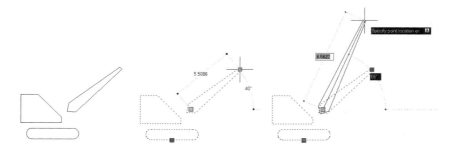

Figure 10–46 *Example of using the Polar Stretch action with the polar parameter*

The polar parameter establishes a vector by setting user-defined distance and angle properties for the block reference. A Move, Scale, Stretch, Polar Stretch, or Array action is associated with a linear parameter. After you invoke the polar parameter from the Authoring palette, AutoCAD prompts:

    Specify base point or ▣: *(specify a start point for the parameter or select an option from the shortcut menu)*
    Specify endpoint: *(specify an endpoint for the parameter)*
    Specify label location: *(specify a location for the label)*

Options for the polar parameter include Name, Label, Chain, Description, Palette, and Value set.

Choosing NAME, LABEL, CHAIN, DESCRIPTION or PALETTE operates the same as described for the point parameter above. Choosing VALUE SET operates the same as described for the linear parameter above.

Refer to the section on Using Actions for a detailed explanation on adding actions.

XY Parameter

Figure 10–47 shows a dynamic block representing a logo on the left with an XY parameter added to it whose X distance is the width from one side to the other and Y distance is the height from top to bottom. In this example, a Scale action has been added to the XY parameter and associated with the objects that make up the logo. The view in the middle shows using the Scale action to make the X distance of the logo fill the width of a rectangular shape. The view on the right shows using the Scale action to make the Y distance of the logo fill the height of a rectangular shape.

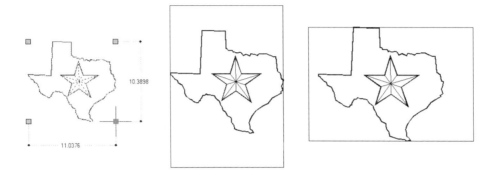

Figure 10-47 *Example of using the Scale action with the XY parameter*

The XY parameter establishes a pair of orthogonal vectors by setting user-defined horizontal and vertical distance properties for the block reference. A Move, Scale, Stretch, or Array action can be associated with an XY parameter. After you invoke the XY parameter from the Authoring palette, AutoCAD prompts:

> Specify base point or ⬇: *(specify a base point or select one of the options from the shortcut menu)*
> Specify endpoint: *(specify an endpoint for the parameter)*

Options for the XY parameter include Name, Label, Chain, Description, Palette, and Value set.

Choosing NAME, LABEL, CHAIN, DESCRIPTION, or PALETTE operates the same as described for the point parameter above. Choosing VALUE SET operates the same as described for the linear parameter above.

Refer to the section on Using Actions for a detailed explanation on adding actions.

Rotation Parameter

Figure 10-48 shows a dynamic block on the left representing a door with jambs with a rotation parameter added to it. In this example, a Rotate action has been added to the rotation parameter and associated with the objects that make up the door. The view in the middle shows the grip being selected for modifying the geometry. The view on the right shows the result of using the Rotate action to change the door's opening angle from 45° to 90°.

Figure 10–48 Example of using the Rotate action with the rotation parameter

The rotation parameter establishes a direction of rotation by setting a user-defined angle property for the block reference. Only a Rotate action can be associated with a rotation parameter. After you invoke the rotation parameter from the Authoring palette, AutoCAD prompts:

> Specify base point or ⤓: *(specify a base point for the parameter or select an option from the shortcut menu)*
> Specify radius of parameter: *(specify a radius for the parameter)*
> Specify default rotation angle or [Base angle] <0>: *(specify a base angle for the parameter)*
> Specify label location: *(specify a location for the label)*

Choosing BASE ANGLE lets you specify a base angle for the parameter and places the grip for the parameter at this angle. AutoCAD prompts:

> Specify base angle <0>: *(specify a base angle for the parameter or press ENTER)*

Options for the rotation parameter include Name, Label, Chain, Description, Palette, and Value set.

Choosing NAME, LABEL, CHAIN, DESCRIPTION or PALETTE operates the same as described for the point parameter above. Choosing VALUE SET operates the same as described for the Linear parameter above.

Refer to the section on Using Actions for a detailed explanation on adding a Rotation action.

Alignment Parameter

Figure 10–49 shows a dynamic block on the left with an alignment parameter at a point midway between the two jambs and aligned with where the door and jambs should be drawn on the hinge side of the wall where it will be installed. The view on the right shows how the block automatically aligns itself during the insertion process as the cursor nears the face of the wall. In this example the door needed to be placed at the midpoint of the angled wall, so the object snap mode MIDPOINT was invoked.

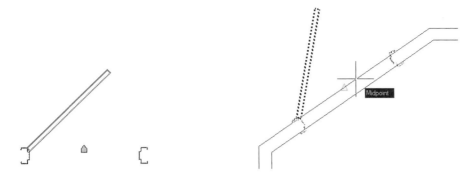

Figure 10–49 *Example of using the alignment parameter*

The alignment parameter establishes an X and Y location and an angle. It affects the entire block and does not require an action associated with it. An alignment parameter allows the block reference to be automatically rotated around the base point to align with other objects in the drawing. It changes the angle property of the block reference. After you invoke the rotation parameter from the Authoring palette, AutoCAD prompts:

> Specify base point of alignment or [Name]: *(Specify a point or enter name)*

When you specify the base point for the alignment parameter, an X is displayed in the Block Editor. When the command is completed, an alignment grip is added at this base point. The block reference automatically rotates about this point to align with another object in the drawing.

The parameter name is displayed in the Properties palette when you select the parameter in the Block Editor. AutoCAD prompts:

> Specify alignment direction or alignment type or ⬇: *(specify an alignment direction or select TYPE from the shortcut menu)*

The alignment direction specifies the direction of the grip and the angle of alignment for the block reference. The TYPE selection allows you to select whether the parameter type is perpendicular or tangent as shown in the following prompt:

> Enter alignment type [Perpendicular/Tangent] <Perpendicular>: *(specify an alignment type)*

Choosing PERPENDICULAR lets you specify that the dynamic block reference aligns perpendicular to objects in a drawing.

Choosing TANGENT lets you specify that the dynamic block reference aligns tangent to objects in a drawing.

Flip Parameter

Figure 10–50 shows a dynamic block on the left representing a logo with a flip parameter added to it. In this example, a Flip action has been added to the flip parameter and associated with the objects that make up the dolphin and the text. The view on the right shows the result of using the Flip action to flip the objects associated with it.

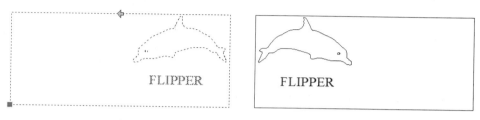

Figure 10–50 *Example of using the Flip action with the flip parameter*

The flip parameter establishes a user-defined flip property for the block reference. A flip parameter flips objects. In the Block Editor, a flip parameter is displayed as a reflection line. Objects can be flipped about this reflection line. A flip parameter displays a value that shows if the block reference has been flipped or not. You associate a Flip action with a flip parameter. After you invoke the flip parameter from the Authoring palette, AutoCAD prompts:

> Specify base point of reflection line or ⬇: *(specify a base point for the reflection line or select an option from the shortcut menu)*
> Specify endpoint of reflection line: *(specify an endpoint for the reflection line)*
> Specify label location: *(specify a location for the label)*

Options for the flip parameter include Name, Label, Description, and Palette.

Choosing NAME, LABEL, DESCRIPTION, or PALETTE operates the same as described for the point parameter above.

Refer to the section on Using Actions for a detailed explanation on adding a Flip action.

Visibility Parameter

Figure 10–51 shows a dynamic block on the left representing an instrument panel with a visibility parameter added to it. In this example, visibility states have been applied to enable the user to select between one set of objects to be visible in one state and another set to be visible in another state. The view on the right shows the result of using the visibility states to hide the objects associated with it and show others.

Figure 10–51 *Example of using the visibility parameter*

The visibility parameter establishes a user-defined visibility property for the block reference. A visibility parameter allows you to create visibility states and to control the visibility of objects in the block. A visibility parameter always applies to the entire block and needs no action associated with it. After you invoke the visibility parameter from the Authoring palette, AutoCAD prompts:

> Specify parameter location or ⬇: *(specify a location for the parameter or select an option from the shortcut menu)*

Options for the visibility parameter include Name, Label, Description, and Palette.

> Choosing NAME, LABEL, DESCRIPTION, or PALETTE operates the same as described for the point parameter above.

Lookup Parameter

Figure 10–52 shows four insertions of a dynamic block representing a raised-face-weld-neck piping flange in which dimensions of selected parts of the geometry are changed to suit different requirements. In this example each of the four flange ratings (150#, 300#, 400#, and 600#) requires a unique combination of flange length and outside diameter.

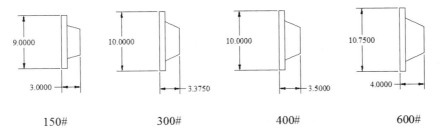

Figure 10–52 *Example of using the lookup parameter*

The lookup parameter establishes user-defined lookup properties for the block reference. A lookup parameter defines a custom property that you can specify or set to evaluate to a value from a list or table you define. You associate a Lookup action with a lookup parameter. Each lookup parameter you add to the block definition can be added as a column in the Property Lookup Table dialog box. After you invoke the lookup parameter from the Authoring palette, AutoCAD prompts:

> Specify parameter location or ⬇: *(specify a location for the parameter or right-click and select an option)*

Options for the lookup parameter include Name, Label, Description, and Palette.

Choosing NAME, LABEL, DESCRIPTION, or PALETTE operates the same as described for the point parameter above.

Refer to the section on Using Actions for a detailed explanation on adding a Lookup action.

Base Parameter

The base point parameter defines the base point for the dynamic block reference in relation to the geometry in the block. This provides a way to control the location of the base point within the block reference when it is edited in a drawing. You do not associate any actions with a base point parameter. The base point parameter is generally included in a selection set of the block definition's actions. After you invoke the base parameter from the Authoring palette, AutoCAD prompts:

> Specify parameter location: *(specify a location)*

Only one base point parameter is allowed in a dynamic block definition. Trying to add another base point parameter causes the following alert to be displayed:

> Base point parameter already exists in block definition

In the Block Editor, when the Display UCS Icon option is turned on, the icon is located at the base point of the block.

ACTION, PARAMETER, AND GRIP PROPERTIES

The font, size, and color of actions, parameters and grip properties can be changed.

The BPARAMETERCOLOR system variable lets you specify the color of the parameter. You can choose BYLAYER, BYBLOCK, a color by its integer from 1 to 255, or a true color specified by three integers each ranging from 1 to 255 in the format: RGB:000,000,000.

The BPARAMETERSIZE system variable lets you set the size of the parameter specified by an integer from 1 to 255 (pixels).

The BPARAMETERFONT system variable lets you specify the font of text in the parameter by responding with any TrueType or SHX font on the system.

The BACTIONCOLOR system variable lets you specify the color of the action. You can choose BYLAYER, BYBLOCK, a color by its integer from 1 to 255, or a true color specified by three integers each ranging from 1 to 255 in the format: RGB:000,000,000.

The BGRIPOBSIZE system variable lets you set the size of the grips specified by an integer from 1 to 255 (pixels).

The BGRIPCOLOR system variable lets you specify the color of the grip. You can choose BYLAYER, BYBLOCK, a color by its integer from 1 to 255, or a true color specified by three integers each ranging from 1 to 255 in the format: RGB:000,000,000.

The BTMARKDISPLAY system variable lets you control if the value set markers are displayed or not. A value set is a range or list of values specified for a parameter. These values can be displayed for the block reference as a drop-down list next to the parameter label under Custom in the Properties palette. When you define a value set for a parameter, the parameter is limited to these values when the block reference is manipulated in a drawing. For example, if you define a linear parameter in a block that represents a door to have a value set of 24, 28, and 36, the door can only be stretched to 24, 28, or 36 units. If the system variable BTMARKDISPLAY is set to 1, then the value set markers are displayed. They are not displayed if it is set to 0.

USING ACTIONS

Actions are associated with parameters to allow the user to manipulate and/or change the values of the geometric properties, specifically locations, sizes, distances, angles, and multiple instances. Actions are accessible while editing a dynamic block in the Block Editor either by invoking the BACTION command from the On-Screen prompt or choosing a specific action from the Action tab of the Authoring palette. Actions include Move, Scale, Stretch, Polar Stretch, Rotate, Flip, Array, and Lookup.

Move Action

Move adds a Move action to the block definition and can be associated with a point, linear, polar, or XY parameter. A **Move** action stipulates that a specified selection set of objects will move when the action is initiated in a dynamic block reference.

In the following example, a block named RECTANGLE is the current block being edited in the Block Editor. A point parameter has already been added to a block definition (see Figure 10–53), so you can choose it as the parameter for the Move action. Invoke the Move action from the Action tab of the Block Authoring palette and select the point parameter defined on the screen. AutoCAD prompts:

Select objects: *(select the rectangle and press* ENTER*)*
Specify action location (or ⬇): *(specify a location for the label or select an option from the shortcut menu)*

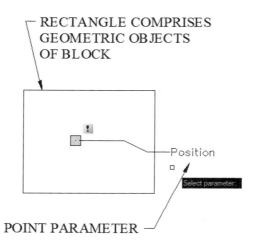

Figure 10–53 *Associating an action with a parameter*

Location is not significant, but when the block definition becomes crowded with other symbols for parameters, actions, and grips, it is advisable to locate the action symbol near the parameter with which is it associated (see Figure 10–54).

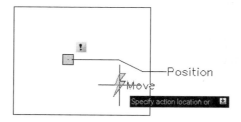

Figure 10–54 *Adding a Move action to a point parameter in the dynamic block definition*

 Note: The names of the action and parameter are not visible in a block reference (a block that has been inserted). Nor is the grip visible until the block has been selected for modifying. And in the Block Editor, when selecting objects to include in an action, if a grip or the name of a parameter or action has been included in the window or crossing for determining the selection set, it has no effect on that grip, action, or parameter.

Instead of specifying an action location, you can choose one of the two options available from the shortcut menu: Multiplier or Offset.

Choose MULTIPLIER to change the associated parameter value by a specified factor when the action is triggered. AutoCAD prompts:

>   Enter distance multiplier <current>: *(enter a numeric value)*

Specify a multiplier factor. For example, if you enter 10 for the multiplier and the user specifies a move that is 3.5 units in a specific direction, the objects will be moved 35 units in the specified direction. Or, if the multiplier factor is 0.25 and the user specifies a distance of 40 units, the selection set associated with that Move action will be moved 10 units.

Choose OFFSET to increase or decrease by a specified number the angle of the associated parameter when the action is triggered. AutoCAD prompts:

>   Enter angle offset <current>: *(enter a numeric value)*

Specify an angle offset. For example, if you specify 90 for the angle offset and the user specifies a move that is vertical (say this is 90° in the current UCS) units for a specific distance, the objects will be moved at a direction of 180° for the specified distance. Or, if the angle offset is specified as -45° and the user specifies an angle of -45°, the selection set associated with that Move action will be moved in a direction of -90°.

The Multiple and Offset options can be applied together to the same Move action.

You can use the grips (by default, the color of the dynamic block grips will be different from the normal grip color) or custom property in the Properties palette to manipulate the block reference. When you manipulate the block reference in a drawing, by moving a grip or changing the value of a custom property in the Properties palette, you change the value of the parameter that defines that custom property in the block. When you change the value of the parameter, it drives the action that is associated with that parameter, which changes the geometry (or a property) of the dynamic block reference.

Without applying the Multiple or Offset option to the Move action, it will operate like the Move option when selecting the normal grip associated with a block, except for one difference. When you use the Move action grip to move the selected geometry of the block, only that geometry moves. Instead of a point parameter, if you select a linear (see Figure 10–55) or polar parameter, AutoCAD prompts:

>   Specify parameter point to associate with action or enter ⏎: *(select one of the parameter points)*

You can select a point or choose START POINT or SECOND POINT. After you specify a point, AutoCAD prompts:

> Select objects: *(select one or more objects to be associated with the Move action and then press ENTER)*
> Specify action location or ⬇: *(specify a location near the selected point or choose MULTIPLIER or OFFSET from the shortcut menu)*

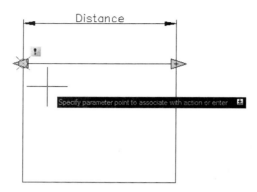

Figure 10–55 *Adding a Move action to a linear parameter in the dynamic block definition*

In Figure 10–55 a linear parameter (which has two points) was selected. Whichever point is selected will become the location for the grip in the block reference that can be selected for initiating the associated Move action. Pressing ENTER automatically associates the Move action with the second point. The grip for the other point will be visible when the block is selected, but it cannot be used to initiate this Move action. The primary difference in using a linear parameter versus a point parameter is that the move is restrained in the direction of the linear parameter. In the example, the linear parameter is horizontal; therefore, when the user applies the Move action associated with it, the selection set of objects can only be moved horizontally.

If a polar parameter (which has two points) is selected with which to associate the Move action, AutoCAD prompts you to select one of the points in the same manner as the linear parameter. Unlike the linear parameter, when the user initiates the Move action by selecting that grip in the block reference, the selection set of objects associated with that Move action can be moved both specified distances and specified angles. The Move action associated with a point of a polar parameter is practically the same as the Move action associated with a point parameter.

When you are being prompted to "Specify parameter point to associate with action or enter ⬇" you can choose either START POINT or SECOND POINT from the shortcut menu to specify the point with which to associate the Move action instead of selecting the point with the pointing device on the screen. If you add a Move action to an XY parameter (see Figure 10–56), AutoCAD prompts:

> Specify parameter point to associate with action or enter ⊡: *(select one of the parameter points or choose one of the options from the shortcut menu)*

Specify a point and AutoCAD prompts:

> Select objects: *(select one or more objects to be associated with the Move action and then press* ENTER*)*
> Specify action location or ⊡: *(specify a location near the selected point or choose one of the options from the shortcut menu)*

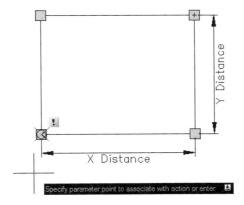

Figure 10–56 *Adding a Move action to an XY parameter in the dynamic block definition*

In Figure 10–56 an XY parameter (which has four points) was selected. The grip for the point on the same X axis as the primary point (referred to as the Xcorner) can be used to initiate a Move action in the direction of the Y axis (referred to as the Ycorner). The grip for the point on the same Y axis as the primary point can be used to initiate a Move action in the direction of the X axis. The grip diagonally opposite the primary point (referred to as the second point) cannot be used to initiate this Move action.

Instead of specifying the parameter point, you can select either BASE POINT, SECOND POINT, XCORNER, or YCORNER from the shortcut menu to specify the point with which to associate the Move action instead of selecting the point with the pointing device on the screen.

 Note: Pressing ENTER automatically associates the Move action with the second point. However, with four points in the XY parameter, it is not always predictable which point AutoCAD might consider the second point. You should note which grip has an X placed on it during the selection process to be sure it is the one you wish to become the Base point for the Move action to be able to work in any direction.

Instead of specifying an action location, you can choose from the shortcut menu the MULTIPLE, OFFSET, or XY option. Multiple and Offset behave similar to the point parameter explained earlier. If you choose the XY option, AutoCAD allows you to specify whether the distance that is applied to the action is the XY parameter's X distance, Y distance, or XY distance from the parameter's base point.

Scale Action

Scale adds a Scale action to the block definition and can only be associated with a linear, polar, or XY parameter. A Scale action stipulates that a specified selection set of objects in a dynamic block reference will be scaled when the action is initiated in a dynamic block reference in a manner similar to the scale command. The Scale action causes the objects to scale when the associated parameter is edited by moving grips or by using the Properties palette.

In a dynamic block definition, you do not associate a Scale action with a key point on the parameter but with an entire parameter.

To add a Scale action, invoke the Scale action from the Action tab of the Block Authoring palette and select the polar parameter defined on the screen. AutoCAD prompts:

>Select objects: *(select one or more objects to be associated with the Scale action and then press* ENTER*)*
>Specify action location or ▼: *(specify a location as shown in Figure 10–57)*

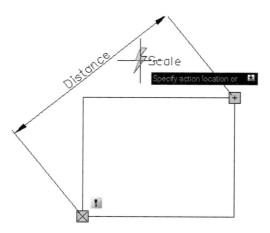

Figure 10–57 *Adding a Scale action to a polar parameter in the dynamic block definition*

Instead you can choose the BASE option from the shortcut menu. AutoCAD prompts:

>Enter base point type [Dependent/Independent]: *(select the base point type)*

Choosing **Dependent** causes the objects in the selection set to scale relative to the base point of the parameter with which the Scale action is associated. If the custom grip is used to scale the block, it scales relative to the lower left corner of the rectangle.

Choosing **Independent** (shown in the Block Editor as an X marker) lets you specify a base point independent of the parameter with which the Scale action is associated. The objects in the selection set will scale relative to this independent base point you specify.

When the block is inserted and the user selects it for modification, the grips for first and second points of the polar parameter will appear. When the cursor is moved to the second point and polar tracking is ON, the distance and angle between the first and second points are displayed (see Figure 10–58).

Note: It is the distance between the first and second point that the Scale action uses as the base distance. If you enter a value, AutoCAD will scale the selected objects an amount determined by the ratio of the value entered divided by the distance between the first and second parameter points. In the example in Figure 10–58, that distance is 15.3832 units. If you enter a value of 20.0, the selected objects will be scaled up so that the new distance between the first and second points is 20.0. If you enter a value of 10.0, the selected objects will be scaled down so that the new distance between the first and second points is 10.0.

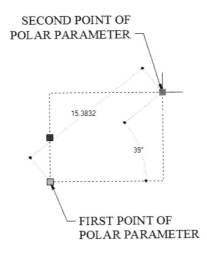

Figure 10–58 *Selecting the second point of a polar parameter to initiate the associated Scale action in a dynamic block reference*

After you have selected the grip at the second point of the polar parameter (with which a Scale action has been associated), you can drag the cursor and AutoCAD displays the distance and angle between the first point and the cursor (see Figure 10–59). The angle has no effect on the Scale action, but if you specify the point as displayed, the distance (21.0503 in the example) will be used as the new distance between the corresponding points on the scaled-up objects and all of the selected objects will be scaled accordingly. In this example the base point of the Scale action is the first point specified when adding the Scale action to the polar parameter.

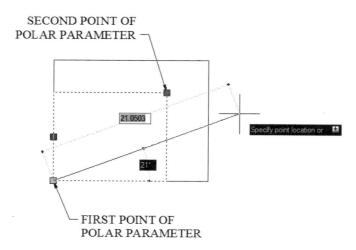

Figure 10–59 *Using the cursor to scale up the objects in a dynamic block reference associated with the Scale action that was added to a polar parameter*

Instead of using the cursor to determine the distance, you can enter a value from the keyboard. That value will be used as the new distance between the corresponding points on the scaled-up objects and all of the selected objects will be scaled proportionately.

To add a Scale action to an XY parameter, invoke the Scale action from the Action tab of the Block Authoring palette and select the XY parameter. The prompts are the same in the selection of linear or polar parameter, except that an additional XY option is provided to *Base* type. AutoCAD prompts when you choose the *XY* option:

Enter XY distance type [X/Y/XY] <XY>: *(select X, Y, or XY distance type)*

Choosing X causes the proportion that the user scales the selected objects in a dynamic block reference to be determined by the movement of the cursor in direction of the X axis. Whichever grip the user selects to initiate the Scale action, the objects will be scaled according to the distance in the X direction that the cursor is from a base point. The base point will be the point opposite the selected grip in the X direction. The direction in which the objects are scaled depends on how the XY parameter was originally added to the object. For example, if the XY parameter were added by first selecting the lower left corner (see Figure 10–60) and then the upper right corner, the base point for certain Scale actions will be the lower left grip. If the XY parameter had been added by first selecting the upper left corner (see Figure 10–61) and then the lower right corner, the base point for certain Scale actions will be the upper left grip.

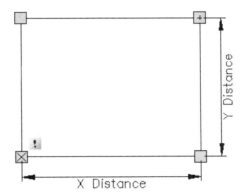

Figure 10–60 *Adding an XY parameter from the lower left corner to the upper right corner of the dynamic block definition*

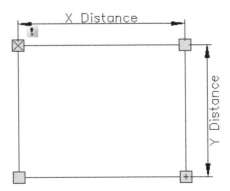

Figure 10–61 *Adding an XY parameter from the upper left corner to the lower right corner of the dynamic block definition*

Choosing Y (see Figure 10–62) causes the proportion that the user scales the selected objects in a dynamic block reference to be determined by the movement of the cursor in the direction of the Y axis. Whichever grip the user selects to initiate the Scale action, the objects will be scaled according to the distance in the Y direction that the cursor is from a base point. The base point will be the point opposite the selected grip in the Y direction. As in choosing X, in choosing Y the direction in which the objects are scaled depends on how the XY parameter was originally added to the object.

Choosing XY causes the proportion that the user scales the selected objects in a dynamic block reference to be determined by the movement of the cursor in any direction.

 Note: In choosing X, Y, or XY, where the user moves the cursor after selecting a grip determines how the objects will be scaled based on the parameter's X, Y, or XY distance respectively. But for entering a value from the keyboard, the scale factor is based only on the X distance. No matter which you choose (X, Y, or XY), when the dynamic block reference is selected and highlighted, and a grip is selected, the only text box that is editable (displayed as dynamic) is the X distance.

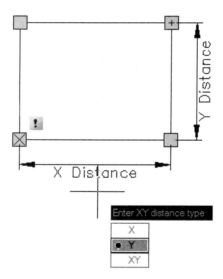

Figure 10–62 *Choosing the Y distance on which to base the Scale action added to an XY parameter*

Choosing the second point of the XY parameter of the dynamic block reference (see Figure 10–63) to initiate the Scale action lets you move the cursor up and down (in the direction of the Y axis) to scale the selected objects larger or smaller in proportion to the Y distance. If you enter a value from the keyboard, however, it will be applied to the X distance. It is the X distance that is highlighted in an editable text box on the screen.

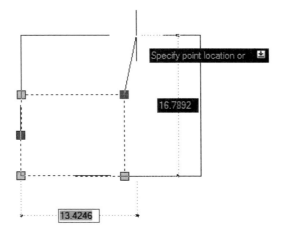

Figure 10–63 *Choosing the second point in the dynamic block reference to initiate the Scale action added to an XY parameter*

Stretch Action

Stretch adds a Stretch action to the block definition and can be associated with a point, linear, polar, or XY parameter. A **Stretch** action stipulates that a specified selection set of objects in a dynamic block reference will be stretched when the action is initiated in a dynamic block reference in a manner similar to the STRETCH command. The Stretch action causes the objects to stretch when the associated parameter is edited by moving grips or by using the Properties palette.

Figure 10–64 shows a typical floor plan of a room with a symbol for a light in the ceiling, connected to a wall switch by a dashed arc, indicating electrical wiring. Drawing the three entities (light, switch, and arc) would normally take three separate operations, even if the light and switch were inserted as blocks. But, with the power of the dynamic block, you can insert a single block and then use the STRETCH command to relocate the light symbol and have the dashed arc stretched to maintain the "connection."

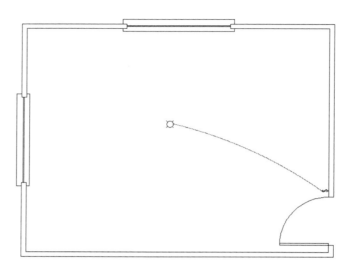

Figure 10–64 *Floor plan of a room with symbols for a light, switch, and a connecting dashed arc*

In the following example, a block named LT&SW (for Light and Switch) is the current block being edited in the Block Editor. A point parameter has already been added to a block definition (see Figure 10–65), so you can choose it as the parameter for the Stretch action. To add the Stretch action, invoke the Stretch action from the Action tab of the Block Authoring palette and select the point parameter (labeled Position) defined on the screen. AutoCAD prompts:

> Specify first corner of stretch frame or ⬇: *(specify the first corner)*
> Specify opposite corner: *(specify the opposite corner)*

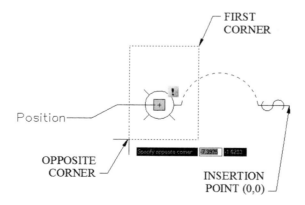

Figure 10–65 *Specifying the first point and opposite point of the stretch frame*

AutoCAD then prompts you to select the objects to be stretched:

Select objects: *(with the crossing selection mode, select the symbol for the light and one end of the dashed arc and then press* ENTER*)*

This causes the light symbol to move and the arc to stretch when the Stretch action is used in a reference of this block in the drawing. AutoCAD then prompts you to locate the action label in the block definition:

Specify action location or ⬇: *(specify a location near the selected point)*

You can now stretch the dashed arc and move the light symbol by selecting the block reference, selecting the grip for the Stretch action, and then moving the cursor. Figure 10–66A shows the dynamic block reference as it is inserted. Figure 10–66B shows the grips highlighted when the reference is selected for modification. Figure 10–66C shows the dynamic stretching of the dashed arc and moving of the light symbol. Figure 10–66D shows the result of applying the Stretch action to a point parameter in the dynamic block.

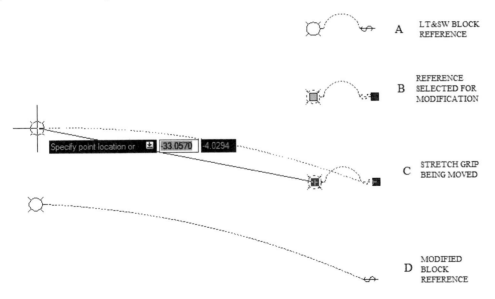

Figure 10–66 *Modifying a block reference with a Stretch action*

The following example (see Figure 10–67, weld-neck flange) shows how to apply a Stretch action to a linear parameter. A linear parameter (with the label Distance) has been added that coincides with the length of the flange. In order for the parameter to align with the length of the flange, it was drawn from right (start point) to left (second point). This example shows how to add a Stretch action to enable the change in length.

To add the Stretch action, invoke the Stretch action from the Action tab of the Block Authoring palette and select the linear parameter (labeled Distance) defined on the screen. AutoCAD prompts for the point to associate with the Stretch action:

> Specify parameter point to associate with action or enter ⬇: *(press* ENTER *to accept the default second point to be associated with the Stretch action)*

If you want to specify the start point, then select START POINT from the shortcut menu.

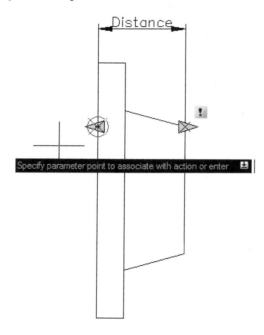

Figure 10–67 *Specifying the second point with the Stretch action added to a linear parameter*

AutoCAD prompts for the corners of the rectangle to determine the stretch frame:

> Specify first corner of stretch frame or ⬇: *(specify the first corner)*
> Specify opposite corner: *(specify the opposite corner)*

The corners in this example need to include the rectangle representing the flange face and only the two lines representing the part of the weld neck that attach to the flange. The line representing the welded end of the weld neck should not be included; otherwise, it will be moved away from the insertion point when a Stretch action is initiated (see Figure 10–68). Then AutoCAD prompts you to select objects. In the example, specify the rectangle that includes the flange and two lines of the weld neck that are attached to the flange. After you select the objects, AutoCAD prompts you to place the action symbol, which can be placed at any convenient location.

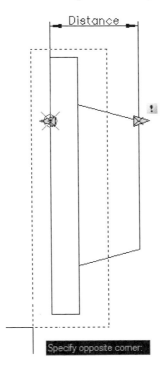

Figure 10–68 *Specifying the first point and opposite point of the stretch frame*

When the block reference is selected for modifying in a drawing, the grips for the Stretch action are represented by arrows when associated with a linear parameter, and the distance that will be affected is highlighted in an editable text box (see Figure 10–69). The cursor in Figure 10–69 is shown being moved upward and to the left, but because this Stretch action has been associated with a linear parameter instead of a point parameter, the only effect cursor movement will have is its change in the direction of the parameter (the block definition's X-axis in this case).

This is where the flange length can best be changed to the desired value. Once the associated grip has been selected (initiating the Stretch action) and the value of the linear parameter is being highlighted, you can enter the desired value from the keyboard and have it applied in the correct direction without moving the insertion point.

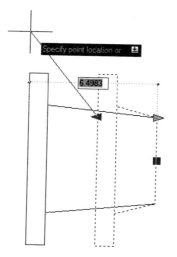

Figure 10-69 *Using the cursor to change the flange length by means of the Stretch action applied to a linear parameter*

Figure 10-70 shows the block reference that has been inserted on a 90° pipe elbow that is angled at 22.50° off horizontal. The block reference has been selected for modifying and the grip associated with the Stretch action has also been selected. When selected, it turned a specified color (default is red). The linear parameter that will be affected by the Stretch action rotates with the block when it is inserted, even though the text box remains horizontal for legibility.

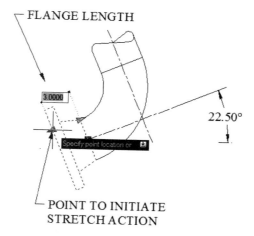

Figure 10-70 *Stretch action initiated in a block reference angled at 22.50°*

Figure 10–71 shows the block reference with its new flange length and the new location of the grip associated with the Stretch action. In the example, the value of 4.5 was entered from the keyboard when the text box was highlighted.

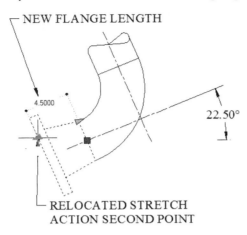

Figure 10–71 *Result of entering new value for flange length when text box was highlighted*

If, instead of a point or linear parameter, you choose an XY parameter for a Stretch action (see Figure 10–72), AutoCAD prompts for the point to associate with the Stretch action:

Specify parameter point to associate with action or enter ⬇: *(press* ENTER *to choose the base point as the point associated with the Stretch action)*

You can choose START POINT, SECOND POINT, XCORNER, or YCORNER from the shortcut menu to specify a point to associate with the XY parameter.

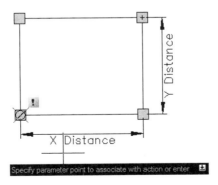

Figure 10–72 *Choosing the base point in the dynamic block reference to initiate the Stretch action added to an XY parameter*

AutoCAD prompts for the corners of the rectangle to determine the stretch frame:

Specify first corner of stretch frame or ⬇: *(specify the first corner)*
Specify opposite corner: *(specify the opposite corner as shown in Figure 10–73)*

Then AutoCAD prompts you to select objects. Select the rectangle in the given example. After you select the objects, AutoCAD prompts you to place the action symbol, which can be placed at any convenient location.

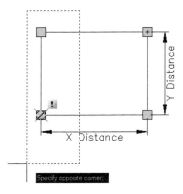

Figure 10–73 *Specifying the first point and opposite point of the stretch frame*

When the block reference has been selected for modifying in a drawing, the grips for the Stretch action are represented by squares when associated with an XY parameter (see Figure 10–74). The cursor in Figure 10–74 is shown being moved downward and to the left and, because the base point had been selected, stretching is allowed in both axes. If the Xcorner or Ycorner is selected, this Stretch action is restricted to only the X or Y direction respectively (see Figures 10–75 and 10–76).

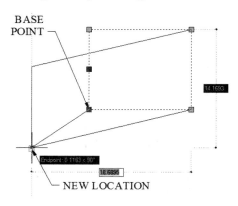

Figure 10–74 *The Stretch action applied to an XY parameter in the dynamic block reference using the base point*

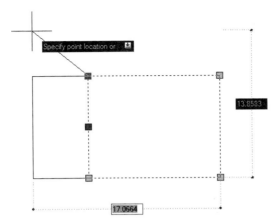

Figure 10-75 *The Stretch action applied to an XY parameter in the dynamic block reference using the Xcorner*

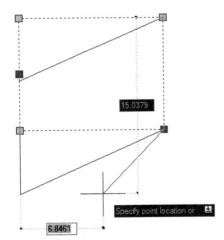

Figure 10-76 *The Stretch action applied to an XY parameter in the dynamic block reference using the Ycorner*

If, instead of a point or linear parameter, you choose a polar parameter for a Stretch action, AutoCAD prompts for the point to associate with the Stretch action.

Applying a Stretch action to a polar parameter is similar to applying it to a linear parameter. When the grip associated with the Stretch action is selected, it acts in the same manner as selecting the base point when associated with an XY parameter; stretching is allowed in both axes.

Polar Stretch Action

Polar Stretch adds a Stretch action to the block definition and can be associated only with a polar parameter. A **Polar Stretch** action stipulates that a specified selection set of objects in a dynamic block reference will be stretched when the action is initiated in a dynamic block reference in a manner similar to the stretch command. The Stretch action causes the objects to stretch when the associated parameter is edited by moving grips or by using the Properties palette.

Figure 10–77 shows a rectangle as a block in the Block Editor with a Polar Stretch action added to a polar parameter (labeled "Distance"). The stretch frame windowed the right end of the rectangle and the rectangle was selected as the object associated with the Polar Stretch action. When this block is saved and inserted, the grip at the upper right of the polar parameter becomes the grip that will be used to initiate the Polar Stretch action.

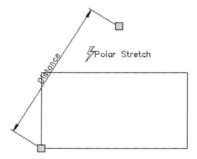

Figure 10–77 *The Polar Stretch action added to a polar parameter in the Block Editor*

Figure 10–78 shows how the rectangle will be changed by using the Polar Stretch action added to the polar parameter. The shape and size on the left are in response to selecting the grip and moving the cursor on a line that is an extension of the polar parameter. The changes on the right reflect moving the cursor at an angle, causing a rotation of the stretched rectangle.

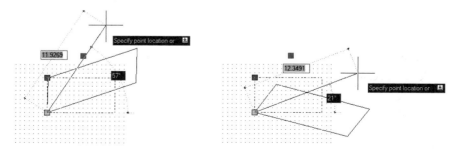

Figure 10–78 *The Polar Stretch action initiated in the dynamic block reference*

Rotate Action

Rotate adds a Rotate action to the block definition and can only be associated with a rotation parameter. A **Rotate** action stipulates that a specified selection set of objects in a dynamic block reference will rotate when the action is initiated in a dynamic block reference in a manner similar to the rotate command. The Rotate action causes the objects to rotate when the associated parameter is edited by moving the cursor or by using the Properties palette.

Figure 10–79 shows a block representing a door with jambs. The block definition has been created with the door open at 45°.

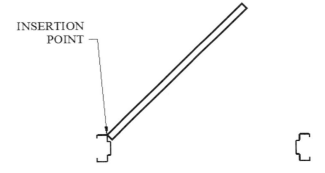

Figure 10–79 *Dynamic block reference showing door and jambs – door open at 45°*

In the Block Editor a rotate parameter is added to the definition (see Figure 10–80) with the base point located at the hinge point of the door (insertion point of the block) and the radius of the parameter set to the door width and the default rotation angle set to 45°. This places a round grip at the end of the door.

To add a Rotate action, invoke the Rotate action from the Action tab of the Block Authoring palette and select the rotate parameter (labeled Angle) defined on the screen. AutoCAD prompts:

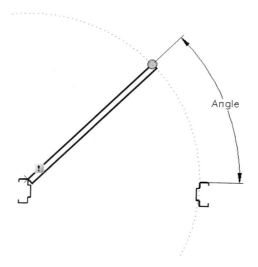

Figure 10–80 *A rotation parameter added with base point at hinge of door*

Select objects: *(select the rectangle that represents the door as shown in Figure 10–81)*

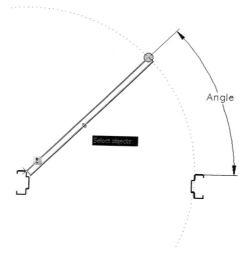

Figure 10–81 *A Rotation action is added and geometry for door selected to be associated with action*

Specify default rotation angle or ⬇: *(press ENTER to accept the default)*

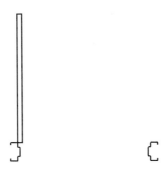

Figure 10–82 *Result of initiating Rotation action and changing property value from 45° to 90°*

Figure 10–82 shows the reference block that is modified with the Rotate action from the default angle of 45 degrees to 90 degrees.

Flip Action

Flip adds a Flip action to the block definition and can only be associated with a flip parameter. A Flip action stipulates that a specified selection set of objects in a dynamic block reference will be flipped about a base line when the action is initiated in a dynamic block reference in a manner similar to the rotate command. The Flip action causes the objects to flip when the associated parameter is edited by moving the cursor or by using the Properties palette.

Figure 10–83 shows a block with a rectangle, logo (dolphin), and text to which a flip parameter has been added in the Block Editor. Adding the flip parameter establishes a base line about which objects associated with a Flip action will be flipped. Note the label "Flip state" and the grip with the alert that an action has not been added.

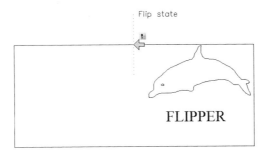

Figure 10–83 *A flip parameter added to a block in the Block Editor*

After the Flip action is added and associated with the logo and text, the block can be inserted. Figure 10–84 shows the result of the grip being selected for initiating the Flip action. Note that the logo graphics has been flipped about the base in both location and orientation but the text is flipped in location only. The text maintains its original orientation so it can be read. Having the name "FLIPPER" in an example of the Flip action is purely coincidental.

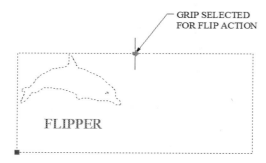

Figure 10–84 *Geometry location and orientation flipped – text location flipped and orientation unchanged*

Array Action

Array adds an Array action to the block definition and can be associated with a linear, polar, or XY parameter. An **Array** action stipulates that a specified selection set of objects in a dynamic block reference will be arrayed when the action is initiated in a dynamic block reference in a manner similar to the array command.

In Figure 10–85 a dynamic block has been created to represent a bolt with threads and a linear parameter has been added, labeled "Distance," and two grips are displayed. To add the Array action, invoke the Array action from the Action tab of the Block Authoring palette and select the linear parameter (labeled Distance) defined on the screen. AutoCAD prompts you to select the objects.

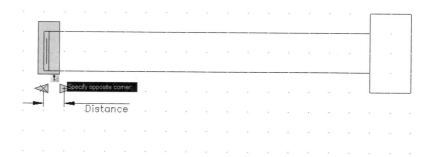

Figure 10–85 *Lines for bolt threads selected by window to be included in selection set of objects associated with an Array action*

Select the two lines near the end of the bolt, which represent the threads. Do not use crossing window to select the objects, because the lines for the outline of the bolt would be included in the selection set, which would cause them to be arrayed also.

AutoCAD prompts you to specify the distance between columns. Because this is a linear parameter, only the column value (and not the row) is required. In Figure 10–86 the distance is specified as the pitch of the threads, which is 0.125.

> Enter the distance between columns (|||): *(specify the distance between columns as **0.125**)*
> Specify action location: *(specify the action location in the block definition)*

The Array action symbol is placed in a convenient location.

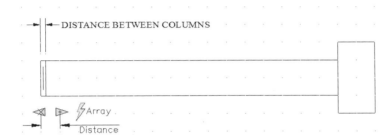

Figure 10–86 *Distance between columns being specified for an Array action*

The linear parameter is the distance that the array will cover when the block is inserted as a reference. When the block is inserted and selected for modifying, the two grips of the linear parameter are displayed (see Figure 10–87). The base grip of the block is also displayed, as usual.

Figure 10–87 *Dynamic block reference selected for modification – grips for Array action displayed*

When the grip for the Array action is selected and moved to the right, the two lines associated with the action are arrayed in accordance with the distance between columns specified when the Array action was added to the block definition. Figure 10–88 shows the grip being moved to increase the linear parameter distance to 4 units.

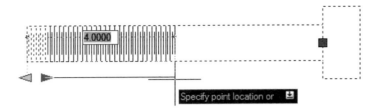

Figure 10–88 *Grip selected and moved to initiate Array action in block reference*

Figure 10–89 shows the block reference modified by using the Array action.

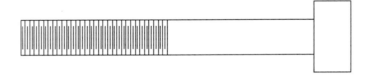

Figure 10–89 *Block Reference after modifying with Array action*

If you select an XY parameter instead of a linear or polar parameter, AutoCAD prompts you to select the objects. Then AutoCAD prompts:

> Enter the distance between rows or specify unit cell (---): *(specify the distance between rows of arrayed objects or enter two values separated by a comma for each of the two points for a unit cell for the arrayed objects)*
> Enter the distance between columns (|||): *(specify the distance between columns of arrayed object)*
> Specify action location: *(specify the action location in the block definition)*

The Array action added to an XY parameter operates in both axes (columns and rows) in the same manner that it acts in one axis for a linear or polar parameter.

Lookup Action

Lookup adds a Lookup action to the block definition and can only be associated with a lookup parameter. When you associate a Lookup action with a lookup parameter in a dynamic block definition, a lookup table is created. A lookup table can be used to assign custom properties and values to a dynamic block.

When you add a Lookup action to a lookup parameter, AutoCAD prompts:

> Specify action location: *(specify the action location in the block definition)*

There is no selection set associated with the Lookup action. AutoCAD displays the Property Lookup Table dialog box (see Figure 10–90).

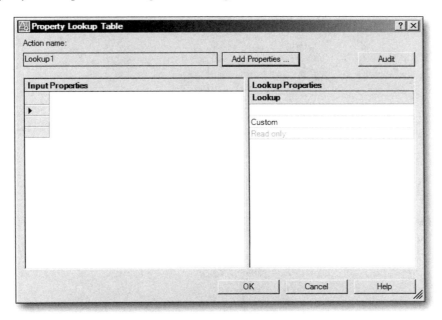

Figure 10–90 *Property Lookup Table dialog box*

The Property Lookup Table dialog box establishes a table for controlling parameter values (lookup properties) in the dynamic block definition. The lookup table also allows the values of lookup parameters to be controlled by the values of other parameters (input properties).

The lookup table assigns property values to the dynamic block reference based on how it is manipulated in a drawing. If Reverse Lookup from the **Custom** list box is selected for a lookup property, the block reference displays a lookup grip. When the lookup grip is clicked in a drawing, a list of lookup properties is displayed. Selecting an option from that list can be used to change the geometry of the dynamic block reference.

Figure 10–91 shows a pictorial view on the left of a piping make-up that includes a gate valve, two 90° ells, and two raised-face-weld-neck flanges. The middle view shows the two-dimensional drawing. The raised-face-weld-neck flange (RFWN) flange can be made into a block as shown on the right for insertion into the drawing when required.

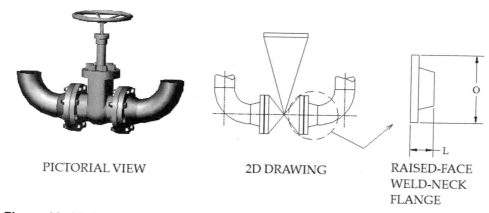

PICTORIAL VIEW   2D DRAWING   RAISED-FACE WELD-NECK FLANGE

Figure 10–91 *Piping make-up that includes two raised-face-weld-neck flanges*

The illustration of the block for the flange (on the right in Figure 10–91) shows two of the dimensions of the flanges that will vary depending on the flange rating (used to determine pressure rating) of the piping specification that has been called out for the piping system. The flange length "L" and the outside diameter "O" are different for each flange rating (see the following table).

PIPE SIZE	FLANGE RATING	FLANGE LENGTH L	FLANGE DIA O
4"	150#	3.000	9.000
4"	300#	3.375	10.000
4"	400#	3.500	10.000
4"	600#	4.000	10.750

Without using the dynamic block feature, it would be necessary to have a separate block drawn for each flange rating. This could require dozens of blocks for various fittings in a large piping drawing.

Figure 10–92 is the block for the flange in the Block Editor with three linear parameters added to the definition. The parameters include FLANGE LENGTH, HALF FLANGE O.D. 1, and HALF FLANGE O.D. 2. The HALF FLANGE O.D. 1 and HALF FLANGE O.D. 2 are flange outside diameters that are to be stretched symmetrically about the center line of the flange. Stretch actions will be added to these three linear parameters.

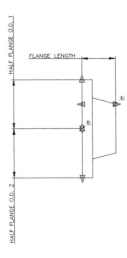

Figure 10–92 *Block for raised-face-weld-neck flange in Block Editor with three linear parameters added*

Figure 10–93 shows adding a Stretch action to the FLANGE LENGTH parameter. The rectangle represented by the dotted line is the stretch frame. All of the objects in the block will be associated with the Stretch action.

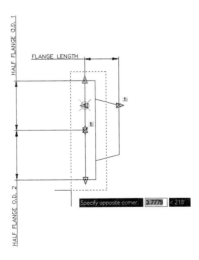

Figure 10–93 *Stretch frame defined for Stretch action added to FLANGE LENGTH linear parameter*

Figure 10–94 shows adding a Stretch action to the HALF FLANGE O.D. 1 parameter. The rectangle represented by the dotted line is the stretch frame. All of the objects in the block will be associated with the Stretch action. Similarly a Stretch action is added to HALF FLANGE O.D. 2.

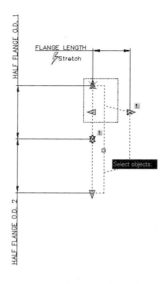

Figure 10–94 *Stretch frame defined and objects selected for Stretch action added to HALF FLANGE O.D. 1 linear parameter*

Figure 10–95 shows the three linear parameters with three Stretch actions added in the Block Editor.

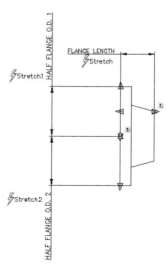

Figure 10–95 *Three linear parameters with Stretch actions added in Block Editor*

Figure 10–96 shows a Lookup action added to a lookup parameter. The location is not critical to any geometry. However, its location determines where it will appear when a reference of the block is selected for modifying.

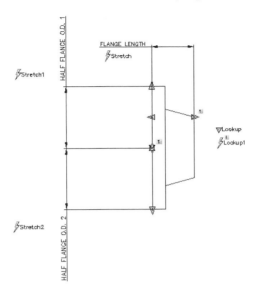

Figure 10–96 *Lookup action added to lookup parameter in Block Editor*

Figure 10–97 shows the Property Lookup Table dialog box being displayed when the Lookup action is added to the lookup parameter.

 Note: If you have closed the Property Lookup Table dialog box and need to reopen it, double-click on the Lookup action.

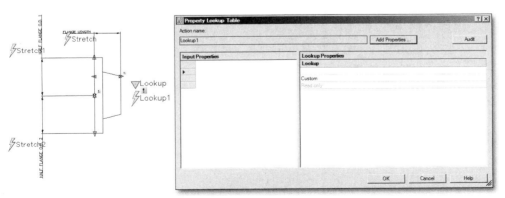

Figure 10–97 *Property Lookup Table dialog box*

Action Name displays the name of the Lookup action associated with the table.

To add properties to the Lookup Table, choose **Add Properties**. AutoCAD displays the Add Parameter Properties dialog box. Choose **Add input properties** and AutoCAD lists the available input properties (see Figure 10–98).

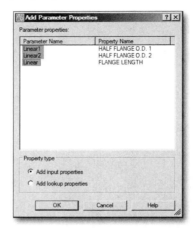

Figure 10–98 *Selecting parameters from the Property Lookup Table dialog box*

Select all three parameters listed, choose **OK**, and AutoCAD returns to the Property Lookup Table dialog box with the selected parameters listed under **Input Properties**. The values for the properties of each flange rating are entered under their respective headings in the Input Properties list and the corresponding flange rating is entered under the Lookup Properties list. Figure 10–99 shows Property Lookup Table with all the values entered in the Input Properties values and corresponding flange rating in the Lookup Properties. You can choose **Audit** and AutoCAD checks that each record is unique. If not, a warning is displayed.

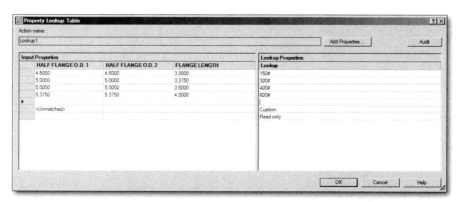

Figure 10–99 *Entering input properties for the parameters and corresponding lookup properties for each flange rating*

In the text box at the bottom of the Input Properties list, select **Allow reverse lookup** (see Figure 10–100) in the **Custom** list box. Selection of **Allow Reverse Lookup** enables the lookup property for a block reference to be set from a drop-down list that is displayed when the lookup grip is clicked in a drawing. Selecting an option from this list changes the block reference to match the corresponding input property values in the table.

Choose **OK** to accept the changes and AutoCAD returns to the Block Editor.

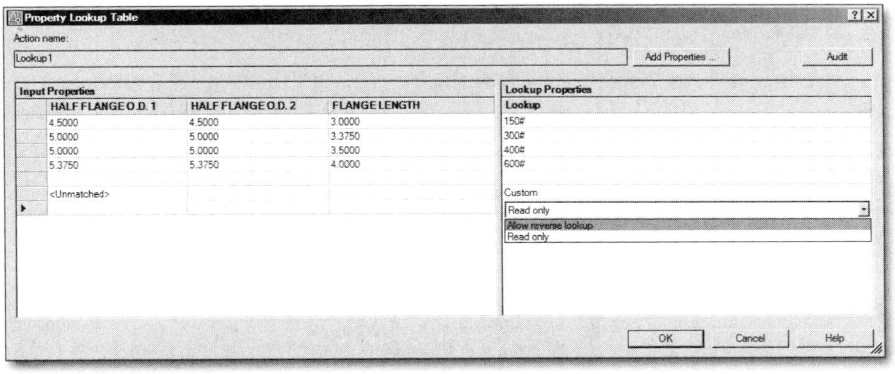

Figure 10–100 *Property Lookup Table with Allow reverse lookup selected in the* **Custom** *list box.*

Close the Block Editor and insert two references of the dynamic block (see Figure 10–101). The dimensions are added to the figure for illustrative purposes. They will not be displayed unless added by using the DIMENSION command. The values for the flange length and flange O.D. have been changed in the block reference on the right in Figure 10–101 by selecting the lookup grip and then choosing *600#* in the list of Lookup Properties in the shortcut menu.

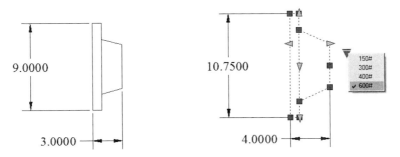

Figure 10–101 *Block reference with flange length and flange o.d. changed with Lookup action*

ADDING VISIBILITY STATES TO DYNAMIC BLOCKS

By adding a visibility parameter in your dynamic block definition, you can create and name visibility states and then specify which geometric objects are invisible for a given visibility state. One block can have multiple visibility states.

The visibility parameter includes a lookup grip. A block reference that contains visibility states will always display the grip. When you select the grip in the block reference, a shortcut menu is displayed listing all the visibility states in the block reference. Selecting one of the states from the list causes the geometry that is visible for that state to be displayed.

An instrument panel (see Figure 10–102) is used to demonstrate the visibility feature in a dynamic block. In this case a single dynamic block definition is created that represents two sets of geometry to suit two separate requirements. One visibility state will display the English version of a speedometer for reading miles per hour. The second visibility state will display the German version of a speedometer for reading kilometers per hour (see Figure 10–102).

Figure 10–102 *Two block references from the same dynamic block definition applying the visibility feature*

The fundamental geometry (panel outline, dial outline, and indicating needle) is created as a block. It is opened in the Block Editor and a visibility parameter is added to the definition. Double-click on the visibility parameter and AutoCAD displays the Visibility States dialog box (see Figure 10–103).

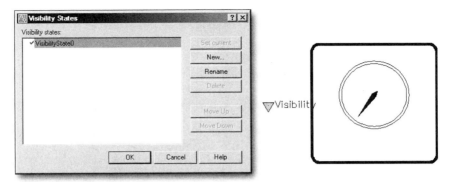

Figure 10–103 *Visibility States dialog box displayed in the Block Editor*

When the dialog box first appears, it contains one visibility state named VisibilityState0. Rename to ENGLISH. Create a new visibility state by choosing **New** and AutoCAD opens the New Visibility State dialog box (see Figure 10–104). Rename it as GERMAN and set it as current. Choose **OK** to close the dialog box.

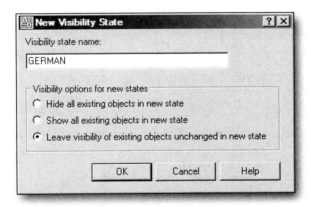

Figure 10–104 *Visibility States dialog box displayed in the Block Editor*

Draw numeral 0 through 110 and the text objects "MPH" and "SPEEDOMETER" (Figure 10–105).

To apply the visibility states, it is necessary to think in reverse. That is, a **visibility** state is used primarily to make selected objects **invisible** when that state is selected during the modifying of the block reference.

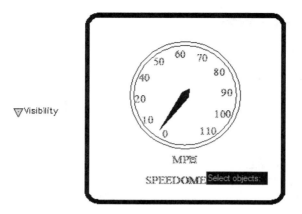

Figure 10–105 *Objects not made invisible in the English Visibility State*

Set English as the current visibility state by selecting English from the list box (see Figure 10–106).

Figure 10–106 *Selecting English from the list box (Block Editor toolbar)*

Numeral 0 through 110 and the text objects "MPH" and "SPEEDOMETER" will disappear. Draw numeral 0 through 140 at the same location and the text objects "KPH" and "GESCHWINDIGKEITSMESSER" (see Figure 10–107).

Figure 10–107 *Objects not made invisible in the German Visibility State*

By switching between English and German from the list box, you can see the corresponding objects disappear. Save the changes and close the Block Editor. When the block reference is selected to change visibility, the grip for the visibility parameter is displayed. To change the geometry and text, select the grip and then select GERMAN from the shortcut menu (see Figure 10–108). Numeral 0 through 140 and the text objects "KPH" and "GESCHWINDIGKEITSMESSER" will appear. To change it back, select the grip and then select ENGLISH from the shortcut menu. Numeral 0 through 110 and the text objects "MPH" and "SPEEDOMETER" will appear. From the shortcut menu of the selected reference block you can control the visibility of objects.

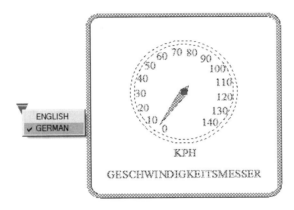

Figure 10–108 *Selecting the Block reference for modifying, selecting the visibility grip, and then selecting* GERMAN *from the shortcut menu*

ADVANCED DYNAMIC BLOCK UTILITIES AND FEATURES

Parameters, actions, and grips on blocks are the fundamental elements that make the blocks dynamic. This section covers the more refined and advanced features available when working with dynamic blocks.

Parameter Sets

The Parameter Sets tab of the Authoring palette contains 20 parameter sets (see Figure 10–109). Parameter sets are combinations of parameters and actions with a preset number of grips that can be quickly applied to a dynamic block and associated with selected geometry without having to go through the process of adding them separately. For example, the Linear Move parameter set adds a linear parameter with one grip and an associated Move action. The Linear Move Pair parameter set adds a linear parameter with two grips and a Move action associated with each grip. Once the configuration is learned, using a parameter set can save time in adding the desired dynamics to a block.

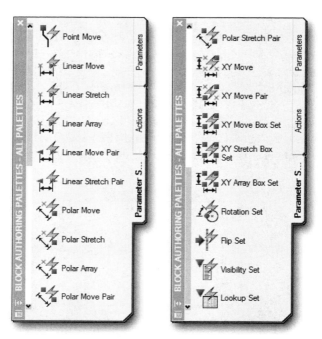

Figure 10–109 *Parameter Sets Tab of the Authoring palette (view on right after scrolling)*

Property Values

The values of a selected object's properties are displayed in the Properties palette. The Block Editor lets you can specify values for properties for a parameter in a dynamic block definition, which can be displayed under Custom properties for the dynamic block reference when it has been inserted into a drawing.

You can specify parameter labels while in the Block Editor so that when the dynamic block reference is selected in a drawing, these properties are listed by their label under Custom in the Properties palette. Parameter labels within the block should be unique.

Other parameter properties may be listed under Custom in the Properties palette when you select the dynamic block reference in a drawing depending on the parameters used. For example, a polar parameter has an angle property that displays in the Properties palette. Depending on how the dynamic block is defined, properties such as size, angle, and position might be displayed.

Geometric properties (such as color, linetype, and lineweight) can also be specified by using the Properties palette. They are listed in the Properties palette under Geometry when you select a parameter in the Block Editor.

How the block reference will function in a drawing is defined by parameter properties such as Value Set properties and Chain actions, whether or not a block can be exploded, and whether the block can be non-uniformly scaled can be specified in the Block Editor.

You can specify whether or not custom properties are displayed for the block reference when it is selected in a drawing.

Properties can be extracted using the Attribute Extraction wizard.

Changing the Insertion Point

The BCYCLEORDER command lets you specify certain grips in the dynamic block definition, which can be used as the insertion point when inserting the block. Figure 10–110 shows a dynamic block definition in the Block Editor with several parameter/action grips.

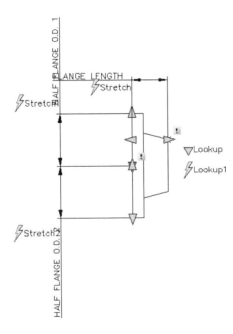

Figure 10–110 *Dynamic block in Block Editor with several action/parameter grips*

While in the Block Editor, type **bcycleorder** and AutoCAD displays the Insertion Cycling Order dialog box (see Figure 10–111). The **Grip cycling list** displays the grips, their parameter name, Type, and Location. If a grip has a check under the Cycling column, it will be available as an insertion point when the block is inserted.

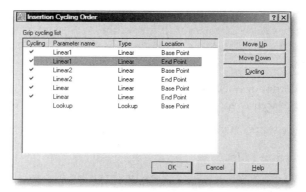

Figure 10–111 *Insertion Cycling Order dialog box showing action/parameter grips*

Choosing **Move Up** or **Move Down** causes the selected grip to move up or down respectively in the list.

Choosing **Cycling** causes the Cycling column for the selected grip to toggle between being checked or unchecked.

Choose **OK** to accept the selections and close the dialog box.

Figure 10–112 shows the dynamic block on the left as it is being inserted with the default insertion point. To cycle between selectable points, press CTRL. The illustration on the right shows the insertion point after cycling through the selectable points (pressing CTRL until the desired point is selected).

Figure 10–112 *Cycling from the default action/parameter grip to the desired one for the insertion point*

Open the Exercise Manual PDF file for Chapter 10 on the accompanying CD for project- and discipline-specific exercises.

If you have the accompanying Exercise Manual, refer to Chapter 10 for project-and discipline-specific exercises.

REVIEW QUESTIONS

1. The maximum number of characters for a block name is:
   a. 8
   b. 16
   c. 23
   d. 31
   e. 255

2. A block is:
   a. a rectangular-shaped figure available for insertion into a drawing
   b. a single element found in a block formation of a building drawn with AutoCAD
   c. one or more objects stored as a single object for later retrieval and insertion
   d. none of the above

3. A block reference cannot be exploded if it:
   a. consists of other blocks (nested)
   b. has a negative scale factor
   c. has been moved
   d. has different *X* and *Y* scale factors
   e. none of the above

4. All of the following can be exploded, except:
   a. block references
   b. associative dimensions
   c. polylines
   d. Block references inserted with MINSERT command
   e. none of the above

5. To return a block reference back to its original objects, use:
   a. EXPLODE
   b. BREAK
   c. CHANGE
   d. UNDO
   e. STRETCH

6. To identify a new insertion point for a drawing file which will be inserted into another drawing, invoke the:

    a. BASE command
    d. WBLOCK command
    b. INSERT command
    e. DEFINE command
    c. BLOCK command

7. MINSERT places multiple copies of an existing block similar to the command:

    a. ARRAY
    d. COPY
    b. MOVE
    e. MIRROR
    c. INSERT

8. If one drawing is to be inserted into another drawing and editing operations are to be performed on the inserted drawing, you must first:

    a. use the PEDIT command
    b. EXPLODE the inserted drawing
    c. UNDO the inserted drawing
    d. nothing, it can be edited directly
    e. none of the above

9. The BASE command:

    a. can be used to move a block reference
    b. is a subcommand of PEDIT
    c. will accept 3D coordinates
    d. allows one to move a dimension baseline
    e. none of the above

10. Attributes are associated with:

    a. objects
    d. layers
    b. block references
    e. shapes
    c. text

11. To merge two drawings, use:

    a. INSERT
    d. BLOCK
    b. MERGE
    e. IGESIN
    c. BIND

12. The DIVIDE command causes AutoCAD to:
    a. divide an object into equal length segments
    b. divide an object into two equal parts
    c. break an object into two objects
    d. all of the above

13. One cannot explode:
    a. polylines containing arcs
    b. blocks containing polylines
    c. dimensions incorporating leaders
    d. block references inserted with different X, Y, and Z scale factors
    e. none of the above

14. The DIVIDE command will:
    a. place points along a line, arc, polyline, or circle
    b. accept 1.5 as segment input
    c. place markers on the selected object and separate it into different segments
    d. divide any object into the equal number of segments

15. The MEASURE command causes AutoCAD to divide an object
    a. into specified length segments
    b. into equal length segments
    c. into two equal parts
    d. all of the above

16. A command used to edit attributes is:
    a. DDATTE
    b. EDIT
    c. EDITATT
    d. ATTFILE
    e. none of the above

17. Attributes are defined as the:
    a. database information displayed as a result of entering the LIST command
    b. X and Y values which can be entered when inserting a block reference
    c. coordinate information of each vertex found along a SPLINE object
    d. none of the above

18. It is possible to force invisible attributes to display on a drawing.

    a. True

    b. False

19. The first step in making a block dynamic is to add action.

    a. True

    b. False

20. The Alignment, Visibility, and Base Point parameters do not have actions associated with them.

    a. True

    b. False

21. The point parameter can be associated with:

    a. Move action   c. Scale action

    b. Stretch action   d. Rotation action

    e. a and b

22. The Rotate parameter associate only with Rotate action.

    a. True

    b. False

23. A dynamic block cannot be associate more than one action.

    a. True

    b. False

24. Move action to the block definition can be associated with _____

    a. Point parameter

    b. Linear parameter

    c. Polar parameter

    d. XY parameter

    e. all of the above

25. Parameter sets are combinations of parameters and actions with a preset number of grips.

    a. True

    b. False

CHAPTER 11

External References and Images

INTRODUCTION

One of the most powerful time-saving features of AutoCAD is the ability to have one drawing (referred to as an external reference) become part of a second drawing while maintaining the integrity and independence of the first one. And if the external reference is changed, those changes will be reflected in the drawing in which it is referenced. This feature is provided by the XREF command, short for external reference.

After completing this chapter, you will be able to do the following:

- Attach and detach reference files
- Change the path for reference files
- Load and unload reference files
- Decide whether to attach or overlay an external reference file
- Clip external reference files
- Control dependent symbols
- Edit external references
- Manage external references
- Use the BIND command to add dependent symbols to the current drawing
- Attach and detach image files

EXTERNAL REFERENCES

Prior to Release 11, existing AutoCAD drawings could be combined in only one way, by using the INSERT command to insert one drawing into another. When one drawing is inserted into another, it is actually a duplicate of the inserted drawing that becomes a part of the drawing into which it is inserted. The data from the inserted drawing is added to the data of the current drawing. Once the duplicate of a drawing is inserted, no link or association remains between the original drawing from which the inserted duplicate came and the drawing it has been inserted into.

The INSERT command and the XREF command now give users a choice of two methods for combining existing drawing files. The external reference (XREF) feature does not obsolete the INSERT feature; users can decide which method is more appropriate for the current application.

When a drawing is externally referenced (instead of inserted as a block), the user can view and object snap to objects in the external reference from the current drawing, but each drawing's data is still stored and maintained in a separate drawing file. The only information in the reference drawing that becomes a permanent part of the current drawing is the name of the reference drawing and its folder path. If necessary, externally referenced files can be scaled, moved, copied, mirrored, or rotated by using the AutoCAD modification commands. You can control the visibility, color, and linetype of the layers belonging to an external drawing file. This lets you control which portions of the external drawing file are displayed, and how. No matter how complex an external reference drawing may be, it is treated as a single object by AutoCAD. If you invoke the MOVE command and select a line in the external reference, for example, the entire external reference moves, not just the line you selected. You cannot explode the externally referenced drawing. A manipulation performed on an external reference will not affect the original drawing file, because an external reference is only an image, however scaled or rotated.

Borders are an excellent example of drawing files that are useful as external reference files. The objects that make up a border will use considerable space in a file, and commonly amount to around 20,000 bytes. If a border is drawn in each drawing file, this would waste a large amount of disk space when you multiply 20,000 bytes by 100 drawing files. If external reference files are used correctly, they can save 2 MB of disk space in this case.

Accuracy and efficient use of drawing time are other important design benefits that are enhanced through external reference files. When an addition or change is made to a drawing file that is being used as an external reference file, all drawings that use the file as an external reference will reflect the modifications. For example, if you alter the title block of a border, all the drawing files that use that border as an external reference file will automatically display the title block revisions. (Can you imagine accessing 100 drawing files to change one small detail?) External reference files will save time and ensure the drawing accuracy required to produce a professional product. Figure 11–1 shows a drawing that externally references four other drawings to illustrate the doorbell detail.

There is a limit of 32,000 external references you can add to a drawing. In practice, this represents an unlimited number. If necessary, you can nest them so that loading one external reference automatically causes another external reference to be loaded. When you attach a drawing file as an external reference file, it is permanently attached until it is detached or bound to the current drawing. When you load the drawing with external references, AutoCAD automatically reloads each external reference drawing file; thus, each external drawing file reflects the latest state of the referenced drawing file.

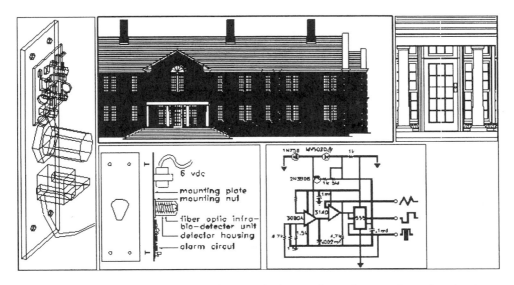

Figure 11–1 *This drawing appears to be a single drawing but references four other drawings*

The XREF command, when combined with the networking capability of AutoCAD, gives the project manager powerful tools for coping with the problems of file management. The project manager can instantaneously sees the work of the departments and designers working on aspects of the contract. If necessary, you can overlay a drawing where appropriate, track the progress, and maintain document integrity. At the same time, departments need not lose control over individual designs and details.

If you need to make changes to an attached external reference file while you are in the host drawing, you can do so by the REFEDIT command. AutoCAD also allows you to open an attached external reference in a separate window. You do not need to browse to and open the xref file. With the XOPEN command, the external reference opens immediately in a new window. You can make necessary changes and save the changes. Immediately the changes will be reflected in the host drawing.

In AutoCAD you can control the display of the external reference file by means of clipping, so you can display only a specific section of the reference file.

EXTERNAL REFERENCES AND DEPENDENT SYMBOLS

The symbols that are carried into a drawing by an external reference are called dependent symbols, because they depend on the external file, not on the current drawing, for their characteristics. The symbols have arbitrary names and include blocks, layers, linetypes, text styles, and dimension styles.

When you attach an external reference drawing, AutoCAD automatically renames the xref's dependent symbols. AutoCAD forms a temporary name for each symbol by combining its original name with the name of the xref itself. The two names are separated by the vertical bar (|) character. Renaming the symbols prevents the xref's objects from taking on the characteristics of existing symbols in the drawing.

For example, you created a drawing called PLAN1 with layers 0, First-fl, Dim, and Text, in addition to blocks Arrow and Monument. If you attach the PLAN1 drawing as an external reference file, the layer First-fl will be renamed as PLAN1|First-fl, Dim as PLAN1|Dim, and Text as PLAN1|Text, as shown in Figure 11–2. Blocks Arrow and Monument will be renamed as PLAN1|Arrow and PLAN1|Monument. The only exceptions to renaming are unambiguous defaults like layer 0 and linetype continuous. The information on layer 0 from the reference file will be placed on the active layer of the current drawing when the drawing is attached as an external reference of the current drawing. It takes on the characteristics of the current drawing.

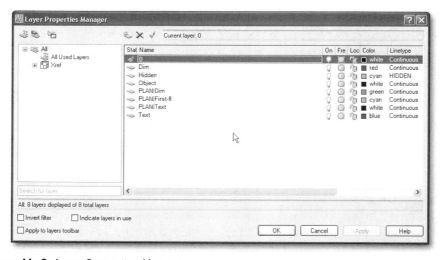

Figure 11–2 *Layer Properties Manager*

This prefixing is carried to nested xrefs. For example, if the external file PLAN1 included an xref named "Title" that has a layer Legend, it would get the symbol name PLAN1|Title|Legend, if PLAN1 were attached to another drawing.

This automatic renaming of an xref's dependent symbols has two benefits:

- It allows you to see at a glance which named objects belong to which external reference file.
- It allows dependent symbols to have the same base name in both the current drawing and an external reference, and coexist without any conflict.

The AutoCAD commands and dialog boxes for manipulating named objects do not let you select an xref's dependent symbols. Usually, dialog boxes display these entries in lighter text.

For example, you cannot insert a block that belongs to an external reference drawing in your current drawing, nor can you make a dependent layer the current layer and begin creating new objects. These types of tasks are made easier by using AutoCAD's DesignCenter, explained in Chapter 12.

You can control the visibility of the layers (ON/OFF, Freeze/Thaw) of an external reference drawing and, if necessary, you can change the color and linetype. When the VISRETAIN system variable is set to 0 (default), any changes you make to these settings apply only to the current drawing session. They are discarded when you end the drawing. If VISRETAIN is set to 1, then the current drawing visibility, color, and linetype for xref dependent layers take precedence. They are saved with the drawing and are preserved during xref reload operations.

There may be times when you want to make your xref data a permanent part of your current drawing. To make an xref drawing a permanent part of the current drawing, use the BIND option of the XREF command. With the BIND option, all layers and other symbols, including the data, become part of the current drawing. This is similar to inserting a drawing via the INSERT command.

If necessary, you can make dependent symbols such as layers, linetypes, text styles, and dim styles part of the current drawing by using the XBIND command instead of binding the whole drawing. This allows you to work with the symbol just as if you had defined it in the current drawing.

ATTACHING AND MANIPULATING XREFS WITH THE XREF MANAGER

The Xref Manager lists the external references attached to the current (or host) drawing. In addition, you can also attach, detach, overlay, bind, reload, unload, rename, and modify paths to external references (xrefs).

Choose Xref Manager from the Reference toolbar (see Figure 11–3) and AutoCAD displays the Xref Manager dialog box (see Figure 11–4).

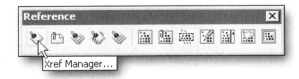

Figure 11–3 *Invoking the* XREF *command from the Reference toolbar*

AutoCAD provides two options for listing attached external reference drawings: the List View and the Tree View. By default, the List View option (see Figure 11–4) displays a list of the attached reference files and their associated data. To sort a column alphabetically, select the column heading. A second click sorts it in reverse order. To resize a column's width, select the separator between columns and drag the pointing device to the right or left.

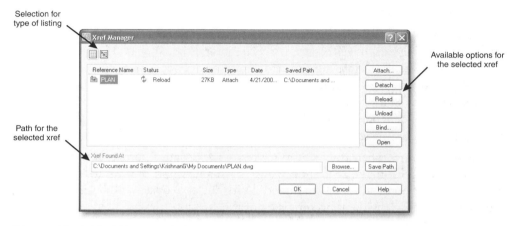

Figure 11–4 *Xref Manager dialog box*

The external reference file's **Reference Name** does not have to be the same as its original file name. To rename the external file, double-click the name or select it and press F2. AutoCAD allows you to change the name. The new name can contain up to 255 characters, including embedded spaces and punctuation.

The **Status** column displays the state of the external reference file, which can be Loaded, Unloaded, Unreferenced, Unresolved, Orphaned, Reload, or Not found. A detailed discussion of these states is provided later in the chapter.

The **Size** column shows the file size of the corresponding external reference drawing. The size is not displayed if the external reference is unloaded, not found, or unresolved.

The **Type** column indicates whether the external reference is an attachment or an overlay.

The **Date** column displays the last date the associated drawing was modified. The date is not displayed if the external reference is unloaded, not found, or unresolved.

The **Saved Path** column shows the saved path of the associated external reference file.

Selecting any field highlights the external file's reference name.

You can also display the information as a Tree View. To do so, choose **Tree View** (see Figure 11–4) at the top left of the dialog box, or press F4. To switch back to the List View, choose **List View** or press F3.

In a Tree View listing, AutoCAD displays a hierarchical representation of the external references in alphabetical order (see Figure 11–5). Tree View shows the level of nesting relationship of the attached external references, whether they are attached or overlaid, whether they are loaded, unloaded, marked for reload or unload, or not found, unresolved, or unreferenced.

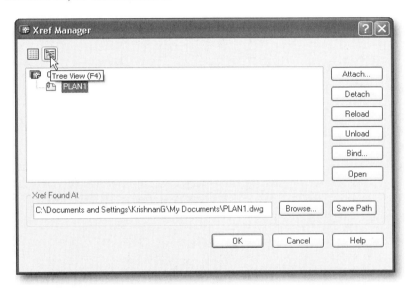

Figure 11–5 *Xref Manager dialog box with Tree View listing*

ATTACHING EXTERNAL REFERENCE DRAWINGS

Choose Attach to attach a drawing as an external reference. Use it when you want to attach a new external reference file or a copy of the external reference file already attached to the current drawing. If you attach a drawing that itself contains an attached external reference, the attached external reference appears in the current drawing. If another person is currently editing the external reference drawing, the drawing attached is based on the most recently saved version.

To attach a drawing as an external reference, choose **Attach** in the Xref Manager and AutoCAD displays the Select Reference File dialog box.

Select the drawing file from the appropriate folder to attach to the current drawing. AutoCAD displays the External Reference dialog box (see Figure 11–6).

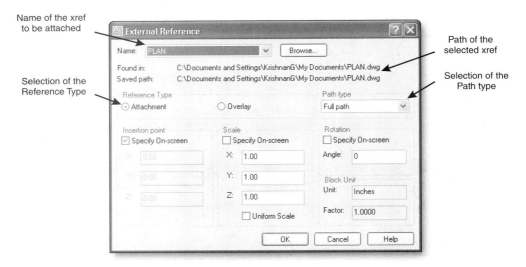

Figure 11–6 *External Reference dialog box*

Once an external reference drawing is attached to the current drawing, the external reference drawing name is added to the list box located next to the **Browse** button. Choose **Browse** to display the Select a Reference dialog box, in which you can select new xrefs for the current drawing.

When an attached external reference file name is selected from the list box, AutoCAD displays its path where the xref was found. If no path was saved for the xref or if the xref is no longer located at the specified path, AutoCAD searches for the xref in the following order:

1. Current folder of the host drawing
2. Project search paths defined on the **Files** tab in the Options dialog box and in the PROJECTNAME system variable
3. Support search paths defined on the **Files** tab in the Options dialog box
4. Start-in folder specified in the Windows® application shortcut

AutoCAD displays the saved path, if any, that is used to locate the xref. This path can be an absolute (fully specified) path, a relative (partially specified) path, or no path.

In the **Reference Type** section of the dialog box, select one of the two available options: Attachment or Overlay.

> Selecting **Attachment** causes the external reference to be included in the drawing when the drawing itself is attached as an external reference to another drawing.

Selecting **Overlay** causes the external reference to not be included in a drawing when the drawing itself is attached as an external reference or overlaid external reference to another drawing.

For example, PLAN-A drawing is attached as an overlaid external reference to PLAN-B. Then PLAN-B is attached as an external reference to PLAN-C. PLAN-A is not seen in PLAN-C because it is overlaid in PLAN-B, not attached as an external reference. But, if PLAN-A is attached as an external reference in PLAN-B, and in turn PLAN-B is attached in PLAN-C, both PLAN-A and PLAN-B will be seen in PLAN-C.

So, the only behavioral difference between overlays and attachments is how nested references are handled. Overlaid external references are designed for data sharing. If necessary, you can change the status from Attachment to Overlay, or vice versa, by double-clicking the field in the **Type** column in the Xref Manager dialog box or by selecting the appropriate radio button in the **Reference Type** section of the External Reference dialog box.

The **Path type** list box specifies whether the saved path to the xref should be set to No path, Full path, or Relative path. When you specify the path type as No path, AutoCAD first looks for the xref in the folder of the host drawing. This option is useful when the xref files are in the same folder as the host drawing. Instead, if you specify the path type as Full path, AutoCAD saves the xref's precise location to the host drawing. This option is the most precise but the least flexible. If you move a project folder, AutoCAD cannot resolve any xrefs that are attached with a full path. And if you specify the path type as Relative path, AutoCAD saves the xref's location relative to the host drawing. If you move a project folder, AutoCAD can resolve xref's attached with a relative path, as long as the xref's location relative to the host drawing has not changed. You must save the current drawing before you can set the path type to Relative path.

In the **Insertion point** section of the External Reference dialog box, you can specify the insertion point of the external reference.

In the **Scale** section of the External Reference dialog box, you can specify the *X* scale factor, *Y* scale factor, and *Z* scale factor of the external reference. Selecting **Uniform Scale** causes the *Y* and *Z* scale factors to be the same as the specified *X* scale factor.

In the **Rotation** section of the External Reference dialog box, you can specify the rotation angle of the external reference.

The **Insertion point**, **Scale**, and **Rotation** features are similar to those in the insertion of a block, explained in Chapter 10.

The **Block Unit** section displays the units (inches, millimeters, etc.) that were used when the selected reference drawing was created and the unit scale factor, calculated based on the reference drawing units value and the current drawing units.

Choose **OK** to attach the selected external reference drawing to the current drawing.

 Note: Once a drawing is attached as a reference file, the Manage Xref's icon is displayed in the status bar that allows you to open the Reference Manager.

DETACHING EXTERNAL REFERENCE DRAWINGS

Choosing **Detach** in the Xref Manager dialog box allows you to detach one or more external reference drawings from the current drawing. Only the external reference drawings attached or overlaid directly to the current drawing can be detached. You cannot detach an external reference drawing referenced by another external reference drawing. If the external reference is currently being displayed as part of the current drawing, it disappears when you detach it.

To detach an external reference drawing from the current drawing, first select the drawing Reference name from the displayed list in the Xref Manager dialog box, and then choose **Detach**. AutoCAD detaches the selected external reference drawing(s) from the current drawing.

RELOADING EXTERNAL REFERENCE DRAWINGS

Choosing **Reload** allows you to update one or more external reference drawings attached to the current drawing. When you open a drawing, it automatically reloads any external references attached. The **Reload** option will reread the external drawing from the external drawing file in case it has been changed during the current AutoCAD session.

To reload external reference drawing(s), first select the drawing name(s) from the displayed list in the Xref Manager dialog box, and then choose **Reload**. AutoCAD reloads the selected external reference drawing(s).

UNLOADING EXTERNAL REFERENCE DRAWINGS

Choosing **Unload** allows you to unload one or more external reference drawings from the current drawing. Unlike the **Detach** option, the **Unload** option merely suppresses the display and regeneration of the external reference definition, to help current session editing and improve performance. This option can also be useful when a series of external reference drawings needs to be viewed during a project on an as-needed basis. Rather than have the referenced files displayed at all times, you can reload the drawing when you require the information.

To unload external reference drawing(s), first select the drawing name(s) from the displayed list in the Xref Manager dialog box, and then choose **Unload**. AutoCAD turns off the display of the selected external reference drawing(s).

The results of Unload and Reload take effect when you close the dialog box.

BINDING EXTERNAL REFERENCE DRAWINGS

Choosing **Bind** allows you to make your external reference drawing data a permanent part of the current drawing. To bind the external reference drawing(s), first select the drawing name(s) from the displayed list in the Xref Manager dialog box, and then choose **Bind**. AutoCAD displays the Bind Xrefs dialog box (see Figure 11–7). Select one of the two available bind types: **Bind** or **Insert**.

Figure 11–7 *Bind Xrefs dialog box*

Choose **Bind** and the external reference drawing becomes an ordinary block in your current drawing. It also adds the dependent symbols to your drawing, letting you use them as you would any other named objects. In the process of binding, AutoCAD renames the dependent symbols. The vertical bar symbol (|) is replaced with three new characters: a $, a number, and another $. The number is assigned by AutoCAD to ensure that the named object will have a unique name.

For example, if you bind an external reference drawing named PLAN1 that has a dependent layer PLAN1|FIRST-FL, AutoCAD will try to rename the layer to PLAN1$0$FIRST-FL. If there is already a layer by that name in the current drawing, then AutoCAD tries to rename the layer to PLAN1$1$FIRST-FL, incrementing the number until there is no duplicate.

If you do not want to bind the entire external reference drawing, but only specific dependent symbols, such as a layer, linetype, block, dimension style, or text style, then you can use the XBIND command, explained later in this chapter, in the section on "Adding Dependent Symbols to the Current Drawing."

Choose **Insert** and the external reference drawing is inserted in the current drawing just like inserting a drawing with the INSERT command. AutoCAD adds the dependent symbols to the current drawing by stripping off the external reference drawing name.

For example, if you insert an external reference drawing named PLAN1 that has a dependent layer PLAN1|FIRST-FL, AutoCAD will rename the layer to FIRST-FL.

If there is already a layer by that name in the current drawing, then the layer FIRST-FL would assume the properties of the layer in the current drawing.

OPENING THE EXTERNAL REFERENCE

If you need to make changes to an attached external reference file while you are in the host drawing, choose **Open** and AutoCAD opens the selected xref for editing in a new window. The new window is displayed after the Xref Manager is closed. You can also open an external reference for editing by invoking the XOPEN command.

CHANGING THE PATH

To change to a different path or file name for the currently selected external reference file, choose **Browse** in the Xref Manager dialog box. AutoCAD displays the Select New Path dialog box, in which you can specify a different path or file name.

SAVING THE PATH

To save the path as it appears in the **Saved Path** field of the currently selected external reference file (and displayed in the **Xref Found At** text box), select **Save Path**. AutoCAD saves the path of the currently selected external reference file.

After making the necessary changes in the Xref Manager dialog box, choose **OK** to accept the changes and close the dialog box.

ADDING DEPENDENT SYMBOLS TO THE CURRENT DRAWING

The XBIND command lets you permanently add a selected subset of external reference-dependent symbols to your current drawing. The dependent symbols include the block, layer, linetype, dimension style, and text style. Once the dependent symbol is added to the current drawing, it behaves as if it were created in the current drawing and is saved with the drawing when you close the drawing session. While adding the dependent symbol to the current drawing, AutoCAD removes the vertical bar symbol (|) from each dependent symbol's name, replacing it with three new characters: a $, a number, and another $ symbol.

For example, you might want to use a block that is defined in an external reference. Instead of binding the entire external reference with the BIND option of the XREF command, it is advisable to use the XBIND command. With the XBIND command, the block and the layers associated with the block will be added to the current drawing. If the block's definition contains a reference to an external reference, AutoCAD binds that xref and all its dependent symbols as well. After binding the necessary dependent symbols, you can detach the external reference file.

Invoke the XBIND command by selecting **Bind** from the Reference toolbar (see Figure 11–8) and AutoCAD displays the Xbind dialog box (see Figure 11–9).

Figure 11–8 *Invoking the* XBIND *command from the Reference toolbar*

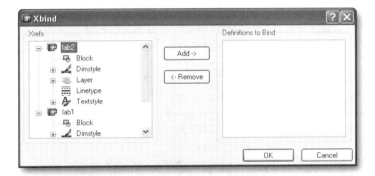

Figure 11–9 *Xbind dialog box*

On the left side of the Xbind dialog box AutoCAD lists the external reference files currently attached to the current drawing. Double-click on the name of the external reference file, and AutoCAD expands the list to show the dependent symbols. Select the dependent symbol from the list and choose **Add**. AutoCAD moves the selected dependent symbol into the **Definitions to Bind** list. If necessary, return it to the **Xrefs** list from the **Definitions to Bind** list by choosing **Remove** after selecting the appropriate dependent symbol.

Choose **OK** to bind the definitions to the current drawing.

CONTROLLING THE DISPLAY OF EXTERNAL REFERENCES

The XCLIP command allows you to control the display of unwanted information by clipping the external reference drawings and blocks. Clipping does not edit or change the external reference or block, it just prevents part of the object from being displayed. The defined clipping boundary can be visible or hidden. You can also define the front and back clipping planes.

The clipping boundary is created coincident with the polyline. Valid boundaries are 2D polylines with straight or spline-curved segments. Polylines with arc segments, or fit-curved polylines, can be used as the definition of the clip boundary, but the clip boundary will be created as a straight segment representation of that polyline. If the polyline has arcs, the clip boundary is created as if it had been decurved prior to being used as a clip boundary. An open polyline is treated as if it were closed.

The XCLIP command can be applied to one or more external references or blocks. If you set the clip boundary to OFF, the entire external reference or block is displayed. If you subsequently set the clip boundary to ON, the clipped drawing is displayed again. If necessary, you can delete the clipping boundary; AutoCAD redisplays the entire external reference or block. In addition, AutoCAD also allows you to generate a polyline from the clipping boundary.

Invoke the XCLIP command by choosing Xref from the Reference toolbar (see Figure 11–10) and AutoCAD prompts:

> Select objects: *(select one or more external references and/or blocks to be included in the clipping and press* ENTER *to complete the selection)*
> Enter clipping option [ON/OFF/Clipdepth/Delete/generate Polyline/New boundary]; *(choose one of the options from the shortcut menu)*

Figure 11–10 *Invoking the* EXTERNAL REFERENCE CLIP *command from the Reference toolbar*

Choose NEW BOUNDARY to define a rectangular or polygonal clip boundary or generate a polygonal clipping boundary from a polyline. AutoCAD prompts:

> Specify clipping boundary:
> [Select polyline/Polygonal/Rectangular]: *(choose one of the three options from the shortcut menu)*

> > Choosing RECTANGULAR (default) allows you to define a rectangular boundary by specifying the opposite corners of a window. The clipping boundary is applied in the current UCS and is independent of the current view.
> >
> > Choosing SELECT POLYLINE defines the boundary by using a selected polyline. The polyline can be open or closed, and can be made of straight-line segments, but cannot intersect itself.
> >
> > Choosing POLYGONAL defines a polygonal boundary by specifying points for the vertices of a polygon.

Once the clipping boundary is defined, AutoCAD displays only the portion of the drawing that is within the clipping boundary and then exits the command.

If you already have a clipping boundary of the selected external reference drawing, and you choose the NEW BOUNDARY option, then AutoCAD prompts:

Delete old boundary(s)? [Yes/No]: *(select one of the two options)*

If you choose YES, the entire reference file is redrawn and the command continues; if you choose NO, the command sequence is terminated.

Note: The display of the boundary border is controlled by the XCLIPFRAME system variable. If it is set to 1 (ON), then AutoCAD displays the boundary border; if it is set to 0 (OFF), then AutoCAD does not display the boundary border.

The ON/OFF option controls the display of the clipped boundary. Choose OFF to display all of the geometry of the external reference or block, ignoring the clipping boundary. Choose ON to display the clipped portion of the external reference or block only.

Choosing CLIPDEPTH sets the front and back clipping planes on an external reference or block. Objects outside the volume defined by the boundary and the specified depth are not displayed.

Choosing DELETE removes the clipping boundary for the selected external reference or block. To turn off the clipping boundary temporarily, use the OFF option explained earlier. The DELETE option erases the clipping boundary and the clipdepth and displays the entire reference file.

Note: The ERASE command cannot be used to delete clipping boundaries.

AutoCAD draws a polyline coincident with the clipping boundary. The polyline assumes the current layer, linetype, and color settings. When you delete the clipping boundary, AutoCAD deletes the polyline. If you need to keep a copy of the polyline, then choose the GENERATE POLYLINE option. AutoCAD makes a copy of the clipping boundary. You can use the PEDIT command to modify the generated polyline, and then redefine the clipping boundary with the new polyline. To see the entire external reference while redefining the boundary, choose OFF to turn off the clipping boundary.

EDITING REFERENCE FILES/XREF EDIT CONTROL

You can edit block references and external references while working in a drawing session by means of the REFEDIT command. This is referred to as in-place reference editing. If you select a reference for editing and it has attached xrefs or block definitions, the nested references and the reference are displayed and available for selection in the Reference Edit dialog box.

Note: You can edit only one reference at a time. Block references inserted with the MINSERT command cannot be edited. You cannot edit a reference file if it is in use by someone else.

You can also display the attribute definitions for editing if the block reference contains attributes. The attributes become visible and their definitions can be edited along with the reference geometry. Attributes of the original reference remain unchanged when the changes are saved back to the block reference. Only subsequent insertions of the block will be affected by the changes.

As mentioned earlier, you can also edit an external reference by opening it in a separate window with the XOPEN command.

Invoke the REFEDIT command by selecting **XRef and Block In-place Editing>Edit Reference In-place** from the Tools menu and AutoCAD prompts:

> Select reference: *(select external references and/or blocks to be included in the editing)*

AutoCAD displays the Reference Edit dialog box (see Figure 11–11).

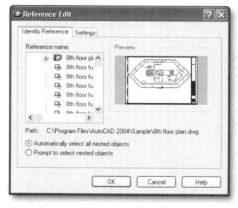

Figure 11–11 *Reference Edit dialog box (Identify Reference tab)*

The **Identify Reference** tab provides visual aids (see Figure 11–11) for identifying the reference to edit and controls how the reference is selected. If you select an object that is part of one or more nested references, the nested references are displayed in the dialog box. Objects selected that belong to any nested references cause all the references to become candidates for editing. Select the specific reference you want to edit by choosing the name of the reference in the **Reference name** list box of the Reference Edit dialog box. This will lock the reference file to prevent other users from opening the file. Only one reference can be edited in place at a time. The path of the selected reference is displayed at the bottom of the dialog box. If the selected reference is a block, no path is displayed.

The **Preview** section of the dialog box displays a preview image of the currently selected reference. The preview image displays the reference as it was last saved in the drawing. The reference preview image is not updated when changes are saved back to the reference.

Choosing **Automatically select all nested objects** controls whether nested objects are included automatically in the reference editing session. If this option is chosen, all the objects in the selected reference will be automatically included in the reference editing session (becoming part of the working set).

Choosing **Prompt to select nested objects** means that nested objects must be selected individually in the reference editing session. If this option is chosen, after you close the Reference Edit dialog box and enter the reference edit state, AutoCAD prompts you to select the specific objects in the reference that you want to edit.

The **Settings** tab (see Figure 11–12) provides options for editing references.

Selecting **Create unique layer, style, and block names** controls whether layers and other named objects extracted from the reference are uniquely altered. If this option is set to ON, named objects in xrefs are altered (names are prefixed with $#$), similar to the way they are altered when you bind xrefs. If it is set to OFF, the names of layers and other named objects remain the same as in the reference drawing. Named objects that are not altered to make them unique assume the properties of those in the current host drawing that share the same name.

Figure 11–12 *Reference Edit dialog box (Settings tab)*

Selecting **Display attribute definitions for editing** controls whether all variable attribute definitions in block references are extracted and displayed during reference editing. If this option is set to ON, the attributes (except constant attributes) are made invisible, and the attribute definitions are available for editing along with the selected reference geometry. When changes are saved back to the block reference, the attributes of the original reference remain unchanged. The new or altered attribute definitions affect only subsequent insertions of the block; the attributes in existing block instances are not affected. Xrefs and block references without definitions are not affected by this option.

Selecting **Lock objects not in working set** locks all objects not in the working set. If this option is set to ON, it will prevent you from accidentally selecting and editing objects in the host drawing while in a reference editing state. The behavior of locked objects is similar to objects on a locked layer. If you try to edit locked objects, they are filtered from the selection set.

Choose **OK** to close the Reference Edit dialog box. AutoCAD prompts to select objects if the Prompt to select nested objects option is selected.

The objects you choose are temporarily extracted for modification in the current drawing and become the *working set*. The working set objects are highlighted so they can be distinguished from other objects. All other objects not selected appear faded. You can now perform modifications on the working set objects.

 Note: Make sure *Reference Edit fading intensity* is set to appropriate setting in the *Display* tab of the Options dialog box.

ADDING/REMOVING OBJECTS FROM THE WORKING SET

If a new object is created while editing a reference, it is usually added to the working set automatically. However, if making changes to objects outside the working set causes a new object to be created, it will not be added to the working set.

Objects removed from the working set are added to the host drawing and removed from the reference when the changes are saved back. Objects created or removed are automatically added to or deleted from the working set. You can tell whether an object is in the working set or not by the way it is displayed on the screen; a faded object is not in the working set. When a reference is being edited, the Refedit toolbar is displayed (see Figure 11–13).

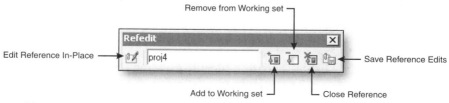

Figure 11–13 *The Refedit toolbar*

To add objects to the working set, select **Add to Working set** from the Refedit toolbar and select objects to be added. You can only select items when the type of space (Model or Paper) is in effect that was in effect when the REFEDIT command was initiated. To remove objects from the working set, select **Remove from Working set** from the Refedit toolbar and select objects to be removed. As mentioned earlier, if you remove objects from the working set and save the changes, the objects are removed from the reference and added to the current drawing. Any changes you make to objects in the current drawing (not in the xref or block) are not discarded. Once the modifications are complete, select **Save Reference Edits** to save the changes to the reference file and to discard the changes, select **Close Reference** from the Refedit toolbar.

MANAGING EXTERNAL REFERENCES

Several tools are available to help in the management and tracking of external references.

One of the tracking mechanisms is an external ASCII log file that is maintained on each drawing that contains external references. This file, which AutoCAD generates and maintains automatically, has the same name as the current drawing with a file extension *.xlg*. You can examine the file with any text editor and/or print it. The log file registers each attach, bind, detach, and reload of each external reference for the current drawing. AutoCAD writes a title block to the log file that contains the name of the current drawing, the date and time, and the operation being performed. Once a log file has been created for a drawing, AutoCAD continues to append to it. The log file is always placed in the same folder as the current drawing. The log file is maintained only if the XREFCTL system variable is set to 1. The default setting for XREFCTL is 0.

External references are also reported in response to the ? option of the -XREF command and the BLOCK command. Because of the external reference feature, the contents of a drawing may now be stored in multiple drawing files. This means that new backup procedures are required to handle drawings linked in external reference partnerships. Three possible solutions are:

1. Make the external reference drawing a permanent part of the current drawing prior to archiving with the BIND option of the XREF command.
2. Modify the path of the current drawing to the external reference drawing so that they are both stored in the same folder, and then archive them together.
3. Archive the folder location of the external reference drawing with the drawing which references it. Tape backup machines do this automatically.

In AutoCAD, a combination of demand loading and the saving of drawings with indexes helps you increase the performance of drawing with external references. In conjunction with the XLOADCTL and INDEXCTL system variables, demand loading provides a method of displaying only those parts of the referenced drawing that are necessary.

The XLOADCTL system variable controls whether demand loading is set to ON or OFF and whether it opens the original drawing or a copy. If XLOADCTL is set to 0, then AutoCAD turns off demand loading, and the entire reference file is loaded. If XLOADCTL is set to 1, then AutoCAD turns on the demand loading, and the reference file is kept open. AutoCAD loads only the objects that are necessary to display on the current drawing. AutoCAD places a lock on all reference drawings that are set for demand loading. Other users can open those reference drawings, but they cannot save changes to them. If XLOADCTL is set to 2, then AutoCAD turns on demand loading and a copy of the reference file is opened. AutoCAD makes a temporary copy of the externally referenced file and demand-loads the temporary file. Other users are allowed to edit the original drawing. When you disable demand loading, AutoCAD reads in the entire reference drawing regardless of layer visibility or clip instances.

The INDEXCTL system variable determines whether layer, spatial, or layer and spatial indexes are created when a drawing file is saved. Using layer and spatial indexes increases performance when AutoCAD is demand-loading external references. If INDEXCTL is set to 0, then indexes are not created; if INDEXCTL is set to 1, a layer index is created. The layer index maintains a list of objects that are on specific layers, and with demand loading it determines which objects need to be read in and displayed. If INDEXCTL is set to 2, then a spatial index is created. The spatial index organizes lists of objects based on their location in three-dimensional space, and it determines which objects lie within the clip boundary and reads only those objects into the current session. If INDEXTCTL is set to 3, both layer and spatial indexes are created and saved with the drawing. If you intend to take full advantage of demand loading, then INDEXCTL should be set to 3.

 Note: If the drawing you are working on is not going to be referenced by another drawing, it is recommended that you set INDEXCTL to 0 (OFF).

AutoCAD provides another system variable, VISRETAIN, to control the visibility of layers in the external reference drawing. If the VISRETAIN system variable is set to 0 (OFF), any changes you make to settings of the external reference drawing's layers, such as ON/OFF, Freeze/Thaw, Color, and Linetype, apply to the current drawing session only. If VISRETAIN is set to 1 (ON), then any changes you make to settings of the external reference drawing's layers take precedence over the external reference layer definition.

Another tool that you can use to manage external references is Autodesk Reference Manager. It provides tools to list referenced files in selected drawings and to modify the saved reference paths without opening the drawing files in AutoCAD. With Reference Manager, drawings containing unresolved references can be easily identified and fixed.

Reference Manager is a stand-alone application that you can access from the Autodesk program group under Programs in the Start menu (Windows).

IMAGES

The IMAGE command provides various options for attaching and detaching a raster or bit-mapped bi-tonal, 8-bit gray, 8 bit color, or 24-bit color image file into the drawing. The image formats that can be inserted into AutoCAD include BMP, TIFF, RLE, DIB, JPG, PCX, FLIC, GEOSPOT, GIF, IG4, IGS, RLC, PCT, PCX, CALS1, PNG, and TGA. More than one image can be displayed in any viewport, and the number and size of images is not limited.

Invoke the IMAGE command by selecting **Image Manager** from the Reference toolbar (see Figure 11–14) and AutoCAD displays the Image Manager dialog box (see Figure 11–15).

Figure 11–14 *Invoking the* IMAGE *command from the Reference toolbar*

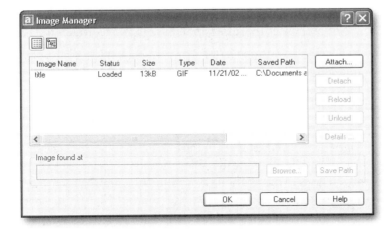

Figure 11–15 *Image Manager dialog box*

The Image Manager dialog box is similar to the Xref Manager dialog box. In the Image Manager dialog box, AutoCAD lists the images attached to the current drawing. The information provided in the list box of the Image Manager dialog box, such as **Image Name**, **Status**, **Size**, **Type**, **Date**, and **Saved Path**, is similar to the information provided in the Xref Manager dialog box. You can switch between List View and Tree View, as in the Xref Manager dialog box, by choosing one of the two buttons at the top left of the Image Manager dialog box.

The **Attach** feature allows you to attach an image object to the current drawing. Choose **Attach** in the Image Manager dialog box and AutoCAD displays the Select Image File dialog box.

Select the image file from the appropriate directory and choose **Open**. AutoCAD displays the Image dialog box (see Figure 11–16).

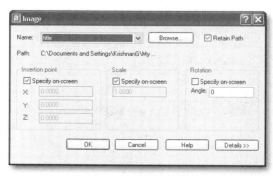

Figure 11–16 *Image dialog box*

To select a different image choose **Browse** to display the Select Image File dialog box. Select the appropriate image file to attach to the drawing.

Once an image file is attached to the current drawing, the image file name is added to the list box located next to the **Browse** button. When an attached image file name is selected from the list box, its path is displayed below.

Selecting **Retain Path** determines whether or not the full path to the image file is saved. If **Retain Path** is set to ON, then the image file path is saved in the drawing database; if it is set to OFF, the name of the image file is saved without a path in the database. AutoCAD searches for the image file in the AutoCAD Support File Search Path and in the paths associated with the **Files** tab of the Options dialog box.

The **Insertion point** section lets you specify the insertion point of the external reference.

The **Scale** section lets you specify *X* scale factor, *Y* scale factor, and *Z* scale factor of the external reference.

The **Rotation** section lets you specify the rotation angle of the external reference.

The **Insertion point**, **Scale**, and **Rotation** features are similar to those in the insertion of a block, explained in Chapter 10.

Choose **Details** to display the **Image Information** for the selected image file. This information includes image resolution in horizontal and vertical units, image size by width and height in pixels, and image size by width and height in the current selected units.

Choosing **Detach** in the Image Manager dialog box removes the selected image definitions from the drawing database and erases all the associated image objects from the drawing and from the display. To detach an image file from the current drawing, first select the image file name from the displayed list in the Image Manager dialog box, and then choose **Detach**. AutoCAD detaches the selected image file from the current drawing.

Choosing **Reload** loads the most recent version of an image. To reload the image file to the current drawing, first select the image file name from the displayed list in the Image Manager dialog box, and then select **Reload**. AutoCAD reloads the selected image file into the current drawing.

Choosing **Unload** unloads the image data from working memory without erasing the image objects from the drawing. It is highly recommended that you unload images that are no longer needed for editing. By unloading the images, you can improve performance by reducing the memory requirement for AutoCAD. To unload the image file from the current drawing, first select the image file name from the displayed list in the Image Manager dialog box, and then select **Unload**. AutoCAD unloads the selected image file from the current drawing.

Choosing **Details** provides detailed information about the selected image, including the image name, saved path, active path, file creation date and time, file size and type, color, color depth, width and height in pixels, resolution, default size in units, and a preview image. To display the detailed information of the selected image file, first select the image file name from the displayed list in the Image Manager dialog box, then choose **Details**. AutoCAD displays the detailed information of the selected image file in the Image File Details dialog box (Figure 11–17).

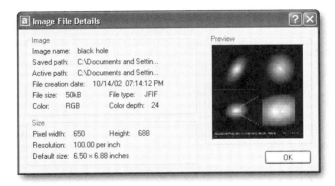

Figure 11–17 *Image File Details dialog box*

Choosing **Browse** in the Image Manager dialog box lets you change to a different path or file name for the currently selected image file. AutoCAD displays the Select Image File dialog box, in which you can select a different path or file name.

Selecting **Save Path** lets you save the path as it appears in the **Saved Path** field of the currently selected image file (and displayed in the **Image found at** text box). AutoCAD saves the path of the currently selected image file.

After making the necessary changes in the Image Manager dialog box, choose **OK** to accept the changes and close the dialog box.

CONTROLLING THE DISPLAY OF THE IMAGE OBJECTS

The IMAGECLIP command allows you to control the display of unwanted information by clipping the image object; this is similar to the use of the XCLIP command for external references and blocks.

Invoke the IMAGECLIP command by selecting **Image** from the Reference toolbar (see Figure 11–18) and AutoCAD prompts:

> Select image to clip: *(select the image to clip)*
> Enter image clipping option [ON/OFF/Delete/New boundary] <New>:
>   *(select one of the options from the shortcut menu)*

Figure 11–18 *Invoking the* IMAGECLIP *command from the Reference toolbar*

Choosing NEW BOUNDARY (default) allows you to define a rectangular or polygonal clip boundary. AutoCAD prompts:

> Enter clipping type ⊥: *(select one of the options from the shortcut menu)*

Choosing RECTANGULAR (default) allows you to define a rectangular boundary by specifying the opposite corners of a window. The rectangle is always drawn parallel to the edges of the image.

Choosing POLYGONAL lets you define a polygonal boundary by specifying points for the vertices of a polygon.

If you already have a clipping boundary of the selected image, and you invoke the NEW BOUNDARY option, then AutoCAD prompts:

> Delete old boundary(s)? [Yes/No] <Yes>: *(select one of the two available options)*

If you choose YES, the entire reference file is redrawn and the command continues; if you choose NO, the command sequence is terminated.

Once the clipping boundary is defined, AutoCAD displays only the portion of the image that is within the clipping boundary and then exits the command.

 Note: The display of the boundary border is controlled by the IMAGEFRAME system variable. If it is set to 1 (ON), then AutoCAD displays the boundary border; if it is set to 0 (OFF), then AutoCAD does not display the boundary border.

The ON/OFF option controls the display of the clipped boundary. The OFF option displays all of the image, ignoring the clipping boundary. The ON option displays the clipped portion of the image only.

ADJUSTING THE IMAGE SETTINGS

The IMAGEADJUST command controls the brightness, contrast, and fade values of the selected image. Invoke the IMAGEADJUST command by selecting **Adjust** from the Reference toolbar (see Figure 11–19) and AutoCAD prompts:

Select image(s): *(select the images, and press* ENTER *to complete selection)*

Figure 11–19 *Invoking the* IMAGEADJUST *command from the Reference toolbar*

AutoCAD displays the Image Adjust dialog box, similar to Figure 11–20.

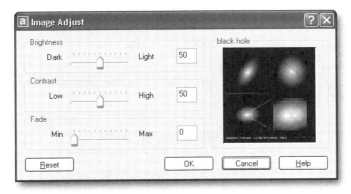

Figure 11–20 *Image Adjust dialog box*

You can adjust the **Brightness**, **Contrast**, and **Fade** within the range of 0 to 100.

Select **Reset** to reset values for the brightness, contrast, and fade parameters to the default settings of 50, 50, and 0, respectively.

ADJUSTING THE DISPLAY QUALITY OF IMAGES

The IMAGEQUALITY command controls the display quality of images. The quality setting affects display performance. A high-quality image takes longer to display. Changing the setting updates the display immediately without causing a regeneration. Images are always plotted using a high-quality display.

Invoke the IMAGEQUALITY command by selecting **Quality** from the Reference toolbar (see Figure 11–21) and AutoCAD prompts:

> Enter image quality setting [High/Draft]: *(select one of the available options from the shortcut menu)*

Figure 11–21 *Invoking the* IMAGEQUALITY *command from the Reference toolbar*

The High option produces a high-quality image on screen, and the Draft option produces a lower-quality image on screen.

CONTROLLING THE TRANSPARENCY OF AN IMAGE

The TRANSPARENCY command controls whether the background pixels in an image are transparent or opaque. Invoke the TRANSPARENCY command from the Reference toolbar (see Figure 11–22) and AutoCAD prompts:

> Select image: *(select the images, and press* ENTER *to complete selection)*
> Enter transparency mode [ON/OFF]: *(select one of the two available options)*

Figure 11–22 *Invoking the* TRANSPARENCY *command from the Reference toolbar*

Choosing ON turns transparency on so that objects beneath the image are visible.

Choosing OFF turns transparency off so that objects beneath the image are not visible.

CONTROLLING THE FRAME OF AN IMAGE

The IMAGEFRAME command controls whether image frames are displayed or hidden from view.

Invoke the IMAGEFRAME command by selecting **Frame** from the Reference toolbar (see Figure 11–23) and AutoCAD prompts:

> Enter image frame setting [0/1/2]: *(select one of the available options from the shortcut menu)*

Figure 11–23 *Invoking the* IMAGEFRAME *command from the Reference toolbar*

Choosing *0* turns display and plotting of frames around images off.

Choosing *1* turns display and plotting of frames around images on.

Choosing *2* turns display of frames on, but plotting of frames off.

Open the Exercise Manual PDF file for Chapter 11 on the accompanying CD for project- and discipline-specific exercises.

If you have the accompanying Exercise Manual, refer to Chapter 11 for project- and discipline-specific exercises.

REVIEW QUESTIONS

1. If an externally referenced drawing named *FLOOR.dwg* contains a block called "TABLE" and is permanently bound to the current drawing, the new name of the block is:

    a. FLOOR0TABLE

    b. FLOOR|TABLE

    c. FLOOR$0TABLE

    d. FLOOR_TABLE

    e. FLOOR$|$TABLE

2. If an externally referenced drawing named *FLOOR.dwg* contains a block called "TABLE," the name of the block is listed as:

    a. FLOOR0TABLE

    b. FLOOR|TABLE

    c. FLOOR$0TABLE

    d. FLOOR_TABLE

    e. FLOOR$|$TABLE

3. The maximum number of files that can be externally referenced into a drawing is:

    a. 32

    b. 1,024

    c. 8,000

    d. 32,000

    e. only limited by memory

4. If you want to retain, from one drawing session to another, any changes you make to the color or visibility of layers in an externally referenced file, the system variable that controls this is:

    a. XREFRET

    b. RETXREF

    c. XREFLAYER

    d. VISRETAIN

    e. these changes cannot be saved from one session to another.

5. XREFs are converted to blocks if you detach them.

   a. True   b. False

6. When detaching XREFs from your drawing, it is acceptable to use wild cards to specify which XREF should be detached.

   a. True   b. False

7. Overlaying XREFs rather than attaching them causes AutoCAD to display the file as a bitmap image, rather than a vector-based image.

   a. True   b. False

8. To make a reference file a permanent part of the current drawing database, use the XREF command with the:

   a. ATTACH option   d. RELOAD option

   b. BIND option   e. PATH option

   c. ? option

9. The XREF command is invoked from which toolbar?

   a. Draw

   b. Modify

   c. External Reference

   d. Any of the above

   e. None of the above

10. The ATTACH option of the XREF command is used to:

    a. bind the external drawing to the current drawing

    b. attach a new external reference file to the current drawing

    c. reload an external reference drawing

    d. all of the above

11. The following are the dependent symbols that can be made a permanent part of your current drawing, except:

    a. Blocks

    b. Dimstyles

    c. Text Styles

    d. Linetypes

    e. Grid and Snap

12. Including an image file in a drawing will incorporate the image similar to the way a drawing file is merged by using:

    a. INSERT

    b. XREF

    c. WBLOCK

13. Which of the following is not a valid file type to use with the IMAGE command?

    a. BMP

    b. TIF

    c. WMF

    d. JPG

    e. GIF

14. Which of the following parameters can be adjusted on a bitmapped image?

    a. Brightness

    b. Contrast

    c. Fade

    d. all of the above

    e. none of the above

CHAPTER 12

AutoCAD DesignCenter

INTRODUCTION

AutoCAD DesignCenter makes it much easier to manage content within your drawing. Content includes blocks, external references, layers, raster images, hatch and gradient fills, linetypes, layouts, text styles, dimension styles, and custom content created by third-party applications. You can now manage content between your drawing and other sources such as other drawings, whether currently open, stored on any drive, or even elsewhere on a network or somewhere on the Internet. The AutoCAD DesignCenter provides a program window with a specialized drawing file-handling section. It allows you to drag and drop content and images into your current drawing or attach a drawing as an external reference. The DesignCenter also gives immediate and direct access to thousands of symbols, manufacturers' product information, and content aggregators' sites through DC Online. Content in the AutoCAD Design-Center can be dragged onto a tool palette for use in the current drawing.

After completing this chapter, you will be able to do the following:

- Open, undock, move, resize, dock, and close the AutoCAD DesignCenter
- Locate drawings, files, and their content in a manner similar to Windows Explorer
- Use the DesignCenter content area and Tree View
- Preview images, drawings, content, and their written descriptions
- Customize and use Autodesk *Favorites* folder
- Manage blocks, layers, xrefs, layouts, dimstyles, textstyles, and raster images
- Manage web-based content and custom content from third-party applications
- Create shortcuts to drawings, folders, drives, the network, and Internet locations
- Browse sources for content and drag and drop (copy) into drawings
- Access symbols and information directly over the Internet

DesignCenter WINDOW

The DesignCenter window allows you to browse, find, preview, and insert content, which includes blocks, hatches, and external references (xrefs). Use the buttons in the toolbar at the top of the DesignCenter for display and access options. You can control the size, location, and appearance of the DesignCenter.

CONTENT

As mentioned above, content includes block definitions, external references, layer names and compositions, raster images, linetypes, layouts, text styles, dimension styles, and custom content created by third-party applications. For example, a layer of one name may have the same properties as a layer of another name. But, because its name is part of its composition, it can be identified as a unique layer. Also, two items of content can have the same name if they are not the same type of content. For example, you can name both a text style and a dimension style "architectural".

Note: You can only view the name of an item of content in the content area along with a raster image if it is a drawing, block, or image and a description if one has been written. The item itself still resides in its container. Through the DesignCenter, you can drag and drop copies of the item's definition into your current drawing, but you cannot edit the item itself from the DesignCenter.

CONTENT TYPE

Content types are only types or categories and not the items themselves. Lines, circles, and other objects are not included in what are considered content types, although a block may be made up of such objects. To view a particular item of content you must first select the name of its content type (under a drawing name) in the Tree View or double-click the name of its content type in the content area (see Figure 12–6 for identifying the Tree View pane and the content area). For example, if you were trying to locate an xref named "1st floor plan", you could search through drawing names and select the content type named "xref" under each drawing name. Under each drawing name there will be a content type named "xref", but under each content type named "xref" there may or may not be any xrefs listed.

CONTAINER

The primary container is the drawing. It contains the blocks, images, linetypes, and other definitions that are most commonly sought to add to your current drawing. A folder can be considered a container because it contains files, and a drawing is a file. An image can also be a file contained in a folder. An image is normally not a container like a drawing. Drawings and images can be dragged and dropped into your current drawing from the content area. So, a drawing can be both content and a container.

Figure 12–1 shows an example of a content (block named Conference Seat), content type (Block), and Container (drawing named *8th floor furniture*).

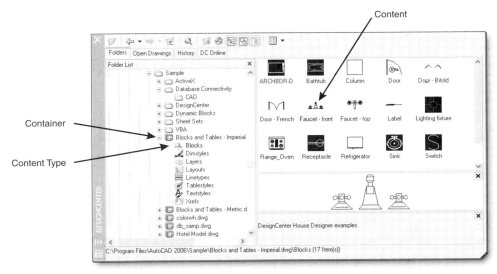

Figure 12–1 *An example showing a content, content type, and container from the DesignCenter window*

OPENING THE DesignCenter WINDOW

The AutoCAD DesignCenter is a window rather than a dialog box. It is like calling up a special program that runs alongside AutoCAD and expedites the tasks of managing files and handling drawing content.

Invoke the AutoCAD DesignCenter command from the Standard toolbar (see Figure 12–2) and AutoCAD displays the AutoCAD DesignCenter window (see Figure 12–3).

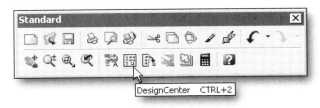

Figure 12–2 *Invoking the AutoCAD DesignCenter from the Standard toolbar*

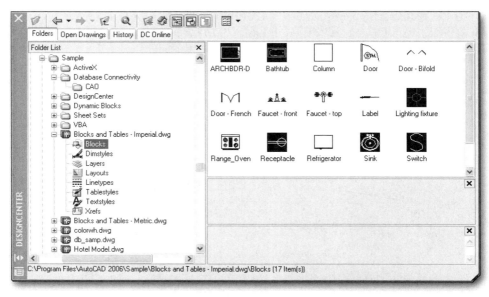

Figure 12–3 *AutoCAD DesignCenter window*

POSITIONING THE DesignCenter WINDOW

The default position of the DesignCenter is docked at the left side of the drawing area. This is where it will be located when you open the DesignCenter window for the first time, as shown in Figure 12–4. However, once the DesignCenter has been repositioned and the drawing session is closed with the DesignCenter window open, the next time you open AutoCAD it will be at its relocated position.

You can undock the DesignCenter by double-clicking on its title bar or border. Or you can drag the window into the drawing area until its drag preview image clears its present position (see Figure 12–5).

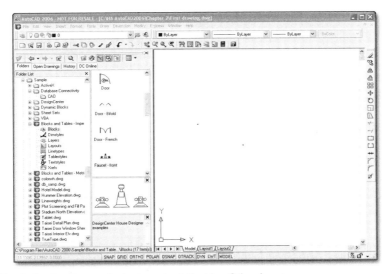

Figure 12–4 *DesignCenter docked on the left side of the drawing area*

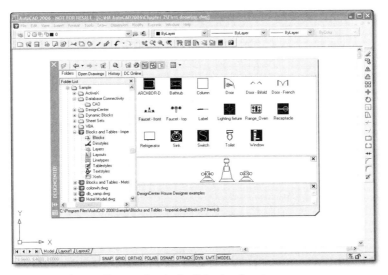

Figure 12–5 *DesignCenter floating (undocked) in the drawing area*

To redock the DesignCenter, double-click the title bar or border or drag the DesignCenter to the left or right side of the drawing area. It cannot be docked at the top or bottom of the screen. When undocked, the DesignCenter can be resized by holding the cursor over an edge or corner of the window and, when the cursor becomes a double arrow holding down the pick button, then dragging the edge or corner with the pointing device.

WORKING WITH THE DesignCenter

The DesignCenter has four tabs: **Folders**, **Open Drawings**, **History**, and **DC Online**. The buttons on the toolbar for the DC Online tab are different from those for the other three tabs.

The **Folders** tab displays a hierarchy of navigational icons, including Networks and computers, Web addresses (URLs), drives, folders, drawings and related support files, Xrefs, layouts, hatch styles, and named objects, including blocks, layers, linetypes, text styles, dimension styles, and plot styles within a drawing. The **Open Drawings** tab displays a list of the drawings that are currently open. The **History** tab displays a list of files opened previously with DesignCenter. The **DC Online** tab provides content from the DesignCenter Online web page including blocks, symbol libraries, manufacturer's content, and online catalogs.

FOLDERS

The DesignCenter, when the **Folders** tab is displayed, consists of five major areas: toolbar, content area, Tree View pane, Preview pane, and Description pane. The content area and toolbar are always displayed (see Figure 12–6). The Tree View, Preview, and Description panes can be turned off and on when desired.

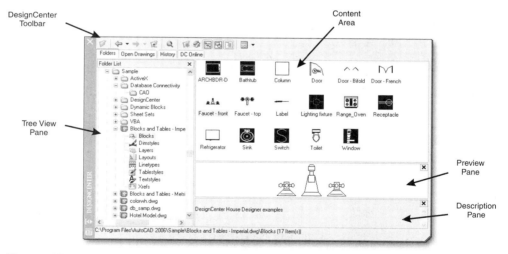

Figure 12–6 *Parts of the DesignCenter window with the Folders tab displayed*

TOOLBAR

The toolbar contains buttons designed especially to help you select and manage the type of content desired within the DesignCenter window (see Figure 12–6). For information about these buttons, see the section "DesignCenter Toolbar."

Content Area

The content area (right pane) is the AutoCAD DesignCenter's primary area for displaying the names and icons representing content (see Figure 12–6).

Tree View

The Tree View or navigation pane is an optional area on the left side of the DesignCenter window for displaying files such as drawings and images and their locations. Choose the Tree View button on the toolbar (see Figure 12–7), to display the Tree View.

Figure 12–7 *Invoking the Tree View from the DesignCenter toolbar*

The Tree View also displays content types. The Tree View displays multiple levels and operates in a manner similar to the Windows Explorer. Figure 12–8 shows the DesignCenter window with the Tree View window display on and off within the **Folders** tab.

 Note: The Tree View can be used to navigate up and down through the hierarchy of networks, drives, folders, files, and content type. But, you cannot drag and drop to, from, or within the Tree View.

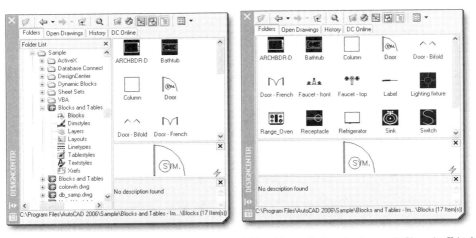

Figure 12–8 *AutoCAD DesignCenter window with Tree View window display on (left) and off (right)*

Preview

The Preview pane displays a raster image of the selected item of content if the item is a drawing, block, or image (see Figure 12–6).

Description

The Description pane displays a written description of the item of content selected on the content area, if a description has been written (see Figure 12–6).

DesignCenter Toolbar

The buttons on the toolbar are used to manage the DesignCenter panes and what is being displayed in them (see Figure 12–9).

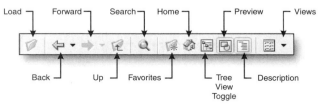

Figure 12–9 *AutoCAD DesignCenter toolbar*

Choosing Load causes the Load dialog box to be displayed (see Figure 12–10). It is similar to the Windows File Manager dialog box. From it you can select a drawing file whose contents will be loaded into the content area.

Figure 12–10 *Load dialog box*

AutoCAD displays a bitmap image in the **Preview** section. In addition, there is a window on the left side of the dialog box displaying quick access icons to folders on your computer: *Desktop* and *My Documents*; icons for **History** (recently opened drawings) and **Favorites**; and locations on the Internet: **Buzzsaw.com**, and **FTP**.

The **Back** and **Forward** icons, when selected, cause the contents in the Tree View and content area to scroll backwards and forwards respectively through the history of viewing activity. The **Up** icon causes the Tree View to display contents of the parent folder.

The **Search** icon provides a means to locate containers, content type, and content in a manner similar to the Windows Find feature. Choosing **Search** causes the Search dialog box to be displayed (see Figure 12–11).

Figure 12–11 *Search dialog box with the Drawings tab displayed*

From the **Look for** list box in the Search dialog box, select the type of content you wish to find. Available options include **Blocks, Dimstyles, Drawings, Drawings and Blocks, Attach Pattern Files, Hatch Patterns Layers, Layouts, Linetypes, Tabstyles, Textstyles,** and **Xrefs**.

From the **In** list box, select the location for searching. Available options include My Computer, local hard drives (C:) [and any others], 3½ Floppy (A:), and any other drives.

Choosing **Browse** causes the Browse for Folder dialog box to be displayed, which has a file manager window (see Figure 12–12). Here you can specify a path by stepping through the levels to the location which you wish to search. When the final location is highlighted, choose **OK**, and the path to this location is displayed in the **In** list box.

Figure 12–12 *Browse For Folder dialog box*

The number of tabs that will be displayed in the Search dialog box depends on the type of content selected in the **Look for** list box. Each of the content types will have a tab that corresponds to the type of content selected. If Drawings is selected, there are two additional tabs: **Date Modified** and **Advanced**.

On the **Drawings** tab (see Figure 12–11), the **Search for the word(s)** text box lets you enter the word(s), such as a drawing name or author, to determine what to search for. The **In the field(s)** text box lets you select a field type to search. Fields include File Name, Title, Subject, Author, and Keywords.

On the **Date Modified** tab (available only when Drawings is selected in the **Look for** text box of the Search dialog box) you can specify a search of drawing files by the date they were modified (see Figure 12–13). The search will include all files that comply with other filters if **All files** is selected. If **Find all files created or modified** is selected, you can limit the search to the range of dates specified by selecting one of the following secondary radio buttons. Selecting **between and** lets you limit the search to drawings modified between the dates entered. Selecting **during the previous months** lets you limit the search to drawings modified during the number of previous months entered. Selecting **during the previous days** lets you limit the search to drawings modified during the number of previous days entered.

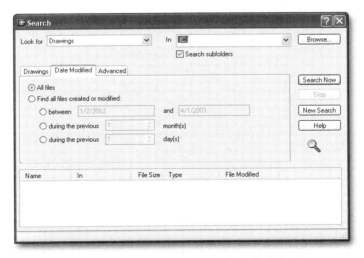

Figure 12–13 Display of the Date Modified tab in the Search dialog box

On the **Advanced** tab (available only when **Drawings** is selected in the **Look For** text box of the Search dialog box), you can specify a search of files by additional parameters (see Figure 12–14). In the **Containing** text box you can specify a file by one of four options: **Block name, Block and drawing description, Attribute tag,** and **Attribute value**. The search will then be limited to items containing the text entered in the **Containing text** text box. The **Size is** text boxes let you limit the search to drawings that are **At least** or **At most** in KBs (size of the drawing file) as the number entered into the second text box.

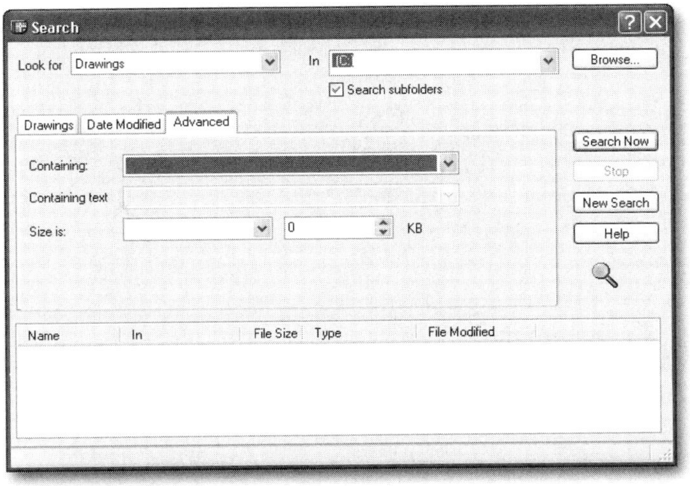

Figure 12–14 Display of the Advanced tab in the Search dialog box

All of the other tabs (**Blocks, Dimstyles, Drawings and Blocks, Hatch Patterns, Layers, Layouts, Linetypes, Textstyles,** and **Xrefs**) have a **Search for the name** text box in which you can enter the name of the item you wish to find. This can be a block name, dimstyle, or one of the other content types.

To the right of the tab section are four buttons: **Search Now, Stop, New Search,** and **Help**.

Once the parameters have been specified, such as the path, content type, and date modified, selecting **Search Now** initiates the search. Select **Stop** to terminate the search. Selecting **New Search** lets you specify new parameters for another search. Selecting **Help** causes the AutoCAD Command Reference (Help) dialog box to be displayed.

Each time a search is performed, the text entered in the **Search for the name** text box or the **Search for the word(s)** text box is saved. If you wish to repeat the same search again, select the down arrow next to the text box and select the text associated with the search you wish to repeat.

Choosing **Favorites** on the DesignCenter toolbar causes the Autodesk Favorites feature to display icons in the content area representing shortcuts to frequently used files. These are files for which you have previously set up shortcut icons to make them quickly accessible. The Tree View displays and highlights the subfolder *Favorites* of the folder Autodesk (see Figure 12–15).

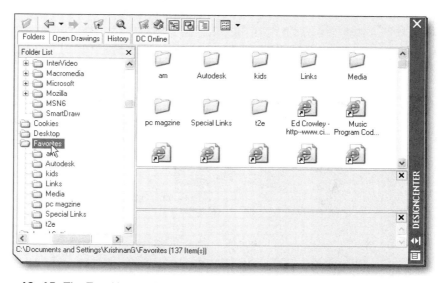

Figure 12–15 *The Tree View with the Autodesk Favorites folder selected*

 Note: Only the shortcuts are in the *Favorites* folder. The files themselves remain in their original locations.

A drawing or image file or a folder can be added to Favorites by right-clicking on the file name or folder name in the content area or Tree View and choosing ADD TO FAVORITES from the shortcut menu. To view and organize items in Favorites, choose ORGANIZE FAVORITES from the shortcut menu. An Autodesk window is displayed for managing items in Favorites.

When the Favorites folder is highlighted in the Tree View, the content area displays the icons, which are shortcuts to other files or folders. Double-click on the icon for the desired folder or file.

After you have double-clicked the desired icon in the content area, the selected file or folder will be displayed in the Tree View and it will be highlighted.

And, in turn, the content area will display the contents of the highlighted folder or file. Figure 12–16 shows the selection of the Favorites folder and corresponding selection.

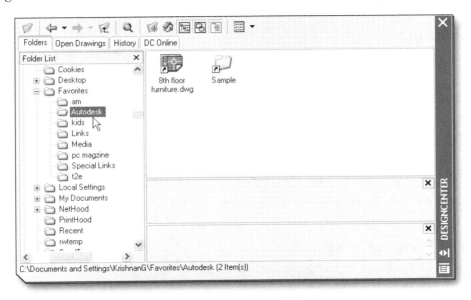

Figure 12–16 *Selection of a subfolder in the Favorites folder*

The buttons for Tree View, Preview, and Description simply cause their respective panes to be displayed or not displayed, depending on their current status.

Choosing **Home** causes the Tree View to display the drives, folders, subfolders, and files that are located on your computer's desktop. The hierarchy is expanded, and the DesignCenter folder is highlighted with its contents displayed in the content area (see Figure 12–17). This is the type of view you normally see when you open Windows Explorer.

From there you can step through the path(s) necessary to the desired location. Select the **Folders** tab whenever you want to return from **Open Drawings** tree view and **History** view (explained later in this section) to display the drives, folders, subfolders, and files that are located on your computer's desktop.

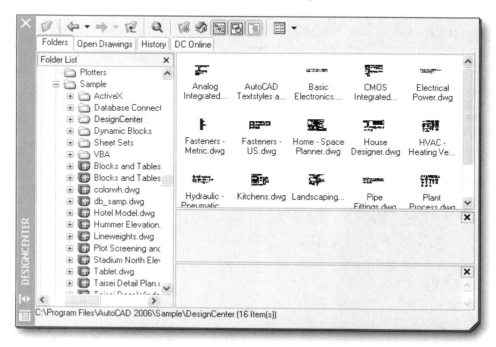

Figure 12–17 *DesignCenter displaying the Desktop with the DesignCenter folder highlighted*

VIEWING CONTENT

The names of content type, such as blocks, linetypes, textstyles, and so on, and containers such as drawings, image files, folders, drives, networks, and Internet locations, can be viewed in the Tree View as well as the content area. Names of content can be viewed in the content area, but not in the Tree View pane. Either pane can be used to move up and down through the path from the drive to the item of content. It is usually quicker to navigate the path in the Tree View because of its ability to display multiple levels of hierarchy. When the container (a folder for drawings/images and a drawing/content type for blocks, images, and other items) appears in the Tree View, select it and then view the name of the items of content (if any) in the content area.

USING THE TREE VIEW

As mentioned earlier, in the Tree View you can view the content type and container, but not the actual content. You can use the Tree View to display the icon for an item of content in the content area. Just doing this will not, however, cause a raster image to be displayed in the Preview panel, as shown in using the content area in this section.

The folder named *Sample* that installs with the AutoCAD program contains numerous drawings which will be used as examples in this section. First, select **Up** as many times as necessary to display the level (in the Tree View pane) that includes the branch leading through the folder named *AutoCAD2006* (or the folder in which AutoCAD has been installed) to the subfolder named *Sample*. This, again, might require going all the way back to a particular drive. Then click the plus sign or double-click the folder name or drive name to display the folders within a folder. Figure 12–18 shows the folder named *Sample* highlighted in the Tree View. Note that the list of content folders is displayed in the content area.

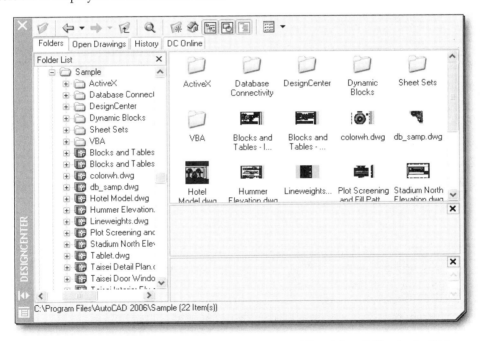

Figure 12–18 *The Tree View showing the drawing named Taisei Detail Plan in the folder named Sample*

Select *Taisei Detail Plan.dwg* in the Tree View. Note that the list of content types is displayed in the content area.

Next, double-click on the drawing named *Taisei Detail Plan.dwg* in the Tree View or select the box with the plus in it to the left of the drawing name. The content types will be listed below the drawing name, as well as in the content area. To display the names of individual items of content, blocks for example, select the content type Blocks in the Tree View (see Figure 12–19).

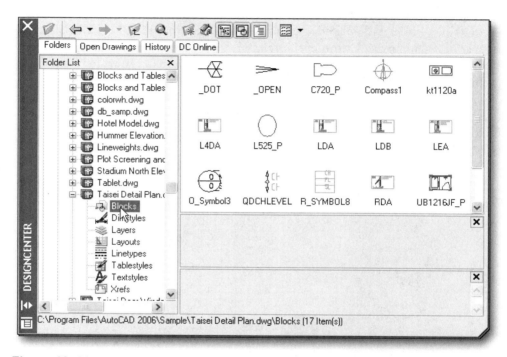

Figure 12–19 *The Tree View with the content type Blocks highlighted*

When one of the block icons displayed in the content area is chosen, AutoCAD displays the corresponding preview image and description (if any) in the Preview pane and Description pane, respectively. Figure 12–20 shows the selection of a block named "C720_ P", and its corresponding preview image and description are shown in the Preview pane and in the Description pane.

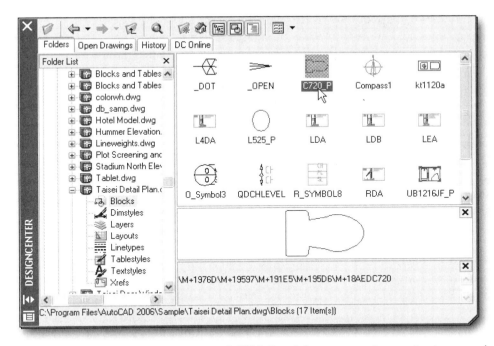

Figure 12-20 *Selection of a block named C720_P and the corresponding preview image and description are shown in the Preview pane and in the Description pane*

USING THE CONTENT AREA

As mentioned earlier, the content area is used for displaying content. It can also display containers and content type. A container is a source of content such as a drawing. In the hierarchy of what can be displayed, containers are one step higher than content types. Content types are one step higher than content. From drawing files and other containers, the next step up is a folder. Then, progressing up through the folders (if there is a hierarchy of folders), next is the drive, which may lead to a network, a Web site on the Internet, a floppy diskette, or a CD-ROM.

Note: The content area displays only items of the same level that are members of a single container one level above them. The content area does not display more than one level at a time. For example, content types such as blocks, xrefs, and layers are members of one drawing. If there are any blocks in the drawing, then they are listed when you have selected that drawing's content type called Blocks. Several drawings may be members of a particular folder. Folders are members of one drive or folders may actually be subfolders of a folder one level up. Remember, in the content area, only one level of the hierarchy is displayed and all items shown are members of the same component/container one level above. For displaying more than one level at a time, use the Tree View.

You can navigate up and down through the hierarchy of drives, folders, drawings, content types, and content in the content area. It is not as easy as in the Tree View, however, because only one level is displayed at a time. If the content is a drawing, block, or image, you can view a raster image of it in the Preview pane. Simply select the item in the content area. Remember that you can enlarge the Preview pane for viewing, even if only temporarily. As mentioned earlier, you can double-click the title bar to dock or undock the DesignCenter.

For example, there is a block named "UB1216JF_P" in the drawing named *Taisei Detail Plan.dwg* in the folder named Sample. To get to the block named "UB1216JF_P" in the content area, as described for the example in the Tree View above, select the Up button as many times as necessary to display the level that includes the branch leading to the drawing named *Taisei Detail Plan.dwg*. This might require going all the way back to a particular drive. Then double-click the sequence of drive and folders on the path to the drawing named *Taisei Detail Plan.dwg* in the content area, and then double-click on *Taisei Detail Plan.dwg* icon. When you double-click the icon for *Taisei Detail Plan.dwg,* the icons for content types will be displayed, including Blocks, Dimstyles, Layers, Layouts, Linetypes, Textstyles, and Xrefs (see Figure 12–21).

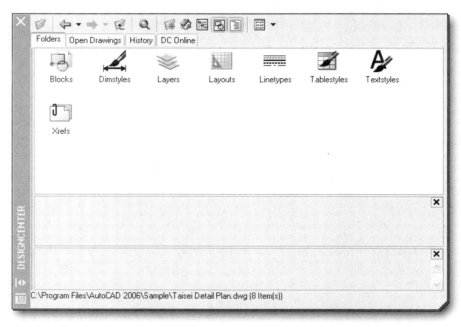

Figure 12–21 *The Content Area with the display of the content types*

AutoCAD also displays the path to the selected drawing at the bottom of the Design-Center window (see Figure 12–21). Select the **Blocks** icon, and AutoCAD displays the icons (when the display is set to Large icons) representing all the blocks in the selected drawing. Select one of the block icons (do not double-click), and AutoCAD displays corresponding raster images and descriptions in the Preview pane and Description pane respectively. Figure 12–22 shows the selection of the block named "UB1216JF_P" and the corresponding display of its raster image and description. You can enlarge the Preview pane to get a bigger picture. Note, however, that it is a raster image, and sharp details are not available in close-up views.

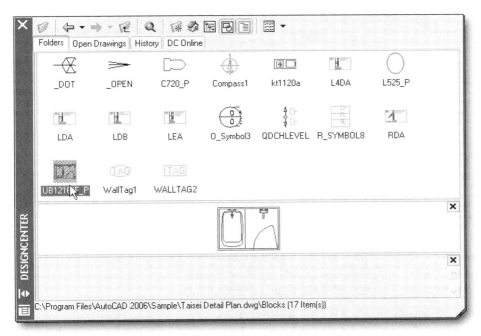

Figure 12–22 *The Preview pane displaying the raster image of the block named "UB1216JF_P"*

VIEWING IMAGES

If the item of content is a bitmap image, its name can be displayed in the content area by selecting the name of its container in the Tree View. For example, there is a subfolder named *Sample* in the folder named *AutoCAD 2006*. In the folder named *Database Connectivity* is another folder named *CAO*. When *CAO* is highlighted in the Tree View, you can choose the bitmap file named *dbcm_query.bmp* in the content area, and the icon resembling a question mark is displayed in the Preview pane (see Figure 12–23). Also, there is a description that can be viewed in the Description pane. All the images in the subfolder have a file extension of *.bmp*. Similarly you can also view any other graphic file type.

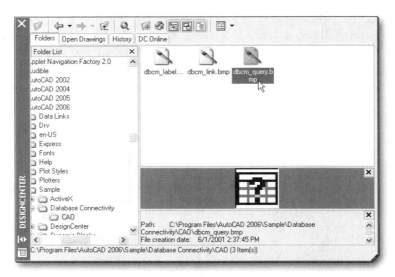

Figure 12–23 *Display of the image named "dbcm_query.bmp" with its raster image in the Preview pane and its corresponding description in the Description pane*

Note: When you have selected the name of a folder in the Tree View, if there are any files in that folder that are viewable, their names will appear in the content area. Some folders may have other files in them, but if they are not viewable, their names will not appear.

You can also load a drawing or folder from the Search dialog box by dragging and dropping it into the content area or Tree View pane. Or you can right-click and select LOAD INTO CONTENT AREA from the shortcut menu.

Note: The Load feature does not load the item selected into your drawing. It only loads its icon into the content area, as shown in the next section on adding content to drawings.

When file manipulations are made to the content area or Tree View, they do not always show up immediately on the screen. If this happens, you can right-click in one of the panes and choose REFRESH from the shortcut menu. The views will be updated.

To open a drawing being displayed in the content area, right-click its icon and choose OPEN IN APPLICATION WINDOW from the shortcut menu. You can also drag and drop the icon into the drawing area. Be sure to drop the icon in an area that is clear of another drawing. If necessary, resize or minimize any other drawing(s) first.

OPEN DRAWINGS

The **Open Drawings** tab of the DesignCenter window displays a list in the Tree View of drawings that are currently open (see Figure 12–24). When you select one of the drawings, its content types will be displayed in the content area. Or you can double-click on the file name, and it will display the content types at one level below the name you double-clicked on. Selecting one of the content types causes content of that type (if any exists in the drawing) to be listed in the content area.

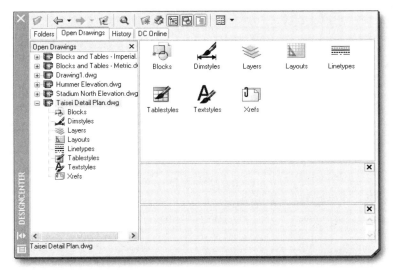

Figure 12–24 *The DesignCenter window with the Open Drawings tab displayed*

HISTORY

The **History** tab of the DesignCenter window displays the items accessed through AutoCAD DesignCenter, including their paths (see Figure 12–25).

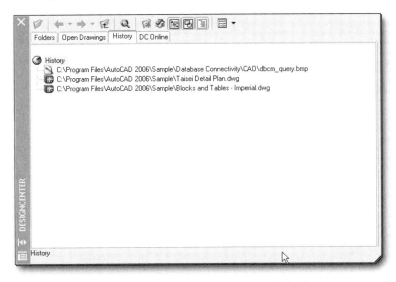

Figure 12–25 *The DesignCenter window with the History tab displayed*

DC ONLINE

The **DC Online** tab of the DesignCenter window, when selected, causes AutoCAD to log on to the DesignCenter Online on the Internet, provided your computer is Internet-ready (see Figure 12–26).

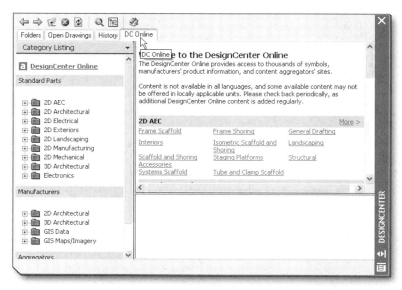

Figure 12–26 *The DesignCenter window with the DC Online tab displayed and Category Listing selected*

There are two panes in the **DC Online** tab of the DesignCenter window. The left pane has four views: **Category Listing**, **Search**, **Settings**, and **Collections**.

Category Listing

In the **Category Listing** view there are three categories: **Standard Parts**, **Manufacturers**, and **Aggregators** (see Figure 12–26). The **Standard Parts** category includes groups of drawings and images that can be used in architectural and engineering design disciplines such as architecture, landscaping, mechanical and GIS. For example, you can select the box to the left of the **2D Architectural** group and then select the box to the left of the **Landscaping** subgroup when expanded. This expands to the list of types of content. If you select **Tables** from this list, the right pane will display thumbnail sketches of drawings or images that are available for download (see Figure 12–27). When you select an image, additional links are displayed in the lower half of the right pane along with a larger sketch of the content.

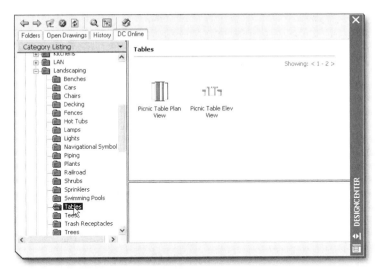

Figure 12–27 *Category Listings with 2D Architectural/Landscaping/Tables groups displayed*

Another group of content related to tables can be found under the **2D Architectural** group by selecting the **Furniture** subgroup and then selecting Tables (see Figure 12–28).

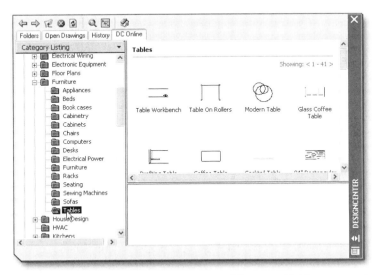

Figure 12–28 *Category Listings with 2D Architectural/Furniture/Tables groups displayed*

Note: If you are looking for drawings, images, or links to drawings and images of a particular type of content (such as tables in the example above), there is usually more than one group or path to a group that includes such drawings, images or links. See the explanation of using the Search view in this section.

The **Manufacturers** category includes groups of Web sites or Internet addresses of manufacturers of products used in architectural and engineering construction. Through these Web sites, drawings and images can be downloaded where available. For example, in the **Manufacturers** group (similar to the selection in the **Category Listing** group), you can select the box to the left of the **2D Architectural** group and then select the box to the left of the **Landscaping** subgroup when expanded. This expands to the list of types of content. If you select **Outdoor Furnishing** from this list, the right pane will display one or more Internet addresses of Web sites from which you can access drawings, images, or other content data that are available for download (see Figure 12–29).

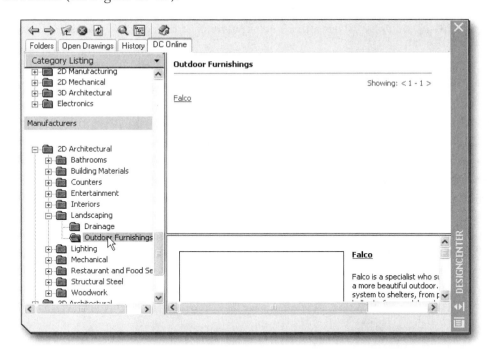

Figure 12–29 *Manufacturers with 2D Architectural/Landscaping/Outdoor Furnishings groups displayed*

The **Aggregators** category contains lists of libraries compiled by commercial catalog providers. Like the items in the **Manufacturers** category, you can access Web sites that contain or lead to drawings and blocks for use in architectural and engineering design and drafting. For example, in the **Aggregators** group (similar to the selection in the **Category Listing** and **Manufacturers** group), you can select the **AEC Aggregators** from this list, and the right pane will display one or more Internet addresses of Web sites from which you can access drawings, images, or other content data that are available for download (see Figure 12–30).

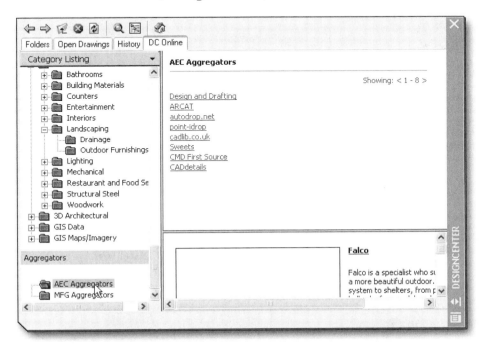

Figure 12–30 *Aggregators groups displayed*

Search

In the Search view there is a text box in which you can enter words or combinations of characters to tell AutoCAD what type of content to search for. You can display details for how to use Boolean and multiple-word search strings by selecting the **Need Help?** link. Figure 12–31 shows an example of entering "table" in the text box and selecting the **Search** button. The right pane shows the result of the search. Selecting one of the items of content in the right pane causes additional links to be displayed in the lower-half of the right pane along with a larger sketch of the content.

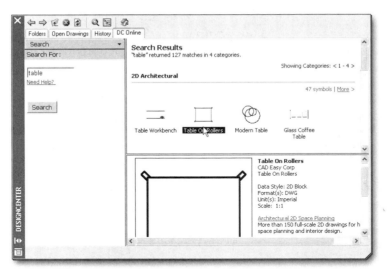

Figure 12–31 *The DesignCenter window with the DC Online tab displayed and Search view selected*

Settings

In the Settings view there are two text boxes: **Number of Categories per page**, and **Number of Items per page** (see Figure 12–32), where you can select from 5, 10, and 20 in the **Number of Categories per page** and from 50, 100, and 200 in the **Number of Items per page** text boxes to specify the **Max Search** numbers for Categories and Items respectively.

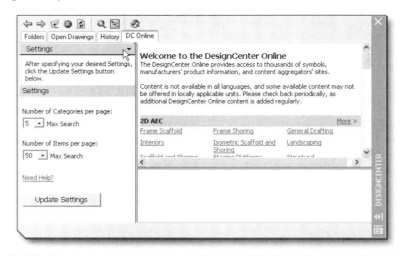

Figure 12–32 *The DesignCenter window with the DC Online tab displayed and Settings view selected*

Collections

In the **Collections** view there is a list of collections for each of the three categories with a check box beside each collection (see Figure 12–33). check a particular collection's check box that you wish to be displayed in the **Category Listing** view. Once you have selected/deselected the desired collections, select **Update Collections**, and AutoCAD will return to the **Category Listing** view displaying the list of collections whose check boxes have checks in them.

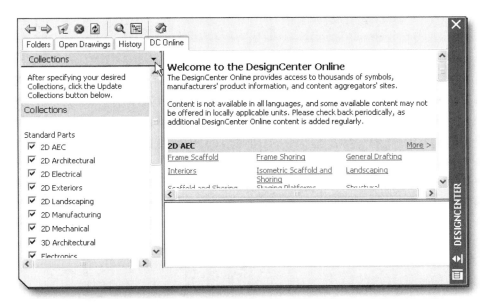

Figure 12–33 *The DesignCenter window with the DC Online tab displayed and Collections view selected*

ADDING CONTENT TO DRAWINGS

As discussed in the introduction to this chapter, blocks, external references, layers, raster images, linetypes, layouts, text styles, dimension styles, and custom content created by third-party applications are the content types that can be added to the current drawing session by using the AutoCAD DesignCenter. Content in the AutoCAD DesignCenter can be dragged into a tool palette for use in the current drawing.

LAYERS, LINETYPES, TEXT STYLES, AND DIMENSION STYLES

A definition of a layer or linetype or a style created for text or dimensions can be dragged into the current drawing from the content area. It will become part of the current drawing as though it had been created in that drawing.

BLOCKS

A block definition can be inserted into the current drawing by dragging its icon from the content area into the drawing area. You cannot, however, do this while a command is active. Only one block definition at a time can be inserted from the content area. Figure 12–34 shows a block in the content area of the AutoCAD DesignCenter being dragged into the tool palette of the current drawing.

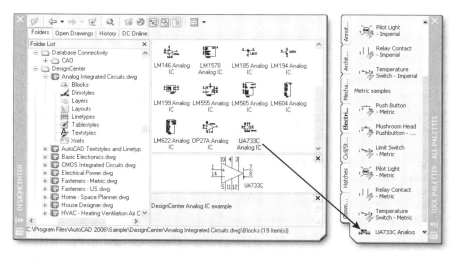

Figure 12–34 *The block named **UA733C Analog IC** being dragged from the DesignCenter window into the tool palette of the current drawing.*

 Note: If you double-click a block and it has nested blocks, the hierarchy is flattened.

There are two methods of inserting blocks from the AutoCAD DesignCenter. One method uses *Autoscaling*, which scales the block reference as needed based on the comparison of the units in the source drawing to the units in the target drawing. Another method is to use the Insert dialog box to specify the insertion point, scale, and rotation.

 Note: When dragging and dropping using the automatic scaling method, the dimension values inside the blocks will not be true.

When you drag a block definition from the content area or the Find dialog box into your drawing, you can release the button on the pointing device (drop) when the block is at the desired location. This is useful when the desired location can be specified with a running object snap mode in effect. The block will be inserted with the default scale and rotation.

To invoke the Insert dialog box, double-click the block icon or right-click the block definition in the content area or Find dialog box and then select INSERT BLOCK from the shortcut menu. In the Insert dialog box, specify the Insertion point, Scale, and Rotation, or select **Specify On-screen**. You can select **Explode** to have the block definition exploded on insertion.

RASTER IMAGES

A raster image such as a digital photo, print screen capture saved in a paint program as a bitmap, or a company logo can be copied into the current drawing by dragging its icon from the content area into the drawing area. Then specify the Insertion point, Scale, and Rotation. Or you can right-click the image icon and choose ATTACH IMAGE from the shortcut menu.

EXTERNAL REFERENCES

To attach an external reference to the current drawing, drag its icon from the content area or Find dialog box with the right button on the pointing device into the drawing area. Then release the button and select ATTACH from the shortcut menu. The Attach Xref dialog box is displayed, from which you can choose between **Attachment** or **Overlay** as the **Reference Type** option. From the Attach Xref dialog box, specify the Insertion point, Scale, and Rotation, or select **Specify On-screen.**

When you copy, insert, or attach content into a drawing that already has an item of the content type with the same name, AutoCAD will display a warning. The item is not added to the drawing. If the item is a block or external reference, AutoCAD checks to determine if the name is already listed in the database. The warning "Duplicate definition of [object][name] ignored" is displayed. If the external reference exists in the drawing, the warning "Xref [name] has already been defined. Using existing definition" is displayed. If the item with the duplicate name is a layer, linetype, or other item that is not a block or xref, the warning "Add [object] operation performed. Duplicate definitions will be ignored" is displayed.

You can exit AutoCAD DesignCenter by either selecting the X in the upper-right corner of the DesignCenter Window or entering **adcclose** at the AutoCAD On-Screen prompt.

REVIEW QUESTIONS

1. What command is used to invoke the AutoCAD DesignCenter?

    a. DSNCEN

    b. DGNCEN

    c. ADCENTER

    d. DGNCTR

2. Which of the following are considered to be drawing content types?

    a. blocks

    b. layers

    c. linetypes

    d. all the above

3. AutoCAD allows items to be directly edited from the DesignCenter.

    a. True

    b. False

4. Which of the following are not drawing content types?

    a. lines

    b. circles

    c. arcs

    d. all of the above

5. Within the DesignCenter a drawing is considered to be a _____?

    a. content

    b. container

    c. folder

    d. a & b

    e. a & c

6. Invoking the AutoCAD DesignCenter command opens the DesignCenter dialog box.

    a. True

    b. False

7. The default position for the DesignCenter is in the lower right corner.

    a. True

    b. False

8. Which of the following areas of the DesignCenter displays the names and icons representing content?

    a. Toolbar

    b. Content area

    c. Tree

    d. Preview pane

    e. Description pane

9. Large Icons, Small Icons, List and Details are four optional modes of displaying content using which of the following buttons?

    a. LOAD

    b. FIND

    c. UP

    d. VIEWS

10. Within the Tree View pane, which button can display the previous items and their paths accessed through the AutoCAD DesignCenter?

    a. Open Drawings

    b. Desktop

    c. History

    d. None of the above

CHAPTER 13

Utility Commands

INTRODUCTION

After completing this chapter, you will be able to do the following:

- Create, customize, and use tool palettes
- Partial Load drawings
- Manage drawing properties
- Manage named objects
- Delete unused named objects
- Use the utility display commands
- Use object properties
- Use X,Y,Z filters
- Use the SHELL command
- Set up a drawing by means of the MVSETUP utility
- Use the Layer Translator
- Use the TIME and AUDIT commands
- Customize AutoCAD settings
- Export and Import data
- Understand CAD Standards
- Use script and slide commands
- Set up and use workspaces
- Use Calculator

TOOL PALETTES

Tool palettes are tabbed areas within the Tool Palettes window that provide an efficient method for organizing, sharing, and placing blocks and hatches, executing a single command or a string of commands, and using custom tools provided by third-party developers. For example, one tool palette is named **Architectural**, with blocks representing doors, windows, and fixtures.

A special facility of tool palettes is the ability to preload a palette with objects with specific properties. When these objects are selected and dragged into the drawing area, they take their properties with them. For instance, if a green polyline on layer 7 is selected and dragged onto a tool palette, then dragging it back into the drawing area invokes the PLINE command, and the resulting polyline will be green and on layer 7, regardless of the current layer or color control. The following objects can be dragged onto a tool palette:

- Blocks
- Dimensions
- External references
- Hatch Patterns and Gradient fills
- Objects such as arcs, circles, lines, and polylines
- Raster images
- AutoCAD commands or strings of commands

The TOOLPALETTES command causes the Tool Palettes window to be displayed. Invoke the TOOLPALETTES command from the Standard toolbar by selecting **Tool Palettes Window** (see Figure 13–1) and AutoCAD displays the Tool Palettes window (see Figure 13–2) with tabs that include: **Command Tools**, Hatches, **Civil/Structural**, **Electrical**, **Mechanical**, **Architectural**, and **Annotation**, which contain commands, hatch patterns, icons representing blocks, and callout/bubble symbols.

Figure 13–1 *Invoking the* TOOL PALETTES *command from the Standard toolbar*

 Note: You can open or close the Tool Palettes window with the CTRL+3 key combination.

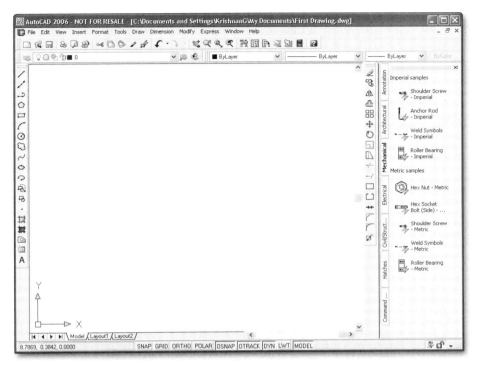

Figure 13–2 *The Tool Palettes window in the docked position with the Mechanical tab displayed*

The default position for the Tool Palettes window is floating on the right side of the screen. When the Tool Palettes window is undocked, it can be docked by double-clicking in the title bar (which may be on the left or right side of the window) or by placing the cursor over the title bar and dragging the window all the way to the side on which you wish to dock it. Its position can be changed by placing the cursor over the double line bar at the top of the window and either double-clicking or dragging the window into the screen area (or across to a docking position on the opposite side of the screen). Double-clicking causes the Tool Palettes window to become undocked and to float in the drawing area (see Figure 13–3).

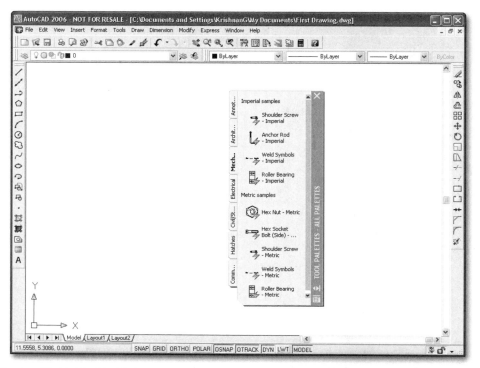

Figure 13–3 *The Tool Palettes window in the floating position*

TOOL PALETTES WINDOW SHORTCUT MENUS

Four different shortcut menus are available for managing the Tool Palettes window, its palettes, and elements on the palettes, depending on the location (tab, tool palette element, open area in tool palette, and tool palette window title bar) of the cursor when you right-click.

The shortcut menu displayed when you right-click the active tab in the Tool Palettes window (see Figure 13–4) includes these options: *MOVE UP, MOVE DOWN, NEW PALETTE, DELETE PALETTE, RENAME PALETTE, VIEW OPTIONS*, and *PASTE* (*DELETE PALETTE, VIEW OPTIONS*, and *PASTE* are not available when an inactive tab is selected).

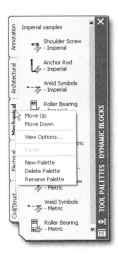

Figure 13–4 *The shortcut menu when right-clicking a Tool Palettes window tab*

Choose MOVE UP or MOVE DOWN to move the selected tab up or down one place respectively in the order of tabs.

Choose NEW PALETTE to create a new palette. AutoCAD will create a new tool palette to which you can add tools. Enter a name or press ENTER to use the default name.

Note: Tool palettes can be used only in the current or later version of AutoCAD in which they were created. For example, you cannot use a tool palette that was created in AutoCAD 2006 in AutoCAD 2005.

Choose RENAME PALETTE to rename the current tool palette.

Choose DELETE PALETTE to remove the current tool palette.

Choose VIEW OPTIONS to display the View Options dialog box (see Figure 13–5) from which you can control the display of tools in the current tool palette or in all tool palettes. The **Image size** slider bar allows you to change the size of the images. The **View style** section lets you set how the tool elements are displayed. Selecting **Icon only** causes the block/hatch pattern icon to be displayed as an image only without text. Selecting **Icon with text** causes the block/hatch pattern icon to be displayed as an image with the descriptive text below it. Selecting **List view** causes the block/hatch pattern icon to be displayed as an image with the descriptive text to its right, allowing for a more compressed listing of the symbols when used with a small image. The **Apply to** list box allows you to choose whether the changes are applied to the **Current Palette** or to **All Palettes**. To exit the View Options dialog box and accept the changes, choose **OK**. To exit without accepting the changes, choose **Cancel**.

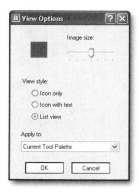

Figure 13–5 *The View Options dialog box*

Choose PASTE to paste a block/hatch pattern shortcut (that has been copied to the Clipboard) on the tool palette. If there is nothing on the Clipboard, or if whatever is on the Clipboard is not a block/hatch pattern shortcut, then the PASTE option is not active and cannot be selected.

The shortcut menu displayed when you right-click one of the elements in a tool palettes window (see Figure 13–6) includes the options CUT, COPY, DELETE, RENAME, UPDATE TOOL IMAGE, BLOCK EDITOR, and PROPERTIES (UPDATE TOOL IMAGE and BLOCK EDITOR are not available if the selected element is a command or hatch pattern).

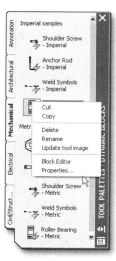

Figure 13–6 *The shortcut menu when right-clicking an element in the Tool Palettes window*

Choose CUT to delete the selected element and place it on the Clipboard.

Choose COPY to copy the selected element and place it on the Clipboard.

Choose DELETE to remove the selected element.

Choose RENAME to rename the selected element.

Choose PROPERTIES to display the Tool Properties dialog box showing the properties of the selected element. Tool properties are explained later in this chapter.

Choose UPDATE TOOL IMAGE to update the selected image of the tool when the definition for a block, xref, or raster image is changed. You must save the drawing before you can update the tool image. The icon for a block, xref, or raster image in a tool palette is not automatically updated if its definition changes.

Choose BLOCK EDITOR to invoke the BEDIT command, explained in Chapter 10, covering Dynamic Blocks.

The shortcut menu that is displayed when you right-click in an open area of the Tool Palettes window (see Figure 13–7) includes the options ALLOW DOCKING, AUTO-HIDE, TRANSPARENCY, VIEW OPTIONS, SORT BY, PASTE, ADD TEXT, ADD SEPARATOR, NEW PALETTE, DELETE PALETTE, RENAME PALETTE, and CUSTOMIZE.

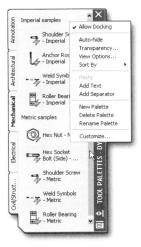

Figure 13–7 *The shortcut menu when right-clicking in an open area of the Tool Palettes window*

The ALLOW DOCKING option, when checked, allows you to drag the Tool Palettes window to one side of the screen and dock it. When ALLOW DOCKING is not checked, the window cannot be docked.

The AUTO-HIDE option, when checked, causes the Tool Palettes window (only when floating) to be hidden, except for the title bar, when the cursor is not over the title bar or the window. To display the window, move the cursor over the title bar.

The TRANSPARENCY option, when selected, causes the Transparency dialog box to be displayed (see Figure 13–8). The **Less-More** slider controls the degree of transparency. Sliding the bar to the left (**Less**) causes the Tool Palettes window (only when floating) to be opaque. The closer the indicator is to the right (**More**), the more transparent the window will become. Selecting **Turn off window transparency** prevents the Tool Palettes window from becoming transparent.

Figure 13–8 *The Transparency dialog box*

The VIEW OPTIONS option, when selected, causes the View Options dialog box to be displayed. It is similar to the one explained earlier.

Choosing SORT BY causes a shortcut menu to be displayed from which you can sort the elements on the tool palette by NAME or TYPE.

The PASTE option functions similarly to the one explained earlier.

Choose ADD TEXT to add descriptive text (at the location of the cursor when right-clicked).

Choose ADD SEPARATOR to create a separating line on the palette at the location of the cursor when right-clicked.

The NEW PALETTE, DELETE PALETTE, and RENAME PALETTE options function similarly to the ones explained earlier.

The CUSTOMIZE option, when selected, causes the Customize dialog box to be displayed (see Figure 13–9).

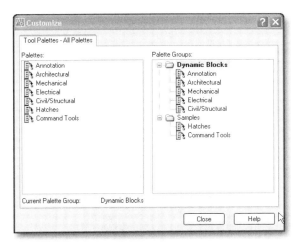

Figure 13-9 *The Customize dialog box*

The Customize dialog box lets you manage the palettes and groups of palettes shown in the Tool Palettes window. It allows you to create, modify, and organize palettes in addition to importing and exporting palette files. You can organize tool palettes into groups and specify which group of tool palettes is displayed. For example, if you have several tool palettes that contain hatch patterns, you can create a group called Hatch Patterns. You can then add all your tool palettes that contain hatch patterns to the Hatch Pattern group. When you set the Hatch Pattern group as the current group, only those tool palettes you've added to the group are displayed.

The **Palettes** section lists all available palettes. Click and drag a palette to move it up or down in the list. From the shortcut menu that appears when you right-click a palette, you can rename, delete, or export the palette. (When you export a palette, it's saved to a file with an *.xtp* extension.) From the shortcut menu that appears when you right-click in the open area of the dialog box in the **Palettes** section, you can create a new palette or import a palette.

The **Palettes Group** section displays the organization of your palettes in a tree view. Click and drag a palette to move it into another group. From the shortcut menu that appears when you right-click a palette group, you can create, delete, rename, export, and import a group, remove a tool from a palette group, and set it current to display the selected group of palettes.

Choosing **Current Palette Group** displays the name of the palette group currently shown.

The shortcut menu displayed when you right-click on the Tool Palette window title bar (see Figure 13–10) includes the options MOVE, SIZE, CLOSE, ALLOW DOCKING, AUTO-HIDE, TRANSPARENCY, NEW PALETTE, RENAME, CUSTOMIZE, DYNAMIC BLOCKS, SAMPLES, and ALL PALETTES.

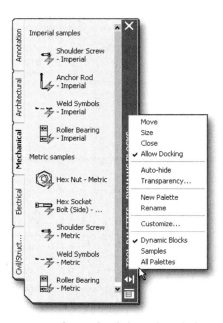

Figure 13–10 *The shortcut menu when right-clicking the title bar of the Tool Palettes window*

Choose MOVE to drag the Tool Palettes window to another location on the screen.

Choose SIZE to drag the title bar edge of the Tool Palettes window to make it wider or narrower. You can make the window longer or shorter (vertically) by placing the cursor on the top or bottom until the double arrow appears and then dragging the edge up or down.

Choose CLOSE to close the Tool Palettes window.

ALLOW DOCKING, AUTO-HIDE, TRANSPARENCY, NEW PALETTE, RENAME, and CUSTOMIZE are explained in the previous section.

The shortcut menu also lists available palette groups and you can switch between the groups.

INSERT BLOCKS/HATCH PATTERNS FROM A TOOL PALETTE

To insert a block from a tool palette, simply place the cursor on the block symbol in the tool palette, press the pick button and drag the symbol into the drawing area. The block will be inserted at the point where the cursor is located when the pick button is released. This procedure is best implemented by using the appropriate OSNAP mode. Another method of inserting a block from a tool palette is to select the block symbol in the tool palette and then select a point in the drawing area for the insertion point.

To draw a hatch pattern that is a tool in a tool palette, place the cursor on the hatch pattern symbol in the tool palette, press the pick button and drag the symbol into the boundary to receive the hatch pattern and release the pick button. Another method of drawing a hatch pattern that is a tool in a tool palette is to select the hatch pattern symbol in the tool palette and then select a point within a boundary in the drawing area.

BLOCK TOOL PROPERTIES

Blocks whose symbols appear in a tool palette are not, as a rule, blocks defined in the current drawing. Usually, they reside as block definitions in another drawing or in some cases they might even be drawing files. As a tool in a tool palette, a block has tool properties. To access the block/drawing tool properties, right-click on the tool's symbol in the tool palette, and select PROPERTIES from the shortcut menu. The Tool Properties dialog box is displayed (see Figure 13–11).

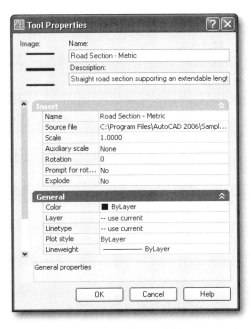

Figure 13–11 *The Tool Properties dialog box for block/drawing tools*

The Tool Properties dialog box controls the properties associated with the selected tool. The Insert properties of a block/drawing tool in a tool palette include **Name, Source file, Scale, Auxiliary Scale, Rotation, Prompt for rotation,** and **Explode**. The General properties include **Color, Layer, Linetype, Plot style,** and **Lineweight**.

The **Name** specifies the name of the block of the selected tool. The **Source file** property edit box lists the path to the drawing file where the block that is a tool in the tool palette resides. Or, if the tool is a drawing file, the edit box lists the path to it.

Note: Any change to the path/file name in the **Source file** property text box will prevent the block/drawing from being inserted.

The **Scale** factor specifies the XYZ scale factor of the block. **Auxiliary scale** overrides the regular scale setting and multiplies your current scale setting by the plot scale or the dimension scale. The **Rotation** specifies the rotation angle of the block. If the **Prompt for rotation** list box is set to **No**, the block will be inserted with the default rotation angle. If it is set to **Yes,** you will be prompted for the angle of rotation. The **Explode** list box determines whether or not the block is exploded when inserted. If the **Explode** list box is set to **No**, the block will be inserted as an unexploded block. If it is set to **Yes**, the separate parts that make up the block will be drawn as separate entities, that is, as a block that is inserted and exploded. The **Color, Layer, Linetype, Plot Style,** and **Lineweight** options specify the override for the color, layer, linetype, plot style, and lineweight respectively.

PATTERN TOOL PROPERTIES

Hatch patterns whose symbols appear in a tool palette have tool properties. The Pattern properties of a hatch pattern tool in a tool palette include **Tool type, Type, Pattern name, Angle, Scale, Auxiliary scale, Spacing,** and **ISO pen width**, and **Double**. The General properties include **Color, Layer, Linetype, Plot style,** and **Lineweight**. To access the pattern tool properties, right-click on the tool's symbol in the tool palette, and select PROPERTIES from the shortcut menu. The Tool Properties dialog box is displayed (see Figure 13–12).

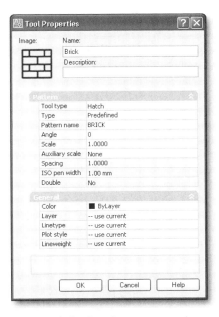

Figure 13–12 *The Tool Properties dialog box for pattern tools*

Tool type specifies whether the tool is a hatch or a gradient. **Type** specifies the pattern type of the hatch. The pattern types include **User-defined**, **Predefined** from one of the hatch patterns listed in the **Pattern** list box, or **Custom** from one of the hatch patterns in the drawing as listed in the **Custom Pattern** list box. **Pattern name** specifies the pattern name of hatch. **Angle, Scale,** and **Spacing** specify the angle, scale and spacing of the hatch respectively. **Auxiliary scale** overrides the regular scale setting and multiplies your current scale setting by the plot scale or the dimension scale. **ISO pen width** specifies the ISO pen width of an ISO hatch pattern. **Double** determines whether the hatch pattern is double. **Color, Layer, Linetype, Plot style**, and **Lineweight** specify the override for the color, layer, linetype, plot style, and lineweight respectively.

CREATING AND POPULATING TOOL PALETTES

A new tool palette can be created from the Tool Palette shortcut menu by selecting the New Tool Palette option as described earlier in this section.

You can also create a new tool palette from the Tree view or Content area of the DesignCenter. From the Tree view or Content area, highlight an item and then right-click to display the shortcut menu. If the selected item is a folder, one of the available options in the shortcut menu is *CREATE TOOL PALETTE OF BLOCKS* (see Figure 13–13). If there are no blocks in the folder, a message will be displayed stating "Folder does not contain any drawing files". If the selected item is a drawing, then one of the available options in the shortcut menu is CREATE TOOL PALETTE. If the drawing does not contain any blocks, a message will be displayed stating "Drawing does not contain any block definitions".

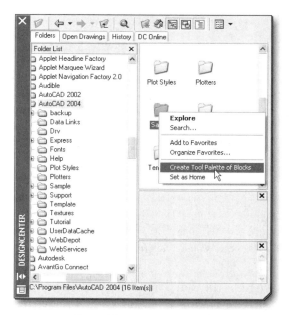

Figure 13–13 *Shortcut menu – DesignCenter content area*

When one of the options to create a tool palette is selected, AutoCAD creates a new tool palette, which will be populated with the drawings from the selected folder or blocks from the selected drawing.

From the Content area, in addition to creating tool palettes by right-clicking a folder or drawing, you can also right-click on a block and select the CREATE TOOL PALETTE option from the shortcut menu. In this case the new tool palette will contain only the selected block. In this case you will be prompted to name the tool palette.

You can also drag and drop a block from a drawing or a drawing from the Content area of the DesignCenter on to the one of the existing palettes.

PARTIAL LOAD

The PARTIALOPEN command allows you to work with just part of a drawing by loading geometry only from specific views or layers. When a drawing is partially open, specified and named objects are loaded. Named objects include blocks, layers, dimension styles, linetypes, layouts, text styles, viewport configurations, UCSs, and views.

Invoke the PARTIALOAD command from the Files menu and AutoCAD displays the Partial Load dialog box (see Figure 13–14).

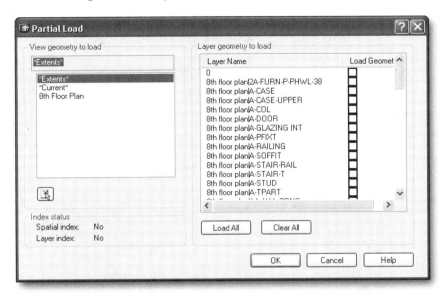

Figure 13–14 *The Partial Load dialog box*

In the Partial Load dialog box, select a view. The default view is Extents. Only geometry from model space views that are saved in the current drawing can be loaded. Select layer(s). No layer geometry is loaded into the drawing if you do not select any layers to load. However, all drawing layers will exist in the drawing. Even if the geometry from a view is specified to load, and no layer is specified to load, no geometry is loaded. When the view(s) and layer(s) have been specified, choose **OK** to partially load the geometry from the specified layers.

 Note: If you invoke the PARTIALOAD command in a drawing that has not been partially opened, AutoCAD will respond with the message "Command not allowed unless the drawing has been partially opened." This response is only displayed at the command line and not shown at the On-Screen prompt.

DRAWING PROPERTIES

Drawing properties allow you to keep track of your drawings by having properties assigned to them. The properties you assign will identify the drawing by its title, author, subject, and keywords for the model or other data. Hyperlink addresses or paths can be stored along with custom properties.

Invoke **Drawing Properties** from the **File** menu and AutoCAD displays the Drawing Properties dialog box (see Figures 13–15).

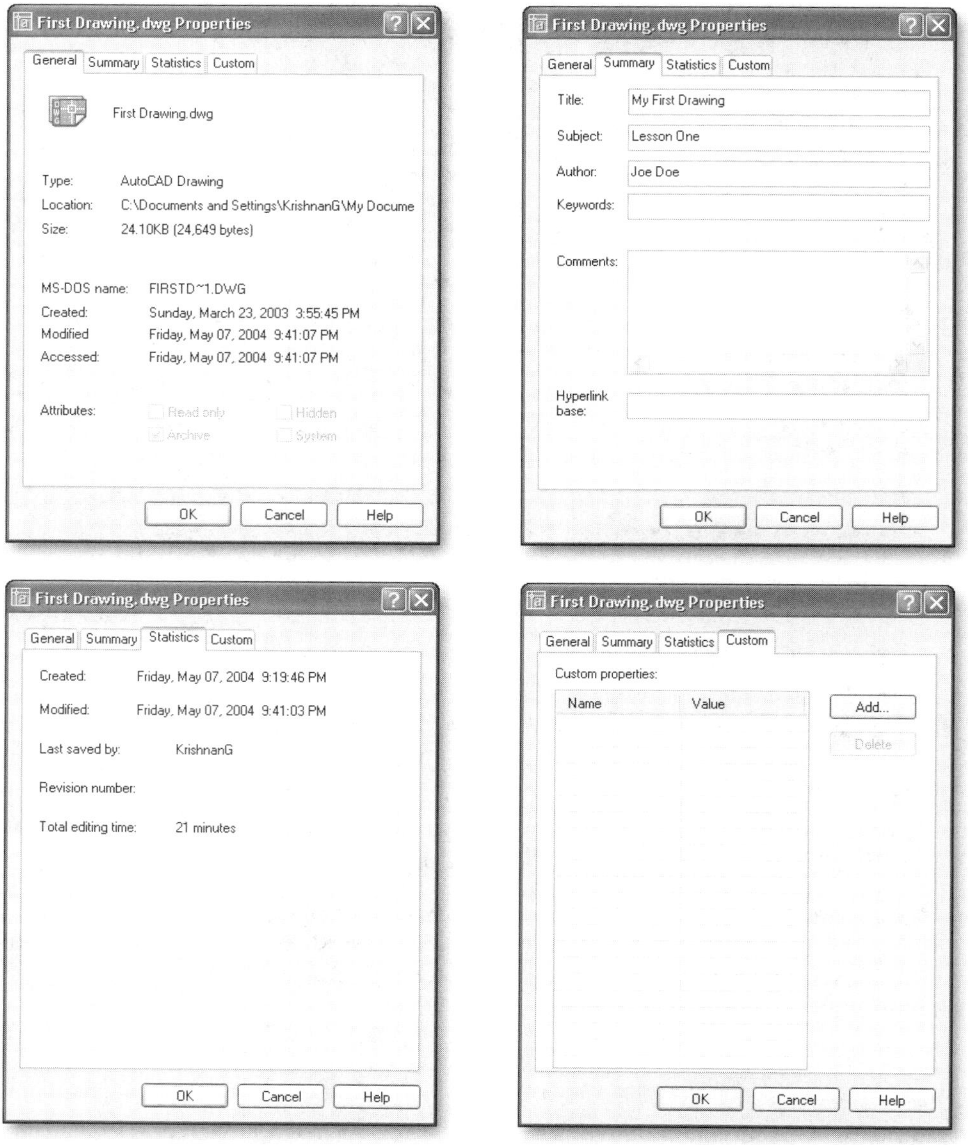

Figure 13–15 *The Drawing Properties dialog box: General tab (top left), Summary tab (top right), Statistics tab (bottom left), Custom tab (bottom right)*

The Drawing Properties dialog box has four tabs: **General**, **Summary**, **Statistics**, and **Custom**.

The **General** tab (see Figure 13–15 top left) displays the drawing type, location, size, and other information. These values come from the operating system. These fields are read-only. However, the attributes options are made available by the operating system if you access file properties through Windows Explorer®.

The **Summary** tab (see Figure 13–15 top right) lets you enter a Title, Subject, Author, Keywords, Comments, and a Hyperlink base for the drawing. **Keywords** for drawings sharing a common property will help in your search. For a **Hyperlink base**, you can specify a path to a folder on a network drive or an Internet address.

The **Statistics** tab (see Figure 13–15 bottom left) displays information such as the dates files were created and last modified. You can search for all files created at a certain time.

The **Custom** tab (see Figure 13–15 bottom right) lets you specify custom properties. Enter the names of the custom fields in the left column, and the value for each custom field in the right column.

Note: Properties entered in the Drawing Properties dialog box are not associated with the drawing until you save the drawing.

QUICKCALC

AutoCAD provides an "on demand" calculator with a full range of mathematical, scientific, and geometric calculations, which can be used to create and use variables and convert units of measurement. This QuickCalc calculator includes shortcut functions that are found in the Geometric Calculator but are more easily accessed in QuickCalc.

Invoke the QUICKCALC command from the Standard toolbar by selecting **QuickCalc** (see Figure 13–16) and AutoCAD displays the QuickCalc palette (see Figure 13–17).

Figure 13–16 *Invoking the* QUICKCALC *command from the Standard toolbar*

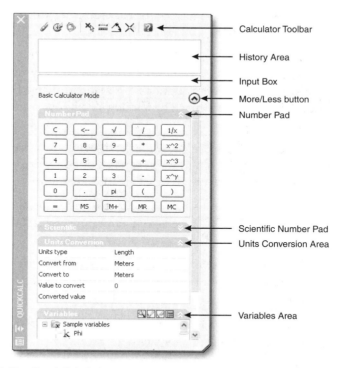

Figure 13–17 *The QuickCalc Palette*

The QuickCalc palette contains the following areas: **Toolbar, History, Input Box, More/Less Button, Number Pad, Scientific, Units Conversion**, and **Variables**.

The Toolbar lets you perform quick calculations of common functions.

Choosing **Clear** clears the **Input** box.

Choosing **Clear History** clears the **History** area.

Choosing **Paste value to command line** pastes the value in the **Input** box to the command line. When QuickCalc is used transparently during a command, this button is replaced by **Apply**.

Choosing **Get Coordinates** lets you calculate the coordinates of a point location that you have selected in the drawing.

Choosing **Distance Between Two Points** lets you calculate the distance between two point locations that you have selected on an object.

Choosing **Angle of Line Defined by Two Points** lets you calculate the angle of two point locations that you have selected on an object.

Choosing **Intersection of Two Lines Defined by Four Points** lets you calculate the intersection of four point locations that you have selected on an object.

Choosing **Help** causes the QuickCalc calculator help to be displayed.

The **History** area shows the expressions that have previously been evaluated. From it you can copy expressions to the Clipboard.

The **Input** area is where expressions are entered and retrieved. Choosing = (equal) evaluates the expression and displays the results in the Input area.

Choosing the **More/Less** icon causes all function areas to be hidden or displayed. Right-clicking allows you to select the individual function areas to hide or display.

The **Number Pad** is a standard calculator keypad where numbers and symbols for arithmetic expressions can be entered. Enter a value or expression and then click the equal (=) sign to evaluate the expression.

The **Scientific** area lets you evaluate trigonometric, logarithmic, exponential, and other expressions commonly associated with scientific and engineering applications.

The **Units Conversion** area lets you convert units of measurement from one unit type to another unit type. Only decimal values without units are accepted.

> The **Units type** box lets you select length, area, volume, and angular values from a list.
>
> The **Convert from** box lists the units of measurement from which to convert.
>
> The **Convert to** box lists the units of measurement to which to convert.
>
> The **Value to convert** box is where you enter the value to convert.
>
> The **Converted value** box is where the converted value is displayed.

The **Variables** area lets you access predefined constants and functions. You can use the **Variables** area to define and store additional constants and functions.

The **Variables Tree** has predefined shortcut functions and user-defined variables stored in tree format. The shortcut functions are common expressions that are a combination of a function and an object snap. Following are the available predefined shortcut functions:

> **Dee** is short for **dist(end,end)** and lets you specify the distance between two endpoints.
>
> **Ille** is short for **ill(end,end,end,end)** and lets you specify the intersection of two lines defined by four endpoints.

Mee is short for **(end+end)/2** and lets you specify the midpoint between two endpoints.

Nee is short for **nor(end,end)** and lets you specify the unit vector in the XY plane and normal to two endpoints.

Rad lets you specify the radius of a selected circle, arc, or polyline arc.

Vee is short for **vec(end,end)** and lets you specify the vector from two endpoints.

Vee1 is short for **vec1(end,end)** and lets you specify the unit vector from two endpoints.

Choosing the **Calculator** icon returns the converted value to the **Input** area.

Choosing **New Variable** opens the Variable Definition dialog box, where you can specify a new variable.

Choosing **Edit Variable** opens the Variable Definition dialog box, where you can make changes to the selected variable.

Choosing **Delete** variable deletes the selected variable.

Choosing **Calculator** returns the selected variable to the Input box.

MANAGING NAMED OBJECTS

The RENAME command allows you to change the names of blocks, dimension styles, layers, linetypes, text styles, views, User Coordinate Systems, or viewport configurations.

Invoke the RENAME command from the **Format** menu, by selecting **Rename**, and AutoCAD displays the Rename dialog box (see Figure 13–18).

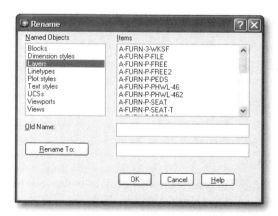

Figure 13–18 *Rename dialog box*

Note: Except for the layer named 0 and the linetype named Continuous, you can change the name of any of the named objects.

In the **Named Objects** list box, select the object name you want to change. The **Items** list box displays the names of all objects that can be renamed. To change the object's name, pick the name in the **Items** list box or enter it into the **Old Name** text box. Enter the new name in the **Rename To** text box, and select **Rename To** to update the object's name in the **Items** list box. To close the dialog box, choose **OK**.

DELETING UNUSED NAMED OBJECTS

The PURGE command is used to selectively delete any unused named objects.

Invoke the PURGE command from the **File** menu by selecting **Drawing Utilities** and then **Purge** from the flyout menu. AutoCAD displays the Purge dialog box (see Figure 13–19).

Figure 13–19 *Purge dialog box with the Items not used in drawing section displayed*

Choosing **View items you can purge** causes the **Items not used in drawing** section list box to list categories of named items, under which are listed the individual named items that have been defined in the drawing but are not currently being used. For example, if a layer has been defined but has nothing drawn on it and it is not the current layer, it can be purged. Or, if the drawing contains a block definition but a reference to that block has not been inserted, then it can be purged. Objects such as lines, circles, and other basic unnamed drawing elements cannot be purged.

Selecting **Confirm each item to be purged** causes AutoCAD to display the Confirm Purge dialog box and ask you to reply by selecting **Yes** or **No** before continuing. Selecting **Purge nested items** causes AutoCAD to purge nested items within any item selected.

Once an individual named item in the list has been selected or a category of items has been selected that has items that can be purged, the **Purge** and **Purge All** buttons at the bottom of the dialog box become operational. Choosing **Purge** purges the selected items and **Purge All** purges all unused items. Choosing **View items you cannot purge** causes the **Items currently used in drawing** list box to list categories of named items, under which are listed the individual named items that have been defined in the drawing but are currently being used (see Figure 13–20). For example, if a layer has something drawn on it or it is the current layer, it cannot be purged. Or, if a reference of a block has been inserted in the drawing, then that block cannot be purged. The text area under the list of items informs you why a selected item cannot be purged. For example, if the Standard Text Style is selected, the message "The default text style, STANDARD, cannot be purged" will be displayed.

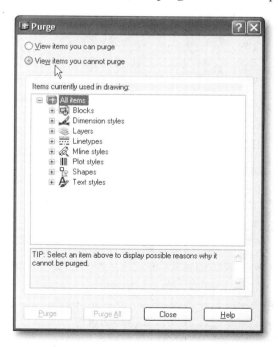

Figure 13–20 *Purge dialog box with the Items currently used in drawing: section displayed*

The PURGE command removes only one level of reference. For instance, if a block has nested blocks, the purge command removes only the most deeply nested block definition. To remove the second-, third-, or deeper-level blocks within blocks, you must repeat the purge command until there are no referenced objects. You can use the PURGE command at any time during a drawing session. Choosing **Purge nested items** removes all unused named objects from the drawing even if they are contained within or referenced by other unused named objects. The Confirm Purge dialog box is displayed, and you can cancel or confirm the items to be purged.

Individual shapes are part of a *.shx* file. They cannot be renamed, but references to those that are not being used can be purged. Views, User Coordinate Systems, and viewport configurations cannot be purged, but the commands that manage them provide options to delete those that are not being used.

COMMAND MODIFIER—MULTIPLE

MULTIPLE is not a command, but when used with another AutoCAD command, it causes automatic recalling of that command when it is completed. You must press ESC or right-click and select CANCEL from the short-cut menu to terminate this repeating process. Here is an example of using this modifier to cause automatic repeating of the ARC command:

> **multiple** (ENTER)
> Enter command name to repeat: **arc** (ENTER)

You can use the MULTIPLE command modifier with any of the draw, modify, and inquiry commands. PLOT, however, will ignore the MULTIPLE command modifier.

UTILITY DISPLAY COMMANDS

The utility display commands include VIEW, REGENAUTO, DRAGMODE, and BLIPMODE.

SAVING VIEWS

The VIEW command allows you to give a name to the display in the current viewport and have it saved as a view. You can recall a view later by using the VIEW command and responding with the name of the view desired. This is useful for moving back quickly to needed areas in the drawing without having to resort to zoom and pan. Naming and saving Views is a significant part of creating a drawing set with the Sheet Set Manager, discussed in Chapter 14.

Invoke the VIEW command by choosing **Named Views** from the **View** menu and AutoCAD displays the View dialog box (see Figure 13–21).

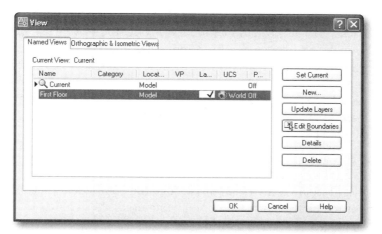

Figure 13–21 *View dialog box with the Named Views tab displayed*

In the **Named Views** tab, AutoCAD lists any saved view(s) in the list box. To create a new view, choose **New**. AutoCAD displays the New View dialog box.

In the New View dialog box, specify a name for the new view in the **View name** box. Specify a category for the named view in the **View category** box. You can select a view category from the list, enter a new category, or leave this option empty. Grouping by category will be helpful in creating drawing sheet sets.

The **Boundary** section lets you select one of the two options to specify the area for the new view.

Choosing **Current display** creates the current display as the new view.

Choosing **Define window** allows you to specify diagonally opposite corners of a window that will define the area for the new view. AutoCAD temporarily closes the dialog boxes so that you can use the pointing device to define the opposite corners of the new view window.

The **Settings** section lets you select where to save the new view.

Choosing **Store Current Layer Settings with View** saves the current layer visibility settings with the new named view.

Selecting **Save UCS with view** saves the coordinate system that is displayed in the UCS name with the new view.

Select one of the available Coordinate Systems (**User** or **World**) from the **UCS name** list box.

Choose **OK** to create a new view and close the New View dialog box.

To restore the selected named view. choose **Set Current** in the View dialog box. You can also restore a named view by double-clicking its name in the list or by right-clicking its name and choosing SET CURRENT on the shortcut menu.

Choosing **Details** causes the View Details dialog box to be displayed, which has information about the view selected in the **Current View** box of the View dialog box. Information includes the area of the view in width and height and twist, target point coordinates, direction point coordinates, front and back offset clipping, perspective lens length, and coordinate system to which the view is relative.

To update the layer information saved with the selected named view to match the layer visibility in the current model space or layout viewport, choose **Update Layers**.

To edit boundaries of the selected view, choose **Edit Boundaries**. AutoCAD displays the selected named view centered and zoomed out and with the rest of the drawing area in a lighter color to show the boundaries of the named view. You can specify opposite corners of a new boundary repeatedly until you press ENTER to accept the results.

To delete the selected named view, choose **Delete**.

After making changes, choose **OK** to close the dialog box and save the changes.

CONTROLLING THE REGENERATION

The REGENAUTO command controls automatic regeneration. When REGENAUTO is set to ON, AutoCAD drawings regenerate automatically. When it is set to OFF, then you may have to regenerate the drawing manually to see the current status of the drawing.

Invoke the REGENAUTO command by entering **regenauto** at the On-Screen prompt and AutoCAD prompts:

> Enter mode *(select an option)*

When you set REGENAUTO to OFF, what you see on the screen may not always represent the current state of the drawing. When changes are made by certain commands, the display will be updated only after you invoke the REGEN command. But the delay caused by constant regenerating can be avoided as long as you are aware of the changing status of the geometry versus what is being displayed. Returning the REGENAUTO setting to ON will cause a regeneration. If a command should require regeneration while REGENAUTO is set to OFF, you will be prompted:

> About to regen, proceed?

Responding with **n** (No) will abort the command.

Regeneration during a transparent command will be delayed until a regeneration is performed at the end of that transparent command. The following message will appear in the text screen window:

REGEN QUEUED

CONTROLLING THE DRAWING OF OBJECTS

The DRAGMODE command controls the way dragged objects are displayed. Certain draw and modify commands display highlighted dynamic (cursor-following) representations of the objects being drawn or edited. This can slow down the drawing process if the objects are very complex. Setting DRAGMODE to OFF turns off this dynamic display while dragging.

Invoke the DRAGMODE command by entering **dragmode** at the On-Screen prompt and AutoCAD prompts:

> Enter new value *(select one of the three available options)*

When DRAGMODE is set to *OFF*, all calls for dragging are ignored. Setting DRAGMODE to *ON* allows dragging by use of the DRAG command modifier. Setting DRAGMODE to *AUTO* (default) causes dragging wherever possible.

CONTROLLING THE DISPLAY OF MARKER BLIPS

The BLIPMODE command controls the display of marker blips. When BLIPMODE is set to *ON*, a small cross mark is displayed when points on the screen are specified with the cursor or by entering their coordinates. After you edit for a while, the drawing can become cluttered with these blips. They have no effect other than visual reference and can be removed at any time by using the REDRAW, REGEN, ZOOM, or PAN commands. Any other command requiring regeneration causes the blips to be removed. When BLIPMODE is set to *OFF*, the blips marks are not displaced.

Invoke the BLIPMODE command by entering **blipmode** at the On-Screen prompt and AutoCAD prompts:

> Enter mode *(select one of the two options)*

CHANGING THE DISPLAY ORDER OF OBJECTS

The DRAWORDER command allows you to change the display order of objects as well as images. This will ensure proper display and plotting output when two or more objects overlay one another. For instance, when a raster image is attached over an existing object, AutoCAD obscures them from view. With the use of the DRAWORDER command, you can make the existing object display over the raster image.

Invoke the DRAWORDER command from the Modify II toolbar, by selectiong **Draw Order** (see Figure 13–22) and AutoCAD prompts:

Select objects: *(select the objects for which you want to change the display order, and press* ENTER *to complete object selection)*
Enter object ordering option *(select one of the options)*
Select reference object: *(select the reference object for changing the order of display)*

Figure 13–22 *Invoking the* DRAWORDER *command from the Modify II toolbar*

When multiple objects are selected for reordering, the relative display order of the objects selected is maintained.

Choosing ABOVE OBJECT moves the selected object(s) above a specified reference object.

Choosing UNDER OBJECT moves selected object(s) below a specified reference object.

Choosing FRONT moves selected object(s) to the front of the drawing order.

Choosing BACK moves selected object(s) to the back of the drawing order.

 Note: The DRAWORDER command terminates when selected object(s) are reordered. The command does not continue to prompt for additional objects to reorder.

OBJECT PROPERTIES

There are three important properties that control the appearance of objects: color, linetype, and lineweight. You can specify the color, linetype, and lineweight for the objects to be drawn with the help of the LAYER command, as explained in Chapter 3. You can do the same thing from their respective list boxes in the Properties toolbar or by invoking the COLOR, LINETYPE, or LINEWEIGHT commands from the **Format** menu.

SETTING AN OBJECT'S COLOR

The Color option menu of the Properties toolbar (see Figure 13–23) allows you to specify a color for the objects to be drawn, separate from the assigned color for the layer. You can select **ByLayer** or **ByBlock**, select one of the primary colors or the **Select Color** options to open the Select Color dialog box (see Figure 13–24) to define the color of objects selecting from the 255 AutoCAD Color Index (ACI) colors, True Colors, and Color Book colors.

Figure 13–23 *Selecting from the Color menu of the Properties toolbar*

The default is set to ByLayer, which causes the objects drawn to assume the color of the layer on which they are drawn. You can also select one of the colors from the chart or one of the color bars below the chart on the **Index Color** tab (Figure 13–24, left) of the Select Color dialog box. All new objects you create are drawn with this color, regardless of the color of the current layer, until you again set the color to **ByLayer** or **ByBlock**. **ByBlock** causes objects to be drawn in white (or black against a white background) until selected for inclusion in a block definition. Subsequent insertion of a block reference that contains objects drawn under the **ByBlock** option causes those objects to assume the color of the current setting of the COLOR command.

Instead of choosing from 256 standard colors, you can also choose colors from the True Color graphic interface located on the **True Color** tab (Figure 13–24, center) with its controls for **Hue, Saturation, Luminance**, and **Color Model** or from standard Color Books (such as Pantone) located on the **Color Books** tab (Figure 13–24, right). True Color and Color Books options make it easier to match colors in your drawing with colors of actual materials.

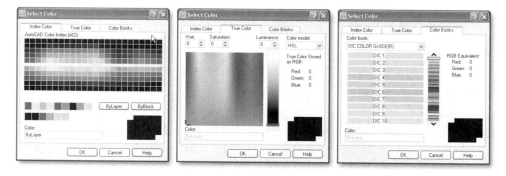

Figure 13–24 *Select Color dialog box: Index Color tab (left), True Color tab (center), Color Books tab (right)*

Note: As noted in Chapter 3, the options to specify colors by both layer and the COLOR command can cause confusion in a large drawing, especially one containing blocks and nested blocks. You are advised not to mix the two methods of specifying colors in the same drawing.

SETTING AN OBJECT'S LINETYPE

The Linetype option menu of the Properties toolbar (see Figure 13–25) allows you to specify a linetype for the objects to be drawn, separate from the assigned linetype for the layer. You can select **ByLayer** or **ByBlock**, select one of the linetypes loaded in the current drawing, or choose the **Other** option to open the Linetype Manager dialog box (see Figure 13–26) to load and set the current linetype.

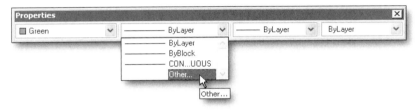

Figure 13–25 *Linetype option menu from the Properties toolbar*

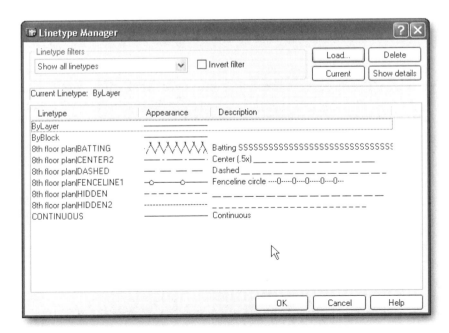

Figure 13–26 *Linetype Manager dialog box*

A linetype must exist in a library file and be loaded before you can apply it to an object or layer. Standard linetypes are in the library file called *acad.lin* and are not loaded with the LAYER command. You must load the linetype before you assign it to a specific layer.

Linetypes are combinations of dashes, dots, and spaces. Customized linetypes permit "out of line" objects in a linetype such as circles, wavy lines, blocks, and skew segments.

Dash, dot, and space combinations eventually repeat themselves. For example, a six-unit-long dash, followed by a dot between two one-unit-long spaces, repeats itself according to the overall length of the line drawn and the LTSCALE setting.

Lines with dashes (not all dots) usually have dashes at both ends. AutoCAD automatically adjusts the lengths of end dashes to reach the endpoints of the adjoining line. Intermediate dashes will be the lengths specified in the definition. If the overall length of the line is not long enough to permit the breaks, the line is drawn continuous.

There is no guarantee that any segments of the line fall at a particular location. For example, when placing a centerline through circle centers, you cannot be sure that the short dashes will be centered on the circle centers as most conventions call for. To achieve this effect, the short and long dashes have to be created by either drawing them individually or by breaking a continuous line to create the spaces between the dashes. This also creates multiple in-line lines instead of one line of a particular linetype. Or you can use the CENTER option of the DIMENSION command to place the desired mark.

Individual linetype names and definitions are stored in one or more files whose extension is *.lin*. The same name may be defined differently in two different files. Selecting the desired one requires proper responses to the prompts in the **Load** option of the LINETYPE command. If you redefine a linetype, loading it with the LINETYPE command will cause objects drawn on layers assigned to that linetype to assume the new definition.

AutoCAD lists the available linetypes for the current drawing and displays the current linetype setting in the Linetype Manager dialog box (see Figure 13–26). By default, it is set to ByLayer. To change the current linetype setting, double-click its name in the list. All new objects you create will be drawn with the selected linetype, regardless of the layer you are working with, until you again set the linetype to ByLayer or ByBlock. ByLayer causes the object drawn to assume the linetypes of the layer on which it is drawn. ByBlock causes objects to be drawn in the Continuous linetype until selected for inclusion in a block definition. Subsequent insertion of a block that contains objects drawn under the ByBlock option will cause those objects to assume the linetype of the block.

To load a linetype explicitly into your drawing, choose **Load** and AutoCAD displays the Load or Reload Linetypes dialog box (see Figure 13–27).

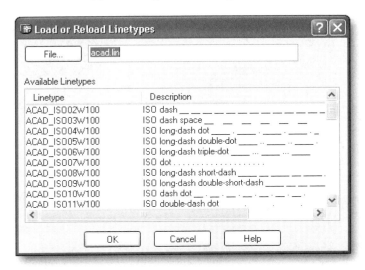

Figure 13–27 *Load or Reload Linetypes dialog box*

By default, AutoCAD lists the available linetypes from the *acad.lin* file. Select the linetype to load from the Available Linetypes list box, and choose **OK**.

If you need to load linetypes from a different file, choose **File** in the Load or Reload Linetypes dialog box. AutoCAD displays the Select Linetype File dialog box. Select the appropriate linetype file and choose **OK**. In turn, AutoCAD lists the available linetypes from the selected linetype file in the Load or Reload Linetypes dialog box. Select the appropriate linetype to load from the **Available Linetypes** list box, and choose **OK**.

To delete a linetype that is currently loaded in the drawing, first select the linetype from the list box in the Linetype Manager dialog box, and then choose **Delete**. You can delete only linetypes that are not referenced in the current drawing. You cannot delete the linetype Continuous, ByLayer, or ByBlock. Deleting is the same as using the PURGE command to purge unused linetypes from the current drawing.

To display additional information of a specific linetype, first select the linetype from the Linetype list box in the Linetype Manager dialog box, and then choose **Show details**. AutoCAD displays an extension of the dialog box, listing additional settings (see Figure 13–28).

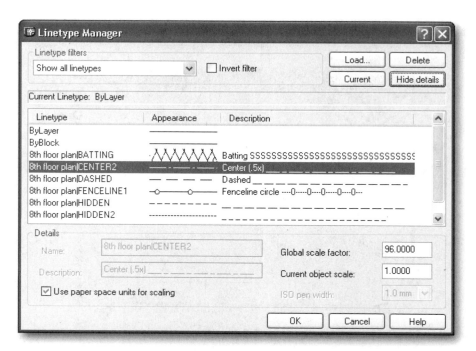

Figure 13-28 *Extended Linetype Manager dialog box*

The **Name** and **Description** edit fields display the selected linetype name and description, respectively.

The **Global scale factor** text field displays the current setting of the LTSCALE factor. The **Current object scale** text field displays the current setting of the CELTSCALE factor. If necessary, you can change the values of the LTSCALE and CELTSCALE system variables.

The **ISO pen width** box sets the linetype scale to one of a list of standard ISO values. The resulting scale is the global scale factor multiplied by the object's scale factor.

Selecting **Use paper space units for scaling** scales linetypes in paper space and model space identically.

After making the necessary changes, choose **OK** to keep the changes and close the Linetype Manager dialog box.

 Note: As noted in Chapter 3, the options to specify linetypes by both the LAYER and the LINETYPE commands can cause confusion in a large drawing, especially one containing blocks and nested blocks. You are advised not to mix the two methods of specifying linetypes in the same drawing.

LINEWEIGHT COMMAND

The Lineweight menu of the Properties toolbar (see Figure 13–29) allows you to specify a lineweight for the objects to be drawn, separate from the assigned lineweight for the layer. You can select **ByLayer** or **ByBlock**, or one of the lineweights listed, or invoke the LINEWEIGHT command and select a lineweight from the listed **Lineweights** in the Lineweight Settings dialog box (see Figure 13–30).

Figure 13–29 *Selecting one of the available lineweights from the menu in the Properties toolbar*

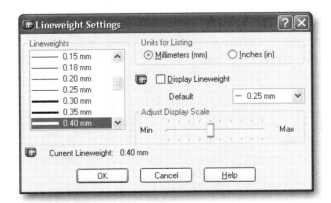

Figure 13–30 *Lineweight Settings dialog box*

AutoCAD displays the current lineweight at the bottom of the dialog box. To change the current lineweight, select one of the available lineweights from the **Lineweights** list box. Bylayer is the default. All new objects you create are drawn with the current lineweight, regardless of which layer is current, until you again set the lineweight to Bylayer, Byblock, or Default. Bylayer causes the objects drawn to assume the lineweight of the layer on which it is drawn. Byblock causes objects to be drawn in the default lineweight until selected for inclusion in a block definition. Subsequent insertion of a block reference that contains objects drawn under the Byblock option causes those objects to assume the Lineweight of the current setting of the LINEWEIGHT command. The Default selection causes the objects to be drawn to the default value as set by the LWDEFAULT system variable and defaults to a value of 0.01 inches or 0.25 mm. You can also set the default value from the Default option menu located on the right side of the dialog box. The lineweight value of 0 plots at the thinnest lineweight available on the specified plotting device and is displayed at one pixel wide in model space. You can use the PROPERTIES command to change the lineweight of the existing objects.

The **Units for Listing** section specifies whether lineweights are displayed in millimeters or inches.

The **Display Lineweight** check box controls whether lineweights are displayed in the current drawing. If it is enabled, lineweights are displayed in model space and paper space. AutoCAD regeneration time increases with lineweights that are represented by more than one pixel. If it is disabled, AutoCAD performance improves. Performance slows down when working with lineweights enabled in a drawing.

The **Adjust Display Scale** controls the display scale of lineweights on the Model tab. On the Model tab, lineweights are displayed in pixels. Lineweights are displayed using a pixel width in proportion to the real-world unit value at which they plot. If you are using a high-resolution monitor, you can adjust the lineweight display scale to better display different lineweight widths. The Lineweight list reflects the current display scale. Objects with lineweight that are displayed with a width of more than one pixel may increase AutoCAD regeneration time. If you want to optimize AutoCAD performance when working in the Model tab, set the lineweight display scale to the minimum value or turn off lineweight display altogether.

Choose **OK** to close the dialog box and keep the changes in the settings.

Note: As noted in Chapter 3, the options to specify lineweights by both the LAYER and the LINEWEIGHT commands can cause confusion in a large drawing, especially one containing blocks and nested blocks. You are advised not to mix the two methods of specifying lineweights in the same drawing.

X, Y, AND Z FILTERS—AN ENHANCEMENT TO OBJECT SNAP

The AutoCAD filters feature allows you to establish a *2D* point by specifying the individual (*X* and *Y*) coordinates one at a time in separate steps. In the case of a *3D* point you can specify the individual (*X*, *Y*, and *Z*) coordinates in three steps. Or you can specify one of the three coordinate values in one step and a point in another step, from which AutoCAD extracts the other two coordinate values for use in the point being established.

The filters feature is used when you are prompted to establish a point, as in the starting point of a line, the center of a circle, drawing a node with the POINT command, or specifying a base point or second point in displacement for the MOVE or COPY command, to mention just a few.

Note: During the application of the filters feature there are steps where you can input either single coordinate values or points, and there are steps where you can input only points. It is necessary to understand these restrictions and options and when one type of input is more desirable than the other.

When selecting points during the use of filters, you need to know which coordinates of the specified point are going to be used for the point being established. It is also essential to know how to combine Object Snap modes with those steps that use point input.

The filters feature is actually an enhancement to either the object snap or the @ (last point) feature. The AutoCAD ability to establish a point by snapping to a point on an existing object is one of the most powerful features in AutoCAD, and being able to have AutoCAD snap to such an existing point and then filter out selected coordinates for use in establishing a new point adds to that power. Therefore, in most cases you will not use the filters feature if it is practical to enter in all of the coordinates from the keyboard, because entering in all the coordinates can be done in a single step. The filters feature is a multistep process, and each step might include substeps, one to specify the coordinate(s) to be filtered out and another to designate the Object Snap mode involved.

Note: The filter prompts will be seen only when you invoke the appropriate commands in the Command window.

FILTERS WITH @

When AutoCAD is prompting for a point, the filters feature is initiated by entering a period followed by the letter designation for the coordinate(s) to be filtered out. For example, if you draw a point starting at (0,0) and use the relative polar coordinate response @3<45 to determine the endpoint, you can use filters to establish another point whose *X* coordinate is the same as the *X* coordinate of the end of the line just drawn. It works for Y and Z coordinates and combinations of XY, XZ, and YZ coordinates also. The following command sequence shows how to apply a filter to a line that needs to be started at a point whose X coordinate is the same as that of the end of the previous line and whose Y coordinate is 1.25. The line will be drawn horizontally 3 units long. The sequence is as follows (the following command sequence is shown when it is invoked in the command window):

> Command: **line**
> Specify first point: **0,0**
> Specify next point or [Undo]: **@3<45**
> Specify next point or [Undo]: (ENTER)
>
> Command: **line** (or ENTER)
> Specify first point: *(hold* SHIFT *and right-click, and from the Point Filters flyout menu choose .x)*
> .X of **end** *(use the endpoint snap mode to select the end of the line)*
> of (need YZ): **0,1.25**
> Specify next point or [Undo]: **@3<0**
> Specify next point or [Undo]:

Entering .x initiates the filters feature. AutoCAD then prompts you to specify a point from which it can extract the X coordinate. The @ (last point) does this. The new line has a starting point whose X coordinate is the same as that of the last point drawn. By using the filters feature to extract the X coordinate, that starting point will be on an imaginary vertical line through the point specified by @ in response to the "of" prompt.

When you initiate filters with a single coordinate (.x in our example) and respond with a point (@), the prompt that follows asks for a point also. From it (the second point specified) AutoCAD extracts the other two coordinates for the new point.

Even though the prompt is for "YZ," the point may be specified in *2D* format as 0,1.25 (the X and Y coordinates), from which AutoCAD takes the second value as the needed Y coordinate. The Z coordinate is assumed to be the elevation of the current coordinate system.

You can use the two-coordinate response to initiate filters. Then specify a point, and all that AutoCAD requires is a single value for the final coordinate. An example of this follows.

> Command: **line**
> Specify first point: **0,0**
> Specify next point or [Undo]: **@3<45**
> Specify next point or [Undo]: (ENTER)

Command: **line** (or ENTER)
Specify first point: *(hold* SHIFT *and right-click, and from the Point Filters flyout menu choose .xz)*
.XZ of **end** *(use the endpoint snap mode to select the end of the line)*
of (need Y): **1.25**
Specify next point or [Undo]: **@3<0**
Specify next point or [Undo]:

You can also specify a point in response to the "of (need Y):" prompt:

of (need Y): **0,1.25** *(or pick a point on the screen)*

In this case, AutoCAD uses the Y coordinate of the point specified as the Y coordinate of the new point.

Remember, it is an individual coordinate in *2D* (one or two coordinates in *3D*) of an existing point that you wish AutoCAD to extract and use for the new point. In most cases you will be object snapping to a point for the response. Otherwise, if you knew the value of the coordinate needed, you would probably enter it at the keyboard.

FILTERS WITH OBJECT SNAP

Without filters, an Object Snap (Osnap) mode establishes a new point to coincide with one on an existing object. With filters, an Osnap mode establishes the selected coordinate(s) of a new point to coincide with the corresponding coordinate(s) of one or two of the coordinates (depending on the filter) on an existing object.

Extracting one or more coordinate values to be applied to corresponding coordinate values of a point that you are being prompted to establish is shown in the following example.

In Figure 13–31, a 2.75" by 7.1875" rectangle has a 0.875"-diameter hole in its center. A board drafter would determine the center of a square or rectangle by drawing diagonals and centering the circle at their intersection. AutoCAD drafters (without filters) could do the same, or they might draw orthogonal lines from the midpoint of a horizontal line and from the midpoint of one of the vertical lines to establish a centering intersection. The following command sequence shows steps in drawing a rectangle with a circle in the center using filters.

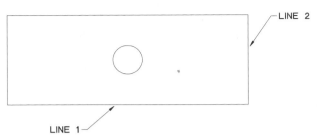

Figure 13–31 *Extracting coordinate values to be applied to corresponding coordinate values*

Command: **line**
Specify first point: *(select a point)*
Specify next point or [Undo]: **@2.75<90**
Specify next point or [Undo]: **@7.1875<0**
Specify next point or [Undo]: **@2.75<270**
Specify next point or [Close/Undo]: **c**

Command: **circle**
Specify center point for circle or [3P/2P/Ttr (tan tan tan radius)]: *(hold SHIFT and right-click, and from the Point Filters flyout menu choose .x)*
of **mid** *(use the midpoint snap mode to select a horizontal line)*
of (need YZ): *(hold SHIFT and right-click, and from the Point Filters flyout menu choose .y)*
of **mid** *(use the midpoint snap mode to select a vertical line)*
of (need Z): *(hold SHIFT and right-click, and from the Point Filters flyout menu choose .z and click any line to place the circle on the same plane)*
Specify radius of circle or [Diameter]<default>: **d** *(choose diameter)*
Specify diameter of circle: **.875**

SHELL COMMAND

The SHELL command allows you to execute operating system programs without leaving AutoCAD. You can execute any operating system program as long as there is sufficient memory to execute.

Invoke the SHELL command by entering **shell** at the On-Screen prompt and AutoCAD prompts:

OS Command: *(invoke one of the available operating system's utility programs)*

When the utility program is finished, AutoCAD takes you back to the On-Screen prompt. If you need to execute more than one operating system program, press ENTER at the "OS command" prompt. AutoCAD responds with the appropriate operating system prompt. You can now enter as many operating system commands as you wish.

When you are finished, you may return to AutoCAD by entering EXIT. It will take you back to the On-Screen prompt.

Note: Do not delete the AutoCAD lock files or temporary files created for the current drawing when you are at the operating system prompt.

SETTING UP A DRAWING

The MVSETUP command is used to control and set up the view(s) of a drawing, including the choice of standard plotted sheet sizes with a border, the scale for plotting on the selected sheet size, and multiple viewports. MVSETUP is an AutoLISP (AutoCAD's embedded programming language) routine that can be customized to insert any type of border and title block.

Options and associated prompts depend on whether the TILEMODE system variable is set to ON (1) or OFF (0). When TILEMODE is set to ON, Tiled Viewports is enabled. When TILEMODE is set to OFF, the Floating Viewports menu item is enabled. Other paper space–related drawing setup options are available.

Invoke the MVSETUP command by entering **mvsetup** at the On-Screen prompt and the AutoCAD prompts that appear depend on whether you are working in model space (Model tab) or paper space (Layout tab). If you are working in model space, then AutoCAD prompts:

> Enable paper space? *(press ENTER to enable paper space, and AutoCAD changes the TILEMODE setting to 0, or choose NO to stay in the TILEMODE setting of 1)*

If you are working in paper space (**Layout** tab), then AutoCAD prompts:

> Enter an option *(select one of the options from the shortcut menu)*

The following is the procedure for setting up the drawing when you are working in model space. AutoCAD prompts:

> Enable paper space? *(choose NO from the shortcut menu)*
> Enter units type *(Select a unit type. Depending on the units selected, AutoCAD lists the available scales. Select one of the available scales, or you can even specify a custom scale factor.)*
> Enter the scale factor: *(specify a scale factor)*
> Enter the paper width: *(specify the paper width at which the drawing will be plotted)*
> Enter the paper height: *(specify the paper height at which the drawing will be plotted)*

AutoCAD sets up the appropriate limits to allow you to draw to full scale and also draws a bounding box enclosing the limits. Draw the drawing to full scale; when you are ready to plot, specify the scale mentioned earlier to plot the drawing.

Following is the procedure for setting up the drawing when you are working in paper space. AutoCAD prompts:

>Enter an option

The options are explained here in the order that is logical to complete the drawing setup.

Choosing TITLE BLOCK allows you to select an appropriate title block. AutoCAD prompts:

>Enter title block option *(select one of the options from the shortcut menu)*

>>Choosing INSERT (default) allows you to insert one of the available standard title blocks. AutoCAD lists the available title blocks (in the text screen) and prompts you to select one of the available title blocks as follows:

>0. None
>1. ISO A4 Size(mm)
>2. ISO A3 Size(mm)
>3. ISO A2 Size(mm)
>4. ISO A1 Size(mm)
>5. ISO A0 Size(mm)
>6. ANSI-V Size(in)
>7. ANSI-A Size(in)
>8. ANSI-B Size(in)
>9. ANSI-C Size(in)
>10. ANSI-D Size(in)
>11. ANSI-E Size(in)
>12. Arch/Engineering (24 x 36in)
>13. Generic D size Sheet (24 x 36in)

>Enter number of title block to load or [Add/Delete/Redisplay]: *(select one of the available title blocks and press* ENTER, *and AutoCAD inserts a border and title block, as shown in Figure 13–32)*

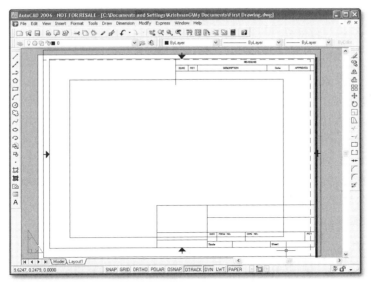

Figure 13–32 *Title block and border for ANSI-B Size*

Choosing ADD allows you to add a title block drawing to the available list.

Choosing DELETE allows you to delete an entry from the available list.

Choosing REDISPLAY redisplays the list of title block options.

Choosing **Create** allows you to establish viewports. AutoCAD prompts:

Enter option *(press ENTER to create viewports, and AutoCAD lists the available viewport layout options)*

Choosing CREATE VIEWPORTS (default) allows you to create multiple viewports to one of the standard layouts. AutoCAD prompts:

Available layout options:

0:   None
1:   Single
2:   Std. Engineering
3:   Array of Viewports

Enter layout number to load or [Redisplay]: *(select one of the available layouts)*

Choosing NONE creates no viewports. Selecting 1 creates a single viewport whose size is determined during subsequent prompts responses. Selecting 2 creates four viewports with preset viewing angles by dividing a specified area into quadrants. The size is determined by responses to subsequent prompts. Selecting 3 creates a matrix of viewports along the *X* and *Y* axes.

Choosing DELETE OBJECTS deletes the existing viewports.

Choosing UNDO reverses operations performed in the current MVSETUP session.

Choosing SCALE VIEWPORTS adjusts the scale factor of the objects displayed in the viewports. The scale factor is specified as a ratio of paper space to model space. For example, 1:48 is 1 paper space unit for 48 model space units (scale for 1/4" = 1'0").

Choosing OPTIONS lets you establish several different environment settings that are associated with your layout. AutoCAD prompts:

Enter an option *(enter an option)*

Choosing LAYER lets you specify a layer for placing the title block.

Choosing LIMITS lets you specify whether or not to reset the limits to drawing extents after the title block has been inserted.

Choosing UNITS lets you specify whether sizes and point locations will be translated to inch or millimeter paper units.

Choosing XREF lets you specify whether the title block is to be inserted or externally referenced.

Choosing ALIGN causes AutoCAD to pan the view in one viewport so that it aligns with a basepoint in another viewport. Whichever viewport the other point moves to becomes the active viewport. AutoCAD prompts:

Enter an option

Choosing ANGLED causes AutoCAD to pan the view in a viewport in a specified direction.

Choosing HORIZONTAL causes AutoCAD to pan the view in one viewport, aligning it horizontally with a basepoint in another viewport.

Choosing VERTICAL ALIGNMENT causes AutoCAD to pan the view in one viewport, aligning it vertically with a basepoint in another viewport.

Choosing ROTATE VIEW causes AutoCAD to rotate the view in a viewport around a basepoint.

Choosing UNDO causes AutoCAD to undo the results of the current MVSETUP command.

LAYER TRANSLATOR

The Layer Translator is used to make selected layers in the current drawing match layers in another drawing or in a CAD standards file.

Invoke the LAYTRANS command from the CAD Standards toolbar, by selecting **Layer Translator** (see Figure 13–33) and AutoCAD displays the Layer Translator dialog box (see Figure 13–34).

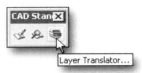

Figure 13–33 *Invoking the* LAYTRANS *command from the CAD Standards toolbar*

Figure 13–34 *Layer Translator dialog box*

The **Translate From** list box lists all the layer(s) in the current drawing that can be changed. A dark colored icon indicates that the layer is referenced in the drawing; a white icon indicates the layer is unreferenced. You can delete unreferenced layers from the drawing by right-clicking in the **Translate From** list box and choosing PURGE LAYERS.

Choose **Load** to load the layers to match from an existing drawing, template file, or CAD Standards file. AutoCAD lists the layers you can translate the current drawing's layers to in the **Translate To** list box. You can also create new layers and assign properties by choosing **New**, which will be added to the list in the **Translate To** list box. You cannot create a new layer with the same name as an existing layer.

Select one or more layers (you can also select layers by using the selection filter) in the **Translate From** list box. Select layers in the **Translate To** list box to map to the selected layer(s). Choose **Map** to map the layers selected in **Translate From** to the layer selected in **Translate To**. Choose **Map same** to map all layers that have the same name in both lists.

AutoCAD lists each layer to be translated and the properties to which the layer will be converted in the **Layer Translation Mappings** section. You can select layers in this list and edit their properties using **Edit**.

Choose **Edit** to open the Edit Layer dialog box (see Figure 13–35), where you can edit the selected translation mapping. You can change the layer's linetype, color, and lineweight. If all drawings involved in translation use plot styles, you can also change the plot style for the mapping.

Choose **Remove** to remove the selected translation mapping from the Layer Translation Mappings list.

Choose **Save** to save the current layer translation mappings to a file for later use. Layer mappings are saved in the *.dwg* or *.dws* file format. You can replace an existing file or create a new file. The Layer Translator creates the referenced layers in the file and stores the layer mappings in each layer. All linetypes used by those layers are also copied into the file.

Choose **Settings** to open the Settings dialog box (see Figure 13–36), where you can customize the process of layer translation.

Figure 13–35 *Edit Layer dialog box*

Figure 13–36 *Settings dialog box*

Selecting **Force object color to ByLayer** causes every object translated to take on the color assigned to its layer. If it is not checked, every object retains its original color.

Selecting **Force object linetype to ByLayer** causes every object translated to take on the linetype assigned to its layer. If it is not checked, every object retains its original linetype.

Selecting **Translate objects in blocks** causes objects nested within blocks (including nested blocks) to be translated. If it is not checked, nested objects in blocks are not translated.

Selecting **Write transaction log option** causes a log file to be created detailing the results of the translation. The log file is assigned the same name as the translated drawing, with a *.log* file name extension and is created in the same folder. If it is not checked, no log file is created.

Selecting **Show layer contents** causes only the layers selected in the Layer Translator dialog box to be displayed in the drawing area. If it is not checked, all layers in the drawing are displayed.

Choose **Translate** to start the translation of the layers you have mapped. If you have not saved the current layer translation mappings, you are prompted to save the mappings before translation begins.

TIME COMMAND

The TIME command displays the current time and date related to your current drawing session. In addition, you can find out how long you have been working in AutoCAD. This command uses the clock in your computer to keep track of the time functions and displays to the nearest millisecond using 24-hour military format.

Invoke the TIME command by selecting **Time** from the **Inquiry** flyout of the **Tools** menu and the following listing is displayed in the text screen followed by a prompt:

```
Current time:           Wednesday, February 16, 2005  11:05:04:406 PM
Times for this drawing:
Created:                Sunday, February 06, 2005  6:55:59:140 AM
Last updated:           Sunday, February 06, 2005  7:25:29:859 AM
Total editing time:     0 days 02:23:51:531
Elapsed timer (on):     0 days 02:23:51:047
Next automatic save in: <no modifications yet>
Enter option [Display/ON/OFF/Reset]: (select one of the options)
```

The first line gives today's date and time.

The third line gives the date and time the current drawing was initially created. The drawing time starts when you initially begin a new drawing. If the drawing was created by means of the WBLOCK command, it is the date and time the command was executed that is displayed here.

The fourth line gives the date and time the drawing was last updated. Initially set to the drawing creation time, this is updated each time you use the END or SAVE command.

The fifth line gives the length of time you are in AutoCAD. This timer is continuously updated by AutoCAD while you are in the program, excluding plotting and printer plot time. This timer cannot be stopped or reset.

The sixth line provides information about the stopwatch timer. You can turn this timer ON or OFF and reset it to zero. This timer is independent of other functions.

The seventh line provides information about when the next automatic save will take place.

Choose DISPLAY to redisplay the time functions, with updated times.

Choose ON to set the stopwatch timer to ON, if it is OFF. By default it is set to ON.

Choose OFF to set the stopwatch timer to OFF and displays the accumulated time.

Choose RESET to reset the stopwatch timer to zero.

To exit the TIME command at the prompt, press ESC or ENTER.

AUDIT COMMAND

The AUDIT command serves as a diagnostic tool to correct any errors or defects in the database of the current drawing. AutoCAD generates an extensive report of the problems, and for every error detected, AutoCAD recommends an action to correct it.

Invoke the AUDIT command by selecting **Audit** from **Drawing Utilities** flyout of the **File** menu and AutoCAD prompts:

Fix any errors detected? ⬇: (choose YES or NO from the shortcut menu)

If you respond with YES, AutoCAD will fix all the errors detected and display an audit report (in the text screen) with detailed information about the errors detected and fixing them. If you answer with NO, AutoCAD will just display a report and will not fix any errors.

In addition, AutoCAD creates an ASCII report file (AUDITCLT system variable should be set to ON) describing the problems and the actions taken. It will save the file in the current directory, using the current drawing's name with the file extension *.adt*. You can use any ASCII editor to display the report file on the screen or print it on the printer, respectively.

Note: If a drawing contains errors that the AUDIT command cannot fix, open the drawing with the RECOVER command to retrieve the drawing and correct its errors.

OBJECT LINKING AND EMBEDDING (OLE)

Object linking and embedding (OLE) is a Microsoft Windows feature that combines various application data into one compound document. AutoCAD has client as well as server capabilities. As a client, AutoCAD permits you to have objects from other Windows applications either embedded in or linked to your drawing.

When an object is inserted into an AutoCAD drawing from an application that supports OLE, the object can maintain a connection with its source file. If you insert an object as an embedded object into AutoCAD (client), it is no longer associated with the source (server). If necessary, you can edit the embedded data from inside the AutoCAD drawing by using the original application. But at the same time, this editing does not change the original file.

If, instead, you insert an object as a linked object into AutoCAD (client), the object remains associated with its source (server). When you edit a linked object in Auto-CAD by using the original application, the original file changes as well as the object inserted into AutoCAD.

Linked or embedded objects appear on the screen in AutoCAD and can be printed or plotted using Windows system drivers.

Let's look at an example of object linking between an AutoCAD (server) drawing and Microsoft Word (client). Figure 13–37 shows a drawing of a desk, a computer, and a chair, that contains various attribute values. We are going to link this drawing to a Microsoft Word document.

From the Edit menu select Copy, and AutoCAD prompts you to select objects. Select the computer, the table, and the chair, and press ENTER. This will copy the selection to the Windows clipboard. Minimize the AutoCAD program.

Instead of selecting specific objects, you can copy the current view into the Windows clipboard by invoking the COPYLINK command from the Edit menu.

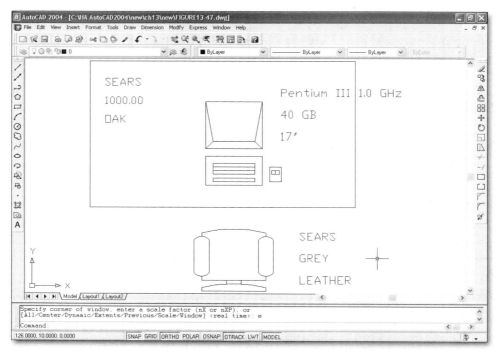

Figure 13–37 *Drawing of a desk, a computer, and a chair with attribute values*

Open the Microsoft Word program from the desktop by double-clicking the Word program icon. The Microsoft Word program is displayed, as shown in Figure 13–38.

From the **Edit** menu in Microsoft Word, select **Paste Special**, and Word displays the Paste Special dialog box shown in Figure 13–39. Choose **Paste Link** to insert the AutoCAD drawing object into Microsoft Word, as shown in Figure 13–40.

Minimize the Word program and maximize AutoCAD or if the AutoCAD program is not open, double-click the drawing image in the Word document, which will launch the AutoCAD program with the image drawing open. Edit the values of the attributes in the computer block to Pentium IV 2.5 GHz, 120.0 GB, 21", which represent a Pentium IV computer with a 120.0 GB hard drive and a 21" monitor.

Switch back to Word, and from the **Edit** menu select **Links**. Word will display the Links dialog box shown in Figure 13–41.

Chapter 13 • *Utility Commands* 697

Figure 13-38 *Microsoft Word program*

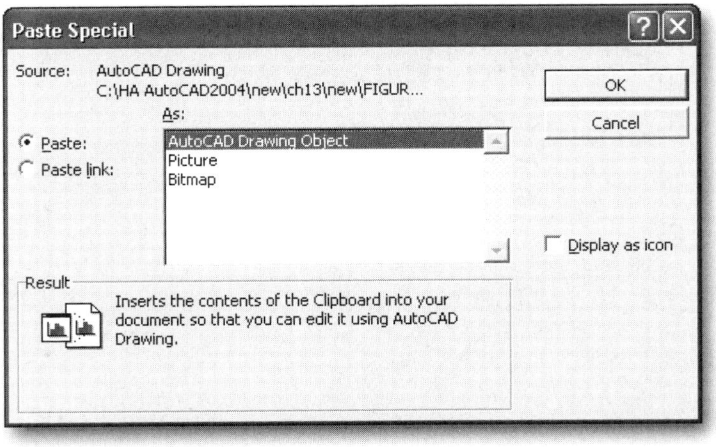

Figure 13-39 *Paste Special dialog box*

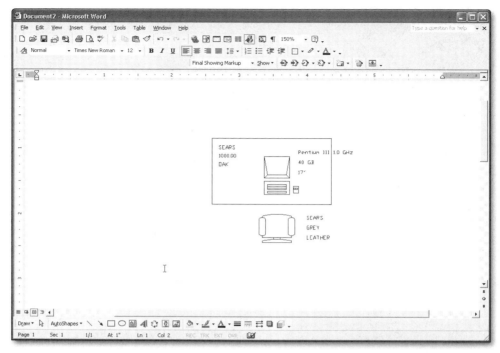

Figure 13–40 *Microsoft Word document with the AutoCAD drawing*

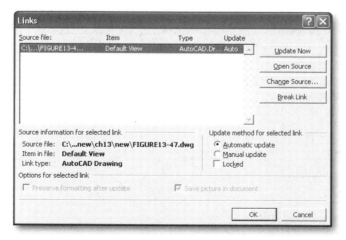

Figure 13–41 *Links dialog box*

Choose **Update Now** and then choose **OK**. The image in the Word document is updated, as shown in Figure 13–42.

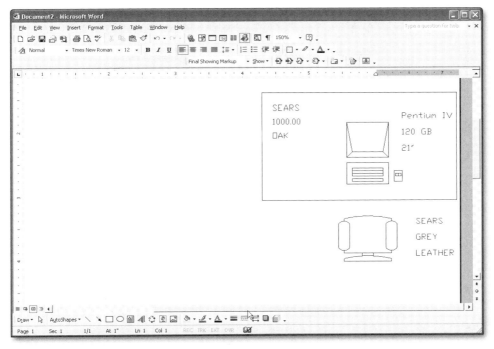

Figure 13-42 *AutoCAD drawing updated in Word as the client*

In our example, AutoCAD is the server and Microsoft Word is the client.

Conversely, you can place a linked object in AutoCAD, where AutoCAD is the client and another application is the server. Let's look at an example in which AutoCAD is the client and Excel is the server.

Start the Excel program and create a spreadsheet. Copy the contents into the Windows clipboard.

From the **Edit** menu in AutoCAD, select **Paste Special**. AutoCAD displays the Paste Special dialog box. Select **Paste Link** and **Microsoft Excel Worksheet** from the list box, and choose **OK**.

The Excel spreadsheet will be linked to the drawing, as shown in Figure 13–43. AutoCAD is now the client and Microsoft Excel is the server.

To edit the spreadsheet, double-click anywhere on the spreadsheet, which in turn will launch Excel with the spreadsheet document open. Any changes made to the spreadsheet will be reflected in the drawing. Figure 13–44 shows the changes that were made in the spreadsheet.

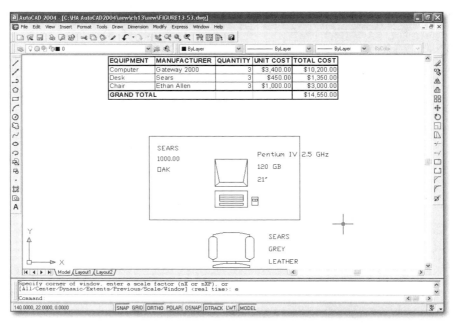

Figure 13-43 *An Excel spreadsheet in the AutoCAD drawing*

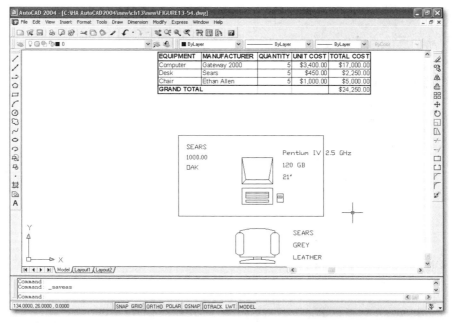

Figure 13-44 *AutoCAD drawing showing the changes made in the Excel spreadsheet*

Here is another example in which an AutoCAD drawing is the client, for both embedding from Word and linking from Excel.

Figure 13–45 shows both an AutoCAD screen and an Excel spreadsheet, in which the spreadsheet is being used for area calculations. The AREA cells are formulas that calculate the product of the corresponding WIDTH and LENGTH cells. In turn, the TOTAL cell is the sum of the AREA cells.

Figure 13–46 shows the same AutoCAD screen and a Word document that was used for typing the GENERAL NOTES, which were in turn embedded into the AutoCAD drawing.

Figure 13–47 shows how the cell value in the WIDTH for ROOM1 has been changed, resulting in changes in the AREA and TOTAL cells. Because this object (the spreadsheet consisting of four columns and seven rows, including the title) was paste linked into the AutoCAD drawing, the linked object automatically reflects the changes.

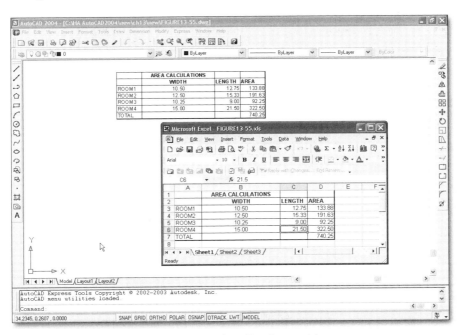

Figure 13–45 *AutoCAD drawing screen and an Excel spreadsheet*

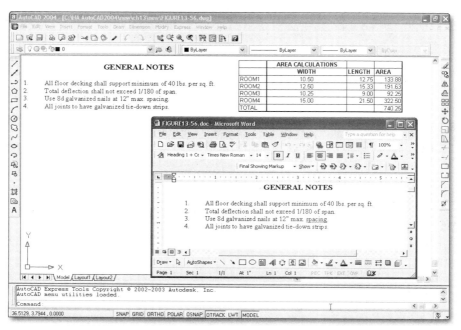

Figure 13-46 *AutoCAD drawing and a Word document*

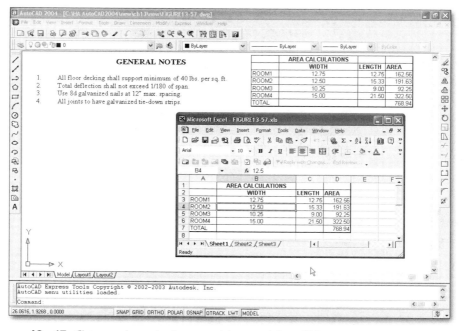

Figure 13-47 *Changes shown in the spreadsheet and AutoCAD screen*

AutoCAD displays the OLE Properties dialog box as shown in Figure 13–48 whenever you insert an OLE object into a drawing.

Figure 13–48 *OLE Properties dialog box*

If necessary, you can change the size of the object by changing the height and width in drawing units, or enter a percentage of the current height or width in the Scale section. You can also change the text size and OLE plot quality. If you need to change the OLE properties after pasting the objects into the current drawing, first select the OLE object, and from the shortcut menu select PROPERTIES or at the On-Screen prompt invoke the OLESCALE command. AutoCAD displays the OLE Properties dialog box.

SECURITY, PASSWORDS, AND ENCRYPTION

Electronic drawing files, like their paper counterparts, often need to have the information they contain protected from unauthorized viewing. AutoCAD provides password and encryption capabilities to achieve this. Also, it might be necessary to determine that the person who last edited and saved the drawing is the person who was supposed to edit and save it. To accomplish this, AutoCAD allows the use of Digital Signatures.

PASSWORDS

AutoCAD password protection makes it possible to prevent a drawing file from being opened without first entering the pre-assigned password.

To assign a password to a drawing, invoke the SECURITYOPTIONS command at the On-Screen prompt. AutoCAD displays the Security Options dialog box (see Figure 13–49).

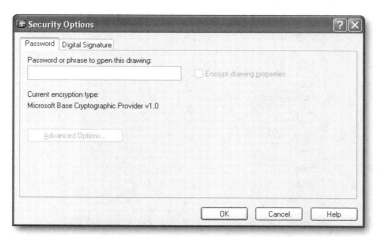

Figure 13–49 *Security Options dialog box with the **Password** tab displayed*

On the **Password** tab of the Security Options dialog box, enter a password in the **Password or phrase to open this drawing** text box. This prevents the drawing from being opened without first entering the password specified. Passwords can be a single word or a phrase and are not case-sensitive.

To view data in a password-protected drawing, open the drawing in a standard way and enter the password in the **Enter password to open drawing** text box in the Password dialog box, as shown in Figure 13–50. Unless the title, author, subject, keywords, or other drawing properties were encrypted when the password was attached, you can view the properties in the Properties dialog box in Windows Explorer.

Figure 13–50 *Password dialog box*

ENCRYPTION

You can encrypt drawing properties, such as the title, author, subject, and keywords, thus requiring a password to view the properties and thumbnail preview of the drawing. If you decide to specify an encryption type and key length, you can select them from the ones available on your computer. On the **Password** tab of the Security Options dialog box, after you have entered a password in the **Password or phrase to open this drawing** text box, set the **Encrypt drawing properties** to ON. Under the **Password or phrase to open this drawing** text box, AutoCAD displays the current encryption type. To change the encryption type, choose **Advanced Options**. AutoCAD displays the Advanced Options dialog box. Select one of the encryption providers listed and specify the key length in the **Choose a key length** text box. The higher the key length, the higher is the protection.

DIGITAL SIGNATURE

AutoCAD provides a means to sign the drawing file electronically. This means that it is possible to verify that a drawing has had a distinct and unique digital signature attached to it when it was last saved. Along with this positive electronic identification, you can also apply a time stamp and comments.

To attach a digital signature to a drawing, you must first obtain a digital ID. This can be done by contacting a certificate authority through a search engine in your Internet browser, using the term "digital certificate." Once a digital ID has been established on your computer, invoke the SECURITYOPTIONS command at the On-Screen prompt. AutoCAD displays the Security Options dialog box (see Figure 13–49). Select the **Digital Signature** tab, and AutoCAD displays various options available for a digital signature. If no valid digital IDs are installed, AutoCAD displays the Valid Digital ID Not Available dialog box from which you can obtain a digital ID by choosing **Get a digital ID,** AutoCAD logs you onto a website where a digital ID can be obtained.

AutoCAD displays a list of digital IDs that you can use to sign files, which includes information about the organization or individual to whom the digital ID was issued, the digital ID vendor who issued the digital ID, and when the digital ID expires. Select one of the available digital IDs and choose **Attach digital signature after saving drawing**. From the **Signature information** section of the **Digital Signature** tab, you can select a time stamp to be attached with the digital ID from the **Get time stamp from** text box. You can also add comments to the digital ID in the **Comment**: text box.

A digital ID has a name, expiration date, serial number, and certain certifying information. The certificate authority that you obtain the digital ID from can provide Low, Medium, and High levels of security.

From the digital signature feature, you can determine whether the file was changed since it was signed, whether the signers are who they claim to be, and if they can be traced. The digital signature is considered invalid if the file was corrupted when the digital signature was attached, it was corrupted in transit, or if the digital signature is no longer valid. In order to maintain validity of the digital signature you must not add a password to the drawing or modify or save it after the digital signature has been attached.

 Note: The digital signature status is displayed when you open a drawing if the SIGWARN system variable is set to ON. If it is set to OFF the signature status is displayed only if the signature is invalid.

In the **Open and Save** tab of the Options dialog box, if **Display Digital Signature Information** is checked, then when you open a drawing that has a digital signature attached, the Digital Signature Contents dialog box is displayed, providing information on the status of the drawing and the signer. In the **Other Fields** list, you can obtain information about the issuer, beginning and expiration dates and serial number of the digital signature.

CUSTOM SETTINGS WITH THE OPTIONS DIALOG BOX

The Options dialog box allows you to customize the AutoCAD settings. AutoCAD allows you to save and restore a set of custom AutoCAD settings called a profile. A profile can include custom settings that are not saved in the drawing, with the exception of pointer and printer driver settings. By default, AutoCAD stores your current settings in a profile named <Unnamed Profile>.

To open the Options dialog box, invoke the OPTIONS command by selecting **Option** from the **Tools** menu and AutoCAD displays the Options dialog box (see Figure 13–51). From the Options dialog box, the user can control various aspects of the AutoCAD environment. The Options dialog box has nine tabs; to make changes to any of the sections, select the corresponding tab from the top of the Options dialog box.

Figure 13–51 *Options dialog box with the Files tab selected*

FILES

The **Files** tab of the Options dialog box (see Figure 13–51), specifies the directory in which AutoCAD searches for support files, driver files, project files, template drawing file location, temporary drawing file location, temporary external reference file location, and texture maps. It also specifies the location of menu, help, log, text editor, and dictionary files.

Choosing **Browse** causes AutoCAD to display the Browse for Folder or Select a File dialog box, depending on what was selected from the list.

Choosing **Add** causes AutoCAD to add a search path for the selected folder.

Choosing **Remove** causes AutoCAD to remove the selected search path or file.

Choosing **Move Up** causes AutoCAD to move the selected search path above the preceding search path.

Choosing **Move Down** causes AutoCAD to move the selected search path below the following search path.

Choosing **Set Current** causes AutoCAD to make the selected project or spelling dictionary current.

DISPLAY

The **Display** tab of the Options dialog box(see Figure 13–52) controls preferences that relate to AutoCAD performance.

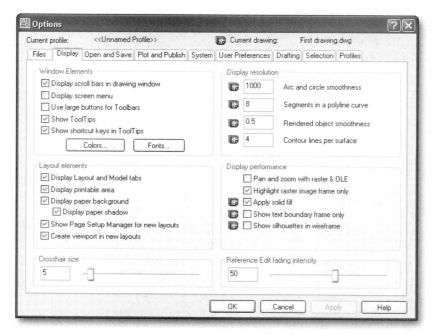

Figure 13–52 *Options dialog box with the Display tab selected*

The **Window Elements** section controls the parameters of the AutoCAD drawing window.

> Selecting **Display scroll bars in drawing window** causes scroll bars at the bottom and right sides of the drawing window to be displayed.
>
> Selecting **Display screen menu** causes the screen menu on the right side of the drawing window to be displayed.
>
> Selecting **Use Large Buttons for Toolbars** displays buttons in a larger format at 32 by 30 pixels. The default display size is 16 by 15 pixels. Selecting **Show ToolTips** causes tooltips to be displayed when the cursor is near a command or option.
>
> Selecting **Show shortcut keys in ToolTips** (available only when **Show ToolTips** is checked) causes the shortcut keys to be included when a tooltip is displayed.

Choosing **Colors** causes AutoCAD to display the AutoCAD Window Colors dialog box, which can be used to set the colors for drawing area, screen menu, text window, and command line.

Choosing **Fonts** causes AutoCAD to display the Graphics Window Font dialog box, which can be used to specify the font AutoCAD uses for the screen menu and command line and in the text window.

The **Display resolution** section lets you set the resolution in the following text boxes:
- Arc and circle smoothness
- Segments in a polyline curve
- Rendered object smoothness
- Contour lines per surface

The **Layout elements** section has check boxes to toggle on and off the following:
- Display Layout and Model tabs
- Display printable area
- Display paper background
- Display paper shadow
- Show page setup dialog for new layouts
- Create viewport in new layouts

The **Display performance** section has check boxes to toggle on and off the following:
- Pan and zoom with raster image
- Highlight raster image frame only
- Apply solid fill
- Show text boundary frame only
- Show silhouettes in wireframe

OPEN AND SAVE

The **Open and Save** tab of the Options dialog box (see Figure 13–53) lets you determine formats and parameters for drawings, external references, and ObjectARX applications as they are opened or saved.

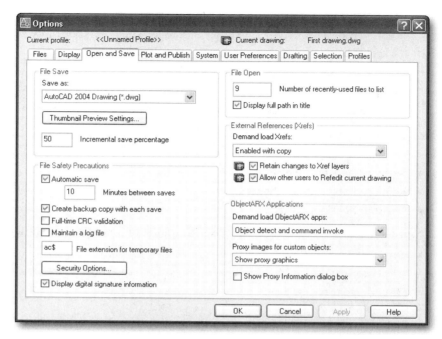

Figure 13-53 *Options dialog box with the **Open And Save** tab selected*

The **File Save** section controls settings related to saving a file.

> The **Save as** text box lets you select a default save format when you invoke the SAVEAS command. Choose from the following formats:
>
> - AutoCAD 2004 Drawing (*.dwg)
> - AutoCAD 2000/LT2000 Drawing (*.dwg)
> - AutoCAD Drawing Template File (*.dwt)
> - AutoCAD 2004 DXF (*.dxf)
> - AutoCAD 2000/LT2000 DXF (*.dxf)
> - AutoCAD R12/LT2 DXF (*.dxf)
>
> Choosing **Thumbnail Preview Settings** causes the Thumbnail Preview Settings dialog box to be displayed, which controls whether thumbnail previews are updated when the drawing is saved.

The **File Open** section shows the number of recently used files to list.

> Selecting **Display full path in title** causes AutoCAD to display the full path in the title.

The **External References (Xrefs)** section controls the settings that relate to editing and loading external references.

The **Demand load Xrefs** text box controls demand loading of xrefs. Demand loading improves performance by loading only the parts of the referenced drawing needed to regenerate the current drawing. You can select **Disabled**, **Enabled**, or **Enabled With Copy**. **Disabled** means that demand loading is not on in the current drawing. Someone else can open and edit an xref file except as it is being read into the current drawing. **Enabled** means that demand loading is ON in the current drawing. No one else can edit an xref file while the current drawing is open. However, someone else can reference the xref file. **Enabled with Copy** means that demand loading is ON in the current drawing. An xref file can still be opened and edited by someone else. AutoCAD only uses a copy of the xref file, treating it as a completely separate file from the original xref.

Selecting **Retain changes to Xref layers** causes the properties of layers in xrefs to be saved as they are changed in the current drawing for reloading later.

Selecting **Allow other users to Refedit current drawing** allows the current drawing to be edited while being referenced by other drawing(s).

The **File Safety Precautions** section helps to detect errors and avoid losing data.

Selecting **Automatic save** with the **Minutes between saves** text box allows you to determine if and for what intervals periodic automatic saves will be performed.

Selecting **Create backup copy with each save** allows you to determine whether a backup copy is created when you save the drawing.

Selecting **Full-time CRC validation** allows you to determine whether a cyclic redundancy check (CRC) is performed when an object is read into the drawing. Cyclic redundancy check is a mechanism for error-checking. If you suspect a hardware problem or AutoCAD error is causing your drawings to be corrupted, check this box.

Selecting **Maintain a log file** allows you to determine whether the contents of the text window are written to a log file. Use the Files tab in the Options dialog box to specify the name and location of the log file.

The **File extension for temporary files** text box allows you to specify an extension for temporary files on a network. The default extension is *.ac$*. **Security Options** and **Display digital signature information** are described in the earlier section "Security, Passwords, and Encryption" in this chapter.

The **ObjectARX Applications** section controls parameters for AutoCAD Runtime Extension applications and proxy graphics.

The **Demand load ObjectARX Apps** text box lets you select when and if a third-party application is demand-loaded when a drawing has custom objects that were created in that application. Choosing **Disable load on demand** turns off demand-loading. Choosing **Custom object detect** demand-loads the source application when you open a drawing that contains custom objects. It does not demand-load the application when you invoke one of the application's commands. Choosing **Command invoke** demand-loads the source application when you invoke the application's command. This setting does not demand-load the application when you open a drawing that contains custom objects. Choosing **Object detect and command invoke** demand-loads the source application when you open a drawing that contains custom objects or when you invoke one of the application's commands.

The **Proxy images for custom objects** text box controls how custom objects in the drawings are displayed. Choosing **Do not show proxy graphics** causes custom objects in drawings not to be displayed. Choosing **Show proxy** causes custom objects in drawings to be displayed. Choosing **Show proxy bounding box** causes a box to be displayed in place of custom objects in drawings. Choosing **Show Proxy Information dialog box** causes a warning to be displayed when you open a drawing that contains custom objects.

PLOT AND PUBLISH

The **Plot and Publish** tab of the Options dialog box (see Figure 13–54), lets you set the parameters for plotting your drawing.

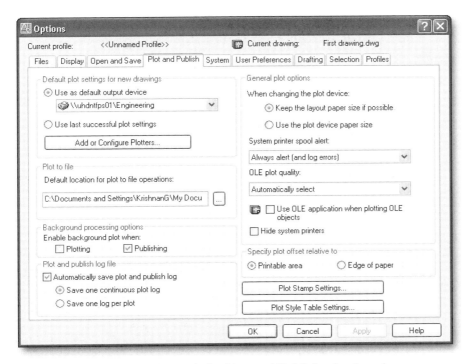

Figure 13-54 *Options dialog box with the **Plot and Publish** tab selected*

The **Default plot settings for new drawings** section determines plotting parameters for new drawings. It will also determine settings for drawings created in earlier releases of AutoCAD prior to AutoCAD 2000 that have never been saved in AutoCAD 2000 format.

Selecting **Use as default output device** causes the device selected in the list box to become the default output device for new drawings and for drawings created in an earlier release of AutoCAD that have never been saved in AutoCAD 2000 format. The list displays all plotter configuration files (*.pcx3*) that are found in the plotter configuration search path. It also displays all system printers configured in the system.

Selecting **Use last successful plot settings** causes the settings of the last successful plot to be used for the current settings.

Choosing **Add or Configure Plotters** lets you add or configure a plotter from the Plotters program window. See Chapter 8 for adding and configuring plotters.

The **General plot options** section lets you set general parameters such as paper size settings, system printer alert parameters, and OLE objects.

Selecting **Keep the layout paper size if possible** applies the paper size in the **Layout Settings** tab in the Page Setup dialog box provided the selected output device is able to plot to this paper size. If it cannot, AutoCAD displays a warning message and uses the paper size specified either in the plotter configuration file (*.pc3*) or in the default system settings if the output device is a system printer.

Selecting **Use the plot device paper size** applies the paper size in either the plotter configuration file (*.pc3*) or in the default system settings if the output device is a system printer.

The **System printer spool alert** list box determines if a warning will be displayed if the plotted drawing is spooled through a system printer because of an input or output port conflict. Selecting **Always alert (and log errors)** causes a warning to be displayed and always logs an error when the plotted drawing spools through a system printer. Selecting **Alert first time only (and log errors)** causes a warning to be displayed once and always logs an error when the plotted drawing spools through a system printer. Selecting **Never alert (and log first error)** causes a warning not to be displayed and logs only the first error when the plotted drawing spools through a system printer. Selecting **Never alert (do not log errors)** causes a warning not to be displayed and does not log an error when the plotted drawing spools through a system printer.

The **OLE plot quality** list box determines plotted OLE objects' quality. Options include **Automatically select, Monochrome (e.g. spreadsheet), Low graphics (e.g. color text & pie charts)**, and **High graphics (e.g. photograph)**.

Selecting **Use OLE application when plotting OLE objects** starts the application that creates the OLE object when you plot a drawing with OLE objects. This will help optimize quality of OLE objects.

Selecting **Hide system printers** causes AutoCAD to Hide system printers.

The **Plot to File** section specifies the default location for plot to file operations. You can enter a location or choose the **Browse** icon to specify a new location.

The **Background processing options** section specifies options for background plotting and publishing. You can use background plotting to start a job you are plotting or publishing and immediately return to work on your drawing while your job is plotted or published as you work.

The **Plot and publish log file** section controls options for saving a plot and publish log file as a comma-separated value (*.csv*) file that can be viewed in a spreadsheet program.

The **Specify plot offset relative to** section specifies whether the offset of the plot area is from the lower-left corner of the printable area or from the edge of the paper.

Choosing **Plot Stamp Settings** opens the Plot Stamp Settings dialog box, which allows you to specify information for the plot stamp.

Choosing **Plot Style Table Settings** opens the Plot Style Table Settings dialog box, which allows you to specify settings for plot style tables.

SYSTEM

The **System** tab of the Options dialog box (see Figure 13–55), has sections for managing the 3D graphics display, pointing devices, dbConnect options, and general options.

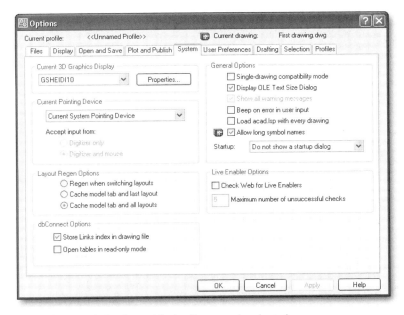

Figure 13–55 *Options dialog box with the System tab selected*

The **Current 3D Graphics Display** section list box allows you to specify the current display from the list.

Choosing **Properties** displays a 3D Graphics System Configuration dialog box for the current 3D graphics display system. In the 3D Graphics System Configuration dialog box, you set options that affect the way objects are displayed and system resources are used in the 3D

Orbit view. The options you set also affect the way objects are shaded with SHADEMODE.

The **Current Pointing Device** section of the **System** tab determines the parameter for the pointing device(s) being used. The text box lists available pointing device drivers from which to choose. Selecting **Current system pointing device** causes the system pointing device to be the current pointing device. Selecting **Wintab Compatible Digitizer ADI 4.2 by Autodesk** causes the Wintab compatible digitizer to be the current pointing device.

The **Accept input from** option determines if the digitizer only is active or both the mouse and digitizer are active. These are determined by choosing either **Digitizer only** or **Digitizer and mouse**.

The **Layout Regen Options** section of the **System** tab lets you specify how AutoCAD updates the display list in the Model and layout tabs. The display list for each tab is updated either by regenerating the drawing when you switch to that tab or by saving the display list to memory and regenerating only the modified objects when you switch to that tab.

Selecting **Regen when switching layouts** causes AutoCAD to regenerate the drawing each time you switch tabs.

Selecting **Cache model tab and last layout** saves the display list to memory for the Model tab and the last layout made current. When checked, it suppresses regenerations when you switch between the two tabs. Regenerations for all other layouts still occur when you switch to those tabs.

Selecting **Cache model tab and all layouts** causes AutoCAD to regenerate the drawing the first time you switch to each tab. For the remainder of the drawing session, when you switch to those tabs, the display list is saved to memory and regenerations are suppressed.

The **dbConnect Options** section of the **System** tab allows you to manage the options associated with database connectivity.

Selecting **Store links index in drawing file** causes AutoCAD to store the database index within the drawing file. When this option is checked, performance during link selection operations is enhanced. When this option is not checked, the drawing file size is decreased and the opening process is enhanced for drawings with database information.

Selecting **Open tables in read-only mode** causes AutoCAD to open database tables in Read-only mode within the drawing file.

The **General Options** section of the **System** tab has check boxes relating to a variety of options for system parameter settings.

> Selecting **Single-drawing compatibility mode** causes AutoCAD to open only one drawing at a time (Single-drawing Interface or SDI). Otherwise AutoCAD can open multiple drawing sessions (Multi-drawing Interface or MDI).
>
> Selecting **Display OLE Text Size Dialog** causes the OLE Properties dialog box to be displayed when inserting OLE objects into AutoCAD drawings.
>
> Selecting **Show all warning messages** causes all dialog boxes that include a Don't Display This Warning Again option to be displayed. Dialog boxes with warning options will be displayed regardless of previous settings specific to each dialog box.
>
> Selecting **Beep on error in user input** causes an alarm beep when AutoCAD detects an invalid entry.
>
> Selecting **Load acad.lsp with every drawing** causes AutoCAD to load the acad.lsp file into every drawing.
>
> Selecting **Allow long symbol names** allows up to 255 characters for named objects.
>
> From the **Startup** list box, you can select one of two possible initial views that might appear when starting AutoCAD: **Show Startup dialog** and **Do not show a startup dialog**. These options are explained in Chapter 1 under "Starting AutoCAD".

The **Live Enabler Options** section of the **System** tab allows you to specify how AutoCAD checks for Object Enablers. Using Object Enablers, you can display and use custom objects in AutoCAD drawings even when the ObjectARX application that created them is unavailable.

> The **Check Web for Live Enablers** check box, when checked, causes AutoCAD to check for Object Enablers on the Autodesk Web site.
>
> The **Maximum number of unsuccessful checks** text box allows you to specify the number of times AutoCAD will continue to check for Object Enablers after unsuccessful attempts.

USER PREFERENCES

The **User Preferences** tab of the Options dialog box (see Figure 13–56), controls options that optimize the way you work in AutoCAD.

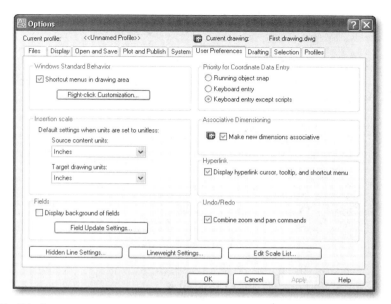

Figure 13–56 *Options dialog box with the User Preferences tab selected*

The **Windows Standard Behavior** section of the **User Preferences** tab lets you apply Windows techniques and methods in AutoCAD.

Selecting **Shortcut menus in drawing area** causes shortcut menus to be displayed when the pointing device is right-clicked. Otherwise, right-clicking is the same as pressing ENTER.

Choosing **Right-click customization** causes the Right-Click Customization dialog box to be displayed. The Right-Click Customization dialog box determines whether right-clicking in the drawing area displays a shortcut menu or is the same as pressing ENTER. This will allow you to have a right-click invoke ENTER while a command is active. You can also disable the following Command shortcut menu options:

Choosing **Turn on time-sensitive right-click** controls right-click behavior. A quick click is the same as pressing ENTER. A longer click displays a shortcut menu. You can set the duration of the longer click in milliseconds.

The **Default Mode** section of the Right-Click Customization dialog box determines the effect of right-clicking when no objects are selected. Selecting **Repeat Last Command** causes right-clicking to be the same as pressing ENTER. Selecting **Shortcut Menu** causes right-clicking to display a shortcut menu when applicable.

The **Edit Mode** section of the Right-Click Customization dialog box determines the effect of right-clicking when one or more objects are selected. Selecting **Repeat Last Command** causes right-clicking to be the same as pressing ENTER. Selecting **Shortcut Menu** causes right-clicking to display the Edit shortcut menu.

The **Command Mode** section of the Right-Click Customization dialog box determines the effect of right-clicking when a command is in progress. Selecting **ENTER** causes right-clicking to be the same as pressing ENTER when a command is in progress. Selecting **Shortcut Menu: always enabled** causes right-clicking to display the Command shortcut menu. Selecting **Shortcut Menu: enabled when command options are present** causes right-clicking to display the Command shortcut menu to be displayed only when options are currently available from the command line. Otherwise, right-clicking is the same as pressing ENTER.

The **Insertion scale** section of the **User Preferences** tab allows you to control the default scale for dragging objects into a drawing using i-drop or DesignCenter.

> The **Source content units** list box allows you to set the units AutoCAD uses for an object being inserted into the current drawing when no insert units are specified with the INSUNITS system variable.

> The **Target drawing units** list box allows you to set the units AutoCAD uses in the current drawing when no insert units are specified with the INSUNITS system variable. (This is stored in the INSUNITSDEFTARGET system variable.) Each of these allow you to choose from the following units: **Inches, Feet, Miles, Millimeters, Centimeters, Meters, Kilometers, Microinches, Mills, Yards, Angstroms, Nanometers, Microns, Decimeters, Decameters, Hectometers, Gigameters, Astronomical Units, Light Years,** and **Parsecs**. If **Unspecified-Unitless** is selected, the object is not scaled when inserted.

The **Fields** section of the **User Preferences** tab sets preferences related to fields.

> Selecting **Display background of fields** displays fields with a light gray background that is not plotted. When this option is cleared, fields are displayed with the same background as any text.

> Choosing **Field Update Settings** displays the Field Update Settings dialog box, which allows you to set the fields that will be updated automatically.

The **Priority for Coordinate Data Entry** section of the **User Preferences** tab determines how input of coordinate data affects AutoCAD's actions.

> Selecting **Running object snap** causes running object snaps to be used at all times instead of specific coordinates.

Selecting **Keyboard entry** causes the coordinates that you enter to be used at all times and overrides running object snaps.

Selecting **Keyboard entry except scripts** causes the specific coordinates that you enter to be used rather than running object snaps, except in scripts.

The **Associative Dimensioning** section of the **User Preferences** tab controls whether new dimensions are associative.

Selecting **Make new dimensions associative** causes new dimensions to be drawn as associative dimensions and will be associated with the objects being dimensioned.

The **Hyperlink** section of the **User Preferences** tab determines display property settings of hyperlinks.

Selecting **Display hyperlink cursor, tooltip, and shortcut menu** causes the hyperlink cursor, tooltip, and shortcut menu to be displayed when the cursor is over an object that contains a hyperlink.

The **Undo/Redo** section of the **User Preferences** tab controls Undo and Redo for ZOOM and PAN.

Selecting **Combine zoom and pan commands** causes these two commands to act together.

Choosing **Hidden Line Settings** causes the Hidden Line Settings dialog box to be displayed (see Figure 13–57). This allows you to change the display properties of hidden lines. These settings are in effect only when the HIDE command is used or when the HIDDEN option of the SHADEMODE command is used.

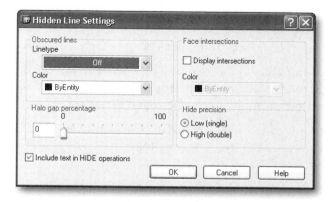

Figure 13–57 *Hidden Line Settings dialog box*

The **Obscured lines** section of the Hidden Line Settings dialog box allows you to specify the linetype and color of obscured lines, which are lines that are made visible by changing its color and linetype. The **Linetype** list box allows you to select from a list of linetypes or select **Off**.

The **Color** list box allows you to select from available colors.

The **Halo gap percentage** section lets you specify the distance to shorten a haloed line at the point where it will be hidden. Moving the slider bar specifies the distance as a percentage of one inch. It is not affected by the zoom level.

Selecting **Include text in HIDE operations** causes text objects created by the TEXT, DTEXT, or MTEXT command to be included during a HIDE command.

The **Face intersections** section controls the display of intersections. Selecting **Display intersections** causes intersection polylines to be displayed. The **Color** drop down allows you specify the color of intersection polylines.

The **Hide precision** section controls the accuracy in the method of creating hides and shades. Selecting **Low (single)** results in low accuracy (and low memory usage). Selecting **High (double)** results in high accuracy.

Choosing **Lineweight Settings** causes the Lineweight Settings dialog box to be displayed. This allows you to set lineweight options.

The **Lineweights** section lets you select lineweights in the list box. Options include **ByLayer**, **ByBlock**, **Default**, and a list of varying lineweights.

The **Units for Listing** lets you choose between **Millimeters** and **Inches**.

Selecting **Display Lineweight** causes lineweights to be displayed in model space and paper space.

The **Adjust Display Scale** slider bar lets you adjust the how wide the selected linewidth will be displayed.

Choosing **Edit Scale List** causes the Edit Scale List dialog box to be displayed, which controls the list of scales available for layout viewports, page layouts, and plotting. Select a scale from the **Scale List** list box and then choose one of the options, which include **Add**, **Edit**, **Move Up**, **Move Down**, **Delete**, and **Reset**.

DRAFTING

The **Drafting** tab of the Options dialog box (see Figure 13–58) lets you customize drafting options in AutoCAD. In this tab are sections for **AutoSnap Settings, AutoSnap Marker Size, AutoSnap Marker Color, AutoTrack Settings, Alignment Point Acquisition, Aperture size, Object Snap Options,** and **Drafting Tooltip Appearance**. The options available in these sections are explained in Chapter 3 in the section on "Drafting Settings."

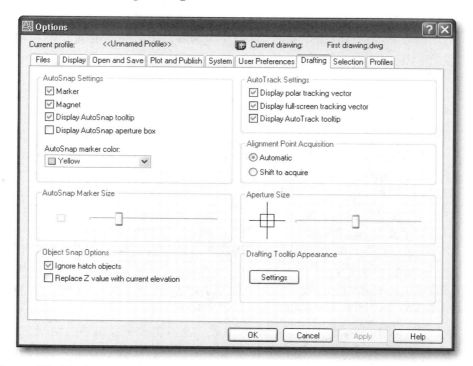

Figure 13–58 *Options dialog box with the Drafting tab selected*

SELECTION

The **Selection** tab of the **Options** dialog box (see Figure 13–59), lets you customize selection options in AutoCAD. In this tab are sections for **Pickbox Size, Selection Preview, Selection Modes, Grip Size,** and **Grips**. The options available for the **Selection Modes** are explained in Chapter 5 in the section "Object Selection Modes." The options available for **Grips** are explained in Chapter 6 in the section "Editing With Grips."

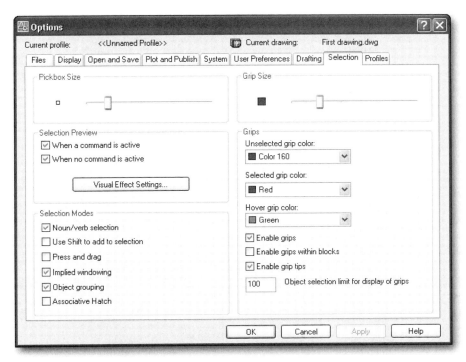

Figure 13–59 *Options dialog box with the **Selection** tab selected*

PROFILES

The **Profiles** tab of the Options dialog box (see Figure 13–60), lets you manage profiles. A profile is a named and saved group of environment settings. This profile can be restored as a group when desired. AutoCAD stores your current options in a profile named **Unnamed Profile**. AutoCAD displays the current profile name, as well as the current drawing name, in the Options dialog box. The profile data is saved in the system registry and can be written to a text file with an *.arg* extension file. AutoCAD organizes essential data and maintains changes in the registry as necessary.

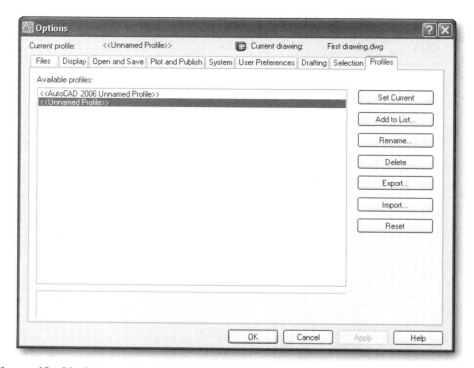

Figure 13–60 *Options dialog box with the Profiles tab selected*

A profile can be exported to or imported from different computers. If changes have been made to your current profile during an AutoCAD session and you want to save them in the *.arg* file, the profile must be exported. After the profile with the current profile name has been exported, AutoCAD updates the *.arg* file with the new settings. Then the profile can be imported again into AutoCAD, thus updating your profile settings.

>   Choosing **Set Current** makes the profile that is highlighted in the **Available profiles** list box the current profile.
>
>   Choosing **Add To List** lets you name and save the current environment settings as a profile.
>
>   Choosing **Rename** lets you rename the highlighted profile.
>
>   Choosing **Delete** lets you delete the highlighted profile.
>
>   Choosing **Export** causes the Export Profiles dialog box to be displayed. This is a file manager dialog box in which the highlighted profile can be saved to the path you specify.

Choosing **Import** causes the Import Profiles dialog box to be displayed. This is a file manager dialog box in which you can select a profile from a saved path to be imported.

Choosing **Reset** causes the highlighted profile to be reset. The default profile name is listed in the description pane at the bottom of the dialog box.

After making changes, choose **Apply** to make the changes effective or choose **OK** to save the settings and close the Options dialog box.

SAVING OBJECTS IN OTHER FILE FORMATS (EXPORTING)

The EXPORT command allows you to save a selected object in other file formats, such as *.bmp*, *.sat*, *.3ds*, and *.wmf*. Invoke the EXPORT command by selecting **Export** from the File menu and AutoCAD displays the Export Data dialog box (see Figure 13–61). In the **Files of type** list box, select the format type in which you wish to export objects. Enter the file name in the **File name** edit box. Select **Save** and AutoCAD prompts:

> Select objects: *(select the objects to export, and press* ENTER *to complete object selection)*

AutoCAD exports the selected objects in the specified file format using the specified file name.

Table 13–1 lists the format types available in AutoCAD for exporting the current drawing.

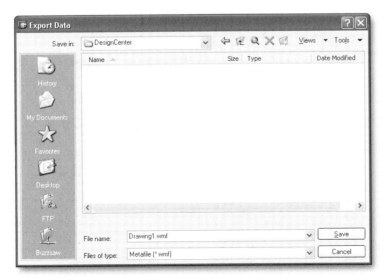

Figure 13–61 *Export Data dialog box*

Table 13-1 *Exportable Format Types*

Format Type	Description
3ds	3D Studio file
bmp	Device-independent bitmap file
dwg	AutoCAD 2000 drawing file (same as invoking the WBLOCK command)
dxx	AutoCAD attribute extract DXF file (same as invoking the ATTEXT command)
eps	Encapsulated PostScript file
sat	ACIS solid-object file
stl	Solid object stereo-lithography file
wmf	Windows metafile

IMPORTING VARIOUS FILE FORMATS

The IMPORT command allows you to import various file formats, such as *.3ds*, *.sat*, and *.wmf*, into AutoCAD. Invoke the IMPORT command from the Insert toolbar (see Figure 13–62) and AutoCAD displays the Import File dialog box (see Figure 13–63).

Figure 13–62 *Invoking the* IMPORT *command from the Insert toolbar*

In the **Files of Type** list box, select the format type you wish to import into AutoCAD. Select the file from the appropriate directory from the list box and choose **Open**. AutoCAD imports the file into the AutoCAD drawing.

Table 13–2 lists the format types available to import into AutoCAD.

Table 13-2 *Format types available to import into AutoCAD*

Format Type	Description
3ds format	Imports a 3D Studio file
sat format	Imports ACIS solid object file
wmf format	Imports Windows Metafile

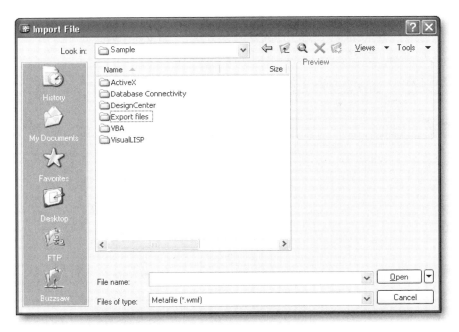

Figure 13–63 *Import File dialog box*

STANDARDS

AutoCAD has a feature that allows you to verify that the layers, dimension styles, linetypes, and text styles of the drawing you are working in conform to an accepted standard, such as a company, trade, or client standard. To utilize this feature, your drawing must have some standard drawing(s) with which it is associated.

You can create a standards file from an existing drawing or you can create a new drawing and save it as a standards file with an extension of *.dws*. Open an existing drawing from which you want to create a standards file, invoke the SAVEAS command, and enter a name for the standards file in the Save Drawing As dialog box. Select AutoCAD Drawing Standards (*.dws*) from the **Files of type** list and then choose **Save**. You can also create a new drawing and set appropriate standards for layers, text styles, dimension styles, and linetypes. Invoke the SAVEAS command, enter a name for the standards file in the Save Drawing As dialog box, select AutoCAD Drawing Standards (*.dws*) from the **Files of type** list, and then choose **Save**.

The STANDARDS command manages the association of standards files with drawings.

Invoke the STANDARDS command by choosing **Configure** from the **CAD Standards** flyout of the **Tools** menu and AutoCAD displays the Configure Standards dialog box (see Figure 13–64).

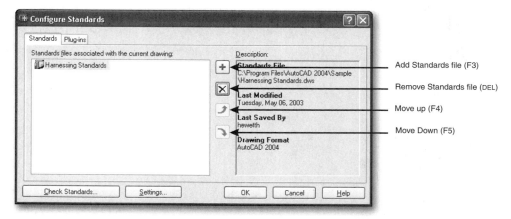

Figure 13–64 *Configure Standards dialog box with the Standards tab selected*

If a drawing has an associated standards file, there will be an Associated Standards File(s) icon on the status bar tray at the bottom right corner of the drawing area.

The Standards files associated with the current drawing section lists all standards (*.dws*) files that are associated with the current drawing. To add a standards file, choose the **Add Standards File** icon. AutoCAD displays the Select Standards File dialog box. Select a standards file from an appropriate folder. To remove a standards file from the current drawing, choose the **Remove Standards File** icon. If conflicts arise between multiple standards in the list (for example, if two standards specify layers of the same name but with different properties), the standard that appears first in the list takes precedence. To change the position of a standards file in the list, select it and choose the **Move Up** or **Move Down** icon.

In the **Plug-ins** tab of the Configure Standards dialog box (see Figure 13–65), the **Plug-ins used when checking standards** section lists the standards plug-ins that are installed on the current system. For the CAD Standards Extension, a standards plug-in is installed for each of the named objects for which standards can be defined (layers, dimension styles, linetypes, and text styles). The **Description** section has descriptions of the Purpose, Version, and Publisher of the plug-in that is highlighted in the **Plug-ins used when checking standards** section.

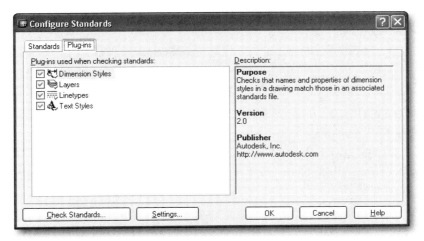

Figure 13–65 *Configure Standards dialog box with the Plug-ins tab selected*

The CHECKSTANDARDS command analyzes the current drawing for standards violations.

Invoke the CHECKSTANDARDS command by choosing **Check** from the **CAD Standards** flyout of the **Tools** menu and AutoCAD displays the Check Standards dialog box (see Figure 13–66).

Figure 13–66 *Check Standards dialog box*

The Check Standards dialog box has sections titled **Problem**, **Replace with**, and **Preview of changes**.

In the **Problem** section there is a description of a nonstandard object in the current drawing. To fix a problem, select a replacement from the **Replace with** list and then choose **Fix**.

The **Replace with** section lists possible replacements for the current standards violation. If a recommended fix is available, it is preceded by a check mark.

The **Preview of changes** section indicates the properties of the nonstandard AutoCAD object that will be changed if the fix currently selected in the **Replace with** list is applied.

Choose **Close** to close the Check Standards dialog box without applying a fix to the standards violation currently displayed in the **Problem** section.

You can use CAD Standards tools to check for violations as you work. You are immediately alerted whenever you create a non-standard named object.

SLIDES AND SCRIPTS

Slides are quickly viewable, non-editable views of a drawing or parts of a drawing. There are two primary uses for slides. One is to have a quick and ready picture to display symbols, objects, or written data for informational purposes only. The other very useful application of slides is to be able to display a series of pictures, organized in a prearranged sequence for a timed slide show. This is a very useful tool for demonstrations to clients or in a showroom. This feature supplements the time-consuming calling up of views required when using the ZOOM, PAN, or other display commands. The "slide show" is implemented through the SCRIPT command (described later in this chapter).

It should be noted that a slide merely masks the current display. Any cursor movement or editor functions employed while a slide is being displayed affects the current drawing under the slide and not the slide itself.

MAKING A SLIDE

The current display can be made into a slide with the MSLIDE command. The current viewport becomes the slide while you are working in model space. The entire display, including all viewports, becomes the slide when using MSLIDE while you are working in paper space. The MSLIDE command takes a picture of the current display and stores it in a file, so be sure it is the correct view.

Invoke the MSLIDE command by typing **mslide** keyboard at the On-Screen prompt and AutoCAD displays the Create Slide File dialog box.

The default is the drawing name, which you can use as the slide file name by pressing ENTER. Or you can type any other name, as long as you are within the limitations of the operating system file-naming conventions. AutoCAD automatically appends the extension *.sld*. Only objects that are visible in the screen drawing area (or in the current viewport when in model space) are made into the slide.

If you plan to show the slide on different systems, you should use a full-screen view with a high-resolution display for creating the slide.

VIEWING A SLIDE

The VSLIDE command displays a slide in the current viewport.

> Invoke the VSLIDE command by typing **vslide** from the On-Screen prompt and AutoCAD displays the Select Slide File dialog box. This is similar to most file-management dialog boxes. Select the slide to display in the current viewport.

SCRIPTS

Of the many means available to enhance AutoCAD through customization, scripts are among the easiest to create. Scripts are similar to the macros that can be created to enhance word processing programs. They permit you to combine a sequence of commands and data into one or two entries. Creating a script, like most enhancements to AutoCAD, requires that you use a text editor to write the script file (with the extension *.scr*), which contains the instructions and data for the SCRIPT command to follow.

Because script files are written for use at a later time, you must anticipate the conditions under which they will be used. Therefore, familiarity with sequences of prompts that will occur and the types of responses required is necessary to have the script function properly. Writing a script is a simple form of programming.

A script text file must be written in ASCII format. That is, it must have no embedded print codes or control characters that are automatically written in files when created with a word processor in the document mode. If you are not using one of the available line editors, be sure you are in the nondocument, programmer, or ASCII mode of your word processor when creating or saving the file. Save the script with the file extension *.scr*.

Each command can occupy a separate line, or you can combine several command/data responses on one line. Each space between commands and data is read as an ENTER, just as pressing the SPACEBAR is read while in AutoCAD. The end of a line of text is also the same as an ENTER.

The following script file contains several commands and data. The commands are GRID, LINE, CIRCLE, LINE, and CIRCLE again. The data are the response ON, coordinates (such as 0,0 and 5,5), and distances, such as radius 3.

```
GRID ON LINE 0,0 5,5  CIRCLE 3,3 3
LINE 0,5 5,0
CIRCLE 5,2.5 1
```

The first line includes the GRID, LINE, and CIRCLE commands, and their responses. Note the two spaces after 5,5; these are required to simulate the double ENTER. Not obvious is the extra space following the 5,0 response in the second LINE command and after the 1 in the third line. This extra space and the invisible CR-LF (carriage-return, linefeed) code that ends every line in a text file combine to simulate pressing the SPACEBAR twice. This is necessary, again, to exit the LINE command.

Some text editors automatically remove blank spaces at the end of text lines. To guard against that, an alternative is to have a blank line indicate the second ENTER, as follows:

```
GRID ON LINE 0,0 5,5  CIRCLE 3,3 3
LINE 0,5 5,0
CIRCLE 5,2.5 1
```

Invoke the SCRIPT command by choosing **Run Script** on the **Tools** menu and AutoCAD displays the Select Script File dialog box. Select the appropriate script file from the list box and choose the Open button. AutoCAD executes the command sequence from the script file.

Using a script to perform a repetitive task is illustrated in the following example. This application also offers some insight on changing the objects in an inserted block with attributes without affecting the attribute values.

Figure 13–67 shows a group of drawings all of which utilize a common block with attribute values in one insertion that are different from the attribute values of those in other drawings. In this case, the border/title block is a block named BRDR. It was originally drawn with the short lines around and outside of the main border line. It was discovered that these lines interfered with the rollers on the plotter and needed to be removed. The BRDR block definition is shown in Figure 13–68. Remember, the inserted block has different attribute values in each drawing, such as drawing number, date, and title.

You want to change objects in the block but maintain the attribute values as they are. There are two approaches to redefinition. One is to find a clear place in the drawing and insert the block with an asterisk (*). This is the same as inserting and exploding the block. Then you make the necessary changes in the objects and make the revised group into a block with the same block name. This redefines all insertions of blocks with that same name in the drawing. In this case there is only one insertion. You must be attentive to how any changes to attributes might affect the already inserted block of that name.

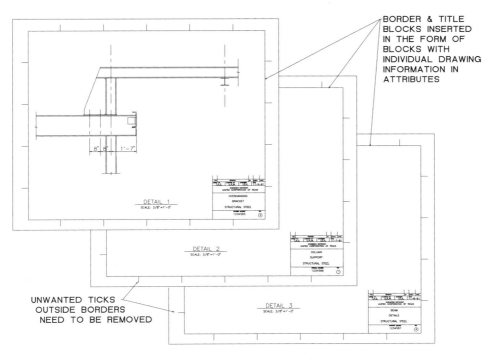

Figure 13-67 *Drawings utilizing common block and attribute values that are different from those of other drawings*

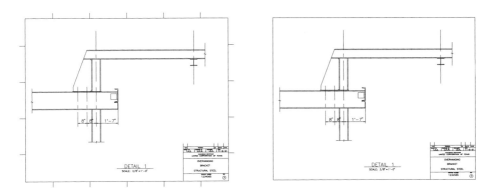

Figure 13-68 *Common block attributes modified: Border removed (left); Ticks removed (right)*

The second method is to use the WBLOCK command to place a copy of the block in a file with the same name. This makes a new and separate drawing of the block. Then you exit the current drawing, call up the newly created drawing, make the required changes in the objects, and end the drawing that was created by the WBLOCK command. Then you re-enter the drawing in which the block objects need to be changed. You now use the INSERT command and respond with "blockname=" and have the block redefined without losing the attribute values. For example, if the block name is BRDR, the sequence would be as follows:

```
Command: insert
Enter block name or [?]: brdr=
Block "b" already exists. Redefine it? [Yes/No] <N>: y
Block BRDR redefined
Regenerating drawing
Specify insertion point or [Scale/X/Y/Z/Rotate/PScale/PX/PY/PZ/
    PRotate]: (press ENTER to cancel the command sequence)
```

The key to this sequence is the equals sign (=) following the block name. This causes AutoCAD to change the definition of the block named BRDR to be that of the drawing named BRDR, although it maintains the attribute values as long as attribute definitions remain unchanged.

If the preceding procedure must be repeated many times, this is where a script file can be employed to automate the process. In the following example, we show how to apply the script to a drawing named PLAN_1. The sequence included an ENTER as it was described to be used while in AutoCAD. This expedited the operation by not actually having the block inserted, but only having its definition brought into the drawing. Because an ENTER during the running of a script file causes the script to terminate, those keystrokes cannot be in the middle of a script; besides, invoking in a SCRIPT command requires using the AutoLISP function "(command)." The script can be written in a file (called BRDRCHNG.SCR for this example) in ASCII format as follows:

```
INSERT
BRDR=
0,0      (the 0,0 is followed by six spaces)
ERASE L (the L is followed by one space)
REDRAW
```

Now, from the On-Screen prompt, you can apply the script to drawing PLAN_1 by responding as follows:

script

There are several important aspects of this sequence:

Lines 1–2 This is where you might enter if you were not in the SCRIPT command and have the definition of the block named BRDR take on that of the drawing BRDR without actually having to continue with the insertion in the drawing.

Line 3 In this case, 0,0 as the insertion point is arbitrary because the inserted block is going to be erased anyway. Special attention is given to the six spaces following the insertion point. These are the same as pressing SPACEBAR or ENTER six times in response to the "X-scale," "Y-scale," and "Rotation angle" prompts, and the number of spaces that follow must correspond to the number of attributes that require responses for values. In this example there were three attributes. Again, the fact that the responses are null is immaterial, because this insertion will not be kept.

Line 4 The ERASE L is self-explanatory, but do not forget that after the "L" you must have another space (press SPACEBAR again) to terminate the object selection process and complete the ERASE command.

Line 5 The REDRAW command is not really required except to show the user for a second time that the changes have been made before the script ends.

Following are the utility commands that may be used within a script file.

The DELAY command causes the script to pause for the number of milliseconds that have been specified by the command. To set the delay in the script for five seconds, you would write:

    DELAY 5000

The RESUME command causes the script to resume running after the user has pressed either or to interrupt the script.

The GRAPHSCR and TEXTSCR commands are used to flip or toggle the screen to the graphics or text mode, respectively, during the running of the script. These two screen toggle commands can be used transparently by preceding them with an apostrophe.

The RSCRIPT command, when placed at the end of a script, causes the script to repeat itself. With this feature you can have a slide show run continuously until terminated by ESC key.

A repeating demonstration can be set up to show some sequences of commands and responses as follows:

    GRID ON
    LIMITS 0,0 24,24
    ZOOM A
    CIRCLE 12,12 4
    DELAY 2000
    COPY L M 12,12 18,12 12,18 6,12 12,6 (an extra space at the end)
    DELAY 5000
    ERASE W 0,0 24,24 (an extra space at the end)

```
DELAY 2000
LIMITS 0,0 12,9
ZOOM A

GRID OFF
TEXT 1,1 .5 0 THAT'S ALL FOLKS!
ERASE L (an extra space at the end)
RSCRIPT
```

This script file utilizes the DELAY and RSCRIPT subcommands. Note the extra spaces where continuation of some actions must be terminated.

The SCRIPT command can be used to show a series of slides, as in the following sequence:

```
VSLIDE SLD_A
VSLIDE *SLD_B
DELAY 5000
VSLIDE
VSLIDE *SLD_C
DELAY 5000
VSLIDE
DELAY 10000
RSCRIPT
```

This script uses the asterisk (*) before the slide name prior to the delay. This causes AutoCAD to load the slide, ready for viewing. Otherwise, there would be a blank screen between slides while the next one is being loaded. The RSCRIPT command repeats the slide show.

WORKSPACES

Any time you are in a drawing session, the combination of settings, content, and arrangements of the menus, toolbars, and dockable windows (such as the Tool Palettes window or DesignCenter) can be saved to a named workspace. Then later, after some of these items have been changed, you can recall the workspace by its name and have the same combination of settings, content, and arrangements that was in effect when you created the workspace. This is especially helpful when your work requires two or more types of workspaces. It is time-consuming to make changes to the drawing environment each time you switch back and forth between different types of work. By naming and saving each combination that makes up the best suitable environment, you can save that time.

Invoke the WORKSPACE command by typing it from the keyboard and AutoCAD prompts:

> Enter workspace option *(choose the desired option)*

Choosing SETCURRENT sets an existing workspace current. AutoCAD prompts:

> Enter name of workspace to make current [?] <current>: *(specify a name or ? to list available workspaces)*

Choosing SAVEAS saves a current environment configuration as a named workspace. AutoCAD prompts:

> Save workspace as <current>: *(specify a name for the new workspace)*

Choosing EDIT opens the Customize User Interface dialog box with the Customize tab displayed where you can modify a workspace.

Choosing RENAME lets you rename a workspace. AutoCAD prompts:

> Enter workspace to rename ⏎ *(specify the name of the workspace to be renamed)*
> Enter new workspace name <current>: *(specify a new name for the workspace)*

Choosing **Delete** lets you delete a workspace. AutoCAD prompts:

> Enter new workspace to delete <current>: *(specify a name or ? to list available workspaces)*
> Workspace <name> deleted.
> Do you really want to delete the workspace "<specified>" ⏎ *(choose Y)*

Choosing **Settings** opens the Workspace Settings dialog box, which controls the display, menu order, and Save settings of a workspace.

Choosing **?** (list workspaces) displays a list of all workspaces defined in the main and enterprise CUI files in the AutoCAD Text Window.

Note: Workspaces differ from Profiles. Workspaces let you change how the menus, toolbars, and dockable windows in the drawing area are displayed. Profiles are a combination of user options, drafting settings, paths, and values. With workspaces, you can easily switch between them during a drawing session and manage them using the Customize User Interface dialog box. Profiles are updated each time an option, setting, or other value is changed. Profiles are managed from the Options dialog box.

REVIEW QUESTIONS

1. All of the following can be renamed using the RENAME command, except:
   a. a current drawing name
   b. named views within the current drawing
   c. block names within the current drawing
   d. text style names within the current drawing

2. The PURGE command can be used:
   a. after an editing session
   b. at the beginning of the editing session
   c. at any time during the editing session
   d. a and b only

3. The most common setting for the current drawing color is:
   a. Red
   b. Bylayer
   c. Byblock
   d. White
   e. none of the above

4. Which of the following are not valid color names in AutoCAD?
   a. Brown
   b. Red
   c. Yellow
   d. Magenta
   e. none of the above (i.e. all are valid)

5. To change the background color for the graphics drawing area, you should use:
   a. COLOR
   b. BGCOLOR
   c. OPTIONS
   d. SETTINGS
   e. CONFIG

6. AutoCAD release 14 can be used to edit a drawing saved as an ACAD2006 drawing.
   a. True
   b. False

7. ACAD2006 can be used to edit a drawing that is saved in an AutoCAD release 14 drawing.
   a. True
   b. False

8. All the following items can be purged from a drawing file, except:

   a. Text styles
   b. Blocks
   c. System variables
   d. Views
   e. Linetypes

9. The following can be deleted with the PURGE command, except:

   a. blocks not referenced in the current drawing
   b. linetypes that are not being used in the current drawing
   c. layer 0, if it is not being used
   d. none of the above (i.e. all can be purged)

10. With OLE (object linking and embedding), you can copy or move information from one application to another while retaining the ability to edit the information in the original application.

    a. True
    b. False

11. If the TIME command is not turned off during lunch break, the TIME command will:

    a. include the lunch break time
    b. exclude the lunch break time
    c. turn itself off after 10 minutes of inactivity
    d. automatically subtract one hour for lunch
    e. none of the above

12. The VIEW command:

    a. serves a purpose similar to the PAN command
    b. will restore previously saved views of your drawing
    c. is normally used on very small drawings
    d. none of the above

13. When using a .X point filter, AutoCAD will request that you complete the point selection process by entering:

    a. An X coordinate
    b. A Y and Z coordinate
    c. An OSNAP mode
    d. Nothing; .X is not a valid filter

14. The geometric calculator function MEE will:
    a. calculate the midpoint of a selected line
    b. calculate the midpoint between the endpoints of any two objects
    c. return the node name for the computer you are working on
    d. average a string of numbers
    e. none of the above

15. To load a script file called *sample.scr*, use:
    a. SCRIPT, then SAMPLE   c. LOAD, SCRIPT, then SAMPLE
    b. LOAD, then SAMPLE   d. none of the above

16. The AutoCAD command used for viewing a slide is:
    a. VSLIDE   d. SLIDE
    b. VIEWSLIDE   e. MSLIDE
    c. SSLD

17. If a REDRAW is performed while viewing a slide:
    a. the command will be ignored
    b. the current slide will be deleted
    c. the current drawing will be displayed
    d. AutoCAD will load the drawing the slide was created from
    e. none of the above

18. A script file is identified by the following extension:
    a. *scr*   d. *spt*
    b. *bak*   e. none of the above
    c. *dwk*

19. A slide file is identified by the following extension:
    a. *sld*   d. *slu*
    b. *scr*   e. none of the above
    c. *sle*

20. Slides can be removed from the display with the command:
    a. ZOOM ALL   d. both a and b
    b. REGEN   e. none of the above
    c. OOPS

21. The SCRIPT command cannot be used to:

    a. insert blocks

    b. create layers

    c. place text

    d. create another script file

    e. none of the above (i.e. all are possible)

22. To cause a script file to execute in an infinite loop, you should place what command at the end of the file?

    a. REPEAT
    b. RSCRIPT
    c. GOTO:START
    d. BEGIN
    e. none of the above

23. If AutoCAD is executing in an infinite script file loop, how can you terminate the loop?

    a. press BACKSPACE
    b. press CTRL+C
    c. press ALT+C
    d. press F1
    e. none of the above

24. What command permits a user to work with just a portion of a drawing by loading geometry from specific views or layer?

    a. PLOAD
    b. PARTOPN
    c. PARTIALOPEN
    d. PLTOPN

25. What command permits additional geometry to be loaded into the current partially loaded drawing?

    a. PLOAD
    b. PARTIAL LOAD
    c. PARTADD
    d. ADDGEO

26. Can the PURGE command be used in partially open drawings?

    a. Yes

    b. No

30. What term can be added to draw, modify, or inquiry commands that will cause them to automatically repeat?

    a. redo
    b. repeat
    c. return
    d. multiple
    e. no such option

31. When one or more objects overlay each other, which command is used to control their order of display?

   a. DWGORDER

   b. DRAWORDER

   c. VIEWORDER

   d. ARRANGE

32. Which command permits you to execute operating system programs without exiting AutoCAD 2006?

   a. RUN

   b. EXOPRG

   c. OPSYS

   d. SHELL

CHAPTER 14

Internet Utilities and Drawing Sets

INTRODUCTION

The Internet is the most important way to convey digital information around the world. You are probably already familiar with the best-known uses of the Internet: e-mail (electronic mail) and surfing the Web (short for "World Wide Web"). E-mail lets users exchange messages and data at very low cost. The Web brings together text, graphics, audio, and video in an easy-to-use format. Other uses of the Internet include FTP (file transfer protocol, for effortless binary-file transfer), Gopher (presents data in a structured, subdirectory-like format), and Usenet, a collection of news groups.

AutoCAD allows you to interact with the Internet in several ways. You can launch a Web browser from within AutoCAD. AutoCAD can create DWF (short for "Design Web Format") files for viewing drawings in two-dimensional format on Web pages. AutoCAD can open and insert drawings from, and save drawings to the Internet.

This chapter covers the AutoCAD feature that makes it possible to manage all the drawings that make up a design project, that is, a drawing set.

After completing this chapter, you will be able to do the following:

- Launch the default Web browser
- Use the Communication Center
- Open drawings from the Internet
- Save drawings to the Internet
- Create and use hyperlinks
- Create and view DWF files
- Etransmit
- Publish to Web
- Collect, sort, create, and manage drawing sheets
- Create layout views automatically
- Automate the numbering of sheets
- Archive a set of drawings
- Publish sets and subsets of drawing sheets

LAUNCHING THE DEFAULT WEB BROWSER

The BROWSER command lets you start a Web browser from within AutoCAD. By default, the BROWSER command uses the Web browser program that is registered in your computer's Windows operating system. The BROWSER command can be used in scripts, toolbar or menu macros, and AutoLISP routines to access the Internet automatically.

Invoke the BROWSER command from the Web toolbar (see Figure 14–1). AutoCAD opens the Web browser (see Figure 14–2), which connects to the default Web site.

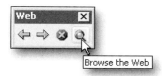

Figure 14–1 *Invoking the* BROWSER *command from the Web toolbar*

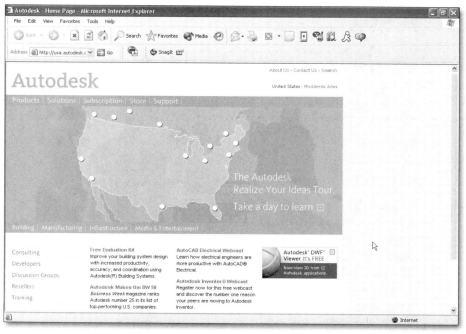

Figure 14–2 *Internet Explorer displaying the Autodesk Web site*

You can specify a default URL in AutoCAD's Options dialog box. Select the **Files** tab, choose **Menu, Help and Miscellaneous File Name**, select **Default Internet Location** (see Figure 14–2) and specify the default URL Web location.

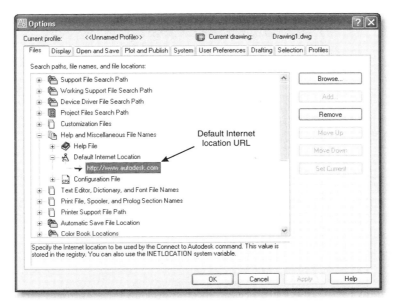

Figure 14-3 *Options dialog box with **Files** tab selected*

COMMUNICATION CENTER

Communication Center resides in the Tray section (lower right corner of the AutoCAD window) and can be customized to offer you just the right amount of information. Here you can choose to be notified about such things as maintenance patches, product support information, subscription information and extension announcements, articles, and tips. Maintenance patches include any program updates and fixes to the existing product. Product support information includes ground-breaking news from the Product Support team at Autodesk. Subscription information and extension announcements include announcements and subscription program news, if you are an Autodesk subscription member. If new information becomes available, a bubble announcement is displayed (see Figure 14-4). If it is selected, the Communication Center window is displayed (see Figure 14-5).

Figure 14-4 *Bubble announcement from the Communication Center*

Figure 14–5 *Communication Center window*

Choose one of the available topics from the list for detailed information.

The Communication Center is an interactive feature that must be connected to the Internet to deliver content and information. Each time the Communication Center is connected, it sends information to Autodesk so that the correct information can be returned. All information is sent anonymously to maintain your privacy. The information that will be sent to Autodesk includes Product Name, Product Release Number, Product Language, and Country.

To customize the Communication Center options, choose **Settings** in the Communication Center window. AutoCAD displays the Configuration Settings dialog box (see Figure 14–6).

Chapter 14 • *Internet Utilities and Drawing Sets* **747**

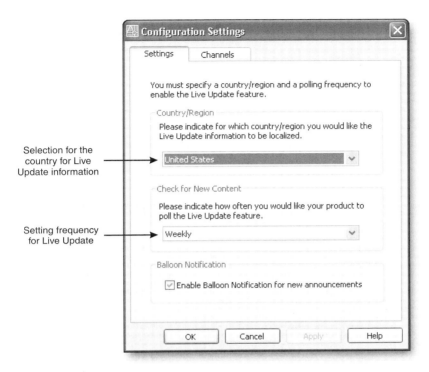

Figure 14–6 *Configuration Settings dialog box with* **Settings** *tab selected*

In the **Settings** tab of the dialog box, choose the **Country/Region** where the software is registered, so that the Communication Center can provide information that is designed specifically for your location. Choose how often you want the Communication Center to synchronize with Autodesk servers to update the software. Set the **Balloon Notification** toggle switch to on or off. When it is set to ON, the Communication Center balloon messages are displayed above the status bar when a new announcement is received. If the balloon notification is disabled in the tray settings, the Balloon Notification setting in the Communication Center is ignored.

In the **Channels** tab of the dialog box (see Figure 14–7) specify the information that you want displayed in the Communication Center.

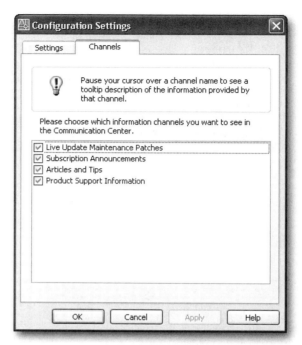

Figure 14–7 *Configuration Settings dialog box with **Channels** tab selected*

In the Configuration Settings dialog box, choose the settings and options that you want to use and then choose **Apply**. Choose **OK** to close the Configuration Settings dialog box and then close the Communication Center window.

OPENING AND SAVING DRAWINGS FROM THE INTERNET

AutoCAD allows you to open and save drawing files from the Internet or an intranet. You can also attach externally referenced drawings stored on the Internet/intranet to drawings stored locally on your system. Whenever you open a drawing file from an Internet or intranet location, it is first downloaded into your computer and opened in the AutoCAD drawing area. Then you can edit the drawing and save it, either locally or back to the Internet or intranet location for which you have appropriate access privileges.

To open an AutoCAD drawing from an Internet/intranet location, invoke the OPEN command from the Standard toolbar. AutoCAD displays the Select File dialog box.

Specify the URL to the file you wish to open in the **File name** text field, and choose **Open** to open the drawing from the specified Internet/intranet location. Be sure to specify the transfer protocol, such as ftp:// or http://, and the file extension, such as *.dwg* or *.dwt*. You can also choose **Search the Web** (located in the toolbar at the top of the dialog) to open the Browse the Web dialog box. From there you can navigate to the Internet location where the file is stored. You can also access the Buzzsaw.com and FTP locations by selecting the appropriate tabs provided on the left side of the dialog box.

To save an AutoCAD drawing to an Internet/intranet location, invoke the SAVEAS command from the File menu. AutoCAD displays the Save Drawing As dialog box.

Specify the URL to the file you wish to save in the **File name** text field and choose **Save** to save the drawing to the specified Internet/intranet location. Be sure to specify the transfer protocol and file extension (such as *.dwg* or *.dwt*). You can also choose **Search the Web** to open the Browse the Web dialog box. From there you can navigate to the Internet location where the file is to be saved.

Note: You can save the drawing to a specific Internet location by using the FTP protocol only.

To attach an xref to a drawing stored on the Internet/Intranet location, invoke the XATTACH command. AutoCAD displays the Select Reference File dialog box.

Specify the URL to the file you wish to attach in the **File name** text field and choose **Open** to attach the drawing as a reference file from the specified Internet/intranet location. Be sure to specify the transfer protocol, such as ftp:// or http://, in the URL. You can also choose **Search the Web** to open the Browse the Web dialog box. From there you can navigate to the Internet location where the file is stored. You can also access the Buzzsaw.com and FTP locations by selecting the appropriate tabs provided on the left side of the dialog box.

WORKING WITH HYPERLINKS

AutoCAD allows you to create hyperlinks that provide jumps to associated files. Hyperlinks provide a simple and powerful way to quickly associate a variety of documents with an AutoCAD drawing. For example, you can create a hyperlink that opens another drawing file from the local drive or network drive, or from an Internet Web site. You can also specify a named location to jump to within a file, such as a view name in an AutoCAD drawing, or a bookmark in a word processing program. You can also attach a URL to jump to a specific Web site. You can attach hyperlinks to any graphical object in an AutoCAD drawing.

AutoCAD allows you to create both *absolute* and *relative* hyperlinks in your AutoCAD drawings. Absolute hyperlinks store the full path to a file location, whereas relative hyperlinks store a partial path to a file location, relative to a default URL or name of the directory you specify using the HYPERLINKBASE system variable.

You can also specify the relative path for a drawing on the **Summary** tab of the Drawing Properties dialog box (see Figure 14–8). The Drawing Properties dialog box is opened from the **File** menu.

Figure 14–8 *Drawing Properties dialog box showing the **Summary** tab*

Whenever you attach a hyperlink to an object, AutoCAD provides cursor feedback as you position the cursor over the object. To activate the hyperlink, first select the object (make sure the PICKFIRST system variable is set to 1). Then, right-click to display the shortcut menu and activate the link from the *HYPERLINK* sub-menu (see Figure 14–9).

Figure 14–9 *HYPERLINK sub-menu on the shortcut menu*

When you create a hyperlink that points to an AutoCAD drawing template (*.dwt*) file, it causes AutoCAD to create a new drawing file when you activate the hyperlink that is based on the selected template, rather than opening the actual template. With this method there is no risk of accidentally overwriting the original template.

When you create a hyperlink that points to an AutoCAD named view and activate the hyperlink, the named view that was created in the model space is restored in the **Model** tab, and the named view that was created in paper space is restored in the **Layout** tab.

To create a hyperlink, invoke the HYPERLINK command from the Insert menu. AutoCAD prompts:

> Select objects: *(select one or more objects to attach the hyperlink and press* ENTER*)*

AutoCAD displays the Insert Hyperlink dialog box (see Figure 14–10).

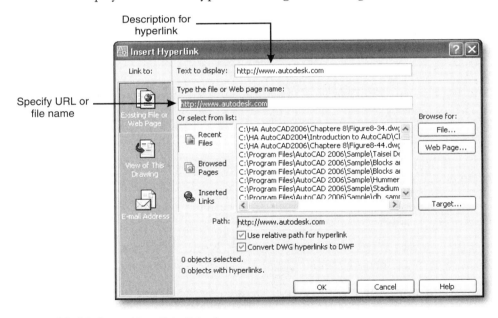

Figure 14–10 *Insert Hyperlink dialog box*

Specify a description for the hyperlink in the **Text to Display** text field. This is useful when the file name or URL is not helpful in identifying the contents of the linked file. Specify the URL or path with the name of the file that you wish to have associated with the selected objects in the **Type the File or Web Page Name** text field. Or choose one of the **Browse for** buttons: **File**, **Web Page**, or **Target**. Choosing **File** opens the Browse the Web – Select Hyperlink dialog box (a standard file selection dialog box). Use the dialog box to navigate to the file that you want associated with the hyperlink. Choosing **Web Page** opens the AutoCAD browser. Use the browser to navigate to a Web page to associate with the hyperlink. Choosing **Target** opens the Select Place in Document dialog box, in which you specify a link to a named location in a drawing. The named location that you select is the initial view that is restored when the hyperlink is executed. You can also select the path, with the name of the file or URL, from the list box categorized from **Recent Files**, **Browsed Pages**, and **Inserted Links**.

The **Path** text field displays the path to the file associated with the hyperlink.

The **Use relative path for hyperlink** check box toggles the use of a relative path for the current drawing. If this option is selected, the full path to the linked file is not stored with the hyperlink. AutoCAD sets the relative path to the value specified by the HYPERLINKBASE system variable or, if this variable is not set, the current drawing path. If this option is not selected, the full path to the associated file is stored with the hyperlink.

The **Convert DWG hyperlinks to DWF** check box specifies that the DWG hyperlink will convert to a DWF file hyperlink when you publish or plot the drawing to a DWF file.

Choosing **Existing File or Web Page** located on the left side of the Insert Hyperlink dialog box displays the options for creating a hyperlink to an existing file or Web page. Choosing **View of This Drawing** allows you to select a named view in the current drawing to link to. Choosing **E-mail Address** specifies an email address to link to. When the hyperlink is executed, a new email is created using the default system email program.

Choose **OK** to create the hyperlink to the selected objects and close the Insert Hyperlink dialog box.

To edit or remove a hyperlink, first select the object that has a hyperlink and from the shortcut menu, choose EDIT HYPERLINK.

AutoCAD displays the Edit Hyperlink dialog box, similar to the Insert Hyperlink dialog box. Make necessary changes in the **Text to display**, and/or **Type the file or Web page name** text fields. To remove the hyperlink, choose **Remove Link**. Choose **OK** to accept the changes and close the Edit Hyperlink dialog box.

AutoCAD allows you to attach hyperlinks to blocks, including nested objects contained within blocks. If the blocks contain any relative hyperlinks, the relative hyperlinks adopt the relative base path of the current drawing when you insert them.

Whenever you attach a hyperlink to a block reference, AutoCAD provides cursor feedback when you position the cursor over the inserted block. To activate the hyperlink, first, select the block reference (make sure the PICKFIRST system variable is set to 1). Then, right-click to display the shortcut menu and activate the hyperlink associated with the currently selected block element from the HYPERLINK sub-menu.

Whenever you include objects that have hyperlinks in a block definition, you can activate the hyperlink from any of the block references. If you attach a hyperlink to a block reference, then you will have the choice of activating the block hyperlink or selected object hyperlink. To remove or edit the hyperlinks of the objects within the block, you must explode the block reference, and then proceed with the removing or editing of hyperlinks. You can remove or edit the hyperlink that was attached to a block reference without exploding the block.

 Note: When a hyperlink is attached to a block reference that already contains an object with a hyperlink, the cursor feedback for that block will only show the block hyperlink. The object hyperlink can still be accessed through the HYPERLINK sub-menu as previously described.

DESIGN WEB FORMAT

Design Web Format™ (DWF) is an open, secure file format developed by Autodesk for the transfer of drawings over networks, including the Internet. DWF files are highly compressed, so they are much smaller (less than half the size of a ".dwg" file) and faster to transmit, enabling the communication of rich design data, without the overhead associated with typical heavy CAD drawings. DWF files are not a replacement for native CAD formats such as DWGs and don't allow editing of the data within the file. The sole purpose of DWF is to allow designers, engineers, developers, and their colleagues to communicate design information and intent to anyone needing to view, review, or print design information.

The latest release of DWF is DWF 6. DWF 6 has been re-designed to enable users to build a complex set of design documents, pages, or layouts in one DWF file. The latest release of Autodesk DWF Viewer® supports viewing and printing of 3D models published from nearly every Autodesk design application. You can also take advantage of new features like print preview, viewing new DWF MapBooks, and accessing block attribute data published from AutoCAD 2006.

To view DWF files, you can use Autodesk DWF Viewer, which is a free, downloadable application (you can download from www.autodesk.com) or Autodesk DWF Composer®, which enables complete round-tripping of markups, annotations, and other changes back into Autodesk® Revit® products and the AutoCAD 2006 family of software products, so you never have to re-enter information.

USING ePLOT TO CREATE DWF FILES

With AutoCAD's ePlot feature, you can generate electronic drawing files that are optimized for either plotting or viewing. The files you create are stored in Design Web Format (DWF). DWF files can be opened, viewed, and plotted by anyone using Autodesk DWF Viewer or Autodesk DWF Composer. With Autodesk DWF Viewer or Autodesk DWF Composer, you can also view DWF files in Microsoft® Internet Explorer 5.01 or later. DWF files support real-time panning and zooming and the display of layers and named views.

With ePlot, you can specify a variety of settings, such as pen assignments, rotation, and paper size, all of which control the appearance of plotted DWF files. With ePlot you can also create DWF files that have rendered images and multiple viewports displayed in the **Layout** tab.

Note: By default, AutoCAD plots all objects with a lineweight of 0.06 inches, even if you haven't specified lineweight values in the Layer Properties Manager. If you want to plot without any lineweight, clear the **Plot Object Lineweights** option from the **Plot Options** section in the Plot dialog box.

AutoCAD provides a preconfigured plotter driver file named DWF6 ePlot to create DWF files. It generates electronic drawing files that are optimized for either printing or viewing. The files you create are stored in Design Web Format (DWF).

DWF files are created in a vector-based format (except for inserted raster image content) and are typically compressed. Compressed DWF files can be opened and transmitted much faster than AutoCAD drawing files. Their vector-based format ensures that precision is maintained.

DWF files are an ideal way to share AutoCAD drawing files with others who don't have AutoCAD. To create a .DWF file, invoke the PLOT command from the Standard toolbar, AutoCAD displays the Plot dialog box (see Figure 14–11).

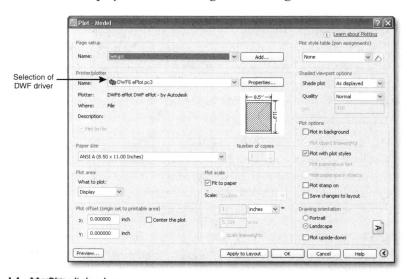

Figure 14–11 *Plot dialog box*

Select DWF6 ePlot.pc3 from the **Name** list box (**Printer/Plotter** configuration section) to create a DWF file.

Make necessary changes to **Paper size** and paper units, **Drawing orientation**, **Plot area**, **Plot scale**, **Plot offset**, **Plot options**, and choose **OK** to create the DWF file. AutoCAD will prompt for the name and location of the DWF file.

AutoCAD allows you to fine-tune custom plotting properties such as resolution, file compression, background color, inclusion of paper boundary, and related settings, when creating DWF files.

To modify custom plotting properties, open the Plot dialog box and select a DWF6 ePlot plotting device and then choose **Properties**. AutoCAD displays the Plotter Configuration Editor dialog box (see Figure 14–12).

Choose the **Device and Document Settings** tab, and then select **Custom Properties** from the tree window (see Figure 14–12). In the **Access Custom Dialog** section, choose **Custom Properties**. AutoCAD displays the DWF6 ePlot Properties dialog box (see Figure 14–13).

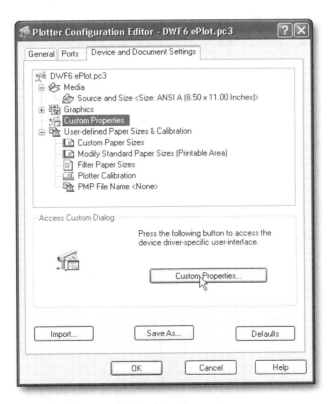

Figure 14–12 *Plotter Configuration Editor dialog box*

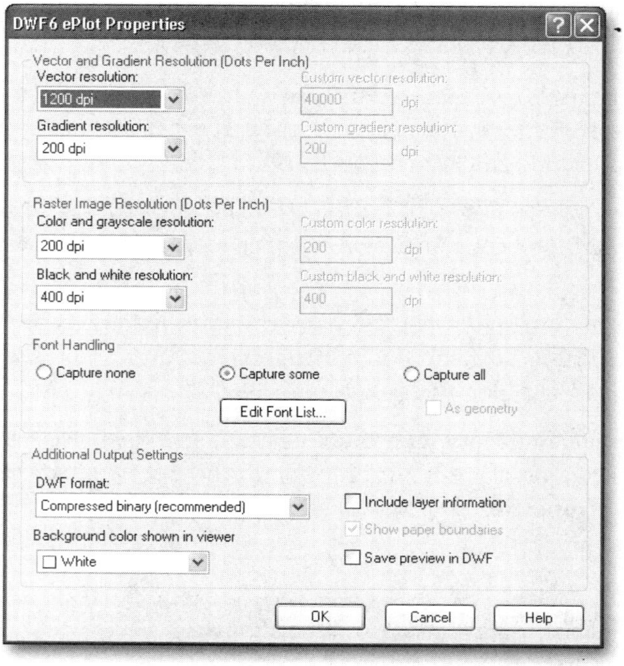

Figure 14-13 *DWF6 ePlot Properties dialog box*

The **Vector and Gradient Resolution (Dots Per Inch)** section, specifies the resolution (in dots per inch) for vector graphics and gradients for DWF files When you set a higher resolution, the file is more precise, but the file size is also larger. Select appropriate vector and gradient resolution from the **Vector resolution** menu and **Gradient resolution** menu, respectively. If there are large numbers of objects in the drawing, then it is recommended that you create a DWF file with high resolution. When you create DWF files intended for plotting, select a resolution to match the output of your plotter or printer.

The **Raster Image Resolution (Dots Per Inch)** section specifies the resolution (in dots per inch) for raster images for DWF files. Select appropriate color and grayscale resolution and black-and-white resolution from the **Color and grayscale resolution** and **Black and white resolution** menus, respectively. When you set a higher resolution, the file is more precise, but the file size is also larger.

The **Font Handling** section specifies the inclusion and handling of fonts in DWF files.

Choosing **Capture none** (all viewer supplied) specifies that no fonts will be included in the DWF file. In order for the fonts used in the source drawing for the DWF file to be visible in the DWF file, the fonts must be present on the DWF viewer's system. If the fonts used to create the DWF file are not present on the viewer's system, other fonts will be substituted.

Choosing **Capture some** (recommended) specifies that fonts used in the source drawing for the DWF file that are selected in the Available True Type Fonts dialog box will be included in the DWF file. The selected fonts do not need to be available on the DWF viewer's system in order for them to appear in the DWF file.

Choose **Edit Font List** to open the Available True Type Fonts dialog box, where you can edit the list of fonts eligible for capture in the DWF file.

Choosing **Capture all** specifies that all fonts used in the drawing will be included in the DWF file. The **As geometry** check box specifies that all fonts used in the drawing will be included as geometry in the DWF file. If you select this option, you should plot your drawing at a scale factor of 1:1 or better to ensure good quality in the output file. This option is only available for DWF files created with the DWF6 ePlot model.

Note: The size of a DWF file can be affected by the font-handling settings, the amount of text, and the number and type of fonts used in the file. If the size of your DWF file seems too large, try changing the font-handling settings.

Choose the compression format for DWF files from the **DWF Format** menu. Selection of **Compressed binary (recommended)** format plots the DWF file in a compressed, binary format; compression does not cause data loss. This is the recommended file format for most DWF files. Selection of **Zipped ASCII encoded 2D stream (advanced)** plots the DWF file in zipped ASCII Encoded 2D Stream (plain text) format. You can use WinZip to unzip the files.

Specify the background color from the **Background color shown in viewer** option menu. In addition, specify toggle settings for **Include layer information, Show paper boundaries**, and **Save preview in DWF.**

After making necessary changes, choose **OK** to close the DWF Properties dialog box. Choose **OK** to save the settings and close the Plotter Configuration Editor dialog box. AutoCAD displays the Changes to a Printer Configuration File dialog box (see Figure 14–24).

Figure 14–14 *Changes to a Printer Configuration File dialog box*

There are two available options. If you want to apply the changes in the settings to the current plot, then select **Apply changes to the current plot only**. If you want to save the settings and apply them to all plots, select **Save changes to the following file** and specify the name of the file in the edit field. Choose **OK** to close the Changes to a Printer Configuration File dialog box.

VIEWING DWF FILES

AutoCAD itself cannot display DWF files, nor can DWF files be converted back to DWG format without using file translation software from a third-party vendor. In order to view a DWF file, you need Autodesk DWF Viewer, which can be obtained free of charge from the Web site: www.autodesk.com or you can purchase Autodesk DWF Composer.

Autodesk DWF viewer enables users to view and print complex 2D and 3D drawings, maps, and models published from Autodesk design applications. Figure 14–15 shows an example of DWF displayed in Autodesk DWF Viewer.

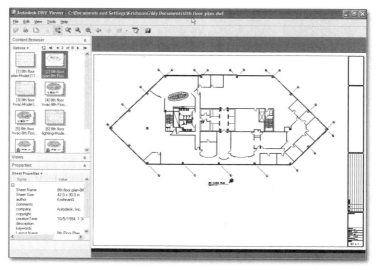

Figure 14–15 *Autodesk DWF Viewer displaying DWF file*

Autodesk DWF Composer improves the efficiency of the entire design review process by enabling the coordination and communication of design information digitally. You can review shared sheet sets faster and more efficiently among project team members. Composer allows you to mark up sheet sets, providing feedback to the project team, while adding redline markups, annotations, and information to the design set. In turn, you can open the revised sheet sets, tracking changes and bring markup information back into the AutoCAD 2006 Markup Set Manager.

PUBLISHING AUTOCAD DRAWINGS TO THE WEB

The Publish to Web Wizard allows you to easily and seamlessly publish AutoCAD drawings to the Web. Drawings are published in HTML format using three pre-defined image types: DWF, JPEG and PNG. The DWF image type translates and publishes specified layouts into DWF, which are easily viewed with either Autodesk DWF Viewer or Autodesk DWF Composer. With the JPEG and PNG image types, you specify a drawing perspective, and AutoCAD translates the specified layout into a JPEG or PNG raster image. Anyone with a standard browser can view JPEG or PNG content. To move your content to the Web or a company's intranet, just specify the server location and configuration, and the content uploads automatically. Once it's posted, updating it is simple and fast.

To publish AutoCAD drawings, invoke the PUBLISHTOWEB command from the File menu, AutoCAD displays the introductory text of the Publish to Web - Begin page (see Figure 14–16). Choose one of the following two options:

- **Create New Web Page** – Allows you to create a new Web page.
- **Edit Existing Web Page** – Allows you to edit an existing page by adding or removing Web pages.

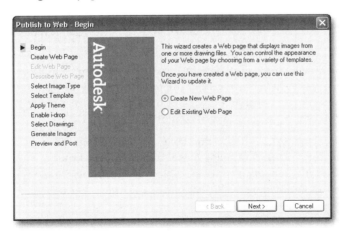

Figure 14–16 *Publish to Web – Begin*

Choose **Next** and AutoCAD displays the Publish to Web – Create Web Page page (see Figure 14–17). Specify the name of your Web page in the first text field and specify the parent directory in your file system where the Web page folder will be created. If desired, provide a description to appear on your Web page in the last text field in the wizard.

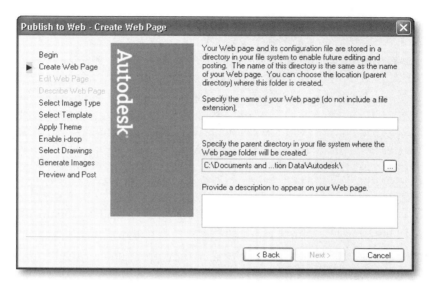

Figure 14–17 *Publish to Web – Create Web Page*

Choose **Next** and AutoCAD displays the Publish to Web – Select Image Type page (see Figure 14–18). Choose one of the following three image types:

- DWF Image type – The DWF template translates and publishes specified DWG files in DWF. DWF files are inserted into your completed Web page in a size that is optimized to display well with most browser settings.

- JPEG Image type – JPEG files are raster-based representations of AutoCAD drawing files. JPEGs are one of the most common formats used on the Web. Due to the compression mechanism used, this format is not suitable for large drawings or drawings that have a lot of text.

- PNG Image type – PNG (Portable Network Graphics) are raster-based representations of drawing files. Most browsers now support the PNG image type, making it more suitable than JPEG for creating images of AutoCAD drawings.

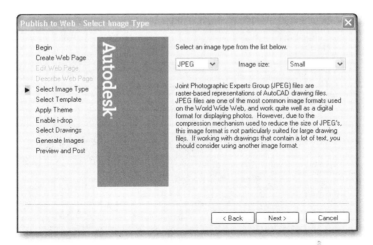

Figure 14–18 *Publish to Web –Select Image Type*

Choose **Next** and AutoCAD displays the Publish to Web – Select Template page (see Figure 14–19). Select one of the available templates. The **Preview** pane demonstrates how the selected template will affect the layout of drawing images in your Web page.

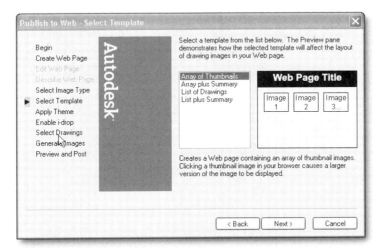

Figure 14–19 *Publish to Web – Select Template*

Choose **Next** and AutoCAD displays the Publish to Web – Apply Theme page (see Figure 14–20). Choose one of the available themes. Themes are preset elements (fonts and colors) that control the appearance of various elements of your completed Web page. The **Preview** pane demonstrates how the selected theme will display the layout of your Web page. The available themes are: Autumn Fields, Classic, Cloudy Sky, Dusky Maize, Ocean Waves, Rainy Day, and Supper Club.

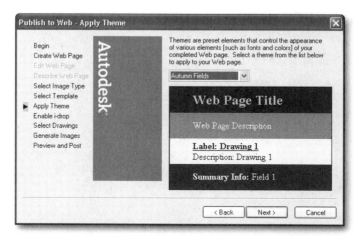

Figure 14–20 *Publish to Web – Apply the Theme*

Choose **Next** and AutoCAD displays the Publish to Web – Enable i-drop page (see Figure 14–21).

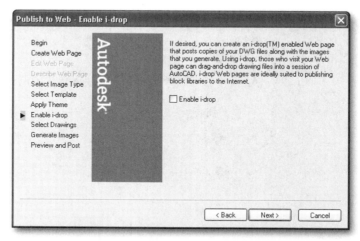

Figure 14–21 *Publish to Web – Enable i-drop*

If desired, you can create an i-drop enabled Web page that posts copies of your DWG files along with the images. Using i-drop, visitors to your Web site can drag and drop drawing files into a session of AutoCAD.

Choose **Next** and AutoCAD displays the Publish to Web – Select Drawings page (see Figure 14–22).

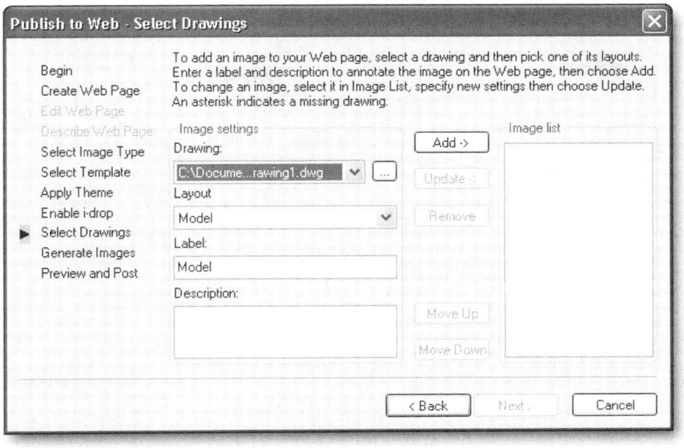

Chapter 14 • *Internet Utilities and Drawing Sets* 763

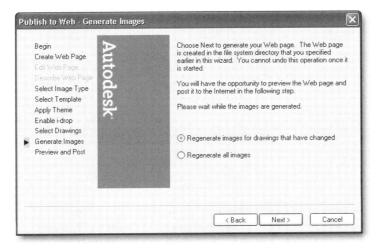

Figure 14–22 *Publish to Web – Select Drawings*

Select a drawing and then choose one of its layouts from the **Layout** menu. Specify a label in the **Label** text box and description in the **Description** text box to annotate the selected image on the Web page. Choose **Add** to add the selected image to the **Image list** box. If necessary, you can change the properties of the selected image and choose **Update** to apply the changes or you can remove it from the selection by choosing **Remove**. Similarly, you can add additional drawings/layouts to the selection. Choosing **Move Up** and **Move Down** allows you to rearrange the selected images.

Choose **Next** and AutoCAD displays the Publish to Web – Generate Images page (see Figure 14–23).

Figure 14–23 *Publish to Web – Generate Images*

Choose one of the following two options:

- **Regenerate images for drawings that have changed** – Generates the images for all the selected drawings that have changed.
- **Regenerate all images** – Generates the images for all the selected drawings.

Choose **Next** and AutoCAD creates the Web pages and stores them in the file directory that you specified earlier in the wizard. You cannot undo this operation once it is started.

When finished, AutoCAD displays the Publish to Web – Preview and Post page (see Figure 14–24). To preview the Web pages, choose **Preview**. AutoCAD opens the default browser and displays the Web pages with appropriate links. To post the Web pages, close the browser, and choose **Post Now**. AutoCAD displays the Posting Web File Handling dialog box. Select the URL where you want to post it and choose the **Save** button. If desired, choose **Send Email** to create and send an email message that includes a hyperlinked URL to its location.

Choose **Finish** to close the Publish to Web Wizard.

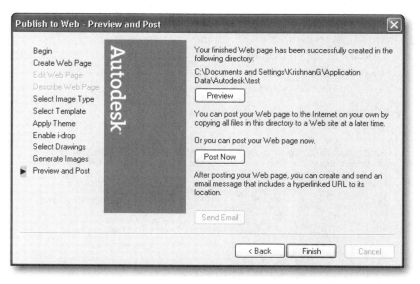

Figure 14–24 *Publish to Web – Preview and Post*

PUBLISH

Publishing allows you to assemble a collection of drawings and plot directly to paper or publish to a DWF (Design Web Format) file. AutoCAD allows you to publish your drawing sets as either a single multi-sheet DWF format file or multiple single-sheet DWF format files, or to plot to the designated plotter in the page setup. You can publish to devices (plotters or files) specified in the page setups for each layout. When using Publish, you have the flexibility to create electronic or paper drawing sets for distribution. The recipients can then view or plot your drawing sets.

You can customize your drawing set for a specific user, and you can add and remove sheets in a drawing set as a project evolves. The PUBLISH command allows you to publish directly to paper or to an intermediate electronic format that can be distributed using email, FTP sites, project Web sites, or CD. You can open DWF files with Autodesk DWF Viewer or Autodesk DWF Composer.

Invoke the PUBLISH command from the Standard toolbar (see Figure 14–25), AutoCAD displays the Publish dialog box (see Figure 14–26).

Figure 14–25 *Invoking the* PUBLISH *command from the Standard toolbar*

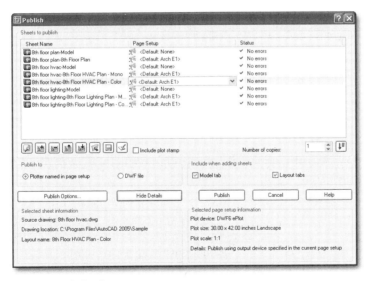

Figure 14–26 *Publish dialog box*

The **Sheets to publish** section lists the drawing sheets to be included for publishing.

- The **Sheet Name** column displays a combination of the drawing name and the layout name, separated by a dash (-). If necessary, you can rename using the RENAME SHEET option available from the shortcut menu. Drawing sheet names must be unique within a single DWF file.

- The **Page Setup** column displays the named page setup for the sheet. You can change the page setup by clicking the page setup name and selecting another page setup from the list. Select **Import** to import page setups from another .dwg file through the Import Page Setups for Publishing dialog box. Only Model tab page setups can be applied to Model tab sheets, and only paper space page setups can be applied to paper space layouts.

- The **Status** column displays the status of the sheet when it is loaded to the list of sheets.

To add sheets to the existing selection, choose the **Add Sheets** icon. AutoCAD displays a standard file selection dialog box, where you can add sheets to the list of drawing sheets. The layout names from those files are extracted, and one sheet is added to the list of drawing sheets for each layout. New drawing sheets are always appended to the end of the current list.

To remove sheets, choose the **Remove Sheets** icon and AutoCAD deletes the currently selected drawing sheet from the list of sheets.

To move the selected drawing sheet up one position in the list, choose the **Move Up** icon. To move the selected drawing sheet down one position in the list, choose the **Move Down** icon.

The **Include when adding sheets** section specifies whether the model and layouts contained in a drawing are added to the sheet list when you add sheets. At least one option must be selected.

> The **Model tab** selection specifies whether the model is included when drawing sheets are added.
>
> The **Layout tab** selection specifies whether all layouts are included when drawing sheets are added.

To save the current selection of the drawing sheets, choose **Save Sheet List**. AutoCAD displays the Save List As dialog box, where you can save the current list of drawings as a *.dsd* (Drawing Set Descriptions) file. These *.dsd* files are used to describe lists of drawing files and selected lists of layouts within those drawing files.

To load a saved list to the current selection, choose **Load Sheet List**. AutoCAD displays a standard file selection dialog box. You can select a *.dsd* file or a *.bp3* (Batch Plot List) file to load. AutoCAD displays the Replace or Append dialog box if a list of drawing sheets is present in the Publish dialog box. You can either replace the existing list of drawing sheets with the new sheets or append the new sheets to the current list.

To include a plot stamp on a specified corner of each drawing and log it to a file, check the **Include plot stamp** check box. To customize the plot stamp settings, open the Plot Stamp dialog box by choosing the **Plot Stamp Settings** icon. AutoCAD displays the Plot Stamp dialog box, in which you can specify the information, such as drawing name and plot scale, that you want applied to the plot stamp.

The **Publish to** section defines how to publish the list of sheets. You can publish to either a multi-sheet *.dwf* file (an electronic drawing set) or to the plotter specified in the page setup (a paper drawing set or a set of plot files).

> Select **Plotter named in page setup** to plot to output devices for each drawing sheet page setup.
>
> Select **DWF file** to publish as a *.dwf* file.

Specify the number of copies to publish in the **Number of copies** box. If the **DWF file** option is selected in the **Publish to** section, the **Number of copies** setting defaults to 1 and cannot be changed.

Choose the **Preview** icon to display the drawing as it will appear when plotted on paper.

To customize publishing, choose **Publish Options**, and AutoCAD opens the Publish Options dialog box (see Figure 14–27), in which you can specify options for publishing.

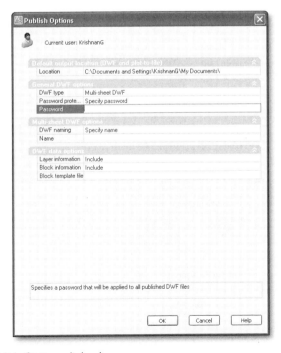

Figure 14–27 *Publish Options dialog box*

The name of the current user or current sheet set is displayed. When the name of the current user is shown, changes made in the dialog box are saved in the current user's profile. When the name of the current sheet set is shown, changes made in the dialog box are saved with the sheet set.

 Note: If you lose or forget the password, it cannot be recovered. Keep a list of passwords and their corresponding *.dwf* file names in a safe place.

Choose **OK** to save the changes and close the Publish Options dialog box.

Choose **Show Details** or **Hide Details** to display or hide the **Selected sheet information** and **Selected page setup information** sections in the Publish dialog box.

Choose **Publish** to publish the selected layouts. AutoCAD begins the publishing operation, creating one or more single-sheet *.dwf* files or a single multi-sheet *.dwf* file, or plotting to a device or file, depending on the option selected in the **Publish to** section and the options selected in the Publish Options dialog box.

If a drawing sheet fails to plot, PUBLISH continues plotting the remaining sheets in the drawing set. A log file is created that contains detailed information, including any errors or warnings encountered during the publishing process. You can stop publishing after a sheet has finished plotting. If you stop publishing a multi-sheet *.dwf* file before it is complete, no output file is generated. After publishing is complete, the **Status** field is updated to show the results.

eTRANSMIT UTILITY

The eTransmit utility allows you to select and bundle together the drawing file and its related files. You can create a transmittal set of files as a compressed self-extracting executable file, as a compressed zip file, or as a set of uncompressed files in a new or existing folder. You can include all the reference files attached to the drawing file, word files, spreadsheet, etc. to be part of the bundle. It is easier to transmit by e-mail one single compressed file consisting of a drawing file and several related files.

To create a transmittal set of a drawing and related files, invoke the ETRANSMIT command from the File menu. AutoCAD displays the Create Transmittal dialog box, shown in Figure 14–28.

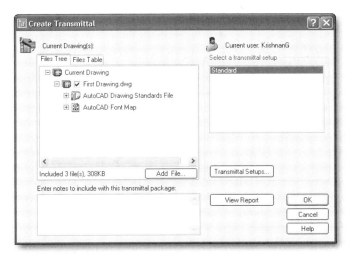

Figure 14–28 *Create Transmittal dialog box*

The **Files Tree** tab lists the files to be included in the transmittal package in a hierarchical tree format. By default, all files associated with the current drawing (such as related external references, plot styles, and fonts) are listed. You can add files to the transmittal package or remove existing files. To add files to the transmittal package, choose **Add File**. AutoCAD opens a standard file selection dialog box, in which you can select additional files to include in the transmittal package.

Enter notes related to a transmittal package in the text area under **Enter notes to be included with this transmittal package**. The notes are included in the transmittal report. You can specify a template of default notes to be included with all your transmittal packages by creating an ASCII text file called *etransmit.txt*. This file must be saved to a location specified by the **Support File Search Path** option on the **Files** tab in the Options dialog box.

Select the transmittal setup from the **Select a transmittal setup** list. AutoCAD lists previously saved transmittal setups. The default transmittal setup is named Standard. To create, modify, and delete transmittal setups, choose **Transmittal Setups**. AutoCAD displays the Transmittal Setup dialog box. You can create a new transmittal setup or modify an existing transmittal setup that specifies the type of transmittal package created.

To view the report that is included with the transmittal package, choose **View Report**. AutoCAD displays report information that includes any transmittal notes that you entered and distribution notes automatically generated by AutoCAD that detail what steps must be taken for the transmittal package to work properly.

Choose **OK** to create the transmittal set and close the Create Transmittal dialog box.

DRAWING SETS

AutoCAD's powerful Drawing Set Management feature allows you to handle sets of drawings for the purpose of plotting, publishing, and otherwise managing and tracking. Chapter 8 explains the use of the Layout/Paper Space feature in AutoCAD to easily produce sheets for plotting. This chapter explains how to apply the Sheet Set Manager feature to collect, organize, and otherwise manage the assortment of layout sheets and views from different drawings so they can be plotted as sets and subsets of deliverables traditionally referred to as "blueprints."

A drawing set is just that, a set of drawings. A small design project might require only two or three sheets, while in a major construction project, a drawing set might consist of hundreds of sheets. And there are often subsets for various disciplines: civil/survey, structural, mechanical, architectural, and electrical. When all of these sheets are scattered throughout one or more offices in one or more locations, and some sheets are just one layout or view in one drawing and other sheets are layouts from many different drawings, it becomes a difficult and almost impossible task to organize, manage, and continually update the sheets individually and as a set and subsets. With the Sheet Set Manager, you can manage drawings as sheet sets. A sheet set is an organized and named collection of sheets from several drawing files. A sheet is a selected layout from a drawing file. You can import a layout from any drawing into a sheet set as a numbered sheet. You can manage, transmit, publish, and archive sheet sets as a unit.

CREATING A NEW SHEET SET

The Create Sheet Set wizard contains a series of pages that step you through the process of creating a new sheet set. You can choose to create a new sheet set from existing drawings, or use an existing sheet set as a template on which to base your new sheet set. The following steps should be performed before creating a sheet set:

- Move the drawing files to be used in the sheet set into a minimum number of folders. This will make sheet set management easier.

- Have only one layout tab in each drawing in the sheet set. This affects access to sheets by multiple users. Only one sheet in each drawing can be open at a time.

- Specify or create a drawing template (*.dwt*) file for use by the sheet set for creating new sheets. This template file is called the sheet creation template and can be specified in the Sheet Set Properties dialog box or the Subset Properties dialog box.

- Create a page setup overrides file. Specify or create another *.dwt* file for storing page setups for plotting and publishing. This file is called the page setup overrides file and can be used to apply a single page setup to each sheet in a sheet set. This will override the individual page setups stored in each drawing.

Invoke the NEWSHEETSET command from the File menu to open the Create Sheet Set wizard (see Figure 14–29).

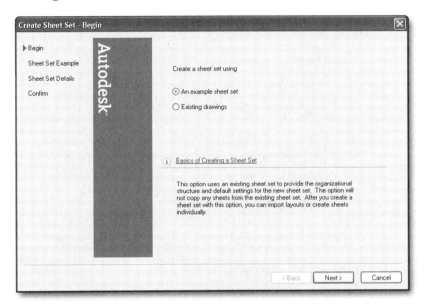

Figure 14–29 *The Begin page of the Create Sheet Set wizard*

On the Begin page of the Create Sheet Set wizard, select a method for creating a sheet set. Two methods are provided: **An example sheet set** and **Existing drawings** (see Figure 14–29).

Choosing **An example sheet set** lets you use a sample sheet set format, structure, and default settings without copying any of the sheets in the sample sheet set to the new set. After you create a sheet set using this option, you can import layouts or create sheets individually.

Choosing **Existing drawings** lets you import layouts from existing drawings in the specified folder(s). The layouts from these drawings are imported into the newly created sheet set automatically.

Creating Sheet Sets from Examples

To create a sheet set from a sample sheet set, select **An example sheet set** in the Create Sheet Set – Begin wizard page and choose **Next**. AutoCAD displays the Create Sheet Set – Sheet Set Example page (see Figure 14–30).

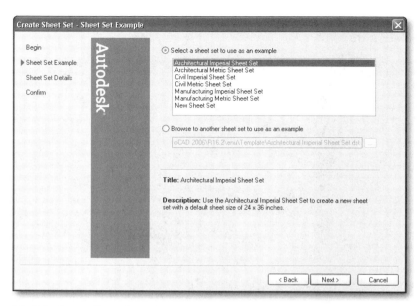

Figure 14–30 *Create Sheet Set – Sheet Set Example page*

AutoCAD displays a list of sample sheet sets in the list box. From here you can choose one to use as the basis of the new sheet set. If you select **Browse to another sheet set to use as an example**, you can use the standard Windows-type browsing mechanism to search folders for a sheet set to use as the basis of a new sheet set. After selecting an existing sheet set from one of the optional methods, choose **Next**. AutoCAD displays the Create Sheet Set – Sheet Set Details page, as shown at left in Figure 14–31.

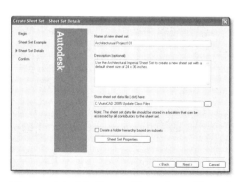
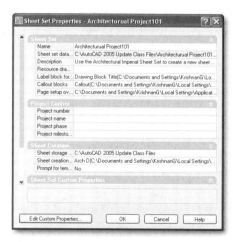

Figure 14–31 *Create Sheet Set – Sheet Set Details page and Sheet Set properties*

Specify the name and description of the newly created sheet set in the **Name of new sheet set** box and **Description (optional)** box, respectively. Specify the folder in which to store the newly created sheet set in the **Store sheet set data file (.dst) here** box. The Sheet Set data file should be stored in a location that can be accessed by all contributors to the sheet set.

Choose **Sheet Set Properties** to view or edit the sheet set properties. AutoCAD displays the Sheet Set Properties dialog box, as shown on the right in Figure 14–31. The Sheet Set Properties dialog box displays information specific to the sheet set selected. The information can be modified in the boxes to the right of individual data descriptions. This includes information such as the path and file name of the sheet set data (*.dst*) file, folder paths that contain drawing files included in the sheet set, and custom properties associated with the sheet set.

The **Sheet Set** section includes the name of the sheet set, location of the sheet set data file, description (if any) for the newly created sheet set, path for the resource drawings, name of the drawing that contains label blocks for views, callout block names associated with the sheet set, and name of the page setup override file.

The **Project Control** section contains information related to the project provided by the user.

The **Sheet Creation** section includes the paths of the folders that contain the drawing files associated with the sheet set, the name of the template for creating a new sheet, and a setting for "prompt for new template."

The **Sheet Set Custom Properties** section contains user-defined properties.

Choose **Edit Custom Properties** to add or remove custom properties associated with a sheet set. Custom properties can be used to store information such as a contract number, the name of the designer, or the release date.

Choose **OK** to return to the Create Sheet Set – Sheet Set Details wizard page and then choose **Next**. AutoCAD displays the Create Sheet Set – Confirm page as shown in Figure 14–32.

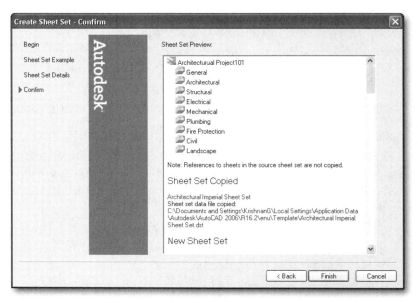

Figure 14–32 *Create Sheet Set – Confirm page*

In the Create Sheet Set – Confirm wizard page, the **Sheet Set Preview** box lists the information about the sheet set for review before you accept it. If the information is acceptable, choose **Finish** to complete the creation of the new sheet set. If any of the information needs to be modified, choose **Back** and make the necessary changes. AutoCAD creates a new sheet set in the form of a sheet set data file with the extension of *.dst* in the specified location with the specified properties.

Creating Sheet Sets from Existing Drawings

To create a sheet set from existing drawings, select **Existing drawings** on the Create Sheet Set – Begin wizard page and choose **Next**. AutoCAD displays the Create Sheet Set – Sheet Set Details page similar to creating a sheet from an example sheet set (at left in Figure 14–31). Specify the name and description of the newly created sheet set in the **Name of new sheet set** box and **Description (optional)** box, respectively. Specify the folder in which to store the newly created sheet set in the **Store sheet set data file (.dst) here** box. Choose **Sheet Set Properties** to view or edit the sheet set properties.

Choose **Next**, and AutoCAD displays the Create Sheet Set – Choose Layouts page (see Figure 14–33).

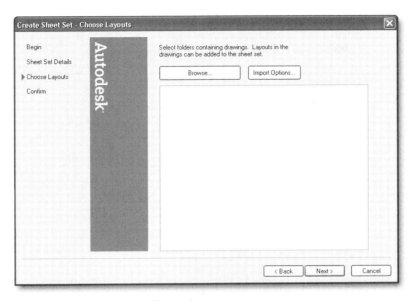

Figure 14–33 *Create Sheet Set – Choose Layouts page*

Choose **Browse** to select a folder or folders to list drawing files from which AutoCAD will import layouts in drawing files to create the sheet set. You can add additional folders containing drawings by choosing **Browse** for each additional folder. Choosing **Import Options** causes the Import Options dialog box to be displayed (see Figure 14–34).

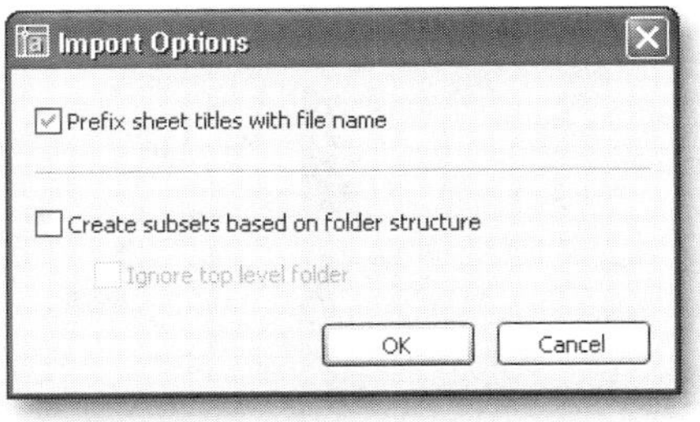

Figure 14–34 *The Import Options dialog box*

Selecting **Prefix sheet titles with file name** causes AutoCAD to automatically add the drawing file name to the beginning of the sheet title.

Selecting **Create subsets based on folder structure** causes AutoCAD to automatically create subsets in the newly created sheet sets based on folder structure.

Choose **OK** to close the dialog box and accept the settings.

On the Create Sheet Set – Choose Layouts page, choose **Next**, and AutoCAD displays the Create Sheet Set – Confirm wizard page, similar to creating a sheet from an example sheet set. The **Sheet Set Preview** box lists the information about the sheet set for review before you accept it. If the information is acceptable, choose **Finish** to complete the creation of the new sheet set. If any of the information needs to be modified, choose **Back** and make the necessary changes. AutoCAD creates a new sheet set in the form of a sheet set data file with the extension of *.dst* in the specified location with the specified properties.

SHEET SET MANAGER

The Sheet Set Manager provides the tools to organize, manage, and update a set of drawings. The Sheet Set Manager not only accesses the drawings associated with a project, but also lets you access the layouts and views that become the plotted sheets making up the final set of plotted drawings. As mentioned earlier, sheet sets are stored in the form of a sheet set data file with the extension of *.dst*. Each sheet in a sheet set is a layout in a drawing (*.dwg*) file. Open the Sheet Set Manager from the Standard toolbar (see Figure 14–35). AutoCAD displays the Sheet Set Manager palette with the **Sheet List** tab displayed as shown at left in Figure 14–36.

Figure 14–35 *Invoking the Sheet Set Manager from the Standard toolbar*

 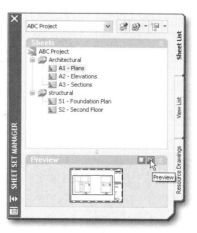

Figure 14–36 *Sheet Set Manager palette with the Sheet List tab displayed at left, Sheet Set Manager palette with the preview image of the selected sheet at right*

Choosing **Sheet Set Control** lists menu options to create a new sheet set, open an existing sheet set, or switch between open sheet sets.

The **Sheet List** tab displays an organized list of all sheets in the sheet set. Each sheet in a sheet set is a specified layout in a drawing file.

The **View List** tab displays an organized list of all sheet views in the sheet set. Only sheet views created with AutoCAD 2005 and later are listed.

The **Resource Drawings** tab displays a list of folders, drawing files, and model space views available for the current sheet set. You can add and remove folder locations to control which drawing files are associated with the current sheet set.

Choosing **Details** displays descriptive information about the currently selected item in the tree view.

Choosing **Preview** displays a thumbnail preview of the currently selected item in the tree view.

VIEWING AND MODIFYING A SHEET SET

The **Sheets** box displays the name of the sheet set that is open (see Figure 14–36 at left), or if no sheet set is open, choose **Open** to open an existing sheet set. To create a new sheet set, choose **New**, which in turn will start the wizard.

The Sheet Set Manager shown at left in Figure 14–36 contains two subsets called Architectural and Structural in the sheet set called ABC Project. The Architectural subset contains three numbered sheets: Plans, Elevations, and Sections. The Structural subset contains two numbered sheets: Foundation Plan and Second Floor. Each of these sheets is a layout in a *.dwg* file. When you select a sheet, information about it is displayed under the **Details** section of the palette. Figure 14–36 at left shows details of the Plans sheet. To view a thumbnail preview of the selected sheet, choose **Preview**. Figure 14–36 at right shows a preview image of the Plans sheet.

You can also open the Sheet Set Properties in the Sheet Set Properties dialog box from the shortcut menu that is displayed when the name of the sheet set is selected.

Instead of using the OPEN command to open a drawing, you can open it using the Sheet Set Manager. Double-click the name of the sheet, and it will open in a new window. When the sheet is open, it will be locked automatically and no other user can open it at the same time. The lock status is indicated in the details section of the selected item (see Figure 14–37). When you close the drawing, the status will be changed to Accessible.

Figure 14–37 *Sheet Set Manager palette with the Details section indicating the status of the selected item*

In a similar manner, you can create a new sheet. Right-click while the cursor is over a sheet set or sheet name in the Sheets list and choose NEW SHEET. AutoCAD displays the New Sheet dialog box, in which you can specify the name and number of the new sheet based on the default template set in the sheet set properties. When you create a new sheet, you create a new layout in a new drawing file, which is stored in the location specified in the sheet set properties. Instead of manually creating new drawing files, you can use the Sheet Set Manager.

To import a layout from an existing drawing, right-click while the cursor is over a sheet set or sheet name in the Sheets list and choose IMPORT LAYOUT AS SHEET. AutoCAD displays the Import Layouts as Sheets dialog box. In the **Select drawing file containing layouts** box, choose **Browse for Drawings** to find a drawing from which to import layouts. In the **Select Drawing** dialog box, select a drawing to import and choose **Open**. You can choose to have AutoCAD prefix sheet titles with the file name. Choose **Import Checked** to import the layout.

With a large sheet set, you will find it necessary to organize sheets and views in the tree view. On the **Sheet List** tab, sheets can be arranged into collections called subsets. To create a new subset, first select the name of the sheet set and choose NEW SUBSET from the shortcut menu. AutoCAD displays the Subset Properties dialog box, where you can create a new sheet subset for organizing the sheets. Figure 14–37 shows two subsets (Architectural and Structural) in the ABC Project sheet set.

If necessary, you can rename and renumber the sheet. To rename and renumber the selected sheet, choose RENAME & RENUMBER from the shortcut menu. AutoCAD displays the Rename & Renumber dialog box (see Figure 14–38), where you can specify the sheet number and title for the selected sheet.

The **Number** box specifies the sheet number of the selected sheet.

The **Sheet title** box specifies the sheet title of the selected sheet.

The **Folder path** box displays the path to the selected sheet.

In the **File rename options** section, you can select **Rename associated drawing file to match sheet title**, and then if desired, select **Prefix sheet number to file name.**

The **Next** option loads the next sheet into this dialog box.

Choose **OK** to close the Rename & Renumber Sheet dialog box.

Figure 14–38 *The Rename & Renumber Sheet dialog box*

Choosing the **Sheet Selections** icon, in the upper right corner of the Sheet List tab of the Sheet Set Manager, causes a menu to be displayed where you can save, manage, and restore sheet selections by name. This makes it easy to specify a group of sheets for publish, transmit, or archive operations. To create a sheet selection, first select several sheets from the sheet list and choose Create from the Sheet Selections menu (see Figure 14–39). AutoCAD displays the New Sheet Selection dialog box, where you can specify a sheet selection name and choose **OK** to create the new sheet selection. To restore the selection, choose the name of the sheet selection from the Sheet Selections menu. Choosing **Manage** causes the Sheet Selections dialog box to be displayed, where you can rename or delete the selected Sheet Selection.

To remove the selected sheet from the sheet set, choose REMOVE SHEET from the shortcut menu. AutoCAD removes the currently selected sheet from the sheet set.

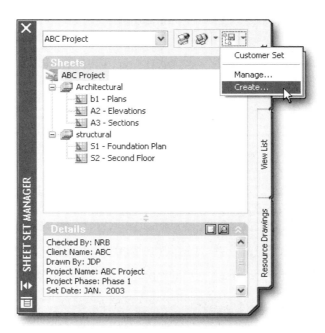

Figure 14-39 *Sheet Set Manager with Create selected from the Sheet Selections menu*

PLACING A VIEW ON A SHEET

The Sheet Set Manager automates and enhances the process for placing a view of a drawing on a sheet. First open the sheet where you want to place the view. To find the view to add to this sheet, select the **Resource Drawings** tab. On this tab you can browse for drawings that contain the views you want to add to your sheet. Any drawings you want to use must be listed at this location. To add a folder that contains drawings to the list, choose the **Add New Location** icon, in the upper right corner of the Resource Drawings tab, and select the folder from the Browse for Folder dialog box. When you select a view, information about it is displayed in the **Details** section of the palette. To view a thumbnail preview of the selected view, choose **Preview**. To place a view, first select the view and from the shortcut menu, select PLACE ON SHEET (see Figure 14–40). Before you place the view, right-click to view or change the scale of the view. Click anywhere on the sheet to insert it. A block label is also inserted. When you place a view on a sheet, AutoCAD attaches the drawing with the named view as an xref. The view is listed as a paper space view on the **View List** tab. From the **View List** tab, you can add a view number to the paper space view. Numbers added to the paper space view are updated when the drawing is regenerated.

Figure 14–40 *Sheet Set Manager - Resource Drawings tab with the shortcut menu to place a view*

Instead of using the OPEN command to open a drawing, you can also open the drawing using the Sheet Set Manager. Double-click the name of the drawing in the **Resource Drawings** tab, and it will open in a new window. When the drawing is open, it will be locked automatically and no other user can open it at the same time. The lock status is indicated in the details section of the selected item. When you close the drawing, the status will be changed to Accessible.

The **View List** tab (see Figure 14–41) displays all named views (also called sheet views) on the layouts in your sheet set. You can use the view list to keep track of all sheet views in the sheet set. You can navigate to any sheet view in the sheet set. You can also link sheet views together for coordination across the sheet set.

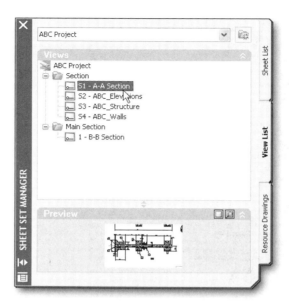

Figure 14–41 *Sheet Set Manager with View List tab*

Sheet views can be created in two ways: You can create a sheet view by creating a named view in paper space on any layout by invoking the VIEW command. In addition, AutoCAD also automatically creates a sheet view whenever you place a view of a drawing on a sheet (for details on placing a view, see the earlier section "Placing a View on a Sheet"). With the Sheet Set Manager, you can apply label and callout blocks to views on sheets.

A callout block refers to other views in the sheet set. It is a symbol that shows, for example, a cross-reference to an elevation, a detail, a section and so on (see Figure 14–42). With the Sheet Set Manager, you can automatically update the information in your label and callout blocks when the reference information changes. For example, when you renumber or rename a view, information on the label and callout blocks is updated when the drawing regenerates. To place a callout block for the selected view, select PLACE CALLOUT BLOCK from the shortcut menu, and choose the type of callout block you want to use (see Figure 14–43). Click anywhere on the sheet to place the callout block. This callout block contains information about the sheet number of the drawing in which the view is saved and the view number (see Figure 14–42).

Figure 14–42 *Example callout symbols*

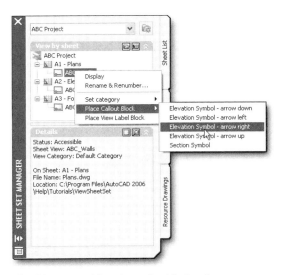

Figure 14–43 *Sheet Set Manager – View List tab with the shortcut menu to place a callout block*

If necessary, you can rename and renumber a view. To rename or renumber the selected view, choose RENAME & RENUMBER from the shortcut menu. AutoCAD displays the Rename & Renumber View dialog box. Specify a view number and view title for the selected view. The **Number** box specifies the view number of the selected view. The **View title** option specifies the view title of the selected view. Choosing **Next** loads the next view into this dialog box. Choose **OK** to close the dialog box.

CREATING A SHEET LIST TABLE

With the Sheet Set Manager you can create a sheet list table and then update it to match changes in your sheet list. Start by opening the sheet in which you want to create the sheet list table. Select the **Sheet List** tab, select the sheet set, and from the shortcut menu, select INSERT SHEET LIST TABLE (see Figure 14–44 at top). AutoCAD displays the Insert Sheet List Table dialog box, as shown at bottom in Figure 14–44. There are several ways to change the appearance of the table before you insert it. For example, when you select **Show Subheader**, the table includes rows displaying subheaders, or the subsets in the sheet set. After making necessary changes in the table, choose **OK**. Click anywhere on the sheet to place the sheet list table. This table lists all the sheets and subsheets in the sheet set (see Figure 14–45). If you remove, add, or make any changes to the sheet number or sheet name, you can update the sheet list table by selecting the table and choosing UPDATE SHEET LIST TABLE from the shortcut menu. AutoCAD updates the sheet list table with the changes.

 Note: If you modify the sheet list table manually, the changes are temporary and are lost when you update the table.

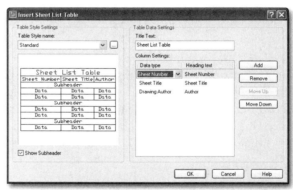

Figure 14–44 *Sheet Set Manager – Sheet List tab with the shortcut menu to insert a sheet list table (top); Insert Sheet List Table dialog box (bottom)*

Figure 14–45 *An example table created from the Sheet Set Manager*

CREATING A TRANSMITTAL PACKAGE

The Sheet Set Manager allows you to eTransmit a sheet set, selected sheets, or a subset. eTransmit packages a set of files for Internet transmittal. In a transmittal package, sheet set data files, xrefs, plot configuration files, font files, and so on are automatically included. To create a transmittal package, first select the sheet set, one or more sheets, or subset you want to include in the transmittal package and select ETRANSMIT from the shortcut menu. Figure 14–46 at top shows the Architectural subset selection and ETRANSMIT selected from the shortcut menu. AutoCAD displays the Create Transmittal dialog box, as shown at bottom in Figure 14–46. The **Sheets** tab lists all the sheets in the transmittal package for the Architectural subset. The **Files Tree** tab lists all the xref, sheet set data, and template files for the transmittal package. Choose **Transmittal Setups** to customize the transmittal setup that will define how your transmittal is packaged. Choose **OK** to create the eTransmit package.

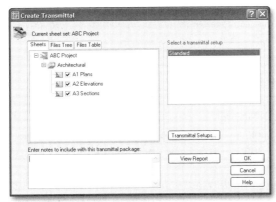

Figure 14–46 *Sheet Set Manager – Sheet List tab with the shortcut menu for eTransmit selection (top); Create Transmittal dialog box (bottom)*

CREATING AN ARCHIVE OF THE SHEET SET

The Sheet Set Manager allows you to archive the selected sheet set. The ARCHIVE option brings together for archiving purposes the files associated with the current sheet set. To create an archive package, first select the sheet set to include in the archive package and select ARCHIVE from the shortcut menu, as shown at left in Figure 14–47. AutoCAD displays the Archive a Sheet Set dialog box, as shown at right in Figure 14–47. The **Sheets** tab lists all the sheets in the archive package for the selected sheet set. The **Files Tree** tab lists all the xref, sheet set data, and template files for the archive package. Choose **Modify Archive Setup** to customize the archive setup that will define how your archive is packaged. If necessary, enter information relative to the archive package in the **Enter notes to include with this archive** box. The information is included in the archive report. Choose **OK** to create the archive package. Be sure that the files to be archived are not open.

Figure 14–47 Sheet Set Manager – Sheet List tab with the shortcut menu for Archive selection; Archive a Sheet Set dialog box

PLOTTING THE SHEET SET AND PUBLISHING TO DWF

The Sheet Set Manager allows you to plot to the default plotter or printer, or publish to specified DWF format a sheet set, selected sheets, or subset. To plot, in the **Sheet List** tab, first select the sheet set, one or more sheets, or subset you want to include. Then select PUBLISH from the shortcut menu and PUBLISH TO PLOTTER from the submenu, as shown in Figure 14–48. AutoCAD will plot the selected sheet(s) to the default plotter. (For detailed information on plotting and its related settings, refer to Chapter 8.) Similarly, to publish to a specified DWF format, first select the sheet set, one or more sheets, or subset you want to include and select PUBLISH from the shortcut menu and PUBLISH TO DWF from the submenu. AutoCAD will create the DWF file of the selected sheet(s).

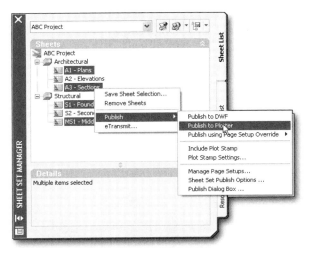

Figure 14–48 *Sheet Set Manager – Sheet List tab with the shortcut menu to Publish to Plotter selection*

REVIEW QUESTIONS

1. The command to invoke the Internet browser in AutoCAD is:

    a. INTERNET

    b. BROWSER

    c. HTTP

    d. WWW

    e. none of the above

2. DWF is short for:

    a. DraWing Format

    b. Design Web Format

    c. DXF Web Format

3. The purpose of DWF files is to view:

    a. 2D drawings on the Internet

    b. 3D drawings on the Internet

    c. 3D drawings in another CAD system

    d. all of the above

4. URL is short for:

    a. Union Region Lengthen

    b. Earl

    c. Useful Resource Line

    d. Uniform Resource Locator

5. Which of the following URLs are valid:

    a. www.autodesk.com

    b. http://www.autodesk.com

    c. all of the above

    d. none of the above

6. What is the purpose of a URL?

    a. to access files on computers, networks, and the Internet

    b. a universal file-naming system for the Internet

    c. to create a link to another file

    d. all of the above

7. FTP is short for:

   a. Forwarding Transfer Protocol

   b. File Transfer Protocol

   c. File Transference Protocol

   d. File Transfer Partition

8. URLs are used in an AutoCAD drawing to browse the Internet.

   a. True   b. False

9. The purpose of URLs is to let you create _____ between files.

   a. backups

   b. links

   c. copies

   d. partitions

10. You can attach a URL to rays and xlines.

    a. True   b. False

11. Compression in the DWF file causes it to take _____ time to transmit over the Internet.

    a. more

    b. less

    c. all of the above

    d. none of the above

12. A "plug-in" lets a Web browser:

    a. plug in to the Internet

    b. display a file format

    c. log in to the Internet

    d. display a URL

13. A Web browser can view DWG drawing files over the Internet.

    a. True   b. False

14. Which AutoCAD command is used to start a Web browser?

    a. WEBSTART   c. BROWSER

    b. WEBDWG   d. LAUNCH

15. Before a URL can be accessed, you must always type the prefix "http://".

    a. True   b. False

16. Which AutoCAD command is used to open an AutoCAD drawing file from an Internet site?

    a. LAUNCH

    b. OPEN

    c. START

    d. LAUNCHFILE

17. Once an Internet AutoCAD drawing file has been modified, it can only be saved back to the Internet site.

    a. True   b. False

18. AutoCAD creates hyperlinks as being _____.

    a. absolute

    b. relative

    c. polar

    d. only a or b

19. Hyperlinks can only be attached to text objects within a drawing.

    a. True   b. False

20. The system variable PICKFIRST must be set to a value of 1 before a hyperlink can be attached.

    a. True   b. False

21. DWF files are compressed up to _____ of the original .DWG file size.

    a. 1/8

    b. 1/4

    c. 1/2

    d. 1/3

CHAPTER 15

AutoCAD 3D

INTRODUCTION

After completing this chapter, you will be able to do the following:

- Define a User Coordinate System
- View in *3D* using the VPOINT, DVIEW, and 3DORBIT commands
- Create *3D* objects
- Use the REGION command
- Use the 3DPOLY and 3DFACE commands
- Create meshes
- Edit in *3D* using the ALIGN, ROTATE3D, MIRROR3D, 3DARRAY, EXTEND, and TRIM commands
- Create solid shapes: solid box, solid cone, solid cylinder, solid sphere, solid torus, and solid wedge
- Create solids from existing *2D* objects and regions
- Create solids by means of revolution
- Create composite solids
- Edit *3D* solids using the CHAMFER, FILLET, SECTION, SLICE, and INTERFERE commands, in addition to Face editing, body editing, and edge editing
- Obtain the mass properties of a solid
- Place a multiview in paper space
- Generate Views in viewports
- Generate profiles

WHAT IS 3D?

In *two-dimensional* drawings you have been working in a single plane with two axes, *X* and *Y*. In *three-dimensional* drawings you work with the *Z* axis, in addition to the *X* and *Y* axes (see Figure 15–1). Plan views, sections, and elevations represent only two dimensions. Isometric, perspective, and axonometric drawings, on the other hand, represent all three dimensions. For example, to create three views of a cube, the cube is drawn as a square with thickness and then viewed along each of the three axes. Drawing in this manner is referred to as extruded *2D*. Only objects that are extrudable can be drawn by this method. Other views are achieved by rotating the viewpoint or the object, just as if you were physically holding the cube. You can get an isometric or perspective view by simply changing the viewpoint.

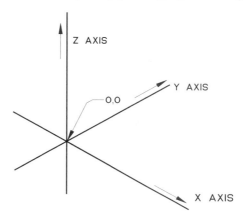

Figure 15–1 *The X, Y, and Z axes for a 3D drawing*

Whether you realize it or not, all the drawings you have done in previous chapters were created by AutoCAD in true *3D*. What this means is that every line, circle, or arc that you have drawn, even if you think you have drawn it in *2D*, is really stored with three coordinates. By default, AutoCAD stores the *Z* value as your current elevation with a thickness of zero. What you think of now as *2D* is really only one of an infinite number of views of your drawing in *3D* space.

Drawing objects in *3D* provides three major advantages:

- An object is drawn once and then can be viewed and plotted from any angle (viewpoint).
- A 3D object holds mathematical information that can be used in engineering analysis such as finite element analysis and computer numerical control (CNC) machinery.
- Shading and rendering enhances the visualization of an object.

There are two major limitations to working in *3D*:

1. Whenever you want to input 3D coordinates whose Z coordinate is different from the current construction plane's elevation, you have to use the keyboard instead of your pointing device. One exception is to Object Snap to an object not in the current construction plane. The input device

(mouse or digitizer) can supply AutoCAD with only two of the three coordinates at a time.

2. Determining where you are in relationship to an object in 3D space is difficult.

COORDINATE SYSTEMS

AutoCAD provides two types of coordinate systems. One is a single fixed coordinate system called the World Coordinate System, and the other is an infinite set of user-defined coordinate systems available through the User Coordinate System.

The World Coordinate System (WCS) is fixed and cannot be changed, as indicated in Chapter 2. In this system (when viewing the origin from 0,0,1), the X axis starts at the point 0,0,0, and values increase as the point moves to the operator's right; the Y axis starts at 0,0,0, and values increase as the point moves to the top of the screen; and finally, the Z axis starts at the 0,0,0 point, and values get larger as it comes toward the user. All drawings from previous chapters are created with reference to the WCS. The WCS is still the basic system in virtually all 2D AutoCAD drawings. However, because of the difficulty in calculating 3D points, the WCS is not suited for many 3D applications.

The User Coordinate System (UCS) allows you to change the location and orientation of the X, Y, and Z axes to reduce the number of calculations needed to create 3D objects. The UCS command lets you redefine the origin of your drawing and establish the positive X and the positive Y axes. New users think of a coordinate system simply as the direction of positive X and positive Y. But once the directions of X and Y are defined, the direction of Z will be defined as well. Thus, the user has only to be concerned with X and Y. As a result, when you are drawing in 2D, you are also somewhere in 3D space. For example, if a sloped roof of a house is drawn in detail using the WCS, each endpoint of every object on the inclined roof plane must be calculated. On the other hand, if the UCS is set to the same plane as the roof, each object can be drawn as if it were in the plan view. You can define any number of UCSs within the fixed WCS and save them, assigning each a user-determined name. But at any given time, only one coordinate system is current and all coordinate input and display is relative to it. This is unlike a rotated snap grid in which direction and coordinates are based on the WCS. If multiple viewports are active, AutoCAD allows you to assign a different UCS to each viewport. Each UCS can have a different origin and orientation for various construction requirements.

RIGHT-HAND RULE

The directions of the X, Y, and Z axes change when the UCS is altered; hence, the positive rotation direction of the axes may become difficult to determine. The right-hand rule helps in determining the rotation direction when changing the UCS or using commands that require object rotation.

To remember the orientation of the axes, do the following:

1. Hold your right hand with the thumb, forefinger, and middle finger pointing at right angles to each other (see Figure 15–2).
2. Consider the thumb to be pointing in the positive direction of the X axis.
3. The forefinger points in the positive direction of the Y axis.
4. The middle finger points in the positive direction of the Z axis.

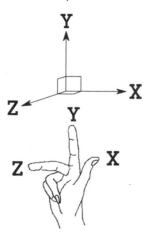

Figure 15–2 *The correct hand position when using the right-hand rule*

SETTING THE DISPLAY OF THE UCS ICON

The UCS icon provides a visual reminder of how the UCS axes are oriented, where the current UCS origin is, and the viewing direction relative to the UCS XY plane. AutoCAD provides two methods of displaying icons: 2D UCS style and 3D UCS style and displays different coordinate system icon in paper space and in model space (see Figure 15–3).

The X and Y axis directions are indicated with arrows labeled appropriately, and the Z axis is indicated by the placement of the icon and in both cases (2D UCS style and 3D UCS style), a plus sign (+) appears at the base of the icon when it is positioned at the origin of the current UCS. With the 2D UCS icon a W is displayed in the Y axis arrow and with the 3D UCS icon, a square is displayed in the XY plane at the origin when the UCS is the same as the world coordinate system. When looking straight up or down on the Z plane, the icon seems flat. When viewed at any other angle, the icon looks skewed. In the case of 2D UCS style, the orientation of the Z axis is indicated further by the presence or absence of a box at the base of the arrows that create the icon. If the box is visible, you are looking down on (from the positive Z side of) the XY plane; if the box is not present, the bottom of the XY plane is being viewed (from the negative Z side of it). With the 3D UCS icon, the Z axis is solid when viewed from above the XY plane and dashed when viewed from below the XY plane. See Figure 15–4 for different orientations of the UCS icon.

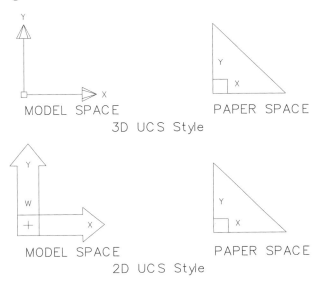

Figure 15–3 *The UCS icon for model space and for paper space*

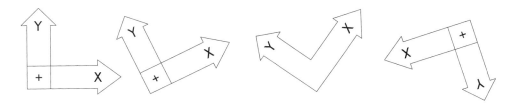

Figure 15–4 *The UCS icon in different orientations*

 Note: If the viewing angle comes within 1 degree of the XY plane, the 2D UCS style icon will change to a "broken pencil," as shown in Figure 15–5. When this icon is showing in a view, it is recommended that you avoid trying to use the cursor to specify points in that view, because the results may be unpredictable. The 3D UCS icon does not use a broken pencil icon.

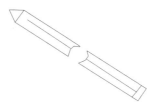

Figure 15–5 *The UCS icon becomes a broken pencil when the viewing angle is within 1 degree of the XY plane*

The display and placement on the origin of the UCS icon is handled by the UCSICON command. Invoke the UCSICON command at the On-Screen prompt and AutoCAD prompts:

Enter an option *(select one of the options)*

Choose ON to set the icon to ON if it is OFF in the current viewport.

Choose OFF to set the icon to OFF if it is ON in the current viewport.

Choose NOORIGIN (default setting) to display the icon at the lower left corner of the viewport, regardless of the location of the UCS origin. This is like parking the icon in the lower left corner.

Choose ORIGIN to display the icon at the origin of the current coordinate system.

 Note: If the origin is off screen, the icon is displayed at the lower left corner of the viewport.

Choose ALL to set whether the options that follow affect all of the viewports or just the current active viewport. This option is issued before each and every option if you want to affect all viewports. For example, to turn ON the icon in all the viewports and display the icon on the origin, the following sequence of prompts is displayed:

ucsicon (ENTER)
Enter an option **a**
Enter an option **on**
ucsicon (ENTER)
Enter an option **a**
Enter an option **or**

Choose PROPERTIES to display the UCS Icon dialog box (see Figure 15–6), in which you can control the style, visibility, and location of the UCS icon.

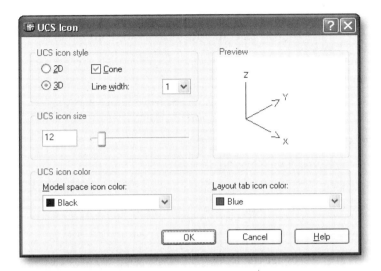

Figure 15-6 *UCS Icon dialog box*

The **UCS icon style** section controls display of either the 2D or the 3D UCS icon and its appearance. The 2D selection displays a 2D icon without a representation of the Z axis, and the 3D selection displays a 3D icon. Setting the **Cone** check box controls the display of the 3D cone arrowheads for the X and Y axes (available only when 3D UCS icon is selected). **Line width** controls the line width of the UCS icon, and you can set to one of the three available selections: 1, 2, or 3 pixels (available only when 3D UCS icon is selected). The Preview section displays a preview of the selected UCS icon style in model space. The UCS icon size section controls the size of the UCS icon as a percentage of viewport size. The default value is 12, and the valid range is from 5 to 95. The UCS icon color section allows you to select the color of the UCS icon in model space viewports and in the Layout tab. Choose **OK** to save the settings and close the UCS Icon dialog box.

 Note: You can also open the UCS Icon dialog box from the View menu (Display > UCS icon > Properties).

DEFINING A NEW UCS

The UCS is the key to almost all 3D operations in AutoCAD, as mentioned earlier. Many commands in AutoCAD are traditionally thought of as 2D commands. They are effective in *3D* because they are always relative to the current UCS. For example, the ROTATE command rotates in only the *X* and *Y* directions; if you want to rotate an object in the Z direction, change your UCS so that X or Y is now in the direction of what was previously Z. Then you could use the ROTATE command.

The UCS command lets you redefine the origin in your drawing. Broadly, you can define the origin by five methods:

- Specify a data point for an origin, specify a new XY plane by providing three data points, or provide a direction for the Z axis.
- Define an origin relative to the orientation of an existing object.
- Define an origin by selecting a face.
- Define an origin by aligning with the current viewing direction.
- Define an origin by rotating the current UCS around one of its axes.

Invoke the UCS command from the UCS toolbar (see Figure 15–7).

Figure 15–7 *Invoking the* UCS *command from the UCS toolbar*

AutoCAD prompts:

> Enter an option *(choose NEW)*
> Specify origin of new UCS or ⬇ *(specify a point to define a new origin or select one of the options)*

Specify a data point to define a new UCS by shifting the origin of the current UCS, leaving the directions of the X, Y, and Z axes unchanged (see Figure 15–9). You can also invoke ORIGIN from the UCS toolbar (see Figure 15–8).

Figure 15–8 *Invoking* ORIGIN *from the UCS toolbar*

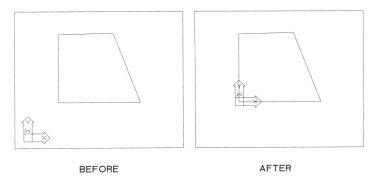

Figure 15-9 *Specifying a new origin point relative to the origin of the current UCS*

Choose ZAXIS to define an origin by specifying a point and the direction for the Z axis. AutoCAD arbitrarily, but consistently, sets the direction of the X and Y axes in relation to the given Z axis. You can also invoke Z AXIS from the UCS toolbar (see Figure 15-10).

Figure 15-10 *Invoking Z AXIS command from the UCS toolbar*

AutoCAD prompts:

> Specify new Origin point <0,0,0>: *(specify a point to define a new origin point)*
> Specify point on positive portion of the Z axis <default>: *(specify the direction for the Z axis)*

If you press ENTER in response to the second prompt, the Z axis of the new coordinate system will be parallel to the previous one. This is similar to using the ORIGIN option.

Choose 3POINT to define three points that include the origin and the directions of the positive X and Y axes. The origin point acts as a base for the UCS rotation, and when a point is selected to define the direction of the positive X axis, the direction of the Y axis is limited because it is always perpendicular to the X axis. When the X and Y axes are defined, the Z axis is automatically placed perpendicular to the XY plane. You can also invoke 3POINT from the UCS toolbar (see Figure 15-11).

Figure 15–11 *Invoking 3POINT from the UCS toolbar*

AutoCAD prompts:

    Specify new Origin point <0,0,0>: *(specify the origin point)*
    Specify Point on positive portion of the X-axis <default>: *(specify a point for the positive X axis)*
    Specify Point on positive-Y portion of the UCS XY plane <default>: *(specify a point for the positive Y axis)*

    Specify a data point and the direction for the positive X and Y axes (see Figure 15–12). The points must not form a straight line. If you press ENTER at the first prompt, the new UCS will have the same origin as the previous UCS. If you press ENTER at the second or third prompt, then that axis will be parallel to the corresponding axis of the previous UCS.

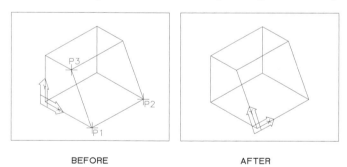

Figure 15–12 *Specifying the direction for the positive X and Y axes using the 3POINT option*

    Choose OBJECT to define a new coordinate system based on a selected object. The actual orientation of the UCS depends on how the object was created. When the object is selected, the UCS origin is placed at the first point used to create the object (in the case of a line, it will be the closest endpoint; for a circle, it will be the center point of the circle); the X axis is determined by the direction from the origin to the second point used to define the object. And the Z axis direction is placed perpendicular to the XY plane in which the object sits. The following table lists the location of the origin and its X axis for different types of objects. You can also invoke Object from the UCS toolbar (Figure 15–13).

Location of the Origin and Its X Axis for Different Types of Objects

Object	Method of UCS Determination
Line	The endpoint nearest the specified point becomes the new UCS origin. The new X axis is chosen so that the line lies in the XZ plane of the new UCS.
Circle	The circle's center becomes the new UCS origin, and the X axis passes through the point specified.
Arc	The arc's center becomes the new UCS origin, and the X axis passes through the endpoint of the arc closest to the pick point.
2D polyline	The polyline's start point becomes the new UCS origin, with the X axis extending from the start point to the next vertex.
Solid	The first point of the solid determines the new UCS origin, and the X axis lies along the line between the first two points.
Dimension	The new UCS origin is the middle point of the dimension text, and the direction of the X axis is parallel to the X axis of the UCS in effect when the dimension was drawn.

Figure 15–13 *Invoking* OBJECT *from the UCS toolbar*

AutoCAD prompts:

Select object to align UCS: *(identify an object to define a new coordinate system)*

Figure 15–14 shows a new coordinate system based on the selected object.

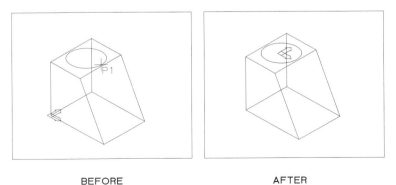

Figure 15–14 *Defining a new coordinate system by the* OBJECT UCS *command*

Choose FACE to align the UCS to the selected face of a solid object. To select a face, click within the boundary of the face or on the edge of the face. The face is highlighted and the UCS *X* axis is aligned with the closest edge of the first face found. You can also invoke FACE from the UCS toolbar (see Figure 15–15).

Figure 15–15 *Invoking* FACE *from the UCS toolbar*

AutoCAD prompts:

> Select face of solid object: *(select the face of the solid object)*
> Enter an option *(select one of the available options)*

>> Choosing NEXT locates the UCS on either the adjacent face or the back face of the selected edge.

>> Choosing XFLIP rotates the UCS 180 degrees around the *X* axis.

>> Choosing YFLIP rotates the UCS 180 degrees around the *Y* axis.

>> Choosing ACCEPT accepts the location. The prompt repeats until you accept a location.

>> Choose VIEW to place the *XY* plane parallel to the screen, and make the *Z* axis perpendicular. The UCS origin remains unchanged. This method is used mainly for labeling text, which should be aligned with the screen rather than with objects. You can also invoke VIEW from the UCS toolbar (see Figure 15–16). AutoCAD prompts:

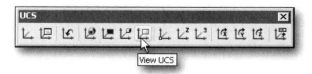

Figure 15–16 *Invoking the* VIEW UCS *command from the UCS toolbar*

>> Choosing x, y, or z rotation lets you define a new coordinate system by rotating the *X*, *Y*, and *Z* axes independently of each other. You can specify the desired angle by selecting two points on the screen, or you can enter the rotation angle from the keyboard. In either case, the new

angle is specified relative to the *X* axis of the current UCS. See Figures 15–17, 15–18, and 15–19 for examples of rotating the UCS around the *X*, *Y*, and *Z* axes, respectively.

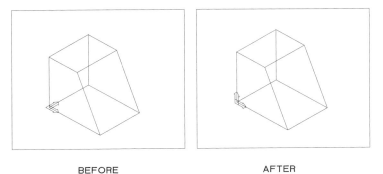

Figure 15–17 *Example of rotating the UCS around the X axis*

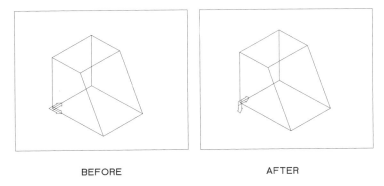

Figure 15–18 *Example of rotating the UCS around the Y axis*

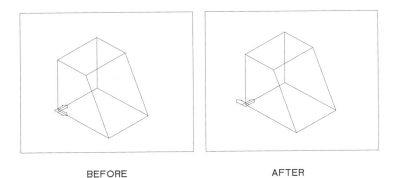

Figure 15–19 *Example of rotating the UCS around the Z axis*

Choose PREVIOUS to restore the previous origin. AutoCAD saves the last 10 coordinate systems in both model space and paper space. You can step back through them by using the PREVIOUS option repeatedly. You can also choose **UCS Previous** from the UCS toolbar.

Figure 15–20 *Invoking PREVIOUS from the UCS toolbar*

Choose RESTORE to restore any previously saved UCS.

 Note: You can also restore a previously saved UCS by invoking the DDUCS command, which in turn displays a dialog box listing the previously saved UCS.

Choose SAVE to save the current UCS under a user-defined name.

Choose DELETE to delete any saved UCS.

Choose ? to list the name of the UCS you specify, the origin, and the X, Y, and Z axes for each saved coordinate system, relative to the current UCS. To list all the UCS names, accept the default, or you can specify wild cards.

Choose WORLD to return the drawing to the WCS. You can also choose **World UCS** from the UCS toolbar (see Figure 15–21).

Figure 15–21 *Invoking WORLD from the UCS toolbar*

SELECTING A PREDEFINED ORTHOGRAPHIC UCS

In addition to various methods explained earlier for defining a new UCS, AutoCAD's UCS dialog box allows you to change the UCS to one of the available standard orthographic settings. In addition, the UCS dialog box lists saved user coordinate systems and allows you to modify UCS icon settings and UCS settings saved with a viewport. Invoke the UCSMAN command to open the UCS dialog box from the UCS toolbar (see Figure 15–22).

Figure 15–22 *Invoking the UCS dialog from the UCS toolbar*

AutoCAD displays the UCS dialog box with the Orthographic UCSs tab selected (see Figure 15–23).

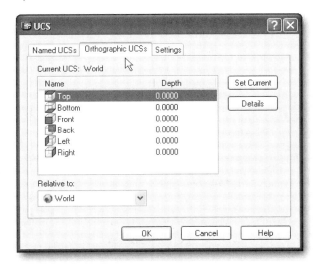

Figure 15–23 *UCS dialog box with the Orthographic UCSs tab selected*

AutoCAD displays the name of the current UCS view at the top of the dialog box. If the UCS setting has not been saved and named, the current UCS reads Unnamed. AutoCAD lists the standard orthographic coordinate systems in the current drawing. The orthographic coordinate systems are defined relative to the UCS specified in the Relative to list box. By default it is set to the World coordinate system. To set the UCS to one of the orthographic coordinate systems, select one of the six listed names and choose **Set Current**, double-click on the listed name, or right-click and select SET CURRENT from the shortcut menu. The **Depth** column lists the distance between the orthographic coordinate system and the parallel plane passing through the origin of the UCS base setting (stored in the UCSBASE system variable). To change the depth, double-click on the Depth field, and AutoCAD displays the Orthographic UCS depth dialog box. Make the necessary changes to the depth and choose **OK** to close the dialog box.

To set the UCS to one of the saved UCSs, select the Named UCSs tab, and AutoCAD lists the saved UCS in the current drawing (see Figure 15–24). Select one of the listed UCS saved names and choose **Set Current**, double-click on the listed name, or right-click and select SET CURRENT from the shortcut menu.

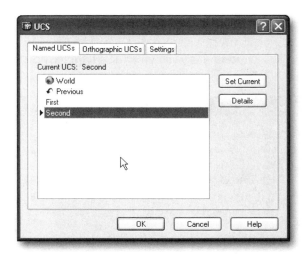

Figure 15–24 *UCS dialog box with the Named UCSs tab selected*

The Settings tab of the UCS dialog box displays and allows you to modify UCS icon settings and UCS settings saved with a viewport(see Figure 15–25).

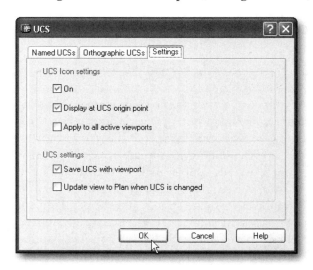

Figure 15–25 *UCS dialog box with the Settings tab selected*

The **UCS icon settings** section of the dialog box specifies the UCS settings for the current viewport. Choose **On** to display the UCS icon in the current viewport. **Display at UCS origin point** controls the display of the UCS icon at the origin of the current coordinate system for the current viewport. If this option is set to off, or if the origin of the coordinate system is not visible in the viewport, the UCS icon is displayed at the lower left corner of the viewport. Choose **Apply all to active viewports** to apply the UCS icon settings to all active viewports in the current drawing.

The **UCS settings** section specifies the UCS settings for the current viewport. Choose **Save UCS with viewport** to save the coordinate system setting with the viewport. If this option is set to off, the viewport reflects the UCS of the viewport that is current. Choose **Update view to Plan when UCS is changed** to restore plan view when the coordinate system in the viewport changes. Plan view is restored when the dialog box is closed and the selected UCS setting is restored.

Choose **OK** to close the UCS dialog box and accept the changes.

You can also set the current UCS to one of the saved UCS or to one of the standard orthographic UCS from the UCS II toolbar menu (see Figure 15–26).

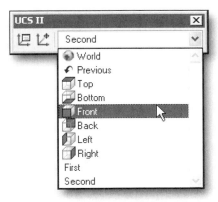

Figure 15–26 UCS II toolbar

VIEWING A DRAWING FROM PLAN VIEW

The PLAN command provides a convenient means of viewing a drawing from plan view. The definition of a plan means that you are at positive Z and looking perpendicularly down on the XY plane, with X to the right and Y pointing up. You can select the plan view of the current UCS, a previously saved UCS, or the WCS.

Invoke the PLAN command by selecting Plan View from the 3D views submenu in the View menu.

> Choose CURRENT UCS to display the plan view of the current UCS.
>
> Choose NAMED UCS to display a plan view of a previously saved UCS. When you select this option, AutoCAD prompts for a name of the UCS.
>
> Choose WORLD to display the plan view of the World Coordinate System.

VIEWING IN 3D

Until now, you have been working on the plan view, or the *XY* plane. You have been looking down at the plan view from a positive distance along the *Z* axis. The direction from which you view your drawing or model is called the viewpoint. You can view a drawing from any point in model space. From your selected viewpoint, you can add objects, modify existing objects, or suppress the hidden lines from the drawing.

The VPOINT, DVIEW, and 3DORBIT commands are used to control viewing of a model from any point in model space.

VIEWING A MODEL BY MEANS OF THE VPOINT COMMAND

To view a model in *3D*, you may have to change the viewpoint. The location of the viewpoint can be controlled by means of the VPOINT command. The default viewpoint is 0,0,1; that is, you are looking at the model from 0,0,1 (on the positive *Z* axis above the model) to 0,0,0 (the origin).

 Note: The value of 1 as the Z coordinate in the point 0,0,1 is not critical. In a parallel view, the Z coordinate can be any positive value (0.5 or even 1000). It establishes the distance from which you are viewing the objects. However, the Z coordinate can affect the appearance of objects drawn in 3D if you are using the PERSPECTIVE option of the 3DORBIT feature (discussed later).

Invoke the VPOINT command from the On-Screen prompt.

AutoCAD prompts:

> Specify a view point or ⬇ *(select one of the available options from the shortcut menu)*

The default method requires you to enter *X*, *Y*, and *Z* coordinates from the keyboard. These coordinates establish the viewpoint. From this viewpoint, you will be looking at the model in space toward the model's origin. For example, a 1,–1,1 setting gives you a –45-degree angle projected in the *XY* plane and 35.264-degree angle above the *XY* plane (top, right, and front views); looking at the model origin (0,0,0). You can set the viewpoint to any *X, Y, Z* location. The following table shows you various viewpoint settings and their results when rotating *3D* objects.

Various Viewpoint Settings for Rotating 3D objects

Viewpoint Setting	Displayed View(s)	Viewpoint Setting	Displayed View(s)
0,0,1	Top	–1,–1,1	Top, Front, Left side
0,0,–1	Bottom	1,1,1	Top, Rear, Right side
0,–1,0	Front	–1,1,1	Top, Rear, Left side
0,1,0	Rear	1,–1,–1	Bottom, Front, Right side
1,0,01	Right side	–1,–1,–1	Bottom, Front, Left side
–1,0,0	Left side	1,1,–1	Bottom, Rear, Right side
1,–1,1	Top, Front, Right side	–1,1,–1	Bottom, Rear, Left side

If instead of entering coordinates, you press ENTER, a compass and axes tripod appear on the screen (see Figure 15–27). The compass, in the upper right of the screen, is a *2D* representation of a globe. The center point of the circle represents the north pole (0,0,1), the inner circle represents the equator, and the outer circle represents the south pole (0,0,–1), as shown in Figure 15–28. A small cross is displayed on the compass. You can move the cross with your pointing device. If the cross is in the inner circle, you are above the equator looking down on your model. If the cross is in the outer circle, you are looking from beneath your drawing, or from the Southern Hemisphere. Move the cross, and the axes tripod rotates to conform to the viewpoint indicated on the compass. When you achieve the desired viewpoint, press the pick button on your pointing device or press ENTER. The drawing regenerates to reflect the new viewpoint position.

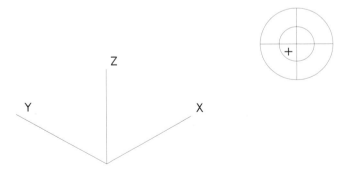

Figure 15–27 *The* VPOINT *command's compass and axes tripod*

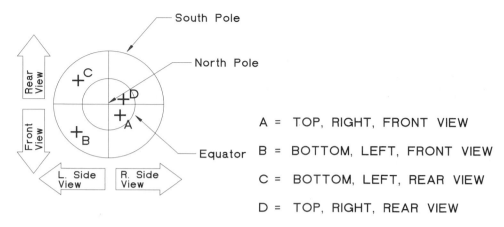

Figure 15–28 *Components of the* VPOINT *command's compass and its poles*

Choose ROTATE from the shortcut menu to specify the location of the viewpoint in terms of two angles. The first angle determines the rotation in the *XY* plane from the *X* axis (0 degrees) clockwise or counterclockwise. The second angle determines the angle from the *XY* plane up or down. When you select the ROTATE option, AutoCAD prompts:

    Enter angle in XY plane from X axis <current>: *(specify the angle in the XY plane from the X axis)*
    Enter angle from XY plane <current>: *(specify the angle from the XY plane)*

AutoCAD regenerates to reflect the new viewpoint position.

AutoCAD provides the Viewpoint Presets dialog box when you invoke the DDVPOINT command. The dialog box lets you set a *3D* viewing direction by specifying an angle from the *X* axis and an angle from the *XY* plane. This is similar to using the ROTATE option of the VPOINT command.

Invoke the DDVPOINT command by selecting Viewpoint Presets from the 3D Views submenu in the View menu.

AutoCAD displays the Viewpoint Presets dialog box (see Figure 15–29). Specify viewing angles from the image tiles, or enter their values in the text boxes. You specify the view direction relative to the current UCS or the WCS, and the viewing angles are updated accordingly. The new angle is indicated by the black arm; the current viewing angle is indicated by the gray arm. By selecting **Set to Plan View**, you can set the viewing angles to display the plan relative to the selected coordinate system.

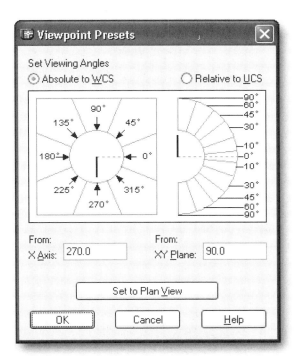

Figure 15-29 *Viewpoint Presets dialog box*

 Note: By default, AutoCAD orients the model to your current viewpoint position in reference to the WCS, not the current UCS. If necessary, you can change the WORLDVIEW system variable from 1 (default) to 0; AutoCAD then orients the model in reference to the UCS for your current viewpoint position. It is recommended that you keep the WORLDVIEW set to 1 (default). Regardless of the WORLDVIEW setting, you are always looking through your viewpoint to the WCS origin.

VIEWING A MODEL BY MEANS OF THE DVIEW COMMAND

The DVIEW command is an enhanced VPOINT command. Here you visually move an object around on the screen, dynamically viewing selected objects as the view changes. The DVIEW command provides either parallel or perspective views, whereas the VPOINT command provides only parallel views. In the case of a parallel view, parallel lines always remain parallel, whereas in perspective view, parallel lines converge from your viewpoint to a vanishing point. Figures 15-30 and 15-31 show parallel and perspective views, respectively, of a model. The viewing direction is the same in each case.

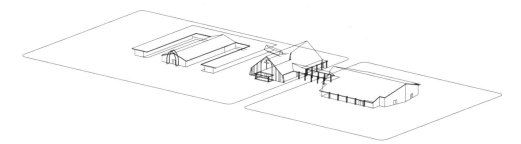

Figure 15–30 *The* DVIEW *command displaying a model with parallel projection*

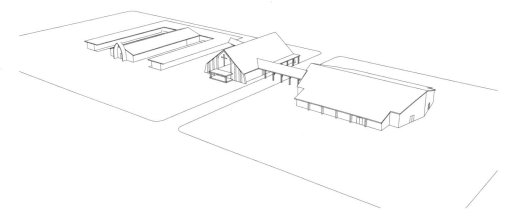

Figure 15–31 *The* DVIEW *command displaying the same model with perspective projection*

Invoke the DVIEW command at the On-Screen prompt.

AutoCAD prompts:

Select objects or <use DVIEWBLOCK>: *(select objects or press* ENTER*)*

All or any part of the objects in the drawing can be selected for viewing during the DVIEW command process. While you are manipulating the view, only the selected items are visible. But once you exit the DVIEW command, all objects in the drawing are visible in the new view created. If your drawing is too large to regenerate quickly in the DVIEW display, small portions can be selected and used to orient the entire drawing. The purpose of this is to save time on slower machines and still give dynamic rotation so that you can quickly and effortlessly adjust the view of your object before you begin working with it.

If you press ENTER in response to the "Select objects:" prompt, AutoCAD provides you with a picture of a *3D* house. Whatever changes you make to the view of the *3D* house under DVIEW will be made to your current drawing when you exit the DVIEW command.

Each time you exit the DVIEW command, AutoCAD performs an unconditional regeneration. No matter what the current setting for REGENAUTO is, AutoCAD automatically performs a regeneration.

After selecting the objects, AutoCAD prompts with the following options:

[CAmera/TArget/Distance/POints/PAn/Zoom/TWist/CLip/Hide/Off/
Undo]: *(select one of the available options)*

Note: These options appear in the command window or in the shortcut menu by right-clicking.

The CAMERA option is one of the six options that adjust what is seen in the view. With the CAMERA option, the drawing is stationary while the camera moves. It can move up and down (above or below) or it can move around the target to the left or right (clockwise or counterclockwise). When you are moving the camera, the target is fixed.

When you select the CAMERA option, AutoCAD prompts:

Specify camera location, or enter angle from XY plane [or Toggle (angle in)] <current>:

You can specify the amount of rotation by moving the cursor in the drawing area. When you move the cursor, you will see the object begin to rotate dynamically, and the AutoCAD status line displays a continuous readout of the new angle. Move the cursor to the desired angle and then press the pick button. Or you can type the desired angle from the keyboard. Either way, you have selected an angle of view above or below the target.

Next, AutoCAD prompts for the desired rotation angle of the camera around the target:

Specify camera location, or enter angle in XY plane from X axis or
[Toggle (angle from)] <current>:

You can move the camera 180 degrees clockwise and 180 degrees counterclockwise around the target. Specify the angle by using the cursor to specify a point on the screen. Or you could type in the desired angle from the keyboard. Either way, you have selected an angle of view around the target clockwise or counterclockwise.

The TOGGLE ANGLE option allows you to move between two angle input modes.

AutoCAD takes you back to the 11-option prompt of the DVIEW command. When the new angle of view is correct, you exit the command sequence by pressing ENTER.

When you exit DVIEW, your entire drawing will rotate to the same angle of view as the few objects that you selected.

The following command sequence shows an example of using the CAMERA option of the DVIEW command.

> **dview**
> Select objects or <use DVIEWBLOCK>: *(select the objects and press* ENTER*)*
> Enter option [CAmera/TArget/Distance/POints/PAn/Zoom/TWist/CLip/Hide/Off/Undo]: **ca**
> Specify camera location, or enter angle from XY plane or [Toggle (angle in)] <current>: **45**
> Specify camera location, or enter angle in the XY plane from X axis or [Toggle (angle from)] <current>: **45**
> Enter option [CAmera/TArget/Distance/POints/PAn/Zoom/TWist/CLip/Hide/Off/Undo]: (ENTER)

The TARGET option is similar to the CAMERA option, but in this case the target is rotated around the camera. The camera remains stationary except for maintaining its lens on the target point. The prompts are similar to those for the CAMERA option. There may seem to be no difference between the CAMERA and TARGET options, but there is a difference in the actual angle of view. For instance, if you elevate the camera 75 degrees above the target, you are then looking at the target from the top down. On the other hand, if you raise the target 75 degrees above the camera, you are then looking at the target from the bottom up. The angles of view are reversed. The real difference comes when you are typing in the angles rather than visually picking them.

The following command sequence shows an example of using the TARGET option of the DVIEW command.

> **dview**
> Select objects or <use DVIEWBLOCK>: *(select the objects and press* ENTER*)*
> Enter option [CAmera/TArget/Distance/POints/PAn/Zoom/TWist/CLip/Hide/Off/Undo]: **ta**
> Specify camera location, or enter angle from XY plane or [Toggle (angle in)] <current>: **75**
> Specify camera location, or enter angle in the XY plane from X axis or [Toggle (angle from)] <current>: **75**
> Enter option [CAmera/TArget/Distance/POints/PAn/Zoom/TWist/CLip/Hide/Off/Undo]: (ENTER)

The DISTANCE option creates a perspective projection from the current view. The only information required for this option is the distance from the camera to the target point. Once AutoCAD knows the distance, it will apply the correct perspective. When perspective viewing is on, a box icon appears on the screen (see Figure 15–32), in place of the UCS icon. Some commands (like ZOOM and PAN) will not work while perspective is on. The perspective view is used just for visual purposes or for plotting.

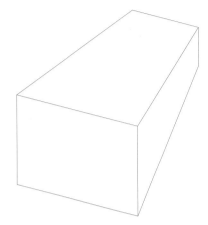

Figure 15–32 *The perspective box icon appears when perspective mode is on*

When you select the DISTANCE option, AutoCAD prompts:

Specify a new camera—target distance <current>:

In addition to the prompt, you also see a horizontal bar at the top of the screen. The bar goes from **0x** to **16x**. These are factor distances times your current distance from the object. Moving the slider cursor toward the right increases the distance between the target and the camera. Moving the slider cursor toward the left reduces the distance between the target and the camera. The current distance is represented by **1x**. For instance, moving the slider cursor to **3x** makes the new distance three times the previous distance. Or you could also type the desired distance in the current linear units from the keyboard.

The following command sequence shows an example of using the DISTANCE option of the DVIEW command.

    dview
    Select objects or <use DVIEWBLOCK>: *(select the objects and press* ENTER*)*
    Enter option
    [CAmera/TArget/Distance/POints/PAn/Zoom/TWist/CLip/Hide/Off/
      Undo]: **d**
    Specify a new camera—target distance <current>: **75**
    Enter option
    [CAmera/TArget/Distance/POints/PAn/Zoom/TWist/CLip/Hide/Off/
      Undo]: (ENTER)

The OFF option turns off the perspective view. The following command sequence shows an example of using the OFF option of the DVIEW command.

dview
Select objects or <use DVIEWBLOCK>: *(select the objects and press* ENTER*)*
Enter option
[CAmera/TArget/Distance/POints/PAn/Zoom/TWist/CLip/Hide/Off/ Undo]: **o**
Enter option
[CAmera/TArget/Distance/POints/PAn/Zoom/TWist/CLip/Hide/Off/ Undo]: (ENTER)

Note: To turn the perspective on again, select the *DISTANCE* option and press ENTER for all the defaults. There is no option called *ON* for turning on the perspective view.

The POINTS option establishes the location of the camera as well as the target points. This gives AutoCAD the basic information needed to create the view. The location of the camera and target points must be specified in a parallel projection. If perspective is set to ON, AutoCAD temporarily turns it off while you specify the new location for camera and target points, and then redisplays the image in perspective.

When you select the option, AutoCAD prompts:

Specify target point: *(specify the target location)*
Specify camera point: *(specify the camera location)*

After the locations are defined, the screen shows the new view immediately.

The following command sequence shows an example of using the POINTS option of the DVIEW command.

dview
Select objects or <use DVIEWBLOCK>: *(select the objects and press* ENTER*)*
Enter option
[CAmera/TArget/Distance/POints/PAn/Zoom/TWist/CLip/Hide/Off/ Undo]: **po**
Specify target point: *(specify a point)*
Specify camera point: *(specify a point)*
Enter option
[CAmera/TArget/Distance/POints/PAn/Zoom/TWist/CLip/Hide/Off/ Undo]: (ENTER)

The PAN option allows you to view a different location of the model by specifying the pan distance and direction. This option is similar to the regular PAN command. The following command sequence shows an example of using the PAN option of the DVIEW command.

```
Command: dview
Select objects or <use DVIEWBLOCK>: (select the objects and press ENTER)
Enter option
[CAmera/TArget/Distance/POints/PAn/Zoom/TWist/CLip/Hide/Off/
    Undo]: pa
Specify displacement base point: (specify a point)
Specify second point: (specify a point)
Enter option
[CAmera/TArget/Distance/POints/PAn/Zoom/TWist/CLip/Hide/Off/
    Undo]: (ENTER)
```

The ZOOM option lets you zoom in on a portion of the model. This option is similar to the regular AutoCAD CENTER option of the ZOOM command, with the center point lying at the center of the current viewport. This option is controlled by a scale factor value.

When you select the ZOOM option, AutoCAD prompts:

```
Specify zoom scale factor <current>: (specify the zoom scale factor)
```

In addition to the prompt, you see a horizontal bar at the top of the screen. The slider bar lets you specify a zoom scale factor, with **1x** being the current zoom level. Any value greater than 1 increases the size of the objects in the view; any decimal value less than 1 decreases the size.

Note: When the perspective is set to ON by the DISTANCE option, the ZOOM option prompts for a lens size rather than a zoom factor, but the effect is similar. The larger the lens size, the closer the object.

The TWIST option rotates or twists the view. It allows you to rotate the image around the line of sight at a given angle from zero, with zero being to the right. The angle is measured counterclockwise.

The following command sequence shows an example of using the TWIST option of the DVIEW command.

```
dview
Select objects or <use DVIEWBLOCK>: (select the objects and press ENTER)
Enter option
[CAmera/TArget/Distance/POints/PAn/Zoom/TWist/CLip/Hide/Off/
    Undo]: tw
Specify view twist angle <current>: (select a point)
Enter option
[CAmera/TArget/Distance/POints/PAn/Zoom/TWist/CLip/Hide/Off/
    Undo]: (ENTER)
```

The CLIP option hides portions of the object in view so that the interior of the object can be seen or parts of the complex object can be more clearly identified.

The CLIP option has three suboptions: Back, Front, and Off.

> The BACK suboption eliminates all parts of the object in view that are located beyond the designated point along the line of sight.
>
> The FRONT suboption eliminates all parts of the object in view that are located between the camera and the front clipping plane.
>
> The OFF suboption turns off front and back clipping.

The following command sequence shows an example of using the CLIP option of the DVIEW command.

```
dview
Select objects or <use DVIEWBLOCK>: (select the objects and press ENTER)
Enter option
[CAmera/TArget/Distance/POints/PAn/Zoom/TWist/CLip/Hide/Off/
    Undo]: cl
Enter clipping option [Back/Front/Off] <current>: b
Specify distance from target or [On/Off] <current>: (specify the distance,
    or turn on and off the previously defined clipping plane)
Enter option
[CAmera/TArget/Distance/POints/PAn/Zoom/TWist/CLip/Hide/Off/
    Undo]: (ENTER)
```

The HIDE option is similar to the regular AutoCAD HIDE command.

The UNDO option will undo the last DVIEW operation. You use it to step back through multiple DVIEW operations.

To exit the UNDO option in the DVIEW command, press ENTER without selecting any of the available options, and AutoCAD returns you to the On-Screen prompt. It is the default option of the DVIEW command.

USING 3DORBIT

The 3DORBIT command allows you to view the model interactively in the current viewport. By using the pointing device you can manipulate the view of the model. You can view the entire model or any object in your model from different points around it.

Invoke the 3DORBIT command from the 3D Orbit toolbar (see Figure 15–33).

Figure 15–33 *Invoking the 3DORBIT command from the 3D Orbit toolbar*

AutoCAD prompts:

Press ESC or ENTER to exit, or right-click to display shortcut-menu.

AutoCAD displays an arcball (see Figure 15–34), a circle divided into four quadrants by smaller circles. While 3DORBIT is active, the point (target) that you are viewing stays stationary and the camera moves around the target. By default, the center of the arcball is the target point.

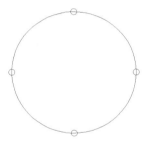

Figure 15–34 *Display of the arcball with a 3D model*

Click and drag the cursor to rotate the view. When you move your cursor over different parts of the arcball, the cursor icon changes.

When you move the cursor inside the arcball, a small sphere encircled by two lines is displayed similar to the first icon (see Figure 15–35). By clicking and dragging, you can manipulate the view freely. You can drag horizontally, vertically, and diagonally.

When you move the cursor outside the arcball a circular arrow around a small sphere is displayed similar to the second icon (see Figure 15–35). Clicking outside the arcball and dragging the cursor around the arcball moves the view around an axis that extends through the center of the arcball, perpendicular to the screen.

When you move the cursor over one of the smaller circles on the left or right of the arcball a horizontal ellipse around a small sphere is displayed similar to the third icon (see Figure 15–35). Clicking and dragging from either of these points rotates the view around the vertical or *Y* axis that extends through the center of the arcball. The *Y* axis is represented on the cursor by a vertical line.

When you move the cursor over one of the smaller circles on the top or bottom of the arcball a vertical ellipse around a small sphere is displayed similar to the fourth icon as shown in Figure 15–35. Clicking and dragging from either of these points rotates the view around the horizontal or *X* axis that extends through the center of the arcball. The *X* axis is represented on the cursor by a horizontal line.

Figure 15–35 *Various cursor icons displayed when the* 3DORBIT *command is invoked*

Panning and Zooming in the 3D Orbit View

You can pan and zoom while you are viewing the model with the 3DORBIT command. It works similarly to the REALTIME PAN and REALTIME ZOOM commands (see Chapter 3 for a detailed explanation). Select the PAN and ZOOM options from the shortcut menu while 3DORBIT is active, from the *3D* Orbit toolbar, or enter **3dpan** and **3dzoom** at the Command prompt to invoke the PAN and ZOOM options of the 3DORBIT command, respectively.

Using Projection Options in the 3D Orbit View

AutoCAD allows you to display a perspective or a parallel projection of the view while 3DORBIT is active. The PERSPECTIVE view option changes the view so that lines parallel to each other converge at a vanishing point. Objects appear to recede into the distance while parts of the objects appear larger and closer to you. The shapes are somewhat distorted when the object is very close. This view correlates more closely to what your eyes see in the real world. The PARALLEL view option changes the view so that two parallel lines never converge at a single point. The shapes in the drawing always remain the same and do not appear distorted when they are closer. You can select PERSPECTIVE and PARALLEL options from the shortcut menu while 3DORBIT is active.

Shading Objects in the 3D Orbit View

AutoCAD allows you to view various Shading Modes while 3DORBIT is active. You can change the way objects are shaded using the different Shading Mode options. The available modes are WIREFRAME, HIDDEN, FLAT SHADED, GOURAUD SHADED, FLAT SHADED – EDGES ON, GOURAUD SHADED – EDGES ON. You can select one of the available shading options from the shortcut menu while 3DORBIT is active. See Chapter 16 for a detailed explanation of various shading options.

 Note: Shading is applied even after exiting the 3DORBIT command. Use SHADEMODE to change the shading mode when 3DORBIT is not active.

Adjusting Clipping Planes in the 3D Orbit View

AutoCAD allows you to set clipping planes for the objects in *3D* orbit view. Objects or parts of the objects that are beyond a clipping plane cannot be seen in the view. Open the Adjust Clipping Planes window by selecting ADJUST CLIPPING PLANES from the MORE flyout of the shortcut menu while 3DORBIT is active, from the *3D* Orbit toolbar, or type 3DCLIP at the Command prompt. In the Adjust Clipping Planes window, there are two clipping planes, front and back. The front and back clipping planes are represented as lines at the top and bottom of the Adjust Clipping Planes window. You can adjust the lines with your pointing device and you can use toolbar buttons or the options from the shortcut menu to choose the clipping plane that you want to adjust.

If clipping planes are ON when you exit the *3D* orbit view, they remain on in the *2D* and *3D* view. The only way to turn off the clipping planes is to remove the check mark from the FRONT CLIPPING ON or BACK CLIPPING ON in the shortcut menu while 3DORBIT is active.

Using Continuous Orbit in the 3D Orbit View

AutoCAD allows you to display the model in a continuous motion. To start the continuous motion, first select the CONTINUOUS ORBIT option from the shortcut menu when 3DORBIT is active. Then click and drag in the direction that you want the continuous orbit to move. Then release the pick button. The orbit continues to move in the direction that you indicated with your pointing device. The speed of the orbit rotation is determined by the speed with which you move the pointing device. If necessary, you can change the direction by clicking and dragging in a new direction. To stop continuous orbit, choose PAN, ZOOM, ORBIT, or ADJUST CLIPPING PLANES from the shortcut menu.

Resetting to Preset Views in the 3D Orbit View

AutoCAD allows you to reset the view to the view that was current when your first entered the *3D* orbit view or set to one of the preset views, such as top, front, side, isometric, and so on. You can choose RESET VIEW or one of the preset views from the shortcut view when 3DORBIT is active.

To exit the 3DORBIT command, press ESC or ENTER.

WORKING WITH MULTIPLE VIEWPORTS IN 3D

As mentioned in Chapter 3, AutoCAD allows you to set multiple viewports to provide different views of your model. For example, you might set up viewports that display top, front, right side, and isometric views. You can do so with the help of the VPOINT, DVIEW, or 3DORBIT commands. To facilitate editing objects in different views, you can define a different UCS for each view. Each time you make a viewport current, you can begin drawing using the same UCS you used the last time that viewport was current. The UCSVP system variable controls the setting for saving the UCS in the current viewport. When UCSVP is set to 1 (default setting) in a viewport, the UCS last used in that viewport is saved with the viewport and is restored when the viewport is made current again. When UCSVP is set to 0 in a viewport, its UCS is always the same as the UCS in the current viewport.

For example, you might set up four viewports: top view, front view, right side view, and isometric view. If you set the UCSVP system variable to 0 in the isometric viewport and to 1 in the top view, front view, and right side view, when you make the front viewport current, the isometric viewport's UCS reflects the UCS front viewport. Likewise, making the top viewport current switches the isometric viewport's UCS to match that of the top viewport.

CREATING 3D OBJECTS

As mentioned earlier, there are several advantages to drawing objects in *3D*, including viewing the model at any angle, automatic generation of standard and auxiliary *2D* views, rendering and hidden-line removal, interference checking, and engineering analysis.

AutoCAD supports three types of *3D* modeling: wireframe, surface, and solid.

The wireframe model consists of only points, lines, and curves that describe the edges of the object. In AutoCAD you can create a wireframe model by positioning *2D* (planar) objects anywhere in *3D* space. In addition, AutoCAD provides additional commands, such as 3DPOLY, for creating a wireframe model.

The surface model is more sophisticated than the wireframe model. It defines not only the edges of a *3D* object but also its surfaces. The AutoCAD surface modeler defines faceted surfaces by using a polygonal mesh. It is possible to create a mesh on a flat or curved surface by locating the boundaries or edges of the surface.

Solid modeling is the easiest type of *3D* modeling. Solids are the unambiguous and informationally complete representation of the shape of a physical object. Fundamentally, solid modeling differs from wireframe or surface modeling in two ways:

- The information is more complete in the solid model.
- The method of construction of the model itself is inherently straightforward.

In wireframe or surface modeling, objects are created by positioning lines or surfaces in *3D* space. In solid modeling, you build the model as you would with building blocks; from beginning to end, you think, draw, and communicate in *3D*. One of the main benefits of solid modeling is its ability to be analyzed. You can calculate the mass properties of a solid object, such as its mass, center of gravity, surface area, and moments of inertia.

Each modeling type uses a different method for constructing *3D* models, and the use of each editing method varies among model types. It is recommended that you not mix modeling methods. It is possible, in AutoCAD, to convert between model types, from solids to surfaces and from surfaces to wireframe; however, you cannot convert from wireframe to surfaces or surfaces to solids.

2D DRAWING COMMANDS IN 3D SPACE

You can use most of the drawing commands discussed in previous chapters with a *Z* coordinate value. But *2D* objects such as polylines, circles, arcs, and solids are constrained on the *XY* plane of the current UCS. For these objects, the *Z* value is accepted only for the third coordinate of a point to set the elevation of the *2D* object above or below the current plane. When you specify a point by using an Object Snap mode, it assumes the *Z* value of the point to which you snapped.

SETTING ELEVATION AND THICKNESS

You can create new objects by first setting up a default elevation (*Z* value). Subsequently, all the objects drawn assume the current elevation as the *Z* value whenever a *3D* point is expected but you supply only the *X* and *Y* values. The current elevation is maintained separately in model space and paper space.

Similarly, you create new objects with extrusion thickness by presetting a value for the thickness. Subsequently, all the objects drawn, such as lines, polylines, arcs, circles, and solids, assume the current thickness and extrude in their *Z* direction. For example, you can draw a cylinder by drawing a circle with preset thickness, or you can draw a cube simply by drawing a square with preset thickness.

Note: Thickness can be positive or negative. Thickness is in the Z axis direction for 2D objects. For 3D objects that can accept thickness, it is always relative to the current UCS. They will appear oblique if they do not lie in or parallel the current UCS. If thickness is added to a line drawn directly in the Z direction, the line appears to extend beyond its endpoint in the positive or negative thickness direction. Text and dimensions ignore the thickness setting.

To set the default elevation and thickness, invoke the ELEV command at the On-Screen prompt.

AutoCAD prompts:

    Specify new default elevation <current>: *(specify the elevation, or press*
        ENTER *to accept the current elevation setting)*
    Specify new default thickness <current>: *(specify the thickness, or press*
        ENTER *to accept the current thickness setting)*

For example, the following are the command sequences to draw a six-sided polygon at zero elevation with a radius of 2.5 units and a height of 4.5 units, and to place a cylinder at the center of the polygon with a radius of 1.0 unit at an elevation of 2.0 units with a height of 7.5 units, (see Figure 15–36).

elev
New current elevation <0.0000>: (ENTER)
New current thickness <0.0000>: **4.5** (ENTER)

polygon (ENTER)
Enter number of sides <default>: **5** (ENTER)
Specify center of polygon or ⬇ (specify a point)
Enter an option (choose INSCRIBED IN CIRCLE)

elev
New current elevation <0.0000>: **2.0**
New current thickness <0.0000>: **7.5**

circle (ENTER)
Specify center point of circle or ⬇ (specify a point)
Specify radius of circle or ⬇ 1.0 (ENTER)

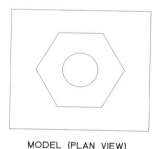
MODEL (PLAN VIEW)

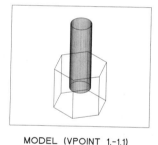
MODEL (VPOINT 1,-1,1)

Figure 15–36 *Specifying a new current elevation and thickness*

You can change the thickness and elevation (Z coordinate) of the existing objects by invoking the PROPERTIES command.

CREATING A REGION OBJECT

Regions are *2D* areas you create from closed shapes or loops. The REGION command creates a region object from a selection set of objects. Closed polylines, lines, curves, circular arcs, circles, elliptical arcs, ellipses, and splines are valid selections. Once you create a region, you can extrude it with the EXTRUDE command to make a *3D* solid. You can also create a composite region with the UNION, SUBTRACTION, and INTERSECTION commands. If necessary, you can hatch a region with the BHATCH command.

AutoCAD converts closed *2D* and planar *3D* polylines in the selection set to separate regions and then converts polylines, lines, and curves that form closed planar loops.

If more than two curves share an endpoint, the resultant region might be arbitrary. Each object retains its layer, linetype, and color. AutoCAD deletes the original objects after converting them to regions and, by default, does not hatch the regions.

Invoke the REGION command from the Draw toolbar (see Figure 15–37).

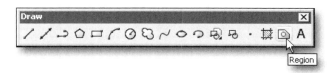

Figure 15–37 *Invoking the* REGION *command from the Draw toolbar*

AutoCAD prompts:

> Select objects: *(select the objects, and press* ENTER *to complete object selection)*

DRAWING 3D POLYLINES

The 3DPOLY command draws polylines with independent *X*, *Y*, and *Z* axis coordinates using the continuous linetype. The 3DPOLY command works similar to the PLINE command, with a few exceptions. Unlike the PLINE command, 3DPOLY draws only straight-line segments without variable width. Editing a *3D* polyline with the PEDIT command is similar to editing a *2D* polyline, except for some options. *3D* polylines cannot be joined, curve-fit with arc segments, or given a width or tangent.

Invoke the 3DPOLY command from the Draw menu:

AutoCAD prompts:

> Specify start point of polyline: *(specify a point)*
> Specify endpoint of line or ⬇: *(specify a point or select an option from the shortcut menu)*
> Specify endpoint of line or ⬇: *(specify a point or select an option)*
> Specify endpoint of line or ⬇: *(specify a point or select one of the options)*

The available options are similar to those for the PLINE command, described in Chapter 4.

CREATING 3D FACES

When you create a *3D* model, it is often necessary to have solid surfaces for hiding and shading. These surfaces are created with the 3DFACE command. The 3DFACE command creates a solid surface, and the command sequence is similar to that for the SOLID command. Unlike the SOLID command, you can give differing *Z* coordinates for the corner points of a face, forming a section of a plane in space. Unlike the SOLID command, a 3DFACE is drawn from corner to corner clockwise or counterclockwise around the object (and it does not draw a "bow tie"). A *3D* face is a plane defined by either three or four points used to represent a surface. It provides a means of controlling which edges of a *3D* face will be visible. You can describe complex, *3D* polygons using multiple *3D* faces, and you can tell AutoCAD which edges you want to be drawn. If you have an object with curved surfaces, then the 3DFACE command is not suitable. One of the mesh commands is more appropriate, as explained later in the chapter.

Invoke the 3DFACE command from the Surfaces toolbar (see Figure 15–38).

Figure 15–38 *Invoking the* 3DFACE *command from the Surfaces toolbar*

AutoCAD prompts:

    Specify first point or ⊥: *(specify the first point)*

Specify the first point, and AutoCAD prompts you for the second, third, and fourth points in sequence. Then AutoCAD closes the face from the fourth point to the first point and prompts for the third point. If you press ENTER in response to the prompt for the third point, AutoCAD closes the *3D* face with four sides, terminates the command, and returns to the On-Screen prompt.

If you want to draw additional faces in one command sequence, the last two points of the first face become the first two points for the second face. And the last two points of the second face become the first two points of the third face, and so on. You have to be very careful in drawing several faces in one command sequence, since AutoCAD does not have an *UNDO* option that works inside the 3DFACE command. A single mistake can cause the entire face to be redrawn. For this reason, it is a good idea to draw *3D* faces one at a time.

For example, the following command sequence demonstrates the placement of *3D* faces(see Figure 15–39).

    3dface
    Specify first point or ⊥: *(select point A1)*
    Specify second point or ⊥: *(select point A2)*
    Specify third point or ⊥: *(select point A3)*
    Specify fourth point or ⊥: *(select point A4)*
    Specify third point or ⊥: *(select point A5)*
    Specify fourth point or ⊥: *(select point A6)*
    Specify third point or ⊥: *(select point A7)*
    Specify fourth point or ⊥: *(select point A8)*

Specify third point or ⬇: *(select point A1)*
Specify fourth point or ⬇: *(select point A2)*
Specify third point or ⬇: (ENTER)

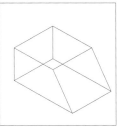
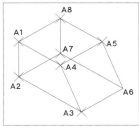
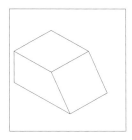

MODEL   MODEL WITH 3D FACES   MODEL–HIDDEN LINES REMOVED

Figure 15–39 *Drawing a 3D object with the* 3DFACE *command*

The surface created in Figure 15–39 required four faces. Some of the edges are overlapping, which is not acceptable when viewing the object. The 3DFACE command allows face edges to be "invisible." To create an invisible edge, the letter **i** must be entered at the prompt for the first point of the edge to be invisible, and then the point can be entered.

The following command sequence shows the placement of *3D* faces for invisible edges (see Figure 15–39).

3dface
Specify first point or ⬇: *(select point A1)*
Specify second point or ⬇: *(select point A8)*
Specify third point or ⬇: *(select point A5)*
Specify fourth point or ⬇: *(select point A4)*
Specify third point or ⬇: *(select point A3)*
Specify fourth point or ⬇: *(select point A6)*
Specify third point or ⬇: *(select point A7)*
Specify fourth point or ⬇: **i** (ENTER)
Specify fourth point or ⬇: *(select point A8)*
Specify third point or ⬇: *(select point A1)*
Specify fourth point or ⬇: *(select point A2)*
Specify third point or ⬇: *(select point A3)*
Specify fourth point or ⬇: *(select point A4)*
Specify third point or ⬇: (ENTER)

The 3DFACE command ignores thickness. The SPLFRAME system variable controls the display of invisible edges in *3D* faces. If SPLFRAME is set to a nonzero value, all invisible edges of *3D* faces are displayed.

Controlling the Visibility of a 3D Face

The EDGE command allows you to change the visibility of *3D* face edges. You can set the edges to ON/OFF.

Invoke the EDGE command from the Surfaces toolbar (see Figure 15–40).

Figure 15–40 *Invoking the* EDGE *command from the Surfaces toolbar*

AutoCAD prompts:

> Specify edge of 3dface to toggle visibility or ⬇: *(select the edges and press* ENTER *to complete selection, or select the* DISPLAY *option from the shortcut menu)*

The *DISPLAY* option highlights invisible edges of *3D* faces so you can change the visibility of the edges. AutoCAD prompts:

> Enter selection method for display of hidden edges

The default option displays all the invisible edges. Once the edges are displayed, AutoCAD allows you to change the status of the visibility.

The *SELECT* option allows you to selectively identify hidden edges to be displayed. Then, if necessary, you can change the status of the visibility.

 Note: All edges are visible regardless of the visibility setting if the SPLFRAME system variable is set to 1 (ON). If necessary, you can also use the PROPERTIES command to modify 3D faces.

CREATING MESHES

A *3D* mesh is a single object. It defines a flat surface or approximates a curved one by placing multiple *3D* faces on the surface of an object. It is a series of lines consisting of columns and rows. AutoCAD lets you determine the spacing between rows (M) and columns (N).

It is possible to create a mesh for a flat or curved surface by locating the boundaries or edges of the surface. Surfaces created in this fashion are called geometry-generated surfaces. Their size and shape depend on the boundaries used to define them and on the specific formula (or command) used to determine the location of the vertices between the boundaries. AutoCAD provides four different commands to create geometry-generated surfaces: RULESURF, REVSURF, TABSURF, and EDGESURF. The differences between these types of meshes depend on the types of objects connecting the surfaces. In addition, AutoCAD provides two additional commands for creating polygon mesh: 3DMESH and PFACE. The key to using meshes effectively is to understand the purpose and requirement of each type of mesh and to select the appropriate one for the given condition.

CREATING A FREE-FORM POLYGON MESH

You can define a free-form *3D* polygon mesh by means of the 3DMESH command. Initially, it prompts you for the number of rows and columns, in terms of mesh M and mesh N, respectively. Then it prompts for the location of each vertex in the mesh. The product of M x N gives the number of vertices for the mesh.

Invoke the 3DMESH command from the Surfaces toolbar (see Figure 15–41).

Figure 15–41 *Invoking the* 3DMESH *command from the Surfaces toolbar*

AutoCAD prompts:

    Enter size of mesh in M direction: *(specify an integer value between 2 and 256)*
    Enter size of mesh in N direction: *(specify an integer value between 2 and 256)*

The points for each vertex must be entered separately, and the M value can be considered the number of lines that will be connected by faces, while the N value is the number of points each line consists of. Vertices may be specified as *2D* or *3D* points, and may be any distance from each other.

The following command sequence creates a simple 5 × 4 polygon mesh. The mesh is created between the first point of the first line, the first point of the second line, and so on (see Figure 15–42).

    3dmesh
    Enter size of mesh in M direction: **5**
    Enter size of mesh in N direction: **4**
    Specify location for vertex (0,0): *(select point A1)*
    Specify location for vertex (0,1): *(select point A2)*
    Specify location for vertex (0,2): *(select point A3)*
    Specify location for vertex (0,3): *(select point A4)*
    Specify location for vertex (1,0): *(select point B1)*
    Specify location for vertex (1,1): *(select point B2)*
    Specify location for vertex (1,2): *(select point B3)*

Specify location for vertex (1,3): *(select point B4)*
Specify location for vertex (2,0): *(select point C1)*
Specify location for vertex (2,1): *(select point C2)*
Specify location for vertex (2,2): *(select point C3)*
Specify location for vertex (2,3): *(select point C4)*
Specify location for vertex (3,0): *(select point D1)*
Specify location for vertex (3,1): *(select point D2)*
Specify location for vertex (3,2): *(select point D3)*
Specify location for vertex (3,3): *(select point D4)*
Specify location for vertex (4,0): *(select point E1)*
Specify location for vertex (4,1): *(select point E2)*
Specify location for vertex (4,2): *(select point E3)*
Specify location for vertex (4,3): *(select point E4)*

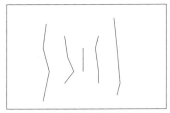

LINE DIAGRAM

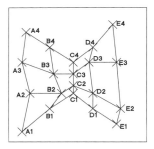 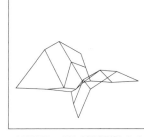

LINE DIAGRAM WITH 3D FACES   MODEL AT VPOINT 1,-1,1

Figure 15–42 *Creating a 3D mesh*

 Note: Specifying a 3D mesh of any size can be time-consuming and tedious. It is preferable to use one of the commands for geometry-generated surfaces, such as RULESURF, REVSURF, TABSURF, or EDGESURF. The 3DMESH command is designed primarily for AutoLISP and ADS applications.

CREATING A 3D POLYFACE MESH

The PFACE command allows you to construct a mesh of any topology you desire. This command is similar to the 3DFACE command, but it creates surfaces with invisible interior divisions. You can specify any number of vertices and *3D* faces, unlike the other meshes. Producing this kind of mesh lets you conveniently avoid creating many unrelated *3D* faces with the same vertices.

AutoCAD first prompts you to pick all the vertex points, and then you can create the faces by entering the vertex numbers that define their edges.

Invoke the PFACE command from the On-Screen prompt. AutoCAD prompts:

> **pface**
> Specify location for vertex 1: *(specify a point)*

One by one, specify all the vertices used in the mesh, keeping track of the vertex numbers shown in the prompts. You can specify the vertices as *2D* or *3D* points and place them at any distance from one another. Press ENTER after specifying all the vertices, and AutoCAD prompts for a vertex number that has to be assigned to each face. You can define any number of vertices for each face, and press ENTER). AutoCAD prompts for the next face. After all the vertex numbers for all the faces are defined, press ENTER, and AutoCAD draws the mesh.

The following command sequence creates a simple polyface for a given six-sided polygon, with a circle of 1" radius drawn at the center of the polygon at a depth of −2 (see Figure 15–43).

> **pface**
> Specify location for vertex 1: *(select point A1)*
> Specify location for vertex 2 or <define faces>: *(select point A2)*
> Specify location for vertex 3 or <define faces>: *(select point A3)*
> Specify location for vertex 4 or <define faces>: *(select point A4)*
> Specify location for vertex 5 or <define faces>: *(select point A5)*
> Specify location for vertex 6 or <define faces>: *(select point A6)*
> Specify location for vertex 7 or <define faces>: (ENTER)
> Face 1, Vertex 1:
> Enter a vertex number or ⏎ **1**
> Face 1, Vertex 2:
> Enter a vertex number or ⏎ **2**
> Face 1, Vertex 3:
> Enter a vertex number or ⏎ **1**
> Face 1, Vertex 4:
> Enter a vertex number or ⏎ **2**
> Face 1, Vertex 5:
> Enter a vertex number or ⏎ **1**
> Face 1, Vertex 6:
> Enter a vertex number or ⏎ **2**
> Face 1, Vertex 7:
> Enter a vertex number or ⏎ (ENTER)
> Face 2, Vertex 1:
> Enter a vertex number or ⏎ (ENTER)

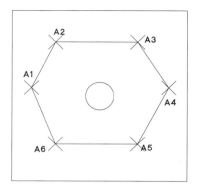 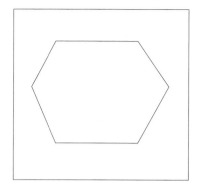

POLYGON WITH PFACE   POLYGON WITH CIRCLE HIDDEN

Figure 15–43 *Creating a polyface for a given six-sided polygon with a circle at the center*

If necessary, you can make an edge of the polyface mesh invisible by entering a negative number for the beginning vertex of the edge. By default, the faces are drawn on the current layer and with the current color. However, you can create the faces in layers and colors different from the original object. You can assign a layer or color by responding to the "Enter a vertex number or ↵" prompt with **L** for layer or **C** for color. AutoCAD then prompts for the name of the layer or color, as appropriate. It will continue with the prompts for vertex numbers. The layer or color you enter is used for the face you are currently defining and for any subsequent faces created.

 Note: Specifying the layer or color within the PFACE command does not change object properties for subsequent commands. Specifying PFACE of any size can be time-consuming and tedious. It is preferable to use one of the commands for geometry-generated surfaces, such as RULESURF, REVSURF, TABSURF, or EDGESURF. The PFACE command is intended primarily for AutoLISP and ADS applications.

CREATING A RULED SURFACE BETWEEN TWO OBJECTS

The RULESURF command creates a polygon mesh between two objects. The two objects can be lines, points, arcs, circles, *2D* polylines, or *3D* polylines. If one object is open, such as a line or an arc, the other must be open too. If one is closed, such as a circle, so must the other be. A point can be used as one object, regardless of whether the other object is open or closed. But only one of the objects can be a point.

RULESURF creates an M × N mesh, with the value of mesh M a constant 2. The value of mesh N can be changed depending on the required number of faces. This can be done with the help of the SURFTAB1 system variable. By default, SURFTAB1 is set to 6.

The following command sequence shows how to change the value of SURFTAB1 from 6 to 20:

surftab1
Enter new value for SURFTAB1 <6>: **20**

Invoke the RULESURF command from the Surfaces toolbar (see Figure 15–44).

Figure 15–44 *Invoking the* RULESURF *command from the Surfaces toolbar*

AutoCAD prompts:

rulesurf
Select first defining curve: *(select the first defining curve)*
Select second defining curve: *(select the second defining curve)*

Identify the two objects to which a mesh has to be created. See Figure 15–45, in which an arc (A1–A2) and a line (A3–A4) were selected and a mesh was created with SURFTAB1 set to 15. Two lines (B1–B2 and B3–B4) were selected and a mesh was created with SURFTAB1 set to 20. A cone was created by drawing a circle at an elevation of 0 and a point (C1) at an elevation of 5, followed by the application of the RULESURF command with SURFTAB1 set to 20.

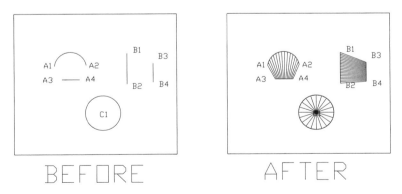

Figure 15–45 *Creating ruled surfaces with the* RULESURF *command*

Note: When you identify the two objects, make sure to select on the same side of the objects, left or right. If you pick the left side of one of the sides and the right side of the other, you will get a bow-tie effect.

CREATING A TABULATED SURFACE

The TABSURF command creates a surface extrusion from an object with a length and direction determined by the direction vector. The object is called the defining curve and can be a line, arc, circle, *2D* polyline, or *3D* polyline. The direction vector can be a line or open polyline. The endpoint of the direction vector nearest the specified point will be swept along the path curve, describing the surface. Once the mesh is created, the direction vector can be deleted. The number of intervals along the path curve is controlled by the SURFTAB1 system variable, similar to the RULESURF command. By default, SURFTAB1 is set to 6.

Invoke the TABSURF command from the Surfaces toolbar (see Figure 15–46).

Figure 15–46 *Invoking the* TABSURF *command from the Surfaces toolbar*

AutoCAD prompts:

> **tabsurf**
> Select object for path curve: *(select the path curve)*
> Select object for direction vector: *(select the direction vector)*

The location at which the direction vector is selected determines the direction of the constructed mesh. The mesh is created in the direction from the selection point to the nearest endpoint of the direction vector. In Figure 15–47, a mesh was created with SURFTAB1 set to 16 by identifying a polyline as the path curve and the line as the direction vector.

 Note: The length of the 3D mesh is the same as that of the direction vector.

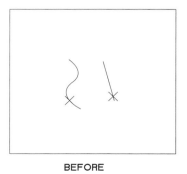
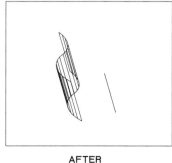

BEFORE　　　　　　　　　　AFTER

Figure 15–47 *Creating a tabulated surface with the* TABSURF *command*

CREATING A REVOLVED SURFACE

The REVSURF command creates a *3D* mesh that follows the path defined by a path curve and is rotated around a centerline. The object used to define the path curve may be an arc, circle, line, *2D* polyline, or *3D* polyline. Complex shapes consisting of lines, arcs, or polylines can be joined into one object using the PEDIT command, and then you can create a single rotated mesh instead of several individual meshes.

The centerline can be a line or polyline that defines the axis around which the faces are constructed. The centerline can be of any length and at any orientation. If necessary, you can erase the centerline after the construction of the mesh. Thus it is recommended that you make the axis longer than the path curve so that it is easy to erase after the rotation.

In the case of REVSURF, both the mesh M size as well as mesh N are controlled by the SURFTAB1 and SURFTAB2 system variables, respectively. The SURFTAB1 value determines how many faces are placed around the rotation axis and can be an integer value between 3 and 1024. The SURFTAB2 determines how many faces are used to simulate the curves created by arcs or circles in the path curve. By default, SURFTAB1 and SURFTAB2 are set to 6.

The following command sequence shows how to change the value of SURFTAB1 from 6 to 20 and that of SURFTAB2 from 6 to 15:

> **surftab1**
> Enter new value for SURFTAB1 <6>: **20**
>
> **surftab2**
> Enter new value for SURFTAB1 <6>: **15**

Invoke the REVSURF command from the Surfaces toolbar (see Figure 15–48).

Figure 15–48 *Invoking the* REVSURF *command from the Surfaces toolbar*

AutoCAD prompts:

> **revsurf**
> Select object to revolve: *(select the path curve)*
> Select object that defines the axis of revolution: *(select the axis of revolution)*
> Specify Start angle<0>: *(specify the start angle, or press* ENTER *to accept the default angle)*
> Specify Included angle (+=ccw,-=cw)<360>: *(specify the included angle, or press* ENTER *to accept the default)*

For the "Specify Start angle:" prompt, it does not matter if you are going to rotate the curve 360 degrees (full circle). If you want to rotate the curve only a certain angle, you must provide the start angle in reference to three o'clock (the default) and then indicate the angle of rotation in the counterclockwise (positive) or clockwise (negative) direction. See Figure 15–49, in which a mesh was created with SURFTAB1 set to 16 and SURFTAB2 set to 12, by identifying a closed polyline as the path curve and the vertical line as the axis of revolution, and then rotated 360 degrees.

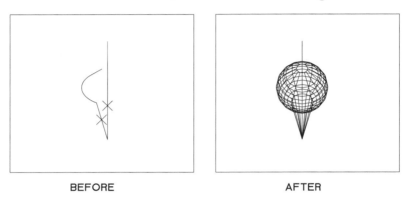

Figure 15–49 *Creating a meshed surface with the* REVSURF *command*

CREATING AN EDGE SURFACE WITH FOUR ADJOINING SIDES

The EDGESURF command allows a mesh to be created with four adjoining sides defining its boundaries. The only requirement for EDGESURF is that the mesh has exactly four sides. The sides can be lines, arcs, or any combination of polylines and polyarcs. Each side must join the adjacent one to create a closed boundary.

In EDGESURF, both the mesh M size and mesh N can be controlled by the SURFTAB1 and SURFTAB2 system variables, respectively, just as in REVSURF.

Invoke the EDGESURF command from the Surfaces toolbar (see Figure 15–50).

Figure 15–50 *Invoking the* EDGESURF *command from the Surfaces toolbar*

    edgesurf
    Select edge 1 for surface edge: *(select the first edge)*
    Select edge 2 for surface edge: *(select the second edge)*

Select edge 3 for surface edge: *(select the third edge)*
Select edge 4 for surface edge: *(select the fourth edge)*

When picking four sides, you must be consistent in picking the beginning of each polyline group. If you pick the beginning of one side and the end of another, the final mesh will cross and look strange. See Figure 15–51, in which a mesh was created with SURFTAB1 set to 25 and SURFTAB2 set to 20 by identifying four sides.

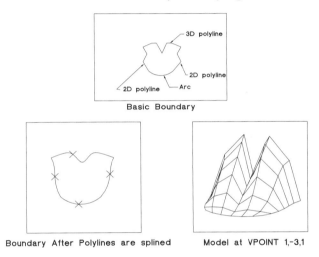

Figure 15–51 *Creating a meshed surface with the* EDGESURF *command*

EDITING POLYMESH SURFACES

As with blocks, polylines, hatch, and dimensioning, you can explode a mesh. When you explode a mesh it separates into individual *3D* faces. Meshes can also be altered by invoking the PEDIT command, similar to editing polylines using the PEDIT command. Most of the options under the PEDIT command can be applied to meshes, except giving width to the edges of the polymesh. For a detailed explanation of the PEDIT command, refer to Chapter 5.

EDITING IN 3D

This section describes how to perform various *3D* editing operations, such as aligning, rotating, mirroring, arraying, extending, and trimming.

ALIGNING OBJECTS

The ALIGN command allows you to translate and rotate objects in *3D* space regardless of the position of the current UCS. Three source points and three destination points define the move. The ALIGN command lets you select the objects to move, and then subsequently prompts for three source points and three destination points.

Invoke the ALIGN command by selecting Align from the 3D Operation submenu in the Modify menu and AutoCAD prompts:

align
Select objects: *(select the objects)*

Select the objects to move and press ENTER. AutoCAD then prompts for three source points and three destination points. Temporary lines are displayed between the matching pairs of source and destination points. If you enter all six points, the move consists of a translation and two rotations based on the six points. The translation moves the 1st source point to the 1st destination point. The first rotation aligns the line defined by the 1st and 2nd source points with the line defined by the 1st and 2nd destination points. The second rotation aligns the plane defined by the three source points with the plane defined by the three destination points.

If instead of entering three pairs of points you enter two pairs of points, the transformation reduces to a translation from the 1st source point to the 1st destination point and a rotation such that the line passing through the two source points aligns with the line passing through the two destination points. The transformation occurs in either *2D* or *3D*, depending on your response to the following prompt:

<2d> or 3d transformation:

If you enter 2d or press ENTER, the rotation is performed in the *XY* plane of the current UCS. If you enter 3d, the rotation is in the plane defined by the two destination points and the 2nd source point.

If you enter only one pair of points, the transformation reduces to a simple translation from the source to the destination point. This is similar to using the AutoCAD regular MOVE command without the dynamic dragging.

ROTATING OBJECTS ABOUT A 3D OBJECT

The ROTATE3D command lets you rotate an object about an arbitrary *3D* axis.

Invoke the ROTATE3D command by selecting Rotate 3D from the 3D Operation submenu in the Modify menu and AutoCAD prompts:

Select objects: *(select object and press* ENTER *when done)*

AutoCAD lists the options for selecting the axis of rotation:

Specify first point on axis or define axis by ⬇: *(specify a point or select one of the options from the shortcut menu)*

Choose *2POINTS* from the shortcut menu to define the axis of rotation. The axis of rotation is the line that passes through the two points, and the positive direction is from the first to the second point.

Choose OBJECT from the shortcut menu to select an object and then derive the axis of rotation based on the type of object selected. Valid objects include line, circle, arc, and pline.

Choose LAST from the shortcut menu to select the last-used axis. If there is no last axis, a message to that effect is displayed and the axis selection prompt is redisplayed.

Choose VIEW from the shortcut menu to base the rotation on a selected view. The axis of rotation is perpendicular to the view direction and passes through the selected point. The positive axis direction is toward the viewer.

Choose X, Y, or Z AXIS option from the shortcut menu to select a point for the axis of rotation parallel to the standard axis of the current UCS and passing through the selected point.

Once you have selected the axis of rotation, AutoCAD prompts:

> Specify by rotation angle or ↵ (specify the rotation angle, or enter r, for reference)

AutoCAD rotates the selected object(s) to the specified rotation angle. The REFERENCE option allows you to specify the current orientation as the reference angle or to show AutoCAD the angle by pointing to the two endpoints of a line to be rotated and then specifying the desired new rotation. AutoCAD automatically calculates the rotation angle and rotates the selected object appropriately.

See Figure 15–52 for an example of rotating a cylinder around the Z axis.

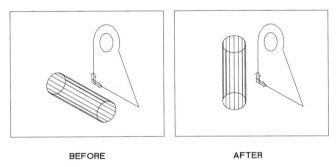

BEFORE   AFTER

Figure 15–52 *Rotating a cylinder about the Z axis with the* ROTATE3D *command*

MIRRORING ABOUT A PLANE

The MIRROR3D command lets you mirror a selected object about a plane.

Invoke the MIRROR3D command by selecting Mirror 3D from the 3D Operation submenu in the Modify menu and AutoCAD prompts:

> Select objects: *(select the objects)*

Select the objects to mirror and press ENTER. AutoCAD then lists the options for selecting the mirroring plane:

Specify first point of mirror plane (3 points) or ⬇: *(select one of the options from the shortcut menu)*
Delete source objects? <No>: *(enter **y** to delete the objects or **n** not to delete the objects)*

Choose 3POINTS from the shortcut menu to define three points. The mirroring plane is the plane that passes through the three selected points.

Choose OBJECT from the shortcut menu to align the mirroring plane with the plane of the object selected. Valid objects include: circle, arc, and pline.

Choose LAST from the shortcut menu to mirror the object based on the last used plane. If there is no previous mirror plane, the On-Screen prompt repeats the default prompt.

Choose ZAXIS from the shortcut menu to define the mirroring plane by two points. The first point is the point at which the object will rotate about and the second defines the angle of rotation on the Z plane's normal (perpendicular to the plane).

Choose VIEW from the shortcut menu to create a mirroring plane perpendicular to the view direction and passing through the selected point.

Choose XY, YZ or XZ from the shortcut menu to select a point to create the mirroring plane parallel to the standard plane of the current UCS and passing through the selected point.

See Figure 15–53 for an example of mirroring a cylinder aligned with the plane of the object selected (pline).

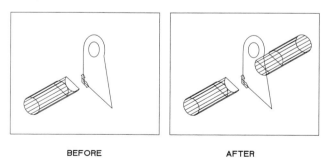

BEFORE AFTER

Figure 15–53 *Mirroring a cylinder about a polyline object*

CREATING A 3D ARRAY

The 3DARRAY command is used to make multiple copies of selected objects in either rectangular or polar array in *3D*. In the rectangular array, specify the number of columns (*X* direction), the number of rows (*Y* direction), the number of levels (*Z* direction), and the spacing between columns, rows, and levels. In the polar array, specify the number of items to array, the angle that the arrayed objects are to fill, the start point and endpoint of the axis about which the objects are to be rotated, and whether or not the objects are rotated about the center of the group.

Invoke the 3DARRAY command by selecting 3D Array from the 3D Operation submenu in the Modify menu and AutoCAD prompts:

> Select objects: *(select the objects to array and press* ENTER*)*
> Enter type of array : *(select one of the two available options)*

To generate a rectangular array, choose RECTANGULAR and AutoCAD prompts:

> Enter the number of rows(---)<1>: *(specify the number of rows or press* ENTER*)*
> Enter the number of columns(|||)<1>: *(specify the number of columns or press* ENTER*)*
> Number of levels(...)<1>: *(specify the number of levels or press* ENTER*)*
> Specify the distance between rows(---)<1>: *(specify a distance)*
> Specify the distance between columns(|||)<1>: *(specify a distance)*
> Specify the distance between levels(...)<1>: *(specify a distance)*

Any combination of whole numbers of columns, rows, and levels may be entered. AutoCAD includes the original object in the number you enter. An array must have at least two columns, two rows, or two levels. Specifying one row requires that more than one column be specified, or vice versa. Specifying one level creates a *2D* array. Column, row, and level spacing can be different from one another. They can be entered separately when prompted, or you can select two points and let AutoCAD measure the spacing. Positive values for spacing generate the array along the positive *X*, *Y*, and *Z* axes. Negative values generate the array along the negative *X*, *Y*, and *Z* axes.

To generate a polar array, choose *POLAR* and AutoCAD prompts:

> Enter the number of items in the array: *(specify the number of items in the array; include the original object)*
> Specify the angle to fill (+=ccw, -=cw) <360>: *(specify an angle, or press* ENTER *for 360 degrees)*
> Rotate arrayed objects? *(enter* **y** *to rotate the objects as they are copied, or enter* **n** *not to rotate the objects as they are copied)*
> Specify center point of array: *(specify a point)*
> Specify second point on axis of rotation: *(specify a point for the axis of rotation)*

EXTENDING AND TRIMMING IN 3D

AutoCAD allows you to extend an object by means of the EXTEND command (explained in Chapter 4) to any object in *3D* space or to trim an object to any other *3D* space by means of the TRIM command (explained in Chapter 4), regardless of whether the objects are on the same plane or parallel to the cutting or boundary edges. Before you select an object to extend or trim in *3D* space, specify one of the three available projection modes: NONE, UCS, or VIEW. The NONE option specifies no projection. AutoCAD extends/trims only objects that intersect with the boundary/cutting edge in *3D* space. The UCS option specifies projection onto the *XY* plane of the current UCS. AutoCAD extends/trims objects that do not intersect with the boundary/cutting objects in *3D* space. The VIEW option specifies projection along the current view direction. The PROJMODE system variable allows you to set one of the available projection modes. You can also set the projection mode by selecting the PROJECT option available in the EXTEND command.

In addition to specifying the projection mode, you have to specify one of the two available options for the edge. The edge determines whether the object is extended/trimmed to another object's implied edge or only to an object that actually intersects it in *3D* space. The available options are EXTEND and NO EXTEND. The EXTEND option extends the boundary/cutting object/edge along its natural path to intersect another object or its implied edge in *3D* space. The NO EXTEND option specifies that the object is extended/trimmed only to a boundary/cutting object/edge that actually intersects it in *3D* space. The EXTEDGE system variable allows you to set one of the available modes. You can also set the edge by selecting the EDGE option available in the EXTEND command.

CREATING SOLID SHAPES

As mentioned earlier, solids are the most informationally complete and least ambiguous of the modeling types. It is easier to edit a complex solid shape than to edit wireframes and meshes.

You create solids from one of the basic solid shapes: box, cone, cylinder, sphere, torus, or wedge. The user-defined solids can be created by extruding or revolving *2D* objects and regions to define a *3D* solid. In addition, you can create more complex solid shapes by combining solids together by performing a Boolean operation—UNION, SUBTRACTION, or INTERSECTION.

Solids can be further modified by filleting and chamfering their edges. AutoCAD provides commands for slicing a solid into two pieces or obtaining a *2D* cross-section of a solid.

Like meshes, solids are displayed as a wireframe until you hide, shade, or render them. AutoCAD provides commands to analyze solids for their mass properties (volume, moments of inertia, center of gravity, etc.). AutoCAD allows you to export data about a solid object to applications such as NC (numerical control) milling and EXTEDGE FEM (finite element method) analysis. If necessary, you can use the AutoCAD EXPLODE command to explode solids into mesh and wireframe objects.

 Note: The ISOLINES system variable controls the number of tessellation lines used to visualize curved portions of the wireframe. The default value for ISOLINES is set to 4.

CREATING A SOLID BOX

The BOX command creates a solid box or cube. The base of the box is defined parallel to the current UCS by default. The solid box can be drawn by one of two options: by providing a center point or a starting corner of the box.

Invoke the BOX command from the Solids toolbar (see Figure 15–54).

Figure 15–54 *Invoking the* BOX *command from the Solids toolbar*

AutoCAD prompts:

> Specify corner of box or ⬇: *(specify a point, or choose* CENTER *from the shortcut menu)*

First, by default, you are prompted for the starting corner of the box. Once you provide the starting corner, the box's dimensions can be entered in one of three ways.

The default option lets you create a box by locating the opposite corner of its base rectangle first, and then its height. The following command sequence defines a box (see Figure 15–55), using the default option:

> **ox**
> Specify corner of box or ⬇: **3,3**
> Specify corner or ⬇: **7,7**
> Specify height: **4**

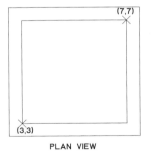
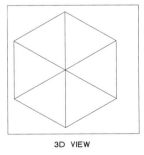

Figure 15–55 *Creating a solid box using the default option of the* BOX *command*

The CUBE option allows you to create a box in which all edges are of equal length. The following command sequence defines a box using the CUBE option:

box
Specify corner of box or ⤓: **3,3**
Specify corner or ⤓: *(choose CUBE from the shortcut menu)*
Specify length: **3**

The LENGTH option lets you create a box by defining its length, width, and height. The following command sequence defines a box using the LENGTH option:

box
Specify corner of box or ⤓: **3,3**
Specify corner or ⤓: *(choose LENGTH from the shortcut menu)*
Specify length: **3**
Specify width: **4**
Specify height: **3**

The CENTER option allows you to create a box by locating its center point. Once you locate the center point, a line rubberbands from this point to help you visualize the size of the rectangle. Then AutoCAD prompts you to define the size of the box by entering one of the following options:

Specify corner or ⤓: *(select one of the options from the shortcut menu)*

 Note: Once you create a box you cannot stretch it or change its size. However, you can extrude the faces of a box with the SOLIDEDIT command.

CREATING A SOLID CONE

The CONE command creates a cone, either round or elliptical. By default, the base of the cone is parallel to the current UCS. Solid cones are symmetrical and come to a point along the Z axis. The solid cone can be drawn two ways: by providing a center point for a circular base or by selecting the elliptical option to draw the base of the cone as an elliptical shape.

Invoke the CONE command from the Solids toolbar (see Figure 15–56).

Figure 15–56 *Invoking the CONE command from the Solids toolbar*

AutoCAD prompts:

Specify center point for base of cone or ⤓: *(specify a point, or choose ELLIPTICAL from the shortcut menu)*

By default, AutoCAD prompts you for the center point of the base of the cone and assumes the base to be a circle. Subsequently, you are prompted for the radius (or enter D, for diameter). Enter the appropriate value and then it prompts for the apex/height of the cone. The height of the cone is the default option, and it allows you to set the height of the cone, not the orientation. The base of the cone is parallel to the current base plane. The Apex option, in contrast, prompts you for a point. In turn, it sets the height and orientation of the cone. For example, the following command sequence lists the steps in drawing a cone (see Figure 15–57), using the default option:

cone
Specify center point for base of cone or ⬇: **5,5**
Specify radius base of cone or ⬇: **3**
Specify height of cone or ⬇: **4**

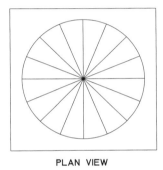 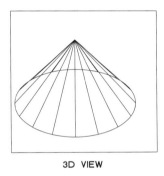

PLAN VIEW   3D VIEW

Figure 15–57 *Creating a solid cone using the default option of the* CONE *command*

Selecting the ELLIPTICAL option indicates that the base of the cone is an ellipse. The prompts are identical to the regular AutoCAD ELLIPSE command. For example, the following command shows steps in drawing a cone using the ELLIPTICAL option:

cone
Specify center point for base of cone or ⬇: *(choose* ELLIPTICAL *from the shortcut menu)*
Specify axis endpoint of ellipse for base of cone or ⬇: **3,3**
Specify second axis endpoint of ellipse for base of cone: **6,6**
Specify length of other axis for base of cone: **5,7**
Specify height of cone or ⬇: **4**

CREATING A SOLID CYLINDER

The CYLINDER command creates a cylinder of equal diameter on each end and similar to an extruded circle or an ellipse. The solid cylinder can be by means of one of two options: by providing a center point for a circular base, or by selecting the elliptical option to draw the base of the cylinder as an elliptical shape.

Invoke the CYLINDER command from the Solids toolbar (see Figure 15–58).

Figure 15–58 *Invoking the* CYLINDER *command from the Solids toolbar*

AutoCAD prompts:

> Specify center point for base of cylinder or ⤓: *(specify a point, or choose* ELLIPTICAL *from the shortcut menu)*

The prompts are identical to those for a cone. For example, the following command sequence lists the steps in drawing a cylinder(see Figure 15–59), using the default option:

cylinder
Specify center point for base of cylinder or ⤓: **5,5**
Specify radius for base of cylinder or ⤓: **3**
Specify height of cylinder or ⤓: **4**

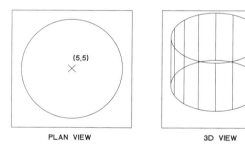

Figure 15–59 *Creating a solid cylinder using the default option of the* CYLINDER *command*

CREATING A SOLID SPHERE

The SPHERE command creates a *3D* body in which all surface points are equidistant from the center. The sphere is drawn in such a way that its central axis is coincident with the *Z* axis of the current UCS.

Invoke the SPHERE command from the Solids toolbar (see Figure 15–60).

Figure 15–60 *Invoking the* SPHERE *command from the Solids toolbar*

AutoCAD prompts:

> Specify center of sphere <0,0,0>: *(specify a point)*

First, AutoCAD prompts for the center point of the sphere; then you can provide the radius or diameter to define a sphere.

For example, the following command sequence shows steps in drawing a sphere (see Figure 15–61):

> **sphere**
> Specify center of sphere <0,0,0>: **5,5**
> Specify radius of sphere or ⤓: **3**

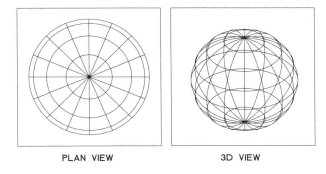

Figure 15–61 *Creating a solid sphere using the default option of the* SPHERE *command*

CREATING A SOLID TORUS

The TORUS command creates a solid with a donut-like shape. If a torus were a wheel, the center point would be the hub. The torus is created lying parallel to and bisected by the *XY* plane of the current UCS.

Invoke TORUS from the Solids toolbar (see Figure 15–62).

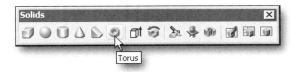

Figure 15–62 *Invoking the* TORUS *command from the Solids toolbar*

AutoCAD prompts:

> Specify center of torus <0,0,0>: *(specify a point)*

AutoCAD prompts for the center point of the torus and then subsequently for the diameter or radius of the torus and the diameter or radius of the tube (see Figure 15–63). You can also draw a torus without a center hole if the radius of the tube is defined as greater than the radius of the torus. A negative torus radius would create a football-shaped solid (if the tube diameter is greater than the absolute value of the specified radius).

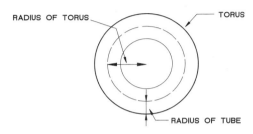

Figure 15–63 *Creating a solid torus with a center hole using the* TORUS *command*

For example, the following command sequence lists the steps in drawing a torus (see Figure 15–64):

torus
Specify center of torus <0,0,0>: **5,5**
Specify radius of torus or ⏷: **3**
Specify radius of tube or ⏷: **0.5**

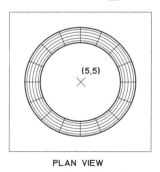 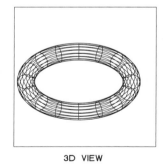

Figure 15–64 *Creating a torus by specifying the baseplane and central axis direction using the* TORUS *command*

CREATING A SOLID WEDGE

The WEDGE command creates a solid like a box that has been a cut in half diagonally along one face. The face of the wedge is always drawn parallel to the current UCS, with the sloped face tapering along the Z axis. The solid wedge can be drawn by one of two options: by providing a center point of the base or by providing starting corner of the box.

Invoke the WEDGE command from the Solids toolbar (see Figure 15–65).

Figure 15-65 *Invoking the* WEDGE *command from the Solids toolbar*

AutoCAD prompts:

> Specify first corner of wedge or ⬇: *(specify a point, or choose* CENTER OF THE WEDGE *from the shortcut menu)*
>
>> By default you are prompted for the starting corner of the box. Once you provide the starting corner, AutoCAD prompts:
>
> Specify corner or ⬇: *(specify the corner of the wedge or select one of the available options)*
>
>> The wedge dimensions can be specified by using one of the three options. The default option lets you create a wedge by locating first the opposite corner of its base rectangle and then its height. The CUBE option allows you to create a wedge in which all edges are of equal length. The LENGTH option lets you create a box by defining its length, width, and height.

The CENTER option allows you to create a wedge by first locating its center point. Once you locate the center point, a line rubber-bands from this point to help you visualize the size of the rectangle. Then AutoCAD prompts you to define the size of the box by entering one of the following options:

> Specify corner or ⬇: *(specify corner of the wedge or select one of the options)*

CREATING SOLIDS FROM EXISTING 2D OBJECTS

The EXTRUDE command creates a unique solid by extruding circles, closed polylines, polygons, ellipses, closed splines, donuts, and regions. Because a polyline can have virtually any shape, the EXTRUDE command allows you to create irregular shapes. In addition, AutoCAD allows you to taper the sides of the extrusion.

 Note: A polyline must contain at least 3 but not more than 500 vertices and none of the segments can cross each other. See Figure 15–66 for examples of shapes that cannot be extruded. If the polyline has width, AutoCAD ignores the width and extrudes from the center of the polyline path. If a selected object has thickness, AutoCAD ignores the thickness.

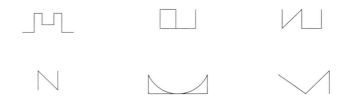

EXAMPLES OF SHAPES THAT CANNOT BE EXTRUDED (SHOWN IN PLAN VIEW)

Figure 15-66 *Shapes (shown in plan view) that cannot be extruded using the* EXTRUDE *command*

Invoke the EXTRUDE command from the Solids toolbar (see Figure 15-67).

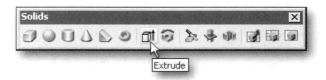

Figure 15-67 *Invoking the* EXTRUDE *command from the Solids toolbar*

AutoCAD prompts:

Select objects: *(select the objects to extrude and press* ENTER *to complete the selection)*

Specify height of extrusion or ⬇: *(specify height of extrusion or choose* PATH *from the shortcut menu)*

The height of extrusion is the distance for extrusion along the positive side of the Z axis of the current UCS. A negative value causes the object to be extruded along the negative axis

The PATH option allows you to select the extrusion path based on a specified curve object. All the profiles of the selected object are extruded along the chosen path to create solids. Lines, circles, arcs, ellipses, elliptical arcs, polylines, or splines can be paths. The path should not lie on the same plane as the profile, nor should it have areas of high curvature. The extruded solid starts from the plane of the profile and ends on a plane perpendicular to the path's endpoint. One of the endpoints of the path should be on the plane of the profile. Otherwise, AutoCAD moves the path to the center of the profile.

Once you specify the height of extrusion and path appropriately, AutoCAD prompts:

Specify angle of taper for extrusion <0>: *(specify the angle)*

Specify an angle between −90 and +90 degrees, or press ENTER or SPACEBAR to accept the default value of 0 degrees. If you specify 0 degrees as the taper angle, AutoCAD extrudes a *2D* object perpendicular to its *2D* plane (see Figure 15–68). Positive angles taper in from the base object; negative angles taper out.

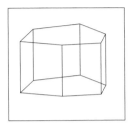 

TAPER ANGLE 0 DEGREES   TAPER ANGLE 15 DEGREES

Figure 15–68 *Creating a solid with the* EXTRUDE *command with 0 degrees and with 15 degrees of taper angle*

 Note: It is possible for a large taper angle or a long extrusion height to cause the object, or portions of the object, to taper to a point before reaching the extrusion height.

CREATING SOLIDS BY MEANS OF REVOLUTION

The REVOLVE command creates a unique solid by revolving or sweeping a closed polyline, polygon, circle, ellipse, closed spline, donut, and region. Polylines that have crossing or self-intersecting segments cannot be revolved. The REVOLVE command is similar to the REVSURF command. The REVSURF command creates a surface of revolution, whereas REVOLVE creates a solid of revolution. The REVOLVE command provides several options for defining the axis of revolution.

Invoke the REVOLVE command from the Solids toolbar (see Figure 15–69).

Figure 15–69 *Invoking the* REVOLVE *command from the Solids toolbar*

AutoCAD prompts:

> Select objects: *(select the objects to revolve and press* ENTER *to complete the selection)*
> Specify start point for axis of revolution or define axis by ⤓ *(specify start point for axis of revolution or select one of the options from the shortcut menu)*

The **Start point of axis option** (default) allows you to specify two points for the start point and the endpoint of the axis, and the positive direction of rotation is based on the right-hand rule.

The OBJECT option allows you select an existing line or single polyline segment that defines the axis about which to revolve the object. The positive axis direction is from the closest to the farthest endpoint of this line.

The X AXIS option uses the positive X axis of the current UCS as the axis of the revolution.

The Y AXIS option uses the positive Y axis of the current UCS as the axis of the revolution.

Once you specify the axis of revolution, AutoCAD prompts:

> Specify angle of revolution <360>: *(Specify the angle for revolution)*

The default angle of revolution is a full circle. You can specify any angle between 0 and 360 degrees.

CREATING COMPOSITE SOLIDS

As mentioned earlier in this chapter, you can create a new composite solid or region by combining two or more solids or regions via Boolean operations. Although the term Boolean implies that only two objects can be operated upon at once, AutoCAD lets you select many solid objects in a single Boolean command. There are three basic Boolean operations that can be performed in AutoCAD: Union, Subtraction, and Intersection

The UNION, SUBTRACT, and INTERSECT commands let you select both the solids and regions in a single use of the commands, but solids are combined with solids, and regions combined only with regions. Also, in the case of regions you can make composite regions only with those that lie in the same plane. This means that a single command creates a maximum of one composite solid, but might create many composite regions.

UNION OPERATION

Union is the process of creating a new composite object from one or more original objects. The union operation joins the original solids or regions in such a way that there is no duplication of volume. Therefore, the total resulting volume can be equal to or less than the sum of the volumes in the original solids or regions. The UNION command performs the union operation.

Invoke the UNION command from the Solids Editing toolbar (see Figure 15–70).

Figure 15–70 *Invoking the* UNION *command from the Solids Editing toolbar*

AutoCAD prompts:

    Select objects: *(select the objects to make one composite object and press* ENTER *to complete the selection)*

You can select more than two objects at once. The objects (solids or regions) can be overlapping, adjacent, or nonadjacent.

For example, the following command sequence shows steps in creating a composite solid by joining two cylinders (see Figure 15–71).

    union
    Select objects: *(select cylinders A and B and press* ENTER*)*

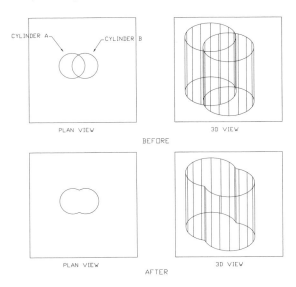

Figure 15–71 *Creating a composite solid by joining two cylinders using the* UNION *command*

SUBTRACTION OPERATION

Subtraction is the process of forming a new composite object by starting with one object and removing from it any volume that it has in common with a second object. In the case of solids, they are created by subtracting the volume of one set of solids from another set. If the entire volume of the second solid is contained in the first solid, then what is left is the first solid minus the volume of the second solid. However, if only part of the volume of the second solid is contained within the first solid, then only the part that is duplicated in the two solids is subtracted. Similarly, in the case of regions, they are created by subtracting the common area of one set of existing regions from another set. The SUBTRACT command performs the subtraction operation.

Invoke the SUBTRACT command from the Solids Editing toolbar (see Figure 15–72).

Figure 15–72 *Invoking the* SUBTRACT *command from the Solids Editing toolbar*

AutoCAD prompts:

> Select solids and regions to subtract from...
> Select objects: *(select the objects from which you will subtract other objects and press* ENTER*)*

You can select one or more objects as source objects. If you select more than one, they are automatically joined. After selecting the source objects, press ENTER or the SPACEBAR, and AutoCAD prompts you to select the objects to subtract from the source object.

> Select solids and regions to subtract...
> Select objects: *(select the objects to subtract and press* ENTER*)*

If necessary, you can select one or more objects to subtract from the source object. If you select several, they are automatically joined before they are subtracted from the source object.

 Note: Objects that are neither solids nor regions are ignored.

For example, the following command sequence shows steps in creating a composite solid by subtracting cylinder B from A (see Figure 15–73).

> **subtract**
> Select objects: *(select cylinder A and press* ENTER*)*
> Objects to subtract from them...
> Select objects: *(select cylinder B and press* ENTER*)*

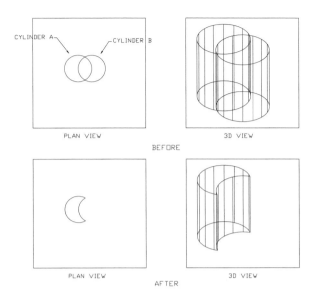

Figure 15–73 *Creating a composite solid by subtracting cylinder B from cylinder A using the* SUBTRACT *command*

INTERSECTION OPERATION

Intersection is the process of forming a composite object from only the volume that is common to two or more original objects. In the case of solids, you can create a new composite solid by calculating the common volume of two or more existing solids. Whereas, in the case of regions, it is done by calculating the overlapping area of two or more existing regions. The INTERSECT command performs the intersection operation.

Invoke the INTERSECT command from the Solids Editing toolbar (see Figure 15–74).

Figure 15–74 *Invoking the* INTERSECT *command from the Solids Editing toolbar*

AutoCAD prompts:

> Select objects: *(select the objects for intersection and press* ENTER *to complete the selection)*

For example, the following command sequence shows the steps in creating a composite solid by intersecting cylinder A with cylinder B (see Figure 15–75):

intersect
Select objects: *(select cylinders A and B and press* ENTER *or the* SPACEBAR*)*

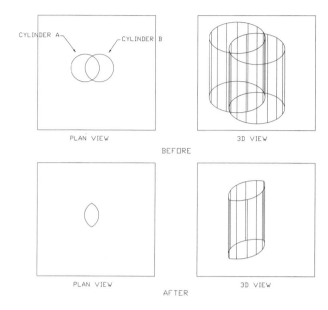

Figure 15–75 *Creating a composite solid by intersecting cylinder A from cylinder B using the* INTERSECT *command*

EDITING 3D SOLIDS

AutoCAD makes the work of creating solids a little easier by providing editing tools, including chamfering or filleting the edges, creating a cross-section through a solid, creating a new solid by cutting the existing solid and removing a specified side, and creating a composite solid from the interference of two or more solids. In addition, AutoCAD provides additional editing tools such as extrude faces, move faces, offset faces, delete faces, rotate faces, taper faces, color faces, copy faces, color and copy edges, imprint, clean, separate solids, shell, and check. If necessary, you can always use the AutoCAD modify and construct commands, such as MOVE, COPY, ROTATE, SCALE, and ARRAY to edit solids.

CHAMFERING SOLIDS

The CHAMFER command (explained in Chapter 4) can also be used to bevel the edges of an existing solid object.

Invoke the CHAMFER command from the Modify toolbar (see Figure 15–76).

Figure 15–76 Invoke the CHAMFER command from the Modify toolbar

AutoCAD prompts:

Select first line or ⬇: (select an edge on a 3D solid)

If you pick an edge that is common to two surfaces, AutoCAD highlights one of the surfaces and prompts:

Base surface selection

Enter surface selection option

If this is the surface you want, press ENTER to accept it. If it is not, enter **N** (for next) to highlight the adjoining surface and then press ENTER. AutoCAD prompts:

Specify base surface chamfer distance <default>: (specify a distance, or press ENTER to accept the default)
Specify other surface chamfer distance <default>: (specify a distance, or press ENTER to accept the default)

Once you provide the chamfer distances, AutoCAD prompts:

Select an edge or ⬇: (select the edges of the highlighted surface you want chamfered, and then press ENTER).

> The LOOP option allows you to select one of the edges on the base surface, and AutoCAD automatically selects all edges on the base surface for chamfering.

The following command sequence draws a chamfer for a solid object (see Figure 15–77).

chamfer
Select first line or ⬇: (select the edge)
Enter surface selection option (ENTER)
Specify base surface chamfer distance <default>: **0.25**
Specify other surface chamfer distance <default>: **0.5**
Select an edge or ⬇: (choose LOOP from the shortcut menu)
Select an edge loop or ⬇ (select an edge and press ENTER)

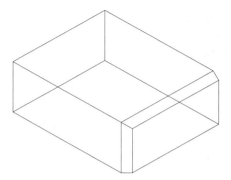

Figure 15-77 *Example of chamfering a solid surface*

FILLETING SOLIDS

The FILLET command (explained in Chapter 4) can also be used to round the edge of an existing solid object.

Invoke the FILLET command from the Modify toolbar (see Figure 15-78).

Figure 15-78 *Invoking the FILLET command from the Modify toolbar*

AutoCAD prompts:

> Select first object or ⤓: *(select an edge in a 3D solid)*

If necessary, you can select multiple edges; but you must select the edges individually after specifying the radius for the fillet. AutoCAD prompts:

> Enter fillet radius : *(specify radius for fillet, or press ENTER to accept the default)*
> Select an edge or ⤓: *(select addition edges and press ENTER to complete the selection)*

The following command sequence fillets a solid object (see Figure 15-79).

> **fillet**
> Select first object or ⤓: *(select the solid edge)*
> Enter fillet radius : **0.5**
> Select an edge or ⤓: (ENTER)

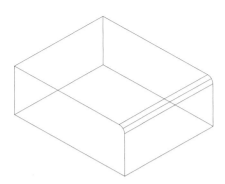

Figure 15-79 *Example of filleting a solid surface*

SECTIONING SOLIDS

The SECTION command creates a cross-section of one or more solids. The cross-section is created as one or more regions. The region is created on the current layer and is inserted at the location of the cross-section. If necessary, you can use the MOVE command to move the cross-section.

Invoke the SECTION command from the Solids toolbar (see Figure 15–80).

Figure 15-80 *Invoking the SECTION command from the Solids toolbar*

AutoCAD prompts:

> Select objects: *(select the objects from which you want the cross-section to be generated)*
> Specify first point on Section plane by ⬇ *(specify one of the three points to define a plane or select one of the options from the shortcut menu)*

The 3POINTS option (default) allows you to define a section plane by locating three points. The first point is the origin, the second point determines the positive direction of the X axis for the section plane, and the third point determines the positive Y axis of the section plane. This option is similar to the 3POINT option of the AutoCAD UCS command.

The OBJECT option aligns the sectioning plane with a circle, ellipse, circular or elliptical arc, 2D spline, or 2D polyline segment.

The ZAXIS option defines the section plane by locating its origin point and a point on the Z axis (normal) to the plane.

The VIEW option aligns the section plane with the viewing plane of the current viewport. Specifying a point defines the location of the sectioning plane.

The XY option aligns the sectioning plane with the *XY* plane of the current UCS. Specifying a point defines the location of the sectioning plane.

The YZ option aligns the sectioning plane with the *XY* plane of the current UCS. Specifying a point defines the location of the sectioning plane.

The ZX option aligns the sectioning plane with the *XY* plane of the current UCS. Specifying a point defines the location of the sectioning plane.

Figure 15–81 shows a hatched cross-section produced with the SECTION command.

Note: The section may be hatched using the hatching techniques described in Chapter 9.

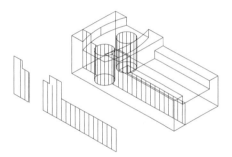

Figure 15–81 *Creating a 2D hatched cross-section using the* SECTION *command*

SLICING SOLIDS

The SLICE command allows you to create a new solid by cutting the existing solid and removing a specified portion. If necessary, you can retain both portions of the sliced solid(s) or just the portion you specify. The sliced solids retain the layer and color of the original solids.

Invoke the SLICE command from the Solids toolbar (see Figure 15–82).

Figure 15–82 *Invoking the* SLICE *command from the Solids toolbar*

AutoCAD prompts:

> Select objects: *(select the objects to create a new solid by slicing and press* ENTER*)*

After selecting the objects, press ENTER and AutoCAD prompts you to define the slice plane:

> Specify first point on slicing plane by ⏎ *(specify one of the three points to define a plane or select one of the options)*

The options are the same as those for the SECTION command explained earlier in this chapter.

After defining the slicing plane, AutoCAD prompts you to indicate which part of the cut solid is to be retained:

> Specify a Point on desired side of the plane or ⏎

By default AutoCAD allows you to select, with your pointing device, the side of the slice that is to be retained in your drawing.

The KEEP BOTH SIDES option allows you to retain both portions of the sliced solids.

Figure 15–83 shows two parts of a solid model that have been cut using the SLICE command and moved apart using the MOVE command.

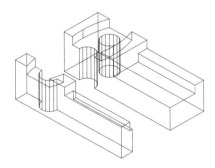

Figure 15–83 *Cutting a solid model into two parts using the* SLICE *command*

SOLID INTERFERENCE

The INTERFERE command checks the interference between two or more solids and creates a composite solid from their common volume.

There are two ways to determine the interference between solids:

- Select two sets of solids. AutoCAD determines the interference between the first and second sets of solids.
- Select one set of solids instead of two. AutoCAD determines the interference between all of the solids in the set. They are checked against each other.

Invoke the INTERFERE command from the Solids toolbar (see Figure 15–84).

Figure 15–84 *Invoking the* INTERFERE *command from the Solids toolbar*

AutoCAD prompts:

Select objects: *(select the first set of solids and press* ENTER*)*
Select objects: *(select the second set of solids or press* ENTER*)*

The second selection set is optional. Press ENTER if you do not want to define the second selection set. If the same solid is included in both the selection sets, it is considered part of the first selection set and ignored in the second selection set. AutoCAD highlights all interfering solids and prompts:

Create interference solids? *(choose an option)*

Entering **y** creates and highlights a new solid on the current layer that is the intersection of the interfering solids. If there are more than two interfering solids, AutoCAD prompts:

Highlight pairs of interfering solids? *(choose an option)*

If you specify **y** for yes, and if there is more than one interfering pair, AutoCAD prompts:

Enter an option *(specify* **x** *or* **n***)*

Pressing ENTER cycles through the interfering pairs of solids, and AutoCAD highlights each interfering pair of solids. Enter **x** to complete the command sequence.

EDITING FACES OF 3D SOLIDS

AutoCAD allows you to edit solid objects by extruding faces, copying faces, offsetting faces, moving faces, rotating faces, tapering faces, coloring faces, and deleting faces.

Extruding Faces

AutoCAD allows you to extrude selected faces of a *3D* solid object to a specified height or along a path. Specifying a positive value extrudes the selected face in its positive direction (usually outward); and a negative value extrudes in the negative direction (usually inward). Tapering the selected face with a positive angle tapers the face inward, and a negative angle tapers the face outward. Tapering the selected face to 0 degrees extrudes the face perpendicular to its plane.

Face extrusion along a path is based on a path curve, such as lines, circles, arcs, ellipses, elliptical arcs, polylines, or splines.

Invoke EXTRUDE FACES from the Solids Editing toolbar (see Figure 15–85).

Figure 15–85 *Invoking EXTRUDE FACES from the Solids Editing toolbar*

AutoCAD prompts:

> Select faces or ⬇: *(select faces and press* ENTER *to complete the selection)*
> Specify height of extrusion or ⬇: *(specify height of extrusion or select path option to extrude along a path)*
> Specify angle of taper for extrusion <0>: *(press* ENTER *to accept the default angle or specify the angle for taper for extrusion)*
> Enter a face editing option *(select the* EXIT *option to exit the face editing)*
> Enter a solids editing option *(select the* EXIT *option to exit solids editing)*

Figure 15–86 shows an example of a solid model in which one of the faces is extruded by a positive value with a 15-degree tapered angle with the EXTRUDE FACES option of the SOLIDEDIT command.

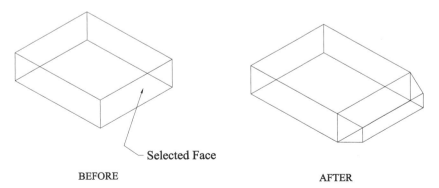

Figure 15–86 *An example in extruding one of the faces in a solid model*

Copying Faces

The COPY FACES option of the SOLIDEDIT command allows you to copy selected faces of a *3D* solid object. AutoCAD copies selected faces as regions or bodies. Prompts are similar to the regular COPY command.

Invoke COPY FACES from the Solids Editing toolbar (see Figure 15–87).

Figure 15–87 *Invoking COPY FACES from the Solids Editing toolbar*

AutoCAD prompts:

> Select faces or ⬇: *(select faces and press* ENTER *to complete the selection)*
> Specify a base point or displacement: *(specify a base point)*
> Specify a second point of displacement: *(specify a second point of displacement or press* ENTER *to consider the original selection point as a base point)*
> Enter a face editing option *(select the* EXIT *option to exit the face editing)*
> Enter a solids editing option *(select the* EXIT *option to exit solids editing)*

Figure 15–88 shows an example of a solid model in which one of the faces is copied by the COPY FACES option of the SOLIDEDIT command.

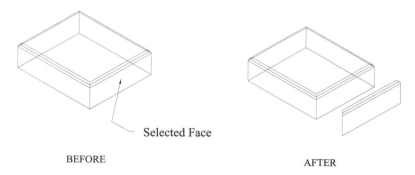

BEFORE   AFTER

Figure 15–88 *An example of copying one of the faces in a solid model*

Offsetting Faces

AutoCAD allows you to uniformly offset selected faces of a *3D* solid object by a specified distance. New faces are created by offsetting existing ones inside or outside at a specified distance from their original positions. Specifying a positive value increases the size or volume of the solid; a negative value decreases the size or volume of the solid.

Figure 15–89 *Invoking* OFFSET FACES *from the Solids Editing toolbar*

AutoCAD prompts:

> Select faces or ⬇: *(select faces and press* ENTER *to complete the selection)*
> Specify the offset distance: *(specify the offset distance)*
> Enter a face editing option *(select the* EXIT *option to exit the face editing)*
> Enter a solids editing option *(select the* EXIT *option to exit solids editing)*

Figure 15–90 shows an example of a solid model in which one of the faces is offset (positive value) by the OFFSET FACES option of the SOLIDEDIT command.

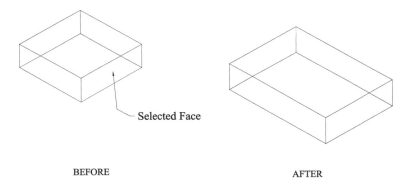

Figure 15–90 *An example of offsetting one of the faces (positive value) in a solid model*

Moving Faces

The MOVE FACES option of the SOLIDEDIT command allows you to move selected faces of a *3D* solid object. You can move holes from one location to another location in a *3D* solid. Prompts are similar to the regular MOVE command.

Invoke MOVE FACES from the Solids Editing toolbar (see Figure 15–91):

Figure 15–91 *Invoking* MOVE FACES *from the Solids Editing toolbar*

AutoCAD prompts:

Select faces or ⬇: *(select faces and press* ENTER *to complete the selection)*
Specify a base point or displacement: *(specify a base point)*
Specify a second point of displacement: *(specify a second point of displacement or press* ENTER *to consider the original selection point as a base point)*
Enter a face editing option *(select the* EXIT *option to exit the face editing)*
Enter a solids editing option *(select the* EXIT *option to exit solids editing)*

Figure 15–92 shows an example of a solid model in which an elliptical cylinder is moved by the MOVE FACES option of the SOLIDEDIT command.

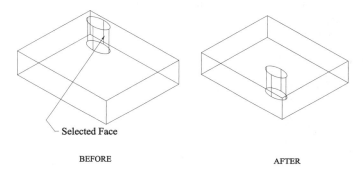

BEFORE   AFTER

Figure 15–92 *An example of moving the elliptical cylinder in a solid model*

Rotating Faces

The ROTATE FACES option of the SOLIDEDIT command allows you to rotate selected faces of a *3D* solid object by choosing a base point to relative or absolute angle. All *3D* faces rotate about a specified axis.

Invoke ROTATE FACES from Solids Editing toolbar.

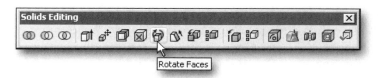

Figure 15–93 *Invoking* ROTATE FACES *from the Solids Editing toolbar*

Select faces or ⬇: *(select faces and press* ENTER *to complete the selection)*
Specify an axis point or ⬇: *(specify an axis point or select one of the available options)*
Enter a face editing option *(select the* EXIT *option to exit the face editing)*
Enter a solids editing option *(select the* EXIT *option to exit solids editing)*

Figure 15–94 shows an example of a solid model in which elliptical cylinder is rotated by the ROTATE FACES option of the SOLIDEDIT command.

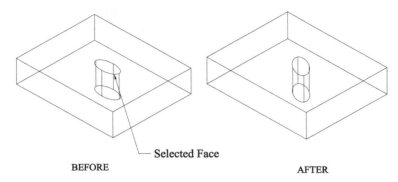

Figure 15–94 *An example of rotating elliptical cylinder of a solid model*

Tapering Faces

The SOLIDEDIT command allows you to taper selected faces of a *3D* solid object with a draft angle along a vector direction. Tapering the selected face with a positive angle tapers the face inward, and a negative angle tapers the face outward.

Invoke TAPER FACES from Solids Editing toolbar (see Figure 15–95).

Figure 15–95 *Invoking TAPER FACES from the Solids Editing toolbar*

AutoCAD prompts:

    Select faces or ⬇: *(select faces and press ENTER to complete the selection)*
    Specify the base point: *(specify the base point)*
    Specify another point along the axis of tapering: *(specify a point to define the axis of tapering)*
    Specify the taper angle: *(specify the taper angle and press ENTER to continue)*
    Enter a face editing option *(select the EXIT option to exit the face editing)*
    Enter a solids editing option *(select the EXIT option to exit solids editing)*

Figure 15–96 shows an example of a solid model in which the cylinder tapered angle is changed by the TAPER FACES option of the SOLIDEDIT command.

Coloring Faces

The COLOR FACES option of the SOLIDEDIT command allows you to change color of selected faces of a *3D* solid object. You can choose a color from the Select Color dialog box. Setting a color on a face overrides the color setting for the layer on which the solid object resides.

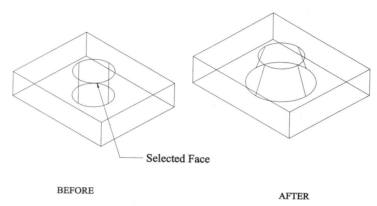

Figure 15–96 *An example of change in a tapered angle of the cylinder in a solid model*

Invoke COLOR FACES from the Solids Editing toolbar (see Figure 15–97).

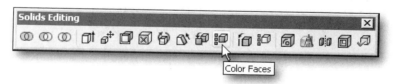

Figure 15–97 *Invoking COLOR FACES from the Solids Editing toolbar*

AutoCAD prompts:

Select faces or ⬇: *(select faces and press* ENTER *to complete the selection)*

AutoCAD displays the Select Color dialog box. Select the color to change for selected faces and choose the **OK** button. AutoCAD prompts:

Enter a face editing option *(select the* EXIT *option to exit the face editing)*
Enter a solids editing option *(select the* EXIT *option to exit solids editing)*

AutoCAD changes the color of the selected faces of the *3D* solid model.

Deleting Faces

The DELETE FACES option of the SOLIDEDIT command allows you to delete selected faces, holes, and fillets of a *3D* solid object.

Invoke DELETE FACES from Solids Editing toolbar (see Figure 15–98).

Figure 15–98 *Invoking* DELETE FACES *from the Solids Editing toolbar*

AutoCAD prompts:

Select faces or ⬇: *(select faces and press* ENTER *to complete the selection)*

 Note: If the object is a simple box, AutoCAD responds with: Modeling Operation Error – Gap cannot be filled. (This informs us that there is no way to join the opposite sides to fill the gap.)

Enter a face editing option *(select the* EXIT *option to exit the face editing)*
Enter a solids editing option *(select the* EXIT *option to exit solids editing)*

AutoCAD deletes the selected faces of the *3D* solid model.

EDITING EDGES OF 3D SOLIDS

AutoCAD allows you to copy individual edges and change color of edges on a *3D* solid object. The edges are copied as lines, arcs, circles, ellipses, or spline objects.

Copying Edges

The COPY EDGES option of the SOLIDEDIT command allows you to copy selected edges of a *3D* solid object. AutoCAD copies selected edges as lines, arcs, circles, ellipses, or splines. Prompts are similar to the regular COPY command.

Invoke COPY EDGES from the Solids Editing toolbar (see Figure 15–99).

Figure 15–99 *Invoking* COPY EDGES *from the Solids Editing toolbar*

AutoCAD prompts:

> Select edges or ⬇: *(select edges and press* ENTER *to complete the selection)*
> Specify a base point or displacement: *(specify a base point)*
> Specify a second point of displacement: *(specify a second point of displacement or press* ENTER *to consider the original selection point as a base point)*
> Enter an edge editing option *(select the* EXIT *option to exit the edge editing)*
> Enter a solids editing option *(select the* EXIT *option to exit solids editing)*

Figure 15–100 shows an example of a solid model in which the edges (right side of the model) are copied by the COPY EDGES option of the SOLIDEDIT command.

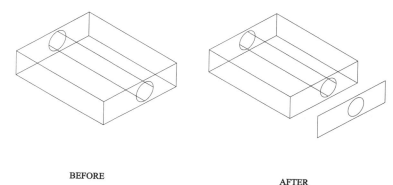

BEFORE   AFTER

Figure 15–100 *An example of copying edges of a solid model*

Coloring Edges

The COLOR EDGES option of the SOLIDEDIT command allows you to change color of selected edges of a *3D* solid object. You can choose a color from the Select Color dialog box. Setting a color on an edge overrides the color setting for the layer on which the solid object resides.

Invoke COLOR EDGES from the Solids Editing toolbar (see Figure 15–101).

Figure 15–101 *Invoking* COLOR EDGES *from the Solids Editing toolbar*

AutoCAD prompts:

> Select edges or ⬇: *(select faces and press* ENTER *to complete the selection)*

AutoCAD displays the Select Color dialog box. Select the color to change for selected faces and choose **OK**. AutoCAD prompts:

> Enter an edge editing option *(select the EXIT option to exit the edge editing)*
> Enter a solids editing option *(select the EXIT option to exit solids editing)*

AutoCAD changes the color of the selected edges of the *3D* solid model.

IMPRINTING SOLIDS

AutoCAD allows you to have an imprint of an object on the selected solid. The object to be imprinted must intersect one or more faces on the selected solid in order for imprinting to be successful. Imprinting is limited to the following objects: arcs, circles, lines, *2D* and *3D* polylines, ellipses, splines, regions, bodies, and *3D* solids.

Invoke IMPRINT from the Solids Editing toolbar (see Figure 15–102).

Figure 15–102 *Invoking* IMPRINT *from the Solids Editing toolbar*

AutoCAD prompts:

> Select a *3D* solid: *(select a 3D solid object)*
> Select an object to imprint: *(select an object to imprint)*
> Delete the source object *(press ENTER to delete the source objects or ENTER y to keep the source object)*
> Select an object to imprint: *(select another object to imprint or press ENTER to complete the selection)*
> Enter a body editing option *(select the EXIT option to exit body editing)*
> Enter a solids editing option *(select the EXIT option to exit solids editing)*

AutoCAD creates an imprint of the selected object.

SEPARATING SOLIDS

AutoCAD separates solids from a composite solid. But it cannot separate solids if the composite *3D* solid object shares a common area or volume. After separation of the *3D* solid, the individual solids retain the layers and colors of the original.

Invoke SEPARATE from the Solids Editing toolbar (see Figure 15–103).

Figure 15–103 *Invoking SEPARATE from the Solids Editing toolbar*

AutoCAD prompts:

    Select a 3D solid: *(select a 3D solid object and press ENTER to complete the selection)*
    Enter a body editing option *(select the EXIT option to exit body editing)*
    Enter a solids editing option *(select the EXIT option to exit solids editing)*

AutoCAD separates the selected composite solid.

SHELLING SOLIDS

AutoCAD creates a shell or a hollow thin wall with a specified thickness from the selected *3D* solid object. AutoCAD creates new faces by offsetting existing ones inside or outside their original positions. AutoCAD treats continuously tangent faces as single faces when offsetting. A positive offset value creates a shell in the positive face direction; a negative value creates a shell in the negative face direction.

Invoke SHELL from the Solids Editing toolbar (see Figure 15–104).

Figure 15–104 *Invoking SHELL from the Solids Editing toolbar*

AutoCAD prompts:

    Select a 3D solid: *(select a 3D solid object)*
    Remove faces or ⬇: *(select faces to be excluded from shelling and press ENTER to complete the selection)*
    Specify the shell offset value: *(specify the shell offset value)*
    Enter a body editing option *(select the EXIT option to exit body editing)*
    Enter a solids editing option *(select the EXIT option to exit solids editing)*

AutoCAD creates a shell with the specified thickness from the *3D* solid object.

CLEANING SOLIDS

AutoCAD allows you to remove edges or vertices if they share the same surface or vertex definition on either side of the edge or vertex. All redundant edges, imprinted as well as used, on the selected *3D* solid object are deleted.

Invoke CLEAN from the Solids Editing toolbar (see Figure 15–105).

Figure 15–105 *Invoking* CLEAN *from the Solids Editing toolbar*

AutoCAD prompts:

>Select a *3D* solid: *(select a 3D solid object and press* ENTER *to complete the selection)*
>Enter a body editing option *(select the* EXIT *option to exit body editing)*
>Enter a solids editing option *(select the* EXIT *option to exit solids editing)*

AutoCAD removes the selected edges or vertices of the selected *3D* model.

CHECKING SOLIDS

AutoCAD checks to see if the selected solid object is a valid *3D* solid object. With a *3D* solid model, you can modify the object without incurring ACIS failure error messages. If the selected solid *3D* model is not valid, you cannot edit the object.

Invoke CHECK from the Solids Editing toolbar (see Figure 15–106).

Figure 15–106 *Invoking* CHECK *from the Solids Editing toolbar*

AutoCAD prompts:

>Select a *3D* solid: *(select a* 3D *solid object and press* ENTER *to complete the selection)*
>Enter a body editing option *(select the* EXIT *option to exit body editing)*
>Enter a solids editing option *(select the* EXIT *option to exit solids editing)*

AutoCAD checks the solid *3D* model and displays with appropriate information about the selected solid.

MASS PROPERTIES OF A SOLID

The MASSPROP command calculates and displays the mass properties of selected solids and regions. The mass properties displayed for solids are mass, volume, bounding box, centroid, moments of inertia, products of inertia, radii of gyration, and principal moments with corresponding principal directions. The mass properties are calculated based on the current UCS.

Invoke the REGION/MASS PROPERTIES command from the Inquiry toolbar (see Figure 15–107).

Figure 15–107 *Invoking the* REGION/MASS PROPERTIES *command from the Inquiry toolbar*

AutoCAD prompts:

>Select objects: *(select the objects and press* ENTER *to complete the selection)*

The MASSPROP command displays the object mass properties of the selected objects in the text window, as shown in Figure 15–108. Then, AutoCAD prompts:

>Write analysis to a file? *(type* **y** *to create a file or* **n** *t to not create a file)*

```
AutoCAD Text Window
Edit
Mass:                    105.66
Volume:                  105.66
Bounding box:     X:  0.00  --  8.00
                  Y:  0.00  --  7.00
                  Z: -3.00  --  2.00
Centroid:         X:  3.55
                  Y:  3.50
                  Z: -1.25
Moments of inertia:  X: 1989.68
                     Y: 2177.49
                     Z: 3640.82
Products of inertia: XY: 1312.68
                     YZ: -461.91
                     ZX: -600.08
Radii of gyration:   X:  4.34
                     Y:  4.54
                     Z:  5.87
Press ENTER to continue:
Principal moments and X-Y-Z directions about centroid:
     I:  497.04 along [0.97 0.00 -0.25]
     J:  681.40 along [0.00 1.00 0.00]
     K: 1048.65 along [0.25 0.00 0.97]
Write to a file ? <N>:
```

Figure 15–108 *Mass properties listing*

If you enter **y**, AutoCAD prompts for a file name and saves the file in ASCII format.

 Note: You can also use the AutoCAD LIST and AREA commands to obtain information about coordinates and areas of solid(s).

HIDING OBJECTS

The HIDE command hides objects (or displays them in different colors) that are behind other objects in the current viewport. Complex models are difficult to read in wireframe form, and become more clear when the model is displayed with hidden lines removed. HIDE considers circles, solids, traces, wide polyline segments, *3D* faces, polygon meshes, and the extruded edges of objects with a thickness to be opaque surfaces hiding objects that lie behind them. The HIDE command remains active only until the next time the display is regenerated. Depending on the complexity of the model, hiding may take from a few seconds to even several minutes.

Invoke the HIDE command from the Render toolbar (see Figure 15–109).

Figure 15–109 *Invoking the* HIDE *command from the Render toolbar*

There are no prompts to be answered. The current viewport goes blank for a period of time, depending on the complexity of the model, and is then redrawn with hidden lines removed temporarily. Hidden-line removal is lost during plotting unless you specify that AutoCAD remove hidden lines in the plotting configuration.

PLACING MULTIVIEWS IN PAPER SPACE

The SOLVIEW command creates untiled viewports (Layout) using orthographic projection to lay out orthographic views and sectional views. View-specific information is saved with each viewport as you create it. The information that is saved is used by the SOLDRAW command, which does the final generation of the drawing view. The SOLVIEW command automatically creates a set of layers that the SOLDRAW command uses to place the visible lines, hidden lines, and section hatching for each view. In addition, the SOLVIEW command creates a specific layer for dimensions that are visible in individual viewports. The SOLVIEW command applies the following conventions in naming the layers:

Layer	Name
For visible lines	VIEW NAME-VIS
For hidden lines	VIEW NAME-HID
For dimensions	VIEW NAME-DIM
For hatch patterns (sections)	VIEW NAME-HAT

The viewport objects are drawn on the VPORTS layer. The SOLVIEW command creates the VPORTS layer if it does not already exist. All the layers created by SOLVIEW are assigned the color white and the linetype continuous. The stored information is deleted and updated when you run the SOLDRAW command, so do not draw any objects on these layers.

Invoke the VIEW from the Solids toolbar (see Figure 15–110).

Figure 15–110 *Invoking the VIEW command from the Solids toolbar*

AutoCAD prompts:

> Enter an option *(select one of the available options)*

Choose UCS to create a profile view relative to a User Coordinate System. AutoCAD prompts:

> Enter an option *(select one of the available options)*

>> Choose NAMED to use the *XY* plane of a named UCS to create a profile view. After prompting for the name of the view, AutoCAD prompts as follows:

>>> Enter view scale <1.0>: *(specify the scale factor for the view to be displayed)*
>>> Specify View center: *(specify the center location for the viewport to be drawn in the paper space and press* ENTER*)*
>>> Specify first corner of viewport: *(specify a point for one corner of the viewport)*
>>> Specify opposite corner of viewport: *(specify a point for the opposite corner of the viewport)*
>>> Enter view name: *(specify a name for the newly created view)*

>> Choose WORLD to use the *XY* plane of the WCS to create a profile view. The

prompts are the same as just described for the Named option.

Choose *?* to list the names of existing User Coordinate Systems. After the list is displayed, press any key to return to the first prompt.

Choose CURRENT (default option) to use the *XY* plane of the current UCS to create a profile view. The prompts are the same as explained earlier for the Named option.

If no untiled viewports exist in your drawing, the UCS option allows you to create an initial viewport from which other views can be created.

Choose ORTHO to create a folded orthographic view from an existing view. AutoCAD prompts:

> Specify side of viewport to project: *(select the one of the edges of a viewport)*
> Specify view center: *(specify the center location for the viewport to be drawn in the paper space and press* ENTER*)*
> Specify first corner of viewport: *(specify a point for one corner of the viewport)*
> Specify opposite corner of viewport: *(specify a point for the opposite corner of the viewport)*
> Enter view name: *(specify a name for the newly created view)*

Choose AUXILIARY to create an auxiliary view from an existing view. An auxiliary view is one that is projected onto a plane perpendicular to one of the orthographic views and inclined in the adjacent view. AutoCAD prompts:

> Specify first point of inclined plane: *(specify a point)*
> Specify second point of incline plane: *(specify a point)*
> Specify side to view from: *(specify a point that determines the side from which you will view the plane)*
> Specify view center: *(specify the center location for the viewport to be drawn in the paper space and press* ENTER*)*
> Specify first corner of viewport: *(specify a point for one corner of the viewport)*
> Specify opposite corner of viewport: *(specify a point for the opposite corner of the viewport)*
> Enter view name: *(specify a name for the newly created view)*

Choose SECTION to create a sectional view of solids with crosshatching. AutoCAD prompts:

> Specify first point of cutting plane: *(specify a point to define the first point of the cutting plane)*
> Specify second point of cutting plane: *(specify a point to define the second point of the cutting plane)*
> Specify side to view from: *(specify a point to define the side of the cutting plane from which to view)*

Enter view scale: *(specify the scale factor for the view to be displayed)*
Specify view center: *(specify the center location for the viewport to be drawn in the paper space and press* ENTER*)*
Specify first corner of viewport: *(specify a point for one corner of the viewport)*
Specify opposite corner of viewport: *(specify a point for the opposite corner of the viewport)*
Enter view name: *(specify a name for the newly created view)*

To exit the command sequence, press ENTER.

GENERATING VIEWS IN VIEWPORTS

The SOLDRAW command generates sections and profiles in viewports that have been created with the SOLVIEW command. Visible and hidden lines representing the silhouette and edges of solids in the viewports are created and then projected to a plane perpendicular to the viewing direction. AutoCAD deletes any existing profiles and sections in the selected viewports, and new ones are generated. In addition, AutoCAD freezes all the layers in each viewport, except those required to display the profile or section.

Invoke the SOLDRAW command from the Solids toolbar (see Figure 15–111).

Figure 15–111 *Invoking the* SOLDRAW *command by selecting Drawing from the Solids toolbar*

AutoCAD prompts:

Select objects: *(select the viewports to be drawn and press* ENTER*)*

AutoCAD generates views in the selected viewports.

GENERATING PROFILES

The SOLPROF command creates a profile image of a solid, including all of its edges, according to the view in the current viewport. The profile image is created from lines, circles, arcs, and/or polylines. SOLPROF will not give correct results in perspective view; it is designed for parallel projections only.

The SOLPROF command will work only when you are working in layout tab and you are in model space.

Invoke SOLPROF from the Solids toolbar by selecting **Setup Profile** (see Figure 15–112).

Figure 15–112 *Invoking Setup Profile from the Solids toolbar*

AutoCAD prompts:

> Select objects: *(select one or more objects)*

Select one or more solids and press ENTER. The next prompt lets you decide the placement of hidden lines of the profile on a separate layer:

> Display hidden profile lines on separate layer?

Enter **Y** or **N**. If you answer **Y** (default option), two block inserts are created—one for the visible lines in the same linetype as the original and the other for hidden lines in the hidden linetype. The visible lines are placed on a layer whose name is PV-(viewport handle of the current viewport). The hidden lines are placed on a layer whose name is PH-(viewport handle or the current viewport). If these layers do not exist, AutoCAD will create them. For example, if you create a profile in a viewport whose handle is 6, then visible lines will be placed on a layer PV-6 and hidden lines on layer PH-6. To control the visibility of the layers, you can turn the appropriate layers on and off.

The next prompt determines whether *2D* or *3D* entities are used to represent the visible and hidden lines of the profile:

> Project profile lines onto a profile?

Enter **Y** or **N**. If you answer **Y** (default option), AutoCAD creates the visible and hidden lines of the profile with *2D* AutoCAD entities; N creates the visible and hidden lines of the profile with *3D* AutoCAD entities.

Finally, AutoCAD asks if you want tangential edges deleted. A tangential edge is an imaginary edge at which two facets meet and are tangent. In most of the drafting applications, the tangential edges are not shown. The prompt sequence is as follows for deleting the tangential edges:

> Delete tangential edges?

Enter **Y** to delete the tangential edges and **N** to retain them.

Open the Exercise Manual PDF file for Chapter 15 on the accompanying CD for project- and discipline-specific exercises.

If you have the accompanying Exercise Manual, refer to Chapter 15 for project- and discipline-specific exercises.

REVIEW QUESTIONS

1. What type of coordinate system does AutoCAD use?

    a. Lagrangian system

    b. Right-hand system

    c. Left-hand system

    d. Maxwellian system

    e. None of the above

2. If the origin of the current coordinate system is off screen, then the UCSICON, if it is set to Origin, will:

    a. appear in the lower left corner of the screen

    b. attempt to be at the origin, and will therefore not be on screen

    c. appear at the center of the screen

    d. force AutoCAD to perform a ZOOM so that it will be on screen

3. Which of the following UCS options will perform a translation only on the coordinate system?

    a. Rotate

    b. Z-axis

    c. Origin

    d. View

    e. none of the above

4. Which of the following UCS options will perform only a rotation of the coordinate system?

    a. Rotate

    b. Z-axis

    c. Origin

    d. View

    e. none of the above

5. Which of the following UCS options will perform both a rotation and a translation of the coordinate system?

   a. Rotate
   b. Z-axis
   c. Origin
   d. View
   e. none of the above

6. The command used to generate a perspective view of a *3D* model is:

   a. VPOINT
   b. PERSPECT
   c. VPORT
   d. DVIEW
   e. none of the above

7. The VPOINT command will allow you to specify the viewing direction in all of the following methods *except*:

   a. 2 points
   b. 2 angles
   c. 1 point
   d. dynamically dragging the X, Y, and Z axes
   e. none of the above (that is, all are valid)

8. Objects can be assigned a negative thickness.

   a. True
   b. False

9. Regions:

   a. are *3D* objects
   b. can be created from any group of objects
   c. can be used in conjunction with Boolean operations
   d. cannot be crosshatched

10. The PFACE command:

    a. generates a mesh
    b. requests the vertexes of the object
    c. can create an opaque (will hide) flat polygon
    d. all of the above

11. To align two objects in *3D* space, invoke the command called:
    a. ALIGN
    b. ROTATE
    c. 3DROTATE
    d. TRANSFORM
    e. 3DALIGN

12. *3D* arrays are limited to rectangular patterns (that is, rows, columns, and layers).
    a. True   b. False

12. The UNION command will allow you to select more than two solid objects concurrently.
    a. True   b. False

14. Which of the following is not a standard solid primitive shape used by AutoCAD?
    a. SPHERE
    b. BOX
    c. DOME
    d. WEDGE
    e. CONE

15. Fillets on solid objects are limited to planar edges.
    a. True   b. False

16. A command which will split a solid object into two solids is:
    a. TRIM
    b. SLICE
    c. CUT
    d. SPLIT
    e. DIVIDE

17. To display a more realistic view of a *3D* object, use:
    a. HIDE
    b. VIEW
    c. DISPLAY3D
    d. MAKE3D
    e. VIEWEDIT

19. To force invisible edges of *3D* faces to display, what system variable should be set to 1?

    a. 3DEGE

    b. 3DFRAME

    c. SPLFRAME

    d. EDGEFACE

    e. EDGEVIEW

20. Advantages of solid modeling include:

    a. creation of objects which are manufacturable

    b. interfaces with Computer Aided Manufacturing (CAM)

    c. analysis of physical properties of objects

    d. all of the above

21. The SECTION command:

    a. creates a region

    b. creates a polyline

    c. crosshatches the area where a plane intersects a solid

    d. is an alias for the HATCH command

    e. will give you a choice of either a or b

22. Which AutoCAD command permits a 3D model to be viewed interactively in the current viewport?

    a. 3DVIEW

    b. 3DPAN

    c. 3DROTATE

    d. 3DORBIT

23. When the cursor is positioned inside the arcball of the 3DORBIT command, the 3D object can be viewed _____?

    a. horizontally

    b. vertically

    c. diagonally

    d. all of the above

24. When the cursor is positioned outside the arcball of the 3DORBIT command, the 3D object can be viewed _____?

    a. horizontally

    b. vertically

    c. diagonally

    d. all of the above

    e. both a and b

25. Which of the following is not an option in the 3DORBIT command?

    a. Pan

    b. Zoom

    c. Tilt

    d. none of the above

26. AutoCAD allows you to display a perspective or a parallel projection of the view while 3DORBIT is active.

    a. True

    b. False

27. AutoCAD allows you to view shaded view while 3DORBIT is active.

    a. True

    b. False

28. Which of the following AutoCAD commands allows you to set clipping planes for the objects in 3D orbit view?

    a. CLIPLANE

    b. 3DCLIP

    c. 3DFACED

    d. CLIPVIEW

29. The Continuous Orbit option of 3DORBIT permits the 3D model to be set into motion, even in shaded mode?

    a. True   b. False

30. AutoCAD allows you to edit solid objects by _____.

    a. extruding faces

    b. copying faces

    c. offsetting faces

    d. moving faces

    e. all the above

31. When offsetting a face of a 3D solid a positive value will increase the volume of the solid.

    a. True   b. False

32. A hole in a 3D solid model can be relocated using the Move Faces option within the SOLIDEDIT command.

    a. True   b. False

33. Colors of individual edges of a 3D solid can be changed using SOLIDEDIT?

    a. True   b. False

34. When an edge is copied using the SOLIDEDIT command, it is copied as an _____.

    a. arc

    b. spline

    c. circle

    d. line

    e. all of the above

35. When a composite 3D solid is separated into individual solids, each solid element will retain the layer and color settings of the original solid.

    a. True   b. False

36. AutoCAD refers to the removal of specified edges or vertices as _____.

    a. extraction

    b. cleaning

    c. removing

    d. erasing

CHAPTER 16

Rendering

INTRODUCTION

Shading or rendering turns your three-dimensional model into a realistic (eye-catching) image. The AutoCAD SHADEMODE command allows you to produce quick, shaded models. However, the AutoCAD RENDER command gives you more control over the appearance of the final image. You can add lights and control lighting in your drawing and define the reflective qualities of surfaces in the drawing, making objects dull or shiny. You can create the rendered image of your 3D model entirely within AutoCAD.

After completing this chapter, you will be able to do the following:

- Render a 3D model
- Create and modify lighting—ambient light, distant light, point light, and spotlight
- Create and modify a scene
- Create and modify a material
- Save and replay an image

SHADING A MODEL

The AutoCAD SHADEMODE command lets you produce a shaded picture of the 3D model in the current viewport. It also provides various shading and wireframe options, and you can even edit shaded objects without regenerating the drawing. By default, AutoCAD uses a single light that is logically placed just over your right shoulder in the current viewport. You can display ambient, point, distant, and spot lights defined with the LIGHT command in SHADEMODE. To display these lights, you must set SHADEMODE to Flat Shaded, Gouraud Shaded, Flat Shaded Edges On, or Gouraud Shaded Edges On.

Invoke the SHADEMODE command from the View menu and AutoCAD prompts:

> Enter option [2D wireframe/3D wireframe/Hidden/Flat/Gouraud/
> fLat+edges/gOuraud+edges]: *(select one of the available options from the shortcut menu)*

Choose 2D WIREFRAME to display the objects using lines and curves to represent the boundaries. Raster and OLE objects, linetypes, and lineweights are visible.

Choose 3D WIREFRAME to display the objects using lines and curves to represent the boundaries. Raster and OLE objects, linetypes, and lineweights are visible. AutoCAD also displays a shaded 3D user coordinate system (UCS) icon.

Choose HIDDEN to display the objects using the 3D wireframe representation and hide the lines representing the back faces. Text objects are ignored unless they are given a thickness. The thickness can be as large a value as you want or as small as 0.000001 units (or any value other than 0).

Choose FLAT to display the objects shaded between the polygon faces. The objects appear flatter and less smooth than Gouraud shaded objects. Materials that have been applied to objects are shown.

Choose GOURAUD to display the shaded objects and smooth the edges between polygon faces, giving the objects a smooth, realistic appearance. Materials that have been applied to objects are shown.

Choose FLAT+EDGES to display the objects as a combination of Flat and Wireframe options. The objects are flat shaded with the wireframe showing through.

Choose GOURAUD+EDGES to display the objects as a combination of Gouraud and Wireframe options. The objects are Gouraud shaded with the wireframe showing through.

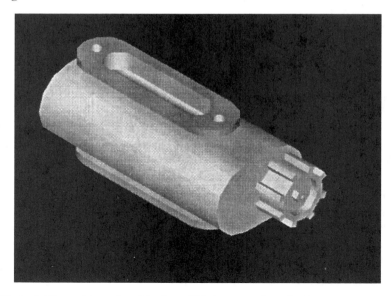

Figure 16-1 *A 3D model rendered by AutoCAD (courtesy Autodesk)*

RENDERING A MODEL

The AutoCAD render facility allows you to create realistic models from your AutoCAD drawings. With the AutoCAD RENDER command you can adjust lighting factors, material finishes, and camera placement. All these options give you a great deal of flexibility. If you wish to take rendering "to the max," you may wish to purchase 3D Studio Max or 3D VIZ to gain the greatest flexibility and photorealism.

The tools available for rendering allow you to adjust the type and quality of the rendering, set up lights and scenes, and save and replay images. But you can always use the RENDER command without any other AutoCAD render setup. By default, RENDER uses the current view if no scene or selection is specified. If there are no lights specified, the RENDER command assumes a default over-the-shoulder distant light source with an intensity of 1.

Invoke the RENDER command from the Render toolbar (see Figure 16–2):

Figure 16–2 *Invoking the* RENDER *command from the Render toolbar*

AutoCAD displays the Render dialog box (seeFigure 16–3).

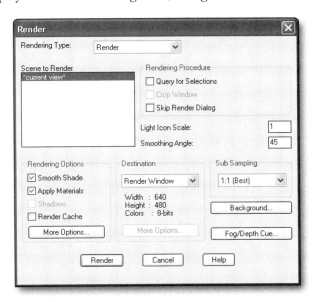

Figure 16–3 *Render dialog box*

The **Rendering Type** list box lists the available rendering types, including AutoCAD Render, Photo Real, and Photo Raytrace. The default type is set to AutoCAD Render.

The **Scene to Render** list box lists the scenes available in the current drawing, including the current view, from which you can select the scene/view for rendering.

The **Rendering Procedure** section of the dialog box sets the default value for rendering. Three settings are available:

> Choose **Query for Selections** to control whether or not to display a prompt to select objects to render.
>
> Choose **Crop Window** to control the prompts for whether or not to pick an area on the screen for rendering.
>
> Choose **Skip Render Dialog** to control whether or not to skip the Render dialog box and render the current view.

Light Icon Scale controls the size of the light blocks inserted in the drawing. The default is set to a scale factor of 1.0. If necessary, you can change the value to some other real number to rescale the blocks. Overhead, Direct, and Sh_spot are the blocks affected by the scale factor.

Smoothing Angle sets the angle at which AutoCAD interprets an edge. The default value is 45 degrees. Angles greater than 45 degrees are considered edges, and those less than 45 degrees as smoothed.

The **Rendering Options** section of the dialog box controls the rendering display. Four main settings plus additional options are available:

> Selecting **Smooth Shade** yields a smooth appearance. Depending on the object's surfaces and the direction of the screen's lights, smooth shading adds a cleaner, more realistic appearance to a rendering.
>
> Selecting **Apply Materials** applies surface materials you have defined and attaches them by color or to specific object(s). If the **Apply Materials** option is not selected, then all objects in the drawing assume the color, ambient, reflection, transparency, refraction, bump map, and roughness attribute values defined for the *GLOBAL* material.
>
> Selecting **Shadows** generates shadows when selected. This option is applicable only when **Photo Real** or **Photo Raytrace** rendering is selected.
>
> Selecting **Render Cache** specifies that rendering information be written to a cache file on the hard disk. As long as the drawing or view is unchanged, the cached file is used for subsequent renderings, eliminating the need for AutoCAD to retessellate.

For fine-tuning the rendering quality, choose **More Options**. AutoCAD displays the Windows Render Options dialog box. The options available vary depending on whether you have selected **Render, Photo Real, Photo Raytrace**, or a third-party application as your rendering type.

The **Destination** section of the dialog box controls the image output setting. Three options are available:

Choose **Viewport** to render to the current viewport.

Choose **File** to render to a file. When the **File** option is selected, you must choose **More Options** to select the file type, set the colors in the output file, and set the PostScript options, if necessary.

Choose **Render Window** to render the model in a window (see Figure 16–4).

In the Render window, select Open from the File menu to open three types of bitmap files: *.bmp*, *.dib*, and *.rle* files. The Save command, also available from the File menu, allows you to save an image to a bitmap file. If desired, you can select Print from the File menu to print the image. Select Copy from the Edit menu to copy an image from the active render window to the clipboard. Select Options from the File menu to display the Windows Render Options dialog box (see Figure 16–5), from which you can select aspect ratios (Size in Pixels) and resolutions (Color Depth) for bitmap images.

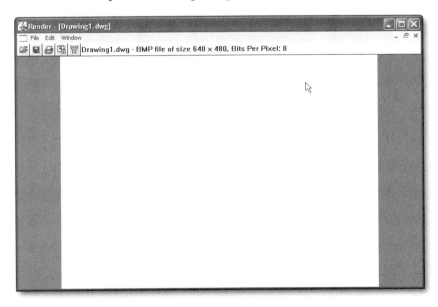

Figure 16–4 *Render window*

The **Sub Sampling** section of the Render dialog box controls the rendering time and image quality without abandoning effects such as shadows by rendering a fraction of all pixels. The available ratios include 1:1 (for best quality) to 8:1 (fastest).

Choose **Background** to set the background for your scene. AutoCAD displays the Background dialog box (see Figure 16–6). Following are the four options available to select the type of background for rendering:

> Choose **Solid** to select a one-color background. Use the controls in the **Colors** section to specify the color.
>
> Choose **Gradient** to specify a two- or three-color gradient background. Use the **Colors** section controls and the **Horizon**, **Height**, and **Rotation** controls on the lower right of the dialog box to define the gradient.
>
> Choose **Image** to use a bitmap (.*bmp*) file for the background. Manipulate the controls in the **Image** and **Environment** sections to define the bitmap.
>
> Choose **Merge** to use the current AutoCAD image as the background.

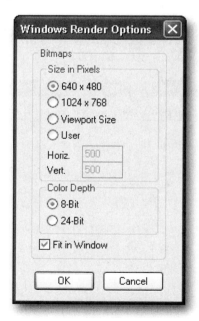

Figure 16–5 *Windows Render Options dialog box*

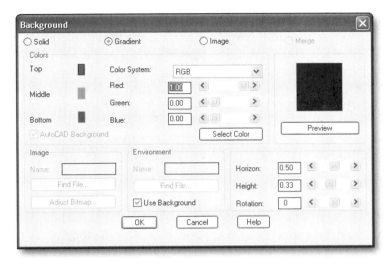

Figure 16–6 *Background dialog box*

Choose **Fog/Depth Cue** in the Render dialog box to set the visual cues for the apparent distance of objects. AutoCAD displays the Fog/Depth Cue dialog box (see Figure 16–7).

Figure 16–7 *Fog/Depth Cue dialog box*

Choose **Enable Fog** to set the fog to ON without affecting the settings in the dialog box. Choose **Fog Background** to apply the fog to the background as well as to the geometry. The **Color Controls** section of the dialog box controls whether AutoCAD uses the red-green-blue color system or the hue-lightness-saturation color system. The **Near Distance** and **Far Distance** controls define where the fog starts and ends. Each value is a percentage of the distance from the camera to the back clipping plane. The **Near Fog Percentage** and **Far Fog Percentage** controls define the percentage of fog at the near and far distances, ranging from 0% fog to 100% fog.

After making all the necessary changes in the Render dialog box, choose **Render**. AutoCAD renders the selected objects in the selected destination window.

SETTING UP LIGHTS

AutoCAD's render facility gives you great control over four types of lights in your renderings:

> **Ambient light** can be thought of as background light that is constant and distributed equally among all objects.

> **Distant light** gives off a fairly straight beam of light that radiates in one direction. Another property of distant light is that its brilliance remains constant, so an object close to the light receives as much light as a distant object.

> **Point light** can be thought of as a ball of light. A point light radiates beams of light in all directions. Point lights also have more natural characteristics. Their brilliance may be diminished as the light moves away from its source. An object that is near a point light appears brighter; an object that is farther away will appear darker.

> **Spotlight** is very much like the kind of spotlight you might be accustomed to seeing at a theater or auditorium. Spotlights produce a cone of light toward a target that you specify.

Invoke the LIGHTS command from the Render toolbar (see Figure 16–8):

Figure 16–8 *Invoking the* LIGHTS *command from the Render toolbar*

Chapter 16 • *Rendering* 895

AutoCAD displays the Lights dialog box (see Figure 16–9).

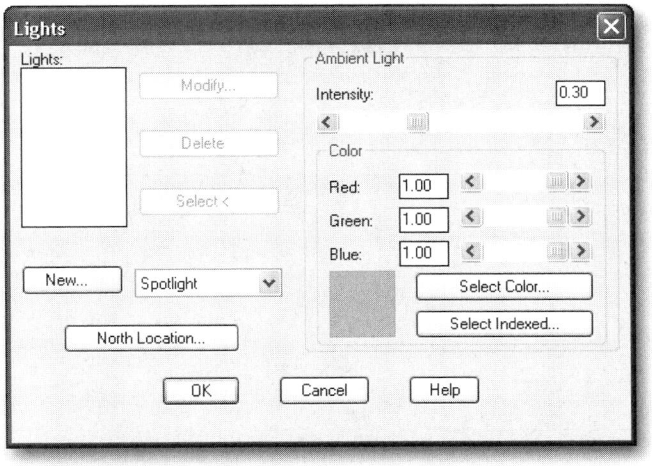

Figure 16–9 *Lights dialog box*

CREATING A NEW LIGHT

In the Lights dialog box, select one of the three available light types from the list box located to the right of the **New** button. Then choose **New**. Depending on the type of light selected, AutoCAD displays the New Point Light dialog box (see Figure 16–10), New Distant Light dialog box (see Figure 16–11), or the New Spotlight dialog box (see Figure 16–12).

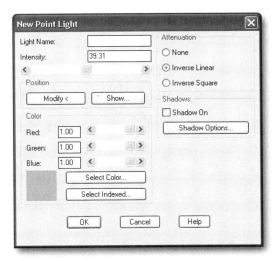

Figure 16–10 *New Point Light dialog box*

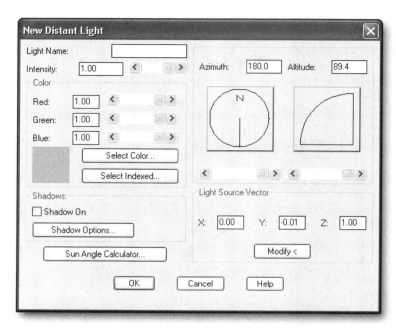

Figure 16–11 *New Distant Light dialog box*

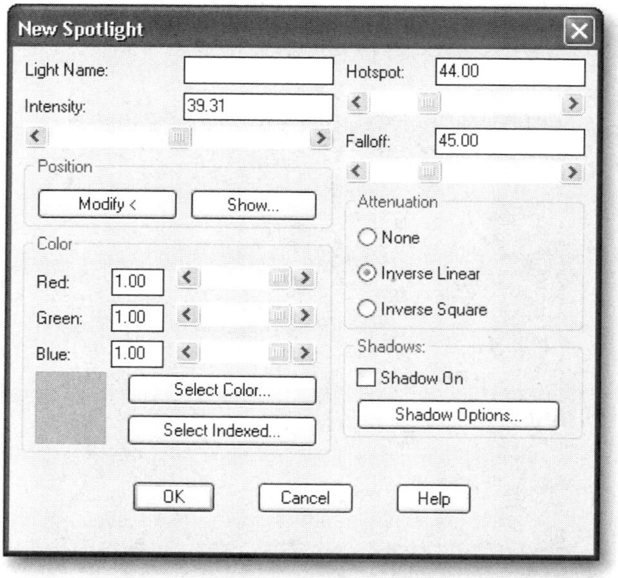

Figure 16–12 *New spotlight dialog box*

Specify the name of the light in the **Light Name** text box. The name must be eight or fewer characters.

The slider bar located below the **Intensity** text box controls the brightness of the light, with 0 turning the light off. Distant light intensity values may range from 0 to 1. Point lights have a more complex intensity setting. This setting can be any real number. The factors that control the maximum intensity are the extents of the drawing and the current rate of falloff. Spotlight intensity factors are the same as for point lights except that the falloff is always inverse linear.

Choose **Modify** and **Show** in the **Position** section to modify or look at the X,Y,Z coordinate location of the light and its target (available only with point lights and spot lights).

The **Color** section controls the current color of the light. To set the color you can either adjust the Red, Green, and Blue slider bars or choose **Select Color** or **Select Indexed** to choose true color or index color respectively.

The setting of the attenuation controls how light diminishes over distance. The attenuation applies to both point light as well as spotlight. Select one of the three radio buttons. Choosing **None** sets no attenuation. Objects far from the point light are as bright as objects close to the light. Choosing **Inverse Linear** decreases the light intensity linearly as the distance increases. For example, an object 10 units away from the light source will be 1/10 as illuminated as an object adjacent to the light source. Choosing **Inverse Square** decreases the light intensity by the inverse of the squared distance. An object 10 units away from the light source receives 1/100 the amount of light of an item adjacent to the light source. This function provides rapid falloff so that a point light can be more localized.

The **Azimuth** and **Altitude** edit boxes in the New Distant Light dialog box specify the position of the distant light by using the site-based coordinates. Azimuth can be set at any value between −180 and 180. Altitude can be set at any value between 0 and 90.

The **Light Source Vector** section in the New Distant Light dialog box displays the coordinates of the light vector that result from the light position you set using Azimuth and Altitude. You can also enter values directly in the edit boxes. AutoCAD updates the corresponding Azimuth and Altitude values.

The **Hotspot** and **Falloff** edit fields in the New Spotlight dialog box specify the angle that defines the brightness cone of light and the full cone of light, respectively.

Once you set the appropriate values in the dialog box, choose **OK** to create a new light and close the dialog box. AutoCAD lists the name of the newly created light in the **Lights** list box of the Lights dialog box.

The **Ambient Light** slider bar allows you to adjust the intensity of the ambient (background) light from a value of 0 (off) to 1 (bright). The **Color** section controls the current color of the ambient light. To set the color you can adjust either the Red, Green, or Blue slider bars or choose **Select Color** or **Select Indexed**. Choose **OK** to accept the changes and close the Lights dialog box.

MODIFYING A LIGHT

In the Lights dialog box, first select the light name to modify from the **Lights** list box and then choose **Modify**. AutoCAD displays the appropriate Modify dialog box, depending on the light type selected. Make the necessary changes, and choose **OK** to accept the changes. You cannot change a light type. For example, you cannot make a point light a distant light. But you can delete the point light and insert a new distant light in the same location.

DELETING A LIGHT

First, select the light name to delete from the **Lights** list box and then choose **Delete**. AutoCAD deletes the selected light.

SELECTING A LIGHT

Choose **Select** to select a light from the screen. AutoCAD temporarily dismisses the dialog box so you can select a light on screen using the pointing device. The Lights dialog box returns, with the selected light highlighted in the **Lights** list.

SETTING UP A SCENE

Inserting and adjusting lights allows you to render images from an unlimited number of viewpoints. If necessary, you can save a certain combination of lights and a particular view as a scene, which you can recall at any time. A scene represents a particular view of the drawing together with one or more lights. Making a scene avoids re-creating a particular set of conditions every time you need to render that image. The VPOINT and DVIEW commands are used to control viewing of a model from any point in model space, and the LIGHT command allows you to add one or more lights to the model or modify lights. The SCENES command allows you to save a scene. You can have an unlimited number of scenes in a drawing.

 Note: The VIEW command allows you to save a view but doesn't save the lights, whereas the scene can include both the view and the light positions.

Invoke the SCENES command from the Render toolbar (see Figure 16-13):

Figure 16-13 *Invoking the* SCENES *command from the Render toolbar*

AutoCAD displays the Scenes dialog box (see Figure 16–14).

Figure 16–14 *Scenes dialog box*

AutoCAD lists the scenes available in the current drawing in the **Scenes:** list box.

CREATING A NEW SCENE

Choose **New** in the Scenes dialog box to add a new scene to the current drawing. AutoCAD displays the New Scene dialog box (see Figure 16–15).

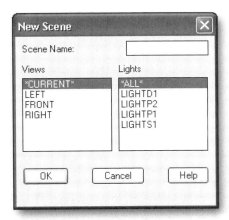

Figure 16–15 *New Scene dialog box*

Specify the name of the scene in the **Scene Name** text box, which may be up to eight characters long.

The **Views** list box displays the list of the views in the current drawing. *CURRENT* is the current view in the active viewport. The active view in the current scene is highlighted. Selecting another view makes it the new view of the scene. You can have only one view in a scene.

The **Lights** list box displays the list of lights in the current drawing. *ALL* represents all the lights in the drawing. When you select the *ALL* option, all the lights in the drawing are added to the scene. The lights in the current scene are highlighted. Holding down the CTRL key while selecting a non-highlighted light adds that light to the scene. Selecting a highlighted light deselects that light and removes it from the scene.

 Note: You can create a scene with no lights, so that in this case the only lighting in the scene is ambient light.

Choose **OK** to create a new scene.

MODIFYING AN EXISTING SCENE

Choose **Modify** in the Scenes dialog box to modify the selected scene. AutoCAD displays the Modify Scene dialog box and allows you to add or delete views and lights or change the name of the scene.

DELETING A SCENE

Choose **Delete** in the Scenes dialog box to delete the selected scene from the current drawing.

MATERIALS

The RMAT command gives you the power to modify the light reflection characteristics of the objects you will render. By modifying these characteristics, you make objects appear rough or shiny. These finish characteristics are stored in the drawing via surface property blocks. The drawing contains one surface property block for each finish you create, an attribute from the name, and AutoCAD color index (ACI) if assigned. You can modify materials by manipulating ambient, diffuse, specular, and roughness factors.

Invoke the RMAT command from the Render toolbar (see Figure 16–16):

Figure 16–16 *Invoking the* MATERIALS *command from the Render toolbar*

AutoCAD displays the Materials dialog box (see Figure 16–17).

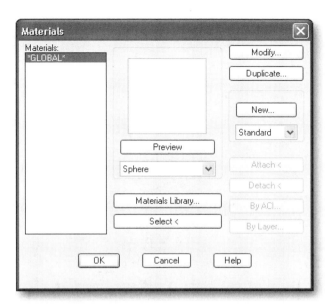

Figure 16–17 *Materials dialog box*

The **Materials** list box lists the available materials. The default for objects with no other material attached is *GLOBAL*.

Choosie **Materials Library** to display the Materials Library dialog box (see Figure 16–18), from which you can select a material.

The Materials Library dialog box allows you to import a predefined material from an *.mli* materials library into the current drawing. The **Current Drawing** list box in the dialog box lists the materials currently in the drawing. The **Current Library** list box lists the materials available in the library file. If necessary, you can preview a sample of the material selected in the list. The sample can be applied to a sphere or square. You can preview only one material at a time. To import materials from the **Current Library** list box into the current drawing, first select the materials you want to import from the **Current Library** list box and then choose **Import**. AutoCAD adds the selected materials to the **Current Drawing** list box. Choose **OK** to close the Materials Library dialog box, and AutoCAD returns control to the Materials dialog box.

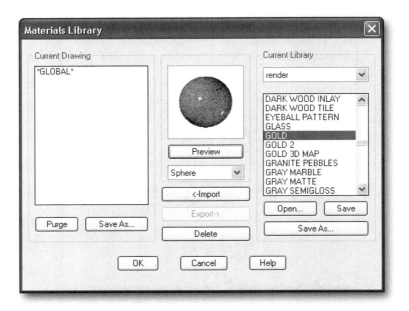

Figure 16-18 *Materials Library dialog box*

Choose **Select** to select an object to display the attached material. AutoCAD temporarily removes the Materials dialog box and displays the graphics area so you can select an object and display the attached material. After you select the object, the Materials dialog box reappears, highlighting the material applied to the object, with the method of attachment displayed at the bottom of the dialog box.

Choose **Modify** to modify a material by displaying the Modify Standard Material dialog box.

Choose **Duplicate** to duplicate a selected material and display the New Standard Material dialog box. Make the necessary changes, and save the material with a new material name.

Choose **New** to create a new material by displaying the New Standard Material dialog box.

Choose **Attach** to display the graphics area so you can select an object and attach the current material to it.

Choose **Detach** to display the graphics area so you can select an object and detach the material from it.

Choose **By ACI** to display the Attach by AutoCAD Color Index (ACI) dialog box, from which you can select the available ACI to attach to a material.

Choose **By Layer** to display the Attach by Layer dialog box, from which you can select a layer by which to attach a material.

Choose **OK** to accept the changes and close the Materials dialog box.

SETTING PREFERENCES FOR RENDERING

The Rendering Preferences dialog box allows you to establish the default settings for rendering. Invoke the RPREF command from the Render toolbar (see Figure 16–19):

Figure 16–19 *Invoking the* RENDER PREFERENCES *command from the Render toolbar*

AutoCAD displays the Rendering Preferences dialog box (see Figure 16–20). The available settings in the Rendering Preferences dialog box are the same as in the Render dialog box. For a detailed explanation of the available settings, refer to the earlier section on "Rendering a Model."

Figure 16–20 *Rendering Preferences dialog box*

SAVING AN IMAGE

You can save the contents of the frame buffer to a *.bmp*, *.tif*, or *.tga* file format by invoking the SAVEIMG command.

Invoke the SAVEIMG command from the Tools menu by choosing Save the Display Image submenu: AutoCAD displays the Save Image dialog box (see Figure 16–21).

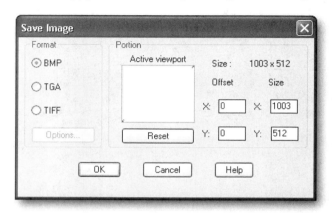

Figure 16–21 *Save Image dialog box*

Select one of the three file formats for the output image. You can save an image in any one of the three industry standard file formats: *.bmp*, *.tga*, or *.tif*.

The **Portion** section specifies the portion of the image to be rendered.

Choose **OK** to save the image. AutoCAD displays the Image File dialog box. Select the appropriate directory in which to save the file, and specify the file name. Choose **Save** to save the image to the given file name.

VIEWING AN IMAGE

The REPLAY command allows you to load a *.bmp*, *.tif*, or *.tga* file into the frame buffer for viewing, converting, or using as a background for a composite rendering.

Invoke the REPLAY command from the Tools menu by choosing View from the Display Image submenu:

AutoCAD displays the Replay File dialog box. Select a file or enter a file name. After you select an image file, choose Open, and an Image Specifications dialog box is displayed (see Figure 16–22).

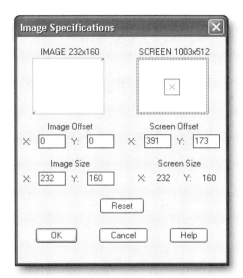

Figure 16–22 *Image Specifications dialog box*

The **Image** tile in the Image Specifications dialog box lets you select a smaller part of the image you want to display. The default size of the image in the image tile reflects the entire display size in pixel measurement, with offset set to 0,0, the lower left corner of the image. To resize the image, specify two points, one for the lower left corner and the other for the upper right corner of the image in the image tile. AutoCAD automatically draws a box to mark the bounds of the reduced image area and updates the values in the *X* and *Y* Image Size coordinates.

The **Screen** tile lets you adjust the offset location of the selected, sized image in relation to your screen. The tile displays the size of your screen or your current viewport. To change the offset, select a point in this tile to offset the center of the image to that point. AutoCAD automatically redraws the image size boundaries to mark the new offsets and updates the values in the *X* and *Y* Screen Offset coordinates.

Choose **Reset** to reset the size and offset values to the original values.

Choose **OK** to close the Image Specifications dialog box.

STATISTICS

The STATISTICS command gives detailed information on your last rendering. This can be useful for diagnosing problems with your drawing.

Invoke the STATISTICS command from the Render toolbar (see Figure 16–23):

Figure 16–23 *Invoking the* STATISTICS *command from the Render toolbar*

AutoCAD displays the Statistics dialog box. If necessary, you can save the information in the Statistics dialog box to a file.

REVIEW QUESTIONS

1. Text is ignored by the SHADEMODE command.

    a. True

    b. False

2. Approximately where is the light source for the SHADE command?

    a. pure ambient light (i.e. all around)

    b. just behind your right shoulder

    c. directly behind the object

    d. in the upper right corner of the screen

3. It is possible to make a slide of a shaded image.

    a. True

    b. False

4. If you save your drawing while a shaded view is displayed, the preview of the drawing will appear shaded.

    a. True

    b. False

5. Which of the following is not a valid type of light you can add to a drawing for the RENDER command?

    a. point

    b. spot

    c. distant

    d. ambient

    e. none of the above (i.e. all are valid)

6. It is possible to make a slide of a rendered image.

    a. True

    b. False

7. If there is a single-point light source in your drawing, objects will be darker if you set the attenuation for that light to:

    a. none

    b. inverse linear

    c. inverse square

    d. both b and c

8. Distant light sources do not attenuate (i.e. get dimmer with distance).
   a. True
   b. False

9. The SCENE command operates much like the VIEW command, but it also saves all the lights you have added to the drawing.
   a. True
   b. False

10. The maximum number of scenes that can be saved in a drawing is:
    a. 8
    b. 64
    c. 256
    d. 32,000
    e. unlimited (except by memory)

11. Material finishes can be assigned based on the color of the object in AutoCAD.
    a. True
    b. False

12. Which of the following file types is not available to the SAVEIMG command?
    a. .bmp
    b. .gif
    c. .tga
    d. .tif

13. By default all lights emit white light.
    a. True
    b. False

14. The smoothing angle is the minimum angle:
    a. between faces which AutoCAD will ignore
    b. between the viewing angle and the face which AutoCAD will display the face
    c. between the incident light and the face which AutoCAD will still reflect the light
    d. none of the above

15. Both 2D and 3D models can be rendered using the SHADEMODE command?

    a. True

    b. False

16. The image option allows which type of files to be used as a background image?

    a. .GIF

    b. .BMP

    c. .JPG

    d. .TIF

17. Which AutoCAD command is used to control the position from which a 3D model is viewed in model space?

    a. VPORTS

    b. VIEWRES

    c. RENDER

    d. VPOINT

18. Which AutoCAD command is used to adjust the light reflection characteristics of an object?

    a. REFL

    b. LGTRFL

    c. LIGHTREF

    d. RMAT

19. Which of the following depicts the point at which a light's value fades to zero?

    a. fade

    b. highlight

    c. falloff

    d. diminish

CHAPTER 17

Customizing AutoCAD

INTRODUCTION

Off the shelf, AutoCAD is extremely powerful. But, like many popular engineering and business software programs, it does not automatically do all things for all users. It does—probably better than any software available—permit users to make changes and additions to the core program to suit individual needs and applications. Word processors offer a feature by which you can save a combination of many keystrokes and invoke them at any time with just one or a combination of two keystrokes or choosing a button on a user-created toolbar. This automation is known as a *macro*. Spreadsheet and database management programs (as well as other types of programs) have their own library of user functions that can be combined and saved as user-named, custom-designed commands. Using these features to make your copy of a generic program unique and more powerful for your particular application is known as *customizing*.

Customizing AutoCAD can include several facets requiring various skill levels. The topics in this chapter include creating custom linetypes, shapes, and hatch patterns. And, with the Customize User Interface (CUI) dialog box, you can customize the interface features used in a drawing session such as workspaces, toolbars, menus, shortcut menus, and keyboard shortcuts.

After completing this chapter, you will be able to customize the following:

- Workspaces
- Toolbars
- Menus
- Shortcut menus
- Keyboard shortcuts
- Mouse buttons

Other customizations included are:

- Loading Partial CUI files
- Creating custom linetypes, hatch patterns, shapes, and fonts

CUSTOMIZING THE USER INTERFACE

To customize workspaces, toolbars, menus, shortcut menus, keyboard shortcuts, or mouse buttons, right-click on any toolbar and choose CUSTOMIZE from the shortcut menu. AutoCAD displays the Customize User Interface (CUI) dialog box (see Figure 17–1).

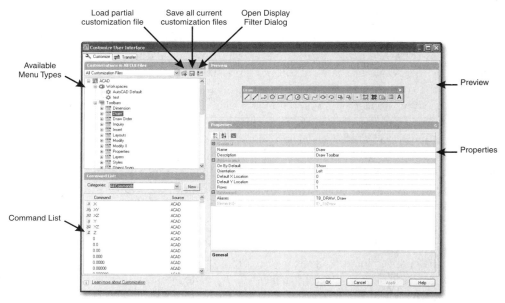

Figure 17–1 *The Customize User Interface (CUI) dialog box with the Customize Tab selected*

Panes shown on this tab include the **Customizations in All CUI Files**, **Preview**, **Command List**, and **Properties** panes. Various panes are displayed, depending on the selection made in the list box, and will be explained along with the node selection with which they are associated.

The buttons to the right of the text box in the **Customizations in All CUI Files** pane include options to **Load partial customization file, Save all current customization files**, and **Open Display Filter Dialog**.

WORKSPACES

The CUI dialog box can be used to edit existing workspaces and create new ones. Application of the workspace concept and using the WORKSPACE command are discussed in Chapter 13.

To customize workspaces, right-click on any toolbar and choose CUSTOMIZE from the shortcut menu. AutoCAD displays the Customize User Interface (CUI) dialog box with the **Customize** tab chosen and the node named **Workspaces** selected in the list of nodes (see Figure 17–2). The **Information** pane, when a node is selected, gives a description of the node (Workspaces in this case) and how it can be used and customized.

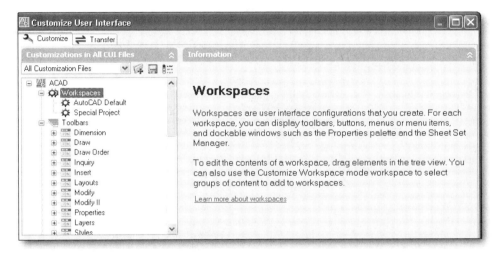

Figure 17–2 *Upper half of the Customize Tab of the Customize User Interface (CUI) dialog box with Workspaces selected and the Information Pane displayed*

In the CUI dialog box, open the **Workspaces** node (double-click or select the + in the square next to the Workspace node), and the named workspaces are displayed in tree mode. If you select the AutoCAD Default workspace (see Figure 17–3), its contents are displayed in the **Workspace Contents** pane. Other panes displayed include the **Command List** and **Properties** panes.

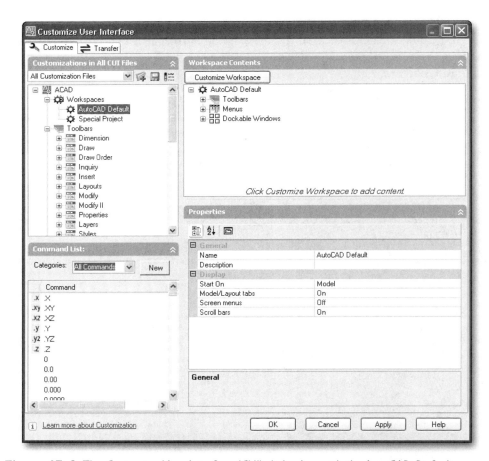

Figure 17–3 *The Customize User Interface (CUI) dialog box with the AutoCAD Default Workspaces selected and the Workspace Contents, Command List, and Properties panes displayed*

  Note: The title of the first pane varies, depending on the selection made in the text box below the title. Titles include **Customizations in All CUI Files**, **Customizations in Main CUI**, **Customizations in custom.cui**, **Customizations in acetmain.cui**, or **Customizations in filename.cui** if *filename.cui* is a Custom User Interface file selected through the Windows search function. The **Customizations in All CUI Files** pane lets you select a node such as Workspaces (see Figure 17–2), a category such as Dimension in the Toolbars node, or a command such as UNDO in the **Menu** node in the Edit toolbar category.

Creating a New Workspace Using the CUI Dialog Box

To create a new workspace, select the Workspace node or one of the individual workspaces in the **Customizations in All CUI Files** pane and right-click to display the shortcut menu. Choose NEW and from the flyout menu choose WORKSPACE.

AutoCAD adds the new workspace to the list and assigns it the name Workspace1. It is in an editable text box where you can change its name. The **Workspace Contents** pane displays the new workspace with a list of empty nodes (Toolbars, Menus, etc.) in a tree view (see Figure 17–4).

Figure 17–4 *New Workspace1 displayed in the CUI Dialog Box Panes*

To populate the workspace with nodes, choose **Customize Workspace** at the top of the **Workspace Contents** pane and then from the list in the **Customizations in All CUI Files** pane, expand (select the adjacent box or double-click) a node such as

Toolbar. AutoCAD lists the available toolbars with empty check boxes beside them. To add a toolbar (for example, Dimension as shown in Figure 17–5) to the newly named workspace, check the box next to the desired toolbar name. The toolbar will appear under the Toolbars in the **Workspace Contents** pane.

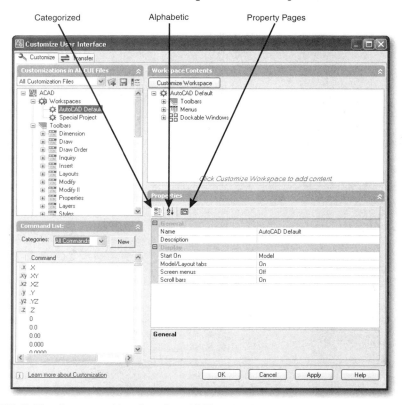

Figure 17–5 *The Customize User Interface (CUI) dialog box with the Dimension toolbar being added to a new Workspace*

The **Properties** pane shows the **Orientation** (floating or docked), **Default X Location**, **Default Y Location** and **Rows** under the **Appearance** property list. These properties can be edited in this pane by selecting the value of the property and then changing its value in place. After the workspace has been named and the nodes arranged, choose **Done** and then **Apply** at the bottom of the CUI dialog box to save the new workspace. Changing the orientation, location, and shape of a toolbar in this manner may not be as exact as it is when you place the toolbar by dragging it and releasing it with the cursor. See the explanation in Chapter 13 of how to save a workspace by using the WORKSPACE command for more exact control of the toolbar location.

Note: At the top of all **Properties** panes are three buttons to sort properties according to categories, sort properties alphabetically, and toggle the display of the tips box below the properties grid. These buttons, when chosen, act in accordance with their respective titles.

To make the new workspace current, right-click on its name in the list in the **Customizations in All CUI Files** pane and choose SET CURRENT. Other options in the shortcut menu include CUSTOMIZE WORKSPACE, DUPLICATE WORKSPACE, RENAME, DELETE, FIND, and REPLACE. The DUPLICATE WORKSPACE option lets you create a new workspace by using the selected one as a basis and making the necessary changes.

TOOLBARS

Toolbars, as explained in Chapter 1, contain the buttons that, when chosen, invoke a command, mode, or custom macro, or display a dialog box or flyout toolbar.

Customizing toolbars can make your drawing tasks easier and more efficient. Frequently used buttons can be consolidated on one toolbar and rarely used buttons can be removed or hidden. You can also specify a text string to be displayed when the cursor is near a button. You can create custom toolbars and flyout toolbars, and create or change the button image associated with a toolbar command. Toolbar buttons that cause flyouts to be displayed have a black triangle in the lower right corner. To create a flyout, you can start from scratch or drag an existing toolbar onto another toolbar.

Creating a New Toolbar Using the CUI Dialog Box

To create a new toolbar, select the Toolbar node or one of the individual toolbars in the **Customizations in All CUI Files** pane of the CUI dialog box.

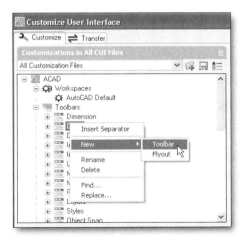

Figure 17–6 *Creating a new toolbar from the flyout menu*

Right-click and choose NEW from the shortcut menu and from the flyout menu choose TOOLBAR (see Figure 17–6). AutoCAD creates a new toolbar in the list with the name Toolbar1 in an editable text box where you can enter a name for the new toolbar. Initially the newly created toolbar has no commands or buttons assigned to it. AutoCAD will not recognize the toolbar if you do not add at least one command to it.

If necessary, you can make changes to the properties of the newly created toolbar. Select the new toolbar in the tree view, and update the Properties pane:

> The **General** section displays and lets you change the **Name** and **Description** of the newly created toolbar.
>
> The **Appearance** section displays and lets you change the status of the **On By Default** mode between **Show** and **Hide**. If you choose **Show**, the selected toolbar will be displayed in all workspaces. The **Orientation** box allows you to choose between **Floating**, **Top**, **Bottom**, **Left**, or **Right**. Specify the default X and Y location in the **Default X Location** and **Default Y Location** boxes respectively. In the **Rows** box, specify the number of rows for an undocked toolbar. In the **Aliases** box, enter an alias for the toolbar.

Adding Commands to a Toolbar Using the CUI Dialog Box

AutoCAD displays a list of commands that are loaded in the program in the **Command List** pane for the selected category. From the **Categories** list box you can select a filter so that only commands in the selected category will be displayed in the main list box below. Categories include **All Commands, ACAD Commands, EXPRESS Commands, Custom Commands, Control Elements, File, Edit, View, Insert, Format, Tools, Draw, Dimension, Modify, Window, Help**, and **Legacy** (see Figure 17–7).

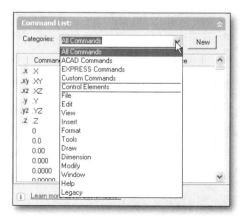

Figure 17–7 *Command categories in the Command List pane*

To add a command to a toolbar, drag the command from the **Command List** pane to a location just below the name of the newly created toolbar in the **Customizations in All CUI Files** pane.

When you choose an individual command that is added to the newly created toolbar in the **Customizations in All CUI Files** pane, AutoCAD displays under the **Properties** pane the lists of properties for the selected command under the headings of **General**, **Macro**, **Advanced,** and **Images** (see Figure 17–8).

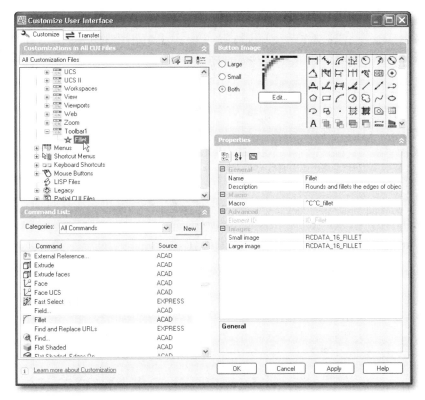

Figure 17–8 *The Customize User Interface (CUI) dialog box with the selected Command Properties*

The **General** section displays the name and description of a user interface element (the command). The name you enter is the label displayed for the element. The description you enter is displayed on the status bar.

The **Macro** section displays the macro assigned to a selected command. You can edit the macro assigned to the button.

The **Advanced** section displays the aliases, DIESEL strings, and element ID that you can define for each user interface element.

The **Images** section is where you can provide an image resource file as the button's icon.

CUSTOM COMMANDS (MACROS)

In addition to placing standard AutoCAD commands on toolbars, you can create custom commands from macro syntax and AutoLISP or combinations of the two. These custom commands can then be added to existing or custom created toolbars.

Creating New Commands Using the CUI Dialog Box

AutoCAD allows you to create a new command from scratch or edit the properties of an existing command. When you create or edit a command, the properties you can define are the command name, description, macro, element ID (for new commands only), and large or small image. When you change any properties of a command in the **Command List** pane, the command is updated for all interface items that reference that command.

To create a new command, choose **New** from the **Command List** pane. AutoCAD creates a new command in the list with the name Command1. In Figure 17–9, three new commands were added to the Furniture toolbar. They were named **Chair**, **Computer**, and **Desk**.

Choose the newly created command and AutoCAD displays under the **Properties** pane the lists of properties for the selected command under the headings of **General**, **Appearance**, **Macro**, **Advanced**, and **Images**.

The **General** section displays the name and description of the selected command. In the **Name** field, enter a name for the command. The name will be displayed as a tooltip or menu name when you select this command. In the **Description** field, enter a description for the command. The description will be displayed on the status bar when the cursor hovers over the menu item or toolbar button.

The **Macro** section displays the macro assigned to a selected command. You can create a macro or edit an existing macro. See the explanation of Menu Macro Syntax later in this chapter. The macro line "^C^C-insert;chair;\;;\\\\" (see Figure 17–9) uses the INSERT command to insert a block named Chair and pauses for you to specify an insertion point and rotation angle on the screen. This command macro requires that a block named Chair be available for insertion or an error message will result. It does not, however, pause for you to specify the X and Y scale factors, but automatically invokes the ENTER key for these prompts.

The **Advanced** section lets you enter an element ID for the command (for new commands only; you cannot modify the element ID of an existing command).

The **Images** section is where you can provide an image resource file as the button's icon.

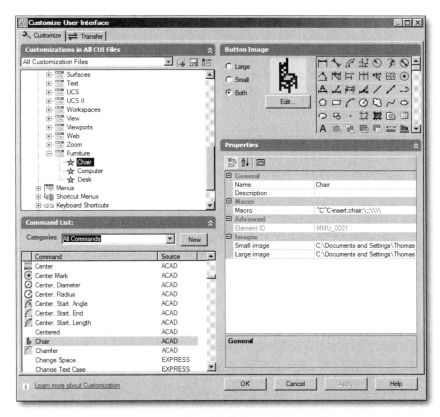

Figure 17-9 *The Customize User Interface (CUI) dialog box with the custom command named Chair highlighted*

In the **Button Image** pane you can select one of the available icon images from the list and if it is not suitable, you can edit it or start from scratch. Choose **Edit** and AutoCAD displays the Button Editor in which you can modify or create button images.

> The **Button Image** window displays a thumbnail of the image as it is edited.
>
> Choosing **Pencil** lets you edit one pixel at a time using the selected color.
>
> Choosing **Line** lets you draw lines using the selected color. Click to set the start point and draw the line, drag, and then release to complete the line.

Choosing **Circle** lets you draw circles using the selected color. Click to set the center and drag to set the radius and then release to complete the circle.

Choosing **Erase** lets you erase pixels. Click and drag over colored pixels to erase them.

The **Color Palette** lets you select the color palette used by the editing tools. Choosing **More** opens the **True Color** tab in the Select Color dialog box. When you select a color, it is displayed in the color swatch.

The **Editing Area** is where you create the button image using the selected editing tool with the specified color.

Choosing **Grid** causes a grid to be displayed in the editing area. Each grid square represents a single pixel.

Choosing **Clear** erases the editing area.

Choosing **Open** opens an existing button image file for editing. Button images are stored as bitmap (*.bmp*) files.

Choosing **Undo** undoes the last action.

Choosing **Save** saves the customized button image.

Choosing **Save As** saves the customized button image using a different name or location.

Choosing **Close** closes the Button Editor dialog box.

Note: Creating a custom command macro (such as Chair in the above example) does not actually create an AutoCAD command named CHAIR that can be entered at the command prompt. The name Chair is merely the word that appears on the button that, when chosen, invokes the INSERT command in a macro that calls for the block named Chair. It is not even significant that the name of the button and the name of the block are the same. In order to create and name a command that can be entered as that name, you must use one of the higher-level programming languages available in AutoCAD such as AutoLISP.

MENUS

Pull-down menus are displayed as a list under a menu bar. In the CUI dialog box, the **Menus** node displays menus defined in all workspaces. Pull-down menus should have one alias in the range of POP1 through POP499. Menus with an alias of POP1 through POP16 are loaded by default when a menu loads. All other menus must be added to a workspace to be displayed.

To edit properties of an existing menu, select the menu name in the **Customizations in All CUI Files** pane and AutoCAD lists the corresponding properties for the selected menu under the headings of **General** and **Advanced**. The **General** section displays the **Name** and **Description** of the selected menu. The **Advanced** section displays the **Alias** name and **Element ID** (read-only). Here you may make changes to the properties of an existing menu.

To edit properties of an individual command listed under an existing menu, first choose the menu name and then choose the name of the command in the **Customizations in All CUI Files** pane (see Figure 17–10). AutoCAD lists the corresponding properties for the selected command in the **Properties** section under the headings of **General**, **Macro**, **Advanced**, and **Images**. Here you may make changes to the properties of an individual command. If desired, you can edit the image of the selected command in the **Button Image** section.

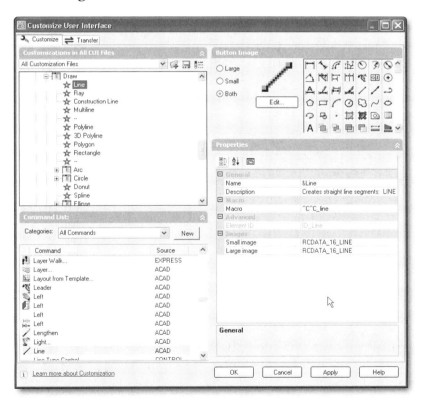

Figure 17–10 *The Customize User Interface (CUI) dialog box with Line selected*

To create a new menu, right-click **Menus** in the **Customizations in All CUI Files** pane and from the shortcut menu, choose MENU from the NEW flyout menu. AutoCAD creates a new menu (named Menu1) and places it at the bottom of the **Menus** tree. Rename it with an appropriate name. Select the new menu in the tree view, and update the **Properties** pane as follows:

> In the **Description** box, enter a description for the menu.
>
> In the **Aliases** box, an alias is automatically assigned to the new menu, based on the number of menus already loaded. For example, if the alias assignment is POP14, thirteen menus are already loaded. Here you can edit the alias.

To add the commands to the newly created menu, from the **Command List** pane, drag the command to a location just below the newly created menu in the **Customizations in All CUI Files** pane. If necessary, make the necessary changes to the properties of the newly added command to the menu.

You can also create a submenu much the same way that you create a menu.

SHORTCUT MENUS

Shortcut Menus are displayed at your cursor location when you right-click. The shortcut menu and the options it provides depend on the pointer location and whether an object is selected or a command is in progress. Context-sensitive shortcut menus display options that are relative to the current command or the selected object. Shortcut menus are referenced by their aliases and are used in specific situations.

To create a new shortcut menu, right-click **Shortcut Menus** in the **Customizations in All CUI Files** pane and from the shortcut menu, choose MENU from the NEW flyout menu. AutoCAD creates a new menu (named ShortcutMenu1) and places it at the bottom of the **Shortcut Menu** tree. Rename it with an appropriate name. Select the new shortcut menu in the tree view, and update the **Properties** pane as follows:

> In the **Description** box, enter a description for the shortcut menu.
>
> In the **Aliases** box, enter additional aliases for this menu. An alias is automatically assigned, and defaults to the next available POP number, based on the number of shortcut menus already loaded in the program.

To add the commands to the newly created shortcut menu, from the **Command List** pane, drag the command to a location just below the newly created menu in the **Customizations in All CUI Files** pane. Here you may make changes to the properties of the newly added command.

You can also create a submenu much the same way that you create a shortcut menu.

KEYBOARD SHORTCUTS

Keyboard shortcuts include shortcut keys and temporary override keys. Keyboard shortcuts are used to assign commands to custom keystroke combinations. You can assign shortcut keys (sometimes called accelerator keys) to commands you use frequently, and temporary override keys to execute a command or change a setting when a key is pressed. Shortcut keys are keys or key combinations that start commands. For example, you can press CTRL+N to create a new drawing file and CTRL+S to save a file, which is the same result as choosing **New** and **Save** from the **File** menu. Temporary override keys are keys that temporarily turn on or turn off one of the drawing aids that are set in the Drafting Settings dialog box (for example, Ortho mode, object snaps, or Polar mode). Shortcut keys can be associated with any command in the command list. You can create new shortcut keys or modify existing shortcut keys.

Temporary override keys execute a command or change a setting only while the key or combination of keys is being pressed. For example, if you are prompted for a second point of a line and the ORTHO mode is ON, you can press and hold the SHIFT key and the ORTHO mode will be temporarily turned OFF. For some of the common overrides there is a temporary override for the left hand (for right-handed mouse operators) and another one for the right hand. For example, to temporarily invoke the Endpoint Object Snap mode with the left hand, the combination of keys to hold down at the same time is SHIFT+E and for the right hand it is SHIFT+P.

Creating a New Shortcut Key Using the CUI Dialog Box

You can create and edit keyboard shortcuts for a selected command in the **Properties** pane. In the CUI dialog box, expand the **Shortcut Keys** node in the **Customizations in All CUI Files** pane. Then drag the command (to be invoked by the new shortcut combination of strokes) from those listed on the **Command List** pane in the CUI dialog box to the list of **Shortcut Keys** in the **Customizations in All CUI Files** pane. In the **Properties** pane, the properties for the newly created shortcut key are displayed.

In the **Key(s)** field, choose the [...] button to open the Shortcut Keys dialog box. In the Shortcut Keys dialog box, in the **Press new shortcut key** field, hold a modifier key (CTRL, ALT, or SHIFT) and press a letter, number, or function key. Valid modifier keys include the following:

Function (Fn) keys containing no modifiers

CTRL+letter, CTRL+number, CTRL+function key

CTRL+ALT+letter, CTRL+ALT+number, CTRL+ALT+function key

SHIFT+CTRL+letter, SHIFT+CTRL+number, SHIFT+CTRL+function key

SHIFT+CTRL+ALT+letter, SHIFT+CTRL+ALT+number, SHIFT+CTRL+ALT+function key

The **Currently Assigned To** section displays any current assignments for the shortcut key. If you do not want to replace the current assignment, use a different shortcut key. Otherwise, choose **Assign**.

 Note: More than one command can share the same shortcut, but only the last command assigned will be active.

Choose **OK** to assign the shortcut key and close the Shortcut Keys dialog box.

Creating a New Temporary Override Key Using the CUI Dialog Box
In the CUI dialog box (see Figure 17–11) select the **Temporary Override** node in the **Customizations in All CUI Files** pane, right-click and choose NEW and from the flyout menu choose TEMPORARY OVERRIDE. AutoCAD creates a new temporary override in the list with the name TemporaryOverride1. Rename it with an appropriate name.

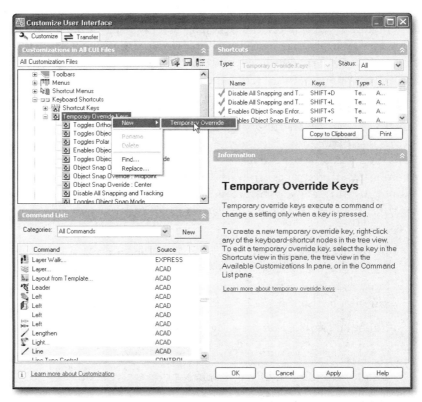

Figure 17–11 *The Customize User Interface (CUI) dialog box creating a new Temporary Override*

The **Properties** pane displays the lists of properties for the new toolbar under the headings of **General** and **Shortcut**.

> The **General** section displays and lets you change the **Name** and **Description** of the new toolbar.
>
> In the **Shortcut** section, click in the **Key(s)** text box and then choose the button on the right. AutoCAD displays the Shortcut Keys dialog box where you can enter the new shortcut key combination after clicking in the **Press new shortcut key** text box. Any current keys assigned to an existing temporary override selected to be edited will be shown in the **Current Keys** list box. Choose **Assign** to assign the selected item to the override or **Remove** to remove the assignment.
>
> In the **Macro 1 (Key Down)** text box, enter the macro code for the new key down mode of the new override or edit the existing override. See the explanation of Menu Macro Syntax later in this chapter.
>
> In the **Macro 2 (Key Up)** text box, enter the macro code for when the temporary override key is released. If no code is defined, releasing the key restores the application to the state that existed before the temporary override was executed.

MOUSE BUTTONS

Mouse buttons control the functions of a Windows system pointing device. You can customize these functions in the Customize User Interface dialog box. For a pointing device with more than two buttons, you can change the functions of the second and third button. The first button on any pointing device cannot be changed in the Customize User Interface dialog box.

Different functions of the pointing device can be created by using various combinations of SHIFT and CTRL keys with clicking. The number of possible functions depends on the number of assignable buttons. The combinations include click, SHIFT+click, CTRL+click, and CTRL+SHIFT+click. The tablet buttons are numbered sequentially.

Adding Commands to Mouse Button Combinations Using the CUI Dialog Box

To add a command to a mouse button combination, drag the command from the **Command List** pane to the name of the mouse button combination in the **Customizations in All CUI Files** pane.

The **Properties** pane displays the lists of properties for the button combination under the headings of **General**, **Macro**, and **Advanced**.

> The **General** section displays and lets you change the **Name** and **Description** of a button combination.

In the **Macro** section text box, enter the macro code for a button combination. See the explanation of Menu Macro Syntax later in this chapter.

The **Advanced** section displays the button combination **Element ID** as read-only.

LEGACY

The **Legacy** node of the **Customizations in All CUI Files** pane lets you make changes in features that are considered obsolete or of minimal usage in the latest version of AutoCAD. These include **Tablet Menus**, **Tablet Buttons**, **Screen Menus**, and **Image Tile Menus**.

PARTIAL CUI FILES

The **Partial CUI Files** node of the **Customizations in All CUI Files** pane lets you load partial CUI files. Partial CUI files are loaded on top of the main CUI file. They allow you to create and modify most interface elements (such as toolbars, menus, and so on) in an external CUI file without having to import the customizations to your main CUI file.

Note: If you define a workspace in a partial CUI file, you need to transfer the contents of the workspace to the main CUI file so that AutoCAD can display the customized workspace. Use the **Transfer** tab in this dialog box to move the workspace.

Loading a Partial CUI File Using the Customize Tab

In the Customize User Interface dialog box, to the right of the drop-down list, choose **Load partial customization file** icon. AutoCAD displays the Open dialog box; choose the partial CUI file you want to open, and choose **Open**.

Note: You need to change the customization group name if the partial CUI file you are attempting to load has the same customization group name as the main CUI file. To create a new customization group name, open the CUI file in the Customize dialog box, select the file name, and right-click to rename it. Choose the main CUI file from the drop-down list in the **Customizations in All CUI Files** pane to check that the file has been loaded into the main CUI file,

In the tree view of the main customization file, click the plus sign (+) next to the **Partial CUI Files** node to expand it. Any partial menus loaded in the main CUI file are displayed.

Choose **OK** to save the changes and close the Customize User Interface dialog box.

If you open the CUI file with a text editor you will see the following message:

```
Warning! Do not edit the contents of this file.
```

If you attempt to edit this file using an XML editor, you could
lose customization and migration functionality. If you need to
change information in the customization file, use the Customize
User Interface dialog box in the product. To access the Customize User Interface dialog box, click the Tools menu > Customize >
Interface, or enter **CUI** on the command line.

Note: If the loaded file configuration is not satisfactory, you can restore the menus and other environmental components by using the MENU command. When prompted to select a CUI file, choose *acad.cui*.

EXTERNAL COMMANDS AND ALIASES

A command alias is an abbreviation that you enter at the On-Screen prompt instead of entering the entire command name. For example, you can enter **l** instead of **line** to start the LINE command. An alias is not the same as a keyboard shortcut, which is a combination of keystrokes, such as CTRL+N for NEW.

An alias can be defined for any AutoCAD command, device driver command, or external command. A command alias is defined in the *acad.pgp* file. You can change existing aliases or add new ones by editing the *acad.pgp* file in an ASCII text editor (such as Notepad). To open the *.pgp* file, choose **Edit Program Parameters (acad.pgp)** from the **Customize** flyout of the **Tools** menu. The file can also contain comment lines preceded by a semicolon (;).

Prior to modifying the *acad.pgp* file, it is recommended that you make a backup copy of the file, such as *xacad.pgp*, so that if you make a mistake, you can restore the original version.

The External Commands are defined at the top of the *acad.pgp* file.

The format for a command line is as follows:

```
<Command name>,<executable>,flags,[*]<Prompt>,    <Return code>
```

The format to define an alias is as follows:

```
<Alias>,*<Full command name>
```

The abbreviation preceding the comma is the character or characters to be entered at the On-Screen prompt. The asterisk (*) must precede the command you wish to be invoked. It may be a standard AutoCAD command name, a custom command name that has been defined in and loaded with AutoLISP or ADS, or a display or machine driver command name. Aliases cannot be used in scripts. You can prefix a command in an alias with the hyphen that causes a command-line version to be used instead of a dialog box, as shown here:

```
BH,  *-BHATCH
```

See Appendix H for list of available aliases from the default *acad.pgp* file.

CUSTOMIZING MENUS WITH MACROS

Selecting an item from a menu or toolbar might execute a command, an AutoLISP routine, or a macro, or cause another menu to be displayed. If you perform an application-specific task on a regular basis that requires multiple steps to accomplish this task, you can place this in a menu macro as shown in the examples of Macro lines (in the Macro text box in the **Properties** pane of the CUI dialog box) earlier in this chapter and have AutoCAD complete all the required procedures in a single step while pausing for input if necessary. Menu macros are similar to script files (files ending with *.scr*). Script files are also capable of executing many commands in sequence but have no decision-making capability and cannot pause for interactive user input.

AutoCAD menu macros can be used in the following kinds of menus:

- Pull-down menus
- Shortcut menus
- Toolbars
- Keyboard shortcuts

MENU MACRO SYNTAX

Following is a partial list of the special charaters you will encounter in menu macros:

Syntax	Description
;	Semicolon; equivalent to pressing ENTER on the keyboard
^M	Caret M; equivalent to pressing ENTER on the keyboard
^I	Caret I; equivalent to pressing TAB on the keyboard
space	A space character; equivalent to pressing ENTER on the keyboard
\	Pause for user input
_	Translates AutoCAD commands and options that follow
=*	Displays the current top-level pull-down, shortcut, or image menu
*^C^C	Repeats a command until another command is chosen
$	Introduces a conditional DIESEL macro expression ($M=)
^B	Repeats a command until another command is chosen
^C	Cancels a command (equivalent to ESC)
^D	Turns Coords on or off (equivalent to CTRL+D)
^E	Sets the next isometric plane (equivalent to CTRL+E)
^G	Turns Grid on or off (equivalent to CTRL+G)
^H	Equivalent to pressing BACKSPACE
^O	Turns Ortho on or off
^P	Turns MENUECHO on or off
^Q	Echoes all prompts, status listings, and input to the printer
^T	Turns tablet on or off (equivalent to CTRL+T)
^V	Changes the current viewport
^Z	Null character that suppresses the automatic addition of SPACEBAR at the end of a command

The following example macro will create a layer called "EL_OFFEQ" (Electrical Office Equipment), assign the color "RED", and make it the current layer:

```
^C^C-LAYER;M;EL-OFFEQ;C;RED;;;
```

This is the equivalent of typing the -LAYER command, selecting the MAKE option, typing **EL-OFFEQ** as the desired layer name, selecting the Color option to assign the color red, and finally, pressing ENTER three times to exit the command. Selecting this macro will execute all of this in one operation. Use the MAKE option of the -LAYER command in case the layer does not yet exist. If the layer does exist it will become the current layer. Note that there is no space after the ^C^C.

CUSTOMIZING HATCH PATTERNS

Certain concepts about hatch patterns should be understood before learning to create one.

1. Hatch patterns are made up of lines or line segment/space combinations. There are no circles or arcs available in hatch patterns as there are in shapes and fonts (which will be covered next).
2. A hatch pattern may be one or more series of repeated parallel lines, repeating dot, or line segment/space combinations. That is, each line in one so-called family is like every other line in that same family. And each line has the same offset and stagger (if it is a segment/space combination), relative to its adjacent sibling, as does every other line.
3. One hatch pattern can contain multiple families of lines. One family of lines may or may not be parallel to other families. With properly specified basepoints, offsets, staggers, segment/space combinations, lengths, and relative angles; you can create a hatch pattern from multiple families of segment/space combinations that will display repeated closed polygons.
4. Each family of lines is drawn with offsets and staggers based on its own specified basepoint and angle.
5. All families of lines in a particular hatch pattern will be located (basepoint), rotated, and scaled as a group. These factors (location, angle of rotation, and scale factor) are determined when the hatch pattern is loaded by the HATCH command and used to fill a closed polygon in a drawing.
6. The pattern can be achieved by specifying parameters in different ways.

Hatch patterns are created by including their definition in a file whose extension is *.pat*. This can be done by using a text editor such as Notepad or a word processor in the non-document (or programmer) mode, which will save the text in ASCII format. Your hatch pattern definition can also be added to the *acad.pat* file. You can also create a new file specifically for a pattern.

Each pattern definition has one header line giving the pattern name and description and a separate specification line describing each family of lines in the pattern.

The header line has the following format:

```
*pattern-name[,description]
```

The pattern name will be the name for which you will be prompted when using the HATCH command. The description is optional and is there so that someone reading the *.pat* file can identify the pattern. The description has no effect, nor will it be displayed while using the HATCH command. The leading asterisk denotes the beginning of a hatch pattern.

The format for a line family is as follows:

```
angle, x-origin, y-origin, delta-x, delta-y [,dash-1, dash-2...]
```

The brackets "[]" denote optional segment/space specifications used for noncontinuous-line families. Note also that any text following a semicolon (;) is for comment only and will be ignored. In all definitions, the angle, origins, and deltas are mandatory (even if their values are zero).

An example of continuous lines that are rotated at 30 degrees and separated by 0.25 units (see Figure 17–12):

```
*P30, 30 degree continuous
30, 0,0, 0,.25
```

The 30 specifies the angle.

The first and second zero specify the coordinates of the origin.

The third zero, though required, is meaningless for continuous lines.

The 0.25 specifies the distance between lines.

A pattern of continuous lines crossing at 60 degrees to each other could be written as follows (see Figure 17–13):

```
*PX60,x-ing @ 60
30, 0,0, 0,.25
330, 0,0, 0,.25
```

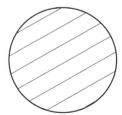

Figure 17–12 *A hatch pattern with continuous lines rotated at 30 degrees and separated by 0.25 units*

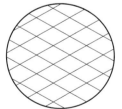

Figure 17–13 *A hatch pattern with continuous lines crossing at 60 degrees to each other and separated by 0.25 units*

A pattern of lines crossing at 90 degrees but having different offsets is as follows (see Figure 17–14):

```
*PX90, x-ing @ 90 w/ 2:1 rectangles
0, 0,0, 0,.25
90, 0,0, 0,.5
```

Note the effect of the delta-Y. It is the amount of offset between lines in one family. Hatch patterns with continuous lines do not require a value (other than zero) for delta-X. Orthogonal continuous lines also do not require values for the X origin unless they are used in a pattern that includes broken lines.

To illustrate the use of a value for the Y origin, two parallel families of lines can be written to define a hatch pattern for steel as follows (see Figure 17–15):

```
*steel
45, 0,0, 0,1
45, 0,.25, 0,1
```

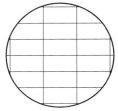
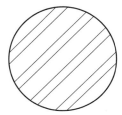

Figure 17–14 *A hatch pattern with lines crossing at 90 degrees and having different offsets*

Figure 17–15 *Defining a hatch pattern for steel*

The folowing three concepts are worthy of notation in this example:

- If the families were not parallel, then specifying origins other than zero would serve no purpose.
- Parallel families of lines should have the same delta-Y offsets. Different offsets would serve little purpose.
- Most important, the delta-Y is at a right angle to the angle of rotation, but the Y origin is in the Y direction of the coordinate system. The steel pattern as written in the example would fill a polygon, as shown in Figure 17–16. Note the dimensions when used with no changes at a scale factor of 1.0 or a rotation angle of zero.

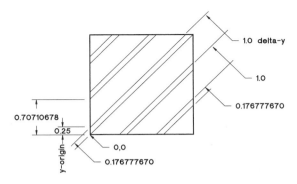

Figure 17–16 *The steel hatch pattern with the delta-Y at a right angle to the angle of rotation and the Y origin in the Y direction of the coordinate system*

CUSTOM HATCH PATTERNS AND TRIGONOMETRY

The dimensions in the hatch pattern resulting from a 0.25 value for the delta-Y of the second line-family definition may not be what you expected (see Figure 17–17). If you wished to have a 0.25 separation between the two line families (see Figure 17–18), then you must either know enough trigonometry/geometry to predict accurate results or else put an additional burden on the user to reply to prompts with the correct responses to achieve those results. For example, you could write the definition as follows:

```
*steel
0, 0,0, 0,1
0, 0,.25, 0,1
```

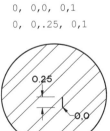
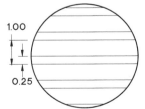

Figure 17–17 *Hatch pattern dimensions resulting from a 0.25 value for the delta Y of the second line-family definition*

Figure 17–18 *Steel hatch pattern defined with a 0.25 separation between the two line families*

In order to use this pattern as shown, the user will have to specify a 45-degree rotation when using it. This will maintain the ratio of 1 to .25 between the offset (delta-Y) and the spacing between families (Y origin). However, if you wish to avoid this inconvenience to the user, but still wish to have the families separated by .25, you can write the definition as follows:

```
*steel
45, 0,0, 0,1
45, 0,.353553391, 0,1
```

The value for the *Y* origin of .353553391 was obtained by dividing .25 by the sine (or cosine) of 45 degrees, which is .70710678. The *X* origin and *Y* origin specify the coordinates of a point. Therefore, setting the origins of any family of continuous lines merely tells AutoCAD that the line must pass through that point. See Figure 17–19 for the trigonometry used.

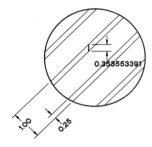

Figure 17–19 *Steel hatch pattern defined with a 45-degree rotation to maintain a 1:25 offset ratio*

For families of lines that have segment/space distances, the point determined by the origins can tell AutoCAD not only that the line passes through that point, but that one of the segments will begin at that point. A dashed pattern can be written as follows (see Figure 17–20):

```
*dashed
0, 0,0, 0,.25, .25,-.25
```

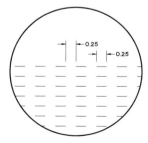

Figure 17–20 *Writing a dashed pattern*

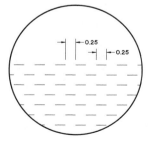

Figure 17–21 *Writing a staggered dashed pattern*

Note that the value of the *X* origin is zero, thus causing the dashes of one line to line up with the dashes of other lines. Staggers can be produced by giving a value to the *X* origin as follows (see Figure 17–21):

```
*dashstagger
0, 0,0, .25,.25, .25,-.25
```

In a manner similar to defining linetypes, you can cause lines in a family to have several lengths of segments and spaces (see Figure 17–22).

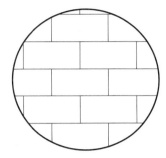

Figure 17–22 *A pattern with several lengths of segments and spaces*

```
*simple
0, 0,0, 0,.5
90, 0,0, 0,1, .5,-.5
```

A similar, but more complex, hatch pattern could be written as follows (see Figure 17–23):

```
*complex
45, 0,0, 0,.5
-45, 0,0, 0,1.414213562, 0,1.41421356
```

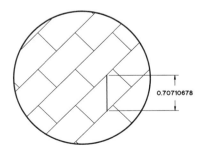

Figure 17–23 *A pattern with more complex hatch patterns*

REPEATING CLOSED POLYGONS

Creating hatch patterns with closed polygons requires planning. For example, a pattern of 45/90/45-degree triangles, as shown in Figure 17–24, should be started by first creating construction lines, as shown in part a of Figure 17–24. Extend the construction lines through points of the object parallel to other lines of the object. A grid pattern will emerge when you place the lines based on distances obtained from the object to form a pattern.

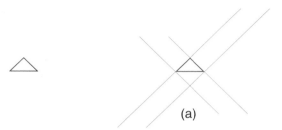

Figure 17–24 *Closed polygons*

It is also helpful to sketch construction lines that are perpendicular to the object lines. This will assist you in specifying segment/space values. In the example, two of the lines are perpendicular to one another, thus making this easier. Figures 17–25 through 17–28 illustrate potential patterns of triangles. Once the pattern is selected, the grid, and some knowledge of trigonometry, will assist you in specifying all of the values in the definition for each line family.

For the following pattern (PA) the horizontal line families can be written as follows (see Figure 17–29):

```
0, 0,0, 0,1, 1,-1
```

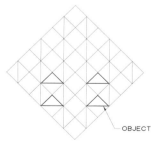

Figure 17–25 *Creating triangular hatch patterns—method #1*

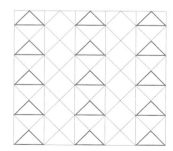

Figure 17–26 *Creating triangular hatch patterns—method #2*

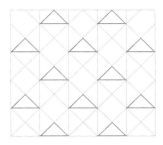

Figure 17–27 *Creating triangular hatch patterns—method #3*

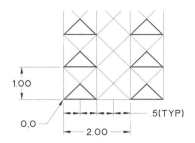

Figure 17–28 *Creating triangular hatch patterns—method #4*

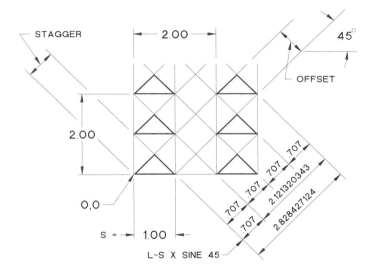

Figure 17–29 *Triangular patterns—Pattern PA*

The specifications for the 45-degree family of lines can be determined by using the following trigonometry:

    sin 45 degrees = 0.70710678
    S = 1
    L = S times sin 45 degrees
    L = 1 times sin 45 degrees = 0.70710678

Note that the trigonometry function is applied to the hypotenuse of the right triangle. In the example, the hypotenuse is 1 unit. A different value would simply produce a proportional result; i.e., a hypotenuse of .5 would produce L = S x 0.70710678 =

0.353553391. The specifications for the 45-degree family of lines could be written as follows:

```
45,     0,0     0.70710678,   0.70710678,   0.70710678,   -2.121320343
angle,  origin, offset,       stagger,      segment,      space
```

For the 135-degree family of lines, the offset, stagger, segment, and space have the same values (absolute) as for the 45-degree family. Only the angle, the *X* origin, and the sign (+ or −) of the offset or stagger may need to be changed.

The 135-degree family of lines could be written as follows:

```
135,  1,0,   -0.70710678,-0.70710678,  0.70710678,-2.121320343
```

Putting the three families of lines together under a header could be written as follows, and as shown in Figure 17–30.

```
*PA,45/90/45 triangles stacked
0,  0,0,  0,1,  1,-1
45, 0,0 0,  0.70710678,0.70710678,  0.70710678,-2.121320343
135, 1,0,  -0.70710678,0.70710678,  0.70710678,-2.121320343
```

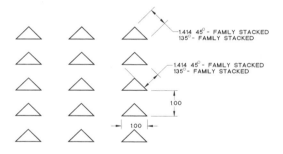

Figure 17–30 *Families of triangular patterns—Pattern PA*

In the preceding statement, "could be written" tells you that there may be other ways to write the definitions. As an exercise, write the descriptions using 225 degrees instead of 45 degrees and 315 instead of 135 for the second and third families, respectively. As a hint, you determine the origin values of each family of lines from the standard coordinate system. But to visualize the offset and stagger, orient the layout grid so that the rotation angle coincides with the zero angle of the coordinate system. Then the signs and the values of delta-*X* and delta-*Y* will be easier to establish along the standard plus for right/up and negatives for left/down directions. Examples of two hatch patterns, PB and HONEYCOMB, follow.

The PB pattern can be written as follows, and as shown in Figure 17–31.

```
*PB, 45/90/45 triangle staggered
0,  0,0,  1,1,  1,-1
45, 0,0,  0,1.414213562,  0.70710678,-0.70710678
135, 1,0, 0,1.414213562,  0.70710678,-0.70710678
```

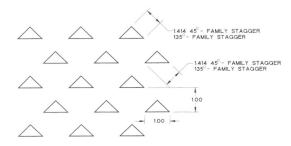

Figure 17–31 *Example of the pattern PB*

Note that this alignment simplifies the definitions of the second and third families of lines over the PA pattern.

The HONEYCOMB pattern can be written as follows, and as shown in Figure 17–32.

```
*HONEYCOMB
90,  0,0,     0.866025399,0.5,  0.577350264,-1.154700538
330, 0,0,     0.866025399,0.5,  0.577350264,-1.154700538
30,  0.5, -0.288675135,  0.866025399, 0.5, 0.577350264,-1.154700538
```

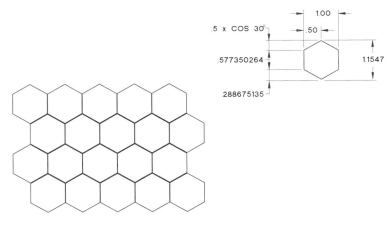

Figure 17–32 *Example of the HONEYCOMB pattern*

CUSTOMIZING SHAPES AND TEXT FONTS

Shapes and fonts are written in the same manner, and both are stored in files with the *.shp* file extension. The *.shp* files must be compiled into *.shx* files. This section covers how to create and save *.shp* files and how to compile *.shp* files into *.shx* files. To compile an *.shp* file into an *.shx* file for shapes or fonts, enter **compile** at the prompt.

From the Select Shape or Font File dialog box, select the file to be compiled. If the file has errors they will be reported; otherwise, the Command window will show:

>   Compilation successful
>   Output file name *.shx* contains nnn bytes

The main difference between shapes and fonts is in the commands used to place them in a drawing. Shapes are drawn by using the SHAPE command, and fonts are drawn using commands that insert text, such as TEXT or DIM. Whether or not an object in an *.shp* or *.shx* file can be used with the SHAPE command or as a font character is partly determined by whether its shape name is written in uppercase or lowercase (explained below).

Each shape or character in a font in an *.shp* or *.shx* file is made up of simplified objects. These objects are simplified lines, arcs, and circles. The reason they are referred to as simplified is because in specifying their directions and distances, you cannot use decimals or architectural units. You must use only integers or integer fractions. For example, if the line distance needs to be equal to 1 divided by the square root of 2 (or .7071068), the fraction 70 divided by 99 (which equals .707070707) is as close as you can get. Rather than call the simplified lines and arcs "objects," we will refer to them as "primitives."

Individual shapes (and font characters) are written and stored in ASCII format. *.shp/.shx* files may contain up to 255 Shape-Characters. Each Shape-Character definition has a header line, as follows:

>   ```
>   *shape number, defbytes, shapename
>   ```

The codes that describe the Shape-Character may take up one or more lines following the header. Most of the simple shapes can be written on one or two lines. The meaning of each item in the header is as follows:

>   The shape number may be from 1 to 255 with no duplications within one file.
>
>   *Defbytes* is the number of bytes used to define the individual Shape-Character, including the required zero that signals the end of a definition. The maximum allowable bytes in a Shape-Character definition is 2,000. Defbytes (the bitcodes) in the definition are separated by commas. You may enclose

pairs of bitcodes within parentheses for clarity of intent, but this does not affect the definition.

The *shapename* should be in uppercase if it is to be used by the SHAPE command. Like a block name is used in the BLOCK command, you enter the shapename when prompted to do so during the SHAPE command. If the shape is a character in a font file, you may make any or all of the shapename characters lowercase, thereby causing the name to be ignored when compiled and stored in memory. It will serve for reference only in the *.shp* file for someone reading that file.

PEN MOVEMENT DISTANCES AND DIRECTIONS

The specifications for pen movement distances and directions (whether the pen is up or down) for drawing the primitives that will make up a Shape-Character are written in bitcodes. Each bitcode is considered one defbyte. Codes 0 through 16 are not Distance-Direction codes, but special instructions-to-AutoCAD codes (to be explained shortly, after the Distance-Direction codes discussion).

Distance-Direction codes have three characters. They begin with a zero. The second character specifies distance. More specifically, it specifies vector length, which may be affected by a scale factor. Vector length and scale factor combine to determine actual distances. The third character specifies direction. There are 16 standard directions available through use of the Distance-Direction bitcode (or defbyte). Vectors 1 unit in length are shown in the 16 standard directions in Figure 17–33.

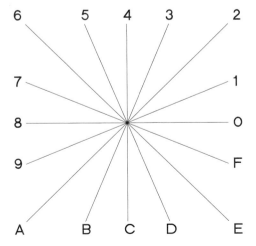

Figure 17–33 *Distance-Direction bitcodes*

Directions 0, 4, 8, and C are equivalent to the default 0, 90, 180, and 270 degrees, respectively. Directions 2, 6, A, and E are 45, 135, 225, and 315 degrees, respectively. But the odd-numbered direction codes are not increments of 22.5 degrees, as you might think. They are directions that coincide with a line whose delta-*X* and delta-*Y* ratio are 1 unit to 2 units. For example, the direction specified by code 1 is equivalent to drawing a line from 0,0 to 1,.5. This equates to approximately 26.56505118 degrees (or the arctangent of 0.5). The direction specified by code 3 equates to 63.434494882 degrees (or the arctangent of 2) and is the same as drawing a line from 0,0 to .5,1.

Distances specified will be measured on the nearest horizontal or vertical axis. For example, 1 unit in the 1 direction specifies a vector that will project 1 unit on the horizontal axis. Three units in the D direction will project 3 units on the vertical axis (downward). So the vector specified as 1 unit in the 1 direction will actually be 1.118033989 units long at an angle of 26.65606118 degrees, and the vector specified as 3 units in the D direction will be 3.354101967 units long at an angle of 296.5650512 degrees. See Figure 17–34 for examples of specifying direction.

To illustrate the codes specifying the Distance-Direction vector, the following example is a definition for a shape called "oddity" that will draw the shape shown in Figure 17–35.

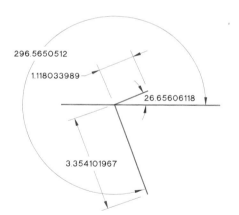

Figure 17–34 *Specified directions*

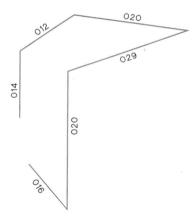

Figure 17–35 *Distance-direction vector specifying code*

The "oddity" shape could could be written as follows:

```
*200,7,ODDITY
014,012,020,029,02C,016,0
```

To draw the shape named "oddity", you would first load the shape file that contains the definition and then use the SHAPE command as follows:

shape
Enter shape name or ⏎ **oddity**
Specify insertion point: *(specify a point)*
Specify height <default>: *(specify a scale factor)*
Specify rotation angle <default>: *(specify a rotation angle)*

 Note: An alternative to the standard Distance-Direction codes is to use codes 8 and 9 to move the pen by paired (delta*X*,delta*Y*) ordinate displacements. This is explained in the following section on "Special Codes."

SPECIAL CODES

Special codes can be written in decimal or hexadecimal. You can specify a special code as 0 through 16 or as 000 through 00E. A three-character defbyte with two leading zeros will be interpreted as a hexadecimal special code. Code 10 is a special code in decimal. However, 010 is equivalent to decimal 16. But more important, it will be interpreted by AutoCAD as a Distance-Direction code with a vector length of 1 and a direction of 0. The hexadecimal equivalent to 10 is 00A. The code functions are as follows.

Code 0: End of Shape The end of each separate shape definition must be marked with the code 0.

Codes 1 and 2: Pen Up and Down The Pen Down (or Draw) mode is on at the beginning of each shape. Code 2 turns the Draw mode off or lifts the pen. This permits moving the pen without drawing. Code 1 turns the Draw mode on.

Note the relationship between the insertion point specified during the SHAPE command and where you wish the object and its primitives to be located. If you wish for AutoCAD to begin drawing a primitive in the shape at a point remote from the insertion point, then you must lift the pen with a code 2 and move the pen (with the proper codes) and then lower the pen with a code 1. Movement of the pen (directed by other codes) after a Pen Down code 1 is what causes AutoCAD to draw primitives in a shape.

Codes 3 and 4: Scale Factors Individual (and groups of) primitives within a shape can be increased or decreased in size by integer factors as follows. Code 3 tells AutoCAD to divide the subsequent vectors by the number that immediately follows the code 3. Code 4 tells AutoCAD to multiply the subsequent vectors by the number that immediately follows the code 4.

Scale factors are cumulative. The advantage of this is that you can specify a scale factor that is the quotient of two integers. A two-thirds scale factor can be achieved by a code 4 followed by a factor of 2 followed by a code 3 followed by a factor of 3. But the effects of scale factor codes must be reversed when they are no longer needed. They do not go away by themselves. Therefore, at the end of the definition (or when you wish to return to normal or other scaling within the definition), the scale factor must be countered. For example, when you wish to return to the normal scale from a two-thirds scale, you must use code 3 followed by a factor of 3 followed by a code 4 followed by a factor of 2. There is no law that states you must always return to normal from a scaled mode. You can, with codes 3 and 4 and the correct factors, change from a two-thirds scale to a one-third scale for drawing additional primitives within the shape. You should *always*, however, return to the normal scale at the end of the definition. A scale factor in effect at the end of one shape will carry over to the next shape.

Codes 5 and 6: Saving and Recalling Locations Each location in a shape definition is specified relative to a previous location. However, once the pen is at a particular location, you can store that location for later use within that shape definition before moving on. This is handy when an object has several primitives starting or ending at the same location. For example, a wheel with spokes would be easier to define by using code 5 to store the center location, drawing a spoke, and then using code 6 to return to the center.

Storing and recalling locations are known as *pushing* and *popping* them, respectively, in a stack. The stack storage is limited to four locations at any one time. The order in which they are popped is the reverse of the order in which they were pushed. Every location pushed must be popped.

More pushes than pops will result in the following Command line error message:

    Position stack overflow in shape nnn

More pops than pushes will result in the following Command line error message:

    Position stack underflow in shape nnn

Code 7: Subshape One shape in an *.shp* or *.shx* file can be included in the definition of another shape in the same file by using the code 7 followed by the inserted shape's number.

Codes 8 and 9: *X-Y* Displacements Normal vector lengths range from 1 to 15 and can be drawn in one of the 16 standard directions unless you use a code 8 or code 9 to specify *X-Y* displacements. A code 8 tells AutoCAD to use the next two bytes as the *X* and *Y* displacements, respectively. For example, 8, (7,−8) tells AutoCAD to move the pen a distance that is 7 in the *X* direction and *8* in the Y direction. The parentheses are optional. After the displacement bytes, specifications revert to normal.

Code 9 Code 9 tells AutoCAD to use all following pairs of bytes as *X-Y* displacements until terminated by a pair of zeros. For example; 9,(7,–8),(14,9), (–17,3),(0,0) tells AutoCAD to use the three pairs of values for displacements for the current mode and then revert to normal after the (0,0) pair.

Code 00A: Octant Arc Code 00A (or 10) tells AutoCAD to use the next two bytes to define an arc. It is referred to as an octant (an increment of 45 degrees) arc. Octant arcs start and end on octant boundaries. Figure 17–36 shows the code numbers for the octants. The specification is written in the following format:

```
10, radius, (-)OSC
```

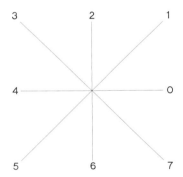

Figure 17–36 *Code numbers for octants*

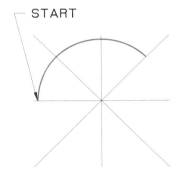

Figure 17–37 *An arc drawn with code 10,(2,–043)*

The radius may range from 1 to 255. The second byte begins with zero and specifies the direction by its sign (clockwise if negative, counterclockwise otherwise), the starting octant (S), and the number of octants it spans (C), which may be written as 0 to 7, with 0 being 8 (a full circle). Figure 17–37 shows an arc drawn with the following codes:

```
10,(2,-043)
```

The arc has a radius of 2, begins at octant arc 4, and turns 135 degrees (3 octants) clockwise.

Code 00B: Fractional Arc Code 00B (11) can be used to specify an arc that begins and ends at points other than the octants. The definition is written as follows:

```
11, start-offset, end-offset, high-radius, low-radius, (-)OSC
```

Start and end offsets specify how far from an octant the arc starts and ends. The high-radius, if not zero, specifies a radius greater than 255. The low-radius is specified in the same manner as the radius in a code 10 arc, as are the starting octant and octants covering specifications in the last byte. The presence of the negative also signifies a clockwise direction.

The units of offset from an octant are a fraction of 1 degree times 45 divided by 256, or approximately .17578125 degrees. For example, if you wish to specify the starting value near 60 degrees, the equation would be:

offset = (60-45)*(256/45) = 85.333333

So the specification value would be 85.

To end the arc at 102 degrees, the equation would be:

offset = (102-90)*(256/45) = 68.2666667

So the specification value would be 68.

To draw an arc with a radius of 2 that starts near 60 degrees and ends near 102 degrees, the specifications would be as follows:

    11,(85,68,0,2,012)

The last byte (012) specifies the starting octant to be 1 (45 degrees) and the ending octant to be 2 (90 degrees).

Codes 00C and 00D: Bulge-Specified Arc Codes 00C and 00D (12 and 13) are used to specify arcs in a different manner from octant codes. Codes 00C and 00D call out bulge factors to be applied to a vector displacement. The effect of using code 00C or 00D involves specifying the endpoints of a flexible line by the *X-Y* displacement method and then specifying the bulge. The bulge determines the distance from the straight line between the endpoints and the extreme point on the arc. The bulge can range from –127 to 127. The maximum or minimum values (127 or –127) define a 180-degree arc (half circle). Smaller values define proportionately smaller-degree arcs. That is, an arc of a given value, say x, will be x times 180 divided by 127 degrees. A bulge value of zero will define a straight line.

Code 00C precedes a single-bulge-defined arcs and 00D precedes multiple arcs. This is similar to the way codes 008 and 009 work on *X-Y* displacement lines. Code 00D, like 009, must be terminated by a 0,0 byte pair. You can specify a series of bulge arcs and lines without exiting the code 00D by using the zero bulge value for the lines.

Code 00E: Flag Vertical Text Command Code 00E (14) is only for dual-orientation text font descriptions, where a font might be used in either horizontal or vertical orientations. When code 00E is encountered in the shape definition, the next code will be ignored if the text is horizontal.

TEXT FONTS

Text fonts are special shape files written for use with AutoCAD text drawing commands. The shape numbers should correspond to ASCII codes for characters. Table 17–1 shows the ASCII codes. Codes 1 through 31 are reserved for special control characters. Only code 10 (line feed) is used in AutoCAD. In order to be used as a font, the file must include a special shape number, 0, to describe the font. Its format is as follows:

```
*0,4,fontname
above, below, modes, 0
```

Note: Codes 1 to 31 are for control characters, only one of which is used in AutoCAD text fonts.

"Above" specifies the number of vector lengths that uppercase letters extend above the baseline, and "below" specifies the number of vector lengths that lowercase letters extend below the baseline. A modes byte value of zero (0) defines a horizontal (normal) mode, and a value of two (2) defines dual-orientation (horizontal or vertical). A value of 2 must be present in order for the special code 00E (14) to operate.

Standard AutoCAD fonts include special shape numbers 127, 128, and 129 for the degrees symbol, plus/minus symbol, and diameter dimensioning symbol, respectively.

The definition of a character from the *txt.shp* file is as follows:

```
*65,21,uca
2,14,8,(-2,-6),1,024,043,04D,02C,2,047,1,040,2,02E,14,8,(-4,
  -3),0
```

Note that the number 65 corresponds to the ASCII character that is an uppercase "A." The name "uca" (for uppercase a) is in lowercase to avoid taking up memory. As an exercise, you can follow the defbytes to see how the character is drawn. The given character definition starts by lifting the pen. A font containing the alphanumeric characters must take into consideration the spaces between characters. This is done by having similar starting and stopping points based on each character's particular width.

Table 17-1 ASCII Codes for Text Fonts

Code	Character	Code	Character	Code	Character
32	space	64	@	96	' left apostrophe
33	!	65	A	97	a
34	" double quote	66	B	98	b
35	#	67	C	99	c
36	$	68	D	100	d
37	%	69	E	101	e
38	&	70	F	102	f
39	' apostrophe	71	G	103	g
40	(	72	H	104	h
41	)	73	I	105	i
42	*	74	J	106	j
43	+	75	K	107	k
44	, comma	76	L	108	l
45	- hyphen	77	M	109	m
46	. period	78	N	110	n
47	/	79	O	111	o
48	0	80	P	112	p
49	1	81	Q	113	q
50	2	82	R	114	r
51	3	83	S	115	s
52	4	84	T	116	t
53	5	85	U	117	u
54	6	86	V	118	v
55	7	87	W	119	w
56	8	88	X	120	x
57	9	89	Y	121	y
58	: colon	90	Z	122	z
59	; semicolon	91	[	123	{
60	<	92	\ backslash	124	| vertical bar
61	=	93	]	125	}
62	>	94	^ caret	126	~ tilde
63	?	95	_ underscore		

CUSTOM LINETYPES

Linetype definitions are stored in files with an *.lin* extension. Approximately 40 standard linetype definitions are stored for use in the *acad.lin* file. The definitions are in ASCII format and can be edited, or you can add new ones of your own by using either a text editor or the CREATE option of the -LINETYPE command. Or you can save new or existing linetype definitions in another *filename.lin* file.

Simple linetypes consist of series of dashes, dots, and spaces. Their definitions are considered the in-line pen-up/pen-down type. Complex linetypes have repeating "out-of-line" objects, such as text and shapes, along with the optional in-line dashes, dots, and spaces. These are used in mapping or surveying drawings for such things as topography lines, fences, utilities, and many other descriptive lines. Instrumentation or control drawings also use many lines with repeating shapes to indicate the purpose of each line.

Each linetype definition in a file comprises two lines. The first line must begin with an asterisk, followed by the linetype name and an optional description, in the following format:

```
*ltname,description
```

The second line gives the alignment and description by using proper codes and symbols, in the following format:

```
alignment,patdesc-1,patdesc-2,...
```

A simple linetype definition for two dashes and a dot, called DDD, could be written as follows:

```
*DDD, _ _ _   _ _ _    .   _ _ _   _ _ _    .   _ _ _   _ _ _
A,.75,-.5,.75,-.5,0,-.5
```

The linetype name is DDD. A graphic description of underscores, spaces, and periods follow. The dashes are given as .75 in length (positive for pen down) separated by spaces −.5 in length (negative for pen up), with the 0 specifying a dot. No character other than the A should be entered for the alignment; it is the only one applicable at this time. This type of alignment causes the lines to begin and end with dashes (except for linetypes with dots only).

The complex linetype definitions include a descriptor (enclosed in square brackets) in addition to the alignment and dash/dot/space specification. A shape descriptor will include the shape name, shape file, and optional transform specification, as follows:

```
[shapename,filename,transform]
```

A text descriptor will include the actual text string (in quotes), the text style, and optional transform specification, as follows:

```
["string",textstyle,transform]
```

Transform specifications (if included) can be one or more of the following:

A=##	absolute rotation	X=##	X offset
R=##	relative rotation	Y=##	Y offset
S=##	scale		

The ## for rotation is in decimal degrees (plus or minus); for scale and offset it is in decimal units.

The following example of an embedded shape in a line for an instrument air line (with repeating circles) could be written as follows:

```
*INSTRAIR, _ _ _   [CIRC]   _ _ _ _   [CIRC]   _ _ _ _
A,2.0,-.5,[CIRC,ctrls.shx],-.5
```

If the ctrls.shx file contains a proper shape description of the desired circle, it will be repeated in the broken line (with spaces on each side) when applied as the INSTRAIR linetype. If the scale of the circle needed to be doubled in order to have the proper appearance, it could be written as follows:

```
*INSTRAIR, _ _ _   [CIRC]   _ _ _ _   [CIRC]   _ _ _ _
A,2.0,-.5,[CIRC,ctrls.shx,S=2],-.5
```

The following example of an embedded text string in a line for a storm sewer (with repeating SS's) could be written as follows:

```
*STRMSWR, _ _ _   SS   _ _ _ _   SS   _ _ _ _
A,3.0,-1.0,["SS",simplex,S=1,R=0,X=0,Y=-0.125],-1.0
```

EXPRESS TOOLS

In addition to the customization tools, AutoCAD has a set of tools under the category of Express tools that will make your job easier in using AutoCAD. You can install AutoCAD Express Tools from the install program provided in the AutoCAD 2006 CD. Before you install the Express tools, AutoCAD must already be installed on your system.

See Appendix I for a list of commands available with a brief description.

CUSTOMIZING AND PROGRAMMING LANGUAGE

There are various other topics with respect to the customization of AutoCAD, some of which are beyond the scope of this book. AutoCAD provides a programming language called AutoLISP (refer to Chapter 18). AutoLISP is a structured programming language similar in a number of ways to other programming languages. A compiled language (AutoCAD supports a number of these, too) is first converted into object code (this is called compiling) or machine language and then linked with various other compiled object code modules (this is called linking) to form an executable file. C/C++ and ARx (AutoCAD Runtime Extension) are examples of compiled programming languages.

REVIEW QUESTIONS

1. The Customize User Interface dialog box can be used to customize:

    a. workspaces

    b. toolbars

    c. menus

    d. shortcuts

    e. all of the above

2. AutoCAD contains no workspaces unless you create one.

    a. True

    b. False

3. A workspace contains:

    a. rows

    b. columns

    c. orientation

    d. docking location

    e. none of the above

4. A toolbar can contain:

    a. rows

    b. columns

    c. orientation

    d. flyouts

    e. all of the above

5. Customization can include:

    a. macro language

    b. AutoLISP functions

    c. DIESEL expressions

    d. script files

    e. all of the above

6. You can modify the element ID of an existing command.

    a. True

    b. False

7. The purpose of the *acad.pgp* file is to:

    a. allow other programs to be accessed while editing a drawing

    b. enable shape files to be compiled

    c. store system configurations

    d. serve as a "file manager" for system variables

    e. none of the above

8. AutoCAD menu files are stored with what type of file extension?

    a. *dwg*

    b. *dxf*

    c. *mnu*

    d. *men*

    e. none of the above

9. When developing screen menus, the information you would like to see displayed in the screen menus should be:

    a. typed in uppercase letters only

    b. enclosed with square brackets "[]"

    c. longer than four characters, but shorter than ten characters

    d. all of the above

10. What file defines external commands and their parameters?

    a. *acad.dwk*

    b. *acad.ext*

    c. *acad.pgp*

    d. *acad.lsp*

    e. *acad.cmd*

11. The system variable used in conjunction with DIESEL to modify the contents of the status line is:

    a. STATUSLINE

    b. MACRO

    c. MODEMACRO

    d. MODESTATUS

12. AutoCAD allows you to create your own icons for use in toolbar menus.

    a. True

    b. False

13. Command aliases are stored in what file?

    a. *acad.ini*

    b. *acad.als*

    c. *acad.pgp*

    d. *acad.mnu*

    e. none of the above

14. The first line of a custom crosshatching definition always begins with:
    a. *
    b. **
    c. ***
    d. !
    e. !!

15. To define a custom linetype with three elements—a 1-unit dash, a 0.5-unit dash, and a dot, all separated by 0.25 unit spaces—what would the definition look like?
    a. A,1,0.25,0.5,0.25,0
    b. A,1,.25,0.5,0.25,0,0.25
    c. A,1,−0.25,0.5,−0.25,0
    d. A,1,−0.25,0.5,−0.25,0,−0.25
    e. A,1,0.5,0,−0.25

16. The bit code which specifies a direction of 12 o'clock in a shape file is:
    a. 0
    b. 4
    c. 8
    d. 12
    e. C

17. When specifying the name of a shape character, in order to conserve memory you should:
    a. use uppercase
    b. use lowercase
    c. preface the name with an *
    d. use a short name
    e. it doesn't matter; all shapes require the same amount of memory regardless of their names

18. A macro permits the user to save and invoke a series of keystrokes with a combination of only one or two keystrokes.
    a. True
    b. False

19. Which of the following can be customized in AutoCAD?
    a. aliases
    b. linetypes
    c. shapes
    d. hatch patterns
    e. all of the above

20. Command aliases can include command options.
    a. True
    b. False

21. Which of the following is the correct format to define an alias using *L* for the LINE command?
    a. Line = L
    b. L = <line>
    c. L, *line
    d. L = *line

22. Which of the following is not a valid AutoCAD menu type?
    a. pull-down
    b. screen
    c. tablet
    d. tooltip
    e. all of the above

23. When you're writing a script file, command aliases cannot be used.
    a. True
    b. False

24. How many pull-down menus does AutoCAD have available to use?
    a. 12
    b. 256
    c. 498
    d. 1024
    e. unlimited

CHAPTER 18

The Tablet and Digitizing

INTRODUCTION

A digitizing tablet (digitizer) can be used to trace geometry on an overlaid paper into a drawing file. It can also be used to invoke commands from an appropriately configured overlay. Using the Wintab driver, the tablet pointer can be used instead of a mouse as a system pointer, to choose menu items and drawing objects, or to interact with the operating system.

The digitizer must first be configured and then calibrated. When configured, areas of the tablet surface are assigned as command input areas and a screen-pointing area. When calibrated, the digitizer can be used to trace geometry from an overlaid paper drawing, map or image into a drawing.

You can alternate between using the tablet uncalibrated, as a system pointer (Tablet mode off) or calibrated, for digitizing a drawing (Tablet mode on) by clicking the **Tablet** button on the status bar.

After completing this chapter, you will be able to do the following:

- Configure the tablet areas
- Calibrate the tablet for digitizing

TABLET OPERATION

When a digitizing tablet, and an overlay have been installed and properly set up to work together, you can use the digitizer's attached puck on the tablet surface to achieve the following results.

>**Normal operation** The most common use of a tablet is to allow the user to move the puck and press the pick button while pointing to one of the various commands or symbols on a preprinted overlay, and be able to invoke that particular command or initiate a program (perhaps written by that user) that will draw the object(s) that that symbol represents. In addition, when the puck is moved within the overlay's designated

screen area, the screen cursor will mimic the puck movement, thereby permitting the user to specify points or select objects on the screen.

Mouse movement Some tablets have an option that causes the puck to emulate mouse-type movement rather than absolute movement. Mouse emulation means that if you pick the puck up off of the tablet surface and move it to another place in the screen area, the screen cursor does not move. The cursor moves only with puck movement while it is on the tablet and in the screen area. Absolute (normal) tablet-puck operation means that, once configured, each point in the tablet's screen area corresponds to only one point on the screen. So, while in the normal mode, if you pick up the puck and put it down at another point in the tablet screen area, the cursor will immediately move to the screen's corresponding point.

Paper copying By switching Tablet mode to ON, you can cause points on the tablet to correspond to drawing coordinates rather than to screen pixel locations as it does in the normal or mouse operations just described. This allows you to fix a drawing (like a map) on the tablet surface, select two points on the map, and specify their coordinate locations on the map, after which the puck movement around the map will cause screen cursor movement to correspond to the same coordinates in the computer-generated drawing. Options and precautions for using this feature (referred to as *digitizing*) are discussed later in this chapter.

TABLET CONFIGURATION

The intended procedure is to have a preprinted template (the overlay) arranged on a sheet that can be fixed to the tablet. Tablet menu areas can then be configured to coincide with the template. Although you could try to configure a bare tablet, it would be difficult to select the required points for rectangular menu areas and also impractical to try and place a template on the tablet after it was configured in such a manner. However, if a tablet has been configured for one template, you can use another template in the same location without reconfiguring as long as the areas are the same. One benefit of this is being able to change from one set of icons/symbols or commands to another set without having to configure again. However, a change in the menu must be made in order to accommodate changes in the template, even if the configuration is the same.

AutoCAD supports a multi-button pointing device and the tablet overlay that is provided with the AutoCAD program package. That overlay is approximately 11" x 11" and has four areas for selecting icon/commands and a screen area, as shown in Figure 18–1. The pointing device (the puck furnished with every tablet) usually has three or more buttons (the setup supports up to a 10-button puck or mouse), and the cursor movement on the screen mimics the puck's movement in the tablet's configured screen area.

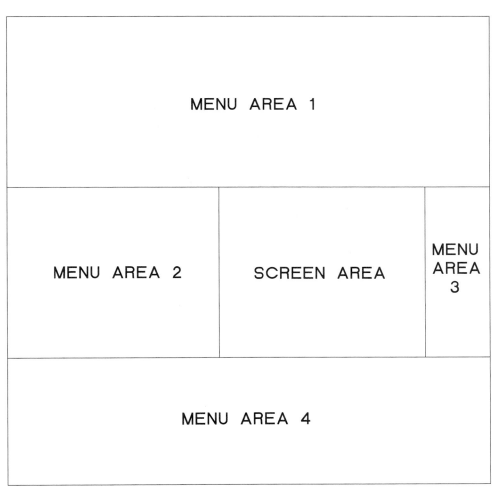

Figure 18–1 *The four screen areas of the overlay*

CUSTOM MENUS

As mentioned in chapter 17, the **Legacy** node of the **Customizations in All CUI Files** pane lets you make changes in Tablet Menus,

The **Tablet Menus** node organizes each of the four tablet menus by column and row. Commands are assigned to a row. Drag a command to assign it to a **Row**. Drag a command to the **Column** node to add the command to the next open row.

You can configure up to four areas of your digitizing tablet as menu areas for command input. The nodes in the Customize User Interface dialog box are labeled **Tablet Menu 1** through **Tablet Menu 4** and define the macros associated with tablet selections.

TABLET COMMAND

The TABLET command is used to switch between digitizing paper drawings and normal command/icon/screen area selections on a configured overlay. The TABLET command is also used to calibrate a paper drawing for digitizing or to configure the overlay to suit the current menu. The TABLET command has four options: ON, OFF, CAL, and CFG. Invoke one of these options from the flyout menu, after selecting Tablet from the Tools menu.

The ON or OFF option turn the Tablet mode on or off, respectively.

The default setting of the Tablet mode is OFF. The OFF setting does not incapacitate the tablet as you might think; rather, it means that you are not going to use the tablet for digitizing (making copies of paper drawings). With the Tablet mode set to OFF you can use the tablet to select command/icons in the areas programmed accordingly. You may also use the puck in the screen area of the tablet to control the screen cursor. In order to digitize paper drawings, you must set the Tablet mode to ON. Most systems have a toggle key to switch the Tablet mode ON and OFF. With many PCs, the toggle is either F10 or CTRL+T.

The CFG option is used to set up the individual tablet menu areas and the screen pointing area. At this time, a preprinted overlay should have been fixed to the tablet. Its menu areas should suit the menu you wish to use. The sequence of prompts is as follows:

> Enter number of tablet menus desired (0-4) <0>:

Select the number of individual menu areas desired (with a limit of 4). The next prompt asks:

> Digitize upper left corner of menu area n:
> Digitize lower left corner of menu area n:
> Digitize lower right corner of menu area n:

The "n" refers to tablet menu areas of the corresponding tablet number in the menu. If the three corners you digitize do not form a right angle (90 degrees), you will be prompted to try again. Individual areas may be skewed on the tablet, with respect to each other, but such an arrangement usually does not provide the most efficient use of total tablet space. Tablet areas should not overlap.

The next prompts are:

> Enter the number of columns for menu area n:
> Enter the number of rows for menu area n:

Enter the numbers from the keyboard. The area will be subdivided into equal rectangles determined by the row and column values you have entered. If the values you enter do not correspond to the overlay row/column values, the results will be unpredictable when you try to use the tablet. Remember also that the overlay must suit the menu being used. After the menus areas have been specified, AutoCAD prompts:

> Do you want to respecify the Fixed Screen Pointing Area? [Yes/No] <N>:

If you respond with y, AutoCAD prompts:

> Digitize lower left corner of Fixed Screen pointing area: *(specify a point with the pick button)*

> Digitize upper right corner of Fixed Screen pointing area: *(specify a point with the pick button)*

AutoCAD then prompts:

> Do you want to specify the Floating Screen Pointing area? [Yes/No] <N>:

If you respond with y, AutoCAD prompts:

> Do you want the Floating Screen Area to be the same size as the Fixed Screen Pointing Area? [Yes/No] <Y>:

If you respond with n, AutoCAD prompts:

> Digitize lower-left corner of the Floating Screen pointing area: *(specify a point with the pick button)*

> Digitize upper-right corner of the Floating Screen pointing area: *(specify a point with the pick button)*

In order to toggle the Floating Screen Area ON and OFF, press F12.

AutoCAD prompts:

> The F12 Key will toggle the Floating Screen Area ON and OFF. Would you also like to specify a button to toggle the Floating Screen Area? [Yes/No] <N>:

If you respond with y, AutoCAD prompts:

> Press any non-pick button on the digitizer puck that you wish to designate as the toggle for the Floating Screen Area.

The standard AutoCAD overlay is installed as follows:

> Command: **tablet**

Option (ON/OFF/CAL/CFG): **cfg**
Enter number of tablet menus desired (0-4) <default>: **4**
Do you want to realign tablet menu areas? <N>: **y** *(if required)*

At this time, digitize areas 1 through 4, as shown in Figure 18–2. The values for columns and rows must be entered as follows:

MENU AREA	COLUMN	ROW
1	25	9
2	11	9
3	9	7
4	25	7

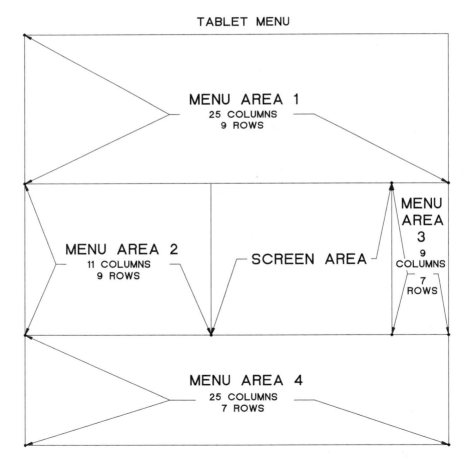

Figure 18–2 *Digitizing the screen areas of the tablet menu*

The CAL option is used to set up the digitizer by selecting points on the overlaid drawing or image and mapping them to their real coordinates. Calibration can be performed in model space or paper space. The CAL option turns on the Tablet mode in the space in which the tablet is calibrated. When the space is changed, the CAL option turns off Tablet mode.

If the tablet has been calibrated already, the last calibration coordinates will still be in effect. If not, or if you wish to change the calibration (necessary when you move the paper drawing on the tablet), invoke the CAL option and respond as follows:

> Digitize point #1: *(digitize the first known point)*

The point you select on the paper drawing must be one whose coordinates you know. The next prompt asks you to enter the actual paper drawing coordinates of the point you just digitized:

> Enter coordinates point for #1: *(enter those known coordinates)*

You are then prompted to digitize and specify coordinates for the second known point:

> Digitize point for #2: *(digitize the second known point)*
> Enter coordinates point #2: *(enter those known coordinates)*
> Digitize point #3 (or press ENTER to end): *(digitize the third known point or press ENTER)*

An example of a drawing that might be digitized is a map, as shown in Figure 18-3.

If, for example, the map in Figure 18-3 has been printed on an 11" x 17" sheet and you wish to digitize it on a 12" x 12" digitizer, you can overlay and digitize on one-half of the map at a time. You may use the coordinates 10560,2640 and 7920,5280 for two calibrating points. But because *X* coordinates increase toward the left, you must consider them as negative values in order to make them increase to the right. Therefore, in calibrating the map, you may use coordinates −10560,2640 and −7920,5280 to calibrate the first half and coordinates −7920,2640 and −5280,5280 for the second half.

The points on the paper should be selected so that the *X* values increase toward the right and the *Y* values increase upward.

Once calibration has been initiated in a particular space (model or paper), turning on the Tablet mode must be done while in that particular space.

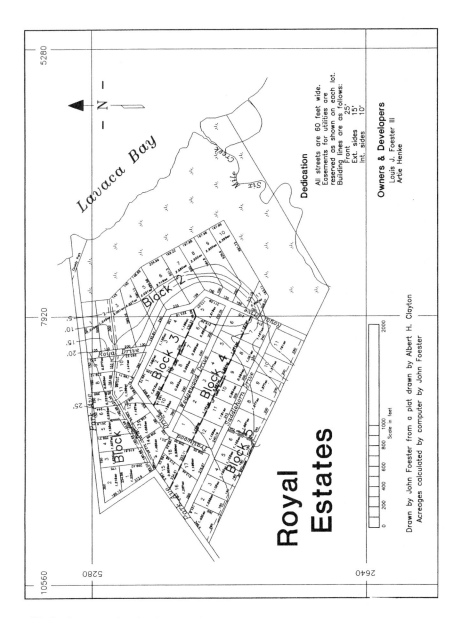

Figure 18–3 *An example of a drawing digitized as a map*

TRANSFORMATION OPTIONS

Tablet calibration can be done by one of several transformation methods. These include Orthogonal, Affine, Projective, and Multiple-Point. The method you choose may depend on the condition of the map/drawing that is to be digitized and the desired accuracy.

If three or more points are entered, AutoCAD uses each of the three transformation types (Orthogonal, Affine, and Projective) to compute the transformation to determine which best fits the calibration points. Entering more than four points can cause computing the best-fitting projective transformation to take a long time. The process can be canceled by pressing ESC. A table is displayed when the computations are complete, showing the number of calibration points with a column for each transformation type.

If no failures of projection transformation have occurred, you are prompted to choose a transformation type:

> Enter transformation type [Orthogonal/Affine/Projective/Repeat table]
> <Repeat>: *(Enter an option or press* ENTER*)*

This prompt includes only transformation types for which the outcome was Success, Exact, or Canceled.

Orthogonal

This option involves two points. It results in uniform scaling and rotation. The translation is arbitrary. Orthogonal (two-point) translation is most suitable for tracing paper drawings that are dimensionally and rotationally accurate (right angles are not skew). It is advisable to use this option for long, narrow applications (a pipeline, for example).

Affine

This option involves three points and can be applied to a paper drawing with scale factors, rotation, and right angle representations that are not to an acceptable accuracy in two dimensions. This can be applied to drawings with parallel lines that are represented parallel, but with an *X* direction scale/*Y* direction scale differential that is out of tolerance. Right angles may not be represented by right angles.

Whether or not you should use the AFFINE option depends on whether or not those lines that should be parallel are represented by parallel lines on the paper drawing. You can check the report displayed (in the form of a table) when you have digitized at least three points. If the RMS error (described in this section) is small, then the AFFINE option should be acceptable.

Projective

This option involves four points, to simulate a translation comparable to a perspective in which points from one plane converge while passing through another plane (skew to the first) to one point of view. This option is applicable to copying paper sheets with irregularities that differ from one area to another (known as rubber-sheeting) and parallel lines that are not always represented as parallel. However, lines do project as lines.

Repeat Table

Redisplays the computed table, which rates the transformation types.

THE CALIBRATION TABLE

If you use three or more points during calibration, AutoCAD computes the range or error (if any) and reports the results and displays information designed to help you determine if the paper copying process is acceptable. An example of a calibration table follows:

	Four Calibration Points		
Transformation type:	Orthogonal	Affine	Projective
Outcome of fit:	Success	Success	Exact
RMS Error:	143'-6.2"	73'-4.1"	
Standard deviation:	62'-2.7"	1'-7.8"	
Largest residual:	193'-1"	74'-1"	
At point:	2	3	
Second largest residual:	177'-9"	7'-7"	
At point:	3	2	

Outcome of Fit

>**Cancelled** Cancelled occurs only with projective transformations. It indicates the fit has been cancelled.

>**Exact** The number of points used was exactly correct and the transformation defined from them was valid.

>**Failure** Points selected, though the correct number, were probably collinear or coincidental.

>**Impossible** Insufficient points were selected for the transformation type under which "Impossible" is reported.

>**Success** The transformation defined was valid and more points were used than were required.

RMS Error

If the transformation is reported as a "Success," then the RMS (root mean square) error is reported. This is the square root of the average of the squares of the distances (called residuals) of each selected point from their respective targets.

Standard Deviation

This reports the standard deviation of all residuals (the distance each point misses its target).

Point(s)/Residual(s)

The two points whose residuals (see "RMS Error" and "Standard Deviation") are largest and second largest are reported along with their respective residual values.

TABLET MODE AND SKETCHING

Sketching while in Tablet mode operates similarly to sketching with a mouse or on a tablet with the Tablet mode off. The difference is that the entire tablet surface is used for digitizing while the Tablet mode is on, making the maximum area available for tracing but making the pull-down menus inaccessible.

EDITING SKETCHES

Once sketched lines have been recorded and the SKETCH command has been terminated, you can use regular editing commands (like COPY, MOVE, ERASE) to edit the individual line segments or sketched polylines (discussed next) just as though they had been drawn by the LINE or PLINE command. In the case of sketched polylines, the PEDIT command can be used for editing.

SKETCHING IN POLYLINES

You can cause AutoCAD to make the created sketch segments into polylines instead of lines by setting the SKPOLY system variable to a nonzero value.

LINETYPES IN SKETCHING

You should use the Continuous linetype while sketching, whether with regular lines or polylines.

REVIEW QUESTIONS

1. What filename extension is used to store the menu command entries for a tablet menu?
   a. .tab
   b. .mnu
   c. .cmd
   d. .pgp
   e. .ini

2. How many different tablet menu areas can be specified on a digitizing pad?
   a. 1
   b. 2
   c. 3
   d. 4
   e. 5

3. When aligning a tablet menu with the digitizing pad, how many points are required?
   a. 1
   b. 2
   c. 3
   d. 4
   e. 5

4. To toggle the digitizing pad between menu functions and paper copying, you press:
   a. F10
   b. F11
   c. F12
   d. CTRL+O
   e. none of the above

5. The minimum number of points required to calibrate a digitizing pad using an Affine calibration is:
   a. 1
   b. 2
   c. 3
   d. 4
   e. 5

6. If you have an isometric drawing on paper and wish to digitize an orthographic view of the top, what type of calibration would you use?
   a. 2 point (Orthogonal)
   b. 3 point (Affine)
   c. 4 point (Projective)
   d. cannot be done

7. If you have photograph of a building and wish to digitize the front elevation, what type of calibration would you use?

   a. 2 point (Orthogonal)

   b. 3 point (Affine)

   c. 4 point (Projective)

   d. cannot be done

8. What does RMS stand for in RMS error?

   a. Real Measure Statistic

   b. Root Mean Square

   c. Radical Motion Setting

   d. ReMainder Sum

   e. none of the above

9. If you select three points in a straight line to calibrate the digitizing pad, what "Outcome of Fit" will AutoCAD report for an Affine fit:

   a. Canceled   d. Impossible

   b. Exact   e. Success

   c. Failure

10. When using the SKETCH command, what system variable determines if lines segments are polylines?

    a. SKLINE   d. LINESEG

    b. SKPOLY   e. POLYGEN

    c. SKTYPE

11. What command will allow you to generate freehand lines when digitizing a paper drawing?

    a. DIGITIZE   d. FREEHAND

    b. TABLET   e. DRAW

    c. SKETCH

12. The option of the TABLET command that allows for the configuration of a tablet menu is:

    a. ON   d. CFG

    b. OFF   e. MENU

    c. CAL

13. The RMS error will always be lower (or equal) for a projective fit versus an orthogonal fit when 6 points are selected.

    a. True b. False

14. AutoCAD's default menu template has how many menu areas?

    a. 1
    b. 2
    c. 3
    d. 4
    e. 5

15. To toggle the floating screen area on and off press which of the following?

    a. F10
    b. F11
    c. F12
    d. CTRL+F
    e. none of the above

16. The three corners that are selected to define the tablet menu area must form a _____ degree angle.

    a. 30
    b. 45
    c. 60
    d. 90

17. Which of the following is used to toggle the tablet mode on and off?

    a. CTRL+B
    b. CTRL+M
    c. CTRL+T
    d. none of the above

18. Which of the following is not a tablet transformation method?

    a. Orthogonal
    b. Affine
    c. Polygonal
    d. Projective
    e. Multi-Point

19. The Projective transformation option allows copying paper with irregularities from one area to another. This type of copying is known as _____.

    a. Rubber banding
    b. Rubber sheeting
    c. Rubber stamping
    d. Applique

CHAPTER 19

Visual LISP

INTRODUCTION

This chapter will cover the fundamental concepts of the Visual LISP programming system, including writing, storing, and loading .LSP files; variables and expressions; lists; custom functions; and file handling.

After completing this chapter, you will be able to do the following:

- Grasp fundamental concepts of the Visual LISP programming system
- Interpret Visual LISP applications written by others
- Establish a basis for more advanced programming

BACKGROUND

In order to support Alonzo Church's lambda calculus operations, John McCarthy invented Lisp in the late 1950s. Internal data structures are ideally represented as lists, which can contain additional nested lists. The symbolic expressions for the lists are sandwiched with balanced parentheses. List processing is the intrinsic functionality of Lisp.

John Walker, the principal Autodesk co-founder, recognized the synergistic potential of Lisp for AutoCAD customization at a time when competitive programs were either offering very simple macro operations or demanding extensive investment in higher-level programming languages for similar operations. The implementation of Lisp for use within AutoCAD with some key constraints is referred to as AutoLISP™ and was introduced with AutoCAD Release 2.1 (later referenced as Release 6).

VLISP

As Autodesk placed emphasis on its ARX™ technology for the development of high-performance custom applications in recent years, developments in AutoLISP remained largely static. The introduction of Visual LISP has brought new life to legacy applications and also allowed users to maintain their investment in AutoLISP code. AutoCAD Visual LISP is referred to as VLISP™ in the remaining sections of this chapter.

INTERFACES

VLISP provides a very rich set of interfaces to the underlying AutoCAD object model. It relies on AutoCAD-specific (namely, *Proteus*) programmable dialog boxes for a graphical user interface when applications are run. (These topics are beyond the scope of this chapter.)

INTEGRATED DEVELOPMENT ENVIRONMENT

VLISP is comprised of a set of tools and resources that are packaged together to aid the programmer in the development of custom applications by simplifying and automating many of the tasks performed by the developer. Collectively these tools are referred to as VLIDE's Integrated Development Environment (IDE) and include such tools as a project window, color-coded text editor, debugger, resource windows. Their primary benefit is to reduce the amount of time required to analyze, design and implement a VLIDE project. Consequently, a user is generally more productive with VLISP than previously was possible.

A TEST DRIVE OF VISUAL LISP

Let us take a test drive to understand the key components of VLISP. In the remaining sections of this chapter, we will explore some simple projects to illustrate the basic components of VLISP.

SIMPLE EXAMPLE

It is customary to build a very simple program to test the installation of an integrated development environment. In this chapter we will begin with such a program that is commonly referred to as a *Hello World* example. We will then develop an additional exercise to demonstrate other key features of VLISP.

LAUNCH VISUAL LISP

VLISP operates under AutoCAD. You can edit programs without running AutoCAD by using an ASCII text editor, but you can only launch VLISP when AutoCAD is running.

Invoke the Visual LISP integrated development environment:

Tools menu	Choose AutoLISP > Visual LISP Editor
Command: prompt	**vlisp** (ENTER) or **vlide** (ENTER)

The Visual LISP window is displayed as a separate window with two basic child windows:

- VLISP console
- Trace (by default it's minimized)

The main areas within the VLISP window are:

- Console window
- Status bar
- Menu bar
- Toolbars
- Trace window

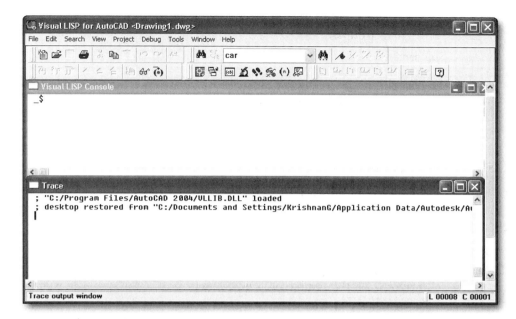

Figure 19–1 *Visual LISP initial window*

Console Window

This child window allows you to enter VLISP commands and is scrollable. You can test VLISP expressions as well as review previous commands or operations in this window. Most of the selections in the menu bar and toolbars have equivalent commands that you can enter in this window.

Status Bar

The current conditions related to your operations are displayed in the bottom left bar of the VLISP window.

Menu Bar

The menu bar operates like the conventional Windows menu bar with a brief description of the selected menu item displayed in the status bar.

Trace Window

This window is minimized during startup. It contains information regarding the VLISP release number and other startup messages. You will also use this window during debug operations.

Toolbars

Toolbars allow you to issue commands using the point-and-click method. Instead of navigating the menu bar or memorizing the command syntax to be entered in the console window you can select a button to activate the command. There are five principal toolbar groups:

- Standard
- Tools
- Search
- Debug
- View

DEVELOPING AND APPLICATION

Our first sample program will simply display the text string "Hello World" at the AutoCAD command prompt. Let us assume that you have launched VLISP using the instructions detailed earlier.

TEXT EDITOR

The VLISP text editor allows you to write AutoLISP code. Since we have not written any code until now for the sample program we will invoke the text editor with a new file. Invoke the VLISP text editor:

File menu	Choose New
Standard toolbar	Choose New (Figure 19–2)

Figure 19–2 *Invoke New file from the Standard toolbar (VLISP)*

On completion of this command, a new child window is displayed with the title <Untitled-0> and the status bar has the message "Edit –no file- (Visual LISP)." You should proceed to enter the following code in this new window:

```
(defun c:hello ( )        ; function definition
  (prompt "Hello World!") ; displays the string at the command prompt
  (princ)                 ;
)
```

Color Codes

As you enter the above code you will notice that certain elements are color-coded to assist you in recognizing the syntax. The color scheme depends on the type of file that is being edited. This file type is implicit from the file type extension in use. Some of the standard color schemes are:

Style	Description
None	No color coding
AutoLISP	LISP syntax color coding for file type extension LSP
C++	C syntax color coding for file type extensions C, CPP, and H
DCL	AutoCAD programmable dialog boxes for file type extension DCL (Dialog Control Language)
SQL	SQL procedure files for file type extension SQL (Structured Query Language)

The code is also automatically indented for clarity. You can change the color codes and the number of spaces indented by the VLIDE IDE yourself to suit your needs more appropriately, but these issues are beyond the scope of this chapter.

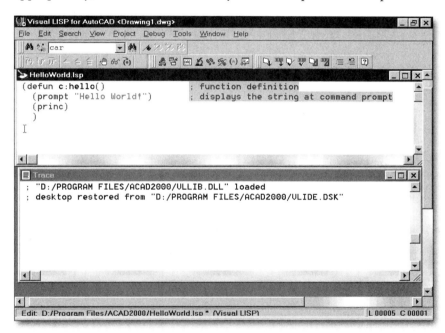

Figure 19-3 *Hello World program*

SAVE YOUR PROGRAM

After you have completed entering the code you should save your code.

Invoke the save command:

File menu	Choose Save As
Standard toolbar	Choose Save File

The Save As dialog box is displayed. For this sample program, enter the name **hello** in the **Filename** text box. Before you select the **Save** button in the dialog box you should ensure that the **Save As Type** text box is set to **Lisp Source File** and the desired path (drive:\folder(s)) is set up. AutoCAD can find source files such as your hello.lsp file most easily in the Acad2004\Support folder.

Note: It is recommended that the students save their work in a directory other than the AutoCAD2004\Support directory. This directory contains many files that are needed by the AutoCAD application. A first-time programmer can easily trash one of the many system files needed by AutoCAD. It is better to create a directory for development aside from the support directory and include the directory in AutoCAD PROJECT FILE/Search PATH Option.)

Once you have successfully saved the source program file; its text window and the status bar will reflect the file name you entered in the text box control field.

LOAD YOUR PROGRAM

Your program must be loaded before it can be run. Make your program window active by selecting anywhere within it.

Invoke the LOAD TEXT IN EDITOR command:

Tools menu	Choose Load Text in Editor
Tools toolbar	Choose Load Active Edit Window

On completion of the load operation the VLISP console window will display a confirmation message. If errors are reported, you must carefully review what you have entered and correct the mistake appropriately. Two very common sources of error messages are unbalanced parentheses (error: malformed list on input) and apostrophes (i.e., for strings — malformed string on input —).

RUN YOUR PROGRAM

You must switch to AutoCAD to run this short program. You can test the execution from the console window. Because this program specifically writes a string to the AutoCAD command prompt window you will not see any output in the console window except the completion code, which is *nil* for this program.

Switch to the AutoCAD window:

Window menu	Choose Activate AutoCAD
View toolbar	Choose Activate AutoCAD

Note: If the AutoCAD window is minimized, this operation will not restore it. You must restore the AutoCAD window to run the program.

Before you can use an AutoLISP application, it must first be loaded into memory. You can use the APPLOAD command or the AutoLISP "load" function to load an application. Loading an AutoLISP application loads the AutoLISP code from the .LSP file into your system's memory. Then, you can invoke the lisp routine from the "Command:" prompt.

Invoke the APPLOAD Command:

Tools menu	Choose AutoLISP > Load
Command: prompt	**appload** (ENTER)

AutoCAD displays Load/Unload Applications dialog box. Select the lisp file from the appropriate folder, and click the **Load** button. AutoCAD loads the selected AutoLISP file into memory and is ready to be used for the current session. AutoCAD lists the loaded applications in the **Loaded Applications** list box. If necessary, you can unload the application by selecting the AutoLISP file name from the **Loaded Applications** list box and click the **Unload** button. Choose the **Close** button to close the dialog box.

You can also drag lisp files from the files list box, or from any application with drag and drop capabilities such as Windows Explorer, into the **Startup Suite**, located in the Load/Unload Applications dialog box. Each time AutoCAD is started the contents of the **Startup Suite** are automatically loaded into memory.

In addition to loading AutoLISP applications into memory by using the APPLOAD command, you can also load AutoLISP applications using the AutoLISP function called "load" as shown in the following example:

Command: **(load "hello")**

Notes: The use of the parentheses distinguishes AutoLISP functions and routines. This is especially important when using a function, such as "load," for which there is an AutoCAD command of the same name.

Do **not** include the .LSP extension. AutoCAD appends it automatically.

You may also specify a path if necessary. For example, if the hello.LSP file is on the A: drive in a directory called LISP, the following response can be used:

Command: **(load "a:/lisp/hello")**

Notice the use of the nonstandard forward slashes to specify the directory path.

After loading the AutoLISP routing into memory, at the AutoCAD command prompt, enter the command **hello**. Immediately following your command you will see the string *Hello World!* If there is only one line in the command history window you may have to increase it to two lines or press F2 to see the results of your program in the AutoCAD text window. You have built and run a simple AutoLISP program using the VLISP IDE.

VLISP FUNDAMENTALS

While you develop simple AutoLISP programs, the entire source code is often retained in a single file. If the programs become more complicated or more than one person is involved during development, then it is customary to divide the program into modular sections. VLISP uses a project file to allow autonomous modular development.

PROJECT FILE

To specify a project file:

| Project menu | Choose New Project |

VLISP responds by displaying a New Project dialog box to specify the folder and name of the project file.

You should select the correct folder where you would like to keep the project file. Enter the file name in the corresponding File name text box and ensure that the file type is set to VL project file.

After the project file name is processed, the Project Properties dialog box (as shown in Figure 19–4) is displayed to allow the selection of the files associated with the project.

Figure 19–4 *Project properties dialog box*

PROJECT WINDOW

After you have selected the files using the appropriate buttons to move the files from one list to the other you should click the **OK** button. The dialog box is dismissed, and a child window listing the project files is displayed, as shown Figure 19–5.

Figure 19–5 *Project window*

The source code for the three files is listed below.

Balloon.lsp

```
; Name    : balloon
;
; Purpose : to place a balloon with leader in a 2D drawing file
;
; Last update:   12/29/89   Rev: 1.0    initial implementation
;     05/19/92     2.0      customer simplification
;     04/01/99     3.0      2000 update
;
; Usage   : balloon
;
; Prompts :
;    Scale/<Insertion point>:
;    Scale factor:
;    End point:
;
;    The author makes no warranty, either expressed or implied,
;    including but not limited to any implied warranties of
;    merchantability or fitness for a particular purpose,
;    regarding these materials and makes such materials
;    available solely on an "as-is" basis.
;
;   Auxiliary functions:
;      MODER MODES
```

```
;
;                             save entry state variables
;
; a list of variables whose entry states have to be saved
;   cv          current variable
;   MLST              save list
;
; Input variables:
;             none
;
; Output variables:
;             none
;
; Local variables:
;   atd     INSERT mode for attributes (restored upon exit)
;   bm      blip mode (restored upon exit)
;   cn      current balloon number
;   itm     current item number
;   la      current layer name (restored upon exit)
;   pt1     insertion point for block
;   pt2     terminating point for leader line
;   qty     quantity (attribute) for block (=1 default)
;   rad     radius of bubble circle
;   sx      scale factor (=1 for prototype)
;
(defun C:balloon ( / cn cw er oe la ln mors pp pt1 pt2 sx zz)
;                           establish local error handler routine
;                           get entry state variables
;
; Internal error handler defined locally
;
   (defun er (s)            ; If an error (such as CTRL-C) occurs
                            ; while this command is active...
      (if (/= s "Function cancelled")
        (if (= s "quit / exit abort")
          (princ)
          (princ (strcat "\nballoon: ERROR! " s))))
      (moder)
   ); defun
  (if *error*                     ; Set our new error handler
     (setq oe *error* *error* er)
     (setq *error* er)
```

```
    ); if
  (modes '("ATTDIA" 0 "ATTREQ" 0 "CMDECHO" 0 "COORDS" 1 "DRAGMODE"
    2 "GRIDMODE" 0
    "ORTHOMODE" 0 "OSMODE" 0 "OSNAPCOORD" 1 "SNAPMODE" 0))
  (setq la (getvar "CLAYER") ln "bomsym")
;                       establish the correct layer
  (command "layer" "m" ln "")
;                       check the scale factors
  (setq cn (getvar "USER2") cw 786 zz (getvar "USERI1"))
  (if (/= zz cw)
    (progn
      (setvar "USERI1" cw)
      (setvar "USERR1" 12)
    ); progn
  ); if
  (if (not cn)
    (setq cn 0)
    )
  (setvar "USERI2" (setq cn (1+ cn)))
;                       collect data for curr invocation of proc
  (setq pp T)
  (while pp
    (initget (+ 1 2 4 32) "Scale")
;   (graphscr)
    (setq pt1 (getpoint "\nScale/<Insertion point>: "))
    (cond ((eq pt1 "Scale") (progn
        (initget (+ 1 2 4))
        (setvar "USERR1" (getreal "Scale factor: "))
      ); progn
   )
  (T (setq pp nil))
    ); cond
  ); while
  (setq pt2 (getpoint pt1 "End point: ")
        sx (getvar "USERR1"))
  (if (or (<= 0.0 sx) (not sx))
    (setq sx (setvar "USERR1" 1.0))
    )
  (setq rad (* 0.25 sx));setq
;                       place block in dwg file
  (command "INSERT" "balloon" pt1 sx sx 0 )
;                       place leader from balloon to part geometry
```

```
      (setq pt1 (polar pt1 (angle pt1 pt2) rad))
      (command "LINE" pt1 pt2 "" "LAYER" "S" la "")
;                           restore entry state variables
      (moder)
;                           restore old error handler
     (if oe
       (setq *error* oe)
       ); if
     (princ)
   );defun
```

Modes.lsp

```
   (defun modes ( a / cv)
     (setq saveList '())
     (repeat (/ (length a) 2)
       (setq cv (car a)
         saveList (append saveList (list (list cv (getvar cv))))
         a (cdr a))
       (setvar cv (car a))
       (setq    a (cdr a)))
     ); defun
```

Moder.lsp

```
   (defun moder ()
     (repeat (length saveList)
       (setvar (caar saveList)(cadar saveList))
       (setq saveList (cdr saveList)))
      ); defun
```

COMPILING VLISP APPLICATION (OR, SO LONG KELVINATOR)

A significant change introduced by VLISP is the ability to compile an AutoLISP application into a stand-alone executable file that is assigned a file type extension of VLX. You can load VLX files just the way you would load an ARX, EXE, or EXP file. A desirable result of this technique is that your investment in your source code is protected to a reasonable extent from casual prying.

MAKING AN APPLICATION

The Make Application wizard will guide you through the steps of creating an application. For a new application the wizard offers two paths for this process—simple or expert, as shown in Figure 19–6.

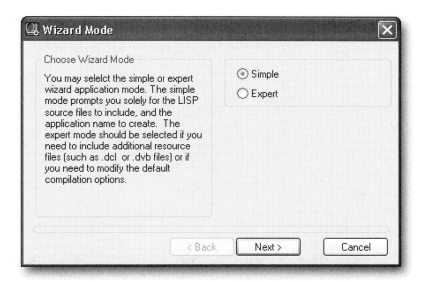

Figure 19–6 *Wizard mode selection*

SIMPLE WIZARD

If you select the **Simple** radio button and click the **Next** button, the following window is displayed for you to specify the home folder and name for the application.

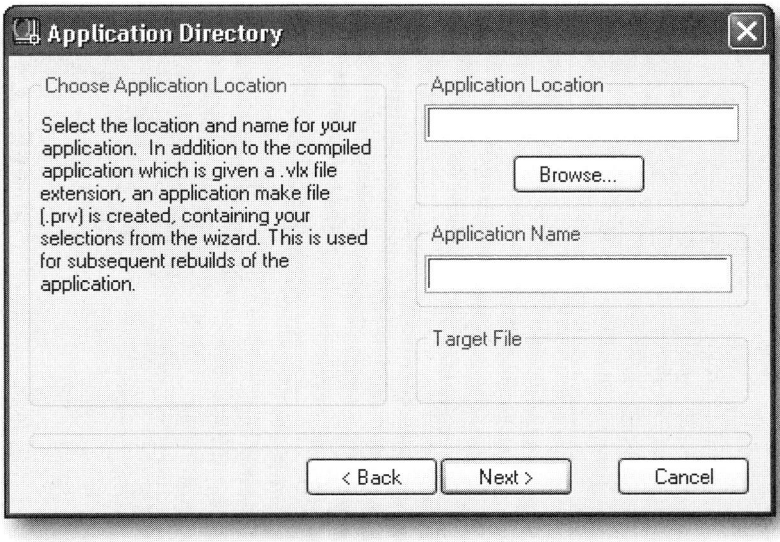

Figure 19–7 *Application directory data*

You can use the **Browse...** button to navigate to the appropriate folder, and the corresponding folder name will be entered in **Application Location** text box. You must enter an application name in the corresponding text box before you click the **Next** button.

PROJECT FILE SELECTIONS

The three types of files that you can select for inclusion in your project are:

- Lisp source
- Compile lisp files
- VLISP project files

After selecting the appropriate files, select the **Next** button to proceed to the final phase of this wizard.

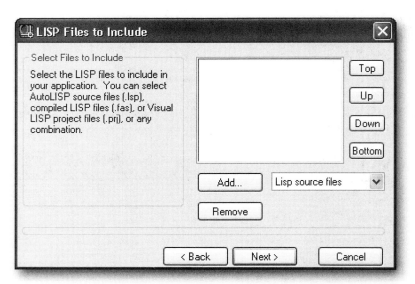

Figure 19-8 *Project file selections*

REVIEW SELECTIONS

If you are ready to build the applications, you should select the corresponding check box. Or, if you need to make changes to the code or add additional files, there is no need to build the application. Select the **Finish** button to complete the processing by the wizard.

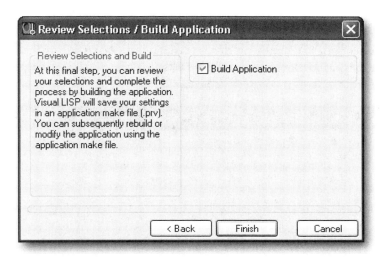

Figure 19–9 *Build Application wizard*

EXPERT WIZARD

If you choose the **Expert** radio button in the wizard, then you will be presented with some additional dialog boxes and options. For example, as shown in Figure 19–10, in addition to the standard Lisp and project files, you can also choose AutoCAD Visual Basic for Applications modules (DVB file type extension) and dialog declaration files (DCL file type extension).

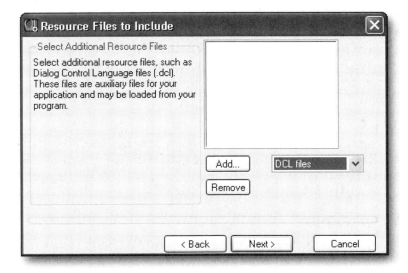

Figure 19–10 *Additional file selections*

You can select different compile and link options, as shown below, particularly to improve the performance of the application. The **Standard** mode is typically used for small projects and generally produces the smallest output file of the two options. The **Optimize and Link** option usually results in a more efficient output file and is strongly recommended for complex projects.

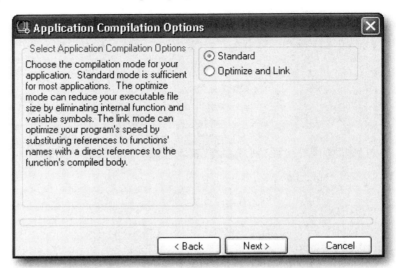

Figure 19-11 *Compilation and link options*

PROJECT DEFINITION

If you are not ready to build or make your application, but still want to proceed along the development path, then you can create a project file. When you are ready to build or make the application you can simply include the project file as a source. In the meantime, you can coordinate your development as a team effort using the project file as the control source.

PROJECT FILE

From the **Project** menu, select **New Project**. When the New Project dialog box is displayed you should navigate to the correct folder before you enter the file name. Note that the file type extension is VLJ for *VL*ISP project file. After you have entered the file name, click the **Save** button to proceed to the next step.

PROJECT PROPERTIES

The Project properties dialog box has two tabs. The **Project Files** tab is used to collect the names of the project files. Once again, you have to be careful that you navigate to the correct folder to select the files.

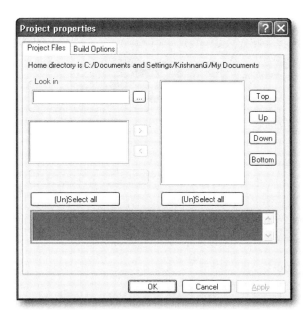

Figure 19–12 *Project Files tab*

The second Project properties tab, **Build Options,** allows you to specify the build options for the project. You can always revise these selections later as appropriate.

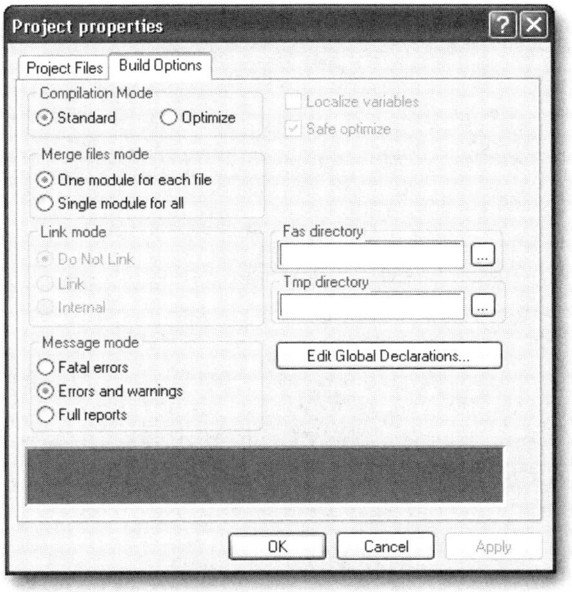

Figure 19–13 *Build Options tab*

VIEW TOOLBAR

As you work in the VLISP IDE you will become familiar with the toolbar buttons.

Figure 19-14 *View toolbar*

The View toolbar contains the most common navigation buttons:

- Activate AutoCAD
- Select Window
- VLISP Console
- Inspect
- Trace
- Symbol Service
- Apropos
- Watch Window

Activate AutoCAD
This button allows you to activate the AutoCAD window. During program development and testing you may have to switch between the VLISP IDE and the AutoCAD window.

Select Window
This button allows you set focus to an existing child window in the VLISP IDE. In addition to the VLISP console and trace windows, other windows include any source code windows.

Visual LISP Console
This button places the cursor at the end of the last record in the console window. You can use this command button to switch to the console command line quickly.

Inspect
The Inspect windows allow you to browse, examine, and modify objects associated with AutoCAD or your program. The VLISP Inspect feature can help you figure out how to process the information stored in the association list. You can scroll through complex objects to see their simpler components.

Trace
The Trace window lists the calling sequence of the programs or procedures that lead to the current stopping point.

Apropos

With this command, VLISP will offer to complete partially entered words from the entries in its symbol table. This feature reduces typing errors.

Watch Window

If you wish to keep track of the values of selected variables, you can place these variables in the watch window. The name and current value of the selected variable is always on display for your inspection.

Note: VLISP provides an interface with the embedded AutoLISP programming language that makes it easier to program and more like other "visual" programming languages, for example Visual Basic and Visual C++. But the structure, syntax, functions, expressions, and symbols that are the body of the program are still unique and specific to AutoLISP. The following sections cover that specific structure and syntax and those functions, expressions, and symbols that must be learned before you can successfully apply AutoLISP through the VLISP interface. Once the basics of AutoLISP are mastered, there is a section on debugging at the end of the chapter that is part of the VLISP interface.

AUTOLISP BASICS

In Version 2.1 of AutoCAD (Release 6, May 1985), Autodesk first introduced AutoLISP, its embedded programming language. It provided on-board computational power for the operator while in AutoCAD. It also permitted true programming routines to be used by way of menu devices, including interactive functions to receive input from the operator from keyboard entries and screen picks for use in the routine. Version 2.18 (January 1986) included a full implementation of user-defined functions and custom commands, adding a whole new world of open architecture to AutoCAD (meaning you can customize the program to suit your needs).

EXPRESSIONS AND VARIABLES

Expressions in AutoLISP should be understood before getting into variables. The simplest application of AutoLISP is to evaluate an equation just by entering it in and pressing ENTER. Of course, you must enter the equation in the proper format, which is somewhat different from ordinary algebraic notation. It involves a format that is unique among those used in other, more popular computer programming languages. For example, if you wish to add 5 and 3, simply enter the following expression:

Command: **(+ 5 3)**

The integer 8 will be displayed in the prompt area. That is, AutoLISP evaluates the expression and returns the integer 8. Throughout this chapter, the word *return* will be used to describe the result of an evaluation. The following expression returns the integer 108:

Command: **(+ 5 3 1 99)**

Note: When AutoCAD sees an open parenthesis (unless responding to a prompt to enter text), it knows that it is entering an AutoLISP expression to be evaluated. The AutoLISP evaluator remains in effect until it encounters the closing parenthesis that is the mate of the first open parenthesis.

AutoLISP uses prefix notation, which means that expressions begin with the operator (after the opening parenthesis, of course). In the preceding examples, the plus sign (+) is the arithmetic operator. The *operator* tells AutoLISP what operation to perform on the items that follow. The items that follow the operator are called *arguments*.

As you can see by the second example, the plus operator can have more than the usual two arguments to which an algebraic plus sign is restricted.

Elements in an AutoLISP expression are separated by one or more spaces. Multiple adjoining spaces (unlike spaces in a menu line) are considered one space in an AutoLISP routine.

Variables as Symbols and Symbols That Should Not Vary

Just as algebra uses letter names for the unknown values in an equation, AutoLISP uses symbols as variables, whose name you may select during the writing of the program. In algebra, for example, a sequence might be written as follows:

Algebra	AutoLISP
a = 3	(setq a 3)
b = 7	(setq b 7)
therefore	
a + b = 10	(+ a b) returns 10
and	
ab = 21	(* a b) returns 21

Assigning Names

There are three situations for assigning names:

Expressions: For example, pi is the name for 3.14159…

Variables: A variable is an expression whose value is not known when originally written into the program. Variables will take on some value after the program has been called into use. The value of the variable is usually determined by some operation on some other value that the user has been prompted to enter while the program is in progress.

Custom-Defined Functions: AutoCAD permits users to create and name a customized function and then use it in AutoCAD in a manner similar to a standard AutoCAD command.

TERMINOLOGY AND FUNDAMENTAL CONCEPTS

Lists, operators, arguments, types, parentheses, the exclamation point, and the concept of the function list comprise the basis of AutoLISP.

Lists: Practically everything in AutoLISP is a list of one sort or another. Functions are usually represented as a list of expressions enclosed in parentheses. For example, (+ 1 2.0) is an AutoLISP function with three elements in it, the "+," the "1," and the "2.0." Other lists can be established by applying a function called list or by applying the single quote, as in '(1 2 3.0 a "b").

Operators and Arguments: Arguments are those items in a function list on which the operator operates. For example, in the function list (+ 1 2), the operator is (+) and the arguments are "1" and "2."

Types: Arguments are classified by their type. The arguments in the example (+ 1 2) are of the type called integer. In the function list (+ 1.0 2.0), the arguments are of the type called real, signifying a real number.

Parentheses: The primary mechanisms for entering and leaving AutoLISP and entering expressions within AutoLISP are the open and close parentheses.

> **Note:** For every open parenthesis [(] there must be a close parenthesis [)]. If Auto-CAD encounters a condition wherein the close parentheses are one fewer than the open ones, the prompt will display the following:
>
>   1>
>
> Most often, this is caused by the need for one close parenthesis. A display of 3> could indicate the need for three closing parentheses. The problem is usually remedied by just entering the specified number of close parentheses. This type of error message can also be caused by having an odd number of double quotation marks inside of an AutoLISP expression, in which case one double quotation mark must be entered, followed by the required number of close parentheses, in order to eliminate the error message in the prompt area. Further examination will usually reveal that the program line needs to be corrected to prevent the message from recurring.

Exclamation Point: A leading exclamation point in response to a prompt is another mechanism for entering AutoLISP. This tells AutoCAD that the symbol following the exclamation point is an AutoLISP variable that has been set equal to some value and that AutoCAD should use the value of the variable as a response to the prompt.

The terms and concepts just described are explained in detail, with examples, in the sections that follow.

Functions: The first function list to be introduced involves three expressions: the operator SETQ, the variable that we have arbitrarily named x, and the real number 2.5. Within this function SETQ will perform a special operation on the variable x and the real number 2.5. SETQ is probably the most common AutoLISP function. It means "set equal." Using SETQ as the first of three expressions will set the second expression equal to the third. We are accustomed to performing this operation by the conventional expression x = 2.5. But as you will see in the next example, in order to write x = 2.5 in AutoLISP you must write in a special format: SET x EQUAL 2.5. It is done by writing three expressions in parentheses, with the first being the operator SETQ, the second being x, and the third being the value to which the variable named x is to be equal, for example:

    Command: **(setq x 2.5)**

Having x = 2.5, or, more properly, to have SET x EQUAL 2.5, is necessary only if you wish to use the value of the real number 2.5 later by just entering its name, x. If, earlier, you have entered (setq x 2.5) at the "Command:" prompt, then any time during that current editing session you could re-enter AutoLISP right from the keyboard by using ! before the name of the value in response to a prompt for some real number. If the prompt is asking for a name of something (like a layer name, which requires responding with something called a string variable), trying to use a variable that has been set equal to a real number will cause an error message. The use of ! is another way to enter AutoLISP directly from the keyboard or in a string of custom menu commands and responses.

For example, if you create a unit block named UB and wish to insert it with a scale factor of 2.5, the sequence of prompts and responses would be as follows:

    Command: **insert**
    Block name (or ?): **ub**
    Insertion point: *(pick an insertion point)*
    X scale factor <1> / Corner / XYZ: **!x**

Note at this point that the use of the name x and the fact that it is being applied to the *X* scale is purely coincidental.

By using the ! in front of the x, you will have the value of x (or 2.5 in this case) used as a response to the prompt asking for the *X* scale.

This example is not very efficient. Entering !x appears to take only two keystrokes, whereas entering 2.5 takes three. True, this would save a keystroke. But because entering the exclamation point requires the SHIFT key, it is a double stroke requiring both hands. Along with the x it is actually less convenient than entering 2.5, which has three one-hand strokes. Additionally, the characters "2," ".," and "5" are all accessible on the AutoCAD tablet overlay, and the ! is not. Entering 2.5 will, however, give the same accuracy as entering 2.50000000000000.

 Note: If x has not previously been set equal to a value, then entering !x will return "nil." Believe it or not, advanced programming does make use of the nil. Normally, variables and expressions retain their names and values only during the current editing session. The names are lost when a drawing is ended. At the beginning of each session they must be reestablished. But even this problem of saving variable values from one session to another can be addressed by a custom program.

Naming and SETQing the Value of an Expression: Suppose the value you wish to enter is 1 divided by the square root of 2. This could be written as 0.7071068 depending on the accuracy desired. Now compare three keystrokes using AutoLISP versus nine or ten from the keyboard or tablet. Not only will a variable having a short name (and having been set equal to that value) save time in entering but it will decrease the probability of errors in both reading and keying in a long string of characters, for example:

Command: **(setq x 0.7071068)**

If the value 0.7071068 is needed in response to a prompt, simply enter !x. Programming begins to appear both expedient and practical.

Note: You may respond with .7071068 outside AutoLISP, but while in AutoLISP you must use some value (even if zero) on both sides of the decimal point. Expressions beginning or ending with a dot (.) have another special meaning.

Before leaving this example, additional power of AutoLISP can be seen in a demonstration of using an expression within an expression within an expression. Two other AutoLISP operators will be introduced at this time. They are the SQRT and the "/" functions. SQRT returns the square root of the argument following (it must be only one argument and must be a nonnegative real number). The "/" function requires two or more arguments and returns the quotient of the first divided by the second, for example, the following returns 1.414213562373... When an expression contains more than two arguments, the result of the previous two arguments is then divided by the next number. For example (/ 16 2 2) would yield a result of 4 (16/2 = 8, 8/2 = 4).

Command: **(setq y (sqrt 2))**

You may now use this value as follows:

Command: **(sqrt 2)**

This does two things. It sets y equal to the expression that follows, which is the square root of 2, and returns 1.414213562.

Having entered this, you may operate on y as follows:

Command: **(/ 1.0 y)**

This expression will return the quotient of 1 divided by the value of y, which was previously set equal to the square root of 2. Therefore, the expression evaluates to 0.7071068.

Using an AutoLISP function in this fashion neither sets values nor gives them names. It is of little use later, but does return the result of the prescribed computation for immediate viewing. Sometimes this is even handier than picking up a hand-held calculator.

A two-step use of AutoLISP could be written as follows:

> Command: **(setq y (sqrt 2))**
> Command: **(setq x (/ 1 y))**

The first step sets y equal to the square root of 2. The second step sets x equal to the inner expression, which uses the / operator (for division) and returns the value of 1 divided by the value of y. This two-step sequence involves four operators (SQRT, /, and SETQ twice). It names y and sets it equal to the square root of 2. It names x and sets it equal to the reciprocal of y.

Instead of this two-step routine, a simpler one-step routine is as follows:

> Command: **(setq x (/ 1 (sqrt 2)))**

So, in one step you can name an expression x and set it equal to the reciprocal of the square root of 2. It allows you then to respond to any prompt for a real number and apply this value by simply entering !x. This saves time and ensures accuracy to 14 decimal places, which is what CAD is all about in the first place—speed and accuracy.

EXAMPLE

At the "Command:" prompt enter the following expression:

> Command: **(setq a 1 b 2 c 99 d pi e 2.5)**

This expression is a function list using the operator SETQ. Note the opening and closing parentheses that distinguish it as an individual expression. Unlike the (+ . . .) function that performs a single operation of as many arguments as you wish to furnish, the (setq . . .) must have pairs of arguments. SETQ operates on each pair of arguments individually. After entering the preceding, you can have AutoLISP evaluate equations by entering them as follows:

> Command: **(setq x (+ a b))** *(returns 3)*
> Command: **(setq y (- a c))** *(returns -98)*
> Command: **(setq z (* b d))** *(returns 6.283185308)*
> Command: **(setq q (/ c e))** *(returns 39.6)*

Perform the following exercise:

> Name and give values to variables corresponding to the following algebraic equations:
>
> $a = 4$
> $b = 7$
> $c = a \times b$
> $d = b^2$ *(with 2 as exponent)*
> $e = \sqrt{a^2 + b^2}$ *(square root of a-squared plus b-squared)*

EXERCISES FROM THE KEYBOARD

Line: After first drawing a random line (with the Snap mode off), establish variables for the points p1 and p2 as follows (see Figure 19–15). Enter **(setq p1 (getpoint))**. Use the Osnap mode Endpoint to pick one end of the line. Enter **(setq p2 (getpoint))**, and use the Osnap mode Endpoint to pick the other endpoint of the line. Do not be alarmed that the prompt area is blank. When the points have been returned, record the *X* and *Y* coordinates on paper for use in the following exercises. For example, if (1.2345 6.7890 0.0000) is returned for p1, enter the following:

Command: **(setq X1 1.2345)**   *(returns 1.2345)*
Command: **(setq Y1 6.7890)**   *(returns 6.7890)*

Figure 19–15 *The line for writing a routine*

 Note: Enter the numbers you obtain, not the 1.2345 and 6.789 from the example!

Write in from the keyboard routines to perform the following:

1. Set *dx* equal to the horizontal distance between p1 and p2.
2. Set *dy* equal to the vertical distance between p1 and p2.
3. Set *d* equal to the distance between p1 and p2.
4. Set *a* equal to the angle between p1 and p2 in radians.

Circle: After first drawing a random circle (with the Snap mode off), set variables to the center c of the circle and to one point p on the circle (see Figure 19–16). Write routines to perform the following:

1. Set *r* equal to the radius of the circle.
2. Set *d* equal to the diameter of the circle.
3. Set *p* equal to the perimeter of the circle.
4. Set *a* equal to the area of the circle.

Arc: Establish endpoints p1 and p2 and the center c of the arc (see Figure 19–17) as in the exercises. Write routines to perform the following:

1. Set *r* equal to the radius of the arc.
2. Set *lc* equal to the chord length of the arc.
3. *(A doozie)* Set *la* equal to the arc length of the arc.

Computation: Write routines to perform the following (see Figure 19–18):

1. Given $a = 1$ and $b = 2$, find c.
2. Given $b = 2$ and $c = 3$, find a.
3. Find the area of the triangle for question 1.
4. Find the area of the triangle for question 2.

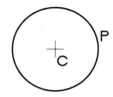

Figure 19–16 *The circle for writing a routine*

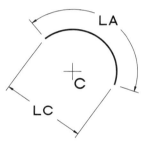

Figure 19–17 *The arc for writing a routine*

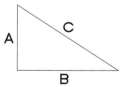

Figure 19–18 *The triangle for writing a routine*

According to Ben Shneiderman in his preface to Dan Friedman's *The Little LISPer*, "The fundamental structure of the LISP programming language was derived from the abstract notions of lambda calculus and recursive function theory by John McCarthy. His goal was to produce a programming language with a powerful notation for defining and transforming functions. Instead of operating on numeric quantities, LISP was designed to manipulate abstract symbols, called atoms, and combinations of symbols, called lists. The expressive power was recognized by a small number of researchers who were primarily concerned with difficult symbolic manipulation problems in artificial intelligence."

Translating an algebraic word problem into a true algebraic expression demands a certain symbol classification as well as list processing. Many algebra students who learn how to solve an algebraic expression (or at least memorize the procedures) once it is in the proper structure reach an impasse when asked to "interpret" the word problem into that proper structure.

Words in a sentence or paragraph have characteristics similar to AutoLISP functions or expressions. Verbs act on subjects like subroutines act on variables and expressions. The conjunction *and* joins phrases like the "+" joins real numbers. A phrase like *out of here* has the subphrase of *here* in it.

LISTS

TYPES OF ARGUMENTS

Although an argument is defined, in part, by *The American Heritage Dictionary* as "a quarrel; contention," it also is defined under the heading of math as "the independent variable of a function." *The AutoCAD Programmer's Reference* consistently describes the variables on whose value the function value depends as arguments.

The List in AutoLISP Is an Argument: When is a list not a list? As mentioned earlier, almost every function in AutoLISP is some form of list. The AutoLISP interpreter operates on lists as they occur in a routine or program. The AutoLISP expression (setq x (/ 1 (sqrt 2))) is a list. To the AutoCAD LISP interpreter it is a list to be evaluated according to the operator SETQ. Likewise, (/ 1 (sqrt 2)) and (sqrt 2) are a sublist and subsublist, respectively, to be evaluated. But none of these would be considered as the type of defined list (known as list to the type function) that some of the AutoLISP function lists, such as (CAR <list>), normally require as their argument. The lists that are acceptable arguments to these certain functions might be looked on as special lists within lists. They must take on a special format in order to satisfy the main function list as the required argument(s).

AutoLISP function lists that require lists as one of their arguments include the CAR, CADR, LAST, LENGTH, NTH, and REVERSE functions.

Whenever an AutoLISP function list requires a list as one of its arguments, you must use one of the following procedures:

1. You may use the function-list that has as its operator the function called list. This function list is written in the format (list <expr>...) as described in the *AutoCAD Programmer's Reference*. Entering in **(list 1.0 2.0)** will return (1.000000 2.000000). But more importantly, it will satisfy the requirements of this special list when needed for data input. Other function lists (such as the one with the operator SQRT) most likely will *not* satisfy the special requirement of an argument that must be of the type called list.

2. If you must respond within AutoLISP with the special form of list required, writing **(list 1.0 2.0)** every time becomes laborious (like writing 3.14159 and

so forth when the use of *pi* will do). Therefore, it is convenient to use the SETQ function and write:

(setq p (list 1.0 2.0))

The name p is strictly arbitrary in this case. Any unique variable name you want to use will suffice. Then whenever your routine requires you to respond with a point whose *X* and *Y* coordinates are 1.0 and 2.0, respectively, you can respond with p within AutoLISP or !p directly from the keyboard while in AutoCAD.

Try this exercise for an illustration of the preceding: The default limits of 0,0 and 12,9 as in the ACAD.DWG file will be helpful. Enter at the "Command:" prompt the following:

Command: **(setq p (list 1.0 2.0))** *((1.000000 2.000000) will be displayed)*
Command: **line**
Specify first point: **(list 6.0 7.0)**
Specify next point or [Undo]: **!p**

CREATING A LIST

As explained earlier in this chapter, the (list <expr>...) function list is often used under restrictive conditions. For example, using it as a response to certain commands requiring a 2D point, the "<expr>..." part of the overall expression must be "<expr> <expr>," where both <expr>s are real numbers.

Remember that an integer can be entered in AutoLISP and that AutoCAD will translate it into a real number under certain conditions. For a 3D point, the form of the response must be "<expr> <expr> <expr>." Using "<real> <real>," "<real> <int>," or "<int> <int>" may be considered proper responses to a prompt for the starting point of a 2D line. This coincides with the type of entry required for a point even when not in AutoLISP.

For example, the following are considered appropriate while in AutoCAD:

Command: **line**
Specify first point: **1.5, .7081068**
   or
Specify first point: **3.1415, 5**
   or
Specify first point: **1, 1**

A similar mixture of reals and integers is evident in the following example from within AutoLISP:

Command: **line**
Specify first point: **(list 1.5 0.7071068)**
   or
Specify first point: **(list pi 5)**

or
Specify first point: **(list 1 1)**

Note two subtle differences (not the parentheses, because they are not subtle): (1) From within AutoLISP, any real number must begin with an integer, not a decimal point. (2) The two expressions comprising the list within each function are separated by a space, as in the last examples (within AutoLISP), instead of a comma, as in the first examples (not in AutoLISP).

A List That Doesn't Look Like a List: Once a symbol (name of your choice) has been SETQed (set equal) to a list by the (setq <sym> (list <expr> <expr>...)) function, that symbol becomes a defined list. It will evaluate to that (<expr> <expr>...) for use as a response within AutoLISP or out of AutoLISP by using the ! prefix. For example, entering (setq p1 (list 2.0 3.0)) returns (2.0000 3.0000). Using the TYPE command, as in entering (type p1), will return "list." It will work for the following response to the LINE command:

Specify first point: **!p1**

But if (setq p1 (list 2.0 "z")) has been entered as the prior routine instead of (setq p 1 (list 2.0 3.0)), the preceding sequence will not accept !p1 as a valid response, even though (type p1) still returns "list." Its argument, (<expr> <expr>..), as a list is not (<real> <real>..), as points require.

Before you even considered programming in AutoLISP, you were using lists. If, in response to the "From point:" prompt, you entered 2,3.75 from the keyboard (or any form of "real,real," "real,integer," or "integer,integer"), then you used a list. Even if you had entered x1,2.5 or 3,y, you would have been using a list. But you would have found out that even though it was in a similar format, the elements of the entry were not of the proper classification. Once you understood that AutoCAD rejected anything except certain classifications of elements, you learned to work using that knowledge (or went back to the drawing board, literally!).

A disguised use of the list input is when you specify a point on the screen. AutoCAD defines that specification as a list in the proper format (*X* coordinate, *Y* coordinate), and enters it for you. It is a list in the acceptable format.

Association List The following is a list, but it is a very special list:

((a 097) (b 098) (c 099) (d 100)...(x 120) (y 121) (z 122))

Each expression in the list similarly has two expressions enclosed in parentheses (making each of them a list), with the first being a lowercase letter and the second an integer.

What makes the last list special is that it satisfies the requirements of an association list. It may be used (like other qualified association lists) as an argument in the AutoLISP function ASSOC. It takes the form (assoc <item> <alist>). Association lists are used exclusively in the creation of blocks and other complex entities.

EXERCISES

Determine if the following evaluation results below satisfy the requirements of the expression of the type list. If not, explain why.

a. 1 , 2
b. 1 2
c. (12)
d. (1 2)
e. (1.0 2)
f. (1.0 2.0)
g. ((1 2))
h. ((1 2)
i. (1 .2)
j. (1.0 0.2)
k. (1.0 2.0 3.0 x)
l. (1.0 2.0x3.0)
m. (* 2 4)
n. (-2 -4)

PAUSING FOR USER INPUT

There are very few AutoCAD commands that begin and end just by entering the command name. Except for REGEN, REDRAW, OOPS, UNDO, and the like, most commands pause for user input before they are completed. You can include this interactive capability in your AutoLISP routines through several input functions. The exercises in this section will introduce two such functions, (getpoint) and (getreal). The (getpoint) function pauses for user input of a point (either from the keyboard or on screen) and returns that point in the form of a list of three reals. The (getreal) function pauses for user input of a real number and returns that real number.

AUTOLISP FUNCTIONS—THE COMMAND FUNCTION

The command function (command <args>...) provides a method of invoking an AutoCAD command from within AutoLISP. It is usually the culmination of the routine. The <args> are written in the same sequence as if entered from the keyboard from AutoCAD. The AutoCAD command name that follows the (command ...) function must be enclosed in quotation marks, as in the following example:

```
(setq p1 (list 1.0 1.0))   ;Sets the variable p1 = 1,1
(setq p2 (list 2.0 2.75))  ;Sets the variable vap2 = 2.0, 2.75
(command "line" p1 p2 "")  ;Constructs a line from p1 to p2
```

When included in an AutoLISP routine, this sequence will draw a line from 1,1 to 2,2.75.

Note that the double-double quotation mark is the equivalent of pressing ENTER while in AutoLISP. It is referred to as the null string (""). It simulates pressing the SPACEBAR.

In the (command ...) function list, the arguments that are AutoCAD entries are enclosed within quotation marks, whereas AutoLISP symbols and expressions are not. One of the notable differences is illustrated in responding to a prompt for an angle. Remember, AutoLISP requires radians, and AutoCAD normally uses degrees (unless the units are set for angle input to be in radians). For this illustration, certain symbols were named and SETQed to values, as shown here:

```
(setq ublkname "UB1")          ;Sets the variable ublkname = UB1
(setq inspt (list 2.0 2.0))    ;Sets the variable inspt = 2.0, 2.0
(setq xscal 2.5)               ;Sets the variable xscale = 2.5
(setq yscal 2.5)               ;Sets the variable yscale = 2.5
 setq ang 90)                  ;Sets the variable ang = 90
```

Then a routine could be written in two different ways with the same results, as follows:

```
(command "insert" ublkname inspt xscal yscal ang)
```

or

```
(command "insert" "UB1" "2,2" "2.5" "" "90")
```

Although angular responses to normal AutoLISP functions must be in radians, this is an angular response while temporarily back in the AutoCAD screen and therefore must be 90, for degrees.

The first program line uses AutoLISP symbols as responses to the INSERT command that the (command ...) function invoked. The second line uses AutoCAD equivalents of the same responses. In the latter, the responses must be enclosed within quotation marks. This use of the quotation marks is different from their use to mark characters as a string type of argument in function lists other than the (command ...) function.

Two subtle lessons can be gleaned from the foregoing:

- The use of quotation marks returns you to AutoCAD types of responses.
- The double-double quotation marks (null string) is used to cause the Y scale factor to default to the X scale factor. This would not be appropriate if the Y scale needed to be different, of course. Using the null string in this manner is equivalent to returning to the AutoCAD screen, pressing the SPACEBAR, and then returning to AutoLISP.

PAUSE SYMBOL

During a (command...) function you can cause a pause if you wish to allow the user input during the particular AutoCAD command that you have called up. This is done by using the pause symbol in lieu of a variable or of a fixed value where a particular response is required. For example, in the preceding program line you could have allowed the user to input a name and an angle:

```
(command "insert" pause inspt xscal yscal pause)
```

or

```
(command "insert" pause "2,2" "2.5" "" pause)
```

SYMBOLOGY USED IN THIS CHAPTER TO DESCRIBE FUNCTIONS

The function descriptions in this section include the operator (+ or -, for example), and the elements (arguments) that must (or may) follow the operator. If the arguments following the operator in the description are enclosed with angle (< >) brackets only, then those arguments must follow the operator and must be of the type specified. An argument in square ([ ]) brackets following the operator (not necessarily immediately) is optional. When an argument is followed by an ellipsis (...) then the operator will accept multiple arguments of the type specified.

ELEMENTARY FUNCTIONS

(+ <number> <number>...) This function returns the sum of the <number>s. There may be any quantity of <number>s.

(- <number> <number>...) This function returns the difference of the <number>s. There may be any quantity of <number>s.

(* <number> <number>...) This function returns the product of any quantity of <number>s.

(\ <number> <number>...) This function returns the quotient of the first <number> divided by the second <number>. If there are more than two arguments, the quotient of the first and second will be divided by the third, and so on.

RULES OF PROMOTION OF AN INTEGER TO A REAL

As noted in the *Reference Manual*, integers may range from -32,768 to 32,767, depending on the platform. Adding or multiplying integers whose sum or product exceeds 32,767 (or is less than -32,768) will not provide an acceptable result.

1. If any argument in one of the functions is entered as an integer and that integer is outside the integer limits, then the result is not usable.

2. If all of the number arguments entered are integers, then the result will be an integer, and if that result is outside the integer limits, the result will be subject to error.

3. If any of the arguments is entered as a real (and none is an integer exceeding the integer limits), then the result will be a real without limits.

MORE ELEMENTARY FUNCTIONS

(1+ <number>) and (1- <number>) These functions are just different methods of writing (+ <number> 1) and (- <number> 1), respectively. They are used primarily for counters in loops. For example:

```
(1+7) returns 8
(1-7) returns 6
```

(abs \<number>) This function returns the absolute value of the \<number>. For example:

```
(abs (- 4 7))   returns 3
```

(ascii \<string>) This function returns the ASCII value of the first character of the \<string>. For example:

```
(ascii "All")   returns 65
(ascii "a")     returns 97
(ascii "B")     returns 66
```

(chr \<number>) This function returns the character (as a one-character string) whose ASCII code is \<number>. For example:

```
(chr 65)    returns "A"
(chr 97)    returns "a"
(chr 100)   returns "d"
```

(eval \<expr>) This function returns the result of evaluating \<expr>, where \<expr> is any LISP expression. For example:

```
(setq z 6)
(setq q 'z)
(eval z)   returns 6
(eval q)   returns 6
```

(exp \<number>) This function returns e raised to the \<number> power (natural antilog). It returns a real. For example:

```
(exp 1.0)    returns 2.718282
(exp -0.2)   returns 0.818730753
```

(expt \<base> \<power>) This function returns \<base> raised to the specified \<power>. If both arguments are integers, the result is an integer; otherwise, the result is a real. For example:

```
(expt 3 4)   returns 81
```

(log \<number>) This function returns the natural log of \<number> as a real. For example:

```
(log 3.74)    returns 1.32175584
(log 1.025)   returns 0.024692613
```

(sqrt <number>) This function returns the square root of <number> as a real. For example:

```
(sqrt 16)    returns 4.000000
(sqrt 2.0)   returns 1.414213562
```

(type <*item*>) This function returns the type of <*item*>, where type is one of the following (as an atom):

REAL	floating point numbers
FILE	file descriptors
STR	strings
INT	integer
SYM	symbol
list	lists (and user functions)
SUBR	internal AutoLISP functions
PICKSET	AutoCAD selection sets
ENAME	AutoCAD entity names
PAGETB	function paging table

TRIGONOMETRY FUNCTIONS

(sin <angle>) This function returns the sine of <angle>, where <angle> is expressed in radians.

(cos <angle>) This function returns the cosine of <angle>, where <angle> is expressed in radians. For example:

```
(cos 1)         returns .540302306
(sin (/ pi 2))  returns 1.00000
```

(atan <num1> [<num2>]) If only <num1> is present, then (atan ...) returns the angle (in radians) whose tangent is <num1>. If <num1> and <num2> are present, then (atan ...) returns the angle whose tangent is the dividend of <num1> divided by <num2>. For example:

```
(atan 0.75)      returns 0.643501109
(atan 1.0 2.0)   returns 0.463647609
```

LIST HANDLING FUNCTIONS

(list <expr>...) This function has expression(s) as its arguments. It is included in this section because it is the function that creates the lists that the other functions in this section require as arguments.

(**list ...**) This function takes any number of expressions and makes them into a list. For example:

```
(list 1 1)           returns (1 1)
```

Remember: A created list is enclosed in parentheses.

A specific application of the (list ...) function is to combine two or three reals in the format required by a function whose argument is <pt>, which is a 2D or 3D point. The point is a special form of a list. For example:

```
(setq p1 (list 1.0 1.0))
(setq p2 (list 2.0 2.0))
```

allows the following:

```
(setq a (angle p1 p2))
```

which returns 0.785398163.

But

```
(setq p1 (list "you" 1.0))   ;Sets the variable p1 to the list "You",1.0
(setq p2 (list 2.0 2.0))     ;Sets the variable p2 to the list 2.0,2.0
(setq a (angle p1 p2))       ;Sets the variable a to the value of the
                             ;angle between list p1 and p2.
```

returns

    error: bad point value

(**angle <pt1> <pt2>**) This function returns the angle (in radians) between the baseline of angle zero and the line from pt1 to pt2.

(**distance <pt1> <pt2>**) This function returns the distance in decimal units from pt1 to pt2. If the units are set to architectural and the distance between two points is 6'3", the (distance ...) function will return 75.000000.

(**polar <pt> <angle> <distance>**) This function returns a point. It can be one of the most useful tools in the AutoLISP arsenal. The arguments must be of the proper type. The <pt> must evaluate to a list of two reals. The <angle> and <distance> are each a real. The value of the <angle> is in radians.

(**osnap <pt> <mode-string>**) This function returns a point. It allows object snapping to a point while in AutoLISP. Like the (polar ...) function, it returns a point in the form of a list of two or three reals. For example, if a circle has been drawn using p1 and p2 in the 2P method as follows:

```
(setq p1 (list 1.0 3.0))     ;Sets the variable p1 to the list 1.0,3.0
(setq p2 (list 4.0 3.0))     ;Setq the variable p2 to the list 4.0,3.0
(command "circle" "2P" p1 p2) ;Creates a cirle between the points p1 and p2
```

then

    (command "line" "0,0" (setq c (osnap p1 "center")) "")

returns (2.5 3.0), or the list or two reals representing the center of the circle as the endpoint of the line.

(inters \<pt1> \<pt2> \<pt3> \<pt4> [\<onseg>]) This function returns a point. It can be used for the following:

1. To determine if two nonparallel lines intersect.
2. If they intersect, the location of that point.
3. If they do not intersect, where they would intersect if one or both were extended until they intersected.

If the optional \<onseg> argument is present and is nil, the lines will be considered infinite in length and the function will return a list of three reals designating that intersection point.

If the optional \<onseg> argument is not present or is not nil, then their intersection must be on both segments in order for a point to be returned; otherwise, the function will return nil.

CAR, CDR, AND COMBINATIONS

CAR and CDR are the primary functions that select and return element(s) of a list. Unlike the (angle ...) and (distance ...) functions, these functions will operate on a list comprised of elements of any type. The elements can be atoms or lists within the list. The atoms can be reals, integers, strings, or symbols. The argument to the CAR and CDR functions can even be a list of mixed types of elements. These were the foundation functions designed to analyze a list of symbols (which is what language is). In AutoLISP, these functions break down points into coordinates. For example:

    (setq p1 '(1.0 2.0)) returns (1.0 2.0)

Then

    (car p1) returns 1.0, a real

But

    (cdr p1) returns (2.0), a list

Note the parentheses enclosing 2.0. If

    (setq L1 (list 1.0 2.0)) and (setq L2 (list L1 3.0))

then

    (car L2) returns (1.0 2.0), a list

Here is a list within a list. Or in the function list (setq L2 (list L1 3.0)), L1 is a list within a list within a function list.

The preceding expression is not used very often in AutoLISP, but its capability is noteworthy.

    (car <list>) returns the first element of <list>.

    (cdr <list>) returns <list> without the first element.

The type of the return of the (car <list>) function depends on the type of the first element.

If

    `(setq L (list 1.0 2.0))`

then

    `(car L)` returns 1.0, a real

If

    `(setq L (list 1 2.0))`

then

    `(car L)` returns 1, an integer

If

    `(setq L (list "1" 2))`

then

    `(car L)` returns "1," a string

If the symbol "a" evaluates to nil and

    `(setq L (list a 2))`

then

    `(car L)` returns the symbol "a"

But if

    `(setq a 1.0)` and `(setq L (list a 2.0))`

then

    `(car L)` returns 1.0, a real

Whereas (car <list>) can return any type of expression, (cdr <list>) always returns a list. The return may not look like a list, but its type will be a list (even if nil). For example:

```
(setq L1 (1.0 2.0))
(cdr L1)
```

returns (2.0), a list. Remember, the parentheses designate the list and (car L1) returns 1.0, a real. So how can the list (2.0) be used as a real? By using the following combination:

```
(car (cdr L1))
```

which is the same as

```
(car (2.0))
```

which returns 2.0, a real. Note how (car <list>) breaks the first expression out of the <list> and returns it evaluated to its type.

Although (car <list>) and (cdr <list>) operate on the same list, (cdr <list>) always returns a list. Consider the following:

```
(setq L1 (list 1.0))   returns (1.0), a list
(car L1)   returns 1.0, a real
(cdr L1)   returns nil
```

This is the null list, or the same as (list ( )).

If

```
(setq L1 (list 1.0 2.0))
```

then

```
(car L1)   returns 1.0
(cdr L1)   returns (2.0)
(car (cdr L1))   returns 2.0
```

A shortcut for this is: (cadr <list>), or

```
(cadr L1)
```

which returns 2.0, a real. CADR is one of several short forms of combinations of CAR and CDR.

 Note: (car <list>) may return any type. (cdr <list>) is a list.

The short forms of CAR and CDR begin with C and end with R. The characters between will be either an A or a D and will determine the sequence of combined CAR(s) and CDR(s). Here are some short forms:

CAAR	(car (car <list>))
CDDR	(cdr (cdr <list>))
CADR	(car (cdr <list>))
CDAR	(cdr (car <list>))

Any form that begins with CA will return an expression.

Any form that begins with CD will return a list.

All CARs and CDRs represented by A's and D's in such forms as CADAR and CDDAR, however deep, must have lists as their individual arguments.

For example: If

```
(setq L1 (list (1.0 2.0) 3.0))   returns ((1.0 2.0) 3.0)
(car L1)   returns (1.0 2.0), a list
(cdr L1)   returns (3.0), also a list
```

(list (1.0 2.0) 3.0) makes a list out of the list (1.0 2.0) and the atom 3.0. That is why (car L1) returned a list, even though 3.0 is a real in the list "L1;" (cdr <list>) always returns a list. The first element of L1 is (1.0 2.0), when it is removed from ((1.0 2.0) 3.0), (3.0) remains. And that is the function of (cdr L1), to return a list with its first element removed.

So

```
(caar L1)   is the same as   (car (car L1))
```

and returns 1.0, a real. But

```
(cdar L1),   which is the same as   (cdr (car L1))
```

returns (2.0), a list. Because 2.0 and 3.0 are not first elements, in order to return their values as reals, they must first be returned as the first elements of a list, as follows:

```
(cadar L1)   means   (car (cdr (car L1)))
```

and returns 2.0, which is the same as

```
(car (cdr (1.0 2.0)))
```

and the same as

```
(car (2.0))
```

And

```
(cadr L1) means (car (cdr L1))
```

and returns 3.0, or

```
(car (3.0))
```

CAR AND CADR MAINLY FOR GRAPHICS

(car ...) and **(cadr ...)** are the primary functions for accessing the *X* and *Y* coordinates of a point in AutoCAD. Remember, if (setq L1 (list 1.0 2.0)) is entered, (cdr L1) returns the not-so-useful list (2.0). But by using (cadr L1), which means **(car** (cdr L1)), in the following manner (cadr L1) returns 2.0, which is no longer a list but a real. For example, the following expression prompts the user to select a point and then returns the Y component of the point selected.

```
(cadr(getpoint))
```

REVIEW

All forms of the CAR-CDR combination that begin with CA will return the first expression. All forms that begin with CD will return a list.

For graphics applications, lists represent the *X* and *Y* coordinates of a 2D point and the *X*, *Y*, and *Z* coordinates of a 3D point. Therefore, the more useful CAR-CDR combination forms are as follows:

In a 2D point:

```
(setq p2d (list 1.0 2.0))
(car p2d)   returns 1.0, the X coordinate
(cadr p2d)  returns 2.0, the Y coordinate
(caddr p2d) returns nil
```

In a 3D point:

```
(setq p3d (list 1.0 2.0 3.0))
(car p3d)   returns 1.0, the X coordinate
(cadr p3d)  returns 2.0, the Y coordinate
(caddr p3d) returns 3.0, the Z coordinate
(cdr p2d)   returns (2.0), a list
(cdr p3d)   returns (2.0 3.0), a list
```

Neither of the last two statements is very useful unless the programmer wishes to project all of the 3D points on the Y-Z plane.

(last <list>) This function returns the last expression in the <list>. Although (last <list>) can be used to return the *Y* coordinate of a 2D point and the *Z* coordinate of a 3D point, it is not recommended for that purpose. Because there might be an erroneous return, it is recommended that (cadr <list>) be used for the *Y* coordinate and (caddr <list>) for the *Z* coordinate.

(cons \<new first element> \<list>) This function returns the \<first new element> and \<list> combined into a new list, as in the following:

```
(setq a 2.0)
(setq b 3.0)
(setq L1 (list a b))
(cons 1.0 L1)  returns (1.0 2.0 3.0)
```

(cons ...) This function will also return what is known as a "dotted pair" when there is an atom in place of the \<list> argument, as in the following:

```
(cons 1.0 2.0)  returns (1.000000 . 2.000000)
```

This special form of a list requires less memory than ordinary lists.

(length \<list>) This function returns the number of elements in \<list> as in the following:

```
(setq L1 (list "you" (1.0 2.0) 3.0))
returns ("you" (1.0 2.0) 3.0)
(length L1)  returns 3
(length (car L1))  returns nil, because (car L1) is an atom
(length (cdr L1))  returns 2
(length (cadr L1))  returns 2
(length (caddr L1))  returns nil
```

If

```
(setq L1 (list "you" (1.0) 2.0))
(length L1)  returns 3
(length (cadr L1))  returns 1
```

(nth \<n> \<list>) This function returns the nth element of \<list>. (Zero is the first element.) For example:

```
(setq L1 (list "you" (1.0 2.0) 3))
(nth 0 L1)  returns "you," a string
(nth 1 L1)  returns (1.0 2.0), a list
(nth 2 L1)  returns 3, an integer
(nth 0 (cadr L1))  returns 1.0, a real
(nth 1 (cadr L1))  returns 2.0, a real
(nth 3 L1)  returns nil
```

(reverse \<list>) This function returns the \<list> with the elements in reverse order, as in the following:

```
(setq L1 (list (1.0 2.0) 3.0))
(reverse L1)  returns (3.0 (1.0 2.0))
(reverse (car L1))  returns (2.0 1.0)
```

TYPE-CHANGING FUNCTIONS

In order for the AutoCAD operator, AutoCAD, and AutoLISP to communicate properly between (and within) themselves, data is constantly exchanged. Of the different data types (as classified by AutoLISP), there are three types that can be stored as one type but need to be communicated as another type: integer, the real, and the string.

Some AutoLISP functions are designed to take data of one type as their argument and return that data as another type. The basic outline below, Figure 19–19, shows the functions and the types that they are designed to translate from and to.

		FROM		
		INTEGER	REAL	STRING
TO	INTEGER	(FIX)	FIX	ATOI
	REAL	FLOAT	(FLOAT)	ATOF
	STRING	ITOA	ANGTOS	
			RTOS	

Figure 19–19 *The basic translation for communication between AutoCAD and AutoLISP*

(angtos \<angle\> [\<mode\> [\<precision\>]]) This function takes \<angle\> input as a real in radians and returns a string in the format determined by \<mode\>. The values of \<mode\> and their corresponding format are as follows:

Angtos Mode	Format
0	degrees
1	degrees/minutes/seconds
2	grads
3	radians
4	surveyor's units

For example: If

```
(setq p1 (list 1.0 1.0))  ;Sets the variable p1 to the list 1.0,1.0
(setq p2 (list 2.0 2.0))  ;Sets the variable p2 to the list 2.0,2.0
(setq a (angle p1 p2))  returns 0.78539816
```

then

```
(angtos a 0) returns "45"
(angtos a 1) returns "45.000000"
(angtos a 2) returns "45d0'0.0000""
(angtos a 3) returns "0.78539816r"
(angtos a 4) returns "N 45d0'0.0000 W""
```

The optional <precision> determines the decimal places to be displayed.

(atof <string>) This function takes a <string> and returns a real. For example:

```
(atof "3.75")  returns 3.750000
(atof "4")  returns 4.000000
```

(atoi <string>) This function takes a <string> and returns an integer. For example:

```
(atoi "3.75")  returns 3
(atoi "4")  returns 4
```

(itoa <int>) This function takes an <integer> and returns a string. For example:

```
(itoa 33)  returns "33"
(itoa -4)  returns "-4"
```

(rtos <number> [<mode> [<precision>]]) This function takes a real input and returns a string in the format determined by <mode>. The values of <mode> and their corresponding format is as follows:

Rtos Mode	Format
1	scientific
2	decimal
3	engineering
4	architectural
5	arbitrary Fractional Units

The optional <precision> determines the decimal places to be displayed.

(fix <number>) This function takes a real or integer and returns an integer. For example:

```
(fix 4)  returns 4
(fix 4.25)  returns 4
```

(float <number>) This function takes a real or integer and returns a real. For example:

```
(float 4)  returns 4.000000
(float 4.25)  returns 4.250000
```

INPUT FUNCTIONS

The input functions cause a program to pause for user input of a particular type and return data in the format of a specified type.

The optional [<prompt>] in all (get ...) functions allows the programmer to display the <prompt> message in the prompt area on the screen. It will be demonstrated in the first (get ...) function description.

Note: The (getvar ...) function is not a function for user input.

Caution: For all (get ...) functions, the user input cannot be in the form of an AutoLISP function. For example, using an exclamation point followed by a variable name in response to a getxxxx function will cause AutoCAD to generate an error.

(getangle [\<pt>] [\<prompt>]) This function will return an angle, in radians, between two points, the first of which may be the optional [\<pt>] in the function. Note the option of either selecting two points or inputting the first point into the AutoLISP function and selecting the second point. This method is used in the (getdist ...) and the (getorient ...) functions and will be referred to in their descriptions. For example:

```
(setq a (getangle "PICK TWO POINTS: "))
```

will pause for the user to input two points, either of which may be entered in at the keyboard (as in 1'2,3'6-1/2 if in the architectural units mode) or picked on the screen. If the response to the preceding (getangle ...) function were 1,1 and 2,2, then the function would return 0.785398. Or if

```
(setq p1 (list 14.0 42.5))
```

then

```
(setq a (getangle p1 "PICK SECOND PT: "))
```

will use p1 for the first point and pause for the user to input the second point from the keyboard or on the screen.

If

```
(setq p1 (list 1 1)) and (setq a (getangle p1 "PICK SECOND PT: "))
```

and 2,2 were entered, it would return 0.785398.

Caution: Unlike the (angle ...) function, the angle returned by the (getangle ...) function is affected by a change in the ANGBASE system variable. If ANGBASE were changed from 0 degrees to 45 degrees, the preceding (getangle p1 ...) function with a response of 2,2 would return 0.000000. (See the upcoming (getorient ...) function description.)

(getcorner \<pt> [\<prompt>]) This function returns a point selected during the pause. As the user places the cursor for selection, a rectangle is displayed with the \<pt> as one corner of the rectangle and the cursor location as the diagonally opposite corner.

(getdist [\<pt>] [\<prompt>]) This function returns the distance between two points in the same manner and with the same options as the (getangle ...) function returns an angle. The return will be a real. If the units are set to architectural a length of 3'-6 1/2" would be returned as 42.500000.

(getint [<prompt>]) This function pauses for an integer input and returns that integer.

(getkword [<prompt>]) This function pauses for user input of a keyword that must correspond to a word on a list set up by the (initget ...) function prior to using the (getkword ...) function. If the response is not appropriate, AutoCAD will retry. This function prevents a program from terminating prematurely due to the wrong type of data being input by mistake and gives the user another chance. It also permits returning a string by just inputting initial letters. For example:

```
(initget 1 "SET Make New")
(setq g (getkword "LAYER CHOICES? (SET, M, or N): "))
```

AutoCAD will reject any response that is not made up of the initial uppercase characters of the options in the list set by (initget ...). The response may also include any and all of the lowercase characters in the string, but nothing in addition to the characters of any of the options. The responses that are valid to the example here are: SET, M, Ma, Mak, Make, N, Ne, and New, with any of the preceding in uppercase.

```
(initget 1 "SET MaKe New")
(getkword)
```

will accept m because the K was preceded by a lowercase a.

```
(initget 1 "SET MAke New")
(getkword)
```

will not accept m, but will accept ma, mak, or make, but not makeup.

(getorient [<pt>] [<prompt>]) This function will pause for the user to input two points and will return an angle, in radians, between two points.

The point selection options are the same as for the (getangle ...) function. Unlike the (getangle ...) function, the (getorient ...) function is not affected by a change in the ANGBASE system variable. It will return an angle determined by the line connecting the two input points. The angle will be measured between that line and the zero East baseline regardless of the ANGBASE setting.

(getpoint [<pt>] [<prompt>]) This function pauses for input of a point (either from the keyboard or a pick on the screen), and returns that point in the form of a list of two reals.

The optional <pt>, if used, will cause a rubber-band line from <pt> to the placement of the cursor until a pick is made.

(getreal [<prompt>]) This function pauses for user input of a real number and returns that real number.

(getstring [<cr>] [<prompt>]) This function pauses for keyboard characters to be entered and returns them as a string. It is not necessary to enclose the input in quotation marks. AutoLISP will do that automatically. The optional <cr>, if present and not nil, will permit the string to have blank spaces. The string must be terminated by pressing ENTER. Otherwise, if <cr> is present and nil, pressing the SPACEBAR will terminate the entry. For example, T in AutoLISP is a predefined constant symbol that contains a non nil value. When this is combined with the getstring function the user is allowed to enter blank spaces. (setq InputString (getstring T ("\nEnter string : "))

(initget [<bits>] [<string>]) This function offers the programmer a one-time control of the user's response to the next (get ...) function, and that (get ...) function only.

This means that anytime control is needed for a (get ...) function, a new (initget ...) function must precede it.

The type of control that is offered by the (initget ...) function is as follows:

> The program can be set up to reject responses of a certain unwanted type or value (without terminating the program) and offer the user a second chance to enter an acceptable response.
>
> The program can be made to accept points outside of the limits even when LIMCHECK is on.
>
> The program can be made to return 3D points rather than 2D points.
>
> Dashed lines can be used when drawing a rubber band or a box.
>
> The program can be made to accept a string when the (get ...) function normally requires a specific type, such as point or real.

The controls offered by using the <bits> option are shown in Figure 19–20. The <bits> may be a sum of whichever values in Figure 19–20 correspond to the controls desired for the next (get ...) function. For example:

```
(setq p1 (list 0 0))
(initget 9)
(setq d (getdist p1 "SECOND POINT: "))
```

The bits in the next to last line are a sum of 1 (rejects null input) and 8 (allows input outside limits). This will allow the second point to be outside the limits even if the LIMITS mode is on. It also will not accept a null return.

BITS VALUE	MEANING
1	REJECTS NULL INPUT
2	REJECTS ZERO VALUES
4	REJECTS NEGATIVE VALUES
8	ALLOWS INPUT OUTSIDE LIMITS
16	RETURNS 3D POINTS RATHER THAN 2D
32	USES DASHED LINES FOR RUBBERBAND/BOX

Figure 19–20 *The controls offered by the <bits> option*

CONDITIONAL AND LOGIC FUNCTIONS

AutoLISP conditional and logic functions allow the user to have a program test certain conditions and proceed according to the result of those tests. Or, by using the WHILE function, a programmer's LOOP situation will allow iteration of a changing variable between the extents of a specified range.

The symbol T is used when an expression is needed that will never evaluate to nil.

(if <testexpr> <thenexpr> [<elseexpr>]) This function evaluates <testexpr>. If <textexpr> does not return nil, the function returns the evaluation of <thenexpr>. If the optional <elseexpr> is present and <testexpr> evaluates to nil, the function returns the evaluation of <elseexpr>. Otherwise, the function returns nil. For example:

```
(setq q (getint "ENTER QUANTITY FROM 1-99: " ))
(if (< q 10)
   (setq c q)
   (setq c (fix (/ q 10)))
)
```

This program will take a number (from user input) and test to see if it is less than 10. If so, it will SETQ the symbol c to that number. If it is 10 or greater, it will divide the number by 10 and SETQ c to the whole number of the result; for instance, 37 becomes 3.7 becomes 3.

(cond (<test1> <result1> ...) ...) This function accepts any number of arguments. The first item in each list is evaluated, and when one returns not nil, the following expressions in that list are evaluated and the function returns the value of the last expression. For example, a routine could be written to return the angle of a line to be only in the first or fourth quadrants, regardless of how it was originally selected. Note the four possibilities shown in Figure 19–21.

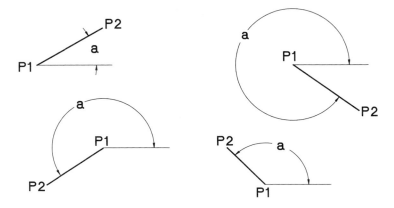

Figure 19–21 *Conditional and logic function return possibilities*

We will not consider the four ortho directions, N, E, S, or W, in this example. If an angle is returned that was determined by the function (setq a (angle p1 p2)), then it could be in one of four quadrants. Then to ensure that no matter which angle was set by p1-p2, the function would return an angle in the first or fourth quadrant. For example:

```
(setq a (angle p1 p2))
(cond
   ((and (> a pi) (< a (* 3 (/ pi 2)))) (setq a (- a pi))
   ((and (> a (/ pi 2)) (< a pi)) (setq a (+ a pi))
)
```

(while <testexpr> <expr>...) This function evaluates <textexpr>, and, if it is not nil, evaluates the following expressions and then repeats the procedure again with <testexpr>.

This repetition continues until <textexpr> evaluates to nil; the function then returns the evaluation of <lastexpr>.

TEST EXPRESSIONS

(if ...), (cond ...), and (while ...) These functions normally use one of the logic functions as <textexpr>. For example, the following program lines were previously entered:

```
(setq a 10.0)
(if (null a) (setq a 6.0))
```

This sequence will not change the value of the symbol a because the evaluation of the expression (null a) is nil. Therefore, the following expression will not be evaluated. Had a not been previously SETQed, the expression (null a) would evaluate to T (for TRUE), and the expression following would be evaluated and would have SETQed the symbol a to 6.0.

(= <atom> <atom> ...) This expression returns T if all of the <atom>s evaluate to the same thing.

(/= <atom1> <atom2>) This expression returns T if <atom1> is "not equal to" <atom2>. It is nil otherwise.

(< <atom> <atom> ...) This expression returns T if each <atom> is "less than" the <atom> to its right. It is nil otherwise.

(<= <atom> <atom> ...) This expression returns T if each <atom> is "less than or equal to" the <atom> to its right. It is nil otherwise.

(> <atom> <atom> ...) This expression returns T if each <atom> is "greater than" the <atom> to its right. It is nil otherwise.

(>= <atom> <atom> ...) This expression returns T if each <atom> is "greater than or equal to" the <atom> to its right. It is nil otherwise.

(and <expr>...) This expression returns T if all <expr>s return T. It is nil otherwise.

(boundp <atom>) This expression returns T if <atom> has a value bound to it. It is nil otherwise.

(eq <expr1> <expr2>) This expression returns T if <expr1> and <expr2> are identical and are bound to the same object. It is nil otherwise.

(equal <expr1> <expr2>) This expression returns T if <expr1> and <expr2> evaluate to the same thing. It is nil otherwise.

(not <expr>) This expression returns T if <expr> is nil. Otherwise, the function returns nil.

(null <item>) This expression returns T if <item> is bound to nil. The function returns nil otherwise.

(or <expr>...) This expression returns T if any of the <expr>s evaluate to something that is not nil. Otherwise the function returns nil.

EXERCISES

1. Identify the atoms in the following expressions:

   ```
   (if (not (null dfr)) (setq dfr (rtos rad)) (setq dfr "0"))
   ```

Write the AutoLISP expression equivalent to each of the following:

2. $a = 1$
3. $b = 2.0$
4. $c = a$
5. $d = 1 + 2$
6. $e = \sqrt{2}$
7. $f = 1 + b^2$
8. $g = 7 + 9 + 3 + 7$
9. $h = (3 + 5 + 6) + 7$
10. $i = (7 + 3) + (4 - (6 + 3))$
11. $j = 3 (7 + 6)$
12. $k = 3 + (7 \times 6)$
13. $l = 5 - (7 + 2)$
14. $m = \sin\ 0.75$
15. $n = \cos\ 1.75$
16. $o = \sin\ (pi\ ,\ 2)$
17. $p =$ absolute value of a
18. $q = x^3$
19. $r = \tan^{-1}((3\ pi) + 4)$

In the rectangle shown in Figure Ex19–1, write the expressions that evaluate to the following:

20. the vertical distance between p1 and p3
21. the distance between p1 and p3
22. the horizontal distance between p2 and p4
23. the distance between p1 and p4
24. p2 in terms of p1 and p3
25. p3 in terms of p2 and p4
26. the sum of the four sides (perimeter)
27. the area enclosed by the four sides
28. the area of a triangle whose vertices are p1, p2, and p3
29. the angle between lines from p1 to p3 and from p1 to p4

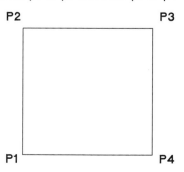

Figure Ex19–1 *The rectangle for writing expressions*

Answer the following:

30. In (setq L1 (list 1.0 2.0)) what is (car L1)?
31. In (setq L2 (list 3.0 4.0)) what is (cdr L2)?
32. In (setq L3 (list 5.0 6.0)) what is (cadr L3)?

For the following expression, write the combination of CAR and CDR that will return the given values from list L5:

    (setq L5 (list 1.0 '(2.0 (3.0 4.0)) 5.0))

33. 1.0
34. 2.0
35. 3.0
36. 4.0
37. 5.0

Give the evaluations of the following expressions:

40. (+ 5 6.0)
41. (- 7 6)
42. (- 8 9)
43. (+ 1 20 300)
44. (- 1 20 300)
45. (- 300 20 1)
46. (- 30000 1.0)
47. (+ 30000 30000)
48. (* 1 2)
49. (* 1.0 2.0)
50. (* 3 4)
51. (* 5.0 6)
52. (* 1 2 3)
53. (* 4 4 6.0)
54. (* 7 8000)
55. (* 8.0 9000)
56. (* 3 pi)
57. (* pi pi)
58. (/ 2.0 1.0)
59. (/ 2.0 1)
60. (/ 2 1)
61. (/ 1 2)
62. (/ 3 2)
63. (/ 3.0 2)
64. (/ 3 2.0)
65. (/ 1.0 2)
66. (/ 40000 2)
67. (/ 2 40000)
68. (abs (+ 4 2))
69. (abs (- 4 2))
70. (ascii "ABC")
71. (ascii "aBC")
72. (chr 66)
73. (chr 98)
74. (expt 3 2)
75. (expt 2 3)
76. (sqrt 9)
77. (sqrt (* 2 8))
78. (type 1)
79. (type pi)
80. (type 3.0)
81. (type T)
82. (sin 0)
83. (cos (/ pi 2))
84. (atan 1)
85. (angle (list 1.0 1.0) (list 2.0 2.0))
86. (angle (list 2.0 1.0) (list 0.0 1.0))
87. (distance (list 1.0 1.0) (list 3.5 1.0))
88. (distance (list 6.0 5.0) (list 2.0 2.0))
89. (polar (list 1.0 1.0) 0 2.0)
90. (polar (list 2.0 3.0) pi 1)
91. (inters (list 0.0 0.0) (list 3.0 3.0)
92. (list 3.0 0.0) (list 0.0 3.0))

CUSTOM COMMANDS AND FUNCTIONS

(defun ...) and (defun C: ...)   This is an AutoLISP function called DEFUN that will create a custom-designed function. Its format is:

>   (defun <sym> <argument list> <expr>...)

The DEFUN function operates in a manner similar to the SETQ function. SETQ names a variable and sets it equal to a value. DEFUN names a new function so that when its name is entered in a routine, it will evaluate to the subsequent expression(s). In addition, it allows variables to be operated on by the function by entering the name of that defined function, followed by a variable or group of variables.

The customizing power of this feature cannot be emphasized enough. What this means to the programmer is that when a routine has been worked out in AutoLISP, the entire routine can be given a name. Then, when the routine needs to be used again, the programmer simply enters the name of the routine. This makes it a user-defined or custom-defined function.

The name of a custom-defined function can also be entered while in AutoCAD, or it can be included in a menu string by entering its name within parentheses.

In addition to having a one-word entry perform the task of many lines of programming, there is an optional feature that makes the custom-defined function even more effective. The user can define the custom function to perform the prescribed task on a dummy variable in the definition and then enter the defined function followed by some real variable later and have the task performed on the real variable. The dummy variables are listed in the <argument list>.

ARGUMENTS AND LOCAL SYMBOLS

Every custom-defined function and command must include the parentheses used to enclose the arguments. Even if there are no arguments or local symbols (which means the opening and closing parentheses would be empty), those opening and closing parentheses must follow the user-designated function name. Remember, however, that arguments (the dummy variables) are not permitted in custom commands using the "C:FNAME" format. The following forms are examples of how function and command definitions may be written:

(defun fname ( )...def exprs...)	No arguments or local symbols
(defun C:FNAME ( )...def exprs...)	No arguments or local symbols
(defun fname (/ x y z)...def exprs...)	Three local symbols only
(defun C:FNAME (/ x y z)...def exprs...)	Three local symbols only
(defun fname (a b)...def exprs...)	Two arguments only
(defun fname (a b / x y z)...def exprs...)	Two arguments and three local symbols

Caution: If the arguments' enclosing parentheses are inadvertently omitted, AutoLISP will take whatever is in the first set of parentheses of the definition expressions and try to use them as arguments and/or local symbols. Naturally, this will not operate as expected and will result in an error message. Therefore, include parentheses as shown, even if empty.

Also, within the <argument list>, optional local symbols may be listed (after the slash) to name symbols to be used within the function definition only, without having any effect on the same symbol name outside the defined function. This form of creating a defined function is for use within AutoLISP and cannot be entered without the enclosing parentheses. An added feature, to be discussed later, will allow the user-defined function to be invoked in the same manner as AutoCAD commands. That is, with the added feature, the user-defined functions can be entered from the keyboard or in the menu without parentheses and then recalled immediately afterward by simply pressing ENTER.

DEFINED FUNCTIONS AND COMMANDS

When one starts to delve into custom programs in many graphic applications, one of the first geometric-trigonometric stumbling blocks to overcome is the use of radians to measure angles. AutoCAD is no exception. Even if the use of radians is second nature to someone, using degrees seems to be first nature with almost all designers/drafters. The main problem is in converting one to the other. This is often necessary because AutoCAD uses degrees for the screen and radians in AutoLISP. Therefore, one of the first AutoLISP routines introduced in articles and books on AutoLISP is one that converts the value of an angle in degrees to its value in radians. The second routine is usually one that converts from radians back to degrees. The usual name for the degrees-to-radians function is the abbreviation dtr. The dtr function is written as follows:

```
(defun dtr (a)          ;Creates the custom defined function dtr with
                        ;one argument
   (* pi (/ a 180.0))   ;Divides the variable a by 180 then multiples
                        ;the quotient by 3.14
)                       ;End user defined function
```

In the format (defun <sym> <argument list> <expr>...), dtr is the symbol (<sym>), a is the only argument in the argument list, and (* pi (/ a 180.0)) is the expression (<expr>).

Though (/ a 180.0) is an expression, it resides within the expression (* pi (/ a 180.0)) and is not considered one of the separate <expr> expressions in the format: (defun <sym> <argument list> <expr>...). The <expr>... indicates there is no limit to the number of expressions possible in a defined function. Although this is a rather minor point, it may help in understanding what an expression is and how AutoLISP looks at them.

WRITING A DEFINED FUNCTION

The dtr function may be entered in from the keyboard while in the AutoCAD screen, as follows:

Command: **(defun dtr (a) (* pi (/ a 180.0)))**

This function may be used and applied to any real number in the following manner:

    Command: **(dtr 90.0)** (returns 1.570796327)

or

    Command: **(dtr 180.0)** (returns 3.141592654)

or

    Command: **(dtr 7.5)** (returns 0.130899694)

Although entering this function from the keyboard will define the dtr function for use at any time during the current editing session, the definition will be lost when the drawing is ended. Although writing this routine is fairly simple, it is still not very economical. There is a better way.

WRITING AND STORING A DEFINED FUNCTION

Usually, any function that needs to be defined in one drawing will be useful enough to warrant saving the definition for instant reuse later in any drawing without having to enter it in repeatedly. Therefore, defined functions may be saved for loading into a drawing. By using a text editor or word processor in the programmer mode, the custom programmer can create a file having the file extension of .LSP and then enter in the routine. The following example is similar to the first example, only with a slight variation in format:

File name: ANGLE.LSP

```
(defun dtr (a)            ;degrees to radians
    (* pi (/ a 180.0))    ;pi times "a" divided by 180
)                         ;leave function definition
```

The indentation is permissible in writing a defined function in a FILENAME.LSP file (unlike the program lines in a FILENAME.MNU file). It is even desirable in order to be able to identify easily the distinct elements and subelements of the function definition. The value of this procedure will become more evident with more elaborate definitions. The closing parenthesis, on a line by itself, is lined up with the opening parenthesis. This is an example of how indentation coincides with the depth of nesting of expressions.

In this last example, remarks have been written to the right, following semicolons. AutoLISP ignores anything on a line following a semicolon.

Function Names The name of the file may or may not be the same as any of the function names it includes. It may even be the same as a function in another file, although that would not be logical. Uniqueness of names is critical within a group of named items, such as files, functions, variables, and drawings in one directory. But in most cases it is not a problem to duplicate names across groups, such as having a

FILENM1.LSP and a FILENM1.DWG, and a FILENM1.SHP at the same time along with defining a custom function as follows:

    (defun filenm1 (...

Duplication in this case is not only acceptable but sometimes desirable in coordinating a group of specially named custom files and functions that are all in a single enhancement program being used for a singular purpose.

Loading a Defined Function As mentioned earlier in this chapter, custom-defined functions must be written into a FILENAME.LSP file in order to be saved and usable later. Once a custom-defined function is stored in that file, the user must load the file in order to use it. For example, the file named ANGLE.LSP has been created to store the custom-defined function named dtr as follows:

File name: ANGLE.LSP

```
(defun dtr (a)           ;degrees to radians
    (* pi (/ a 180.0))   ;pi times a divided by 180
)                        ;leave function definition
```

To make dtr usable, the user must invoke the AutoLISP function named LOAD. It has the format of:

(load <filename>)

with <filename> being a string without the extension of .LSP. Therefore, the file name must be enclosed within quotations, as follows:

(load "angle")

Caution: Be aware of the distinction between the AutoLISP load function and the AutoCAD load command. The AutoCAD LOAD command named LOAD is used to load FILENAME.SHP (shape) files and is entered without the parentheses, in the following manner:

Command: **load**

The AutoLISP function named LOAD" is used to load FILENAME.LSP (AutoLISP) files and is entered with the parentheses, in the following manner:

Command: **(load "angle")**

The preceding sequence will make the custom-defined function named dtr usable in the current editing session. The user must be careful when trying to load .LSP files from some other drive or directory. If the file named ANGLE.LSP is stored on the directory named LISPFILE, the format would be as follows:

Command: **(load "/lispfile/angle")**

Special note should be made of the forward slash versus the backslash. The file name is a string, and a leading backslash in a string is itself a special control character used in conjunction with other code characters for specific operations on the string. Thus, in order to have a backslash read as a backslash in a string, there must be a double backslash. Therefore, the forward slash is recommended in designating a directory path in PC-DOS/MS-DOS.

Drive specifications can also be included in the AutoLISP LOAD function format as follows:

Command: **(load "a:angle")**

or

Command: **(load "a:/lispfile/angle")**

The last entry may be used if the ANGLE.LSP file is in the LISPFILE directory on the A: drive.

If the custom programmer has created custom-defined functions in a file called ACAD.LSP and that file is accessible when a drawing is edited, AutoCAD will "load" that file automatically. Only functions that might be used in all drawings should be included in that file if it is created.

SIN and COS, but not TAN! Sometimes designers/drafters wish to determine some distances in the all-powerful right triangle (from which trigonometry is derived) by using the opposite and adjacent sides instead of the hypotenuse. So why doesn't AutoLISP have the TAN (tangent) function in addition to the SIN and COS? First of all, an angle is defined in *Webster's New Collegiate Dictionary* as "a measure of the amount of turning necessary to bring one line or plane into coincidence with or parallel to another." Two special cases exist in computing the tangent of an angle, one of which causes a problem that the sine and cosine do not cause. When two lines exist, it is assumed that they must have length and therefore have a nonzero value. When considering the sine or cosine of an angle, the hypotenuse is always one of the existing nonzero lines. In the sine and cosine of any angle, the hypotenuse is always the divisor. So, even if the opposite side (in the case of the sine) or the adjacent side (in the case of the cosine) turn out to be zero, the worst that can happen is that the function will result in zero. In the case of the tangent, where the divisor is the adjacent side, if it is zero the result approaches infinity and is therefore not valid for use within the program. This occurs, of course, at 90 degrees.

The other special case is where the angle is zero and the tangent is zero. Zero may be a valid entry, whereas infinity cannot be. This problem does not occur in the sine and cosine functions because the results of either range from 1 to 0 to -1 and back to 1. All results within this range are valid for use within programs.

An approach to arriving at the tangent is possible by using existing AutoLISP functions named SIN and COS. The tangent can be expressed as the quotient of the sine divided by the cosine. But here again is the possibility of the divisor being zero because the cosine of 90 degrees is just that. If used for specific purposes where it is known that the value of the adjacent side involved is nonzero, then a custom-defined function that will return the tangent of an angle (along with a radians-to-degrees function) might be included in the ANGLE.LSP file, as follows:

File name: ANGLE.LSP

```
(defun dtr (a)              ;degrees to radians
   (* pi (/ a 180.0))        ;pi times a divided by 180
)                            ;leave function definition
(defun rtd (b)              ;radians to degrees
   (* 180.0 (/ b pi))  ;180 times b divided by pi
)                            ;leave function definition
(defun tan (c)              ;tangent
   (/ (sin c) (cos c)) ;sine divided by cosine
)                            ;leave function definition
```

Defined AutoLISP Commands AutoCAD programming features allow the creation of custom-defined AutoLISP commands with two characteristics similar to regular AutoCAD commands.

- AutoLISP commands can be entered from the keyboard or within a menu string by just entering the name of the custom command without having to enclose it in parentheses.

- If a particular AutoLISP command was the last command used, then typing « will recall that same command for immediate use.

Note: AutoLISP commands cannot operate on external variables (arguments) of an <argument list>. However, local symbols are allowed.

Writing an AutoLISP Function The format used in creating an AutoLISP command is as follows:

(defun C:CMDNAME </ local symbols> <expr>...)

where CMDNAME can be a name of your choice and can duplicate the name of an existing AutoCAD command or function, if your purpose is to override that command with a definition of your own.

WHAT IS THE DATABASE?

The file that stores information about the drawing has tables and Block sections that include data associated with linetype, layer, style (for text), view, UCS, viewport, and blocks. These are accessible through the (tblnext...) and (tblsearch...) functions.

The Entities section contains data associated individually with each entity in the drawing. There are functions that point to the entity data and that can use or manipulate that data.

ENTITY NAMES, ENTITY DATA, AND SELECTION SETS

Information is continually being updated during an editing session. An ongoing record is being kept for each new, modified, or deleted entity. The information or data concerning each entity is stored at some particular location in the drawing file. (This location is not to be confused with its *X*, *Y*, and *Z* coordinates.)

AutoLISP gains access to an entity's data through its entity name, which points to the location of the entity's data within the drawing file. The entity name is the address of that entity's data. You must grasp the concepts of entity name and entity data and be able to distinguish between them.

ENTITY NAME FUNCTIONS

Entity name functions evaluate to an item whose data type is called AutoCAD entity name. An entity name can be used as a response when an AutoCAD command prompts you to "Select object:."

The **(entnext [<ename>])** function, if performed with no arguments, returns the entity name of the first nondeleted entity in the database. If (entnext...) is performed with an entity name (which we will call en1 for illustration purposes) as the argument, it returns the entity name of the first nondeleted entity that follows en1 in the database.

WALKING THROUGH THE DATA BASE

Certain functions can search through the database. The **(entlast)** function returns the entity name of the last nondeleted main entity in the database. It may be used to return the entity name of a new entity that has just been added by a previous (command...) function.

The **(entsel [<prompt>])** function returns a list with the entity name as the first element and the point by which the entity was selected as the second element. Note that the second element (a point) is a list itself. For example:

    Command: **line**
    From point: **1,1**
    to point: **4,4**

to point: ENTER
Command: **(setq es (entsel "Select an Entity: "))**
Select an Entity: **2.5,2.5** *(returns (<Entity name: 60000018> (2.5 2.5 0.0)))*

While the evaluation of (entsel...) is a list and not an AutoCAD entity name, its first element is an entity name, obtainable by using the following:

Command: **(car es)** *(returns <Entity name: 60000018>, an entity name)*

Note that the address of this entity is displayed in hexadecimal form, 60000018, and is different for each different entity name. It will also differ from one editing session to the next for the same entity.

Point data is also obtainable by:

Command: **(cadr es)** *(returns (2.5 2.5 0.0),a list)*

And the *X* coordinate is obtainable by:

Command: **(caadr es)** *(returns 2.5, a real)*

Note that

Command: **(cdr es)**

returns ((2.5 2.5 0.0)), which is a list whose only element is also the list (2.5 2.5 0.0).

The (handent <handle>) function returns the entity name associated with the specified handle. This more advanced concept will not be covered within this section. It is noteworthy, however, that the (handent...) function assists in the problem of entities changing their names from one editing session to the next.

ENTITY DATA FUNCTIONS

Entity data functions have entity names as their arguments.

ENTGET Function The primary entity data function is (entget <ename>). This function returns a list of entity data describing <ename> in that special format called an association list.

Data Types for Database Functions Before continuing with detailed descriptions of the entity data functions and even before introducing selection sets, it would be convenient to discuss certain data types that AutoLISP has set aside especially for use when accessing the database. Some of the concepts used in this discussion of data types will not be described in detail until later in this section. Therefore, the reader will probably wish to refer back here later for review after studying those concepts.

In a manner similar to integers, reals, strings, and lists, the special data types called AutoCAD selection sets and AutoCAD entity names can be used as arguments required by certain AutoLISP functions. It should be noted that the special functions designed to operate on these data types can operate on them only. For example; having performed the following sequence:

```
(setq en1 (entnext))
(setq ss1 (ssget "w" '(0 0) '(12 9))
(setq od1 (list 1.0 2.0))
(setq od2 "HELLO")
```

the following are valid entries:

```
(setq ed1 (entget en1)) and (setq en2 (ssname ss1 0))
```

But the following are not valid:

```
(setq ed2 (entget od1)) and (setq ed3 (entget od2))
```

The first two are valid entries because en1 is an AutoCAD entity name, which is the required data type for an argument to the (entget...) function. The second two are not valid entries because od1 and od2 are other data types and so are not valid as arguments. Od1 is a list and od2 is a string.

Similarly, ss1 is a special data type called an AutoCAD selection set and is valid as an argument to the (ssname...) function.

Further study will show that the (ssname...) function returns a value in the form of that special data type called an AutoCAD entity name. Therefore, if the preceding entries had been performed, then the following is a valid entry:

```
(setq ed4 (entget (ssname ss1 0)))
```

returning the data (as an association list) about the first entity in the drawing file that is included in a window whose corners are 0,0 and 12,9. Even if no entities had been found, and the operation had returned nil, the operation would still have been proper, having had the required data type as an argument to the (entget...) function.

Back to Entget For an example let's say a newly created drawing had the following as the first entries:

Command:
Specify first point: **1,1**
Specify next point or [Undo]: **4,4**
Specify next point or [Undo]: (ENTER)

Command: **line**
Specify first point: **1,4**
Specify next point or [Undo]: **4,1**
Specify next point or [Undo]: (ENTER)

then

```
(setq L1 (entget (entnext)))
```

returns

((-1 . <Entity name: 60000018>) (0 . "LINE") (8 . "0")
(10 1.0 1.0 0.0) (11 4.0 4.0 0.0) (210 0.0 0.0 1.0))

and

```
(setq L2 (entget (entlast))  or (setq L2 (entget (entnext L1))
```

returns

((-1 . <Entity name: 60000030>) (0 . "LINE") (8 . "0")
(10 1.0 4.0 0.0) (11 4.0 1.0 0.0) (210 0.0 0.0 1.0))

Also note the syntax in the following example:

Having entered

```
(setq en1 (entnext))
(setq en2 (entlast))
```

then

```
(setq L1 (entget en1))
```

returns the same as

```
(setq L1 (entget (entnext)))
```

and

```
(setq L2 (entget en2))
```

returns the same as

```
(setq L2 (entget (entlast)))  or (setq L2 (entget (entnext L1)))
```

This is to emphasize that the entity names en1 and en2 are only addresses (or pointers) to where the data associated with the entities are located within the drawing file. The data list itself is gotten through the (entget...) function, which must have that address or entity name as its argument.

THE ASSOCIATION LIST

Once the concepts of <ename>s and <elist>s are understood and the throes of creating those first routines to extract them are survived, the custom programmer must now deal with how to make use of the results.

The (assoc <item> <alist>) function is the primary mechanism used to extract specific data associated with a selected entity. To store data in an organized, efficient, and economical fashion, each common group of data is assigned what is called a group code. That group code is simply an integer. For example, the starting point for a line is a group code 10 and its layer is a group code 8.

Using the first entity in the previous example, you may then enter the following sequence:

Command: **(setq ed1 (entget (entnext)))**

This returns the association list we saw earlier, which we will display as follows:

```
((-1 . <Entity name:60000018>)
  (0 . "LINE")
  (8 . "0")
  (10 1.0 1.0 0.0)
  (11 4.0 4.0 0.0)
  (210 0.0 0.0 1.0)
)
```

Dotted Pairs The first and last parentheses enclose the association list. Each of the other matched pairs of parentheses encloses lists, with each list comprising a specific group code integer as the first element followed by its associated value. In the case of the group code 8, its value is the string "0," which is the name of the layer that this entity is on. Likewise, the value of the group code 0 is the entity type, which is the string "LINE." Note that these two sublists have a period between the group code and its associated value (separated by spaces). This is a special list, called a dotted pair, that requires less memory in storage. For economy, AutoCAD uses these dotted pairs where feasible. Some group codes, like the starting point, may be in the more common form of the list, such as (10 1.0 1.0 0.0) with the group code as the first element followed by the X, Y, and Z coordinates, respectively.

Group codes are broken down as shown in Tables 19–1 and 19–2.

Table 19–1 *Group Codes and Their Respective Value*

GROUP CODE RANGE	FOLLOWING VALUE
0–9	String
10–59	Floating-point
60–79	Integer
210–239	Floating-point
999	Comment (String)

Table 19-2 *Group Codes and Their Respective Value Type*

GROUP CODE RANGE	VALUE TYPE
0	Identifies the start of an entity, table entry, or file separator
1	The primary text value for an entity
2	A name, attribute tag, block name, etc.
3–4	Other textual or name values
5	Entity handle expressed as a hexadecimal string
6	Linetype name (fixed)
7	Text style name (fixed)
8	Layer name (fixed)
9	Variable name identifier (used only in the Header section of the DXF file)
10	Primary point (start point of a line or text entity, center of a circle, etc.)
11–18	Other points
39	This entity's thickness if nonzero (fixed)
40–48	Floating-point values (text height, scale, etc.)
49	Repeated value—multiple 49 groups may appear in one entity for variable-length tables (such as the dash lengths in the Ltype table)
50–58	Angles
62	Color number (fixed)
66	"Entities follow" flag (fixed)
70-78	Integer values, such as repeat counts, modes.
210, 220, 230	X, Y, and Z components of extrusion direction.
999	Comments

EXTRACTING DATA FROM A LIST

```
((-1 . <Entity name:60000018>)
   (0 . "LINE")
   (8 . "0")
   (10 1.0 1.0 0.0)
   (11 4.0 4.0 0.0)
   (210 0.0 0.0 1.0)
)
```

If the preceding list had been SETQed to the variable named ed1, examples of the (assoc...) function would be as follows:

Command: **(assoc 0 ed1)** *(returns (0 . "LINE"))*

Command: **(assoc 8 ed1)** *(returns (8 . "0"))*

Command: **(assoc 10 ed1)** *(returns (10 1.0 1.0 0.0))*

These results are lists. The first two are dotted pairs. To extract data from these, the following may be entered:

Command: **(cdr (assoc 0 ed1))** *(returns "LINE," a string)*

Command: **(cdr (assoc 8 ed1))** *(returns "0," also a string)*

But the following returns (1.0 1.0 0.0),a list:

Command: **(cdr (assoc 10 ed1))**

Therefore, in order to extract the *X* coordinate, enter:

Command: **(cadr (assoc 10 ed1))**

which returns 1.0, a real. Note how CDR returns the second element of a dotted pair as an atom, whereas it requires CADR to return the second element of an ordinary list as an atom. CDR applied to an ordinary list returns a list (unless it is applied to a single-element list).

OTHER ENTITY DATA FUNCTIONS

The **(entdel <ename>)** function deletes the entity specified by <ename> from the drawing or restores the entity previously deleted during the current editing session.

The **(entmod <elist>)** updates the database information for the entity specified. Using this function in conjunction with the AutoLISP (subst...) function is a convenient way to make specific changes to selected entities.

The **(entupd <ename>)** function is used for more advanced handling of block and polyline subentities. The reader is referred to the *AutoLISP Programmer's Reference* for use of this function.

SELECTION SETS

A selection set is a collection of entity names. It is the programmer's equivalent to a group of entities selected by one or more of the optional methods of selecting objects from the screen; except that, by using the selection set functions, certain entities not visible on the screen may even be included in the group. It should be emphasized that a selection set comprises the entity names of the entities in the group. The selection set can be used in response to any AutoCAD command where selection by "Last" is valid. It also is a valid argument to selection set functions that supply entity names to the (entget) function, which then returns entity data in the form of an association list.

The **(ssget [<mode>] [<pt1> [<pt2>]])** function returns the selection set and prompts something like <Selection set: 1>. This indicates that the selection set contains one or more entities. If no objects meet the selection method, then nil will be returned.

The <mode>, if included, determines by what method the selection process is made. The "W," "L," "C," and "P" correspond to the window, last, crossing, and previous selection modes, respectively. Examples are as follows:

(ssget)	Asks the user for entity selection with one or more standard options
(ssget "W" '(2 2) '(8 9))	Selects the entities inside the window from 2,2 to 8,9
(ssget "L")	Selects the last entity added to the database
(ssget "C" '(0 0) '(5 3))	Selects the entities crossing the box from 0,0 to 5,3
(ssget "P")	Selects the most recently selected objects
(ssget '(1.0 1.0))	Selects the entity passing through the point 1,1
(ssget "X" <filter-list>)	Selects the entities matching the "filter-list"

SSGET Filters The **(ssget "X" <filter-list>)** function provides a method of scanning the entire drawing file and selecting the entities having certain values associated with specified group codes. This function is used in conjunction with the (cons...) function. Note that while the (ssget...) function returns a group of entity names, it scans a group of association lists (the drawing file, that is). So the filter will be constructed in a manner that can be tested against entity data. Examples are as follows:

`(ssget "X"(list(cons 0 "CIRCLE")))`	Returns a selection set consisting of all the circles in the drawing
`(ssget "X"(list(cons 8 "0")))`	Returns all entities on layer "0"
`(ssget "X"(list(cons 0 "CIRCLE") (cons 8 "0")))`	Returns all circles on layer "0"

Note that the (ssget "X"...) function selects only from main entities. Special methods must be used to gain data about subentities, such as attributes and polyline vertices. For example, (setq entities (ssget "x")) sets the variable entities to a selection set consisting of all the main objects contained with the current drawing.

The **(sslength <ss>)** function returns an integer containing the number entities in selection set <ss>. For example, the following returns 1:

```
(setq sset (ssget "L"))
(sslength sset)
```

The **(ssname <ss> <index>)** function returns the entity name of the <index>th element of selection set <ss>. The first element begins with number 0. For example, let's say that (setq sset (ssget)) results in five items:

```
(setq en1 (ssname sset 0))   (returns the first entity)
(setq en3 (ssname sset 2))   (returns the third entity)
```

The **(ssadd [<ename> [<ss>]])** function without arguments constructs a selection set with no members. If performed with a single-entity-name argument, it constructs a new selection set containing that single entity name. If performed with an entity name and a selection set, it adds the named entity to the selection set.

The **(ssdel <ename> <ss>)** function deletes entity name <ename> from selection set <ss>.

The **(ssmemb <ename> <ss>)** function tests whether entity name <ename> is a member of selection set <ss>.

ADVANCED ASSOCIATION LIST AND (SCANNING) FUNCTIONS

Examples of **(assoc <item> <alist>)** have been presented previously in this section. It will now be used in our custom-designed command.

The **(subst <newitem> <olditem> <list>)** function searches <list> for <olditem>, and returns a copy of <list> with <newitem> substituted in every place where <olditem> occurred.

Two sample routines are listed next. The TSAVE routine redefines the REDRAW command and then uses a time-checking part to cause the newly defined REDRAW command to invoke a SAVE command after a predetermined time has passed. The PARAB routine draws a parabola.

Saved in the file named TSAVE.LSP:

```
;-Autosave-
(defun S::STARTUP ( )
          (command "undefine" "redraw")
          (setq savetime 0.25)
)

(defun C:TSAVE ( )
          (setq temptime (getreal "ENTER INTERVAL OF TIME IN MINUTES
```

```
                     FOR SAVING: "))
            (setq savetime (/ temptime 60.0))
    )
(defun C:REDRAW ( )
        (if (null cdate1)
            (setq cdate1 (decihr (getvar "cdate"))))
        (setq cdate2 (decihr (getvar "cdate")))
        (if (or (> (- cdate2 cdate1) savetime)
                (> cdate1 cdate2)
                )
            (progn
                (setq tempex (getvar "expert"))
                (setvar "expert" 2)
                        (command "save" "c:backup")
                (setq cdate1 cdate2)
                (setvar "expert" tempex)
                )
        )
            (command ".redraw")
)
(defun decihr (dt / hms dh dm ds hd)
        (setq hms (* (- dt (fix dt)) 10000.0))
        (setq dh (/ (fix hms) 100.0))
        (setq dm (/ (- dh (fix dh)) 0.6))
        (setq ds (/ (- hms (fix hms)) 36.0))
        (setq hd (+ (fix dh) dm ds))
)
```

Saved in the file named PARAB.LSP:

---------------- PARABOLA ----------------------------

```
(defun c:parab ( )
        (setq point1 (getpoint "ENTER POINT: " ))
        (setq number (getint "ENTER ITERATIONS: "))
        (setq i (getreal "ENTER INCREMENTS: "))
        (setq p1 point1 p (car point1) counter 0)
        (while (< counter number)
            (setq p2
                (list
                    (+ (car point1) i)
                    (+ (cadr point1) (* 2.0 (sqrt (* p i))))
                )
            )
        (command "line" p1 p2 "")
        (setq p1 p2 counter (1+ counter) i (+ i i))
        )
)
```

DEBUG YOUR PROGRAM

As you develop an integrated suite of procedures for your applications, you will undoubtedly be confronted with the situation in which the applications are not producing the intended results. You may no longer find it practical to visually inspect your source code to locate the cause. You may prefer to use the debugging techniques in VLISP. Using the sample application developed earlier, this section illustrates the basic debug operations.

LOAD PROJECT

If the project needs to be loaded you may do so by selecting from the Visual LISP for AutoCAD window:

| Project menu | Choose Open Project |

Or from the AutoCAD window select:

| Tools menu | Choose Macro > Load Project |

You may use the BROWSE command button to navigate to the folder when the project file is located and then select it. If the selection is successful, then a project window will pop up and list the names of the files associated with the project.

These files can be loaded in a single step by selecting:

| Project menu | Choose Load Project Source Files |

The VLISP console window will display the results of this selection as shown below.

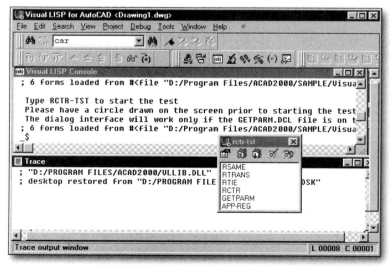

Figure 19–22 Load project source files

SET BREAK POINT

In order to debug the source code you will need to step through the statements as the program executes. Since it may be impractical, especially for the larger or more iterative programs, to patiently step through each statement, you can insert break points in your source code. If the program reaches a break point statement during a run, execution is halted. In this state or condition you can examine the values of program variables, list the call sequence, and perform other debug operations.

To set a break point, place the cursor at the position where you would like to halt the execution. Then select:

| Debug menu | Choose Toggle Breakpoint |

The break point is identified with an inverse video attribute (*viz.*, default is red foreground, white background). If you attempt to set the break point again at the same position, you will notice that the command operates as a toggle switch. In other words, selecting it again disables the break point. If at least one break point is still set then the **Break on Error** option in the **Debug** pop-up menu is set with a check mark to indicate the current status.

DEBUG TOOLBAR

All the buttons in the Debug toolbar are active when a break point is reached. The key buttons are:

- Step into
- Step over
- Step out
- Continue
- Quit
- Reset
- Toggle breakpoint
- Add Watch
- Last Break

The View toolbar contains additional buttons that may be useful during debugging operations.

Step Into

When you click this button, the next expression in the code is evaluated and the program halts once again. If the expression is nested then the first expression at next level of the nest is evaluated.

Step Over

You should use this button to proceed to the closing parenthesis of the current expression. You can avoid having to step through nested expressions with this button

Step Out

This button will allow you to proceed to end of the current function before halting. If you do not want to step through the remainder of the code in the current function, you can use this button.

Continue

This button allows you to resume the execution of the program. The program will subsequently halt at the next break point, if any, or any other program stop condition.

Quit

This button terminates the current break loop. The execution of the program is returned to the next higher loop level.

Reset

This button terminates all the loops. The execution of the program is terminated and control is returned to the console window.

Add Watch

This button allows you to examine the values of the selected variables in dynamic mode. As the values of the variables change during the execution of the program the current values are displayed in the watch window.

AUTOCAD DEVELOPMENT SYSTEM (ADS)

The AutoCAD Development System (ADS) is a programming interface that, like AutoLISP, permits you to write or use applications for AutoCAD in high-level languages, such as C. The applications are loaded in a similar manner to AutoLISP. AutoLISP is more appropriate for smaller applications, whereas ADS makes use of an extensive, powerful, and more complex library of programming functions. ADS, because of the large library, demands more of your system. ADS is not stand-alone, but is tied to AutoLISP; therefore, it is not a good substitute for simpler tasks that can be so easily implemented with AutoLISP.

REVIEW QUESTIONS

1. To load the LISP file SETUP into the current AutoCAD drawing, enter the following at the "Command:" prompt:
   a. SETUP
   b. LOAD SETUP
   c. (LOAD SETUP)
   d. (LOAD "SETUP")
   e. none of the above

2. The command used to easily select and load AutoLISP and ADS routines is:
   a. APPLOAD
   b. LOAD
   c. DBLIST
   d. ADS
   e. none of the above

3. What does the (CADR L) function do?
   a. returns the first element of list L
   b. returns the second element of list L
   c. returns the third element of list L
   d. none of the above

4. What does the (atan N) function do?
   a. returns the angle, in degrees, whose tangent is N
   b. draws a line tangent to the last object drawn
   c. returns the tangent of the angle N, where N is in radians
   d. none of the above

5. What does the function (/ A B) do?
   a. returns the quotient of A divided by B
   b. returns the quotient of B divided by A
   c. returns the remainder of A divided by B
   d. returns the remainder of B divided by A

6. In (+ a 7) the + is called the:
   a. operand
   b. operator
   c. symbol
   d. none of the above

7. The following are acceptable AutoLISP data types, **except**:
   a. real
   b. integer
   c. string
   d. storage
   e. symbol

8. If (setq a 7), what does (list 4 a) return?
   a. (4 7)
   b. (7 4)
   c. (11)
   d. an error message
   e. none of the above

9. What does (/ 3 7) return?
   a. 10
   b. 0
   c. 0.4285714285714
   d. 4
   e. none of the above

10. The expression to add 7 to 3 is:
    a. (7 + 3)
    b. (+ 7 3)
    c. (7 3 +)
    d. none of the above

11. The expression (* 10 .5) returns:
    a. 5
    b. 20
    c. an error
    d. (5)

12. The statement (setq pi 3) will:
    a. redefine the symbol PI
    b. give an error message
    c. be ignored by AutoCAD because PI is a predefined symbol
    d. none of the above

13. To associate a LISP file with a menu file, so that it will be automatically loaded when the menu file is loaded you must use the same file name for both, with only the extensions varying. The proper extension for the LISP file is:
    a. MLP
    b. MNL
    c. LSP
    d. LSM
    e. LLP

14. When working in AutoCAD and you get the prompt "1>", it means:
    a. you have one minute until the network shuts down
    b. you need to specify one more argument for the previous function
    c. you need to enter one more close parenthesis ")"

APPENDIX a

Hardware Requirements

INTRODUCTION

The configuration of your CAD system is a combination of the hardware and software you have assembled to create your system. There are countless PC configurations available on the market. The goal for a new computer user should be to assemble a PC workstation that does not block future software and hardware upgrades. This section lists the essential hardware required to run AutoCAD 2006.

When you install AutoCAD 2006 it automatically configures itself to the various drivers already set up in the Windows operating system. The Options dialog box can be used to custom configure your installation, as explained in Chapter 13.

RECOMMENDED CONFIGURATION

Following is the minimum configuration required by Autodesk for effective use of AutoCAD 2006 software:

1. Intel® Pentium® III processor or later, 800 MHz.
2. Microsoft® Windows® XP (Professional with Service Pack 1 or 2, Home Edition with Service Pack 1 or 2, or Tablet PC Edition) or Windows 2000 Professional with Service Pack 4 and Microsoft Internet Explorer 6.0 with Service Pack 1 or later.
3. 512 MB RAM, 500 MB free disk space for installation
4. Video 1024 x 768 VGA video display recommended, with Windows-supported display adapter.
5. Peripherals: Mouse, trackball, or compatible pointing device, CD ROM drive

Listing of AutoCAD Commands (Menu and Command Prompt)

Command Aliases	Explanation	Options	Toolbar	Pull-Down Menu
'About	Displays a dialog box with the AutoCAD version and serial numbers, a scrolling window with the text of the acad.msg file, and other information			Help
Acisin	Imports an ACIS file			Insert
Acisout	Exports AutoCAD solid objects to an ACIS file			Export
Adcclose	Closes AutoCAD DesignCenter			TYPE IN
Adcenter	Manages content		Standard	
Adcnavigate	Directs the Desktop in AutoCAD DesignCenter to the file name, directory location, or network path you specify			TYPE IN
Align	Aligns objects with other objects in 2D and 3D	Specify either one, two, or three pairs of source points and definition points to align the selected objects	Modify	TYPE IN
Ameconvert	Converts AME solid models to AutoCAD solid objects			TYPE IN
Aperture	Regulates the size of the object snap target box	Select (1–50 pixels) to increase or reduce size of box		TYPE IN
Appload	Loads AuotLISP, ADS, and ARX applications			Tools
Arc (A)	Draws an arc of any size	A Included angle C Center point D Starting direction E Endpoint L Length of chord R Radius S Start point ENTER Continues arc from endpoint of line or arc	Draw	Draw
Archive	Packages the current sheet set files to be archived			TYPE IN

Command Aliases	Explanation	Options		Toolbar	Pull-Down Menu
Area	Calculates the area of a polygon, pline, or circle	A S O	Sets add mode Sets subtract mode Calculates area of the circles, ellipses, splines, polylines, polygons, regions, and solids	Object Properties	Tools
Array	Copies selected objects in circular or rectangular pattern	P R	Polar (circular) arrays about a center point Rectangular arrays objects in horizontal rows and vertical columns	Modify I	Modify
Arx	Loads, unloads, and provides information about ObjectARX applications				TYPE IN
Assistclose	Closes the Quick Help window			Standard	Help
Assist	Opens the Quick Help window			Standard	Help
Attachurl	Attaches hyperlinks to objects or areas in a drawing	Select area or Object			TYPE IN
Attdef	Creates an attribute definition that assigns (tags) textual information to a block	I C V P	Regulates visibility Regulates constant/variable mode Regulates verify mode Regulates preset mode		TYPE IN
Attdisp	Regulates the visibility of attributes in the drawing	ON OFF N	Makes all attribute tags visible Makes all attributes invisible Normal visibility set individually		View
Attedit	Permits the editing of attributes				Modify
Attext	Extracts attribute information from a drawing	C D S E	CDF comma-delimited format DXF format SDF format Select objects		TYPE IN
Attredef	Redefines a block and updates associated attributes				TYPE IN
Attsync	Updates all instances of specified block with current attributes defined for the block	? Name Select	Lists all blocks in drawing Lets you enter block name Lets you select on screen	Modify II	TYPE IN
Audit	Performs drawing integrity check while in AutoCAD	Y N	Fixes errors encountered Reports, but does not fix, errors		File
Background	Sets up the background for your scene	Solid Gradient Image Merge Colors Preview	Selects one color background Specifies a two- or three-color gradient background Uses a bitmap file for the background Uses the current AutoCAD® image as the background Sets color for a solid or gradient background Displays a preview of	Render	

Command Aliases	Explanation	Options		Toolbar	Pull-Down Menu
			the current Background settings		
		Image	Specifies the image file name including BMP, JPG, PCX, TGA, & TIFF		
		Environment	Defines an environment for creating reflection and refraction effects on objects with reflective, raytraced materials		
		Horizon	Represents the percentage of unrotated height		
		Height	Represents a percentage of the second color in a three-color gradient. Use the box or scroll bar to set the value		
		Rotation	Sets an angle at which you can rotate a gradient background		
Baction	Adds an action to a dynamic block definition	Array	Adds an array action to the current dynamic block		TYPE IN
		Move	Adds a move action to the current dynamic block		
		Scale	Adds a move action tp tje current dynamic block		
		Stretch	Adds a stretch action tp tje current dynamic block		
		Polar Stretch	Adds a polar stretch action to the current dynamic block		
		Rotate	Adds a rotate action tp tje current dynamic block		
		Flip	Adds a flip action tp tje current dynamic block		
		Lookup	Adds a lookup action tp tje current dynamic block		
Bactionset	Specifies the selection set of objects associated with an action in a dynamic block definition				TYPE IN
Bactiontool	Adds an action to a dynamic block definition	See options for Baction above			TYPE IN
Base	Defines the origin point for insertion of one drawing into another				Draw
Bassociate	Associates an action with a parameter in a dynamic block definition				TYPE IN
Battmann	Displays Block Attribute Manager			Modify II	Modify
Battorder	Specifies the order of attributes for a block				TYPE IN
Bauthorpaletteclose	Closes the Block Authoring Palettes window in the Block Editor				TYPE IN

Command Aliases	Explanation	Options		Toolbar	Pull-Down Menu
Bauthorpalette	Opens the Block Authoring Palettes window in the Block Editor				TYPE IN
Bclose	Closes the Block Editor				TYPE IN
Bcycleorder	Changes the cycling order of grips for a dynamic block reference				TYPE IN
Bedit	Opens the Edit Block Definition dialog box and then the Block Editor				TYPE IN
Bgripset	Creates, deletes, or resets grips associated with a parameter				TYPE IN
Bhatch	Fills an automatically defined boundary with a hatch pattern through the use of dialog boxes; also allows previewing and repeated adjustments without starting over each time			Draw	Draw
Blipmode	Turns blip markers on and off				TYPE IN
Block	Makes a compound object from one or more objects	?	Lists names of defined blocks a group of objects	Draw	Draw
Blockicon	Generates preview images for blocks displayed in DesignCenter				File
Blocklookuptable	Generates preview images for blocks displayed in DesignCenter				TYPE IN
Bmpout	Exports selected objects to bitmap format				
Boundary	Creates a polyline of a closed boundary				Draw
Box	Creates a three-dimensional solid box			Solids	Solids
Bparameter	Adds a parameter with grips to a dynamic block definition				TYPE IN
Bparameter	Adds a parameter with grips to a dynamic block definition	Alignment	Adds an alignment parameter to the current dynamic block		TYPE IN
		Base	Adds a base parameter to the current dynamic block		
		Point	Adds a point parameter to the current dynamic block		
		Linear	Adds a linear parameter to the current dynamic block		
		Polar	Adds a polar parameter to the current dynamic block		
		XY	Adds an XY parameter to the current dynamic block		
		Rotation	Adds a rotate parameter to the current dynamic block		
		Flip	Adds a flip parameter to the current dynamic block		
		Visibility	Adds a visibility parameter to the current dynamic block		
		Lookup	Adds a lookup parameter to the current dynamic block		
Break	Breaks out (erases) part of an object or splits it into parts	F	Allows you to reselect the first point again	Modify I	Modify
Browser	Launches the Web browser				

Appendix B • *Listing of AutoCAD Commands (Menu and Command Prompt)*

Command Aliases	Explanation	Options		Toolbar	Pull-Down Menu
Bsaveas	Saves a copy of the current block definition under a new name			Block Editor	
Bsave	Saves the current block definition			Block Editor	
Bvhide	Makes objects invisible in the current visibility state or all visibility states in a dynamic block definition			Block Editor	
Bvshow	Makes objects visible in the current visibility state or all visibility states in a dynamic block definition			Block Editor	
Bvstate	Creates, sets, or deletes a visibility state in a dynamic block				TYPE IN
Cal	Evaluates mathematical and geometric expressions				TYPE IN
Camera	Sets the camera and target location to change the view of objects			View	
Chamfer	Makes a chamfer at the intersection of two lines	D	Sets chamfer distance	Modify I	Modify
		P	Chamfers all intersections of a pline figure		
		A	Sets the chamfer distances using a chamfer distance for the first line and an angle for the second		
		T	Controls whether AutoCAD trims the selected edges to the chamfer line endpoints		
		M	Controls whether AutoCAD uses two distances or a distance and an angle to create the chamfer		
Change	Makes changes in the location, size, orientation, and other properties of selected objects; is very helpful for editing text	P	Changes properties of objects		TYPE IN
		E	Elevation		
		C	Color		
		LA	Layer		
		LT	Linetype		
		T	Thickness		
		LW	Lineweight		
		PL	Plotstyle		
Checkstandards	Checks for standards violations			Tools	Tools
Chprop	Makes changes in the properties of selected objects	C	Color		TYPE IN
		LA	Layer		
		LT	Linetype		
		T	Thickness		
		LW	Lineweight		
		PL	Plotstyle		
Circle (C)	Draws a circle of any size; default is center point and radius	2P	Two endpoints on diameter	Draw	Draw
		3P	Three points on circle.		
		D	Enters circle diameter		
		TTR	Tangent, Tangent, Radius		
		R	Enter radius		
Cleanscreenoff	Restores display of toolbars and dockable windows (excluding the command line)				View
Cleanscreenon	Clears the screen of toolbars and dockable windows (excluding the command line)				View

Command Aliases	Explanation	Options		Toolbar	Pull-Down Menu
Close	Closes the current drawing				File
Closeall	Closes all open drawings				TYPE IN
Color	Sets color for objects by name or number; also sets color to be by block or layer BYBLOCK BYLAYER	number name 1. Red 3. Green 5. Blue 7. White	Sets color by number Sets a color by name Retains color of block Uses color of layer 2. Yellow 4. Cyan 6. Magenta		TYPE IN
Commandline	Displays the command line				Tools
Commandlinehide	Hides the command line				Tools
Compile	Compiles shape files and PostScript font files into SHX files				TYPE IN
Cone	Creates a 3D solid cone			Solids	Solids
Convert	Converts associative hatches and 2D polylines to optimized format				
Convertctb	Converts color-dependent plot style table				TYPE IN
Convertpstyles	Converts drawing to color dependent plot styles				TYPE IN
Copy (CP)	Makes one or more copies of selected objects	M	Makes more than one copy of the selected object	Modify I	Modify
Copybase	Copies objects with a specified base point				Edit
Copyclip	Copies selected objects to the Windows clipboard				
Copyhist	Text in command line history is copied to the clipboard				
Copylink	Current view is copied to the clipboard for OLE applications				
CUI	Manages customized user interface elements such as workspaces, toolbars, menus, shortcut menus and keyboard shortcuts				Tools
CUIExport	Exports customized settings from acad.cui to an enterprise or partial CUI file				Tools
CUIImport	Imports customized settings from an enterprise or partial CUI file to acad.cui				Tools
CUILoad	Loads a CUI file				TYPE IN
CUIUnload	Unloads a CUI file				TYPE IN
Customize	Customizes toolbars, buttons, and shortcut keys				TYPE IN
Cutclip	Cuts and copies selected objects from the drawing to the clipboard				
Cylinder	Creates a 3D solid cylinder			Solids	Solids
Dbclose	Closes the dbConnect Manager				Tools
Dbconnect	Provides interface to external database tables				Tools
Dbclkedit	Controls double-click behavior				Tools
Dblist	Makes a listing of every object in the drawing database			Inquiry	Tools
Ddedit	Edits text and attribute definitions			Modify II	

Appendix B • Listing of AutoCAD Commands (Menu and Command Prompt)

Command Aliases	Explanation	Options		Toolbar	Pull-Down Menu
'Ddptype	Specifies point object display mode and sizes				
Ddvpoint	Sets the 3D viewing direction				View
Delay	Provides timed pause within a script				TYPE IN
Detachurl	Removes hyperlinks in a drawing				TYPE IN
Dim	Accesses the dimensioning mode				TYPE IN
Dimxx	See the end of this appendix for Dimensioning Commands				
Dist	Determines the distance between two points			Inquiry	Tool
Divide	Places markers along selected objects, dividing them into a specified number of parts	B	Sets a specified block as a marker		Draw
Doughnut (Donut)	Draws a solid circle or a ring with a specified inside and outside diameter			Draw	Draw
Dragmode	Allows control of the dynamic specification (dragging) feature for all appropriate commands	ON	Honors drag requests when applicable		TYPE IN
		OFF	Ignores drag requests		
		A	Sets Auto mode: drags whenever possible		
Drawingrecovery	Displays a list of drawing files that can be recovered after a program or system failure				File
Drawingrecoveryhide	Closes the Drawing Recovery Manager				TYPE IN
Draworder	Changes the display order of images and objects				
Dsettings	Specifies settings for Snap mode, grid, and polar and object snap tracking				Tools
Dsviewer	Opens the Aerial View window				View (Aerial View)
Dview (DV)	Defines parallel or visual perspective views dynamically	CA	Selects the camera angle relative to the target		View
		CL	Sets front and back clipping planes		
		D	Sets camera-to-target distance, turns on perspective		
		H	Removes hidden lines on the selection set		
		OFF	Turns perspective off		
		PA	Pans the drawing across the screen		
		PO	Specifies the camera and target points		
		TA	Rotates the target point about the camera		
		TW	Twists the view around your line of sight		
		U	Undoes a Dview subcommand		
		X	Exits the Dview command		
		Z	Zooms in/out, or sets lens length		
Dwgprops	Sets and displays the properties of the current drawing				File
Dxbin	Imports specially coded binary files				Insert
Eattedit	Edits attributes in a block reference			Modify II	Modify

Command Aliases	Explanation	Options		Toolbar	Pull-Down Menu
Eattext	Exports attribute to external file			Modify II	Tools
Edge	Changes the visibility of three-dimensional face edges			Surfaces	
Edgesurf	Creates a three dimensional polygon mesh			Surfaces	
Elev	Sets the elevation and extrusion thickness for entities to be drawn in 3D drawings				TYPE IN
Ellipse	Draws ellipses using any of several methods	C	Selects center point	Draw	Draw
		R	Selects rotation rather than second axis		
		I	Draws isometric circle in current isoplane		
		A	Creates an elliptical arc		
Erase (E)	Deletes objects from the drawing			Modify I	Modify
Etransmit	Creates a set with a transmittal drawing and associated files				TYPE IN
Explode	Changes a block or polyline back into its original objects			Modify I	Modify
Export	Saves objects to other file formats				
Extend	Extends a line, arc, or polyline to meet another object	U	Undoes last extension	Modify I	Modify
		P	Specifies the Projection mode AutoCAD uses when extending objects		
		E	Extends the object to another object's implied edge, or only to an object that actually intersects it in 3D space		
Extrude	Creates unique solid primitives by extruding existing 2D objects			Solids	Draw
Field	Creates a multiline text object with a field that can be updated automatically as the field value changes				Insert
Fill	Determines if solids, traces, and wide polylines are automatically filled	ON	Solids, traces, and wide polylines filled		TYPE IN
		OFF	Solids, traces, and wide polylines outlined		
Fillet	Constructs an arc of specified radius between two lines, arcs, or circles	P	Fillets an entire polyline; sets fillet radius	Modify I	Modify
		R	Sets fillet radius		
		T	Controls whether AutoCAD trims the selected edges to the fillet arc endpoints		
Filter	Creates lists to select objects based on properties				TYPE IN
Find	Finds, replaces, selects, or zooms to specified text			Stanadard	
Fog	Provides visual cues for the apparent distance of objects			Render	

Appendix B • Listing of AutoCAD Commands (Menu and Command Prompt)

Command Aliases	Explanation	Options		Toolbar	Pull-Down Menu
Gotourl	Opens URL associated with selected object				TYPE IN
Gradient	Fills an enclosed area or selected objects with a gradient fill			Draw	
Graphscr F2	Flips to the graphics display on single-screen systems; used in command scripts and menus				TYPE IN
Grid F7 On/Off toggle	Displays a grid of dots, at desired spacing, on the screen	ON OFF S A number number X	Turns grid on Turns grid off Locks grid spacing to snap resolution Sets grid aspect (differing X–Y spacings) Sets grid spacing (0 = use snap spacing) Sets spacing to multiple of snap spacing	Status bar	
Group	Creates a named selection set of objects			Standard	Tools
Hatch	Creates crosshatching and patternfilling	name U ?	Uses hatch pattern name from library file Uses simple user-defined hatch pattern Lists selected names of available hatch patterns		TYPE IN
		NAME and U can be followed by a comma and a hatch style from the following list:			
		I N O	Ignores internal structure Normal style: turns hatch lines off and on when internal structure is encountered Hatches outermost portion only		
Hatchedit	Modifies an existing associative hatch block			Modify II	Modify
'Help or '?	Displays a list of valid commands and data entry options or obtains help for a specific command or prompt	To get a set of Help modes, use ESC and F2 for flip screen		Standard	Help
Hide	Regenerates a 3D visualization with hidden lines removed			Render	View
HIsettings	Displays Hidden Line Settings dialog box to set the display properties of hidden lines				TYPE IN
Hyperlink	Attaches a hyperlink to a graphical object or modifies an existing hyperlink			Standard	
Hyperlinkoptions	Controls the visibility of the hyperlink cursor and the display of hyperlink tooltips				TYPE IN
Id	Displays the coordinates of a point selected on the drawing				Tools
Imagexxx	Commands used for modifying and displaying images to the clipboard				

Command Aliases	Explanation	Options		Toolbar	Pull-Down Menu
Import	Imports various file formats into AutoCAD				
Insert	Inserts a copy of a block or Wblock complete drawing into the current drawing	fname	Loads fname as block	Draw	Draw
		fname=f	Creates block fname from file f		
		*name	Retains individual part objects		
		C	(as reply to X scale prompt) Specifies scale via two points (Corner specification of scale)		
		XYZ	(as reply to X scale prompt) Readies Insert for X,Y, and Z scales		
		~	Displays a File dialog box		
		?	Lists names of defined blocks		
Insertobj	Inserts embedded or linked objects				
Interfere	Finds the interference of two or more solids and creates a composite solid from their common volume			Solids	Solids
Intersect	Creates composite solids or regions from the intersection of two or more solids or regions			Modify II	Modify
Isoplane CTRL + E	Changes the location of the isometric crosshairs to left, right, and top plane	L	Left plane		TYPE IN
		R	Right plane		
		T	Top plane		
		ENTER	Toggle to next plane		
Join	Joins objects to form a single object			Modify I	Modify
Jpgout	Displays the Create Raster File dialog box, creates a JPEG file from selected objects				TYPE IN
Justifytext	Changes text justification point			Text	Modify
Layer (LA)	Allows for the creation of drawing layers and the assigning of color and linetype properties	C	Sets layers to color selected	Object Properties	Format
		F	Freezes layers		
		LT	Sets specified layers to linetype		
		M	Makes a layer the current layer, creating it if necessary		
		N	Creates new layers		
		ON	Turns on layers		
		OFF	Turns off layers		
		S	Sets current layer to existing layer		
		T	Thaws layers		
		?	Lists specified layers and their associated colors, linetypes, and visibility		
		L	Lock		
		U	Unlock		
		LW	Changes the lineweight associated with a layer		
		PS	Sets the plot style assigned to a layer		
Layerp	Undoes last changes made to layer settings				TYPE IN
Layerpmode	Toggles layer change tracking on and off				TYPE IN
Layout	Creates a new layout and renames, copies, saves, or deletes an existing layout	Copy	Copies a layout	Layout	
		Delete	Deletes a layout		

Command Aliases	Explanation	Options		Toolbar	Pull-Down Menu
		New	Creates a new layout tab		
		Template	Creates a new template based on an existing layout in a template (DWT) or drawing (DWG) file		
		Rename	Renames a layout		
		Save	Saves a layout		
		Set	Makes a layout current		
		?-List Layouts	Lists all the layouts defined in the drawing		
Layoutwizard	Starts the Layout wizard, in which you can designate page and plot settings for a new layout				TYPE IN
Laytrans	Changes layers to specified layer standards			CAD Standards	Tools
Leader	Draws a line from an object to, and including an annotation				
Lengthen	Lengthens an object			Modify I	Modify
Light	Manages lights & lighting in model space			Render	View
Limits	Sets up the drawing size	2 points	Sets lower left/upper right drawing limits		Format
		ON	Enables limits checking		
		OFF	Disables limits checking		
Line (L)	Draws straight lines of any length	ENTER	(as reply to "From point:") Starts at end of previous line or arc	Draw	Draw
		C	(as reply to "To point:") Closes polygon		
		U	(as reply to "To point:") Undoes segment		
Linetype	Defines, loads, and sets the linetype	?	Lists a linetype library		TYPE IN
		C	Creates a linetype definition		
		L	Loads a linetype definition		
		S	Sets current object linetype; set suboptions:		
		name	Sets object linetype name		
		BYBLOCK	Sets floating object linetype		
		BYLAYER	Uses layer's linetype for objects		
		?	Lists specified loaded linetypes		
List	Provides database information for objects that are selected			Inquiry	Tools
Load	Loads a file of user-defined shapes to be used with the SHAPE command	?	Lists the names of loaded shape files		TYPE IN
Logfileoff	Closes the log file opened by LOGFILEON				TYPE IN

Command Aliases	Explanation	Options		Toolbar	Pull-Down Menu
Logfileon	Writes the text window contents to a file				TYPE IN
Lsedit	Edits a landscape object			Render	
Lslib	Maintains libraries of landscape objects			Render	
Lsnew	Adds realistic landscape items, such as trees and bushes, to your drawings			Render	
Ltscale	Regulates the scale factor to be applied to all linetypes within the drawing				TYPE IN
Lweight	Sets the current lineweight, lineweight display options, and lineweight units				Format
Markupclose	Closes the Markup Set Manager			Standard	Tools
Markup	Displays the details of markups and allows you to change their status			Standard	Tools
Massprop	Calculates and displays the mass properties of regions or solids			Inquiry	Tools
Matchcell	Applies the properties of a selected table cell to other table cells				TYPE IN
Matchprop	Causes properties of one object to be assigned to selected objects				
Matlib	Imports and exports materials to and from a library of materials				Tools
Measure	Inserts markers at measured distances along a selected object	B	Uses specified block as marker	Inquiry	Tools
Menu	Loads a menu into the menu areas (screen, pull-down, tablet, and button)				Tools
Menuload	Loads partial menu files				TYPE IN
Menuunload	Unloads partial menu files				TYPE IN
Minsert	Inserts multiple copies of a block in a rectangular array	fname	Loads fname and forms a rectangular array of the resulting block	Draw	Draw
		fname=f	Creates block fname from file f and forms a rectangular array		
		?	Lists names of defined blocks		
		C	(as reply to X scale prompt) Specifies scale via two points (Corner specification of scale)		
		XYZ	(as reply to X scale prompt) Readies Multiple Insert for X,Y, and Z scales		
		~	Displays a File dialog box		
		S	Sets the scale factor for the X, Y, and Z axes		
		PD	Sets the scale factor for the X, Y, and Z axes to control the display of the block as it is dragged into position		
		PR	Sets the rotation angle of the block as it is dragged into position		

Command Aliases	Explanation	Options		Toolbar	Pull-Down Menu
Mirror	Reflects selected objects about a user-specified axis, vertical, horizontal, or inclined			Modify I	Modify
Mirror3D	Reflects selected objects about a plane				Modify
Mledit	Edits multiple parallel lines			Modify II	Modify
Mline	Draws multiple parallel lines				
Mlstyle	Defines a style for multiple parallel lines				Format
Model	Switches from a layout tab to the Model tab and makes it current				TYPE IN
Move (M)	Moves selected objects to another location in the drawing			Modify I	Modify
Mredo	Reverses action of multiple UNDO commands			Standard	
Mslide	Creates a slide file of the current model viewport or the current layout				TYPE IN
Mspace (MS)	Switches to model space from paper space			Status bar	View
Mtext	Creates paragraph text			Draw	Draw
Multiple	Allows the next command to repeat until canceled				TYPE IN
Mview	Sets up and controls viewports	ON	Turns selected viewport(s) on; causes model to be regenerated in the selected viewport(s)		View
		OFF	Turns selected viewport(s) off; causes model not to be displayed in the selected viewport(s)		
		Hideplot	Causes hidden lines to be removed in selected viewport(s) during paper space plotting		
		Fit	Creates a single viewport to fit the current paper space view		
		2	Creates two viewports in specified area or to fit the current paper space view		
		4	Creates four equal viewports in specified area or to fit the current paper space view		
		Restore	Translates viewport configurations saved with the VPORTS command into individual viewport objects in paper space		
		<point>	Creates a new viewport within the area specified by two points		
		O	Specifies a closed polyline, ellipse, spline, region, or circle to convert into a viewport		

Command Aliases	Explanation	Options		Toolbar	Pull-Down Menu
		P	Creates an irregularly shaped viewport using specified points		
		3	Creates three viewports in specified area or to fit the current paper space view		
Mvsetup	Sets up the specifications of a drawing				TYPE IN
Netload	Loads a .NET application				TYPE IN
New	Creates a new drawing			Standard	File
Newsheetset	Creates a new sheet set				Insert
Offset	Reproduces curves or lines parallel to the one selected	number T	Specifies offset distance Through: allows specification of a point through which the offset curve is to pass	Modify I	Modify
Olelinks	Manages existing OLE links				Edit
Olelscale	Controls the size, scale, and other properties of a selected OLE object				Shortcut
Oops	Recalls last set of objects previously erased				TYPE IN
Open	Opens an existing drawing			Standard	File
Opendwfmarkup	Opens a DWF file that contains markups				File
Opensheetset	Opens a selected sheet set				File
Options	Customizes the AutoCAD settings				Tools
Ortho F8	Restricts the cursor to vertical or horizontal use	ON OFF	Forces cursor to horizontal or vertical use Does not constrain cursor movement	Status bar	
Osnap	Allows for selection of precise points on existing objects	CEN END INS INT MID NEA NOD NON PER QUA QUI TAN EXT APP PAR	Center of arc or circle Closest endpoint of arc or line Insertion point of text/block/shape Intersection of line/arc/circle Midpoint of arc or line Nearest point of arc/circle/line/point Node (point) None (off) Perpendicular to arc/line/circle Quadrant point of arc or circle Quick mode (first find, not closest) Tangent to arc or circle Snaps to the extension point of an object Apparent Intersection includes two separate snap modes: Snaps to an extension in parallel with an object	Object Snap	Tools
Pagesetup	Specifies the layout page, plotting device, paper size, and settings for each new layout				Layout

Appendix B • Listing of AutoCAD Commands (Menu and Command Prompt)

Command Aliases	Explanation	Options		Toolbar	Pull-Down Menu
'Pan (P)	Moves the display window			Standard	View
Partiaload	Loads additional geometry into a partially opened drawing				File
Partialopen	Loads geometry from a selected view or layer into a drawing				TYPE IN
Pastehyperlink	Inserts data from the Clipboard as a hyperlink				Edit
Pasteblock	Pastes a copied block into a new drawing				Edit
Pasteclip	Inserts clipboard data				
Pasteorig	Pastes a copied object in a new drawing using the coordinates from the original drawing				Edit
Pastespec	Specifies fomat of data imported from the clipboard				
Pcinwizard	Displays a wizard to import PCP and PC2 configuration file plot settings into the Model tab or current layout				Tools
Pedit (2D)	Permits editing of 2D polylines	C	Closes to start point	Modify II	Modify
		D	Decurves, or returns a spline curve to its control frame		
		F	Fits curve to polyline		
		J	Joins to polyline		
		O	Opens a closed polyline		
		S	Uses the polyline vertices as the frame for a spline curve (type set by SPLINETYPE)		
		U	Undoes one editing operation		
		W	Sets uniform width for polyline		
		X	Exits PEDIT command during vertex editing		
		E	Edits vertices during vertex editing		
		B	Sets first vertex for Break		
		G	Go (performs Break or Straighten operation)		
		I	Inserts new vertex after current one		
		M	Moves current vertex		
		N	Makes next vertex current		
		P	Makes previous vertex current		
		R	Regenerates the polyline		
		S	Sets first vertex for Straighten		
		T	Sets tangent direction for current vertex		
		W	Sets new width for following segment		
		X	Exits vertex editing, or cancels Break/Straighten		
		L	Generates the linetype in a continuous pattern through the vertices of the polyline		

Command Aliases	Explanation	Options		Toolbar	Pull-Down Menu
Pedit (3D)	Allows editing of 3D polylines	C	Closes to start point.	Modify II	Modify
		D	Decurves, or returns a spline curve to its control frame		
		O	Opens a closed polyline		
		S	Uses the polyline vertices as the frame for a spline curve (type set by SPLINETYPE)		
		U	Undoes one editing operation		
		X	Exits PEDIT command		
		E	Edits vertices		
		During vertex editing:			
		B	Sets first vertex for Break		
		G	Go (performs Break or Straighten operation)		
		I	Inserts new vertex after current one		
		M	Moves current vertex		
		N	Makes next vertex current		
		P	Makes previous vertex current		
		R	Regenerates the polyline		
		S	Sets first vertex for Straighten		
		X	Exits vertex editing, or cancels Break/Straighten		
Pedit (mesh)	Allows editing of 3D polygon meshes	D	Desmoothes-restores original mesh	Modify II	Modify
		M	Opens (or closes) the mesh in the M direction		
		N	Opens (or closes) the mesh in the N direction		
		S	Fits a smooth surface as defined by SURFTYPE		
		U	Undoes one editing operation		
		X	Exits PEDIT command		
		E	Edits mesh vertices during vertex editing		
		D	Moves down to previous vertex in M direction		
		L	Moves left to previous vertex in N direction		
		M	Repositions the marked vertex		
		N	Moves to next vertex		
		P	Moves to previous vertex		
		R	Moves right to next vertex in N direction		
		RE	Redisplays the polygon mesh		
		U	Moves up to next vertex in M direction		
		X	Exits vertex editing during vertex editing		
Pface	Creates a 3D mesh of arbitrary complexity and surface characteristics				TYPE IN

Command Aliases	Explanation	Options		Toolbar	Pull-Down Menu
Plan	Puts the display in plan view (Vpoint 0,0,1) relative to either the current UCS, a specified UCS, or the WCS	C	Establishes a plan view of the current UCS		View
		U	Establishes a plan view of the specified UCS		
		W	Establishes a plan view of the WCS		
Pline (PL)	Draws 2D polylines	H	Sets new half-width	Draw	Draw
		U	Undoes previous segment		
		W	Sets new line width		
		ENTER	Exits PLINE command		
		C	Closes with straight segment		
		L	Segment length (continues previous segment)		
		A	Switches to arc mode		
		In arc mode:			
		A	Included angle		
		CE	Center point		
		CL	Closes with arc segment		
		D	Starting direction		
		L	Chord length, or switches to line mode		
		R	Radius		
		S	Second point of three-point arc		
Plot (Print)	Plots a drawing to a plotting device or a file			Standard	File
Plotstamp	Places a plot stamp on a drawing corner and logs it to a file				TYPE IN
Plotstyle	Sets the current plot style for new objects, or the assigned plot style for selected objects				TYPE IN
Plottermanager	Displays the Plotter Manager, where you can launch the Add-a-Plotter wizard and the Plotter configuration editor				File
Pngout	Displays the Create Raster file dialog box, creates a Portable Network Graphics file from selected objects				TYPE IN
Point	Draws a single point on the drawing			Draw	Draw
Polygon	Creates regular polygons with the specified number of sides indicated	E	Specifies polygon by showing one edge	Draw	Draw
		C	Circumscribes around circle		
		I	Inscribes within circle		
Preview	Displays plotted view of drawing				
Properties	Controls properties of existing objects			Standard	
Propertiesclose	Closes the Properties window				TYPE IN
Psetupin	Imports a user-defined page setup into a new drawing layout				TYPE IN
Pspace (PS)	Switches to paper space			Status bar	View

Command Aliases	Explanation	Options		Toolbar	Pull-Down Menu
Publish	Displays the Publish Drawing Sheets dialog box to begin publishing the current drawing sheets to DWF file or plotter			Standard	
Publishtoweb	Creates HTML pages including images of drawings				TYPE IN
Purge	Removes unused Blocks, text styles, layers, linetypes, and dimension styles from the drawing	A B D LA SH ST LT	Purges all unused named objects Purges unused blocks Purges unused dimstyles Purges unused layers Purges unused shape files Purges unused text styles Purges linetypes		File
Qclose	Closes QuickCalc				TYPE IN
Qdim	Quickly creates a dimension	Continuous Staggered Baseline Ordinate Radius Diameter Datum Point Edit	Creates a series of continued dimensions Creates a series of staggered dimensions Creates a series of baseline dimensions Creates a series of ordinate dimensions Creates a series of radius dimensions Creates a series of diameter dimensions Sets a new datum point for baseline and ordinate dimensions Edits a series of dimensions	Dimension	
Qleader	Quickly creates a leader and leader annotation				Dimension
Qnew	Starts a new drawing from the current default drawing template file				Standard
Qsave	Saves the drawing without requesting a file name			Standard	File
Qselect	Quickly creates selection sets based on filtering criteria				Tools
Qtext	Enables text objects to be identified without drawing the test detail	ON OFF	Quick text mode on. Quick text mode off		TYPE IN
Quickcalc	Opens the QuickCalc calculator			Standard	Tools
Quit	Exit AutoCAD				File
Ray	Creates a semi-infinite line			Draw	
Recover	Attempts to recover damaged or corrupted drawings				File
Rectang	Draws a rectangular polyline			Draw	Draw
Redefine	Restores a built-in command deleted by UNDEFINE				TYPE IN
Redo	Reverses the previous command if it was U or Undo			Standard	Edit

Command Aliases	Explanation	Options		Toolbar	Pull-Down Menu
'Redraw (R)	Refreshes or cleans up the current viewport			Standard	View
'Redrawall	Redraws all viewports			Standard	View
Refclose	Saves back or discards changes made during in-place editing of a reference (an xref or a block)				Modify
Refedit	Selects a reference for editing			Refedit	
Refset	Adds or removes objects from a working set during in-place editing of a reference (an xref or a block)	Add Remove	Adds objects to the working set Removes objects from the working set		Modify
Regen	Regenerates the current viewport				TYPE IN
Regenall	Regenerates all viewports				TYPE IN
Regenauto	Controls automatic regeneration performed by other commands	ON OFF	Allows automatic regenerations Prevents automatic regenerations		TYPE IN
Region	Creates a region object from a selection set of existing objects			Draw	Draw
Reinit	Allows the I/O ports, digitizer, display, plotter, and PGP file to be reinitialized				TYPE IN
Rename	Changes the names associated with text styles, layers, linetypes, blocks, views, UCSs, viewport configurations, and dimension styles	B D LA LT S U VI VP	Renames block. Renames dimension style Renames layer Renames linetype Renames text style Renames UCS Renames view Renames viewport configuration		TYPE IN
Render	Creates a realistically shaded image of a 3D wireframe or solid model			Render	View
Rendscr	Redisplays the last rendering created with the RENDER command				TYPE IN
Replay	Displays a GIF, TGA, or TIFF image				Tools
Resetblock	Resets one or more dynamic block references to the default values of the block definition				TYPE IN
'Resume	Resumes an interrupted command script				TYPE IN
Revcloud	Creates a revision cloud				Draw
Revolve	Creates a solid by revolving a 2D object about an axis			Solids	Draw
Revsurf	Creates a 3D polygon mesh approximating a surface of revolution, by rotating a curve around a selected axis	Surfaces		Draw	
Rmat	Manages rendering materials			Render	View
Rmlin	Inserts markups from an RML file into a drawing			Insert	
Rotate	Rotates existing objects to the angle selected	R	Rotates with respect to reference angles	Modify I	Modify
Rotate3D	Rotates about a 3D axis				Modify

Command Aliases	Explanation	Options		Toolbar	Pull-Down Menu
Rpref	Sets rendering preferences			Render	View
Rscript	Restarts a command script from the beginning				TYPE IN
Rulesurf	Creates a ruled surface between two curves	Surfaces		Draw	
Save	Updates the current drawing file without exiting the Drawing Editor			Standard	File
Saveas	Same as SAVE, but also renames the current drawing				File
Saveimg	Saves a rendered image to a file				Tools
Scale	Changes the size of existing objects to the selected scale factor	R	Resizes with respect to reference size	Modify I	Modify
Scaletext	Changes size of text objects			Text	Modify
Scene	Manages scenes in model space			Render	View
Script	Executes command sequence from script file				Tools
Section	Uses the intersection of a plane and solids to create a region			Solids	Draw
Securityoptions	Lets you add security settings to drawing				TYPE IN
Select	Groups objects into a selection set for use in subsequent commands				TYPE IN
Setidrophandler	Displays the Set Default i-drop Content Type dialog box where you can set default type of i-drop content for current application				TYPE IN
Setuv	Maps materials onto objects			Render	
'Setvar	Allows you to display or change the value of system variables	?	Lists specified system variables		TYPE IN
Shademode	Shades the objects in the current viewport	2D Wireframe	Displays the objects using lines and curves to represent the boundaries		View
		3D Wireframe	Displays the objects using lines and curves to represent the boundaries		
		Hidden	Displays the objects using 3D wireframe representation and hides lines representing back faces		
		Flat Shaded	Shades the objects between the polygon faces		
		Gouraud Shaded	Shades the objects and smooths the edges between polygon faces		
		Flat Shaded, Edges On	Combines the Flat Shaded and Wireframe options		
		Gouraud Shaded, Edges On	Combines the Gouraud Shaded and Wireframe options		

Appendix B • *Listing of AutoCAD Commands (Menu and Command Prompt)*

Command Aliases	Explanation	Options		Toolbar	Pull-Down Menu
Shape	Draws predefined shapes	?	Lists available shape names		TYPE IN
Sheetset	Opens the Sheet Set Manager			Standard	Tools
Sheetsethide	Closes the Sheet Set Manager				Tools
Shell	Allows access to other programs while running AutoCAD				TYPE IN
Showmat	Lists material type and method of attachment for the selected object				
Sigvalidate	Displays the Validate Digital Signatures dialog box				TYPE IN
Sketch	Allows freehand sketching	C	Connect: restarts sketch at endpoint	Draw	Draw
		E	Erases (backs up over) temporary lines		
		P	Raises/lowers sketching pen		
		Q	Discards temporary lines, remains in SKETCH		
		R	Records temporary lines, remains in SKETCH		
		X	Records temporary lines, exits SKETCH; draws line to current point		
Slice	Slices a set of solids with a plane			Solids	Draw
Snap F9	Allows for precision alignment of points	number	Sets snap resolution	Status bar	
		ON	Aligns designated points		
		OFF	Does not align designated points		
		A	Sets aspect (differing X–Y spacing)		
		R	Rotates snap grid		
		S	Selects style, standard or isometric		
Soldraw	Generates profiles and sections in viewports created with SOLVIEW			Solids	Draw
Solidedit	Edits faces and edges of 3D solid objects				Modify
Solid	Creates filled-in polygons			Draw	Draw
Solprof	Creates profile images of three-dimensional solids in paper space			Solids	Draw
Solview	Creates layout viewports using orthographic projection to lay out multi- and sectional view drawings of 3D solids and body objects			Solids	Draw
Spacetrans	Converts length values between model space and paper space			Text	
Spell	Checks the spelling in a drawing			Standard	Tools
Sphere	Creates a 3D solid sphere			Solids	Draw
Spline	Creates a quadratic or cubic spline (NURBS) curve			Draw	Draw
Splinedit	Edits a spline object			Modify II	Modify

Command Aliases	Explanation	Options		Toolbar	Pull-Down Menu
Standards	Manages association of standards files with AutoCAD drawings			CAD Standards	Tools
Stats	Displays rendering statistics			Render	View
Status	Displays drawing setup				Tools
Stlout	Stores a solid in an ASCII or binary file				TYPE IN
Stretch	Allows you to move a portion of a drawing while retaining connections to other parts of the drawing			Modify I	Modify
Style	Sets up named text styles, with various combinations of font, mirroring, obliquing, and horizontal scaling	?	Lists specified currently defined text style		Format
Stylesmanager	Displays the Plot Style Manager				File
Syswindows	Arranges windows				
Table	Creates an empty table object in a drawing			Draw	Draw
Tabledit	Edits text in a table cell				TYPE IN
Tableexport	Exports data from a table object in CSV file format				TYPE IN
Tablestyle	Defines a new table style			Styles	Format
Tablet	Allows for configuration of a tablet menu or digitizing of an existing drawing	ON OFF CAL	Turns tablet mode on Turns tablet mode off Calibrates tablet for use in the current space		Tools
Tabsurf	Creates a polygon mesh approximating a general tabulated suface defined by a path and a direction vector			Surfaces	Draw
Taskbar	Controls how drawings are displayed on the Windows taskbar				TYPE IN
Text	Enters text on the drawing	J S A C F M R BL BC BR ML MC MR TL TC TR	Prompts for justification options Lists or selects text style Aligns text between two points, with style-specified width factor; AutoCAD computes appropriate height Centers text horizontally Fits text between two points, with specified height; AutoCAD computes an appropriate width factor centers text horizontally and vertically Right-justifies text Bottom left Bottom center Bottom right Middle left Middle center Middle right Top left Top center Top right	Draw	Draw

Command Aliases	Explanation	Options		Toolbar	Pull-Down Menu
'Textscr F2	Flips to the text display on singlescreen systems; used in command scripts and menus				TYPE IN
Texttofront	Brings text and dimensions in front of all other objects in the drawing				Tools
Tifout	Displays the Create Raster File dialog box Creates a TIFF file from selected objects				TYPE IN
Time	Indicates total elapsed time for each drawing	D ON OFF R	Displays current times Starts user elapsed timer Stops user elapsed timer Resets user elapsed timer		Tools
Tinsert	Inserts a block in a table cell				Shortcut
Tolerance	Creates geometric tolerances			Dimension	Draw
Toolbar	Customizes, hides, and displays toolbars				View
Toolpalettes	Opens Tool Palettes window				Tools
Toolpalettesclose	Closes the Tool Palettes window				Tools
Torus	Creates a donut-shaped solid			Solids	Draw
Trace	Creates solid lines of specified width			Draw	Draw
Transparency	Determines transparency of opacity of bitonal image background				
Traysettings	Displays the Tray Settings dialog box				TYPE IN
Treestat	Displays information on the drawing's current spatial index, such as the number and depth of nodes in the drawing's database; use this information with the TREEDEPTH system variable setting to fine-tune performances for large drawings				TYPE IN
Trim	Deletes portions of selected entities that cross a selected boundary edge	U	Undoes last trim operation	Modify I	Modify
U	Reverses the effect of the previous command			Standard	Edit
UCS	Defines or modifies the current User Coordinate System	D E O P R S V W X Y	Deletes one or more saved coordinate systems Sets a UCS with the same extrusion direction as that of the selected object Shifts the origin of the current coordinate system Restores the previous UCS Restores a previously saved UCS Saves the current UCS Establishes a new UCS whose Z Axis is parallel to the current viewing direction Sets the current UCS equal to the WCS Rotates the current UCS around the X axis Rotates the current UCS around the Y axis	Standard	View

Command Aliases	Explanation	Options		Toolbar	Pull-Down Menu
		Z	Rotates the current UCS around the Z axis		
		ZA	Defines a UCS using an origin point and a point on the positive portion of the Z axis		
		3	Defines a UCS using an origin point, a point on the positive portion of the X axis, and a point on the positive Y portion of the X plane		
		?	Lists specified saved coordinate systems		
Ucsicon	Controls visibility and placement of the UCS icon, which indicates the origin and orientation of the current UCS; the options normally affect only the current viewport	A	Changes settings in all active viewports		Tools
		N	Displays the icon at the lower-left corner of the viewport		
		OR	Displays the icon at the origin of the current UCS if possible		
		ON	Enables the coordinate system icon		
Ucsman	Manages defined user coordinate systems			UCS	
Undefine	Deletes the definition of a built-in AutoCAD command				TYPE IN
Undo	Reverses the effect of multiple commands, and provides control over the Undo facility	number	Undoes the number most recent commands	Standard	Edit
		A	Auto: controls treatment of menu items as Undo groups		
		B	Back: undoes back to previous Undo mark		
		C	Control: enables/disables the Undo mark		
		E	End: terminates an Undo group		
		G	Group: begins sequence to be treated as one command		
		M	Mark: places marker in Undo file (for back)		
Union	Combines regions or solids by addition			Solids Editing	Modify
Units	Selects coordinate and angle display formats and precision				Format
Updatefield	Manually updates fields in selected objects in the drawing Tools menu: Update Fields				Tools
Updatethumbsnow	Manually updates thumbnail previews for sheets, sheet views, and model space views in the Sheet Set Manager				TYPE IN
U	Reverses the most recent operation			Standard	
Vbaide	Displays the Visual Basic Editor				Tools
Vbaload	Loads a global VBA project into the current AutoCAD session				Tools
Vbaman	Loads, unloads, saves, creates, embeds, and extracts VBA projects				Tools

Command Aliases	Explanation	Options		Toolbar	Pull-Down Menu
Vbarun	Runs a VBA macro				Tools
Vbastmt	Executes a VBA statement on the AutoCAD command line				TYPE IN
Vbaunload	Unloads a global VBA project				TYPE IN
'View	Saves the current graphics display and space as a named view, or restores a saved view and space to the display	D	Deletes named view		View
		R	Restores named view to screen		
		S	Saves current display as named view		
		W	Saves specified window as named view		
		?	Lists specified named views		
Viewports or Vports	Divides the AutoCAD graphics display into multiple viewports, each of which can contain a different view of the current drawing	D	Deletes a saved viewport configuration		View
		J	Joins (merges) two viewports		
		R	Restores a saved viewport configuration		
		S	Saves the current viewport configuration		
		S1	Displays a single viewport filling the entire graphics area		
		2	Divides the current viewport into viewports		
		3	Divides the current viewport into three viewports		
		4	Divides the current viewport into four viewports		
		?	Lists the current and saved viewport configurations		
Viewplotdetails	Displays information about completed plot and publish jobs				File
Viewres	Adjusts the precision and speed of circle and arc drawing on the monitor				Format
Vlisp	Displays the Visual LISP interactive development environment (IDE)				Tools
Vpclip	Clips viewport objects	Object	Specifies an object to act as a clipping boundary		TYPE IN
		Polygonal	Draws a clipping boundary		
		Delete	Deletes the clipping boundary of a selected viewport		
Vplayer	Sets viewport visibility for new and existing layers	?	Lists layers frozen in a selected viewport		Type In
		Freeze	Freezes specified layers in selected viewport(s)		
		Thaw	Thaws specified layers in selected viewport(s)		
		Reset	Resets specified layers to their default visibility		
		Newfz	Creates new layers that are frozen in all viewports		

Command Aliases	Explanation	Options		Toolbar	Pull-Down Menu
		Vpvisdfit	Sets the default viewport visibility for existing layers		
Vpmax	Expands current layout viewport for editing				Status Bar
Vpmin	Restores current layout viewport				Status Bar
Vpoint	Selects the viewpoint for a 3D visualization	R	Selects viewpoint via two rotation angles		View
		ENTER	Selects viewpoint via compass and axes tripod		
		x,y,z	Specifies viewpoint		
Vports	Creates multiple viewports in model space or paper space			Layouts	View
Vslide	Displays an image slide file in the current viewport				TYPE IN
Vtoptions	Displays a change in view as a smooth transition				TYPE IN
Wblock	Creates a block as a separate drawing	name	Writes specified block definition		File
		=	Block name same as file name		
		*	Writes entire drawing		
		ENTER	Writes selected objects		
Wedge	Creates a 3D solid with a tapered sloping face			Solids	Draw
Whohas	Displays ownership information for opened drawing files				TYPE IN
Wmxxx	Controls windows metafiles				
Workspace	Creates, modifies, and saves workspaces and makes a workspace current				TYPE IN
Wssave	Saves a workspaces				TYPE IN
Wssettings	Creates, modifies, and saves workspaces				TYPE IN
Xattach	Attaches an external reference				
Xbind	Permanently adds a selected subset of an external reference's dependent symbols to your drawing	Block	Adds a Block.	Reference	Modify
		Dimstyle	Adds a dimstyle		
		Layer	Adds a layer		
		Ltype	Adds a linetype		
		Style	Adds a style		
Xclip	Defines and external reference				
Xline	Creates an infinite line			Draw	Draw
Xopen	Opens xref to which selected objects belongs				TYPE IN
Xplode	Breaks a compound object into its component objects				TYPE IN
Xref	Allows you to work with other AutoCAD drawings without adding them permanently to your drawing and without altering their contents	Attach	Attaches a new Xref or a copy of an Xref that you have already attached	Reference	Insert
		Bind	Makes an Xref a permanent part of your drawing		
		Detach	Removes an Xref from your drawing		

Command Aliases	Explanation	Options		Toolbar	Pull-Down Menu
		Path	Allows you to view and edit the file name AutoCAD uses when loading a particular Xref		
		Reload	Updates one or more contents Xrefs at any time, without leaving and re-entering the Drawing Editor		
		?	Lists Xrefs in your drawing and the drawing associated with each one		
'Zoom (Z)	Enlarges or reduces the display area of a drawing	number	Multiplier from original scale	Standard	View
		numberX	Multiplier from current scale		
		number XP	Scale relative to paper space		
		A	All		
		C	Center		
		D	Dynamic Pan Zoom		
		E	Extents ("drawing uses")		
		L	Lower left corner		
		P	Previous		
		V	Virtual screen maximum		
		W	Window		

DIMENSIONING COMMANDS

Command	Explanation
Dimaligned	Aligns dimension parallel with objects
Dimangular	Draws an arc to show the angle between two nonparallel lines or three specified points
Dimarc	Creates an arc length
Dimbaseline	Continues a linear dimension from the baseline (first extension line) of the previous or selected dimension
Dimcenter	Draws a circle/arc center mark or centerlines
Dimcontinue	Continues a linear dimension from the second extension line of the previous dimension
Dimdiameter	Dimensions the diameter of a circle or arc
Dimdisassociate	Disassociates dimension with object(s)
Dimedit	Edits dimensions
Dimlinear	Creates linear dimensions
Dimjogged	Creates jogged radius dimensions
Dimlinear	Creates linear dimensions
Dimordinate	Creates ordinate point associative dimensions
Dimoverride	Overrides a subset of the dimension variable settings associated with selected dimension objects
Dimradius	Dimensions the radius of a circle or arc, with an optional center mark or centerlines
Dimreassociate	Reassociates dimension with object(s)
Dimregen	Updates the locations of all associative dimensions
Dimstyle	Displays the Dimension Style Manager window
Dimtedit	Allows repositioning and rotation of text items in an associative dimension without affecting other dimension subentities
Leader	Draws a line with an arrowhead placement of dimension text

DIMENSIONING VARIABLES

Name	Description	Type	Default
DIMADEC	Controls number of places of precision displayed for angular dimension text		
DIMALT	Alternate units	Switch	Off
DIMALTD	Alternate units decimal places	Integer	2
DIMALTF	Alternate units scale factor	Scale	25.4
DIMALTRND	Alternate units rounding value		0.0000
DIMALTTD	Alternate units tolerance value	Integer	2
DIMALTTZ	Toggles suppression of zeros for tolerance values	Integer	0
DIMALTU	Sets unit format for alternate units	Integer	2
DIMALTZ	Toggles suppression of zeros for alternate values	Integer	0
DIMAPOST	Alternate units text suffix	String	None
DIMASO	Controls the associativity of dimesion objects	String	On
DIMASSOC	Obsolete dimension variable (used in AutoCAD 2002 and earlier)		
DIMASZ	Arrow size	Distance	0.18
DIMATFIT	Arrow and text fit		3
DIMAUNIT	Angle format	Integer	0
DIMAZIN	Zero suppression in angles	Integer	0
DIMBLK	Arrow block	String	None
DIMBLK1	Separate arrow block 1	String	None
DIMBLK2	Separate arrow block 2	String	None
DIMCEN	Center mark size	Distance	0.09
DIMCLRD	Dimension line color	Color number	BYBLOCK
DIMCLRE	Extension line color	Color number	BYBLOCK
DIMCLRT	Dimension text color	Color number	BYBLOCK
DIMDEC	Decimal place for tolerance values	Integer	4
DIMDLE	Dimension Line extension	Distance	0.0
DIMDLI	Dimension line increment	Distance	0.38
DIMDSEP	Decimal separator		.

Name	Description	Type	Default
DIMEXE	Extension line extension	Distance	0.18
DIMEXO	Extension line offset	Distance	0.0625
DIMFIT	Preserves integrity of scripts	Integer	3
DIMFRAC	Fraction format		0
DIMGAP	Dimension line gap	Distance	0.09
DIMJUST	Controls horizontal text position	Imteger	0
DIMLDRBLK	Leader arrowhead block name		ClosedFilled
DIMLFAC	Length factor	Scale	1.0
DIMLIM	Limits dimensioning	Switch	Off
DIMLUNIT	Linear unit format		2
DIMLWD	Dimension line and leader lineweight		-2
DIMLWE	Extension line lineweight		-2
DIMPOST	Dimension text suffix	String	None
DIMRND	Rounding value	Scaled distance	0.0
DIMSAH	Separate arrow blocks	Switch	Off
DIMSCALE	Dimension feature scale factor	Switch	1.0
DIMSD1	Suppresses first dimension line	Switch	Off
DIMSD2	Suppresses second dimension line	Switch	Off
DIMSE1	Suppresses extension line 1	Switch	Off
DIMSE2	Suppresses extension line 2	Switch	Off
DIMSHO	Lists dimension styles	Switch	On
DIMSOXD	Suppresses outside dimension lines	Switch	Off
DIMSTYLE	Shows current dimension style	String	
DIMTAD	Text above dimension line	Switch	Off
DIMTDEC	Tolerance values	Integer	4
DIMTFAC	Tolerance text scale factor	Scale	1.0
DIMTIH	Text inside horizontal	Switch	On
DIMTIX	Text inside extension lines	Switch	Off
DIMTM	Minus tolerance value	Scaled distance	0.0
DIMTMOVE	Text movement		0
DIMTOFL	Text outside, force line inside	Switch	Off
DIMTOH	Text outside horizontal	Switch	On

Name	Description	Type	Default
DIMTOL	Tolerance dimensioning	Switch	Off
DIMTOLJ	Tolerance dimensioning justification	Integer	1
DIMTP	Plus tolerance value	Scaled distance	0.0
DIMTSZ	Tick size	Distance	0.0
DIMTVP	Text vertical position	Scale	0.0
DIMTXSTY	Dimension text style	String	Standard
DIMTXT	Text size	Distance	0.18
DIMTZIN	Controls suppression of zeros in tolerance values		
DIMUNIT	Sets unit format	Integer	2
DIMUPT	User-positioned text	Switch	Off
DIMZIN	Zero suppression	Integer	0

The following table lists the Dimension commands as entered at the "Command:" prompt and their equivalent command entered at the "Dim:" prompt.

Dimension Commands entered at the "Command:" prompt	Equivalent Dimension Commands entered at the "Dim:" prompt
DIMALIGNED	ALIGNED
DIMANGULAR	ANGULAR
DIMBASELINE	BASELINE
DIMCENTER	CENTER
DIMCONTINUE	CONTINUE
DIMDIAMETER	DIAMETER
DIMEDIT Home	HOMETEXT
DIMLINEAR Horizontal	HORIZONTAL
LEADER	LEADER
DIMEDIT New	NEWTEXT
DIMEDIT Oblique	OBLIQUE
DIMORDINATE	ORDINATE
DIMOVERRIDE	OVERRIDE
DIMRADIUS	RADIUS
- DIMSTYLE Restore	RESTORE
DIMLINEAR Rotated	ROTATED
- DIMSTYLE Save	SAVE
- DIMSTYLE Status	STATUS
DIMTEDIT	TEDIT
DIMEDIT Rotate	TROTATE
- DIMSTYLE Apply	UPDATE
- DIMSTYLE Variables	VARIABLES
DIMLINEAR Vertical	VERTICAL

OBJECT SELECTION

Object Selection Option	Meaning
point	Selects one object that crosses the small pick box. If no object crosses the pick box and Auto mode has been selected, this designated point is taken as the first corner of a Crossing or Window box
Multiple	Allows selection of multiple objects using a single search of the drawing. The search is not performed until you give a null response to the "Select objects:" prompt
Window	Selects all objects that lie entirely within a window
WPolygon	Selects objects that lie entirely within a polygon shaped selection area
Crossing	Selects all objects that lie within or cross a window
CPolygon	Selects all objects that lie within and crossing a polygon-shaped selection area
Fence	Selects all objects that cross a selection fence line
BOX	Prompts for two points. If the second point is to the right of the first point, selects all objects inside the box (like "Window"); otherwise, selects all objects within or crossing the box (like "Crossing")
AUto	Accepts a point, which can select an object using the small pick box; if the point you pick is in an empty area, it is taken as the first corner of a BOX (see above)
ALL	Selects all entities in the drawing except entities on frozen or locked layers
Last	Selects the most recently drawn object that is currently visible
Previous	Selects the previous selection set
Add	Establishes Add mode to add following objects to the selection set
Remove	Sets Remove mode to remove following objects from the selection set
SIngle	Sets single selection mode; as soon as one object (or one group of objects via Window/Crossing box) is selected, the selection set is considered complete and the editing command uses it without further user interaction
Undo	Undoes (removes objects last added)

APPENDIX C

AutoCAD Toolbars

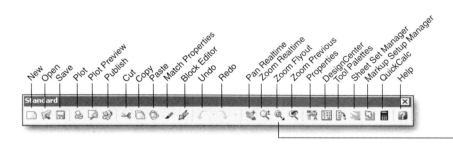

Standard toolbar callouts: New, Open, Save, Plot, Plot Preview, Publish, Cut, Copy, Paste, Match Properties, Block Editor, Undo, Redo, Pan Realtime, Zoom Realtime, Zoom Flyout, Zoom Previous, Properties, DesignCenter, Tool Palettes, Sheet Set Manager, Markup Setup Manager, QuickCalc, Help

Object Snap

- Temporary Tracking Point
- Snap From
- Snap to Endpoint
- Snap to Midpoint
- Snap to Intersection
- Snap to Apparent Intersection
- Snap to Extension
- Snap to Center
- Snap to Quadrant
- Snap to Tangent
- Snap to Perpendicular
- Snap to Parallel
- Snap to Insert
- Snap to Node
- Snap to Nearest
- Snap to None
- Object Snap Settings

UCS

- UCS
- Display UCS Dialog
- UCS Previous
- World UCS
- Object UCS
- Face UCS
- View UCS
- Origin UCS
- Z Axis Vector UCS
- 3-Point UCS
- X Axis Rotate UCS
- Y Axis Rotate UCS
- Z Axis Rotate UCS
- Apply UCS

View

- Named Views
- Top View
- Bottom View
- Left View
- Right View
- Front View
- Back View
- SW Isometric View
- SE Isometric View
- NE Isometric View
- NW Isometric View
- Camera

Zoom

- Zoom Window
- Zoom Dynamic
- Zoom Scale
- Zoom Center
- Zoom In
- Zoom Out
- Zoom All
- Zoom Extents

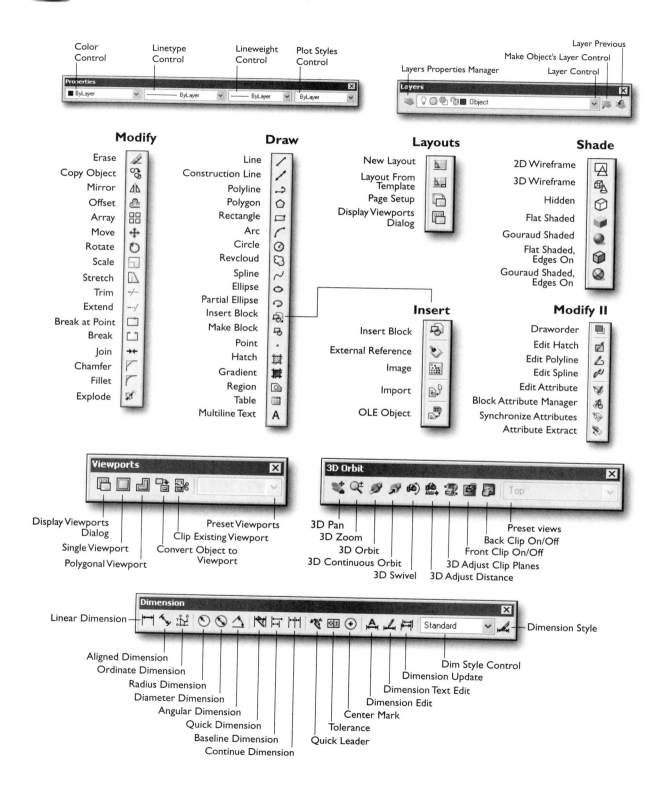

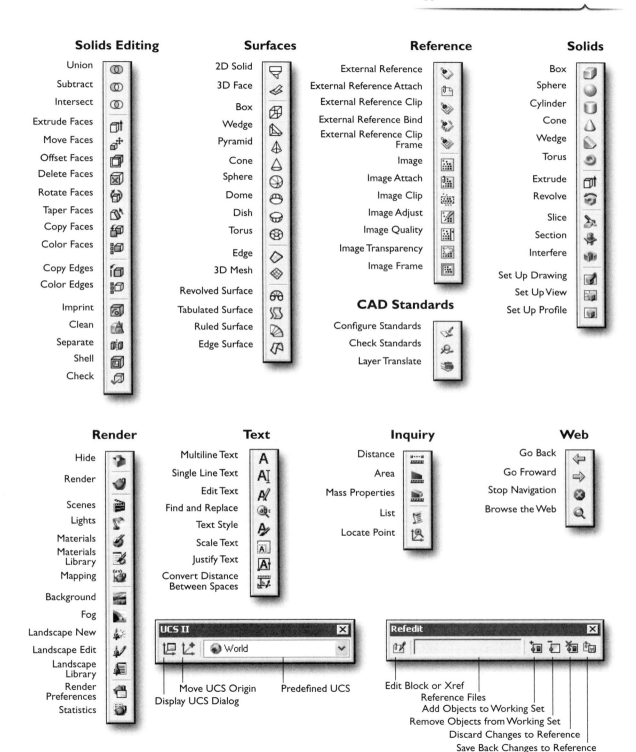

APPENDIX d

System Variables

This is a complete listing of AutoCAD system variables. Each variable has an associated type: integer, real, point, or text string. These variables can be examined and changed (unless read-only) by means of the SETVAR command and AutoLISP (getvar and setvar) functions. Many of the system variables are saved across editing sessions; as indicated in the table, some are saved in the drawing itself, while others are saved in the AutoCAD general configuration file, *ACAD.CFG*

Variable	Default Setting	Type	Saved In	Explanation
ACADLSPASDOC	0	Integer	Registry	Controls whether AutoCAD loads the acad.lsp file into every drawing or just the first drawing opened in an AutoCAD session
ACADPREFIX	" "	String	Read-only	The directory path, if any specified by the ACAD environment variable, with path separators appended if necessary (read-only)
ACADVER		String		This is the AutoCAD version number, which can have values only like "15.0" (read-only). Note that this differs from the DXF file $ACADVER header variable, which contains the drawing database level number.
ACISOUTVER	40	Integer		Controls the ACIS version of SAT files created using the ACISOUTcommand
ADCSTATE	Varies	Integer	Not Saved	Determines if DesignCenter is active or not 0 - DesignCenter not active 1 - DesignCenter active
AFLAGS	0	Integer		Attribute flags bit-code for ATTDEF command (sum of the following): 0 = No Attribute Mode Selected 1 = Invisible 2 = Constant 4 = Verify 8 = Preset
ANGBASE	0	Real	Drawing	Angle 0 direction (with respect to the current UCS)
ANGDIR	0	Integer	Drawing	1= clockwise angles, 0 = counterclockwise (with respect to the current UCS)
APBOX	0	Integer	Registry	Turns Autosnap aperture box on or off 0 - Aperture box not displayed 1 - Apetrure box not displayed
APERTURE	10	Integer	Config	Object Snap target height, in pixels (default value = 10)

Reprinted with permission of Autodesk Inc.

Variable	Default Setting	Type	Saved In	Explanation
APBOX				Sets AutoSnap aperture box to ON or OFF
APERTURE	10	Integer	Registry	Sets object snap target height, in pixels
AREA		Real		True area computed by Area, List, or Dblist (read-only)
ASSISTSTATE	0	Integer	Not-Saved	Indicates whether the Info palette that displays Quick Help is active or not.
ATTDIA	0	Integer	Registry	1 causes the INSERT command to use a dialog box for entry of attribute values; 0 to issue prompts
ATTMODE	1	Integer	Drawing	Attribute display mode (0 = OFF, 1 = normal, 2 = ON)
ATTREQ	1	Integer	Registry	0 assumes defaults for the values of all attributes during insert of blocks; 1 enables prompts (or dialog box) for attribute values, as selected by ATTDIA
AUDITCTL	0	Integer	Config	Controls whether an .adt log file (audit report file) is created 0 = Disables (or prevents) the writing of adt log files 1 = Enables the writing of .adt log files by the AUDIT command
AUNITS	0	Integer	Drawing	Angular units mode (0 = decimal degrees, 1 = degrees/minutes/seconds, 2 = grads, 3 = radians, 4 = surveyor's units)
AUPREC	0	Integer	Drawing	Angular units decimal places
AUTOSNAP	63	Integer	Registry	Controls the display of the AutoSnap marker and Snap-Tips and sets the AutoSnap magnet to ON or OFF
BACKGROUNDPLOT	2	Integer	Registry	Controls whether background plotting is turned on or off for plotting and publishing. By default, background plotting is turned off for plotting and on for publishing. 1 = Plot Foreground, Publish Foreground 2 = Plot Background, Publish Foreground 3 = Plot Foreground, Publish Background 4 = Plot Background, Publish Background
BACKZ	0.0000	Real	Drawing	Back clipping plane offset for the current viewport, in drawing units. Meaningful only if the back clipping bit in VIEWMODE is on. The distance of the back clipping plane from the camera point can be found by subtracting BACKZ from the camera-to-target distance (read-only)
BACTIONCOLOR	7	String	Registry	Sets the text color of actions in the Block Editor. Valid values include BYLAYER, BYBLOCK, and an integer from 1 to 255.
BDEPENDINCYHIGHLIGHT	1	Integer	Registry	Controls whether or not dependent objects are dependency highlighted when a parameter, action, or grip is selected in the Block Editor. 1 = Specifies that dependent objects are not highlighted 2 = Specifies that dependent objects are highlighted
BGRIPCOLOR	141	Integer	Registry	Sets the color of grips in the Block Editor. Valid values include BYLAYER, BYBLOCK, and an integer from 1 to 255.
BGRIPOBJSIZE	8	Integer	Registry	Sets the display size of custom grips in the Block Editor relative to the screen display. Valid values include an integer from 1 to 255.
BINDTYPE	0	Integer		Controls how xref names are handled when binding xrefs or editing xrefs in-place

Variable	Default Setting	Type	Saved In	Explanation
BLIPMODE	0	Integer	Registry	Marker blips ON if 1, OFF if 0
BLOCKEDITLOCK	0	Integer	Registry	Disallows opening of the Block Editor and editing of dynamic block definitions. When BLOCKEDITLOCK is set to 1, double-clicking a dynamic block in a drawing opens the Reference Edit dialog box. If the block contains attributes, double-clicking the block reference opens the Enhanced Attribute Editor 0 = Specifies that the Block Editor can be opened 1 = Specifies that the Block Editor cannot be opened
BLOCKEDITOR	0	Integer	Not-saved	Reflects whether or not the Block Editor is open 0 = Indicates that the Block Editor is not open 1 = Indicates that the Block Editor is open
BPARAMETERCOLOR	7	String	Registry	Sets the text color of parameters in the Block Editor. Valid values include BYLAYER, BYBLOCK, and an integer from 1 to 255.
BPARAMETERFONT	"Simplex"	String	Registry	Sets the font of parameters and actions in the Block Editor.
BPARAMETERSIZE	12	String	Registry	Sets the size of parameter text and features in the Block Editor relative to the screen display. Valid values include an integer from 1 to 255.
BTMARKDISPLAY	1	Integer	Registry	Controls whether or not value set markers are displayed for dynamic block references. 0 = Specifies that value set markers are not displayed 1 = Specifies that value set markers are displayed
BVMODE	0	Integer	Not-saved	Controls how objects that are made invisible for the current visibility state are displayed in the Block Editor. 0 = Specifies that hidden objects are not visible 1 = Specifies that hidden objects are visible but dimmed
CALCINUT	1	Integer	Registry	Controls whether mathematical expressions and global constants are evaluated in text and numeric entry boxes of windows and dialog boxes 0 = Expressions are not evaluated 1 = Expressions are evaluated after you press the END key
CDATE		Real		Calendar date/time (read-only)
CECOLOR	"BY-LAYER"	String	Drawing	Current object color (read-only)
CELTSCALE	1	Real	Drawing	Sets the current global linetype scale for objects
CELTYPE	"BY-LAYER"	String	Drawing	Current object linetype (read-only)
CELWEIGHT	-1	Integer		Sets the lineweight of new objects
CENTERMT	0	Integer	User-set'gs	Controls how grips stretch multiline text that is centered horizontally. CENTERMT does not apply to stretching multiline text by using the ruler in the In-Place Text Editor. 0 = When you move a corner grip in centered multiline text, the center grip moves in the same direction, and the grip on the opposite side remains in place 1 = When you move a corner grip in centered multiline text, the center grip stays in place, and both sets of side grips move in the direction of the stretch
CHAMFERA	0.5000	Real	Drawing	First chamfer distance

Variable	Default Setting	Type	Saved In	Explanation
CHAMFERB	1.0000	Real	Drawing	Second chamfer distance
CHAMFERC	1.0000	Real	Drawing	Sets the chamfer length
CHAMFERD	0.0000	Real	Drawing	Sets the chamfer angle
CHAMMODE	0	Integer		Sets the input method by which AutoCAD creates chamfers 0 = Requires two chamfer distances 1 = Requires one chamfer length and an angle
CIRCLERAD	0.0000	Real		Sets the default circle radius; to specify no default, enter 0 (zero)
CLAYER	"0"	String	Drawing	Sets the current layer (read-only)
CLISTATE	1	Integer	Not-saved	Stores a value that indicates whether the command window is hidden or displayed. 0 = Hidden 1 = Displayed
CMDACTIVE		Integer		Bitcode that indicates whether an ordinary command, transparent command, script, or dialog box is active (read-only). It is the sum of the following: 1 = Ordinary command is active 2 = Ordinary command and a transparent command are active 4 = Script is active 8 = Dialog box is active 16 = Autolist active (only visible to an object arcs-defined)
CMDDIA	1	Integer	Registry	Controls the display of dialog boxes for some commands. 0 = Off 1 = On
CMDECHO	1	Integer		When the AutoLISP (command) function is used, prompts and input are echoed if this variable is 1 but not if it is 0
CMDINPUTHISTORYMAX	20	Integer	Registry	Sets the maximum number of previous input values that are stored for a prompt in a command.
CMDNAMES		String		Displays in English the name of the command (and transparent command) that is currently active, for example; LINE'ZOOM indicates that the ZOOM command is being used transparently during the LINE command
CMLJUST	0	Integer	Drawing	Specifies multiline justification 0 = Top 1 = Middle 2 = Bottom
CMSCALE	1.0000	Real	Drawing	Controls the overall width of a multiline
CMSTYLE	"Standard"	String	Drawing	Sets the name of the multiline style that AutoCAD uses to draw the multiline
COMPASS	0	Integer		Controls whether the 3D compass is on or off in the current viewport
COORDS	1	Integer	Registry	If 0, coordinate display is updated on point picks only; if 1, display of absolute coordinates is continuously updated; if 2, distance and angle from last point are displayed when a distance or angle is requested
CPLOTSTYLE	"ByLayer"		Drawing	Controls the current plot style for new objects

Variable	Default Setting	Type	Saved In	Explanation
CPROFILE		String	Registry	Stores the name of the current profile (read-only)
CROSSINGAREACOLOR	3 (green)	Integer	Registry	Controls the color of the selection area during crossing selection. The valid range is 1 to 255.
CTAB		String	Drawing	(Read-Only) Returns the name of the current (model or layout) tab in the drawing. Provides a means for the user to determine which tab is active
CTABLESTYLE	"Standard"	String	Drawing	Sets the name of the current table style
CURSORSIZE				Sets the size of the crosshairs as a percentage of screen size
CVPORT	2	Integer	Drawing	The identification number of the current viewport
DATE		Real		Julian date/time (read-only)
DBCSTATE	0	Integer	Drawing	Stores active or nonactive state of dbConnect Manager 0 - dbConnect Mgr not displayed 1 - dbConnect Mgr displayed
DBMOD		Integer		Bitcode that indicates the drawing modification status (read-only); it is the sum of the following: 1 = Entity database modified 2 = Symbol table modified 4 = Database variable modified 8 = Window modified 16 = View modified
DCTCUST		String	Registry	Displays the current custom spelling dictionary path and file name
DCTMAIN		String	Registry	Displays the current main spelling dictionary file name
DEFLPLSTYLE	""	String	Registry	Specifies the default plot style for new layers
DEFPLSTYLE	"ByLayer"		Registry	Specifies the default plot style for new objects
DELOBJ	1	Integer	Registry	Controls whether objects used to create other objects are retained or deleted from the drawing database 0 = Objects are retained 1 = Objects are deleted
DEMANDLOAD	3	Integer	Registry	Determines demand loading of a third-party application if a drawing contains custom objects created in that application
DIASTAT		Integer		Dialog box exit status: if 0, the most recent dialog box was exited via CANCEL; if 1, the most recent dialog box was exited by pressing OK (read-only)
DIMxxx		Assorted	Drawing	All the dimensioning variables are also accessible as system variables (see Dimensioning Variables, Appendix B)
DISPSILH	0	Integer	Drawing	Sets display of silhouette curves of body objects in wireframe mode
DISTANCE		Real		Distance computed by DIST command (read-only)
DONUTID	0.5000	Real		Default donut inside diameter; can be zero
DONUTOD	1.0000	Real		Default donut outside diameter; must be nonzero. If DONUTID is larger than DONUTOD, the two values are swapped by the next command
DRAGMODE	2	Integer	Registry	0 = no dragging, 1 = on if requested, 2 = auto
DRAGP1	10	Integer	Registry	Regeneration-drag input sampling rate
DRAGP2	25	Integer	Registry	Fast-drag input sampling rate

Variable	Default Setting	Type	Saved In	Explanation
DRAWORDERCTL	3	Integer	Drawing	Controls the display order of overlapping objects. Use this setting to improve the speed of editing operations in large drawings. The commands that are affected by inheritance are BREAK, FILLET, HATCH, HATCHEDIT, EXPLODE, TRIM, JOIN, PEDIT, and OFFSET. 0 = Turns off the default draw order of overlapping objects 1 = Turns on the default draw order of objects 2 = Turns on draw order inheritance 3 = Provides full draw order display
DRSTATE	Varies	Integer	Not-saved	Determines whether the Drawing Recovery window is active or not. 0 = Drawing Recovery window is not active 1 = Drawing Recovery window is active
DTEXTED	0	Integer	Registry	Specifies the user interface for editing single-line text. 0 = Displays the In-Place Text Editor 1 = Displays the Edit Text dialog box
DWGCHECK	0	Integer	Registry	Determines whether a drawing was last edited by a product other than AutoCAD
DWGCODEPAGE		String	Drawing	Drawing code page: This variable is set to the system code page when a new drawing is created, but otherwise AutoCAD doesn't maintain it. It should reflect the code page of the drawing and you can set it to any of the values used by the SYSCODEPAGE system variable or to "undefined." It is saved in the header DWGNAME String Drawing name as entered by the user; if the user specified a drive/directory prefix, it is included as well (read-only)
DWGNAME	"Drawing.dwg"	String	Not-saved	Stores the drawing name as entered by the user
DWGPREFIX		String		Drive/directory prefix for drawing (read-only)
DWGTITLED		Integer		Bitcode that indicates whether the current drawing has been named (read-only) 0 = The drawing hasn't been named 1 = The drawing has been named
DYNDIVIS	1	Integer	User-set'gs	Controls how many dynamic dimensions are displayed during grip stretch editing 0 = Only the first dynamic dimension in the cycle order 1 = Only the first two dynamic dimensions in the cycle order 2 = All dynamic dimensions, as controlled by the DYNDIGRIP system variable
DYNMODE	3	Integer	User-set'gs	When DYNMODE is set to a negative value, the Dynamic Input features are not visible, but the setting is stored 0 = All off 1 = Only pointer input on 2 = Only dimensional input on 3 = All on
DYNPICOORDS	0	Integer	User-set'gs	Controls whether pointer input uses relative or absolute format for coordinates 0 = Relative 1 = Ablolute

Variable	Default Setting	Type	Saved In	Explanation
DYNPIFORMAT	0	Integer	User-set'gs	Controls whether pointer input uses polar or Cartesian format for coordinates. This setting applies only to a second or next point. 0 = Polar 1 = Cartesian
DYNPIVIS	1	Integer	User-set'gs	Controls when pointer input is displayed 0 = Only when you type at a prompt for a point 1 = Automatically at a prompt for a point 2 = Always
DYNPROMPT	1	Integer	User-set'gs	Controls display of prompts in Dynamic Input tooltips 0 = Off 1 = On
DYNTOOLTIPS	1	Integer	User-set'gs	Controls which tooltips are affected by tooltip appearance settings 0 = Only Dynamic Input value fields 1 = All drafting tooltips
EDGEMODE	0	Integer	Registry	Sets how TRIM and EXTEND determine cutting and boundary edges
ELEVATION	0.0000	Real	Drawing	Current 3D elevation, relative to the current UCS for the current space
ENTERPRISEMENU	"."	String	Registry	Stores the CUI file name (if defined), including the path for the file name
ERRNO	0	Integer	Not Saved	Shows number of error code when AutoLisp function calls an error
EXPERT	0	Integer		Controls the issuance of certain "are you sure?" prompts: 0 = Issues all prompts normally 1 = Suppresses "About to regen, proceed?" and "Really want to turn the current layer off?" 2 = Suppresses the preceding prompts and Block's "Block already defined. Redefine it?" and Save/Wblock's "A drawing with this name already exists. Overwrite it?" 3 = Suppresses the preceding prompts and those issued by linetype if you try to load a linetype that is already loaded or create a new linetype in a file that already defines it 4 = Suppresses the preceding prompts and those issued by "Ucs Save" and "Vports Save" if the name you supply already exists 5 = Suppresses the preceding prompts and those issued by "Dim Save" and "Dim Override" if the dimension style name you supply already exists (the entries are redefined) When a prompt is suppressed, EXPERT, the operation in question, is performed as though you had responded Y to the prompt. In the future, values greater than 5 may be used to suppress additional safety prompts. The setting of EXPERT can affect scripts, menu macros, AutoLISP, and the command functions. The default value is 0.
EXPLMODE	0	Integer		Determines whether EXPLODE supports nonuniformly scaled (NUS) blocks

Variable	Default Setting	Type	Saved In	Explanation
EXTMAX		3D point	Drawing	Upper right drawing uses extents. Expands outward as new objects are drawn; shrinks only by ZOOM All or ZOOM Extents. Reported in World coordinates for the current space (read-only)
EXTMIN		3D point	Drawing	Lower left drawing uses extents. Expands outward as new objects are drawn; shrinks only by ZOOM All or ZOOM Extents. Reported in World coordinates for the current space (read-only)
EXTNAMES	1	Integer	Drawing	Sets the parameters for named object names (such as linetypes and layers) stored in symbol tables
FACETRATIO	0	Integer		Controls the aspect ratio of faceting for cylindrical and conic ACIS solids
FACETRES	0.5	Real	Drawing	Adjusts smoothness of shaded and hidden-line-removed objects
FIELDDISPLAY	1	Integer	Registry	Controls whether fields are displayed with a gray background. The background is not plotted. 0 = Fields are displayed with no background 1 = Fields are displayed with a gray background
FIELDEVAL	31	Integer	Drawing	Controls how fields are updated. The setting is stored as a bitcode using the sum of the following values. 0 = Not updated 1 = Updated on open 2 = Updated on save 4 = Updated on plot 8 = Updated on use of ETRANSMIT 16 = Updated on regeneration
FILEDIA	1	Integer	Registry	1 = Use file dialog boxes if possible; 0 = do not use File dialog boxes unless requested via ~ (tilde)
FILLETRAD	0.5000	Real	Drawing	Fillet radius
FILLMODE	1	Integer	Drawing	Fill mode ON if 1, OFF if 0
FONTALT	"Simplex.shp"	String	Registry	Specifies alternate font
FONTMAP	"acad.fmp"	String	Registry	Specifies font mapping file
FRONTZ	0.0000	Real	Drawing	Front clipping plane offset for the current viewport, in drawing units. Meaningful only if the front clipping bit in VIEWMODE is ON and the "Front clip not at eye" bit is also ON. The distance of the front clipping bit from the camera point can be found by subtracting FRONTZ from the camera-to-target distance (read-only)
FULLOPEN		Integer		Indicates whether the current drawing is partially open (read-only)
FULLPLOTPATH	1	Integer	Registry	Controls whether the full path of the drawing file is sent to the plot spooler. 0 = Sends the drawing file name only 1 = Sends the full path of the drawing file
GRIDMODE	0	Integer	Drawing	1 = Grid on for current viewport, X and Y
GRIDUNIT	0.5000,0.5000	2D point	Drawing	Grid spacing for current viewport, X and Y
GRIPBLOCK	0	Integer	Registry	Controls the assignment of grips in blocks 0 = Assigns grip only to the insertion point of the block 1 = Assigns grips to entities within the block

Variable	Default Setting	Type	Saved In	Explanation
GRIPCOLOR	5 (1–255)	Integer	Registry	Color of nonselected grips; drawn as a box outline
GRIPDYNCOLOR	140 (1–255)	Integer	Registry	Controls the color of custom grips for dynamic blocks
GRIPHOT	1 (1–255)	Integer	Registry	Color of selected grips; drawn as a filled box
GRIPHOVER	3	Integer	Registry	Controls fill color of grip
GRIPOBJLIMIT	100	Integer	Registry	Suppresses number of grips displayed when initial selection set contains more objects than number specified
GRIPS	1	Integer	Registry	Allows the use of selection set grips for the Stretch, Move, Rotate, Scale, and Mirror modes 0 = Disables grips 1 = Enables grips
GRIPSIZE	3 (1–255)	Integer	Registry	The size in pixels of the box drawn to display the grip
GRIPTIPS	1	Integer	Registry	Controls display of grips when cursor hovers on custom objects that support grip tips 0 - Grip tips not displayed 1 - Grip tips displayed
HALOGAP	0	Integer	Drawing	Specifies the distance to shorten a haloed line
HANDLES	1	Integer	Drawing	If 0, entity handles are disabled; if 1, handles are on (read-only)
HIDEPRECISION	0	Integer		Controls the accuracy of hides and shades
HIDETEXT	On	Switch	Drawing	Determines whether text objects created by TEXT, DTEXT, or MTEXT are processed during HIDE
HIGHLIGHT	1	Integer		Object selection highlighting ON if 1, OFF if 0
HPANG	0.0000	Real		Default hatch pattern angle
HPASSOC	1	Integer	Registry	Controls whether hatch patterns and gradient fills are associative 0 - Patterns/fills not associated 1 - Patterns/fills associated
HPBOUND	1			Controls BHATCH and BOUNDARY object types
HPDOUBLE	0	Integer		Default hatch pattern doubling for "U" user defined patterns 0 = Disables doubling 1 = Enables doubling
HPDRAWORDER	3	Integer	Not-saved	Controls the draw order of hatches and fills. Stores the Draw Order setting from the Hatch and Fill Dialog Box. 0 = None. The hatch or fill is not assigned a draw order 1 = Send to back. The hatch or fill is sent to the back of all other objects 2 = Bring to front. The hatch or fill is brought to the front of all other objects 3 = Send behind boundary. The hatch or fill is sent behind the hatch boundary 4 = Bring in front of boundary. The hatch or fill is brought in front of the hatch boundary

Variable	Default Setting	Type	Saved In	Explanation
HGAPTOL	0	Real	Registry	Treats a set of objects that almost enclose an area as a closed hatch boundary
HPINHERIT	0	Integer	Drawing	Controls the hatch origin of the resulting hatch when using Inherit Properties in HATCH and HATCHEDIT 0 = The hatch origin is taken from HPORIGIN 1 = The hatch origin is taken from the source hatch object
HPNAME	"ANSI31"	String		Default hatch pattern name. Up to 34 characters no spaces allowed. Returns " " if there is no default. Enter . (period) to set no default
HPOBJWARNING	10000	Integer	Registry	Sets the number of hatch boundary objects that can be selected before displaying a warning message
HPORIGIN	0,0	2D-Point	Drawing	Sets the hatch origin point for new hatch objects relative to the current user coordinate system
HPORIGINMODE	0	Integer	Registry	Controls how HATCH determines the default hatch origin point. 0 = Hatch origins are set using HPORIGIN 1 = Hatch origins are set using the bottom-left corner of the rectangular extents of the hatch boundaries 2 = Hatch origins are set using the bottom-right corner of the rectangular extents of the hatch boundaries 3 = Hatch origins are set using the top-right corner of the rectangular extents of the hatch boundaries 4 = Hatch origins are set using the top-left corner of the rectangular extents of the hatch boundaries 5 = Hatch origins are set using the center of the rectangular extents of the hatch boundaries
HPSCALE	1.0000	Real		Default hatch pattern scale factor; must be nonzero
HPSEPARATE	0	Integer	Registry	Controls whether HATCH creates a single hatch object or separate hatch objects when operating on several closed boundaries 0 = A single hatch object is created 1 = Separate hatch objects are created
HPSPACE	1.0000	Real		Default hatch pattern line spacing for "U" user defined simple patterns; must be nonzero
HYPERLINKBASE	""	String	Drawing	Specifies the path used for all relative hyperlinks in the drawing
IMAGEHLT	0	Integer	Registry	Controls whether the entire raster image or only the raster image frame is hightlighted
INDEXCTL	0	Integer	Drawing	Controls whether layer and spatial indexes are created and saved in drawing files
INETLOCATION	www.acad.com/acaduser			Saves the Browser location used by the Internet
INPUTHISTORYMODE	15	Integer	Registry	Controls the content and location of the display of a history of user input. 0 = No history of recent input is displayed 1 = History of recent input is displayed at the command line or in a dynamic prompt tooltip with access through Up Arrow and Down Arrow keys 2 = History of recent input for the current command is displayed in the shortcut menu

Variable	Default Setting	Type	Saved In	Explanation
				4 = History of recent input for all commands in the current session is displayed in the shortcut menu
				8 = Markers for recent input of point locations are displayed in the drawing
INSBASE	0.0000, 0.0000, 0.0000	3D point	Drawing	Insertion basepoint (set by BASE command) expressed in UCS coordinates for the current space
INSNAME	" "	String		Default block name for DDINSERT or INSERT. The name must conform to symbol-naming conventions. Returns " " if there is no default. Enter . (period) to set no default.
INSUNITS	0	Integer	Registry	When you drag a block from AutoCAD DesignCenter, specifies a drawing units value
INSUNITSDEFSOURCE	0	Integer	Registry	Sets source content units value
INSUNITSDEFTARGET	0	Integer	Registry	Sets target drawing units value
INTELLIGENTUPDATE	20	Integer	Registry	Controls the graphics refresh rate. The default value is 20 frames per second
INTERSECTIONCOLOR	257	Integer	Drawing	Specifies color of intersection polyline
INTERSECTIONDISPLAY	Off	Switch	Drawing	Specifies display of intersection polylines 0 - intersection plines not displayed 1 - intersection plines displayed
ISAVEBAK	1	Integer	Registry	Optimizes the speed of periodic saves, especially for large drawings in Windows
ISAVEPERCENT	50	Integer	Registry	Specifies the amount of wasted space tolerated in a drawing file
ISOLINES	4	Integer	Drawing	Specifies the number of iso lines first surface on the object
LASTANGLE	0	Real		The end angle of the last arc entered, relative to the XY plane of the current UCS for the current space (read-only)
LASTPOINT	0.0000, 0.0000, 0.0000,	3D point		The last point entered, expressed in UCS coordinates for the current space; referenced by @ during keyboard entry
LASTPROMPT	" "	String		Saves the last string echoed to the command line (read-only)
LAYERFILTERALERT	2	Integer	Registry	Deletes excessive layer filters to improve performance. When a drawing has 100 or more layer filters, and the number of layer filters exceeds the number of layers. 0 = Off 1 = When the Layer Manager is opened, deletes all layer filters; no message is displayed 2 = When the Layer Manager is opened, displays a message that states the problem, recommends deleting all filters, and offers a choice: "Do you want to delete all layer filters now?" 3 = When the drawing is opened, displays a message that states the problem and offers to display a dialog box where you can choose which filters to delete
LAYOUTREGENCTL		Integer		Specifies how the display list is updated in the Model tab and layout tabs
LENSLENGTH		Real	Drawing	Length of the lens (in millimeters) used in perspective viewing, for current viewport (read-only)

Variable	Default Setting	Type	Saved In	Explanation
LIMCHECK	0	Integer	Drawing	Limits checking for the current space: ON if 1, OFF if 0
LIMMAX	12,000, 9,000	2D point	Drawing	Upper right drawing limits for the current space, expressed in World coordinates
LIMMIN	0.0000, 0.0000,	2D point	Drawing	Lower left drawing limits for the current space, expressed in World coordinates
LISPRINT				Detrmines whether names and values of AutoLISP-defined functions and variables are preserved when you open a new drawing
LOCALE	"enu"	String		Displays the ISO language code of the current AutoCAD version
LOCALROOTPREFIX	"pathname"	String	Registry	Stores full path to the root folder where local customizable files installed
LOCKUI	0	Integer	Registry	Locks the position and size of toolbars and windows such as DesignCenter and Properties palette. Locked toolbars and windows can still be opened and closed and items can be added and deleted. To unlock them temporarily, hold down CTRL. 0 = Toolbars and windows not locked 1 = Docked toolbars locked 2 = Docked windows locked 4 = Floating toolbars locked 8 = Floating windows locked
LOGFILEMODE	0	Integer	Registry	Determines whether the contents of the text window are written to a log file
LOGFILENAME				Determines the path for the log file (read-only)
LOGFILEPATH		String	Registry	Specifies the path for the log files for all drawings in a session
LOGINNAME		String		Displays the user's name as configured or input when AutoCAD is loaded (read-only)
LTSCALE	1.000	Real	Drawing	Sets global linetype scale factor
LUNITS	2	Integer	Drawing	Linear units units mode (1 = scientific, 2 = decimal, 3 = engineering, 4 = architectural, 5 = fractional)
LUPREC	4	Integer	Drawing	Linear units decimal places
LWDEFAULT	25	Enum	Registry	Sets the value for the default lineweight
LWDISPLAY	0	Integer	Drawing	Controls whether the lineweight is displayed in the Model or Layout tab
LWUNITS	1	Integer	Registry	Controls whether lineweight units are displayed in inches or millimeters
MAXACTVP	64	Integer	Drawing	Maximum number of viewports to regenerate at one time (read-only)
MAXSORT	200	Integer	Config	Maximum number of symbol/file names to be sorted by listing commands; if the total number of items exceeds this number, then none of the items are sorted (default value is 200)
MBUTTONPAN	1	Integer	Registry	Controls the behavior of the third button or wheel on the pointing device

Variable	Default Setting	Type	Saved In	Explanation
MEASUREINIT		Integer		Controls which hatch pattern and linetype files an existing drawing uses when it's opened
MEASUREMENT	0	Integer	Drawing	Sets drawing units as English or metric
MENUCTL	1	Integer	Config	Controls the page switching of the screen menu 0 = Screen menu doesn't switch pages in response to keyboard command entry 1 = Screen menu switches pages in response to keyboard command entry
MENUECHO	0	Integer		Menu echo/prompt control bits (sum of the following): 1 = Suppresses echo of menu items (^P in a menu item toggles echoing) 2 = Suppresses printing of system prompts during menu 4 = Disables ^P toggle of menu echoing 8 = Displays input/output strings The default value is 0 (all menu items and style prompts are displayed)
MENUNAME	"Acad"	Integer	Drawing	The name of the currently loaded menu file; includes a drive/path prefix if you entered it (read-only)
MIRRTEXT	1	Integer	Drawing	Mirror reflects text if nonzero, retains text direction if 0
MODEMACRO		String		Allows you to display a text string in the status line, such as the name of the current drawing, time/date stamp, or special modes. You can use MODEMACRO to display a simple string of text, or use special text strings written in the DIESEL macro language to have AutoCAD evaluate the macro from time to time and base the status line on user-selected conditions
MSOLESCALE	1.0	Real	Drawing	Controls the size of an OLE object with text that is pasted into model space
MTEXTED	"Internal"	String	Config	Sets the name of the program to use for editing mtext objects
MTEXTFIXED	0	Integer	Registry	Controls apperence of Multiline Text Editor 0 - Editor/text drawing size/location 1 - Editor/text size/location fixed
MTJIGSTRING	"abc"	String	Registry	Sets content of sample text at cursor location when MTEXT command is started
MYDOCUMENTSPREFIX	"pathname"	String	Registry	Stores full path to the My Documents folder for user currently logged-on
NOMUTT	0	Short		Suppresses the message display (muttering) when it wouldn't normally be suppressed
OBSCUREDCOLOR	0	Integer	Drawing	Specifies the color of obscured lines
OBSCUREDLTYPE	0	Integer	Drawing	Specifies the linetype of obscured lines
OFFSETDIST	1.0000	Real		Sets the default offset distance; if you enter a negative value, it defaults to Through mode
OFFSETGAPTYPE	0	Integer	Registry	Controls how to offset polylines when a gap is created as a result of offsetting the individual polyline segments
OLEHIDE	0	Integer	Registry	Controls the display of OLE objects in AutoCAD
OLEQUALITY	1	Integer	Registry	Controls the default quality level for embedded OLE objects

Variable	Default Setting	Type	Saved In	Explanation
OLESTARTUP	0	Integer	Drawing	Controls whether the source application of an embedded OLE object loads when plotting
ORTHOMODE	0	Integer	Drawing	Ortho mode ON if 1, OFF if 0
OSMODE	0	Integer	Registry	Object Snap modes bitcode (sum of the following): 1 = Endpoint 128 = Perpendicular 2 = Midpoint 256 = Tangent 4 = Center 512 = Nearest 8 = Node 1024 = Quick 16 = Quadrant 2048 = Appint 32 = Intersection 4096 = Extension 64 = Insertion 8192 = Parallel
OSNAPCOORD	2	Integer	Registry	Controls whether coordinates entered on the command line override running object snaps
OSNAPHATCH	0	Integer	Registry	Controls whether object snaps ignore hatch objects 0 = Object snaps ignore hatch objects 1 = Object snaps treat hatch objects the same as other objects
OSNAPZ	0	Integer	Not-saved	Controls whether object snaps are automatically projected onto a plane parallel to the XY plane of the current UCS at the current elevation 0 = Osnap uses the Z-value of the specified point 1 = Osnap substitutes the Z-value of the specified point with the elevation (ELEV) set for the current UCS
PALETTEOPAQUE	0	Integer	Registry	Controls whether windows can be opaque. Transparency is: 0 - turned on by user 1 - turned off by user 2 - unavailable though turned on by user 3 - unavailable and turned off by user
PAPERUPDATE	0	Integer	Registry	Controls the display of a warning dialog when attempting to print a layout with a paper size different from the paper size specified by the default for the plotter configuration file
PDMODE	0	Integer	Drawing	Point entity display mode
PDSIZE	0.0000	Real	Drawing	Point entity display size
PEDITACCEPT	0	Integer	Registry	Suppresses display of prompt "Object Selected Is Not a Polyline" 0 - turned on by user 1 - prompt suppressed
PELLIPSE	0	Integer	Drawing	Controls type of ellipse created 0 - turned on by user 1 - turned off by user
PERIMETER		Real		Perimeter computed by Area, List, or Dblist (read-only)
PFACEMAX	4	Integer		Maximum number of vertices per face (read-only)
PICKADD	1	Integer	Config	Controls additive selection of objects 0 = Disables PICKADD. The most recently selected objects, either by an individual pick or windowing, become the selection set. Previously selected objects are removed from the selection set. You can add more objects to the selection set, however, by holding down SHIFT while selecting

Variable	Default Setting	Type	Saved In	Explanation
				1 = Enables PICKADD. Each object you select, either individually or by windowing, is added to the current selection set. To remove objects from the selection set, hold down SHIFT while selecting
PICKAUTO	1	Integer	Config	Controls automatic windowing when the "Select objects:" prompt appears 0 = Disables PICKAUTO 1 = Allows you to draw a selection window (both window and crossing window) automatically at the "Select objects:" prompt
PICKBOX	3	Integer	Config	Object selection target height, in pixels
PICKDRAG	0	Integer	Config	Controls the method of drawing a selection window 0 = You draw the selection window by clicking the mouse at one corner and then at the other corner 1 = You draw the selection window by clicking at one corner, holding down the mouse button, dragging, and releasing the mouse button at the other corner
PICKFIRST	1	Integer	Config	Controls the method of object selection so that you can select objects first and then use an edit/inquiry command 0 = Disables PICKFIRST 1 = Enables PICKFIRST
PICKSTYLE	1	Integer	Registry	Controls group selection and associative hatch selection
PLATFORM		String		Read-only message that indicates which version of AutoCAD is in use
PLINEGEN	1	Integer	Drawing	Sets the linetype pattern generation around the vertices of a 2D polyline. When set to 1, PLINEGEN causes the netype to be generated in a continuous pattern around the vertices of the polyline. When set to 0, polylines are generated with the linetype to start and end with a dash at each vertex. PLINEGEN doesn't apply to polylines with tapered segments
PLINETYPE	2	Integer	Registry	Determines whether AutoCAD uses optimized 2D patterns
PLINEWID	0.0000	Real	Drawing	Default polyline width; it can be zero
PLOTROTMODE	1	Integer	Registry	Controls orientation of plots
PLQUIET	0	Integer	Registry	Controls the display of optional dialog boxes and nonfatal errors for batch plotting and scripts
POLARADDANG	null	String	Registry	Contains user-defined polar angles
POLARANG	90.0000	Real	Registry	Sets the polar angle increment
POLARDIST	0.000	Real	Registry	Sets the snap increment when the SNAPSTYL system variable is set to 1 (polar snap)
POLARMODE	1	Integer	Registry	Controls settings for polar and object snap tracking
POLYSIDES	4	Integer		Default number of sides for the POLYGON command; the range is 3–1024
POPUPS	1	Integer		1 if the currently configured display driver supports dialog boxes, the menu bar, pull-down menus, and icon menus; 0 if these advanced user interface features are not available (read-only)

Variable	Default Setting	Type	Saved In	Explanation
PREVIEWEFFECT	2	Integer	Registry	Specifies the visual effect used for previewing selection of objects. 0 = Dashed lines (the default display for selected objects) 1 = Thickened lines 2 = Dashed and thickened lines
PREVIEWFILTER	7	Integer	Registry	Excludes specified object types from selection previewing. The setting is stored as a bitcode using the sum of the following values: 0 = Excludes nothing 1 = Excludes objects on locked layers 2 = Excludes objects in xrefs 4 = Excludes tables 8 = Excludes multiline text objects 16 = excludes hatch objects 32 = Excludes objects in groups
PRODUCT	"AutoCAD"			Returns the product name (read-only)
PROGRAM	"acad"			Returns the program name (read-only)
PROJECTNAME	""	String	Drawing	Saves the current project name
PROJMODE	1	Integer	Config	Sets the current Projection mode for Trim or Extend operations
PROXYGRAPHICS	1	Integer	Drawing	Determines whether images of proxy objects are saved in the drawing
PROXYNOTICE	1	Integer	Registry	Displays a notice when you open a drawing containing custom objects created by an application that is not present
PROXYSHOW	1	Integer	Registry	Controls the display of proxy objects in a drawing
PROXYWEBSEARCH		Integer		Specifies how AutoCAD checks for Object Enablers
PSLTSCALE	1	Integer	Drawing	Controls paper space linetype scaling 0 = No special linetype scaling 1 = Viewport scaling governs linetype scaling
PSTYLEMODE	0	Integer	Drawing	Indicates whether the current drawing is in a Color-Dependent or Named Plot Style mode (read-only)
PSTYLEPOLICY	1	Integer	Registry	Controls whether an object's color property is associated with its plot style
PSVPSCALE	0	Real		Sets the view scale factor for all newly created viewports
PUCSBASE	""	String	Drawing	Stores the name of the UCS that defines the origin and orientation of orthographic UCS settings in paper space only
QCSTATE	Varies	Integer	Not-saved	Determines whether the QuickCalc calculator is active or not. 0 = Not active 1 = Active
QTEXTMODE	0	Integer	Drawing	Quick text mode ON if 1, OFF if 0
RASTERDPI	300	Integer	Registry	Controls paper size and plot scaling when changing from dimensional to dimensionless output devices, or vice versa. Converts millimeters or inches to pixels, or vice versa. Accepts an integer between 100 and 32,767 as a valid value.
RASTERPREVIEW	1	Integer	Drawing	Controls whether drawing preview images are saved with the drawing

Variable	Default Setting	Type	Saved In	Explanation
RECOVERYMODE	2	Integer	Registry	Controls whether drawing recovery information is recorded after a system failure. 0 = Recovery information is not recorded, the Drawing Recovery window does not display automatically after a system failure, and any recovery information in the system registry is removed 1 = Recovery information is recorded, but the Drawing Recovery window does not display automatically after a system failure 2 = Recovery information is recorded, and the Drawing Recovery window displays automatically in the next session after a system failure
REFEDITNAME	""	String		Indicates whether a drawing is in a reference-editing state and stores the reference file name (read-only)
REGENMODE	1	Integer	Drawing	Regenauto ON if 1, OFF if 0
RE-INIT		Integer		Reinitializes the I/O ports, digitizer, display, plotter, and acad.pgp file using the following bit codes. To specify more than one reinitialization, enter the sum of their values, for example, 3 to specify both digitizer port (1) and plotter port (2) reinitialization 1 = Digitizer port reinitialization 2 = Plotter port reinitialization 4 = Digitizer reinitialization 8 = Display reinitialization 16 = PGP file reinitialization (reload)
REMEMBERFOLDERS		Integer		Controls the default path for the Look In or Save In option in standard file selection dialog boxes
REPORTERROR	1	Integer	Registry	Controls sending of error report 0 - Error Report msg not displayed 1 - Error Report msg displayed
ROAMABLEROOTPREFIX	"pathname"	String	Registry	Stores root folder full path where roamable customizable files were installed
RTDISPLAY	1	Integer	Registry	Controls the display of raster images during Realtime ZOOM
SAVEFILE	""	String	Config	Current auto-save file name (read-only)
SAVEFILEPATH	"C\TEMP\"	String	Registry	Specifies the path to the directory for all automatic save files for the AutoCAD session.
SAVENAME		String		The file name you save the drawing to (read-only)
SAVETIME	120	Integer	Config	Automatic save interval, in minutes (or 0 to disable automatic saves). The SAVETIME timer starts as soon as you make a change to a drawing, and is reset and restarts by a manual SAVE, SAVEAS, or QSAVE. The current drawing is saved to auto.sv$
SCREENBOXES		Integer		The number of boxes in the screen menu area of the graphics area. If the screen menu is disabled configured off), SCREENBOXES is zero. On platforms that permit the AutoCAD graphics window to be resized or the screen menu to be econfigured during an editing session, the value of this variable might change during the editing session (read-only)

Variable	Default Setting	Type	Saved In	Explanation
SCREENMODE		Integer		A (read-only) bit code indicating the graphics/text state of the AutoCAD display. It is the sum of the following bit values: 0 = Text screen is displayed 1 = Graphics mode is displayed 2 = Dual-screen display configuration
SCREENSIZE		2D point		Current viewpoint size in pixels, X and Y (read-only)
SDI	0	Integer	Registry	Controls whether AutoCAD runs in single- or multiple-document interface
SECTIONAREA	1	Integer	Registry	Controls the display of effects for selection areas. Selection areas are created by the Window, Crossing, WPolygon, and CPolygon options of SELECT. 0 - Off 1 - On
SECTIONAREAOPACITY	25	Integer	Registry	Controls the transparency of the selection area during window and crossing selection. The valid range is 0 to 100. The lower the setting, the more transparent the area. A value of 100 makes the area opaque
SECTIONPREVIEW	3	Integer	Registry	Controls the display of selection previewing. Objects are highlighted when the pickbox cursor rolls over them. This selection previewing indicates that the object would be selected if you clicked. 0 = Off 1 = On when no commands are active 2 = On when a command prompts for object selection
SHADEDGE	3	Integer	Drawing	0 = Faces shaded, edges not highlighted 1 = Faces shaded, edges drawn in background color 2 = Faces not filled, edges in object color 3 = Faces in entity color, edges in background color
SHADEDIF	70	Integer	Drawing	Ratio of ambient to diffuse light (in percentage of ambient light)
SHORTCUTMENU	11	Integer	Registry	Controls whether Default, Edit, and Command mode shortcut menus are available in the drawing area
SHOWLAYERUSAGE	0	Integer	Registry	Displays icons in the Layer Properties Manager to indicate whether layers are in use. Setting this system variable to Off improves performance in the Layer Properties Manager 0 = Off 1 = On
SHPNAME	" "	String		Default shape name; must conform to symbol-naming conventions. If no default is set, it returns a " ". Enter . (period) to set no default.
SIGWARN	1	Integer	Registry	Controls display of warning when file with attached digital signature opened 0 - Warning not displayed 1 - Warning displayed
SKETCHINC	0.1000	Real	Drawing	Sketch record increment
SKPOLY	0	Integer	Drawing	Sketch generates lines if 0, polylines if 1
SNAPANG	0	Real	Drawing	Snap/Grid rotation angle (UCS-relative) for the current viewport

Variable	Default Setting	Type	Saved In	Explanation
SNAPBASE	0.0000, 0.0000,	2D point	Drawing	Snap/Grid origin point for the current viewport (in UCS XY coordinates)
SNAPISOPAIR	0	Integer	Drawing	Current isometric plane (0 = left, 1 = top, 2 = right) for the current viewport
SNAPMODE	0	Integer	Drawing	1 = Snap on for current viewport; 0 = Snap off
SNAPSTYL	0	Integer	Drawing	Snap style for current viewport (0 = standard, 1 = isometric)
SNAPTYPE	0	Integer	Registry	Sets the snap style for the current viewport
SNAPUNIT	0.5000, 0.5000,	2D point	Drawing	Snap spacing for current viewport, X and Y
SOLIDCHECK	1	Integer		Turns the solid validation on and off for the current AutoCAD session
SPLFRAME	0	Integer	Drawing	If = 1: − the control polygon for spline-fit polylines is to be displayed − only the defining mesh of a surface-fit polygon mesh is displayed (the fit surface is not displayed) − invisible edges of 3D faces are displayed If = 0: − does not display the control polygon for spline-fit polylines − displays the fit surface of a polygon mesh, not the defining mesh − does not display the invisible edges of 3D faces
SPLINESEGS	8	Integer	Drawing	The number of line segments to be generated for each spline patch
SPLINETYPE	6	Integer	Drawing	Type of spline curve to be generated by PEDIT Spline. The valid values are: 5 = Quadratic B-spline 6 = Cubic B-spline
SSFOUND	""	String	Not-saved	Displays the sheet set path and file name if a search for a sheet set is successful
SSLOCATE	1	Integer	User-Set'gs	Locates and opens the sheet set associated with a drawing when the drawing is opened.: 0 = Does not open a drawing's sheet set with the drawing 1 = Opens a drawing's sheet set with the drawing
SSMAUTOOPEN	1	Integer	User-Set'gs	Displays the Sheet Set Manager when a drawing associated with a sheet is opened 0 = Does not open the Sheet Set Manager automatically 1 = Opens the Sheet Set Manager automatically
SSMPOLLTIME	60	Integer	Registry	Controls the time interval between automatic refreshes of the status data in a sheet set.
SSMSHEETSTATUS	2	Integer	Registry	Controls how the status data in a sheet set is refreshed. 0 = Do not automatically refresh the status data in a sheet set 1 = Refresh the status data when the sheet set is loaded or updated 2 = In addition, refresh the status data at a time interval set by SSMPOLLTIME

Variable	Default Setting	Type	Saved In	Explanation
SSMSTATE	Varies	Integer	Not-saved	Determines whether the Sheet Set Manager window is active or not. 0 = Not active 1 = Active
STANDARDSVIOLATION	2	Integer	Registry	Controls notification of standards violation 0 - Notification turned off 1 - Alert is displayed 2 - Displays icon in status bar
STARTUP	0	Integer	Registry	Displays Create New Drawing dialog box at startup or use of NEW command 0 - Displays dialog box 1 - Displays Create New Dwg diabox
SURFTAB1	6	Integer	Drawing	Number of tabulations to be generated for Rulesurf and Tabsurf; also mesh density in the M direction for Resurf and Edgesurf
SURFTAB2	6	Integer	Drawing	Mesh density in the N direction for Revsurf and Edgesurf
SURFTYPE	6	Integer	Drawing	Type of surface fitting to be performed by PEDIT Smooth. The valid values are: 5 = Quadratic B-spline surface 6 = Cubic B-spline surface 8 = Bezier surface
SURFU	6	Integer	Drawing	Surface density in the M direction
SURFV	6	Integer	Drawing	Surface density in the N direction
SYSCODEPAGE		String	Drawing	Indicates the system code page specified in acad.xmf (read-only)
TABLEINDICATOR	1	Integer	User-Set'gs	Controls the display of row numbers and column letters when the In-Place Text Editor is open for editing a table cell. 0 = Off 1 = On
TABMODE	0	Integer		Controls the use of tablet mode 0 = Disables tablet mode 1 = Enables tablet mode
TARGET	0.0000, 0.0000, 0.0000,	3D point	Drawing	Location (in UCS coordinates) of the target (look-at) point for the current viewport (read-only)
TBCUSTOMIZE	1	Integer	Registry	Controls whether toolbars can be customized. 0 = Disables CUSTOMIZE and TOOLBAR and customization from the toolbar shortcut menu 1 = Enables toolbar customization
TDCREATE		Real	Drawing	Time and date of drawing creation (read-only)
TDINDWG		Real	Drawing	Total editing time (read-only)
TDUCREATE		Real	Drawing	Stores the universal time and date the drawing was created
TDUPDATE		Real	Drawing	Time and date of last update/save (read-only)
TDSURTIMER		Real	Drawing	User elapsed timer (read-only)
TDUUPDATE		Real	Drawing	Stores the universal time and date of the last update/save

Variable	Default Setting	Type	Saved In	Explanation
TEMPOVERRIDES	1	Integer	Registry	Turns temporary override keys on and off. A temporary override key is a key that you can hold down to temporarily turn on or turn off one of the drawing aids that are set in the Drafting Settings dialog box; for example, Ortho mode, object snaps, or Polar mode.. 0 = Off 1 = On
TEMPPREFIX	" "	String		Contains the directory name (if any) configured for placement of temporary files, with a path separator appended if necessary (read-only)
TEXTEVAL	0	Integer		If = 0, all responses to prompts for text strings and attribute values are taken literally. If = 1, text starting with "(" or "!" is evaluated as an AutoLISP expression, as for nontextual input. Note: The DTEXT command takes all input literally, regardless of the setting of TEXTEVAL
TEXTFILL	1	Integer	Registry	Controls the filling of Bitstream, TrueType, and Adobe Type 1 fonts
TEXTQLTY	50	Integer		Sets the resolution of Bitstream, TrueType, and Adobe Type 1 fonts
TEXTSIZE	0.2000	Real	Drawing	The default height for new text objects drawn with the current text style (meaningless if the style has a fixed height)
TEXTSTYLE	"STANDARD"	String	Drawing	Contains the name of the current text style (read-only)
THICKNESS	0.0000	Real	Drawing	Current 3D thickness
TILEMODE	1	Integer	Drawing	1 = Release 10 compatibility mode (uses Vports) 0 = Enables paper space and viewport entities (uses MVIEW)
TOOLTIPMERGE	0	Integer	User-Set'gs	Combines drafting tooltips into a single tooltip. The appearance of the merged tooltip is controlled by the settings in the Tooltip Appearance dialog box. 0 = Off 1 = On
TOOLTIPS	1	Integer	Registry	Controls the display of Tool Tips
TPSTATE	Varies	Integer	Not saved	Controls if Tool Palettes window is active or not 0 - Tools Palettes window not active 1 - Tools Palettes window active
TRACEWID	0.0500	Real	Drawing	Default trace width
TRACKPATH	0	Integer	Registry	Controls the display of polar and object snap tracking alignment paths
TRAYICONS	1	Integer	Registry	Controls if tray is displayed in the status bar 0 - Tray not displayed 1 - Tray displayed
TRAYNOTIFY	1	Integer	Registry	Controls if service notification is displayed in the status bar 0 - Notifications not displayed 1 - Notifications displayed
TRAYTIMEOUT	5	Integer	Registry	Controls time service notification is displayed

Variable	Default Setting	Type	Saved In	Explanation
TREEDEPTH	3020	Integer	Drawing	A 4-digit (maximum) code that specifies the number of times the tree-structured spatial index may divide into branches, hence affecting the speed in which AutoCAD searches the database before completing an action. The first two digits refer to the depth of the model space nodes, and the second two digits refer to the depth of paper space nodes. Use a positive setting for 3D drawings and a negative setting for 2D drawings
TREEMAX	10000000	Integer	Registry	Limits memory consumption during drawing regeneration
TRIMMODE	1	Integer	Registry	Controls whether AutoCAD trims selected edges for chamfers and fillets
TSPACEFAC	1	Real		Controls the multiline text line spacing distance measured as a factor of text height
TSPACETYPE	1	Integer		Controls the type of line spacing used in multiline text
TSTACKALIGN	1	Integer	Drawing	Controls the vertical alignment of stacked text
TSTACKSIZE	70	Integer	Drawing	Controls the percentage of stacked text fraction height relative to selected text's current height
UCSAXISANG	90	Integer	Registry	Stores the default angle when rotating the UCS around one of its axes using the X, Y, or Z options of the UCS command
UCSBASE	"World"	String	Drawing	Stores the name of the UCS that defines the origin and orientation of orthographic UCS settings
UCSFOLLOW	0	Integer	Drawing	The setting is maintained separately for both spaces and can be accessed in either space, but the setting is ignored while in paper space (it is always treated as if set to 0)
UCSICON	3	Integer	Drawing	The coordinate system icon bitcode for the current viewport (sum of the following): 1 = On — icon display enabled 2 = Origin — if icon display is enabled, the icon floats to the UCS origin if possible 3 = On and displayed at origin
UCSNAME	" "	String	Drawing	Name of the current coordinate system for the current space; returns a null string if the current UCS is unnamed (read-only)
UCSORG	0.0000, 0.0000, 0.0000,	3D point	Drawing	The origin point of the current coordinate system for the currentspace; this value is always returned in World coordinates (read-only)
UCSORTHO	1	Integer	Registry	Determines whether the related orthographicUCS setting is restored automatically when an orthographic view is restored
UCSVIEW	1	Integer	Registry	Determines whether the current UCS is saved with a named view
UCSVP	1	Integer	Drawing	Determines whether the UCS in active viewports remains fixed or changes to reflect the UCS of the currently active viewport
UCSXDIR	1.0000, 0.0000, 0.0000,	3D point	Drawing	The X direction of the current UCS for the current space (read-only)

Variable	Default Setting	Type	Saved In	Explanation
UCSYDIR	0.0000, 1.0000, 0.0000,	3D point	Drawing	The Y direction of the current UCS for the current space (read-only)
UNDOCTL	1	Integer		A (read-only) code indicating the state of the UNDO feature; it is the sum of the following values: 0 = undo is turned off 1 = Set if UNDO is enabled 2 = Set if only one command can be undone 4 = Set if Auto-group mode is enabled 8 = Set if a group is currently active
UNDOMARKS		Integer	Not Saved	Stores number of UNDO marks
UNITMODE	0	Integer	Drawing	0 = Displays fractional, feet and inches, and surveyor's angles as previously 1 = Displays fractional, feet and inches, and surveyor's angles in input format
UPDATETHUMBNAIL	15	Integer	Drawing	Controls updating of the thumbnail previews in the Sheet Set Manager 0 = Does not update thumbnail previews for sheet views, model space views, or sheets 1 = Updates model view thumbnail previews 2 = Updates sheet view thumbnail previews 4 = Updates sheet thumbnail previews 8 = Updates thumbnail previews when sheets or views are created, modified, or restored 16 = Updates thumbnail previews when the drawing is saved
USERI1-5	0	Integer	Drawing	Saves and recalls integer values
USERR1-5	0.0000	Real	Drawing	Saves and recalls real numbers
USERS1-5	" "	String		Saves and recalls text string data
VIEWCTR		3D point	Drawing	Center of view in current viewport, expressed in UCS coordinates (read-only)
VIEWDIR	3D Vector	3D point	Drawing	The current viewport's viewing direction expressed in World coordinates; this describes the camera point as a 3D offset from the TARGET point (read-only)
VIEWMODE	0	Integer	Drawing	Viewing mode bitcode for the current viewport (read-only); the value is the sum of the following: 0 = turned off 1 = Perspective view active 2 = Front clipping on 4 = Back clipping on 8 = UCS follow mode 16 = Front clip not at eye. If On, the front clip distance (FRONTZ) determines the front clipping plane. If Off, FRONZ is ignored and the front clipping is set to pass through the camera point (i.e., vectors behind the camera are not displayed). This flag is ignored if the front clipping bit (2) is off
VIEWSIZE		Real	Drawing	Height of view in current viewport, expressed in drawing units (read-only)

Variable	Default Setting	Type	Saved In	Explanation
VIEWTWIST	0	Real	Drawing	View twist angle for the current viewport (read-only)
VISRETAIN	1	Integer	Drawing	If = 0, the current drawing's On/Off, Freeze/Thaw, color, and linetype settings for Xref-dependent layers take precedence over the Xref's layer definition; if = 1, these settings don't take precedence
VPMAXIMIZEDSTATE	0	Integer	Not-saved	Stores a value that indicates whether the viewport is maximized. The maximized viewport state is canceled if you start the PLOT command. 0 = Not maximized 1 = Maximized
VSMAX		3D point		The upper right corner of the current viewport's virtual screen, expressed in UCS coordinates (read-only)
VSMIN		3D point		The lower left corner of the current viewport's virtual screen, expressed in UCS coordinates (read-only)
vtenable	3	Integer	Not-saved	Controls when smooth view transitions are used. Smooth view transitions can be on or off for panning and zooming, for changes of view angle, or for scripts. The valid range is 0 to 7. Setting / For pan/zoom / For rotation / For scripts 0 / Off / Off / Off 1 / On / Off / Off 2 / Off / On / Off 3 / On / On / Off 4 / Off / Off / On 5 / On / Off / On 6 / Off / On / On 7 / On / On / On
VTDURATION	750	Integer	Registry	Sets the duration of a smooth view transition, in milliseconds. The valid range is 0 to 5000
VTFPS	7.0	Real	Registry	Sets the minimum speed of a smooth view transition, in frames per second. When a smooth view transition cannot maintain this speed, an instant transition is used. The valid range is 1.0 to 30.0
WHIPARC	0	Integer	Registry	Controls whether the display of circles and arcs is smooth
WHIPTHREAD		Integer		Controls whether to use an additional processor
WMFBKGND		Integer		Controls whether the background display of AutoCAD objects is transparent in other applications
WMFFOREGND		Integer		Controls the assignment of the foreground color of AutoCAD objects in other applications
WINDOWAREACOLOR	5 (Blue)	Integer	Registry	Controls the color of the transparent selection area during window selection. The valid range is 1 to 255.
WORLDUCS	1	Integer		If = 1, the current UCS is the same as the WCS; if = 0, it is not (read-only)
WORLDVIEW	1	Integer	Drawing	Dview and Vpoint command input is relative to the current UCS. If this variable is set to 1, the current UCS is changed to the WCS for the duration of a DVIEW or VPOINT command. Default value = 1

Variable	Default Setting	Type	Saved In	Explanation
WRITESTAT	1	Integer	Read-only	Indicates whether a drawing file is read-only or can be written to, for developers who need to determine write status through AutoLISP
WSCURRENT	Acad default	String	Not-saved	Returns the current workspace name in the command line interface and sets a workspace to current.
XCLIPFRAME	0	Integer	Drawing	Sets visibilty of Xref clipping boundaries
XEDIT	1	Integer	Drawing	Controls whether the current drawing can be edited in-place when being referenced by another drawing
XFADECTL	50	Integer	Registry	Controls the fading intensity for references being edited in-place
XLOADCTL	1	Integer	Registry	Turns XREF demand loading on/off
XLOADPATH				Creates a path for storing temporary copies of demand-loaded Xref files
XREFCTL	0	Integer	Config	Controls whether .xlg files (external reference log files) are written 0 = Xref log (.xlg) files not written 1 = Xref log (.xlg) files written
XREFNOTIFY	2	Integer	Registry	Controls the notification for updated or missing xrefs. 0 = Disables xref notification 1 = Enables xref notification. Notifies you that xrefs are attached to the current drawing by displaying the xref icon in the lower-right corner of the application window (the notification area of the status bar tray). When you open a drawing, alerts you to missing xrefs by displaying the xref icon with a yellow alert symbol (!). 2 = Enables xref notification and balloon messages. Displays the xref icon as in 1 above. Also displays balloon messages in the same area when xrefs are modified. The number of minutes between checking for modified xrefs is controlled by the system registry variable XNOTIFYTIME.
XREFTYPE	0	Integer	Registry	Controls the default reference type when attaching or overlaying an external reference. 0 = Attachment is the default 1 = Overlay is the default
ZOOMFACTOR	10	Integer	Registry	Controls the incremental change in zoom with each IntelliMouse wheel action, whether forward or backward

APPENDIX e

Hatch and Fill Patterns

INTRODUCTION

AutoCAD supports two types of hatch patterns: vector patterns and PostScript fill patterns. Vector patterns are made of straight lines and dots; they are defined in the *acad.pat* pattern file. You can create custom hatch patterns or purchase patterns created by third-party vendors. You place a hatch pattern with the HATCH and BHATCH commands.

PostScript fill patterns are made via PostScript PDL (page description language); they are defined in the *acad.psf* file. To create a custom fill pattern, you need to know PostScript programming. You place a fill pattern with the PSFILL command.

Over 60 hatch patterns and 12 PostScript fills are supplied with the AutoCAD package. Some are shown on the following pages.

HATCH PATTERNS

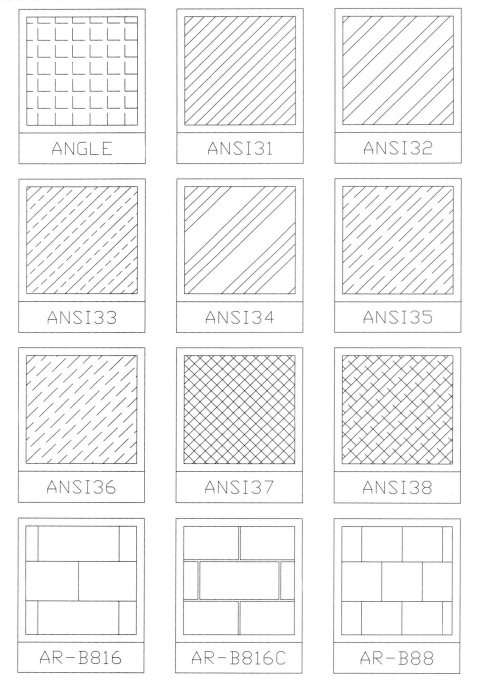

Appendix E • *Hatch and Fill Patterns* 1111

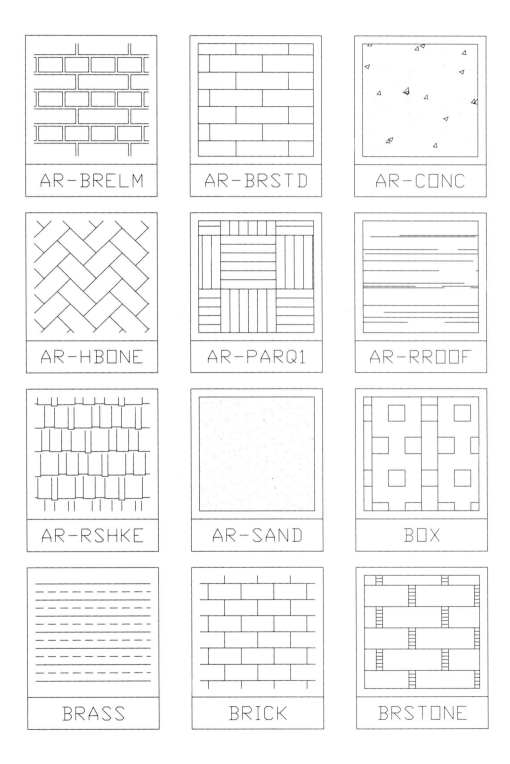

Appendix E • *Hatch and Fill Patterns*

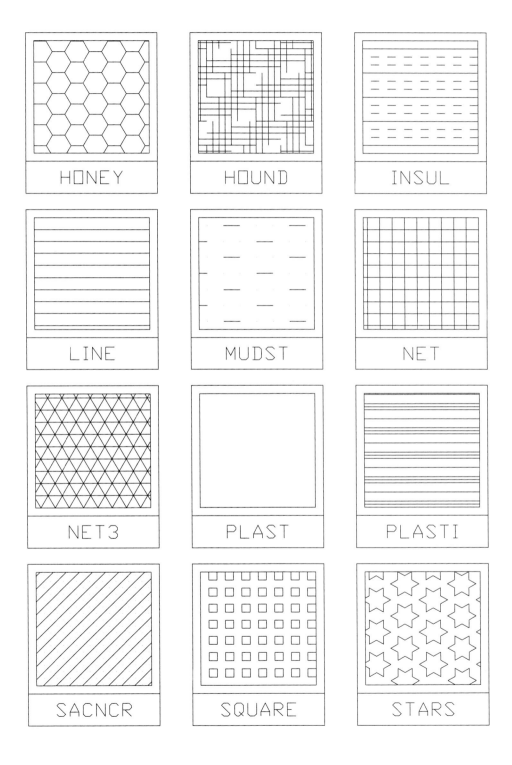

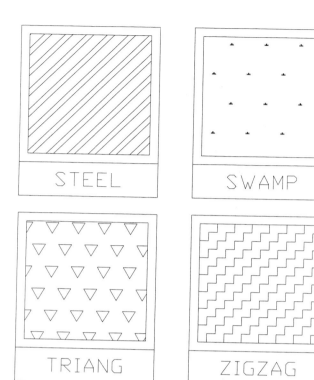
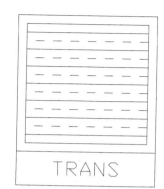

POSTSCRIPT FILL PATERNS

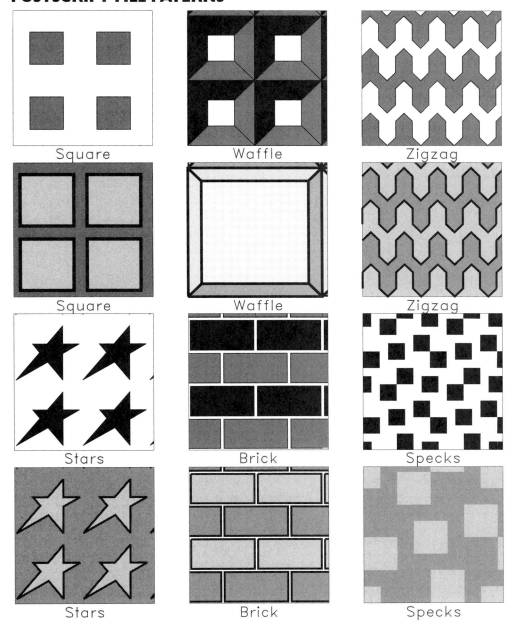

APPENDIX F

Fonts

INTRODUCTION

AutoCAD works with two types of text fonts: the original SHX-format font files and PFB PostScript font files. The AutoCAD package includes 17 SHX text fonts, five SHX symbol fonts, and 16 PFB text fonts. Some of the available fonts are shown on the following pages.

USING SHX AND PFB FILES

Other SHX font files are available from third-party developers. In addition, AutoCAD can use PostScript fonts from any source, many of which are included free with other software packages. Postscript fonts are usually stored in the *Psfonts* folder.

AutoCAD does not store text fonts in the drawing file. Instead, the DWG file references SHX font definition files stored elsewhere on the hard drive. Thus, if you receive a drawing from another AutoCAD system, you might have to tell AutoCAD where to find the font files on your system.

PostScript fonts placed in an AutoCAD drawing have two anomilies: the fonts are unfilled and they are drawn 30% too small. To compensate for the reduced size, specify a text height 50% larger.

STANDARD TEXT FONTS

FAST FONTS

TXT	ABCDEFGHIJKLMNOPQRSTUVWXYZ 1234567890
MONOTXT	ABCDEFGHIJKLMNOPQRSTUVWXYZ 1234567890

SIMPLEX FONTS

ROMANS	ABCDEFGHIJKLMNOPQRSTUVWXYZ 1234567890
SCRIPTS	ABCDEFGHIJKLMNOPQRSTUVWXYZ 1234567890
GREEKS	ABXΔEΦΓHIϑKΛMNOΠΘPΣTΥΝΩΞΨZ 1234567890

DUPLEX FONTS

ROMAND	ABCDEFGHIJKLMNOPQRSTUVWXYZ 1234567890

COMPLEX FONTS

ROMANC	ABCDEFGHIJKLMNOPQRSTUVWXYZ 1234567890
ITALICC	ABCDEFGHIJKLMNOPQRSTUVWXYZ 1234567890
SCRIPTC	ABCDEFGHIJKLMNOPQRSTUVWXYZ 1234567890
GREEKC	ABXΔEΦΓHIϑKΛMNOΠΘPΣTΥΝΩΞΨZ 1234567890

TRIPLEX FONTS

ROMANT	ABCDEFGHIJKLMNOPQRSTUVWXYZ 1234567890
ITALICT	ABCDEFGHIJKLMNOPQRSTUVWXYZ 1234567890

STANDARD TEXT FONTS

GOTHIC FONTS

GOTHICE ABCDEFGHIJKLMNOPQRSTUVWXYZ 1234567890

GOTHICG ABCDEFGHIJKLMNOPQRSTUVWXYZ 1234567890

GOTHICI ABCDEFGHIJKLMNOPQRSTUVWXYZ 1234567890

SYMBOL FONTS

A B C D E F G H I J K L M N O P Q R S T U V W X Y Z
a b c d e f g h i j k l m n o p q r s t u v w x y z

SYASTRO

SYMAP

SYMATH

SYMETEO

SYMUSIC

STANDARD TEXT FONTS

POSTSCRIPT FONTS

CIBT.PFB	ABCDEFGHIJKLMNOPQRSTUVWXYZ 1234567890
COBT.PFB	ABCDEFGHIJKLMNOPQRSTUVWXYZ 1234567890
EUR.PFB	ABCDEFGHIJKLMNOPQRSTUVWXYZ 1234567890
EURO.PFB	ABCDEFGHIJKLMNOPQRSTUVWXYZ 1234567890
PAR.PFB	ABCDEFGHIJKLMNOPQRSTUVWXYZ 1234567890
ROM.PFB	ABCDEFGHIJKLMNOPQRSTUVWXYZ 1234567890
ROMB.PFB	ABCDEFGHIJKLMNOPQRSTUVWXYZ 1234567890
ROMI.PFB	ABCDEFGHIJKLMNOPQRSTUVWXYZ 1234567890
SAS.PFB	ABCDEFGHIJKLMNOPQRSTUVWXYZ 1234567890
SASB.PFB	ABCDEFGHIJKLMNOPQRSTUVWXYZ 1234567890
SASBO.PFB	ABCDEFGHIJKLMNOPQRSTUVWXYZ 1234567890
SASO.PFB	ABCDEFGHIJKLMNOPQRSTUVWXYZ 1234567890
SUF.PFB	ABCDEFGHIJKLMNOPQRSTUVWXYZ 1234567890
TE.PFB	ABCDEFGHIJKLMNOPQRSTUVWXYZ 1234567890
TEB.PFB	ABCDEFGHIJKLMNOPQRSTUVWXYZ 1234567890

CYRILLIC FONTS

CYRILLIC	АБВГДЕЖЗИЙКЛМНОПРСТУФХЦЧШЩ 1234567890
CYRILTLC	АБЧДЕФГХИЩКЛМНОПЦРСТУВШЖЙЗ 1234567890

APPENDIX g

Linetypes and Lineweights

INTRODUCTION

Linetypes are defined by the *acad.lin* file. In addition to the continuous linetype, the AutoCAD program comes with the 45 linetypes. Some of them are listed on the next page. You can add custom linetypes to the *acad.lin* file.

Before you can use a linetype in a drawing, it must be loaded by means of the LINETYPE command. Set the linetype scale with the LTSCALE command; set independent linetype scaling in paper space with the PSLTSCALE system variable; control the generation of linetype along a polyline with the PLINEGEN system variable. The following tables list the standard lineweights and linetypes provided by AutoCAD.

AutoCAD Standard Lineweights

mm	inch		ISO
0.00			
0.05	.002		
0.09	.003		
0.13	.005		
0.15	.006		
0.18	.007		X
0.20	.008		
0.25	.010		X
0.30	.012		
0.35	.014		X
0.40	.016		
0.50	.020		X
0.53	.021		
0.60	.024		
0.70	.028		X
0.80	.031		
0.90	.035		
1.00	.039		X
1.06	.042		
1.20	.047		
1.40	.056		X
1.58	.062		
2.00	.078		X
2.11	.083		

STANDARD LINETYPES

BORDER	—— —— — — —— —— — — —— ——
BORDER2	— — — — — — — — — — — — —
BORDERX2	———— ———— ———— ————
CENTER	———— —— ———— —— ————
CENTER2	—— - —— - —— - —— - ——
CENTERX2	———————— ———— ————————
DASHDOT	—— · —— · —— · —— · —— ·
DASHDOT2	— · — · — · — · — · — · —
DASHDOTX2	———————— · ———————— ·
DASHED	—— —— —— —— —— —— —— ——
DASHED2	— — — — — — — — — — — —
DASHEDX2	———— ———— ———— ————
DIVIDE	—— · · —— · · —— · · ——
DIVIDE2	— · · — · · — · · — · · —
DIVIDEX2	———— · · ———— · · ————
DOT	· · · · · · · · · · · · · · ·
DOT2	· · · · · · · · · · · · · · ·
DOTX2	· · · · · · · · · · · · · · ·
HIDDEN	— — — — — — — — — — — —
HIDDEN2	- - - - - - - - - - - - - - -
HIDDENX2	—— —— —— —— —— —— ——
PHANTOM	———— —— —— ———— —— ——
PHANTOM2	——— — — ——— — — ———
PHANTOMX2	———————— ———— ———— ————

APPENDIX h

Command Aliases

ALIAS NAME	COMMAND NAME
3A	3DARRAY
3DO	3DORBIT
3F	3DFACE
3P	3DPOLY
A	ARC
ADC	ADCENTER
AA	AREA
AL	ALIGN
AP	APPLOAD
AR	ARRAY
ATT	ATTDEF
-ATT	-ATTDEF
ATE	ATTEDIT
-ATE	-ATTEDIT
ATTE	-ATTEDIT
B	BLOCK
-B	-BLOCK
BH	BHATCH
BO	BOUNDARY
-BO	-BOUNDARY
BR	BREAK
C	CIRCLE
CH	PROPERTIES

ALIAS NAME	COMMAND NAME
-CH	CHANGE
CHA	CHAMFER
COL	COLOR
COLOUR	COLOR
CO	COPY
D	DIMSTYLE
DAL	DIMALIGNED
DAN	DIMANGULAR
DBA	DIMBASELINE
DBC	DBCONNECT
DCE	DIMCENTER
DCO	DIMCONTINUE
DDI	DIMDIAMETER
DED	DIMEDIT
DI	DIST
DIV	DIVIDE
DLI	DIMLINEAR
DO	DONUT
DOR	DIMORDINATE
DOV	DIMOVERRIDE
DR	DRAWORDER
DRA	DIMRADIUS
DS	DSETTINGS
DST	DIMSTYLE
DT	DTEXT
DV	DVIEW
E	ERASE
ED	DDEDIT
EL	ELLIPSE
EX	EXTEND
EXIT	QUIT

Appendix H • *Command Aliases*

ALIAS NAME	COMMAND NAME
EXP	EXPORT
EXT	EXTRUDE
F	FILLET
FI	FILTER
G	GROUP
-G	-GROUP
GR	DDGRIPS
H	BHATCH
-H	HATCH
HE	HATCHEDIT
HI	HIDE
I	INSERT
-I	-INSERT
IAD	IMAGEADJUST
IAT	IMAGEATTACH
ICL	IMAGECLIP
IM	IMAGE
-IM	-IMAGE
IMP	IMPORT
IN	INTERSECT
INF	INTERFERE
IO	INSERTOBJ
L	LINE
LA	LAYER
-LA	-LAYER
LE	QLEADER
LEN	LENGTHEN
LI	LIST
LINEWEIGHT	LWEIGHT
LO	-LAYOUT
LS	LIST

ALIAS NAME	COMMAND NAME
LT	LINETYPE
-LT	-LINETYPE
LTYPE	LINETYPE
-LTYPE	-LINETYPE
LTS	LTSCALE
LW	LWEIGHT
M	MOVE
MA	MATCHPROP
ME	MEASURE
MI	MIRROR
ML	MLINE
MO	PROPERTIES
MS	MSPACE
MT	MTEXT
MV	MVIEW
O	OFFSET
OP	OPTIONS
ORBIT	3DORBIT
OS	OSNAP
-OS	-OSNAP
P	PAN
-P	-PAN
PA	PASTESPEC
PARTIALOPEN	-PARTIALOPEN
PE	PEDIT
PL	PLINE
PO	POINT
POL	POLYGON
PR	OPTIONS
PRCLOSE	PROPERTIESCLOSE
PROPS	PROPERTIES

ALIAS NAME	COMMAND NAME
PRE	PREVIEW
PRINT	PLOT
PS	PSPACE
PU	PURGE
R	REDRAW
RA	REDRAWALL
RE	REGEN
REA	REGENALL
REC	RECTANGLE
REG	REGION
REN	RENAME
-REN	-RENAME
REV	REVOLVE
RM	DDRMODES
RO	ROTATE
RPR	RPREF
RR	RENDER
S	STRETCH
SC	SCALE
SCR	SCRIPT
SE	DSETTINGS
SEC	SECTION
SET	SETVAR
SHA	SHADE
SL	SLICE
SN	SNAP
SO	SOLID
SP	SPELL
SPL	SPLINE
SPE	SPLINEDIT
ST	STYLE

ALIAS NAME	COMMAND NAME
SU	SUBTRACT
T	MTEXT
-T	-MTEXT
TA	TABLET
TH	THICKNESS
TI	TILEMODE
TO	TOOLBAR
TOL	TOLERANCE
TOR	TORUS
TR	TRIM
UC	DDUCS
UCP	DDUCSP
UN	UNITS
-UN	-UNITS
UNI	UNION
V	VIEW
-V	-VIEW
VP	DDVPOINT
-VP	VPOINT
W	WBLOCK
-W	-WBLOCK
WE	WEDGE
X	EXPLODE
XA	XATTACH
XB	XBIND
-XB	-XBIND
XC	XCLIP
XL	XLINE
XR	XREF
-XR	-XREF
Z	ZOOM

APPENDIX i

Express Tools

The following commands are included with the Express Tools package:

Command	Description
AliasEdit	Edits the aliases stored in *acad.pgp*.
AlignSpace	Aligns model space objects, whether in different viewports or with objects in paper space.
ArcText	Places text along an arc.
AttIn	Imports attribute data.
AttOut	Quickly extracts attributes in tab-delimited format.
BExtend	Extends open objects to objects in blocks and xrefs.
BlockReplace	Replaces all inserts of one block with another.
BlockToXref	Convert blocks to xrefs.
BreakLine	Creates the break-line symbol.
BScale	Scales blocks from their insertion points.
BTrim	Trims to objects nested in blocks and external references.
Burst	Explodes blocks, converts attributes to text.
ChSpace	Moves objects between model and paper space.
ChUrls	**DdEdit**-like editor for hyperlinks (URL addresses).
ClipIt	Adds arcs, circles, and polylines to the **XClip** command.
CopyM	**Copy** command with repeat, divide, measure, and array options.
CopyToLayer	Copies objects to other layers.
DimEx	Exports dimension styles to an ASCII file.

DimIm	Imports dimension style files created with **DimEx**.
EtBug	Sends bug reports to Autodesk.
ExOffset	Adds options to the **Offset** command.
ExPlan	Adds options to the **Plan** command.
FS	Selects objects that touch the selected object.
FullScreen	Toggles between full-screen and regular window.
GetSel	Selects objects based on layer and type.
ImageApp	Specifies the external image editor.
ImageEdit	Launches the image editor to edit selected images.
LayCur	Changes the layer of selected objects to the current layer.
LayDel	Deletes layers from drawings — permanently.
LayFrz	Freezes the layers of selected objects.
LayIso	Isolates layers of selected objects (all other layers are frozen).
LayLck	Locks the layers of selected objects.
LayMch	Changes the layer of selected objects to that of a selected object.
LayMrg	Merges two layers; removes the first layer from the drawing.
LayOff	Turns off layers of selected objects.
LayOn	Turns on all layers.
LayoutMerge	Places objects from layouts onto one layout.
LayThw	Thaws all layers.
LayUlk	Unlocks layer of selected object.
LayVpi	Isolates object's layer in viewport.
LayWalk	Isolates each layer in sequential order.
LMan	Saves and restores layer settings.
Lsp	AutoLISP function searching utility.
LspSurf	LISP file viewer.
MkLtype	Creates linetypes from selected objects.
MkShape	Creates shapes from selected objects.

Command	Description
MoCoRo	Moves, copies, rotates, and scales objects.
MoveBak	Moves *.bak* files to specified folders.
MStretch	Stretches with multiple selection windows.
NCopy	Copies objects nested inside blocks and xrefs.
Plt2Dwg	Imports HPGL files into the drawings.
Propulate	Updates, lists, and clears drawing properties.
PsBScale	Sets and updates the scale of blocks relative to paper space.
PsTScale	Sets text height relative to paper space.
QlAttach	Associate leaders to annotation objects.
QlAttach	Set Associates leaders with annotations.
QlDetach	Set Dissassociates leaders from annotations.
QQuit	Closes all drawings, and then exits AutoCAD.
ReDir	Changes paths for xrefs, images, shapes, and fonts.
RepUrls	Replaces hyperlinks.
Revert	Closes the drawing, and re-opens the original.
RText	Inserts and edits remote text objects.
RtUcs	Changes UCSs in real time.
SaveAll	Saves all drawings.
ShowUrls	Lists URLs in a dialog box.
Shp2blk	Converts from a shape definition to a block definition.
SuperHatch	Uses images, blocks, external references, or wipeouts as hatch patterns.
SysvDlg	Launches an editor for system variables.
TCase	Changes text between Sentence, lower, UPPER, Title, and toggle CASE.
TCircle	Surrounds text and multiline text with circles, slots, and rectangles.
TCount	Prefixes text with sequential numbers.
TextFit	Fits text between points.
TextMask	Places masks behind selected text.
TextUnmask	Removes masks from behind text.

TFrames	Toggles the frames surrounding images and wipeouts.
TJust	Justifies text created with the **MText** and **AttDef** commands.
TOrient	Re-orients text, multiline text, and block attributes.
TScale	Scales text, multiline text, attributes, and attribute definitions.
TSpaceInvaders	Finds and selects text with overlapping objects.
Txt2Mtxt	Converts single-line to multiline text.
TxtExp	Explodes selected text into polylines.
VpScale	Lists the scale of selected viewports.
VpSynch	Synchronizes viewports with a master viewport.
XData	Attaches xdata to objects.
XdList	Lists xdata attached to objects.
XList	Displays properties of objects nested in blocks and xref.

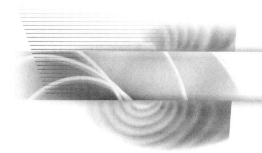

INDEX

Note: AutoCAD command names are in SMALL CAPS; command options are in italics; AutoLISP functions are in (parentheses).

∅ (diameter symbol)
  control characters for, 202
  tolerances with, 366, 367
! (exclamation point), expressions with, 991, 992, 999
" (double quote)
  expressions with, 991, 1000, 1001
  unit input with, 36–37
(pound sign)
  coordinate entry with, 73, 133
  as wild card, 166
$ (dollar sign)
  layer names with, 150, 597, 598
  menu macros with, 931
  xref objects with, 603
%% (percent signs), control characters with, 201–202
' (apostrophe)
  commands with, 19, 26
  unit input with, 36–37
( ) (parentheses), expressions in, 977, 990, 991, 992, 1022–1023
* (asterisk)
  block definition with, 732
  commands with, 929–930
  coordinates with, 72
  hatch names with, 932
  layout/page setups with, 438–439
  linetype definition with, 951
  menu macros with, 931
  as operator, 990, 994, 1002
  slide names with, 736
  as wild card, 166, 167
VARIES properties, 314
+ (plus sign) as operator, 990, 991, 1002
, (comma), coordinate entry with, 71, 72, 132

- (hyphen)
  commands with, 20, 930
  layer names with, 150
  unit input with, 36
-XREF, 605
... (ellipsis)
  arguments with, 1002
  menu commands with, 15, 20
. (period)
  expressions with, 993, 1032
  filtering points with, 683, 684
  SKETCH, 254, 255
  as wild card, 166
/ (forward slash)
  directory path with, 977, 1026
  as operator, 993–994, 997, 1002
  stacking text with, 198
  test expressions with, 1018
; (semicolon)
  comments following, 933, 1024
  menu macros with, 931
</<= test expressions, 1018
< > (angle brackets)
  arguments in, 1002
  plotter settings in, 458
<bits>, (initget), 1016
< (left angle bracket), coordinates with, 72, 133
= (equal sign)
  block redefinition with, 734
  calculations with, 667
  menu macros with, 931
  as test expression, 1018
>/>= test expressions, 1018, 1019
? (option)
  BLOCK/-XREF, 605

1133

UCS, 804
UCS, SOLVIEW, 877
WORKSPACE, 737
? (question mark)
   help via, 26, 27
   as wild card, 166, 167
@ ("at" symbol)
   control characters for, 202
   coordinates with, 71, 72, 73, 133
   filters with, 683–685
   split points with, 241
   as wild card, 166
[ ] (brackets)
   arguments in, 1002
   hatch definition with, 933
   wild card patterns with, 166, 167
\ (backslash)
   menu macros with, 931
   as operator, 1002, 1026
^ (caret), stacking text with, 198
_ (underscore)
   layer names with, 150
   menu macros with, 931
| (vertical bar), layer names with, 590, 597, 598
~ (tilde) as wild card, 166, 167
' (single/reverse quote)
   expressions with, 991
   as wild card, 166
− (minus sign) as operator, 994
↓ (down arrow), on-screen options via, 16, 58
± (plus/minus symbol)
   control characters for, 202
   tolerances with, 363, 364
° (degree symbol), control characters for, 202
(1+)/(1−) functions, 1002
2D environment
   3D vs., 791–792
   plotting in, 404
   viewport setup for, 146, 149, 426
2D objects. see objects
2D UCS icons, 794–795, 797
2D WIREFRAME SHADEMODE, 888

2POINTS, ROTATE3D, 838–839
3DARRAY, 840–841
3DCLIP, 821
3D environment, 791–793
   coordinate systems for, 67–70, 793–808
   creating objects in, 822–828
   dimensions for, 333–335
   editing in, 837–842, 856–873
   generating profiles in, 878–879
   hiding objects in, 875
   mass properties in, 874–875
   placing/generating views in, 875–878
   plotting in, 404
   viewing in, 808–822
   viewport setup for, 146, 147, 149, 426
3DFACE, 825–828
3D Graphics Display settings, 715–716
3DMESH, 829–830
3D objects. see models; objects, 3D; solids
3DORBIT, 808, 818–821
3DORBIT toolbar, 819
3DPAN/3DZOOM, 820
3D points
   filters for, 683
   functions for listing, 1010
3DPOLY, 825
.3ds files, exporting/importing, 726
3D UCS icons, 794–795, 797
3D WIREFRAME SHADEMODE, 888
3POINTS
   MIRROR3D, 840
   SECTION, 859
3POINT UCS, 799–800

A

(abs), 1003
absolute coordinates
   overrides for, 73, 133
   Pointer Input settings for, 133
   rectangular, 70–71
absolute hyperlinks, 492, 749
Absolute Polar Angle measurement, 125, 126
acad.dwt, system variables in, 47

acad.pgp, editing, 929–930
actions, dynamic block
   adding, 522, 523–525
      Array, 566–568
      Authoring palette for, 530
      Flip, 565–566
      Lookup, 568–575
      Move, 543–548
      Parameter Sets for, 527, 579–580
      parameters with, 526–527, 531–532
      Polar Stretch, 562
      properties of, 542–543
      Rotate, 563–565
      Scale, 548–553
      Stretch, 553–561
      using, 525–526, 543
Activate AutoCAD, 976–977, 988
Adaptive linetype scale, 451
ADCCLOSE, 645
Add
   to group, 316, 317–318
   multiline elements, 294
   Parameter Properties, 574–575
   Pick points, hatch boundary, 472–474, 478
   Plot Style, 450
   Select objects, hatch boundary, 472, 474–476
   Standards File, 728
   Vertex, multiline, 286, 290
   Watch, debug operation, 1040
   to Working set, 604, 605
ADD
   AREA, 321–322
   TO FAVORITES, 629
   FIT DATA, SPLINEDIT, 300
   to selection set, 259
   SEPARATOR/TEXT, tool palette, 655, 656
Add-A-Plotter wizard, 455–456
Add Plot Style Table wizard, 408, 446–448
ADJUST CLIPPING PLANES, 3DORBIT, 821
ADS (AutoCAD Development System), 1040
Advanced Options, plot stamp, 412–414
Advanced Setup wizard, 40, 43–45

Aerial View, 144–145
affine tablet transformation, 965, 966
Aggregators, DC Online, 640–641
AIDIMFLIPARROW, 372
Alert icon, parameter, 523, 534
algebraic expressions, 990, 994, 997
aliases, command
   defining, 929–930
   menu, 922, 923, 924
   toolbar, 919
ALIGN
   MVSETUP, 690
   TEXT, 194–195
Align/aligning
   attribute tag, 504, 506
   cell text, 216
   dimension text, 381, 384
   objects (ALIGN), 837–838
ALIGNED, DIMLINEAR, 345
aligned dimensioning, 345–346
alignment parameter, 531, 538–539
ALL
   LOCK LOCATION, 12
   selection by, 258
   UCSICON, 796
   UNDO, 168
ALLOW DOCKING, Tool Palettes window, 655
Alternate Units, dimension
   style settings for, 388, 390
   tolerances for, 363, 364–365
altitude, distant light, 896, 897
ALT key, 19, 925
ambient light, 894, 895, 897–898
(and), 1019
ANGBASE system variable, 1014, 1015
ANGLE
   (ANG), XLINE, 178
   ARC, PLINE, 190
   CHAMFER, 236
   DIMLINEAR, 341
   DIMTEDIT, 371
   LENGTHEN, 263
(angle), 1005, 1014
angle direction, setting, 44–45

angle measurement
  base angle for, 35–36
  settings for, 43–44
angles
  3D viewing, 810–811
  arcs from, 80–82
  computing tangent of, 1026–1027
  constraining leader line, 361
  degree-radian conversions for, 1023–1025, 1027
  extrusion taper, 850–851
  gradient fill, 472
  hatch pattern, 468, 469, 470, 661
  multiline endcap, 295, 296
  polar array, 224
  polar coordinate, 72
  Polar Tracking, 124–127
  QuickCalc for, 666
  render smoothing, 889, 890
  rotation
    array, 222, 223
    block insertion, 494, 495
    object, 265, 266
    Snap, 102–103
    text, 192
  Scale action point, 549–550
(angtos), 1012
angular dimensioning, 351–354, 387, 389–390
angular units, setting
  UNITS for, 33, 34–35
  wizard for, 43–44
annotation. see also text
  attributes for, 498
  dimension, 336
  leader, 359–360
  plot scale for, 402–403
aperture box, Display AutoSnap, 113
APERTURE system variable, 324
Apparent Intersection Object Snap, 114, 116–117, 324
applications
  AutoLISP
    compiling, 982
    debugging, 1038–1040
    developing/loading, 974–977
    making, 982–986
  ObjectARX, options for, 710, 712
APPLOAD, 977
Apropos, VLISP, 989
ARC
  ELLIPSE, 184
  PLINE, 189–190
  Polygonal Viewport, 425
Arc-Arc Continuation ARC, 86–89
arcball, 3DORBIT, 819
architectural units
  dimensioning with, 387
  input format for, 36–37
  setting, 34, 41, 42
archiving
  sheet sets (ARCHIVE), 786
  xref files, 605
ARC LENGTH, REVCLOUD, 303
arc length dimensioning, 350–351, 378, 380
arcs
  continuing lines from, 57–58
  cross marks for, 368
  custom, codes for, 947–949
  dimensioning of
    angular, 351, 352
    arc length, 350–351
    baseline, 356
    diameter, 349–350
    linear, 344–345
    radius, 348–349
  drawing (ARC), 78–89
  multiline start/end cap, 294, 295–296
  offsetting, 226, 228–229
  polylines from, 269–270
  UCS origin/X axis for, 801
area
  calculating (AREA), 320–322
  selection, visual effects for, 256, 257
  set up drawing, 42–43, 45
AREA, RECTANGLE, 61–62
.arg (profile) files, 723–724
arguments, AutoLISP, 990, 991
  (command) and, 1000

for custom functions, 1022–1023
syntax for, 994, 1002
types of, 991, 997–998, 1012–1013
Array action, 531, 566–568
arrays
    3D, 840–841
    creating (ARRAY), 221–226
    viewport, 442, 443
arrowheads
    dimension line, 336, 337
        flip, 372, 373
        style settings for, 378–380, 384, 385
    leader line, 361
    solid, creating, 252
    types of, 379
(ascii), 1003
ASCII files/format
    for audit report, 695
    for scripts, 731, 734
    for xref log, 605
ASCII text font codes, 949, 950
Ask Me, AutoCAD 2006 Help, 28
ASPECT
    GRID, 108–109
    SNAP, 101–102
ASSIST, 25
(assoc), 999, 1032, 1034
association list, 1031–1034
    advanced functions for, 1036
    expressions in, 999
associative dimensions, 337–338
    disassociating/reassociating, 371–372
    exploding, 339, 496
    grip editing, 371
    hatching over, 475
    options for, 718, 720
associative fill, 465, 476
associative hatch, 260, 262, 476
(atan), 1004
(atoi)/(atof), 1012, 1013
Attach/attaching
    images, 607, 608, 645
    materials, 901, 902
    xref drawings, 588, 593–596, 645, 749

Attachment
    multiline leader text, 362
    xref type, 594, 595
ATTDEF, 498, 500, 503–506
ATTDISP, 500, 506
ATTEDIT, 500, 503, 506–507
attenuation, light, 895, 896, 897
ATTEXT, 498, 500–501, 508, 511
ATTMODE system variable, 506
ATTREDEF, 515
ATTREQ system variable, 506
Attribute Extraction wizard, 511–515
attributes, 498
    blocks with
        editing (REFEDIT), 602, 603–604
        exploding, 497
        inserting, 501–503, 506
        managing, 515–519
        redefining, 515, 732–735
    commands for, 500–503
    components of, 499
    controlling visibility of, 499, 500, 506
    defining, 498–499, 503–506
    editing, 205, 503, 506–508
    extracting, 500, 508–515
    hatching around, 475
    plotting, 499
AUDIT, 694–695
Audit parameter properties, 574
AUPREC system variable, 325
Authoring palette, block, 522, 528–530
    Actions in, 543
    Parameter Sets in, 527, 579–580
AUTO
    DRAGMODE, 674
    selection by, 259
    UNDO, 169
AutoCAD
    Activate, 976, 988
    AutoLISP and, 971, 989
    command input for, 18–25
    customizing (see customizing)
    data translation for, 998–999, 1012–1013
    exiting, 51

help for, 25–28, 628
running programs from, 976–977
screen of, 4–18
starting, 1–4
VLISP and, 971–972
AutoCAD Color Index (ACI), 155, 676, 901, 902
AutoCAD Command Reference (Help), 628
AutoCAD DesignCenter. see DesignCenter
AutoCAD Development System (ADS), 1040
AutoCAD 2006 Help window, 26–28
AUTOCAPS, MTEXT, 204
Autodesk
  AutoLISP/VLISP development by, 971, 989
  Communication Center of, 5–6, 745–746
  Design Web Format of, 753
  DWF Composer/Viewer of, 753–754, 758–759
  Plotter Manager of, 454, 455
  Reference Manager of, 606
  Web site of, 744
AUTO-HIDE Tool Palettes window, 655, 656
AutoLISP. see also VLISP
  ADS and, 1040
  applications in
    compiling, 982
    debugging, 1038–1040
    developing/loading, 974–977
    making, 982–986
  custom commands in, 1027
  customizing and, 952
  development of, 971, 989
  expressions/variables in, 989–990
  functions in (see functions)
  lists in, 997–1000
  syntax color codes for, 975
  terminology/concepts for, 991–994
  VLISP interface for, 989
AUTO-LIST, MTEXT, 199
Automatic save, 710, 711
Autoscaling blocks, 644
AutoSnap Settings, 112–113

AutoTrack Tooltip, Display, 125
Auto Viewport, Aerial View, 145
AUXILIARY, SOLVIEW, 877
axes, coordinate
  3D, 792–795
  orientation of, 65–66, 68–69
  rotating UCS around, 802–803
  specifying points on, 70, 71
axis end points, ellipses from, 182–184
axis of revolution, 851–852
azimuth, distant light, 896, 897

B

BACK, UNDO, 168, 169
background
  Display field, 718, 719
  opaque, 203
  plot in, 406, 409
  plot/publish in, 713, 714
  rendering, 889, 892–893
backup copy, Create, 710, 711
backup procedures, xref file, 605
Backwards text, 212
BACTION, 543
BACTIONCOLOR system variable, 543
Balloon Notification for updates, 745, 747
BASE, 497–498
BASE
  linear parameter, 535
  Scale action, 548–550
base angle, angle measurement, 35–36
BASE ANGLE, XY parameter, 538
baseline dimensioning, 354–356, 376, 377
base point
  alignment parameter, 539
  arraying from, 225
  block insertion, 490–491
  copying from, 220, 221
  drawing insertion, 497–498
  MIRROR grip mode, 309
  Move action, 547
  moving from, 237, 238, 308
  Rotate action, 563, 564

rotating from, 265–266, 308
scaling from, 268, 309
Stretch action, 559–560
stretching from, 265, 307
base point parameter, 531, 542
basic lateral tolerance, 364
BATTMAN, 515–519
BCYCLEORDER, 581–582
BEGIN UNDO, 169
BGRIPCOLOR/BGRIPOBSIZE system variables, 543
Bind
   dependent symbols, 591, 598–599
   xref drawings, 591, 593, 597–598
BISECT XLINE, 179
bitmap images
   rendering, 891, 892, 893
   viewing in DesignCenter, 635–636
BLIPMODE, 674
BLIPMODE system variable, 253
.blk (attribute extraction template) files, 513, 515
BLOCK, 487
   creating with, 488, 489–493
      GROUP vs., 315
Block Attribute Manager, 515–519
Block Definition dialog, 489–493, 522, 527
Block Editor
   creating dynamic blocks in, 522–525
   making blocks dynamic in, 527–530
   tool palettes option for, 654, 655
   Visibility States in, 577–579
Block Editor toolbar, 528, 529, 530
block references/blocks, 487. see also dynamic blocks
   associative dimensions and, 338
   with attributes (see also attributes)
      inserting, 501–503, 506
      managing, 515–519
      redefining, 515, 732–735
   callout, placing on sheets, 782–783
   CHANGE properties of, 315
   clipping, 599–601
   common uses of, 489
   creating/defining (BLOCK), 488, 489–493

   custom arrow, 379–380
   divide using, 520
   editing, 601–605
   exploding, 496–497
   grips on/in, 304, 305, 306
   hatching, 465, 475
   hyperlinks to, 752–753
   inserting, 493–495
      from DesignCenter, 644–645
      scale for, 35
      into table, 216
   leader annotation with, 360
   measure using, 521
   nested, 495–496, 497
   object properties for, 677, 680, 682
   purging, 670, 671
   tool palettes for, 12
      creating, 661–662
      inserting from, 14, 659
      properties of, 659–660
   Translate objects in, 693
   viewing in DesignCenter, 632–633, 634–635
Bold text, 198
Boolean operations
   for composite solids, 842, 852–856
   for filtering selection, 312
borders
   hide viewport, 433–434
   IMAGECLIP boundary, 611
   insert title block with, 688–689
   table cell, 216, 218
   XCLIP boundary, 601
   xref files for, 588
boundaries
   clipping
      image object, 610–611
      viewport, 430
      xref/block, 599–601
   hatch/gradient
      boundary set vs., 478–479
      defining, 465–467, 472–476
      Remove/Recreate, 475, 482–483
      Retain, 478

layout, 418, 429–430
view, 672, 673
wipeout, 301–302
boundary set, hatching, 478–479
bounding box, text, 195
adjusting width of, 202
editing in, 206
specifying corners of, 196–197
(boundp), 1019
bow-tie, solid, 251–252
BOX, selection by, 259
box, solid (BOX), 843–844
BPARAMETERCOLOR system variable, 542
BPARAMETERFONT system variable, 543
BPARAMETERSIZE system variable, 542
BREAK
objects, 240–241, 255
PICKFIRST and, 261
break point, set debug, 1039
BREAK VERTEX, PEDIT, 271
brightness, image, 611–612
"broken pencil" UCS icon, 795–796
Browse
for Folder, 625–626
for image file, 607, 608, 610
for template file, 46
BROWSER, 744–745
Browse the Web dialog, 748–749
BTMARKDISPLAY system variable, 543
Build Options, Project, 987
bulge-specified arc, code for, 948–949
bulleted lists, 199
button images, editing, 921–922
Buzzsaw.com quick access icon, 31, 49, 624
ByBlock/ByLayer settings
dimension/extension line, 376–377
dimension text color, 381, 382
object property, 676, 677, 678, 682
plot style, 453
ByLayer
attach materials, 901, 903
Force object color/linetype to, 693

C

cache
Model tab/layouts, 715, 716
render, 889, 890
(cadr), 997, 1008, 1010, 1034
CAD Standards
Configure/Check, 727–730
status bar icon for, 5, 6
CAD Standards toolbar, 691
CAL, TABLET, 960, 963
calculator, QuickCalc, 665–668
calibration, tablet, 960
procedure for, 963–964
table for, 966–967
transformation options for, 964–966
CALLIGRAPHY STYLE, REVCLOUD, 304
callout blocks, placing on sheets, 782–783
CAMERA, DVIEW, 813–814
caps, multiline, 294–296
CAPS LOCK, 204
(car), 997, 1006–1010
Cartesian coordinate system, 68–69
origin point for, 64
Pointer Input settings for, 132
specifying points for, 70–72
cascading submenus, 15
Category Listing, DC Online, 638–641
(cdr), 1006–1010, 1034
cells, table
editing text in, 215–216
properties of, 218, 219
CELTSCALE system variable, 680
CENTER
ARC, PLINE, 190
BOX, 844
DIMTEDIT, 371
ELLIPSE, 183–184
WEDGE, 849
Center, Start, Angle (C,S,A) ARC, 81
Center, Start, End (C,S,E) ARC, 80
Center, Start, Length (C,S,L) ARC, 85
Center-Diameter CIRCLE, 75–76
Centered gradient fill, 472

center marks, circle, 337
   drawing, 368
   style settings for, 378, 380
Center Object Snap, 114, 115–116, 324
center point, polar array, 223, 224
Center-Radius CIRCLE, 74–75
CENTER TEXT, 193
CFG, TABLET, 960–962
C++ files, color codes for, 975
CHAIN, point parameter, 533
CHAMFER, RECTANGLE, 61
chamfering (CHAMFER)
   between objects, 234–236
   object selection for, 255–256
   PICKFIRST and, 261
   solids, 856–858
Change
   Dictionaries, 209
   Group, 316, 317–319
CHANGE CASE, MTEXT, 204
CHANGE DESCRIPTION, layer, 159
CHANGE object properties, 149, 314–315
Character Map, 201
characters, special
   ASCII codes for, 949, 950
   Character Map for, 201
   control characters for, 201–202
   defining, 949
   menu macro, 930–931
   SKETCH option, 254–255
CHARACTER SET, MTEXT, 205
checking solids (CHECK), 873
Check Spelling dialog, 208–209
CHECKSTANDARDS, 729–730
chord length, arcs from, 83–85
(chr), 1003
Church, Alonzo, 971
circles
   center marks for, 337, 368, 378, 380
   CHANGE properties of, 315
   concentric, 226–229
   dimensioning of
      angular, 351, 352–353
      baseline, 356

      diameter, 349–350
      linear, 344
      radius, 348–349
   dividing, 519
   drawing (CIRCLE), 74–78
   grip locations on, 306
   isometric, ellipses from, 184–186
   polygons drawn in/about, 180–181
   solid-filled, 249–250
   UCS origin/X axis for, 801
CIRCUMSCRIBED ABOUT CIRCLE POLYGON, 181
cleaning solids (CLEAN), 873
CLEAR dimension overrides, 392
client/server in OLE, 695
CLIP, DVIEW, 818
CLIPDEPTH, XCLIP, 601
Clip existing viewport, 429–430
clipping
   image objects, 610–611
   xrefs/blocks, 599–601
clipping planes, adjusting, 821
CLOSE
   Block Editor, 525
   LINE, 58, 59
   MLINE, 284
   PEDIT, 270–271
   PLINE, 187
   Polygonal Viewport, 425
   SPLINE, 297, 298
   SPLINEDIT, 300
   Tool Palettes window, 658
   WIPEOUT polyline, 302
CLOSE/CLOSEALL drawings, 50
Close/closing
   Layer Properties Manager, 163–164
   Reference, 604, 605
   toolbars, 10–11
   Tool Palettes window, 650
Closed Cross multiline, 286, 287
Closed Tee multiline, 286, 288
Collections, DC Online, 643
COLOR, 675
Color Books
   for layers, 155, 156

for objects, 676
color codes, program syntax, 974–975
color-dependent plot styles, 445–446
    changing object/layer, 453
    creating/modifying tables for, 446–452
coloring
    edges (COLOR EDGES), 870–871
    faces (COLOR FACES), 868
colors
    action, 543
    AutoSnap marker, 113
    block object, 488, 495–496
    block/pattern tool, 659, 660, 661
    dimension/extension line, 376–377
    dimension text, 381, 382
    Display options for, 708, 709
    gradient fill, 464–465, 471–472
    grip, 304, 305, 543, 545
    hidden line, 720, 721
    layer, 150, 155–156, 591
    light, 895, 896, 897–898
    matching, 276
    Model/Layout tooltip, 135
    multiline, 294, 296
    object, 675–677, 693
    parameter, 542
    plot style, 450
    render background, 892
    table border, 218
    text, 196, 198
    UCS icon, 797
    Window/Crossing selection, 257
columns
    Array action, 567, 568
    rectangular array, 222–223
    table, 214, 215, 216
COMBINE, UNDO/REDO, 168
COMBINE PARAGRAPHS, MTEXT, 205
(command), 1000–1001
command history, 17, 19
command line, 17
COMMANDLINE, 16
Command List, CUI dialog, 912, 914, 918

"Command:" prompt
    AutoCAD, 17, 19
    OS command, 686–687
commands, 18
    2D, 3D operations with, 797–798, 823
    attribute, 500
    custom, 920–922
    custom AutoLISP, 1022–1023, 1027
    defining aliases for, 929–930
    dimensioning, 334, 339
    enhancers for, 118
    help for, 25–28, 628
    input methods for, 19–25
        from command window, 16–18
        with dynamic input, 15–16
        from pull-down menus, 14–15
        from toolbars, 7–8
        from tool palettes, 14
    Inquiry, 319–323, 693–694, 874–875
    invoking from AutoLISP, 1000–1001
    menu, adding/editing, 923, 924
    mouse button, adding, 927
    MULTIPLE modifier for, 671
    non-undoable, 169
    operating system, 686–687
    repeating, 18, 19, 56, 718–719
    right-click customization for, 718–719
    scripts for, 731–732
    shortcut key, assigning, 925–926
    tablet menu, assigning, 959–960
    terminating, 18
    toolbar, adding, 918–920
    transparent, 19, 26, 168, 674
    undoing/redoing, 167–170
    using in Block Editor, 528
    utility display, 671–674
Command window, 16–18
    entering commands in, 19
    shortcut menu for, 23
Communication Center, 5–6, 745–748
compass, VPOINT, 809–810
compiled programming languages, 952
compiling
    .shp (shape) files, 942

VLISP applications, 982
composite solids, 842
   creating, 852–856
   separating solids from, 871–872
compressed files
   DWF, 754, 757
   transmittal set, 768
concentric circles, 226–229
(cond), 1017–1018
conditional functions, 1017–1019
cone, solid (CONE), 844–845
Configure/configuration
   Communication Center, 746–748
   drawing environment, 2–3
   plotters, 406, 454–458, 713
   Standards, 728–729
   tablet, 958–959, 960–962
Connect (C), SKETCH, 254, 255
(cons), 1011
console window, VLISP, 973, 988
Constant attribute mode, 504, 517
construction lines, drawing, 177–179
container, concept of, 618, 619, 633
content, 618, 619
   adding to drawings, 643–645
   DC Online, 638–643
   searching for, 625–628, 641–642
   viewing, 630–635
content area, DesignCenter
   location of, 622, 623
   using, 630, 633–635
Contents, AutoCAD 2006 Help, 28
content type, 618, 619, 633
Continue
   debug operation, 1040
   LINE, 57–58
continue dimensioning, 356–357
CONTINUOUS ORBIT, 3DORBIT, 821
contrast, image, 611–612
CONTROL, UNDO, 168
control characters
   ASCII codes for, 949
   symbol, 201–202
control points, spline, 299, 301

Convert/converting
   to Block, 490, 491
   degrees to radians, 1023–1025, 1027
   distance between spaces, 191
   DWG hyperlinks to DWF, 751, 752
   Object to Viewport, 425–426
   units, QuickCalc for, 667
CONVERTPSTYLES, 446
coordinates
   3D, 792–793
   absolute rectangular, 70–71
   display of, 6, 73–74
   drawing limit, 37–39
   entry of, 5
      format for, 70–72
      locking values during, 16
      overrides for, 73, 133
      priority for, 718, 719–720
   filters for, 683–686
   functions for returning, 1010
   line end point, 56, 57
   locating point, 322–323
   plotting, 400–401
   Pointer Input settings for, 132–133
   QuickCalc for, 666
   relative rectangular/polar, 71–72
   viewpoint, 808–809
coordinate systems, 63–67. see also User
      Coordinate Systems; World
      Coordinate System
   3D, 793–808
      defining new, 797–804
      plan view for, 807–808
      predefined orthographic, 804–807
      right-hand rule for, 793–794
      UCS icon display for, 794–797
   types of, 67–70, 793
copies
   backup, 710, 711
   mirror, 229–231
   multiple, pattern of, 221–226, 840–841
COPY
   MTEXT, 203
   tool palette element, 654, 655

while grip editing, 307, 308, 309, 310
Copy/copying
    edges, 869–870
    faces, 864
    Object, leader annotation, 360
    object properties, 276–277
    objects (COPY), 219–221
COPYLINK, 696
Corner Joint multiline, 286, 290
(cos), 1004, 1026–1027
CPOLYGON (CP), selection by, 258
CRC (cyclic redundancy check), 710, 711
Create New Drawing dialog, 39–47
Create sheet selection, 779–780
Create Sheet Set wizard, 770–776
CREATE TOOL PALETTE OF BLOCKS,
    DesignCenter, 661–662
Create Transmittal dialog, 768–769, 785
CREATE VIEWPORTS, MVSETUP, 689–690
CROSSING, selection by, 90–91
    dynamic block, 525
    MULTIPLE modifier for, 258–260
    visual effects for, 257
cross marks, circle, 368
cross-sections, 859–860
CTRL key
    assigning commands to, 925
    cycling through grips with, 582
    cycling through objects with, 89
    setting mouse button as, 927
CUBE
    BOX, 843–844
    WEDGE, 849
current layer
    concept of, 149–150
    setting, 153, 154, 164
Current Plot Style dialog, 453–454
CURRENT UCS
    PLAN, 808
    SOLVIEW, 876–877
current viewport, 147
Current VP Freeze, 435
cursor
    3D Orbiting with, 819–820

display modes for, 5
dynamic input near, 15, 16, 131
CURSOR COORDINATE VALUES, 6
cursor menu, 20–21. see also shortcut menus
curves
    parallel, 226–229
    spline, 297–301
custom
    AutoLISP commands/functions, 990,
        1022–1027
    Drawing Properties, 664, 665
    hatch patterns, 469
    Properties
        DWF6 ePlot, 755–757
        sheet set, 773
customization, right-click, 718–719
Customizations in All CUI Files, CUI
    dialog, 912, 914
CUSTOMIZE
    toolbars, 8
    tool palettes, 655, 656–657
    user interface, 912
Customize User Interface dialog, 911
    displaying, 8
    tablet menus in, 959–960
    using, 912–929
customizing, 911–912
    AutoCAD settings, 706–725
    Express tools for, 952
    external commands/aliases, 929–930
    hatch patterns, 932–941
    linetypes, 951–952
    menus with macros, 930–931
    programming languages and, 952
    shapes/text fonts, 942–950
    user interface
        commands in, 920–922
        keyboard shortcuts in, 925–927
        Legacy node for, 928
        menus in, 922–924
        mouse buttons in, 927–928
        Partial CUI Files for, 928–929
        tablet menus in, 959–960
        toolbars in, 917–920

workspaces in, 913–917
CUT
   MTEXT, 203
   tool palette element, 654, 655
Cut Single/All multilines, 286, 291–292
cutting edges, trim to, 238, 239
cyclic redundancy check (CRC), 710, 711
Cycling, action/parameter grip, 581–582
cylinder, solid (CYLINDER), 845–846
cylindrical coordinate system, 67–68

D

dashed patterns, 936–937, 951–952
Dash lines for object selection, 256
database
   connectivity options for, 715, 716
   extracting attributes to, 498, 511–515
   functions for manipulating, 1028–1037
data types
   argument, 991, 997–998
   for database functions, 1029–1030
   functions for changing, 1012–1013
date, setting, 693
datum dimensions, 346–348
dbConnect Options, 715, 716
DBLIST, 320
DCL files, color codes for, 975
DC Online, 622
   Category Listing in, 638–641
   Collections in, 643
   Search in, 641–642
   Settings for, 642
DDATTE, 506
DDCHPROP, 314
DDEDIT, 205–206
DDPTYPE, 253
DDUCS, 804
DDVPOINT, 810–811
debugging programs, 1038–1040
Debug toolbar, 1039–1040
decimal code, shape definition, 945
decimal units
   dimension text, 387, 388

setting, 34, 41, 42
DECURVE, PEDIT, 274
default attribute value, 500
   editing, 517
   setting, 504, 505
defbytes, Shape-Character, 942–943
Deferred Perpendicular snap mode, 114–115
Deferred Tangent snap mode, 115
DEFINE ATTRIBUTES, 504–506
(defun...)/(defun C...), 1022–1027, 1036–1037
degree-radian conversions, 1023–1025, 1027
DELAY script, 735
DELETE
   FIT DATA, SPLINEDIT, 300
   OBJECTS, MVSETUP, 690
   PALETTE, 653
   table elements, 216
   tool palette elements, 654, 655
   UCS, 804
   VPCLIP boundary, 430
   WORKSPACE, 737
   XCLIP boundary, 601
Delete/deleting
   faces, 869
   filters/filter lists, 312, 313
   layer filters, 162–163
   layers, 150, 153
   layer states, 160
   lights, 898
   linetypes, 679
   multiline styles/elements, 294, 297
   named objects, 669–671
   named views, 672, 673
   plot styles, 450
   scenes, 899, 900
   table styles, 217
   tangential edges, 879
   text styles, 212
   Vertex, multiline, 286, 291
DELTA, LENGTHEN, 263
demand loading
   ObjectARX applications, 710, 712

xref files, 605–606, 710, 711
Dependent Scale action base point, 549
dependent symbols
    adding, 597, 598–599
    xrefs and, 589–591
Description
    layer, 159
    multiline style, 294
DESCRIPTION, point parameter, 533
Description pane, DesignCenter, 624
    location of, 622
    viewing text in, 632–633, 635–636
DesignCenter, 617
    components of, 622–624
    content in
        adding to drawings, 643–645
        creating tool palettes from, 480, 661–662
        viewing, 630–635
    DC Online in, 638–643
    exiting, 645
    Folders in, 622
    History in, 637
    Open Drawings in, 636–637
    toolbar of, 624–630
    viewing images in, 635–636
    window of, 618–619
        opening, 619–620
        positioning, 620–621
DesignCenter toolbar, 622, 623, 624–630
Design Web Format, 753
    Convert DWG hyperlinks to, 751, 752
    create files in, 753–758
    publish sheet sets to, 764–768, 786–787
    publish to Web in, 759, 760–761
    view files in, 758–759
Desktop quick access icon, 31, 49, 624
Detach
    image files, 607, 609
    materials, 901, 902
    xref drawings, 593, 596
Details
    image file, 607, 609
    sheet/sheet set, 772–773, 777–778

Show linetype, 679–680
View, 672, 673
Developer Help, 28
deviation lateral tolerance, 364
dialog boxes
    invoking commands from, 20
    menu commands displaying, 15
    shortcut menus in, 23
diameter
    circles from, 75–76
    control characters for, 202
    tolerance values with, 366, 367
DIAMETER, ELLIPSE, 186
diameter dimensioning, 349–350
Dictionaries, Change, 209
digital signatures, 703
    options for, 710, 712
    settings for, 705–706
    status bar icon for, 5, 6
digitizing, 958
digitizing tablet, 24, 957
    calibration of, 963, 964–967
    configuration of, 958–959
    custom menus for, 959–960
    operation of, 957–958
    sketching with, 967
    System settings for, 715, 716
    TABLET options for, 960–963
DIMALIGNED, 345–346
DIMANGULAR, 351–354
DIMARC, 350–351
DIMBASELINE, 354–356
DIMCENTER, 368
DIMCONTINUE, 356–357
DIMDIAMETER, 349–350
DIMDISASSOCIATE, 371
DIMEDIT, 369–370
dimensioning system variables, 336, 373–374
    DIMASSOC, 337, 338, 339, 374, 475
    DIMASZ, 402
    DIMBLK, 379
    DIMDLI, 355, 373
    DIMEXO, 374

DIMSCALE, 373, 386, 402, 444
   overriding, 391–392
Dimension Input, dynamic
   Enable, 131
   settings for, 133–135
dimension lines, 336, 337
   drawing, 340–341
   style settings for, 376–378
DIMENSIONS, RECTANGLE, 62
dimensions/dimensioning, 333–336
   aligned, 345–346
   angular, 351–354
   arc length, 350–351
   associative, 337–338
      exploding, 339, 496
      options for, 718, 720
   baseline, 354–356
   center marks for, 368
   commands for, 334, 339
   continue, 356–357
   diameter, 349–350
   disassociating/reassociating, 371–372
   dynamic block, 575
   editing, 369–373
   exploded, 339
   graphics area and, 3
   hatching over, 475
   layer visibility for, 434–435
   leaders for, 358–362
   linear, 339–345
   in model/paper space, 444–445
   non-associative, 339
   oblique, 368–369
   ordinate, 346–348
   overriding features of, 391–392
   plotting, DIMSCALE for, 402, 444
   quick, 358
   radius, 348–349
   terminology for, 336–337
   text for (see text, dimension)
   tolerances for, 362–368
   UCS origin/X axis for, 801
   updating, 392
  Dimension Style Manager, 374–391

dimension styles, 373–391
   adding from DesignCenter, 643
   creating, 375–376
   Manager for, 374–375
   matching, 276, 277
   modifying, 376–391
   system variables for, 336, 373–374
Dimension Text Edit, 370–371
Dimension toolbar, 339
Dimension Update, 392
DIMLINEAR, 339–343
DIMORDINATE, 346–348
DIMOVERRIDE, 391–392
DIMRADIUS, 348–349
DIMREASSOCIATE, 371–372
DIMREGEN, 372
DIMSTYLE, 363, 373–391, 392
DIM STYLE flyout, 373
DIMTEDIT, 370–371
DIM TEXT POSITION flyout, 372, 373
Direct Distance, drawing using, 121–123
direction
   3D viewing, 810–811
   alignment parameter, 539
   angle, 44–45
   drawing arcs using, 80, 83
   UCS axes, 793–795
DIRECTION, ARC, PLINE, 190
Direction Control, base angle, 35–36
directory path, application, 976, 977, 1026
display. see also visibility control
   3D Graphics, 715–716
   attribute, 499, 500, 506
   controlling, commands for, 136–145
   Coordinates, 73–74
   image, 610–611, 612
   lineweight, 682, 721
   options for, 708–709
   parameter value set marker, 543
   set plot area to, 407
   text, 210
   UCS icon, 794–797
   utility commands for, 671–674
   xref, 599–601

DISPLAY
  EDGE, 3D face, 828
  TIME, 694
Display/displaying. see also viewing
  alternate units, 390
  AutoSnap tooltip/aperture box, 113
  AutoTrack tooltip, 125
  information about objects, 319–323
  joints, multiline, 296–297
  screen menu, 24, 708
  Viewports Dialog, 426
DISPLAY LOCKED, viewport, 430
display order, changing, 674–675
display scroll bar, 4
DISPLAY VIEWPORT OBJECTS, 429, 430
distance
  direct, drawing using, 121–123
  between points (DIST), 256, 323
  QuickCalc for, 666
  Scale action point, 549–552
DISTANCE
  CHAMFER, 235
  DVIEW, 814–815, 816, 817
(distance), 1005
Distance and Angle coordinate display, 74
Distance-Direction codes, pen movement, 943–945
distant light, 894, 895–898
Dither, plot style, 450
dividing objects (DIVIDE), 255, 519–520
docked toolbars, closing, 10–11
docking/undocking
  DesignCenter window, 620–621
  Properties palette, 314
  toolbars, 8–9
  Tool Palettes window, 13, 651–652, 655
DONUT (DOUGHNUT), 249–250
dotted pairs, association list, 1032–1033, 1034
double-clicking
  docking/undocking by, 13, 314, 620, 651
  switching spaces by, 404
DOWN-ARROW key, on-screen options via, 16, 58

drafting
  key elements of, 3
  options for, 722
Drafting Settings, 99
  Direct Distance option for, 121–123
  Dynamic Input, 131–136
  Grid, 105–110
  Object Snap, 111–118
  Ortho, 110–111
  Polar/Object Snap Tracking, 123–130
  Snap, 99–105
  Tracking option for, 118–121
dragging
  DesignCenter content, 618, 636, 643–645
  DesignCenter window, 620–621
  docking/undocking by, 314, 651
  Insertion scale for, 718, 719
  to/from tool palettes, 14, 479, 480, 650, 659
  toolbars, 8–9
  Tool Palettes window, 13
DRAGMODE, 674
drawing environment
  setting up, 33–39
  startup configuration of, 2–3
drawing limits. see limits
drawing orientation, 406, 409, 442, 443
Drawing Properties dialog, 663–665, 749–750
drawings
  2D vs. 3D, 791–793
  adding content to, 643–645
  adding dependent symbols to, 598–599
  beginning new, options for, 28–30, 39–47
  changing limits for, 38–39
  command categories for, 18
  as container or content, 618
  creating sheet sets from, 771, 774–776
  digitizing, 963–964
  dimensioning, 333–335
  eTransmit, 768–769, 785
  extracting attributes from, 508–515
  hyperlinks to, 492–493, 749–750

inserting into other, 493, 497–498, 587–588
layout of, 415–421
loading into DesignCenter, 624, 636
multiple, working with, 48, 50
opening
   in DesignCenter, 636–637
   existing, 30–33
   from Internet, 748–749
Partial Open/Partial Load, 32–33, 662–663
plan view of, 807–808
properties of, 663–665
publishing
   to DWF/plotter, 764–768
   to Web, 759–764
purging, 669–671
saving
   commands for, 49–50
   file formats for, 32, 49
   to Internet, 748–749
searching for content in, 625–627
setting up, 687–690
standards for, 727–730
startup with existing, 4
xref, 587–589
   attaching/detaching, 593–596
   binding, 597–598
   listing attached, 592
   reloading/unloading, 596–597
drawing sets, 770
   publishing, 764–768
   Sheet Set Manager for, 776–777
   sheet sets in
      archiving, 786
      creating, 770–776
      placing views on, 780–783
      plotting/publishing, 786–787
      sheet list table for, 783–784
      transmittal package of, 785
      viewing/modifying, 777–780
Drawing Units dialog, 33–37. see also units
DRAWORDER, 302, 674–675
Draw Order, hatch/fill, 476

Draw toolbar, 9–10, 56
.dsd (Drawing Set Descriptions) files, 766
.dst (sheet set data) files, 773, 774, 776
DSVIEWER, 144–145
(dtr), defining, 1023–1025
Duplicate materials, 901, 902
DUPLICATE WORKSPACE, 917
DVIEW, 811–818, 898
DWF. see Design Web Format
DWF6 ePlot device, 754, 755–757
.dwg files. see drawings
.dws (standards) files, 6, 727–728
.dwt files. see template files
DYNAMIC, LENGTHEN, 264
dynamic blocks, 487, 521–522
   actions for, 543–575
   adding visibility states to, 576–579
   advanced utilities/features for, 579–582
   creating/editing, 527–530
   creating simple, 522–525
   parameters for, 531–542
   properties of, 542–543
   using dynamics of, 525–527
Dynamic coordinate display, 73
dynamic horizontal/vertical dimensioning, 340, 342
dynamic input, 15–16
   coordinate entry overrides for, 73, 133
   settings for, 131–136
Dynamic Prompts, 131–132
Dynamic Update, Aerial View, 145
DYNPICOORDS system variable, 73
DYN toggle, 16, 131

E

EATTEDIT, 506–508
EATTEXT, 508, 511–515
EDGE, 828
EDGE
   EXTEND, 242, 842
   POLYGON, 181–182
   TRIM, 239

edges
    3D face, visibility of, 827, 828
    chamfering, 856–858
    coloring, 870–871
    copying, 869–870
    extending/trimming, 842
    filleting, 858–859
    tangential, deleting, 879
edge surface (EDGESURF), 836–837
EDIT
    VERTEX, PEDIT, 271–274
    WORKSPACE, 737
Edit/editing. see also Modify/modifying
    in 3D, 837–842
    3D solids, 856–873
    attributes, 500, 506–508, 516–518
    Block Definition, 522, 528
    Boundaries, view, 672, 673
    button images, 921–922
    content in DesignCenter, 618
    dimensions/dimension text, 369–373
    with grips, 304–310
    hatches/gradients, 481–483
    Hyperlink, 752
    Item, selected filter, 312
    layer states, 160
    Layer translation mapping, 692
    Lineweights, 452
    Links, 696, 698
    menu/menu command properties, 923
    multilines, 286–292
    objects, shortcut menu for, 22
    partial CUI files, 928
    Plot Style Table, 408
    plotter configuration, 456–458
    Polyline, 186, 269–274
    polymesh surfaces, 837
    Program Parameters, 929–930
    reference files, 589, 601–605
    Scale List, 718, 721
    sheet set properties, 773
    sketches, 967
    spline curves, 299–301
    text, 191, 205–210
    multiline, 197
    table cell, 215–216
    Variable, 668
Electrical tool palette, 12–13
Element ID, new command, 920
elements, multiline style, 284, 292, 294
elevation
    setting (ELEV), 823–824
    Z values for, 792
ELEVATION, RECTANGLE, 61
ellipses, drawing (ELLIPSE), 182–186
ELLIPTICAL, CONE, 845
embedding objects, linking and, 695–703
Enable grips, 304, 305
encryption, 703, 705
End of Shape code, 945
Endpoint Object Snap, 111, 114–115, 324
end points
    axis, drawing ellipses from, 182–184
    line, specifying, 56, 57
engineering units
    dimensioning with, 387
    input format for, 36–37
    setting, 34, 41, 42
Enhanced Attribute Editor, 507–508
(entdel), 1034
ENTER
    command entry with, 18, 19
    in script files, 731, 732, 734
    text entry with, 192
(entget), 1029–1031
entity data, 1028
    association list for, 1031–1034
    extracting, 1033–1034
    functions for, 1029–1031, 1034
entity names, 1028
    functions for, 1028–1029
    selection sets for, 1034–1036
(entlast), 1028
(entmod), 1034
(entnext), 1028
(entsel), 1028–1029
(entupd), 1034
(eq)/(equal), 1019

ERASE, OFFSET, 227
Erase/erasing
   floating viewports, 423–424
   objects (ERASE), 92–93, 94
   parts of objects, 240–241
   SKETCH, 254, 255
errors
   AUDIT, 694–695
   AutoLISP syntax, 976, 991
   beep on, 715, 717
   stack storage, 946
   system printer spool, 714
   tablet calibration, 966, 967
eTransmit (ETRANSMIT), 768–769, 785
"ET" toolbars, 10
(eval), 1003
examples, create sheet sets from, 771–774
Excel spreadsheets, OLE with, 699–702
EXIT
   grip mode, 307
   PEDIT, 274
   SPLINEDIT, 301
exiting
   AutoCAD (EXIT), 51
   DesignCenter, 645
eXit (X), SKETCH, 254, 255
(exp)/(expt), 1003
exploded dimensions, 337, 339
exploding (EXPLODE), 496–497
   associative dimensions, 339
   blocks
      Allow, 490, 491
      on insertion, 494, 495
      with nested elements, 497
      tool properties for, 659, 660
   changes caused by, 496–497
   groups, 319
   hatch patterns, 464
   meshes, 837
   solids, 842
   xref drawings, 588
Export/exporting
   layer state, 161
   objects (EXPORT), 725–726

   palettes, 657
   Profiles, 724
   solid object data, 842
   table text, 216
expressions, AutoLISP, 989–990
   names for, 990, 992–993
   naming/SETQing value of, 993–994
   writing simple, 991–993
EXPRESS toolbars, 10
Express tools, 952
EXTEDGE system variable, 842
EXTEND
   in 3D, 841–842
   objects, 241–243
   PICKFIRST and, 261
   TRIM and, 239
EXTEND/NO EXTEND edges, EXTEND/TRIM, 842
extension lines, 336, 337
   origin points for, 339–340
   style settings for, 376–377, 378
Extension Object Snap, 114, 117, 324
extents, plotting to, 407
external commands, defining aliases for, 929–930
External Reference Manager, 5, 6, 591–598
external references, 587–589
   attaching, 593–596, 645, 749
   binding, 597–599
   dependent symbols and, 589–591
   detaching, 593, 596
   display control for, 599–601
   editing, 601–605
   exploding, 497, 588
   load/edit options for, 710, 711
   managing, 605–606
   opening, 598
   reloading/unloading, 596–597
   Xref Manager for, 5, 6, 591–598
EXTNAMES system variable, 490
extracting attributes, 498, 500, 508–515
EXTRUDE, 849–851
extruded 2D, 792
extruding faces, 862–863
extrusion, set thickness for, 823

F

FACE, UCS, 802
faces, 3D
  coloring, 868
  controlling visibility of, 868
  copying, 864
  creating, 825–828
  deleting, 869
  extruding, 862–863
  intersection display for, 720, 721
  moving, 865–866
  offsetting, 864–865
  rotating, 866–867
  tapering, 867, 868
fade, adjust image, 611–612
falloff, spotlight, 896, 897
Favorites
  Autodesk, 624, 628–629
  quick access icon for, 31, 49, 624
Feature Control Frame, 365–367
feedback, on-screen, 15–16
feet-and-inches format, 34, 36–37
FENCE (F), selection by, 258
fields
  database, 511
  Insert text, 196, 200–202
  options for, 718, 719
file formats
  exporting/importing, 725–726
  program, color codes for, 974–975
  render output image, 904
  SAVEAS options for, 32, 49, 710
files
  attribute extraction, 513–515
  finding, 32
  hyperlinks to, 491–493, 749–753
  names for, 49–50
  Open/Save settings for, 710–712
  options for, 707
  plot to, 406, 713, 714
  render to, 891
  security for, 710, 712, 759–762
  Select, 30–33

FILL, 484
fill color
  dimension text, 381, 382
  multiline, 296
filled solids, hatching, 476
FILLET, RECTANGLE, 61
filleting (FILLET)
  between objects, 231–234
  object selection for, 255–256
  PICKFIRST and, 261
  solids, 858–859
FILLMODE system variable, 250, 464
fill style, plot, 452
filters
  layer, 153, 161–163
  selection set (FILTER), 311–313
  (ssget), 1035
  X,Y,Z, 683–686
Filter Tree, Layer Properties Manager, 153
Find
  files, 32
  text (FIND), 191, 206–207
FIND AND REPLACE MTEXT, 203–204
FIT DATA, SPLINEDIT, 299–300
Fit/*FIT*
  dimension, 384–386
  PEDIT, 274
  TEXT, 195
FIT TOLERANCE, SPLINE, 297, 299
(fix), 1012, 1013
Flag Vertical Text Command code, 949
FLAT/FLAT+EDGES SHADEMODE, 888
Flip action, 531, 540, 565–566
FLIP ARROW, 372, 373
flip parameter, 531, 540, 565–566
(float), 1012, 1013
floating toolbars, 8–10
floating viewports. see viewports, floating
flyouts, toolbar button
  customizing, 917
  displaying, 7
Fog/Depth Cue, render, 889, 893–894
folders
  Browse for, 625–626

as container, 618
DesignCenter, 622
quick access icons for, 31, 49, 624
viewing content in, 631–634, 636
fonts. see text fonts
Format menu commands, 18
forms, geometric tolerancing for, 365, 367
Form View, Plot Style Table Editor, 449, 450–452
fractional arc, code for, 947–948
fractional units
dimensioning with, 381, 382, 387, 388
input format for, 36
setting, 34, 41, 42
frames
dimension text, 381, 382, 383
image, 613
leader annotation, 360
stretch, 554, 556, 560, 571–572
FRAMES, WIPEOUT, 302
freezing layers, 154, 164, 435–436
Friedman, Dan, 996
From Object Snap mode, 118
FTP
file transfer using, 749
quick access icon for, 31, 49, 624
function keys, assigning commands to, 925
function lists
argument types for, 997–998
writing, 991–993
functions
command, 1000–1001
conditional/logic, 1017–1019
custom, 990, 1022–1027
database, 1028–1037
advanced association list, 1036
entity data, 1029–1031, 1034
entity name, 1028–1029
selection set, 1034–1036
elementary/trigonometry, 1002–1004
input, 1000, 1001, 1013–1016
list handling, 1004–1011
predefined shortcut, 667–668
symbology for, 1002
type-changing, 1012–1013
user input pause for, 1000, 1001
writing expressions for, 991–993

G

gap tolerance, hatch, 479
GENERATE POLYLINE, XCLIP, 601
geometric characteristic symbols, 365–367
geometric figures, constructing, 55–63, 74–89, 249–255
geometric tolerances, 362, 365–368
geometric values display, 131–136
(get...) functions, 1000, 1013–1016
GOURAUD/GOURAUD+EDGES SHADEMODE, 888
gradient fill, 464–465
defining boundary for, 465–467
editing, 481–483
HATCH settings for, 467–468, 471–479
Gradient render background, 892, 893
gradient resolution, DWF, 756
graphics screen, 1–3
components of, 4–18
drawing layout in, 3
dynamic input in, 15–16, 131–136
switching to (GRAPHSCR), 319, 735
graphics window, 4, 5
Display options for, 708–709
docking toolbars in, 9
shortcut menus in, 21–22
Grayscale, plot style, 450
GRID, 99, 105–106
changing spacing of, 107–108
LIMITS and, 37
ON/OFF settings for, 105, 106–107
options for, 108–110
rotating, 102–104
Grid Snap, 105
grips
action/parameter, 521
cycling order for, 581–582
location/visibility of, 525, 531, 544
properties of, 542–543
using, 525–526, 527

editing with
   dimension, 371
   Dimension Input settings for, 134
   modes for, 306–310
   settings for, 304–305
   viewport, 429
  locations of, 306
  pickbox display for, 305–306
grip tips, Enable, 304, 305
group codes, association list, 1032–1033
Group Filters, layer, 153, 161, 163
Group/grouping
  Object, mode for, 260, 262
  objects (GROUP), 315–319
  Palettes, 657

H

HALFWIDTH PLINE, 189
halo gap percentage, 720, 721
(handent), 1029
Hatch and Gradient dialog, 467–479
Hatch Pattern Palette, 469
hatch patterns, 463–464
  Associative, enabling, 260, 262
  controlling visibility of, 484
  cross-sections with, 860
  customizing, 932–941
  defining boundary for, 465–467
  editing (HATCHEDIT), 481–483
  exploding, 496
  HATCH settings for, 467–471, 472–479
  matching, 276, 277
  tool palettes for, 12
    inserting from, 659
    properties of, 660–661
    using, 14, 479–481
height, extrusion, 850, 851
HEIGHT, MTEXT, 197. see also text height
help, 25–28
  Command Reference, 628
  Info Palette, 25–26
  MTEXT, 203
  traditional (HELP), 26–28

hexadecimal code, shape definition, 945
hidden lines
  profile, displaying, 879
  removal of, 875
  settings for, 718, 720–721
HIDDEN SHADEMODE, 720–721, 888
HIDE, DVIEW, 818
Hide/hiding
  Help window, 27
  Hidden Line Settings for, 720–721
  objects (HIDE), 875
  paperspace objects, 406, 409
  Tool Palettes window, 655, 656
highlighting
  group members, 316, 317, 318
  text, 197
History
  command, 17, 19
  DesignCenter, 622, 637
  quick access icon for, 31, 49, 624
  QuickCalc, 666, 667
Home
  AutoCAD 2006 Help, 28
  DesignCenter, 624, 629
HOME
  DIMEDIT, 369
  DIMTEDIT, 371
honeycomb pattern, 941
HORIZONTAL
  DIMLINEAR, 341, 343–345
  (HOR), XLINE, 178
horizontal dimensions, 334–335
  drawing, 341, 343–345
  dynamic, 340, 342
Horizontal dimension text, 381, 383, 384
hotspot, spotlight, 896, 897
Hover grip color, 304, 305
HYPERLINK, 751–752
HYPERLINKBASE system variable, 492, 749, 752
hyperlinks, 749–753
  in blocks, 490, 491–493
  Field text display for, 201
  options for, 718, 720
HYPERLINK sub-menu, 750, 753

I

ID, 256, 322–323
i-drop, enabling, 762
(if), 1017, 1018
Ignore island display, 473, 475, 477
IMAGEFRAME system variable, 611
Image Manager, 607–610
images, 607–613
   adjusting settings of (IMAGEADJUST), 611–612
   attaching from DesignCenter, 645
   as content, 618
   display of (IMAGECLIP), 610–611
   DWF file, resolution of, 756
   frame of (IMAGEFRAME), 613
   Manager for (IMAGE), 607–610
   new command button, 919, 920, 921–922
   publishing to Web, 759–764
   quality of (IMAGEQUALITY), 612
   render background, 892, 893
   saving, 904
   tool palette icon, 653–654, 655
   transparency of, 612–613
   viewing, 635–636, 904–905
Image Specifications dialog, 905
Imperial units
   input format for, 36–37
   setting, 34
   start drawing with, 40, 47
Implied windowing, 260, 262
Import/importing
   file formats (IMPORT), 726–727
   layer state, 161
   layouts as sheets, 770, 774–775, 778
   materials, 901, 902
   page setup, 440–441, 766
   Profiles, 724, 725
IMPORT LAYOUT AS SHEET, 778
IMPORT TEXT, MTEXT, 203
imprinting solids (IMPRINT), 871
INCREMENT, parameter value, 535
Increment angle, Polar Tracking, 124–127
indents, paragraph, 196, 202
Independent Scale action base point, 549
Index, AutoCAD 2006 Help, 28
Index Color (ACI)
   for layers, 155
   for materials, 901, 902
   for objects, 676
INDEXCTL system variable, 606
indexes, layer/spatial, 606
Info Palette, 25–26
Information Pane, CUI dialog, 913
Inherit Properties, hatch, 476–477
(initget), 1015, 1016
input functions, 1000, 1013–1016
Input Properties, parameter, 569, 574–575
Inquiry commands/toolbar, 319–323, 693–694, 874–875
INSCRIBED IN CIRCLE POLYGON, 180
INSERT
   macro for, 920, 921, 922
   XREF vs., 587–588
INSERT
   FIELD, MTEXT, 200–202
   SHEET LIST TABLE, 783, 784
   TITLE BLOCK, MVSETUP, 688–689
   VERTEX, PEDIT, 271–272
Insert/inserting
   block references, 488, 493–495
      with attributes, 501–503, 506, 732–735
      base points for, 490–491
      from DesignCenter, 644–645
      nested, 496
      scale for, 35
      into table, 216
      from tool palette, 14, 659
   drawings into other, 493, 497–498, 587–588
   Hyperlink, 491–493, 751–752
   Layout(s), 421
   Symbol, 196
   Table, 213–215
   xref drawings, 597–598
Insertion Cycling Order, parameter/action grip, 581–582
Insertion Object Snap, 114, 118, 324

insertion point
   attribute, 504, 505
   block, 494
   dynamic block, changing, 581–582
   image, 608
   table, 215
   xref, 594, 595
insertion scale
   block, 35
   options for, 718, 719
Insert toolbar, 726
Inspect (windows), VLISP, 988
integers
   as argument type, 991
   (list) input for, 998–999
   promoting to real, 1002
   system variables as, 323–325
   translating, 1012–1013
Integrated Development Environment (VLIDE), 972
Intellimouse, 145
intensity, light, 895, 897
interfaces
   ADS programming, 1040
   VLISP, 971–972, 989
interference, solid (INTERFERE), 861–862
Internet, 743
   Communication Center on, 5–6, 745–748
   Design Web Format for, 753–759
   eTransmit via, 768–769, 785
   hyperlinks to, 491–493, 749–753
   opening/saving drawings from, 748–749
   publish drawing sets for, 764–768
   publish drawings to, 759–764
   searching for content on, 638–643
   Web browser for, 744–745
Internet Explorer, 744
(inters), 1006
Intersection Object Snap, 114–115, 324
intersection operation (INTERSECT), 852, 855–856
Invert layer filter, 163
Invisible attribute mode, 504, 517
invisible edges, 3D face, 827, 828

Island Detection, hatch, 473, 477
ISOLINES system variable, 842
ISOMETRIC CIRCLES (ISOCIRCLES), ELLIPSE, 184–186
Isometric Snap, 104–105, 185, 186
ISO pen width
   hatch pattern, 468, 470, 661
   linetype, 680
Italic text, 198
(itoa), 1012, 1013

J

jog, radius dimension, 380–381
JOIN, PEDIT, 270, 271
joining objects (JOIN), 275
joints, display multiline, 296–297
JPEG images, publishing, 759, 760–761
justifying
   leader text, 360
   multiline, 284
   multiline text, 196, 197, 199
   text (JUSTIFYTEXT), 191, 192–194, 208

K

KEEP BOTH SIDES, SLICE, 861
keyboard
   command entry from, 18, 19
   coordinate entry from, 718, 720
   custom shortcut keys for, 925–927

L

LABEL, point parameter, 533
labels
   action, 555, 557
   parameter, 523, 580
   sheet view, 782
Landscape orientation, 409, 442, 443
.las (Layer State) files, 159
LAST
   MIRROR3D, 840
   ROTATE3D, 839

selection by, 91
(last), 1010
lateral tolerances, 362, 363–365
LAYER
   MVSETUP, 690
   OFFSET, 227
layer 0
   block objects on, 488, 495–496, 497
   properties of, 152
layer index, 606
Layer Previous (LAYERP), 164–165
Layer Properties Manager, 152–164
   plot style settings in, 454
   shortcut menu in, 23
   viewport visibility controls in, 434–436
layers
   adding from DesignCenter, 643
   block object, 488, 496
   block/pattern tool, 659, 660, 661
   colors for, 155–156
   creating/modifying, 149–152
      Layers toolbar for, 164
      Manager for, 152–164
   creating new, 153–154
   description for, 159
   filters for listing, 161–163
   freezing/thawing, 154, 435–436
   linetype for, 156–157
   lineweight for, 158
   locked, Exclude Objects on, 257
   locking/unlocking, 154–155
   matching object, 276, 277
   new view, 672, 673
   Partial Load, 662–663
   Partial Open, 32–33
   plot status for, 159
   plot style for, 158–159, 445, 452–454
   purging, 670
   saving properties of, 159–161
   setting current, 154, 164
   SOLVIEW and, 875–876
   translating, 691–693
   undoing settings for, 164–165
   Viewports, turning off, 424, 433–434
   visibility control for, 148, 150, 154, 434–436
   xref-dependent
      editing, 603
      options for, 710, 711
      symbol names for, 590
      visibility control for, 588, 591, 606
Layer States Manager, 153, 159–161
Layers toolbar, 9, 150, 152
   Apply filters to, 163
   changing layer status from, 164
Layer Translator (LAYTRANS), 691–693
LAYOUT, 421
layouts, 404, 414–415. see also paper space
   default, 417–418
   Display options for, 708, 709
   graphics area for, 2, 3
   hyperlinks to, 492
   importing as sheets, 770, 774–775, 778
   page setup for, 437–441
   planning for, 397–403, 415–416
   Plot and Publish options for, 713, 714
   plotting from, 436–437
   Regen Options for, 715, 716
   save changes to, 406, 409
   scale dimensions to, 384, 386, 389
   set plot area to, 407
   setting up, 417–421
      Layout Wizard for, 441–444
      New Layout for, 420–421
      template for, 418–419
   tooltip settings for, 135
   viewports for, 421–424
      centering objects in, 432–433
      controlling layer visibility in, 434–436
      creating, 424–428
      hiding borders of, 433–434
      modifying, 429–431
      placing multiviews in, 875–878
      scaling views in, 431–432
Layouts toolbar, 148, 420
Layout tab(s), 4, 15
   Include when adding sheets, 765, 766
   shortcut menu for, 23

Layout Wizard, 441–444
leaders, 336, 337
   arrowheads for, 378, 379
   Quick (QLEADER), 358–362
   text placement for, 384, 385–386
LEARN ABOUT MTEXT, 203
Least Material Condition (LMC), 366
left-justified text, 192
Legacy node, Customize User Interface, 928, 959–960
LENGTH
   ARC, PLINE, 190
   Polygonal Viewport, 425
   WEDGE, 849
(length), 997, 1011
lengthening objects (LENGTHEN), 262–264
length of chord, arcs from, 83–85
lettered lists, 199
library files
   linetype, 152, 678, 679
   materials, 901–902
lights, 894
   default, 889
   icon scale for, 889, 890
   scenes with, 898, 899–900
   setting up (LIGHTS), 894–898
   SHADEMODE for, 887
limits
   grid spacing and, 105, 106
   plot area for, 398, 399–401, 407
   setting, 37–39, 42–43, 45
LIMITS, 37–39
LIMITS, MVSETUP, 690
limits lateral tolerance, 363, 364
LINE, ARC, PLINE, 190
Line-Arc Continuation ARC, 86–89
linear dimensioning, 339–343
   options for, 334–335, 340–343
   primary units for, 387–388
   by selecting object, 343–345
linear parameter, actions with, 531, 534–535
   Array, 566–568
   Move, 545–546
   Stretch, 526, 531–532, 555–559, 570–572
linear units, setting, 33, 34
line families, hatch pattern, 932–934
lines. see also multilines; polylines
   CHANGE properties of, 315
   construction, 177–179
   dimension/extension, 336, 337, 339–341, 376–378
   dimensioning of
      aligned, 345–346
      angular, 351, 353–354
      continue, 357
      linear, 339–345
   drawing (LINE), 56–60
   filtering points on, 684
   grip locations on, 306
   multiline start/end cap, 294, 295
   parallel
      drawing, 226–229, 283–286
      editing, 286–292
   plot end/join styles for, 451–452
   polylines from, 269–270
   rubber band, 57
   sketching, 254–255
   UCS origin/X axis for, 801
   wide, 62–63, 250
LINE SPACING, MTEXT, 197, 209–210
LINETYPE, 675, 678
Linetype Manager, 677–680
linetypes
   adding from DesignCenter, 643
   block object, 488, 495–496
   block/pattern tool, 659, 660, 661
   custom, 951–952
   dimension/extension line, 376, 377
   layer, 151–152, 156–157, 591
   loading, 157
   matching, 276, 277
   multiline element, 294
   object, 675, 677–680, 693
   obscured line, 720, 721
   plot style, 451
   polyline vertex, 274
   scaling, 165–166, 402, 444

in sketching, 967
LINEWEIGHT, 99, 151, 675, 681
lineweights
   block object, 488, 495–496
   block/pattern tool, 659, 660, 661
   dimension/extension line, 376, 377
   layer, 150–151, 158
   matching, 276, 277
   object, 675, 681–682
   options for, 718, 721
   plot style, 451, 452
   plotting, 407–408, 409
   plotting without, 754
   table border, 218
linking objects, embedding and, 695–703
.lin (linetype) files, 152, 678, 679, 951
LISP
   color codes for, 975
   development of, 971, 996
LIST, 319–320
(list), 997–999, 1004–1005
list/listing
   attached xrefs, 592
   layers, 153, 161–163
   UCS names, 804, 877
   workspaces, 737
LIST parameter values, 535
lists
   AutoLISP, 971, 991
      association, 999, 1031–1034, 1036
      creating, 998–999
      function, 991–993, 997–998
      handling, functions for, 1004–1011
      types of arguments for, 997–998
   text, 196, 199
List View, Xref Manager, 592, 593
Live Enabler Options, 715, 717
Live Update, 747
LMC (Least Material Condition), 366
(load), 977, 1025–1026
Load/loading
   AutoLISP application, 977
   into content area, 624, 636
   defined functions, 1025–1026

layers to translate, 692
linetypes, 157, 678, 679
multiline style, 297
ObjectARX applications, 710, 712
Partial, 662–663
partial CUI file, 912, 928–929
plot stamp parameter file, 412
program, 976
Project, 1038
Sheet List, 766
xref files, 588, 605–606, 710, 711
Load or Reload Linetypes dialog, 157, 679
LOAD TEXT IN EDITOR, 976
Load/Unload Applications dialog, 977
local symbols, custom command/function, 1022–1023
Locate Point, 322–323
location
   geometric tolerancing for, 367
   shape, saving/recalling codes for, 946
LOCK LOCATION, toolbar/window, 11–12
Lock/locking
   attribute in block, 504, 506
   Info Palette help, 26
   layers, 154–155, 164
   objects not in working set, 604
   reference files, 602
   toolbars/windows, 6, 11–12
   values during input, 16
   viewport display, 430
lock status, sheet set, 777, 778, 781
(log), 1003
log files
   ASCII, tracking xrefs with, 605
   layer translation, 693
   options for, 710, 711, 713, 715
   plot stamp, 414
logic functions, 1017–1019
Lookup action, 531, 542, 568–575
lookup parameter, 531, 541–542, 568–575
LOOP, CHAMFER, 857
LOWERCASE MTEXT, 196, 199, 201, 204
.LSP files, color codes for, 975
LTSCALE, 152, 165–166

LTSCALE system variable, 402, 444, 680
LTYPE GEN, PEDIT, 274
LWDEFAULT system variable, 682
LWT toggle, 151

M

macros, custom command, 911, 920–922
   menu, 930–931
   mouse button, 928
   temporary override key, 927
   toolbar, 919
magnet, snap point, 113
maintenance patches, 745
Make Application wizard, 982–986
Make Block, 489–493, 522
Make Object's Layer Current, 164
Manage sheet selection, 779, 780
Manufacturers, DC Online, 640
mappings, layer translation, 691, 692
MARK, UNDO, 168, 169
marker blips, display of, 674
markers, Osnap, 112–113
mass properties (MASSPROP), 842, 874–875
MATCH CELL, table, 216
matching properties (MATCHPROP), 276–277
material conditions of datum symbols, 366, 367
materials, rendering
   Apply, 889, 890
   settings for (MATERIALS), 900–903
Materials Library, 901–902
MAXACTVP system variable, 147
MAXIMIZE VIEWPORT, 430, 431
Maximum Materials Condition (MMC), 366, 367
MBUTTONPAN system variable, 145
McCarthy, John, 971, 996
MEASUREINIT system variable, 40
Measurement Scale, dimension, 387, 389
measuring objects (MEASURE), 255, 520–521
MENU, 20, 929
menu bars
   AutoCAD, 4, 14–15
   VLISP, 973
menus. see also shortcut menus
   command entry from, 19–20
   customizing, 922–924
   side screen, 24–25, 708
   tablet
      configuring, 958–959, 960–962
      customizing, 959–960
Merged Cross multiline, 286, 288
Merged Tee multiline, 286, 289
meshes, 828–837
   creating, commands for, 828–829
   edge, 836–837
   editing/exploding, 496, 837
   free-form polygon, 829–830
   polyface, 830–832
   revolved, 835–836
   ruled polygon, 832–833
   tabulated, 834
METHOD, CHAMFER, 236
metric units, start drawing with, 40, 47
Microsoft Word/Excel, OLE with, 695–702
MIDDLE TEXT, 193
Midpoint Object Snap, 114–115, 324
Midway Between Two Points Object Snap, 118
MINIMIZE VIEWPORT, 430
MINSERT, 497, 601
mirror copy (MIRROR), 229–231
MIRROR grip mode, 307, 309
mirroring about plane (MIRROR3D), 839–840
MIRRTEXT system variable, 230–231
miters, offsetting, 227–229
MLEDIT, 283, 286–292
.mli (Materials Library) files, 901
MLINE, 283–286
.mln (multiline) files, 297
MLSTYLE, 283, 292–297
MMC (Maximum Materials Condition), 366, 367
models, 822–823. see also solids
   rendering, 889–894
      lights for, 894–898

materials for, 900–903
preferences for, 903
saving/viewing images for, 904–905
scenes for, 898–900
statistics on, 905–906
shading, 887–888
viewing
3DORBIT for, 818–821
DVIEW for, 811–818
multiple viewports for, 821–822
VPOINT for, 808–811
model space, 404
changing text to/from, 208
dimensioning in, 444–445
drawing limits for, 37–39
drawing setup for, 687–688
floating viewports in, 422
layout of objects in, 415–416, 417
linetype scaling for, 444
lineweight display in, 151
plotting from, 404–414, 416
switching to/from, 404, 418–419
tooltip settings for, 135
UCS icons for, 795
Model tab, 4, 15
Include when adding sheets, 765, 766
Regen Options for, 715, 716
shortcut menu for, 23
Modify Dimension Style dialog settings, 376–391
Alternate Units, 390
dimension variables and, 336, 374
Fit, 384–386
Lines, 376–378
Primary Units, 386–390
Symbols and Arrows, 378–381
Text, 381–384
Tolerances, 363–365, 390–391
Modify II toolbar, 270, 299
Modify/modifying. see also Edit/editing
block references, 487
commands for, 18
dimension styles, 376–391
floating viewports, 421, 429–431

lights, 898
materials, 901, 902
multiline styles, 297
named group, 315–316
objects, 92–94, 236–243, 262–277
page setups, 439–441
plot style tables, 448–452
scenes, 899, 900
sheet sets, 777–780
tables, 216
table styles, 219
xrefs, 588
Modify toolbar, 9, 92, 219
mouse
controlling display with, 145
customizing buttons on, 927–928
invoking menus with, 20, 22
options for, 715, 716
puck used as, 958
MOVE
objects, 236–238
SLICE and, 861
MOVE
FACES, SOLIDEDIT, 865–866
FIT DATA, SPLINEDIT, 300
grip mode, 307, 308
Tool Palettes window, 658
VERTEX, PEDIT, 272
VERTEX, SPLINEDIT, 301
Move action, 543–548
adding, 524–525
parameters with, 531
point parameter with, 525–526, 532
MSLIDE, 730–731
MTEXT, 195–205
leader, 359
LEARN ABOUT, 203
STYLE option for, 210
MTEXT, DIMLINEAR, 341
mtp/m2p Object Snap mode, 118
multilines
associative dimensions and, 338
drawing, 283–286
editing, 286–292

exploding, 496
multiline styles, 283
   creating/modifying, 292–297
   properties controlled by, 284
multiline text, 191
   creating, 195–205
   editing, 205–210
   leader, 358–360, 362
Multiline Text Editor, 196
Multiline Text Formatting toolbar, 206
MULTIPLE
   CHAMFER, 236
   FILLET, 234
   Move action, 545, 548
   object selection by, 258
   PEDIT, 270
MULTIPLE command modifier, 671
multiple copies, creating, 221–226
multiple drawings, 48
   saving/closing, 50
   turning off mode for, 715, 717
MULTIPLE POINT, POINT, 253
multiple viewports. see viewports
MULTIPLIER, Move action, 545
multiviews
   drawing setup for, 687–690
   generating, 878
   generating profiles for, 878–879
   placing, 875–878
MVSETUP, 687–690
My Computer, configure plotter on, 456
My Documents quick access icon, 31, 49, 624

N

NAME, point parameter, 533
named group, modifying, 315–316
named objects
   deleting unused, 669–671
   managing, 668–669
   Partial Load, 662–663
   wild cards and, 166–167
   xref, editing, 590–591, 603
named plot styles, 446

assigning to layers, 158
changing object/layer, 453
creating/modifying tables for, 446–452
NAMED UCS
   PLAN, 808
   SOLVIEW, 876
Named UCSs, Set Current, 806
Named Viewports, 149
named views
   hyperlinks to, 751
   listing/creating, 671–673
names
   action/parameter, visibility of, 544
   block, 490, 495
   custom hatch pattern, 932
   expression/variable, 990, 992–993
   file, 49–50
   function, 990, 1024–1025
   group, 316, 317, 318
   layer, 150, 153–154, 875–876
   layout, 420
   new command, 920, 921
   plot style, 450
   shape, 943
   sheet, 765–766
   text style, 210, 211
   xref/dependent symbol, 590, 597, 598
Nearest Object Snap, 114, 118, 324
nested blocks, 495–496
   exploding, 497
   object properties for, 677, 680, 682
   purging, 671
nested objects
   exploding, 497
   purging, 670
   xref, editing, 602, 603
nested xrefs, 588, 590
Network Plotter Server, configure plotter on, 456
New
   boundary set, 478
   command, 920–921
   Dimension Style, 375–376
   File, VLISP program, 974

Group, 316, 317
Group Filter, layer, 153, 163
Layer, 153
layer state, 160
Layout, 420
Light, 895–898
materials, 901, 902
Multiline Style, 293–297
Page Setup, 440
Plot Style Table, 408
Project, 978, 986
Property Filter, layer, 153, 162–163
Scene, 899–900
Table Style, 217–219
Text Style, 211
Variable, 668
View, 672
Viewports, 148–149, 426–428
Visibility State, 577

NEW
   DIMEDIT, 370
   MENU, 923–924
   PALETTE, 653, 661
   SHEET, 778
   SUBSET, 778–779
   TEMPORARY OVERRIDE, 926–927
   TOOLBAR, 917–918
   UCS, 798–799
   WORKSPACE, 915

NEW BOUNDARY
   IMAGECLIP, 610
   XCLIP, 600–601

NEW drawing, 29–30, 39–40
New Features Workshop, 28, 203
NEWSHEETSET, 771–776
New VP Freeze, 436
Node Object Snap, 114, 118, 324
non-associative dimensions, 339
   DIMASSOC setting for, 337
   hatching over, 475
   reassociating, 338
non-associative hatch, 476

NONE
   EXTEND/TRIM, 842
   UNDO, 168

None Object Snap mode, 114, 118, 324
nonuniform rational B-spline (NURBS), 297
NOORIGIN, UCSICON, 796
Normal island display, 473, 477, 482
Normal plot style, 158, 450, 453
(not), 1019
notes, transmittal package, 769
Noun/Verb selection, 260, 261
(nth), 997, 1011
(null), 1019
null string, 1000, 1001
numbering lists, 199
NURBS (nonuniform rational B-spline), 297

O

OBJECT
   AREA, 320–321
   MIRROR3D, 840
   REVCLOUD, 303–304
   REVOLVE, 852
   ROTATE3D, 839
   SECTION, 859
   SPLINE, 299
   UCS, 799–801

ObjectARX applications, 710, 712
Object Enablers, 715, 717
Object Grouping, 260, 262, 316–319
object linking and embedding (OLE), 695–703
   options for, 713, 714, 715, 717
   properties for, 703
objects, 55. see also models; solids
   3D
      2D drawing commands for, 823
      aligning, 837–838
      arraying, 840–841
      elevation/thickness for, 823–824
      extending/trimming, 841–842
      face, 825–828
      mesh, 828–837
      mirroring, 839–840
      modeling types for, 822–823

polyline, 825
region, 824–825
rotating, viewpoints for, 809
rotating objects about, 838–839
shading in 3D orbit view, 820
arraying, 221–226
associative dimensioning of, 337–338
base points for, 225
block, 488, 495–496
in Block Editor screen, 529
boundary set, 478–479
chamfering between, 234–236
clipping boundary, 430
Convert to Viewport, 425–426
copying, 219–221
creating from existing, 219
creating solids from, 849–851
custom, proxy images for, 710, 712
cutting edge, 238
cycling through, 89
display order for, 674–675
dividing, 519–520
dragging onto tool palettes, 650
DRAGMODE for, 674
drawing, 3
  Direct Distance for, 121–123
  making layer current for, 149–150, 164
  Tracking for, 118–121
elevation/thickness for, 823–824
erasing, 92–93
erasing parts of (BREAK), 240–241
excluding from selection, 257
exploding, 496–497
exporting, 725–726
extending, 241–243
filleting between, 231–234
grip editing, 304–310
grip locations on, 306
grouping, 315–319
hatching boundary, 465
hiding paperspace, 406, 409
information about, 319–323
joining, 275
lengthening, 262–264

measuring, 520–521
mirror copies of, 229–231
model space
  centering views of, 400–401, 432–433
  layout of, 415–419
  scaling views of, 398–400, 431–432
modifying, commands for, 22, 92, 236, 262
moving, 236–238
named
  deleting unused, 669–671
  managing, 668–669
  Partial Load, 662–663
  wild cards and, 166–167
  xref, editing, 590–591, 603
offsetting, 226–229
OLE, 703, 713, 714, 715, 717
origin/X axis locations for, 801
primitive (see shapes)
properties of, 675–682
  changing, 313–315
  matching, 276–277
  plot style, 445, 452–454
restoring erased, 94
rotating, 265–267
ruled surfaces between, 832–833
scaling, 70, 267–269
stretching, 264–265
trimming, 238–240
wipeout, effect of, 302
xref working set, 604–605
object selection, 89–91, 257–260
  for attribute extraction, 513
  for block definition, 490, 491
  for changing properties, 314
  for dynamic block, 523–525
  for extending/trimming, 242
  filter tool for, 311–313
  grip display limit for, 304, 305
  for hatch boundary, 466–467, 472, 474–476, 478–479
  linear dimensioning by, 343–345
  modes for, 260–262
  options for, 722–723

Quick Select for, 310–311
visual effects for, 256–257
Object Snap, 99
   cursor menu for, 20–21
   Markers/Tooltips for, 112–113
   modes for, 111–112, 114–118
   ON/OFF settings for, 105, 111
   ordinate dimensioning with, 346
   OSMODE values for, 324
   in paper space, 422
   running, 111, 718, 719
   shortcut functions for, 667–668
   X,Y,Z filters with, 683, 685–686
Object Snap toolbar, 111, 114
Object Snap Tracking, 99, 123, 127–130
OBLIQUE, DIMEDIT, 369, 370
Oblique Angle text, 196, 202, 213
oblique dimensioning (OBLIQUE), 368–369
octant arc, code for, 947
OFF, DVIEW, 816
offset
   array row/column, 222–223
   from dim line, 381, 384
   extension line, 336, 376, 378
   faces (OFFSET FACES), 864–865
   hatch pattern, 468, 470–471
   multiline element, 285–286, 294
   objects (OFFSET), 226–229
   plot, 406, 408, 713, 715
   plot stamp, 413
OFFSET
   Move action, 545, 548
   XLINE, 179
OLE (object linking and embedding), 695–703
   options for, 713, 714, 715, 717
   properties for, 703
OLESCALE, 703
ONE, UNDO, 168
online resources, 28
ON/OFF
   BLIPMODE, 674
   Drafting Settings command, 99, 105
   DRAGMODE, 674

Dynamic Input, 131
   GRID, 106–107
   IMAGECLIP boundary, 611
   layer plotting, 159
   layer visibility, 150, 154, 164
   LIMITS, 37–38
   Lineweight, 151
   Object Snap, 105, 111
   Object Snap Tracking, 127–128
   ORTHO, 110–111
   Polar Tracking, 123–124, 125
   REGENAUTO, 673
   SNAP, 100, 323–324
   TABLET, 960
   TIME, 694
   UCSICON, 796
   XCLIP boundary, 601
on-screen prompt
   area add/subtract operations and, 321
   drawing lines using, 56–57, 58
   dynamic input via, 15–16
   SKETCH and, 254
OOPS, 94
opacity, selection area, 257
OPAQUE BACKGROUND, MTEXT, 203
OPEN, 30–33
OPEN
   IN APPLICATION WINDOW, 636
   PEDIT, 271
   SPLINEDIT, 300, 301
Open Cross multiline, 286, 287
Open Drawings, DesignCenter, 622, 636–637
Open/opening
   in block editor, 522, 527
   DesignCenter window, 619–620
   Display Filter Dialog, CUI, 912
   drawings, 30–33
      from Internet, 748–749
      from Startup dialog, 47
   File, options for, 709–712
   image file, 608
   partial CUI file, 928
   Plot Style Table Editor, 448

sheet set, 777
toolbars, 10–11
Tool Palettes window, 650
xrefs, 589, 598
Open Read-Only, 32
Open Tee multiline, 286, 289
operating system programs, executing, 686–687
operators
AutoLISP syntax for, 990, 991, 1002
Boolean, filter selection with, 311
logical, Quick Select with, 312
writing expressions using, 991–993
optimized polyline, 186
OPTIONS, MVSETUP, 690
Options dialog settings (OPTIONS), 706
AutoSnap Settings, 112–113
Default Internet Location, 744–745
digital signature, 706
Display, 708–709
Drafting, 722
Files, 707
Grips, 304–305
layout display, 420
Open and Save, 709–712
Plot and Publish, 712–715
Plot Style Table, 446
Profiles, 723–725
Selection, 260–262, 722–723
Single-drawing compatibility mode, 48
System, 715–717
User Preferences, 717–721
Options toolbar, MTEXT, 196, 199–202
display/SHOW, 198, 203
editing with, 206
(or), 1019
Order Group, 318
ordinate dimensioning, 346–348
ORGANIZE FAVORITES, 629
orientation
coordinate system, 65–66, 68–69, 793–795
drawing, 406, 409, 442, 443
gradient fill, 472
plot stamp, 413

text, Flip action and, 566
origin
default location of, 70–71
Hatch, 468, 470–471, 479
icon for, 64
planes relative to, 68–69
points relative to, 65–66
reference points for, 400–401
UCS, defining, 793, 798–804
ORIGIN
UCS, 798–799
UCSICON, 796
ORTHO, 99, 110–111
ON/OFF settings for, 105, 110
ordinate dimensioning and, 347
Polar Tracking and, 124
ORTHO, SOLVIEW, 877
orthogonal tablet transformation, 965, 966
orthographic UCS, selecting, 804–807
"OS Command:" prompt, 686–687
OSMODE system variable, 324
(osnap), 1005–1006
Osnap Settings, 111. see also Object Snap
outcome of fit, tablet calibration, 966
Outer island display, 473, 477
overlay, tablet
configuring, 958–959, 960–962
using, 957–958
Overlay xref type, 594, 595
OVERLINE MTEXT, 196, 201
overrides
coordinate entry, 73, 133
dimension feature, 391–392
temporary, keys for, 925, 926–927
overrides file, page setup, 770

P

Page Setup
layout, 418, 420, 436, 437–441
model space plot, 406
sheet/sheet set, 765, 766, 770
Page Setup Manager, 420, 437–441
PALETTE, point parameter, 534

PAN
    options for, 718, 720
    Realtime, 143–144
    shortcut menu for, 137
PAN
    3DORBIT, 820
    DVIEW, 816–817
panning
    in Aerial view, 144
    with Intellimouse, 145
paper size
    layout, 442, 443
    model space plot, 406, 407
    options for, 714
paper space, 404. see also layouts
    changing text to/from, 208
    dimensioning in, 444–445
    drawing setup for, 687, 688–690
    lineweight display in, 151
    placing multiviews in, 875–878
    Plot last, 406, 409
    plotting from, 414–437
    scaling to
        dimension, 386, 389
        hatch pattern, 468, 470
        linetype, 444, 680
        view, 431–432
    switching to/from, 404, 418–419
    tab(s) for, 4, 15
    UCS icons for, 795
parabola, routine for drawing, 1036, 1037
paragraphs
    combining, 205
    indents for, 196, 202
PARALLEL, 3DORBIT, 820
parallel lines
    drawing, 226–229, 283–286
    editing, 286–292
Parallel Object Snap, 114, 117, 324
parallel projection, 811–812
PARAMETER, elliptical arc, 184
parameters, dynamic block
    actions for
        adding, 523–525

    types of, 527, 531–532, 543
    adding, 522–523
    alignment, 538–539
    Authoring palette for, 530
    base, 542
    flip, 540
    linear, 534–535
    lookup, 541–542
    lookup/input properties of, 569, 574–575
    point, 532–534
    polar, 536
    properties of, 542–543
    rotation, 537–538
    visibility, 540–541
    XY, 536–537
Parameter Sets, 527, 530, 579–580
PARTIAL Arc Length dimension, 351
Partial CUI Files, loading, 912, 928–929
Partial Load (PARTIALOAD), 662–663
Partial Open (PARTIALOPEN), 32–33,
    662–663, 671
Partial Open Read-Only, 32
passwords, 703, 704–705
PASTE
    MTEXT, 203
    on tool palette, 654
Paste QuickCalc values, 666
Paste Special/Paste Link, 696, 697, 699
path
    image file, 608, 610
    program directory, 976, 977
    relative hyperlink, 749–750, 751, 752
    xref file
        changing/saving, 593, 598
        saved, set type of, 592, 594
PATH, EXTRUDE, 850
.pat (pattern) files, 463, 932
patterns, wild-card, 167. see also arrays;
        hatch patterns
pause for user input, 1000
pause symbol, 1001
PC3 (plotter configuration) files, 454–458
PDMODE system variable, 253
PDSIZE system variable, 253–254

PEDIT, 269–274
   meshes, 837
   optimized polylines, 186
pen
   Distance-Direction codes for, 943–945
   plot stamp, 411
   plot style, 451
Pen Up/Down
   codes for, 945
   SKETCH, 254, 255
PERCENT, LENGTHEN, 263
perimeter, calculating, 320–322
PERPENDICULAR alignment parameter, 538
Perpendicular Object Snap, 114–115, 324
PERSPECTIVE, 3DORBIT, 808, 820
perspective projection, 811–812, 814–816, 817
PFACE, 830–832
.pgp (Program Parameters) file, 929–930
PICKADD system variable, 261, 314
PICKAUTO system variable, 262
pickbox
   display of, 305–306
   size of, 260, 261
PICKDRAG system variable, 262
PICKFIRST system variable, 261, 306
pick points, hatch boundary, 466–467, 472–474, 478
PLACE CALLOUT BLOCK, 782, 783
PLACE ON SHEET, 780, 781
planes
   2D drawing, 63, 66, 68–69
   clipping, adjusting, 821
   mirroring about, 839–840
plan view
   set viewing angles to, 810–811
   viewing from (PLAN), 807–808
PLINE, 186–190
   3DPOLY vs., 825
   FILLMODE settings for, 250
plot area
   computing, 398, 399
   selecting, 406, 407–408
Plot offset, 406, 408, 713, 715

Plot options, 406, 409
PLOT (Plot dialog)
   ePlot, 754–755
   layout, 436–437
   model space, 404–414
Plot Preview, 410
plot scale, 398–399
   computing, 399–401
   layout, 416
   model space, 406, 407
   viewport, changing, 431–432
plot size, computing, 398, 399–401
plot stamp
   options for, 713, 715
   publish drawing set with, 765, 767
   settings for, 406, 409, 410–414
plot status, layer, 159
plot styles, 445–446
   assigning to layers, 158–159
   block/pattern tool, 659, 660, 661
   changing object/layer, 452–454
   matching, 276, 277
   Plot with, 406, 409
Plot Style Table Editor, 448–452
plot style tables, 445–446
   creating, 446–448
   modifying, 448–452
   options for, 406, 408, 713, 715
Plotter Configuration Editor, 456–458, 755–757
plotters
   configuring, 454–458
   DWF6 ePlot, 754, 755–757
   HP, plot area for, 399
   publish to, 765, 767, 786–787
   settings for, 406, 442, 443, 713
Plotters window explorer, 455
plotting
   annotation scaling for, 402–403
   attributes, 499
   configuring plotters for, 454–458
   construction lines, 177
   dimension scaling for, 389, 402, 444–445
   DWF ePlot feature for, 753–758

floating viewports for, 421–431
  centering objects in, 432–433
  controlling layer visibility in, 434–436
  hiding borders of, 433–434
  scaling views in, 431–432
layer visibility for, 150
layouts for
  Layout Wizard for, 441–444
  page setup for, 437–441
  planning, 70, 397–401, 415–416
  Plot settings for, 436–437
  setting up, 417–421
linetype scaling for, 402, 444
model/paper space and, 404
from model space, 404–414, 416
options for, 712–715
from paper space, 414–437
plot style tables for, 445–454
publish drawing sets for, 764–768
sheet sets, 786–787
plug-ins for checking standards, 728–729
PNG images, publishing, 759, 760–761
Pointer Input, settings for, 131, 132–133
pointing devices. see also mouse
  options for, 715, 716
  placing points with, 57
  tablet, 24, 957–959
point light, 894, 895, 897–898
point parameter, 532–534
  actions with, 531
    Move, 526, 543–545
    Stretch, 554–555
  adding, 522–523
points. see also base point; insertion point
  coordinate, 63–67 (see also coordinates)
    functions for listing, 1010
    specifying, 70–72
  Direct Distance option for, 121–123
  drawing (POINT), 252–254
  ellipses from, 182–184
  hatch patterns from, 466–467, 472–474, 478
  limits expressed as, 37
  line end, 56, 57

Locate, 322–323
measure distance between, 323
QuickCalc for, 666
snap between, 118
system variables as, 323, 325
tablet calibration, 964–967
tracking, 118–121
X,Y,Z filters for, 683–686
POINTS, DVIEW, 816
Point Style icon menu, 253
(polar), 1005
POLAR, 3DARRAY, 841
polar arrays
  2D, 221–226
  3D, 840–841
polar coordinates
  Pointer Input settings for, 132, 133
  relative, 72
polar parameter, actions with, 531
  Move, 546–547
  Polar Stretch, 536, 562
  Scale, 548–550
  Stretch, 561
Polar Snap, 105, 127
Polar Stretch action, 531, 536, 562
Polar Tracking, 99
  changing angles for, 124–127
  Object Snap Tracking with, 128–130
  ON/OFF settings for, 105, 123–124
  ORTHO and, 110
polyface mesh
  creating, 830–832
  exploding, 496
POLYGONAL
  IMAGECLIP, 610
  VCLIP, 430
  XCLIP, 600
Polygonal Viewport, 425
polygon mesh
  exploding, 496
  free-form, 829–830
  between objects, 832–833
polygons
  closing, 59

drawing (POLYGON), 179–182
elevation/thickness for, 823–824
finding area of, 320–321
object selection by, 257–258
repeating closed, 938–941
solid-filled, 250–252
POLYLINE
  CHAMFER, 235–236
  FILLET, 233–234
  WIPEOUT, 302
polyline arcs
  closing, 270, 271
  drawing, 189–190
  exploding, 496
polylines
  3D, drawing, 825
  clip boundary, 599–601
  dividing closed, 519–520
  drawing, 186–190
  exploding, 496, 497
  extruding, 849
  FILLMODE settings for, 250
  grip locations on, 306
  hatch boundary, 466–467
  matching properties of, 276, 277
  modifying, 269–274
  offsetting, 226, 227–229
  sketching in, 967
  UCS origin/X axis for, 801
polymesh surfaces, editing, 837
Portrait orientation, 409, 442, 443
PostScript fonts, 210–211
precision
  dimension unit, 372–373, 387, 388, 390
  Hide, 720, 721
  lateral tolerance, 363, 364
  unit/angle display, 33, 34, 43
Preferences, Rendering, 903
prefixes
  command, 19, 20, 26, 929–930
  dimension text, 387, 388
prefix notation, AutoLISP, 990
Preset attribute mode, 504, 505, 517
Press and drag, 260, 261–262

Preview
  array, 222, 223, 224
  attribute extraction, 514
  Customize User Interface, 912
  dimension style, 375
  drawing file, 30, 31
  hatch pattern, 479
  layer filter, 162
  New Viewports, 148, 149
  Plot, 409–410
  plot stamp, 411
  reference for editing, 602–603
  Sheet Set, 774, 776, 777
  table style, 217, 219
  template file, 30, 46
  text style, 213
  Thumbnail, 710
  tooltip appearance, 135
  viewport configuration, 426, 427
Preview pane, DesignCenter, 622, 623
  toolbar button for, 624
  viewing images in, 632–633, 634–636
PREVIOUS
  selection by, 91, 311, 315
  UCS, 804
Primary Units, dimension
  style settings for, 386–390
  tolerances for, 364
primitive objects. see shapes
printers, system
  configuring, 454, 456
  options for, 713, 714
  selecting, 406
product support, 745
profiles (image), generating, 878–879
Profiles (settings), 706
  options for, 723–725
  workspaces vs., 737
profile view, creating, 876–877
programming languages
  ADS interface for, 1040
  color codes for, 974–975
  customizing and, 952
  development of, 971

"visual," 989
Program Parameters, Edit, 929–930
programs
   compiling/making applications for, 982–986
   debugging, 1038–1040
   developing, 974–977
     project file for, 978
     project window for, 979
     source code for, 978, 979–982
     text editor for, 974–975
   loading/running, 976–977
   saving, 975–976
PROJECT
   EXTEND, 243, 842
   TRIM, 239
project definition, 986–987
project files
   build application from, 984–985
   Load for debugging, 1038
   source code for, 979–982
   specifying, 978, 986–987
projective tablet transformation, 965–966
Project window, 979
PROJMODE system variable, 842
prompt, attribute, 500
   editing, 517
   settings for, 504, 505
properties
   3D graphics display, 715–716
   attribute, 517–519
   Customize User Interface, 912, 914, 917
     menu/command, 923, 924
     mouse button, 927–928
     new command, 920–921
     shortcut key, 925–926
     temporary override key, 927
     toolbar/command, 918, 919–920
     workspace, 916–917
   Drawing, 663–665, 749–750
   DWF6 ePlot, 755–757
   dynamic block, 521
     action/parameter/grip, 542–543
     parameter, 534, 569, 574–575, 580–581

   Inherit hatch, 476–477
   layer, setting
     Layers toolbar for, 150–152, 164
     Manager for, 152–164
     saving, 159–161
   mass, 842, 874–875
   matching, 276–277
   multiline, 284, 294
   object, 675–682
     block, 488, 495–496
     hatch/fill, 483
     OLE, 703, 717
     selected, changing, 313–315
     selection based on, 311
     tool, 650
   plot style, 445, 452–454
   plotter configuration, 406
   Project, 978, 986–987
   Sheet Set, 772, 773
   table cell, 218–219
   UCS icon, 796–797
PROPERTIES, 313–315
PROPERTIES
   tool, 654, 655
     block, 659–660
     pattern, 479, 481, 660–661
UCSICON, 796–797
Properties palette
   dynamic block parameter, 534, 580–581
   hatch/fill object, 483
   selected object, 313–315
Properties toolbar, 4, 8, 9, 676
Property Filters, layer, 153, 161, 162–163
Property Lookup Table, 542, 569, 573–575
Property Settings dialog, 276–277
proxy images, custom object, 710, 712
PSLTSCALE system variable, 444
publishing
   to DWF/plotter (PUBLISH), 764–768
   options for, 712–715, 767–768
   sheet sets, 786–787
   to Web (PUBLISHTOWEB), 759–764
Publish to Web wizard, 759–764
puck, 24, 957–959

pull-down menus. see menus
PURGE, 669–671
PURGE FIT DATA, SPLINEDIT, 299–300
PURGE LAYERS, LAYTRANS, 691
"pushing and popping" locations, 946

Q

QDIM, 358
QLEADER (Quick Leader), 358–362
QNew, 29
QSAVE, 49
QTEXT, 210
Quadrant Object Snap, 114, 115–116, 324
quadrilateral area, drawing, 251
Quality, adjusting image, 612
QuickCalc (QUICKCALC), 665–668
quick dimensioning, 358
Quick Object Snap, 324
Quick Select (QSELECT), 310–311, 314, 491
Quick Setup wizard, 40, 41–43
Quit
   debug operation, 1040
   SKETCH, 254, 255
QUIT, 51

R

radian-degree conversions, 1023–1025, 1027
radius
   arcs from, 83, 85–86
   circles from, 74–75, 78
   polygons from, 180–181
RADIUS
   ARC, PLINE, 190
   FILLET, 231–233
radius dimensioning, 348–349, 380–381
raster images
   adding from DesignCenter, 645
   DWF, resolution of, 756
RAY, 177, 179
real numbers
   as argument type, 991

   (list) input for, 998–999
   promoting integers to, 1002
   set variables equal to, 991–993
   system variables as, 323, 325
   translating, 1012–1013
Realtime
   PAN, 143–144
   ZOOM, 136–137, 145
Record (R), SKETCH, 254, 255
records, database, 511
RECOVER, 695
Recreate hatch boundary, 475, 482, 483
rectangles, drawing (RECTANGLE), 60–62
RECTANGULAR
   3DARRAY, 841
   IMAGECLIP, 610
   XCLIP, 600
rectangular arrays
   2D, 221–225
   3D, 840–841
rectangular coordinates, 68–69
   absolute, 70–71
   origin point for, 64
   Pointer Input settings for, 132–133
   relative, 71–72
redefining blocks, 493, 515, 732–735
Redo
   MTEXT, 198, 203
   ZOOM/PAN, 718, 720
REDO, 167, 170
REDRAW, 145
   point display after, 253
   routine for redefining, 1036–1037
REFEDIT, 589, 601–605
Refedit toolbar, 604–605
REFERENCE
   ROTATE, 266–267
   ROTATE3D, 839
   ROTATE grip mode, 308
   SCALE, 268–269
   SCALE grip mode, 309
   XLINE, 178
Reference Edit dialog, 602–604
Reference Manager, Autodesk, 606

reference points, plotting, 400–401
Reference toolbar, 591, 607
REFINE, SPLINEDIT, 301
reflection line, Flip action, 540, 565
REFRESH DesignCenter views, 636
refreshing images, 145
Regardless of Feature Size (RFS), 366, 367
REGEN, PEDIT, 272
REGENAUTO, 145, 506, 673–674
REGEN (regeneration), 145
    ATTDISP and, 506
    controlling, 673–674
    FILL and, 484
    layout, options for, 715, 716
    QTEXT and, 210
REGION/MASS PROPERTIES, 874–875
regions
    composite, 852
    creating, (REGION), 824
relative coordinates, 71–72
    overrides for entry of, 73, 133
    Pointer Input settings for, 133
relative hyperlinks, 492, 749–750, 752
Relative Polar Angle measurement, 127
Reload
    image file, 607, 609
    xref drawing, 588, 593, 596
Remove
    attribute from block, 518
    hatch boundary, 475, 482
    objects from group, 316, 317
    Standards File, 728
    from Working set, 604, 605
REMOVE
    ALL PROPERTY OVERRIDES, table, 216
    FORMATTING, MTEXT, 205
    selected objects, 258, 259–260
    SHEET, 779
Rename
    group, 316, 318
    layer filter, 162–163
    multiline style, 297
    text style, 211
RENAME, 668–669

RENAME
    PALETTE, 653
    palette element, 654, 655
    SHEET, 765–766
    WORKSPACE, 737
RENAME AND RENUMBER
    Sheet, 779
    View, 783
RENDER, 887, 889–894
rendering, 887
    lights for, 894–898
    materials for, 900–903
    preferences for, 903
    RENDER settings for, 889–894
    saving/viewing images of, 904–905
    scenes for, 898–900
    shading for, 887–888
    statistics on, 905–906
RENDER PREFERENCES, 903
Render toolbar, 875, 889
Render Window, 891
Re-Order group, 316, 318
repeating
    closed polygons, 938–941
    commands, 18, 19, 56
    scripts, 735–736
Repeat Last Command, 718–719
Repeat table, tablet calibration, 966
Replace text, Find and, 203–204, 206–207
REPLAY, 904–905
report, transmittal package, 769
Reset
    debug operation, 1040
    profile, 724, 725
RESET TIME, 694
RESET VIEW, 3DORBIT, 821
residuals, tablet calibration, 966, 967
resizing
    DesignCenter window, 621
    floating toolbars, 9–10
resolution
    DWF file, 756
    options for, 708, 709
    shaded viewport, 408

Resource Drawings, Sheet Set Manager, 777, 780, 781
Restore layer state, 160
RESTORE UCS, 804
RESUME script, 735
Retain Path, image file, 608
(reverse), 997, 1011
REVERSE, SPLINEDIT, 301
Reverse Lookup, 569, 575
Reverse Order, group, 318
Revision Cloud (REVCLOUD), 302–304
REVOLVE, 851–852
revolved surface (REVSURF), 835–836
RFS (Regardless of Feature Size), 366, 367
RIGHT
    DIMTEDIT, 371
    TEXT, 194
Right-click customization, 718–719. see also shortcut menus
right-hand rule, UCS, 793–794
RMAT, 900–903
RMS error, tablet calibration, 966, 967
root menu, 25
ROTATE
    DIMEDIT, 370
    grip mode, 307, 308
    SNAP, 102–104
    VPOINT, 810
Rotate action, 531, 537–538, 563–565
ROTATE3D, 838–839
rotated dimensions, 341, 342, 344–345
ROTATION
    ELLIPSE, 183
    MTEXT, 197
    RECTANGLE, 62
rotation parameter, 531, 537–538, 563–565
Rotation/rotating
    3D objects, viewpoints for, 809–810
    arrays, 222, 223, 224, 226
    attribute text, 499, 502–503
    block, 494, 495, 659, 660
    faces (ROTATE FACES), 866–867
    image, 608
    objects about 3D objects, 838–839

objects (ROTATE), 265–267
right-hand rule for, 793–794
text, 192
UCS, 802–803
xref, 594, 595
Round off dimensions, 387, 388
rows
    Array action, 568
    rectangular array, 222–223
    table, 214, 215, 216
RPREF, 903
RSCRIPT, 735–736
(rtos), 1012, 1013
rubber band line, 57
ruled surface (RULESURF), 832–833
Ruler, text, 196, 202
    display/SHOW, 198, 203
    editing with, 206
Running Osnap, 111, 114, 719
runouts, geometric tolerancing for, 367
run program, 976–977
Run Script, 732

S

safety precautions, file, 710, 711–712
SAVE, 49–50
SAVEAS
    drawing file, 49, 50, 749
    file formats for, 32
    options for, 710
    standards file, 727
SAVEAS WORKSPACE, 737
Save/saving
    all current customization files, 912
    attribute extraction template, 515
    drawings, 49–50
    drawings from Internet, 748–749
    image file path, 610
    images (SAVEIMG), 904
    layer properties, 159–161
    layer translation mappings, 692
    multiline style, 297
    options for, 709–712

programs, 975–976
Reference Edits, 604, 605
shape location, code for, 946
Sheet List, 766
text, 196
views, 671–674
xref file path, 593, 598
SAVETIME system variable, 49
SAVE UCS, 804
SCALE, 267–269
SCALE
   grip mode, 307, 309
   MLINE, 284
   VIEWPORTS, MVSETUP, 690
Scale action, 531, 536–537, 548–553
scale factors
   linetype, 152, 165–166
   shape, codes for, 945
   Zoom Scale, 140
Scale List, Edit, 718, 721
Scale/scaling
   annotations/symbols, 402–403
   attributes, 499, 501–503
   block
      Autoscaling for, 644
      inserted, 490, 491, 494–495
      tool properties for, 659, 660
   dimension, plotted, 402, 444–445
   dimension style, 384, 386, 387, 389
   hatch pattern, 468, 470, 661
   image, 608
   Insertion, 35, 718, 719
   linetype, 152, 165–166, 680
      matching, 276, 277
      plotted, 402, 444, 451
   lineweight, 407–408, 682, 721
   objects, 70, 267–269
   OLE object, 703
   plot, 398–399
      computing, 399–401
      layout, 416
      model space, 406, 407
   text, 191, 402–403
   views to paper space, 431–432

   xref, 594, 595
scenes
   selecting, 889, 890
   setting up (SCENE), 898–900
scientific units
   dimensioning with, 387
   setting, 34, 41, 42
Scratch, Start from, 47
screen, AutoCAD. see graphics screen
Screening, plot color, 451
screen menu, Display, 24–25, 708
script files, menu macros vs., 930
scripts (SCRIPT), 730
   creating/running, 731–735
   utility commands for, 735–736
scroll bars, 4, 708
Search
   AutoCAD 2006 Help, 28
   for content, 624, 625–628
   DC Online, 641–642
   the Web, 748–749
SECOND POINT
   Move action, 546–547
   Stretch action, 559
SECOND PT, ARC, PLINE, 190
SECTION, SOLVIEW, 877–878
sectioning solids (SECTION), 859–860
security (SECURITYOPTIONS), 703–706, 710, 712
Select
   Color, 155–156, 675–676
   Filter, 312
   light, 898
   Linetype, 157
   Plot Style, 158
   template, 29–30, 45–46
selectable group, 315, 316, 317
SELECT EDGE, 3D face, 828
Select file dialog, 30–33
Selection, options for, 722–723
selection sets. see also object selection
   changing objects in, 313–315
   clearing grips on, 306
   creating, 89–91, 257–260

FILTER for, 311–313
  modes for, 260–262
  Quick Select for, 310–311
 dynamic block, 523–525, 531–532
 entity name, 1028, 1034–1036
 naming/grouping, 315–319
SELECT POLYLINE, XCLIP, 600
Select Window, VLISP, 988
separating solids (SEPARATE), 871–872
Set Current
 layer, 153, 154
 page setup, 439
 UCS, 805, 806, 807
 view, 672, 673
SET CURRENT WORKSPACE, 737, 917
SET MTEXT WIDTH, 202
(setq), 991–996, 997–998, 999
Settings, DC Online, 642
Setup
 Drawing, 878
 Profile, 878–879
 View, 875–878
SETVAR, 36, 323–324
Shade and Tint, gradient fill, 471, 472
shaded viewports, plotting, 406, 408
shading
 hidden lines for, 720–721
 models (SHADEMODE), 887–888
 objects in 3D orbit view, 820
 smooth, 889, 890
shadows, rendering, 889, 890
SHAPE, 942, 944–945
Shape-Character codes, 942–943
shapes
 customizing, 942–949
 hatching, 475
 line-embedded, 951–952
 non-extrudable, 850
 purging, 671
Sheet List, Sheet Set Manager, 776–780
sheet list table, 783–784
Sheet Selections, create/manage, 779–780
Sheet Set Manager, 770, 776–787
sheet sets/sheets, 770
 archiving, 786
 creating, 770–776, 778
 extracting attributes from, 513
 placing views on, 780–783
 plotting/publishing, 764–768, 786–787
 sheet list table for, 783–784
 transmittal package of, 785
 viewing/modifying, 777–780
shelling solids (SHELL), 872
SHELL (OS command), 686–687
SHIFT key
 assigning commands to, 925
 EXTEND/TRIM with, 239, 242
 selecting grips with, 306
 setting mouse button as, 927
 Use to add to selection, 260, 261
Shneiderman, Ben, 996
shortcut keys
 customizing, 925–926
 Display options for, 708
shortcut menus
 cursor, 20–21
 customizing, 924
 editing dimensions using, 372–373
 grip mode, 307
 hatching options in, 474
 invoking commands from, 21–23
 right-click customization for, 718–719
 status bar, 6
 symbol, 201
 Text Editor, 198, 203–205
 toolbar, 8, 10–12
 Tool Palettes window, 652–658
 viewport options in, 431
 ZOOM/PAN in, 137
Show
 Help window, 27
 Startup dialog, 715, 717
SHOW TOOLBAR/OPTIONS/RULER, MTEXT, 203
.shp (shape) files, 942
.shx (font) files, 210, 671, 942
side screen menu, 24–25, 708
SIGNWARN system variable, 706
(sin), 1004, 1026–1027

SINGLE, selection by, 260
Single-drawing compatibility mode
   CLOSE/CLOSEALL and, 50
   setting, 48, 715, 717
Single Line Text
   creating, 191–195
   editing, 205–210
Single Viewport, 424–425
SIZE, Tool Palettes window, 658
sketching, Tablet mode and, 967
sketch lines (SKETCH), 254–255
SKPOLY system variable, 967
slicing solids (SLICE), 860–861
slides
   making, 730–731
   scripts for, 731–736
   showing series of, 736
   viewing, 731
Smoothing Angle, render, 889, 890
SNAP, 99–100. see also Object Snap
   changing spacing of, 101
   ON/OFF settings for, 100, 105, 323–324
   options for, 101–105
   Polar, 127
   using GRID with, 105, 109–110
SNAPISOPAIR system variable, 324
SNAPMODE system variable, 323–324
SNAPUNIT system variable, 325
SOLDRAW, 875, 878
SOLID
   3DFACE vs., 825–826
   drawing polygons with, 250–252
   FILLMODE settings for, 250
SOLIDEDIT
   edges, 869–871
   faces, 862–869
solid-filled circles (DONUT), 249–250
solid-filled polygons, 250–252
solid models, 822–823. see also models
solids, 822–823
   creating, 842–852
      box/cube, 843–844
      composite, 852–856
      cone, 844–845

      cylinder, 845–846
      by extrusion, 849–851
      by revolution, 851–852
      sphere, 845–847
      torus, 847–848
      wedge, 848–849
   editing, 856–873
      chamfering, 856–858
      checking, 873
      cleaning, 873
      edges of, 869–871
      faces of, 862–869
      filleting, 858–859
      imprinting, 871
      interference checking for, 861–862
      separating, 871–872
      shelling, 872
      slicing, 860–861
   hatching filled, 476
   hiding, 875
   mass properties of, 874–875
   multiviews of
      generating, 878
      generating profiles for, 878–879
      placing, 875–878
   UCS origin/X axis for, 801
Solids Editing toolbar, 853, 863
Solids toolbar, 843, 859, 876
SOLPROF, 878–879
SOLVIEW, 875–878
sorting
   attached xrefs, 592
   CUI properties, 917
SORT tool palette elements, 655, 656
source code, project file, 978–982
source file, block tool, 659, 660
SPACEBAR, 18, 19
space character(s)
   in AutoLISP expressions, 990, 999
   in menu macros, 931
   in script files, 731, 732, 735
SPACETRANS text, 208
spacing
   grid, 105, 106, 107–108

hatch pattern, 468, 469, 470, 661
snap, 101–102
text line, 209–210
spatial index, 606
special characters. see characters, special; symbols
special codes, shape definition, 945–949
spell-checking text (SPELL), 208–209
sphere, solid (SPHERE), 846–847
spherical coordinate system, 67
SPLFRAME system variable, 827, 828
SPLINE, PEDIT, 274
spline curves
   drawing (SPLINE), 297–299
   editing (SPLINEDIT), 299–301
   leader line, 361
splitting objects, 240–241
spool alert, system printer, 713, 714
spotlight, 894, 895–898
spreadsheets, OLE with, 699–702
SQL files, color codes for, 975
(sqrt), 993–994, 997, 1004
(ss...) functions, 1034–1036
Stack MTEXT, 198
stack storage errors, 946
staggered patterns, 936–937, 941
standard deviation, tablet calibration, 966, 967
Standard Parts, DC Online, 638–640
Standards, CAD
   configure/check (STANDARDS), 727–730
   status bar icon for, 5, 6
Standard Snap style, 104
Standard toolbar
   AutoCAD, 4, 7, 8, 9
   VLISP, 974
Start, Center, Angle (S,C,A) ARC, 80–81
Start, Center, End (S,C,E) ARC, 79–80
Start, Center, Length of Chord (S,C,L) ARC, 84–85
Start, End, Angle (S,E,A) ARC, 81–82
Start, End, Direction (S,E,D) ARC, 83
Start, End, Radius (S,E,R) ARC, 85–86
Start from Scratch, 47

starting AutoCAD, 1–4
START POINT
   Move action, 546
   Stretch action, 556, 559
Start point of axis, REVOLVE, 852
Startup dialog, displaying, 47, 715, 717
STARTUP system variable, 29, 30, 39–40, 47
Startup window, 1–3
Static coordinate display, 73
Statistics
   Drawing Properties, 664, 665
   rendering (STATISTICS), 905–906
status bar, 4, 5–6
   coordinates display in, 74
   drafting settings on, 105, 131, 151
   shortcut menu for, 23
   VLISP, 973
status bar tray, 4, 5–6
steel hatch pattern, 934–936
Step into/over/out debug operation, 1039–1040
stopwatch timer, 694
STRAIGHTEN VERTEX, PEDIT, 272–273
Stretch action, 553–561
   linear parameter with, 526, 531–532, 534, 570–572
   parameters with, 531
   Polar, 562
stretch frames, 554, 556, 560, 571–572
STRETCH grip mode, 307
stretching objects (STRETCH), 264–265
   associative dimensions and, 338
   associative hatch and, 476
strings
   null, 1000, 1001
   system variables as, 323, 325
   text, 191
   translating, 1012–1013
STYLE
   MLINE, 286
   MTEXT, 197, 210
   REVCLOUD, 304
   SNAP, 104–105
   TEXT, 195, 210

STYLE, text, 210
Styles toolbar, 217
Sub Sampling, rendering, 889, 892
subsets, sheet, 778–779
subshape code, 946
(subst), 1036
SUBTRACT AREA, 321–322
subtraction operation (SUBTRACT), 852, 853–855
suffixes, dimension text, 387, 388
Summary, Drawing Properties, 664, 665, 750
surface models, 822–823
surfaces, meshed. see also faces, 3D
   creating, 828–829
      edge, 836–837
      free-form polygon, 829–830
      revolved, 835–836
      ruled, 832–833
      tabulated, 834
   editing, 837
Surfaces toolbar, 251, 826
SURFTAB system variables, 832–833, 834, 835
SYMBOL, MTEXT, 201–202
symbols
   Action, 525
   ASCII codes for, 949, 950
   for AutoLISP functions, 1002
   callout, 782
   Character Map for, 201
   control characters for, 201–202
   dimension style, 378–381
   Insert, 196
   local, arguments and, 1022–1023
   plotting, scale for, 402–403
   tolerance, 362, 365–368
   variables as, 990
   xref-dependent, 589–591, 597, 598–599
symmetrical lateral tolerance, 363, 364
Sync block with attribute, 516
System, options for, 715–717
system variables, 47, 323–325. see also dimensioning system variables

T

T, expressions returning, 1017, 1018–1019
TAB key, locking input values with, 16
tables
   creating/modifying (TABLE), 213–216
   extracting attributes to, 513–515
   sheet list, 783–784
Table Style Manager, 217
table styles
   creating/modifying (TABLESTYLE), 216–219
   matching properties of, 276, 277
   selecting, 214, 215
TABLET, 960–963. see also digitizing tablet
Tablet mode
   ON/OFF toggle for, 957, 960
   sketching in, 967
Table View, Plot Style Table Editor, 449–450
tabs, setting, 202
tabulated surface (TABSURF), 834
tags, attribute, 499
   editing, 507–508, 517
   extraction process and, 500, 508, 511
   settings for, 504, 505
(tan), defining, 1026–1027
tangencies
   offsetting, 227–229
   spline, 297, 298–299
TANGENT
   alignment parameter, 538
   VERTEX, PEDIT, 273
tangent, functions for, 1026–1027
Tangent, Tangent, Radius (TTR) CIRCLE, 78
tangential edges, deleting, 879
Tangent Object Snap, 114, 115–116, 324
TANGENTS, FIT DATA, SPLINEDIT, 300
taper angle, extrusion, 850–851
tapering
   faces (TAPER FACES), 867, 868
   polyline width, 188–189
TARGET, DVIEW, 814
target, object selection, 89

(tblnext...)/(tblsearch), 1028
template, tablet
   configuring, 958–959, 960–962
   using, 957–958
template files
   acad.dwt, 47
   attribute extraction, 513, 515
   creating layouts from, 417, 418–421
   drawing settings in, 40
   hyperlinks to, 493, 751
   inserting blocks from, 493
   saving drawings as, 49, 50
   sheet creation, 770
   starting drawings with, 29–30, 45–46
   Web page, 761
temporary files, options for, 710, 712
temporary override keys, creating, 925, 926–927
tessellation lines, controlling, 842
test expressions, AutoLISP, 1017, 1018–1019
text
   adding to tool palette, 655, 656
   attribute, 498
      display of, 499
      modifying, 517
      settings for, 504, 505
   CHANGE properties of, 315
   changing spaces for (SPACETRANS), 208
   controlling display of, 210
   dimension, 336, 337
      changing, 341
      editing, 369–373
      leaders for, 358–362
      radius dimensioning, 348
      style settings for, 381–384, 385–386
   editing, 205–210
   finding/replacing, 206–207
   Flip action and, 566
   freezing layer for, 436
   hatching around, 475
   in HIDE operations, 720, 721
   justifying, 192–194, 208
   line-embedded, 951–952
   line spacing for, 209–210
   mirrored, 230–231
   multiline (MTEXT), 195–205
   plot stamp, 413
   plotting, scale for, 402–403
   single-line (TEXT), 191–195
   spell-checking, 208–209
   table cell, 215–216
   tolerance, 362, 363, 364
TEXT, 191–195, 205, 210
TEXT, DIMLINEAR, 341
Text Editor
   Multiline, 196
   shortcut menu for, 198, 203–205
   VLISP, 974–975
text files, script, 731
text fonts
   ASCII codes for, 950
   customizing, 942, 949–950
   Display options for, 708, 709
   DWF, setting, 756–757
   MTEXT, 196, 198
   parameter, 543
   for text styles, 210–211, 212
Text Formatting toolbar, 196, 197–198, 203
text height
   dimension, 381, 382
   multiline text, 196, 198
   plotted, text size for, 403
   single-line text, 191–192
   style settings for, 212
   tolerance, 363, 364
text screen (TEXTSCR), 319, 735
text styles, 210–213
   adding from DesignCenter, 643
   dimension, 381, 382
   matching, 276, 277
   selecting, 191, 196
Text toolbar, 191, 205
thawing layers, 154, 164, 435
themes, Web page, 761–762
Thicken lines for object selection, 256
thickness
   matching, 276, 277

setting, 823–824
Z values for, 792
THICKNESS, RECTANGLE, 61
Three-Point
   ARC, 78–79
   CIRCLE, 76–77
THROUGH
   OFFSET, 227
   XLINE, 179
Thumbnail Preview Settings, 710
tiled viewports, 146, 147–148
   creating, 148–149
   enabling, 687
TILEMODE system variable
   layer visibility and, 150
   multiple viewports and, 147, 687
   switching spaces with, 404
TIME, 693–694
title bar, 4, 7
TITLE BLOCK, MVSETUP, 688–689
title blocks
   selecting, 442, 443
   from template, 30, 419, 423
TOGGLE ANGLE, DVIEW camera, 813
Toggle Breakpoint, 1039
toggle buttons, status bar, 6
TOLERANCE, FIT DATA, SPLINEDIT, 300
tolerances, 362
   geometric (TOLERANCE), 365–368
   hatch gap, 479
   lateral, 363–365
   leader annotation, 360
   spline fit, 297, 299
   style settings for, 390–391
TOOLBAR, 11
toolbars, 4, 7–8
   adding to workspace, 916
   customizing, 917–920
   Display options for, 708
   docking/undocking, 8–9
   invoking commands from, 19
   locking/unlocking, 11–12
   opening/closing, 10–11

   resizing, 9–10
   VLISP, 973–974
tool palettes, 12–14, 649–650
   creating/populating, 661–662
   hatch pattern, 479–481
   inserting block/pattern to/from, 480, 644, 659
   tool properties for, 655
      block, 659–660
      pattern, 479, 481, 660–661
TOOLPALETTES, 13, 650
Tool Palettes window, 12–14
   docking/undocking, 651–652
   opening/closing, 650
   shortcut menus of, 652–658
Tool Properties
   block, 659–660
   displaying, 655
   pattern, 479, 481, 660–661
tooltips, 7
   Appearance settings for, 135–136
   AutoTrack, Display, 125
   dimension input, 133–135
   Display options for, 708
   Osnap, 112–113
   pointer input, 131–133
tool tray, 4, 5
torus, solid (TORUS), 847–848
TOTAL, LENGTHEN, 263
TRACE, 62–63, 250
trace, hatching, 476
trace window, VLISP, 973, 988
Tracking
   commands for, 99
   drawing objects using, 118–121
   Object Snap, 127–130
   Polar, 110, 123–127
   xref, tools for, 605
TRACKING, MTEXT, 196, 202
Transfer workspace, 928
transformation options, tablet calibration, 964–967
Translate layers, 691–693
transmittal package, 768–769, 785

transparency
- image (TRANSPARENCY), 612–613
- tooltip, 136
- window selection area, 257

TRANSPARENCY, Tool Palettes window, 655, 656

transparent commands
- invoking, 19, 26
- regen during, 674
- undoing, 168

TRAY SETTINGS, 6

Tree View
- Customize User Interface, 912, 913
- DesignCenter, 622
  - toggle for, 623, 624
  - using, 630, 631–633
- Xref Manager, 592, 593

triangles, solid, 252

triangular hatch patterns, 938–940

trigonometry
- custom hatch patterns and, 935–937
- functions for, 1004, 1026–1027

TRIM
- in 3D, 841–842
- EXTEND with, 242
- group, 316
- objects, 238–240
- PICKFIRST and, 261

TRIM
- CHAMFER, 236
- FILLET, 234

TRIMMODE system variable, 231, 234, 236

True Color
- for layers, 155, 156
- for objects, 676

TrueType fonts, 210–211

TWIST, DVIEW, 817

two-point arcs, 80–86

Two-Point CIRCLE, 77

TXT.SHX font, 210

(type), 1004

TYPE, SNAP, 105

type-changing functions, 1012–1013

U

U (command), 167–168, 170

UCS, 69, 793, 798–804. see also User Coordinate Systems

UCS
- EXTEND/TRIM, 842
- SOLVIEW, 876–877

UCS toolbar, 798

UCS II toolbar, 807

UCS icon
- display of, 794–797
  - for base point parameter, 542
  - for current viewport, 806–807
- examples of, 70

UCSICON, 796–797

UCSMAN (UCS dialog), 804–807

UCSVP system variable, 822

Underline text, 198

Undo
- layer settings, 164–165
- text edit, 196, 198
- ZOOM/PAN, options for, 718, 720

UNDO, 167, 168–169, 170

UNDO
- 3DFACE, 826
- COPY, 220
- DVIEW, 818
- EXTEND, 242
- LENGTHEN, 264
- LINE, 58, 59–60
- MLEDIT, 292
- MLINE, 284
- MTEXT, 203
- MVSETUP, 690
- PEDIT, 274
- PLINE, 187
- Polygonal Viewport, 425
- selection, 259
- SPLINEDIT, 301
- TRIM, 239
- WIPEOUT, 302

undocking. see docking/undocking

union operation (UNION), 852–853
UNITMODE system variable, 36
units
   block insertion, 490, 491, 494, 495
   converting, 666, 667
   dimensioning, 386–390
   drawing
      graphics screen and, 3
      Start from Scratch for, 47
      UNITS for, 33–37
      wizards for, 41–42, 43
   insertion, 35, 718, 719
   lateral tolerance, 363, 364–365
   lineweight, 682, 721
   plot stamp, 414
UNITS, MVSETUP, 690
Unload
   image file, 607, 609
   xref drawings, 593, 596–597
unlocking
   layers, 154–155, 164
   toolbars/windows, 11–12
Update
   Dimension, 392
   Layers for new view, 672, 673
updates, product, 745, 747
UPDATE SHEET LIST TABLE, 783
UPDATE TOOL IMAGE, 654, 655
UPPERCASE MTEXT, 196, 199, 201, 204
upside-down
   plot, 409
   plot stamp, 413
   text, 212
User Coordinate Systems, 69, 793
   3D space and, 793
   defining new, 797–804
   for multiple viewports, 821–822
   plan view for, 807–808
   predefined orthographic, 804–807
   profile view relative to, 876–877
   right-hand rule for, 793–794
   Save with view, 672
   UCS icon display for, 70, 794–797
   Viewpoint Presets for, 810–811

User Defined Fields, plot stamp, 412
User-defined hatch patterns, 469
user interface, customizing
   commands in, 920–922
   keyboard shortcuts in, 925–927
   Legacy node for, 928
   menus in, 922–924
   mouse buttons in, 927–928
   Partial CUI Files for, 928–929
   tablet menus in, 959–960
   toolbars in, 917–920
   workspaces in, 913–917
User Preferences, options for, 717–721
Use Shift to add to selection, 260, 261

V

Validate Digital Signatures icon, 5, 6
values, attribute, 499–500
   default, 500
   editing, 506–508
   extracting, 500, 508–515
   modes for, 504–505
VALUE SET, linear parameter, 535
value set, parameter, 543
variables, AutoLISP, 989–990
   names for, 990, 992–993
   writing expressions with, 991–993
Variables/Variables Tree, QuickCalc, 666, 667–668
vector resolution, DWF, 756
Verify attribute mode, 504–505, 517
Vertical
   dimension text, 381, 382
   text, 212
VERTICAL
   DIMLINEAR, 341, 343–345
   (VER), XLINE, 178
vertical dimensions, 334–335
   drawing, 341, 342, 343–345
   dynamic, 340, 342
vertices
   add/delete multiline, 286, 290–291
   edit polyline, 271–274

move spline, 301
polyface mesh, 831–832
VIEW, 875–878, 898
VIEW
   EXTEND/TRIM, 842
   MIRROR3D, 840
   ROTATE3D, 839
   SECTION, 860
   UCS, 802–803
viewing. see also Display/displaying
   in 3D, 808–822
      3DORBIT for, 818–821
      DVIEW for, 811–818
      multiple viewports for, 821–822
      VPOINT for, 808–811
   content, 618, 630–635
   DWF files, 758–759
   images, 635–636, 904–905
   from plan view, 807–808
   sheet sets, 777–780
   slides, 731
View List, Sheet Set Manager, 777, 780, 781–783
View menu commands, 18
VIEW OPTIONS, Tool Palettes window, 653–654
Viewpoint Presets dialog, 810–811
viewpoints
   changing, 792, 808–811
   for rotating 3D objects, 809
viewports
   Auto Aerial View, 145
   Convert Object to, 425–426
   Create in new layouts option for, 420
   default layout, 417–418
   floating, 404, 414, 421–424
      centering objects in, 432–433
      controlling layer visibility in, 434–436
      creating, 424–428, 442, 443
      enabling, 687
      hiding borders of, 433–434
      modifying, 429–431
      scaling objects in, 431–432
   matching properties of, 276, 277

   multiple, creating
      for 3D views, 821–822
      for drawing setup, 689–690
      floating, 426–428, 443
      tiled, 146–149
   multiviews in
      generating, 878
      generating profiles for, 878–879
      placing, 875–878
   render to, 891
   shaded, plotting, 406, 408
   TIDEMODE setting for, 147, 687
   UCS settings for, 806–807
Viewport Scale Control, 431–432
Viewports dialog, 148–149, 426–428
Viewports toolbar, 424–426
views
   3D drawing, 791–792
   Aerial View, 144–145
   centering, 400–401
   creating scenes with, 898, 899
   hyperlinks to, 492, 751
   model vs. layout, 417–418
   multiple
      drawing setup for, 687–690
      generating, 878
      generating profiles for, 878–879
      placing, 875–878
   parallel vs. perspective, 811–812
   Partial Load, 662–663
   Partial Open, 32–33
   plan, 807–808, 810–811
   saving (VIEW), 671–673
   scaling, 431–432
   set plot area to, 407
   sheet, placing, 780–783
View toolbar, VLISP, 988–989
virtual pen #, plot style, 451
visibility control. see also display
   3D face edge, 828
   attribute, 499, 500, 506
   hatch pattern, 464, 484
   layer, 150, 154, 164
      floating viewport, 434–436

tiled viewport, 148
xref-dependent, 588, 591, 606
pointer input, 133
visibility parameter, 531, 540–541, 576–579
visibility states, 540–541, 576–579
VISRETAIN system variable, 591, 606
Visual Effect Settings, 256–257
Visual LISP Console, 988
VLIDE, 972
VLISP, 971. see also AutoLISP
   compiling/making applications in, 982–986
   components of, 972–974
   interfaces/IDE of, 971–972, 989
   launching, 972
   programs in
      debugging, 1038–1040
      developing, 974–977
      loading, 976
      project file for, 978
      project window for, 979
      running, 976–977
      saving, 975–976
      source code for, 978, 979–982
      text editor for, 974–975
   project definition in, 986–987
   View toolbar of, 988–989
vlisp/vlide commands, 972
VLJ (VLISP project) files, 986
VLX (compiled application) files, 982
VPCLIP, 429–430
VPMAX, 429, 430
VPOINT, 808–811, 898
VPORTS, 148–149
VSLIDE, 731
VTENABLE system variable, 136

W

Walker, John, 971
warning messages
   block definition, 493
   partial CUI file edit, 928–929
   printer spooling, 714

Show all, 715, 717
Xref duplicate definition, 645
Watch Window, VLISP, 989, 1040
WBLOCK, 733–734
WCS. see World Coordinate System
Web browser, launching, 744–745
Web sites
   access content on, 640–641
   Autodesk, 744
   hyperlinks to, 751, 752
   publish to, 759–764
   search for files on, 748–749
Web toolbar, 744
wedge, solid (WEDGE), 848–849
Weld All multilines, 286, 292
(while), 1017, 1018
wide lines, drawing, 62–63
width
   leader text, 359, 360
   text bounding box, 202
WIDTH
   MTEXT, 197
   PEDIT, 271
   PLINE, 187–189
   RECTANGLE, 61
   VERTEX, PEDIT, 273, 274
Width Factor, text character, 196, 202, 213
wild cards, 154, 166–167
WINDOW, selection by, 89–90, 93
   dynamic block, 525
   hatch boundary, 466
   MULTIPLE modifier for, 258–260
   visual effects for, 257
windowing, Implied, 260, 262
window(s)
   define view, 672
   locking/unlocking, 6, 11–12
   set plot area to, 407
Windows Render Options dialog, 891, 892
Wintab Compatible Digitizer, 715, 716
WIPEOUT, 301–302
wireframe models, 822–823
wizards
   Add-A-Plotter, 455–456

Add Plot Style Table, 408, 446–448
Advanced/Quick Setup, 40–45
Attribute Extraction, 511–515
Create Sheet Set, 770–776
Layout, 441–444
Make Application, 982–986
Publish to Web, 759–764
Word documents, OLE with, 695–699, 701–702
working set, REFEDIT, 604–605
workspaces
  customizing, 913–917, 928
  drawing environment vs., 3
  options for (WORKSPACE), 736–737
WORLD
  PLAN, 808
  SOLVIEW, 876–877
  UCS, 804
World Coordinate System, 69
  3D space and, 793
  axes for, 68–69
  plan view for, 807–808
  referencing points to, 72
  returning to, 804
  Viewpoint Presets for, 810–811
WORLDVIEW system variable, 811
WPOLYGON (WP), selection by, 257–258
WYSIWYG, plotting in, 404, 414

X

XATTACH, 749
X AXIS
  REVOLVE, 852
  ROTATE3D, 839
X axis for types of objects, 801
XBIND, 591, 597, 598–599
XCLIP, 599–601
XCLIPFRAME system variable, 601
XCORNER
  Move action, 547
  Stretch action, 559, 560–561
Xdatum dimension, 346, 347–348
X distance, Scale action, 550–551, 552

X filters, 683–686
XFLIP, UCS, 802
.xlg (xref log) file, 605
XLINE, 177–179
XLOADCTL system variable, 605–606
XOPEN, 589, 602
x operator, expressions with, 994
XREF. see also external references
  attaching/manipulating with, 591–598
  features of, 587–589
XREF, MVSETUP, 690
XREFCTL system variable, 605
Xref filter, layer, 162
Xref Manager, 5, 6, 591–598
XY
  Move action, 548
  SECTION, 860
X-Y displacement codes, 946–947
XY distance, Scale action, 552
XY parameter, actions with, 531
  Array, 568
  Move, 547–548
  Scale, 550–553
  Stretch, 536–537, 559–561
XY/XZ, MIRROR3D, 840
X/Y/Z axes
  3D orientation of, 792, 793–795
  dimensions aligned with, 340
  location of, 66, 68–69
  rotating UCS around, 802–803
  specifying points on, 70, 71

Y

Y axis. see X/Y/Z axes
Y AXIS
  REVOLVE, 852
  ROTATE3D, 839
YCORNER
  Move action, 547
  Stretch action, 559, 560–561
Ydatum dimension, 346, 347–348
Y distance, Scale action, 551–552
Y filters, 683–686

YFLIP, UCS, 802
YZ
   MIRROR3D, 840
   SECTION, 860

Z

Z axis. see X/Y/Z axes
ZAXIS
   MIRROR3D, 840
   ROTATE3D, 839
   SECTION, 860
   UCS, 799
Z coordinates
   2D vs. 3D object, 792–793, 823
   filters for, 683–686
   setting, 823–824
   viewpoint and, 808
zero suppression
   dimension text, 387, 389, 390
   lateral tolerance, 363, 364
zipped files
   DWF, 757
   transmittal set of, 768

ZOOM, 136–143
   All, 37, 38, 138–139
   Center, 141
   Dynamic, 141–142
   Extents, 139, 177
   flyout menu for, 7
   Object, 139–140
   Previous, 143
   Realtime, 136–137, 145
   Scale, 140
   User Preferences for, 718, 720
   Window, 137–138
   XP, 431
ZOOM
   3DORBIT, 820
   DVIEW, 817
ZOOMFACTOR system variable, 145
zooming
   in Aerial View, 144–145
   with Intellimouse, 145
Zoom toolbar, 138
ZX, SECTION, 860

IMPORTANT-READ CAREFULLY: This End User License Agreement ("Agreement") sets forth the conditions by which Delmar Learning, a division of Thomson Learning Inc. ("Thomson") will make electronic access to the Thomson Delmar Learning-owned licensed content and associated media, software, documentation, printed materials and electronic documentation contained in this package and/or made available to you via this product (the "Licensed Content"), available to you (the "End User"). BY CLICKING THE "I ACCEPT" BUTTON AND/OR OPENING THIS PACKAGE, YOU ACKNOWLEDGE THAT YOU HAVE READ ALL OF THE TERMS AND CONDITIONS, AND THAT YOU AGREE TO BE BOUND BY ITS TERMS CONDITIONS AND ALL APPLICABLE LAWS AND REGULATIONS GOVERNING THE USE OF THE LICENSED CONTENT.

1.0 SCOPE OF LICENSE

1.1 Licensed Content. The Licensed Content may contain portions of modifiable content ("Modifiable Content") and content which may not be modified or otherwise altered by the End User ("Non-Modifiable Content"). For purposes of this Agreement, Modifiable Content and Non-Modifiable Content may be collectively referred to herein as the "Licensed Content." All Licensed Content shall be considered Non-Modifiable Content, unless such Licensed Content is presented to the End User in a modifiable format and it is clearly indicated that modification of the Licensed Content is permitted.

1.2 Subject to the End User's compliance with the terms and conditions of this Agreement, Thomson Delmar Learning hereby grants the End User, a nontransferable, non-exclusive, limited right to access and view a single copy of the Licensed Content on a single personal computer system for noncommercial, internal, personal use only. The End User shall not (i) reproduce, copy, modify (except in the case of Modifiable Content), distribute, display, transfer, sublicense, prepare derivative work(s) based on, sell, exchange, barter or transfer, rent, lease, loan, resell, or in any other manner exploit the Licensed Content; (ii) remove, obscure or alter any notice of Thomson Delmar Learning's intellectual property rights present on or in the License Content, including, but not limited to, copyright, trademark and/or patent notices; or (iii) disassemble, decompile, translate, reverse engineer or otherwise reduce the Licensed Content.

2.0 TERMINATION

2.1 Thomson Delmar Learning may at any time (without prejudice to its other rights or remedies) immediately terminate this Agreement and/or suspend access to some or all of the Licensed Content, in the event that the End User does not comply with any of the terms and conditions of this Agreement. In the event of such termination by Thomson Delmar Learning, the End User shall immediately return any and all copies of the Licensed Content to Thomson Delmar Learning.

3.0 PROPRIETARY RIGHTS

3.1 The End User acknowledges that Thomson Delmar Learning owns all right, title and interest, including, but not limited to all copyright rights therein, in and to the Licensed Content, and that the End User shall not take any action inconsistent with such ownership. The Licensed Content is protected by U.S., Canadian and other applicable copyright laws and by international treaties, including the Berne Convention and the Universal Copyright Convention. Nothing contained in this Agreement shall be construed as granting the End User any ownership rights in or to the Licensed Content.

3.2 Thomson Delmar Learning reserves the right at any time to withdraw from the Licensed Content any item or part of an item for which it no longer retains the right to publish, or which it has reasonable grounds to believe infringes copyright or is defamatory, unlawful or otherwise objectionable.

4.0 PROTECTION AND SECURITY

4.1 The End User shall use its best efforts and take all reasonable steps to safeguard its copy of the Licensed Content to ensure that no unauthorized reproduction, publication, disclosure, modification or distribution of the Licensed Content, in whole or in part, is made. To the extent that the End User becomes aware of any such unauthorized use of the Licensed Content, the End User shall immediately notify Delmar Learning. Notification of such violations may be made by sending an Email to delmarhelp@thomson.com.

5.0 MISUSE OF THE LICENSED PRODUCT

5.1 In the event that the End User uses the Licensed Content in violation of this Agreement, Thomson Delmar Learning shall have the option of electing liquidated damages, which shall include all profits generated by the End User's use of the Licensed Content plus interest computed at the maximum rate permitted by law and all legal fees and other expenses incurred by Thomson Delmar Learning in enforcing its rights, plus penalties.

6.0 FEDERAL GOVERNMENT CLIENTS

6.1 Except as expressly authorized by Delmar Learning, Federal Government clients obtain only the rights specified in this Agreement and no other rights. The Government acknowledges that (i) all software and related documentation incorporated in the Licensed Content is existing commercial computer software within the meaning of FAR 27.405(b)(2); and (2) all other data delivered in whatever form, is limited rights data within the meaning of FAR 27.401. The restrictions in this section are acceptable as consistent with the Government's need for software and other data under this Agreement.

7.0 DISCLAIMER OF WARRANTIES AND LIABILITIES

7.1 Although Thomson Delmar Learning believes the Licensed Content to be reliable, Thomson Delmar Learning does not guarantee or warrant (i) any information or materials contained in or produced by the Licensed Content, (ii) the accuracy, completeness or reliability of the Licensed Content, or (iii) that the Licensed Content is free from errors or other material defects. THE LICENSED PRODUCT IS PROVIDED "AS IS," WITHOUT ANY WARRANTY OF ANY KIND AND THOMSON DELMAR LEARNING DISCLAIMS ANY AND ALL WARRANTIES, EXPRESSED OR IMPLIED, INCLUDING, WITHOUT LIMITATION, WARRANTIES OF MERCHANTABILITY OR FITNESS OF A PARTICULAR PURPOSE. IN NO EVENT SHALL THOMSON DELMAR LEARNING BE LIABLE FOR: INDIRECT, SPECIAL, PUNITIVE OR CONSEQUENTIAL DAMAGES INCLUDING FOR LOST PROFITS, LOST DATA, OR OTHERWISE. IN NO EVENT SHALL DELMAR LEARNING'S AGGREGATE LIABILITY HEREUNDER, WHETHER ARISING IN CONTRACT, TORT, STRICT LIABILITY OR OTHERWISE, EXCEED THE AMOUNT OF FEES PAID BY THE END USER HEREUNDER FOR THE LICENSE OF THE LICENSED CONTENT.

8.0 GENERAL

8.1 Entire Agreement. This Agreement shall constitute the entire Agreement between the Parties and supercedes all prior Agreements and understandings oral or written relating to the subject matter hereof.

8.2 Enhancements/Modifications of Licensed Content. From time to time, and in Delmar Learning's sole discretion, Thomson Thomson Delmar Learning may advise the End User of updates, upgrades, enhancements and/or improvements to the Licensed Content, and may permit the End User to access and use, subject to the terms and conditions of this Agreement, such modifications, upon payment of prices as may be established by Delmar Learning.

8.3 No Export. The End User shall use the Licensed Content solely in the United States and shall not transfer or export, directly or indirectly, the Licensed Content outside the United States.

8.4 Severability. If any provision of this Agreement is invalid, illegal, or unenforceable under any applicable statute or rule of law, the provision shall be deemed omitted to the extent that it is invalid, illegal, or unenforceable. In such a case, the remainder of the Agreement shall be construed in a manner as to give greatest effect to the original intention of the parties hereto.

8.5 Waiver. The waiver of any right or failure of either party to exercise in any respect any right provided in this Agreement in any instance shall not be deemed to be a waiver of such right in the future or a waiver of any other right under this Agreement.

8.6 Choice of Law/Venue. This Agreement shall be interpreted, construed, and governed by and in accordance with the laws of the State of New York, applicable to contracts executed and to be wholly preformed therein, without regard to its principles governing conflicts of law. Each party agrees that any proceeding arising out of or relating to this Agreement or the breach or threatened breach of this Agreement may be commenced and prosecuted in a court in the State and County of New York. Each party consents and submits to the non-exclusive personal jurisdiction of any court in the State and County of New York in respect of any such proceeding.

8.7 Acknowledgment. By opening this package and/or by accessing the Licensed Content on this Website, THE END USER ACKNOWLEDGES THAT IT HAS READ THIS AGREEMENT, UNDERSTANDS IT, AND AGREES TO BE BOUND BY ITS TERMS AND CONDITIONS. IF YOU DO NOT ACCEPT THESE TERMS AND CONDITIONS, YOU MUST NOT ACCESS THE LICENSED CONTENT AND RETURN THE LICENSED PRODUCT TO THOMSON DELMAR LEARNING (WITHIN 30 CALENDAR DAYS OF THE END USER'S PURCHASE) WITH PROOF OF PAYMENT ACCEPTABLE TO DELMAR LEARNING, FOR A CREDIT OR A REFUND. Should the End User have any questions/comments regarding this Agreement, please contact Thomson Delmar Learning at delmarhelp@thomson.com.